MECHANISMS OF DNA DAMAGE AND REPAIR

Implications for Carcinogenesis and Risk Assessment

Edited by

Michael G. Simic

National Bureau of Standards
Gaithersburg, Maryland
and Chemistry Department
American University
Washington, D.C.

Lawrence Grossman

Johns Hopkins University
Baltimore, Maryland

and

Arthur C. Upton

New York University Medical Center
New York, New York

Publications Coordinator
Caprice M. Chappas

Assistant Editor
David S. Bergtold

PLENUM PRESS • NEW YORK AND LONDON

Library of Congress Cataloging in Publication Data

Mechanisms of DNA damage and repair.

(Basic life sciences: v. 38)

Based on papers presented at a conference held at the National Bureau of Standards in Gaithersburg, Md. on June 2–7, 1985, and sponsored by the bureau and other agencies.

Includes bibliographies and index.

1. Carcinogenesis–Congresses. 2. Deoxyribonucleic acid–Decay–Congresses. 3. Deoxyribonucleic acid repair–Congresses. 4. Health risk assessment–Congresses. I. Simic, Michael G. II. Grossman, Lawrence, 1924– . III. Upton, Arthur C. IV. United States. National Bureau of Standards. V. Series. [DNLM: 1. Carcinogens–congresses. 2. DNA Repair–congresses. 3. Mutagens–congresses. 4. Probability–congresses. 5. Radiation Injuries–congresses. W3 BA255 v.38/QZ 202 M48545 1985]

RC268.5.M434 1986 616.94′4071 86-9363

ISBN 978-1-4615-9464-2 ISBN 978-1-4615-9462-8 (eBook)

DOI 10.1007/978-1-4615-9462-8

Proceedings of a symposium on Mechanisms of DNA Damage and Repair: Implications for Carcinogenesis and Risk Assessment, held June 2–7, 1985 in Gaithersburg, Maryland

A Division of Plenum Publishing Corporation
233 Spring Street, New York, N.Y. 10013

DNA is the beginning and the end.

SPONSORS

National Bureau of Standards

International Life Sciences Institute

National Cancer Institute

National Institute of Environmental Health Sciences

National Institute of General Medical Sciences

Department of Energy

U. S. Army Medical R&D Command

American Cyanamid Company

American Industrial Health Council

Shell Development Company

Diamond Shamrock

Dow Chemical U. S. A.

E. I. du Pont de Nemours & Company

Hoffman-LaRoche

The Upjohn Company

Bristol-Myers

Burroughs Wellcome Company

Merck Sharp and Dohme

PREFACE

This book is based on the papers presented at the conference on "Mechanisms of DNA Damage and Repair: Implications for Carcinogenesis and Risk Assessment," held at the National Bureau of Standards on June 2-7, 1985. This volume deals with mechanisms of DNA damage and repair at the molecular level; consequences of unrepaired or misrepaired damage, with major emphasis on carcinogenesis; drugs which bind selectively to altered and potentially damaging DNA sequences; and potential utilization of DNA damage as an endpoint for assessing risks of UV light, ionizing radiations, chemicals, drugs, and hazardous agents in foods.

Because the induction of mutations by radiation and genotoxic chemicals has been observed to follow one-hit kinetics in some instances, it is generally assumed that any level of exposure to a DNA-damaging agent may increase the risk of genetic disease or cancer in an exposed population. At the same time, however, there is evidence that although the DNA of living cells is continually damaged by natural background radiation, free radicals, and other naturally occurring processes, most of the damage is normally repaired.

Knowledge of the intricate relationships between the toxicity of hazardous agents and the level of exposure, possible synergistic effects of exposure to other agents, damage to DNA and its repair or misrepair, repair capacity of the affected cells, and the factors influencing the processes of damage and repair at various stages in their evolution are essential for understanding molecular carcinogenesis and for developing scientifically based risk assessment. The need for further knowledge of these processes is pressing in view of the wide use of genotoxicity assays as screening tests, the uncertainties involved in interpreting such tests in our present state of knowledge, and the health and economic implications of positive or negative test results. The conference was organized primarily to address this need and to highlight opportunities for research in related fields.

The articles include the latest unpublished observations by the authors. Only 59 of the 159 papers presented at the conference could be included in a book intended to provide didactic material for a general audience. These up-to-date reviews make this book particularly useful to research scientists in cancer etiology, cancer therapy, and cancer prevention. The book should also be of interest to biochemists, molecular biologists, radiation chemists and biologists, free-radical chemists, food chemists, and toxicologists. In addition, it is useful background material for those engaged in risk assessment and regulatory decision-making.

Michael G. Simic

Larry Grossman

Arthur D. Upton

ACKNOWLEDGEMENTS

Contributions and assistance of the Organizing Committee: Donald R. Johnson, NBS; Curt Reimann, NBS; Sanford A. Miller, FDA; David G. Longfellow, NCI; Enrico Clementi, IBM; Michael Kssha, Florida State University; Peter A. Cerutti, Swiss Institute for Experimental Cancer Research; George Scholes, University of Newcastle on Tyne; J. Martin Brown, Stanford University; E. Larry Powers, University of Texas; and the Session Chairmen: Dietrich Schulte-Frohlinde, Max Planck Institute; J. William Lown, University of Alberta; Jean-Bernhard LePecq, Institute Gustav-Roussy; Kurt W. Kohn, NCI; John F. Ward, University of California San Diego; Tomas Lindahl, Imperial Cancer Research Fund; Franklin Hutchinson, Yale University; Steven R. Tannenbaum, MIT; Peter N. Magee, Temple University; Bernard D. Goldstein, Rutgers University; W. Gary Flamm, FDA; and Nancy L. Oleinick, Case Western Reserve University.

Special thanks go to Ernest Ambler, Director of NBS; Raymond G. Kammer, Deputy Director of NBS; Sanford A. Miller, Director, Center for Food Safety and Applied Nutrition; and Richard H. Adamson, Director of Cancer Etiology; for their interest, encouragement and assistance.

Technical and management assistance of Janet N. Hauber, Conference Coordinator, Karen A. Taylor, ILSI and the NBS staff: Dene Hocken, Gloria Wiersma, Francis Weygand, Sara R. Torrence, Kathy D. Kilmer and Kathy C. Stang has been greatly appreciated.

Lastly, appreciation and acknowledgement go to those directly involved in putting this book together: Caprice Chappas, Publications Coordinator; David Bergtold, Assistant Editor; Eleanore Cordisco, Denise Sgroi, and Martha Grey, word processing.

CONTENTS

INTRODUCTION

DNA REPAIR AND CONSEQUENCES

RISK ASSESSMENT

INTRODUCTION TO MECHANISMS OF DNA DAMAGE AND REPAIR

Michael G. Simic
National Bureau of Standards
Gaithersburg, MD 20899, USA

INTRODUCTION

From a pessimistic point of view, biological systems are surrounded by an unfriendly medium, the oxygen atmosphere.[1] They are also exposed to damaging solar[2] and cosmic radiations.[3] Food, the basic source of life-sustaining bio-energy, contains a variety of natural and cooking-induced carcinogens.[4] In addition to these natural exposures, technological developments have resulted in man-made hazards from pollution.

From an optimistic point of view, bio-systems have a remarkable capacity for repair of damage[5,6] and inhibition (scavenging) of damaging agents.[6] The inhibitory and repair systems, which are essential for survival of a species, have evolved[7] to deal with natural hazards. Hazards created by humans may induce damage that cannot be repaired by existing systems or may inflict damage that overburdens those repair systems.

DAMAGE

Damage to bio-systems is diverse. It may involve specific organ systems, organs, tissues, or molecular complexes such as the genetic material. The scope of this book is limited to DNA damage because it is apparent that many disorders can be traced to DNA damage. The types of damage to DNA and associated proteins are shown schematically in Figure 1.

Certain types of damage can be induced by classical chemical reactions such as alkylation of DNA bases (base alteration) by alkylating agents. Intra-strand crosslinks (e.g., pyrimidine dimers) and inter-strand crosslinks (e.g., psoralen-type bridging) can be mediated by light.[2] The greatest variety of damage, however, is induced by free radicals.[8] Many damaging agents, including some beneficial drugs, are capable of generating free radicals. Because many scientists are unfamiliar with free radicals and their role in biology and medicine, a short overview of the free radical processes associated with DNA damage and repair is presented here.

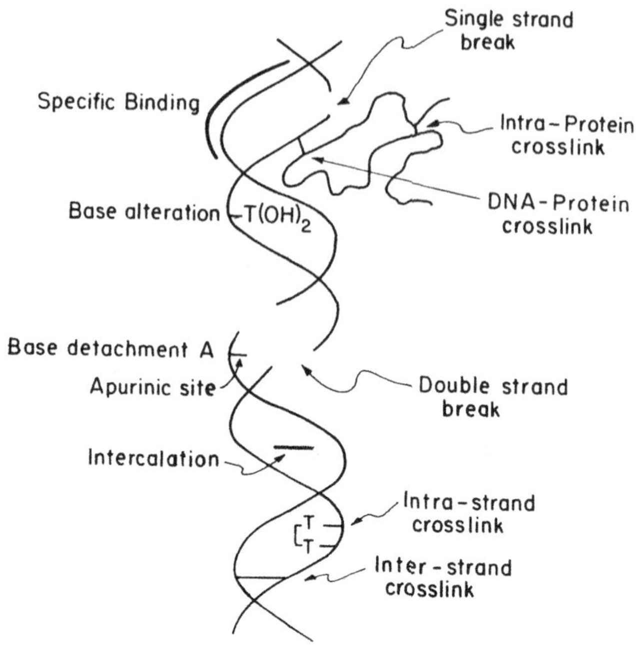

Fig. 1. Schematic presentation of the types of damage to a DNA - protein complex.

FREE RADICALS

The concept of a radical, -R, i.e., a group of atoms, was introduced by organic chemists in the last century to explain how recognizable groups could be easily transferred from molecule to molecule in many organic reactions. The existence of radicals in free form was demonstrated later, at the turn of the century.[9] Hence, such groups of atoms were named "free radicals."

The current, broader definition of a free radical is a specie with an unpaired electron that is generated by the loss of an electron or a gain of an extra electron. For example, when ionized, a molecule loses an electron. When a molecule reacts with a free radical which already has an unpaired electron, it also becomes a free radical because it acquires an unpaired electron. For convenience, the unpaired electron is usually represented graphically by an arbitrarily positioned dot, •R.

A free radical may not be "free" in the usual sense of the word. Free radicals attached to an immobile matrix such as DNA or protein have restricted mobility. In large bio-molecules, a free radical can be associated with a subunit, e.g., a thymine free radical may be associated with DNA. In the broadest sense of the term, this is a DNA radical. In many respects, however, the radical reflects properties of the thymine radical. To emphasize this feature, the radical is represented as DNA-T.

Ionizing radiations — x-rays, γ-rays, high energy electrons, and many other high energy photons and particles — invariably generate ions,[10] which may lead to the formation of a variety of daughter free radicals. For example, when ionizing radiation interacts with water, •OH radical is generated via

$$H_2O \rightsquigarrow H_2O^+ + e^- \quad (1)$$

$$H_2O^+ + H_2O \rightarrow \cdot OH + H_3O^+ \quad (2)$$

Other molecules dissolved in water (solutes) may also become ionized. The fraction of ionization of a solute is closely proportional to the electron fraction of the solute.[11] In bio-systems, ionization and initial distribution of free radicals are roughly proportional to the mass fraction of the components. This mode of generation of free radicals on bio-molecules (Bio) is called direct action.

Since most bio-molecules are surrounded by an environment that consists of up to 90% water, •OH radicals are generated in the vicinity of Bio. By virtue of their extreme reactivity,[12] some •OH radicals react with Bio. This mode of generating free radicals on bio-molecules is referred to as indirect action. In summary,

$$Bio \rightsquigarrow \cdot Bio^+ + \cdot Bio \qquad \text{Direct Action} \quad (3)$$

$$Bio + \cdot OH \rightarrow \cdot Bio\text{-}OH \qquad \text{Indirect Action} \quad (4)$$

Sometimes both direct and indirect actions generate identical radicals. In other cases, a specific radical may be formed by only one of these modes. Both modes contribute to final biological effects and consequences, but generally they cannot be easily distinguished.

Ignoring the distinction between direct and indirect effects of radiation, the interaction of ionizing radiations with DNA and its consequences are shown in Scheme 1,

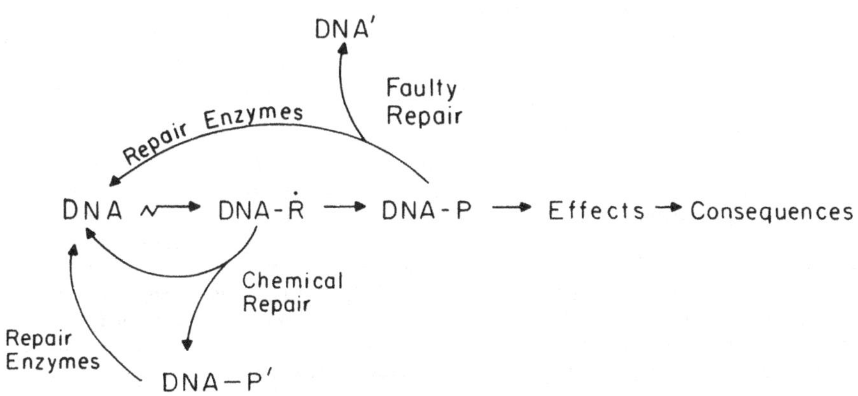

Scheme 1. The processes associated with the interaction of ionizing radiations with DNA in cells, where DNA-Ṙ stands for DNA radicals, and DNA-P for products resulting from these radicals.

Other free radical generating processes can be also represented by Scheme 1. For example, multiphotonic ionization[13] encountered in pulse-laser photobiology, 1 MHz ultrasound used in medical diagnostics,[14] endogenous Fe(II) + peroxides or hydroperoxide reaction (Haber-Weiss-Fenton reactions),[15] and various other chemical reactions induced by drugs, enzymes and chemicals lead to the formation of free radicals. The Haber-Weiss-Fenton reactions are shown on page four.

$$H_2O_2 + Fe(II) \rightarrow \cdot OH + OH^- + Fe(III) \quad (5)$$

$$ROOH + Fe(II) \rightarrow RO\cdot + OH^- + Fe(III) \quad (6)$$

These reactions can be recycled by ascorbate and other electron donors,

$$Fe(III) + AH^- \longrightarrow Fe(II) + A^{\bar{\cdot}} + H^+ \quad (7)$$

The alkoxy radicals, RO•, generated by reaction (6), react similarly to •OH. The RO• radicals are less reactive than •OH, however, and their damaging potential is not yet fully assessed.

PULSE RADIOLYSIS

In general, free radicals are highly reactive and disappear rapidly to give stable final products. Hence, free radicals are transient intermediates. Because of their transient nature and short lifetime, free radical processes must be studied with fast, time-resolved, physicochemical monitoring techniques.[10] These techniques, when coupled to pulsed high energy electron acccelerators (2 to 10 MeV), form a system called pulse radiolysis. In pulse radiolysis, the energy of electrons must be sufficiently high to penetrate the reaction cell (5 mm to 5 cm of water) so that the generated free radicals are distributed uniformly, and must be of short duration (1 ns to 1 μs) so that free radicals accumulate for study before they disappear. The rapid generation of free radicals in relatively high concentrations (10^{-7} to 10^{-4} M) distinguishes pulse radiolysis from other methods of free radical generation. For example, peroxy radicals generated during classical autoxidation processes never reach sufficient concentrations for direct monitoring. It would take days or even weeks at room temperature or moderately elevated temperatures (< 100°C) to produce peroxy radicals in measurable quantities.[16] However, a much faster decay rate than formation rate of these radicals prevents their accumulation. In contrast, peroxy radicals can be produced by pulse radiolysis in less than 1 μs, which is much faster than their rate of disappearance (~ 10 μs to 1 s).

During its 25 years of existence,[10] pulse radiolysis has contributed to major breakthroughs in the kinetic and mechanistic understanding of free radical processes in many fields of science. It must be recognized, however, that investigation of free radical transients alone does not always provide sufficient and/or conclusive information for complete understanding of a free radical process. Quantitative measurements of the products resulting from free radical reactions are not only important, but also are often necessary for reliable and unequivocal characterization of the free radical transients involved.

Pulse radiolysis cannot be applied directly to the study of free radical processes in biology because of its inherent limitations. It requires optically transparent systems for spectroscopic investigations of transients, and simplified model systems for unequivocal mechanistic considerations. Because no other satisfactory technique exists for direct monitoring of free radicals in living organisms, investigations of free radical processes in vivo must rely on the use of model systems and measurement of the products of free radical reactions at present.

SINGLE vs. MULTIPLE DAMAGE

In both model and cellular systems, differences exist between free radicals generated by ionizing radiation and those generated by other modes. The microscopic distribution of ions generated by radiation is not uniform, despite a homogeneous macroscopic appearance. Most ionization events are distinct, being separated from each other by about 500 nm. A fraction of the total ionization events appears in clusters ~ 1 to 2 nm in diameter, containing 2 to 4 ions. These clusters of relatively dense ionization events are called spurs.[17] The diameter of a spur is comparable to the thickness of DNA (2 mm) or the distance between DNA and DNA-bound protein. Therefore, two free radicals could be generated simultaneously on two opposite strands of DNA in close proximity to each other. At biologically relevant doses (< 1 kGy), the probability of having two closely spaced, distinct ionizations is very small. For that reason, the spur effect has been used to explain the mechanisms of DNA double strand breaks[18] and DNA-protein crosslinks,[19] which require two closely spaced radicals.

In model systems consisting of dilute aqueous solutions of DNA, a spur has virtually no effect on the solute. The relative lack of interaction between spurs and DNA in these systems is due to the low probability of a spur and DNA overlapping. Consequently, only water free radicals are formed in those spurs. These radicals then react with each other to recombine or to give molecular products (H_2 and H_2O_2).

The probability of having a distinct ionization of DNA in dilute aqueous solutions is also low (ionization fraction ≈ mass fraction).[11] Under these conditions, most solute damage is caused by water radicals generated by distinct ionization events in water (indirect action of radiation). Two distinct DNA radicals or a DNA radical and an unbound protein radical, separated from each other by ~ 500 nm, can diffuse towards each other, combine, and create a crosslink. In the presence of air, the distance between two oxygen molecules in aqueous solutions is roughly 20 nm (1 atm, 20°C). Because oxygen reacts very efficiently with most free radicals [$k(\cdot R + O_2) = 10^8 - 10^9\ M^{-1}\ s^{-1}$], and because the distance between dissolved oxygen and radicals is much less than the distance between radicals, the radicals will react predominantly with oxygen rather than with each other. Hence, in model systems the crosslinking of solute free radicals is prevented by oxygen.[19] The mechanisms of crosslinking in dilute aqueous model systems are also applicable to cellular conditions if the differences between spatial distribution and mobilities of free radicals are taken into consideration.

In the nucleus of a cell, the concentration of DNA is considerably higher than in model systems because DNA and proteins are tightly packed. As a result, the probability of an overlap between a spur and DNA, as well as a spur and areas of contact between DNA and proteins, is considerably higher. In a spur, when one radical is generated on DNA (either on a base or deoxyribose) and another radical is generated on protein (particularly on aromatic aminoacid residues), it is obvious that these two radicals would be much closer to each other (~ 1 nm) than to an oxygen molecule (~ 20 nm). Since the reactivity of these radicals with each other ($e.g., k(\dot{T} + \dot{Phe}) \sim 10^9\ M^{-1}\ s^{-1}$) is comparable to their reactivity with oxygen, the crosslinking in cells should not be critically affected by the presence of oxygen. It is not surprising, therefore, that irradiation of cells produces DNA-protein crosslinks both in the presence and in the absence of oxygen.[20]

Chemically generated free radicals in vivo should be evenly distributed and could be compared to the indirect action of radiation. However, if the chemical generator of free radicals is attached to DNA (e.g., copper[21]) and its activity recycles, these events could be compared to spurs. Multiply damaged sites can be also generated by polyfunctional chemicals such as psoralene,[22] which crosslinks two opposite DNA strands, and bifunctional alkylating agents.[5]

Free radicals in model systems and in cells eventually disappear. The radicals may react with each other by disproportionation, in which single molecular products are formed, or by combination, in which cross-links are formed. They can also be oxidized or reduced by endogenous and exogenous redox agents, depending on their nature. In the presence of oxygen, reaction of these radicals with oxygen becomes prominent, with the extent of the reaction depending on parameters such as relative reactivities, concentrations, and mobilities. For example,

$$\text{DNA-R} + O_2 \rightarrow \text{DNA-ROO}\bullet \rightarrow \text{DNA-P}_{ox} \qquad (8)$$

where DNA-P_{ox} represents a variety of oxidation products, some of them specific to autoxidation processes.

REPAIR

The free radical state in Scheme 1, DNA-$\dot{R}$, may be reversed back to normal DNA for some kinds of damage. This repair may be accomplished by various H atom and electron donors such as ascorbate, sulfhydryls, and antioxidants.[6] This kind of repair is called chemical repair, and is not discussed comprehensively in this book. More detailed information about chemical repair and repair agents may be obtained elsewhere.[6]

Once final products (designated as DNA-P in Scheme 1) are formed within DNA, they are amenable to enzymatic repair.[5] Because mechanisms of enzymatic repair of damaged DNA are described in Section 2 of this book, they are not discussed here. It is of interest, however, to speculate briefly on the possible interaction of chemical repair and enzymatic repair. Chemical repair of free radicals may lead to complete restitution of the original material by either endogenous or exogenous repair agents (radioprotectors). Chemical repair can also lead to formation of a variety of different products, DNA-P' (Scheme 1). These products may be amenable to a less faulty enzymatic repair than enzymatic repair of DNA-P, which would be reflected in decreased lethality and mutation in cellular systems. In view of these considerations, both enzymatic and chemical repair must be considered in the determination of the repair capacity of a cell.

Free radical scavenging and chemical repair of bio-molecules may also help to maintain the integrity of the enzymatic repair system. At present, very little is known about the damage, prevention of damage, and repair of the repair enzymes themselves. If distributed uniformly, reactive free radicals such as •OH have little chance to react with repair enzymes for two reasons. First, repair enzymes are only a small fraction of the total material in the nucleus. Second, •OH radicals react indiscriminantly with the bio-material, and disappear in the vicinity of their origin. The probability of a spur and a repair enzyme overlapping is also relatively low. Damage to repair enzymes can be expected only by the more selective free radicals[23] such as peroxy radicals, ROO•. These radicals are much less reactive than •OH, yet they react[24] with Trp, Tyr, Met — the active amino acids in many enzymes. It should be noted that the ubiquitous superoxide radical $O_2^{\bar{\bullet}}$, is much less reactive than ROO• radicals.[13] Reactions of either $O_2^{\bar{\bullet}}$ or ROO• with DNA have not been observed.

CONSEQUENCES

Effects of deleterious agents on DNA (Fig. 1), if not repaired by an error-free process, may lead to biological consequences (Scheme 1). Some possible consequences in cells are lethality (reproductive inactivation), mutation, activation of oncogenes (carcinogenic transformation), and senescense. In higher organisms, the consequences could be debilitating diseases, cancer, genetic diseases, and aging. It is of great interest to establish the chain of causality for damaging processes in Scheme 1, i.e., which particular radical (or damaging agents) is responsible for the formation of a particular product, and in turn, which particular product is responsible for a particular effect and its consequences. An understanding of the causality of these processes from the early events to the final consequences is of paramount importance to the advancement of preventive medicine.

It would be interesting to correlate damage to the repair system with its consequences. For example, such a damage may significantly increase when drastic environmental changes occur, which could increase physiological stress, and could result in an increased mutation rate. It is conceivable that an increased mutation rate due to damage to the repair system may be responsible for sudden evolutionary changes.

ACKNOWLEDGMENT

The many useful editorial suggestions of Ms. Karen A. Taylor and discussions with Dr. John F. Ward are greatly appreciated. The technical assistance of Mrs. Gloria Wiersma is kindly acknowledged. Partial support of ILSI is gratefully acknowledged.

REFERENCES

1. J. D. Balantine, "Pathology of Oxygen Toxicity," Academic Press, New York (1982).
2. C. Giese, "Living with our Sun's Ultraviolet Rays," Plenum Press, New York (1976).
3. D. J. Pizzarello, ed., "Radiation Biology," CRC Press, Boca Raton, Florida (1985) p. 312.
4. B. N. Ames, Dietary Carcinogens and Anticarcinogens, Science 221: 1256 (1983).
5. E. C. Friedberg, "DNA Repair," W. H. Freeman, New York (1985).
6. O. F. Nygaard and M. G. Simic, eds., "Radioprotectors and Anti-carcinogens," Academic Press, New York (1983).
7. A. C. Wilson, The Molecular Basis of Evolution, Scientific American 253: 164 (1985).
8. B. Halliwell and J.M.C. Gutteridge, "Free Radicals in Biology and Medicine," Clarendon Press, Oxford (1985).
9. A. R. Forrester, J. M. Hay, and R. H. Thomson, "Organic Chemistry of Stable Free Radicals," Academic Press, New York (1968).
10. M. Z. Hoffman, ed., "Radiation Chemistry: State-of-the-Art Symposium," J. Chem Ed. 58: 83 (1981).
11. M. Inokuti, Radiation Physics as a Basis of Radiation Chemistry and Biology, Appl. Atomic Coll. Phys. 4: 179 (1983).
12. Farhataziz and A. B. Ross, "Selected Specific Rates of Reactions of Transients from Water in Aqueous Solutions," NSRDS-NBS 59, Dept. of Commerce (1977).
13. R. V. Bensasson, E. J. Land, and T. G. Truscott, "Flash Photolysis and Pulse Radiolysis," Pergamon Press, New York (1983),
14. K. Makimo, M. M. Mossoba, and P. Riesz, Formation of •OH and •H in Aqueous Solution by Ultrasound Using Clinical Equipment, Radiat. Res. 96: 416 (1983).

15. W. Bors, M. Saran, and D. Tait, eds., "Oxygen Radicals in Chemistry and Biology," deGruyter, Berlin-New York (1984).
16. M. G. Simic and M. Karel, eds., "Autoxidation Processes in Food and Biology," Plenum Press, New York (1984).
17. M. Mozumder and J. L. Magee, Model of Tracks of Ionizing Radiation for Radical Reaction Mechanisms, Radiat. Res. 28:203 (1966)
18. J. F. Ward, Some Biochemical Consequences of the Spatial Distribution of Ionizing Radiation-Produced Free Radicals, Radiat. Res. 86: 185 (1981).
19. M. G. Simic and M. Dizdaroglu, Formation of Radiation-Induced Crosslinks between Thymine and Tyrosine: Possible Model for Crosslinking of DNA and Proteins by Ionizing Radiation, Biochem. 24: 233 (1985).
20. G. T. Bowden, M. Kasunic, and A. E. Cress, Thermal Enhancement of X-Ray Induced DNA Crosslinking, Radiat. Res. 89: 203 (1982). See also, review by N. L. Oleinick in this book.
21. G. Czapski, J. Arontovitch, A. Samuni, and M. Chevion, The Sensitization of the Toxicity of Superoxide and Vitamin C by Copper and Iron, p. 111, in "Oxy Radicals and Their Scavenger Systems," Vol. I, G. Cohen and R. A. Greenwald, eds., Elsevier Biomedical, New York (1983).
See also, J. F. Ward, W. F. Blakely, and E. I. Joner, Mammalian Cells Are Not Killed by DNA Single Strand Breaks Caused by OH from H_2O_2, Radiat. Res. 103: 383 (1985).
22. P. S. Song and K. J. Tapleg, Jr., Photochemistry and Photobiology of Psoralenes, Photochem. Photobiol. 29: 1177 (1979).
23. M. G. Simic and E.P.L. Hunter, Free Radicals and Antioxidants, p. 449, in "Radioprotectors and Anticarcinogens," O. F. Nygaard and M. G. Simic, eds., Academic Press, New York (1983).
24. C. A. Gee, K. J. Kittridge, and R. L. Willson, Peroxy Free Radicals, Enzymes, and Radiation Damage, Brit. J. Radiology 58: 251 (1985).

HISTORY OF RADIATION BIOLOGY FROM A PERSONAL POINT OF VIEW*

Alexander Hollaender

Council for Research Planning
in Biological Sciences, Inc.
at Associated Universities, Inc.
Washington, D.C. 20036

It is exactly 50 years ago, in 1934, that I was asked by Warren Weaver, then in charge of the Natural Sciences Department at the Rockefeller Foundation, to make a survey of the biological effects of radiation. To emphasize the quantitative aspects was quite natural for me, since I came from a background in physical chemistry into biology. I evaluated about 5,000 papers, not all in-depth, of course, but I gave each paper a definite rating as to its quantitative aspects from my personal point of view. This served, for the Foundation, as a basis for their support of radiation research. No other source of support for research of this type was available during the Depression.

The early 1930s were an exciting time for those who were interested in modern physics, and I hardly need to mention many of the important scientific developments by Niels Bohr or by Drs. Heisenberg, Pauli, Schroedinger, Dirac, Wigner, and many others, especially the Berkeley group who considerably influenced those of us who had come into biology from physics or physical chemistry. I had many interesting discussions with Warren Weaver on this most important development. He emphasized time and again that the biological effects of radiation would become very important, especially since physical and chemical approaches can be used extensively. It was Weaver who first coined the term "molecular biology." After the experience of showing that some radiation work published by a Russian investigator--an excellent histologist but a poor experimenter--could not be repeated, my own work concentrated on the quantitative effects of shorter ultraviolet on bacteria. We recognized in 1936 the recovery of bacteria after ultraviolet irradiation by incubation at 18°C, that was especially pronounced after the effectiveness of 2,650 Å wavelength that was mostly absorbed by nucleic acid. This led 10 years later to the recognition of photoreactivation by Kelner and by Dulbecco.

Coming into biology straight from physical chemistry and, in my case, molecular structure, the study of microorganisms, especially *Escherichia coli*, had a great appeal. The techniques for producing and handling

---

* Excerpted and expanded from an address given on occasion of the 1983 Enrico Fermi award presentation.

bacteria in large numbers of great uniformity and the methods for easy recognition were most attractive to the newcomers to biology. My guess is that this was also the appeal to W.W. Coblentz who, in 1924, published a classical paper: "A Radiometric Investigation of the Germicidal Action of Ultra-violet Radiation".[1] Coblentz's pioneering work in this area and in so many border fields between physics and biology had a most profound influence on many of us, and formed the basis for our friendship when I moved to Washington in 1937.

Another very important report, published in 1929, was by F.L. Gates: "A Study of the Bactericidal Action of Ultraviolet Light. I. The Reaction to Monochromatic Radiations".[2] This paper showed that the wavelength most effective in killing bacteria is around 2,600 Å, and is most absorbed by nucleic acids.

A study that developed a more quantitative approach was by Hollaender and Claus in 1936: "The Bactericidal Effect of Ultraviolet Radiation on Escherichia coli in Liquid Suspensions".[3] A similar study [4] on the effects of ultraviolet radiation on tobacco mosaic virus was published a few years later, showing that the wavelength most effective is 2,650 Å, the wavelength absorbed by nucleic acid. But there was also a smaller maximum found at around 2,800 Å, indicating that tobacco mosaic is a nuclear protein. This was before Stanley, who had crystallized tobacco mosaic, reported that nucleic acid was a major part of tobacco mosaic.

A good picture of the status of genetics, or better, its background, can be obtained by visualizing its problems in the 1930s. In September of each year, many of us who were very much interested in the field of genetics and its border sciences met in Woods Hole, Massachusetts to discuss the position of our field. These meetings were referred to as "gene conferences." Thomas Hunt Morgan was in residence in Woods Hole during the summer. Many geneticists came to these meetings, including L. Stadler, George Beadle, B.A. Bridges, M. Demerec, B.P. Kaufmann, C.L. Huskins, Curt Stern, A.H. Sturdevant, Barbara McClintock, T. Dobzhansky, Alma Howard, and others. The status of the different aspects of genetics was discussed but a major topic was always: what is the gene? Is it a protein or is it a nucleic acid? The dominant feeling was that it must be a protein. These meetings usually included a very pleasant "clam bake" (lobsters, clams, corn, etc., steamed in seaweed) on one of the islands close to Woods Hole.

Our discussions convinced me that one could possibly get evidence of the chemical character of the gene by determining the wavelength dependence of mutation production. The absorption of nucleic acid has a strong maximum of 2,600 Å, whereas proteins absorbed around 2,800 Å.

As soon as I joined the National Institutes of Health (NIH) in 1937, I started an extensive investigation. I had my own quartz monochrometer, thermopile, and galvanometer. In order to measure effectively, I had to irradiate organisms that had little or no absorption in the cell wall at the wavelengths we were interested in. I worked in cooperation with C.W. Emmons, a very well-known mycologist at the NIH. The organism used was Trichophyton mentagrophytes, an organism with colorless spores and with no known sexual stage. By then, I had developed a very simple radiation technique which gave me good quantitative results (monochromatic radiation, exposure in liquid suspension at control temperature, carefully stirred, and so forth).

The spores, as I mentioned, were colorless and had no known sexual stage. The colonies on agar plates were typical and permitted easy recognition of mutations by color, shape, structure of colony, specific aspects of hyphae, and a few other points which could be monitored for many generations. The percentage of mutations appearing on the plates could be increased under carefully controlled conditions in up to 20% of the surviving spores. These were easily recognized and gave quantitative results. We repeated the experiment four times and got identical results. Very similar results were obtained (and published in 1939) by Knapp and Schreiber working with Sphaerocarpus.[5] At the meeting of the American Association for the Advancement of Science in December of 1938 in Richmond, Virginia, we presented the results. The main person who questioned our results was L.J. Stadler who had just published a paper[6] with F.M. Uber, a physical chemist, showing that the wavelength most effective for mutation production in pollen grain of maize was closer to 2,800 Å, the wavelength most absorbed by proteins.

I explained to Stadler that Uber had missed taking into account the absorption of the yellow pigment in the wall of the pollen grains. Stadler went home, and he and Uber repeated the experiments, taking into account the yellow pigment. Their results checked with ours quite closely. Stadler and Uber did not publish their results until about 1942,[7] three years after our announcement.

At the International Genetics Congress in Edinburgh in 1939, I put up a demonstration showing the high effectiveness of 2650 Å on mutation production. Herman Muller brought Dr. Timofeeff-Ressovsky, another well-known geneticist, to show him our results. Timofeeff-Ressovsky said he didn't believe that the crucial compound in the gene could be a nucleic acid. It was thought that there was not enough information available from the nucleic acid to steer the function of the gene.

The scientific community, as I mentioned, at that time believed that the nucleic acid could not contain enough information to explain the function of the genes. I should not be too disappointed about this since a few years later Avery, of the Rockefeller Institute, and his co-workers removed the outside mucous of a strain of Pneumococcus and transferred it to another strain, changing the recipient to the strain which supplied the mucous. An analysis of the transferred material showed that this was mainly nucleic acid. Even this very striking result, which is more direct than the one we produced, did not convince others that the nucleic acid represents the genetic material. Not until the 1950s, when Watson and Crick came out with their finding of the double helix, was this popular misconception corrected.

Naturally, I was very interested in finding out what was in the nucleic acid that gave it the power to direct all genetic information, a concept of which I was already convinced at that time. A very clean, very viscous solution of DNA from Jessie Greenstein at the National Cancer Institute (NCI) was irradiated with ultraviolet (2537 Å); we showed that this suspension became quite depolymerized. Similar results were obtained with X-rays. This finding was published in cooperation with Greenstein and Taylor.[8]

In the late thirties and especially in the early forties, I approached many problems taking advantage of my experience in mutation production in dermatophytes. We produced mutations in Penicillium notatum for increased yield of Penicillin (some in cooperation with Demerec at Cold Spring Harbor). This approach was quite successful and was the source of strains of Penicillium mutations that led to the industrial production of Penicillin. A similar study[9] with Aspergillus terreus by Hollaender, Raper, and Coghill led to mutations with increased production of itaconic acid, compound very much needed as a basis for polymer production during the forties. Similar work[10,11] was done with Neurospora crassa in cooperation with Demerec and Sansome producing mutations with a variety of chemical growth requirements which formed the basis for the classical work of Beadle and Tatum.

Other studies, taking advantage of many aspects in specific fields, led to the inactivation of pin worm eggs (with Jones)[12] and the destruction of the jaundice factor in human serum (with Oliphant),[13] very much needed during the landing in Normandy. Another significant study[14] was on the effect of ultraviolet on the cornea of the mouse with Friedenwald of Johns Hopkins. Extensive work[15,16] in cooperation with Kaufmann on the effect of ultraviolet on Drosophila sperm was somewhat less successful.

Since I had been working intensively on the wavelengths from 1850 to 3100 Å, I thought I would like to try the very long ultraviolet, from 3100 up to 4000 Å, just below the visible region. Very interesting findings[17] were made in these higher parameters which were entirely different from the effects of the lower wavelengths. There was hardly any O_2 at 2537 Å. Oxygen is very important in the long ultraviolet effects, but not on the short ultraviolet spectrum. This is a classical paper which is often quoted.

In 1946, I was asked by the then Surgeon General, Dr. Thomas Perrin, to see what I could do in Oak Ridge to take advantage of the new developments in atomic energy. I had a meeting with Charles Thomas, the head of Monsanto, the operator of the Oak Ridge National Laboratory (ORNL). This was before 1948 when Union Carbide took over. I saw this as an opportunity to do pioneering work in an area in which I was very interested: the mechanism of radiation effects and the implications to man.

I was told that I would thereby lose my civil service status and all the privileges that go with it, and go on a payroll of an industrial company that had comparatively miserable retirement benefits. But I thought it was wise to take the opportunity to set up a laboratory of the type that I would like to see developed.

Dr. Eugene Wigner, the Research Director at ORNL at that time, asked me what I would like to do, and I told him that I would write a proposal for him. I knew definitely what I wanted to do. I wanted to emphasize genetics and nucleic acids, and the study of microorganisms and their genetic make-up, and also do some physiological and biochemical work. As a biophysicist, I have always been interested mostly in quantitative approaches. The outline that I prepared was accepted, and I am still very proud of it because it really showed that one could visualize (this was in 1946) what was coming up in the area of biology.

Clark Center, the manager for Carbide, helped me to get started. He had great respect for basic biological research. There were a great many others, too numerous to mention here. I had many discussions with Clarence Larson and Alvin Weinberg and others. The important thing was to establish quarters for the Laboratory. Center offered me 500,000 square feet of buildings which were originally designed to house facilities where uranium 235 was separated by chemical means. The membrane process at K25 turned out to be successful. There was no need for the chemical process for which these buildings were designed, and so the buildings were filled with very beautiful equipment which was, alas, of no use for biological work. We received adequate funds from the Atomic Energy Commission (AEC) to clean out the buildings and to establish laboratories. But, how could we bring good scientists to an isolated place like Oak Ridge? Good work was already being done in Oak Ridge during the war in regard to protecting workers against radiation damage. This shows that reactors can be handled in an intelligent and careful way if people are trained to understand what they are doing, if they are motivated.

How do you create basic research in an area where mostly applied work is being done with no tradition for basic studies? I first tried to bring in my old friends with whom I had been associated throughout the country, but this did not work because very few had the confidence to come down to Oak Ridge which, at that time, was a highly restricted area. The city was just putting in sidewalks and the town did not make a very modern impression. So, I decided to bring in young investigators. As a matter of fact, it was better than bringing in my old friends who were settled in their own thinking and whose areas of research were really not the directions in which I wanted our research to go. I had to judge people on the basis of my own impressions and I did not necessarily go by recommendations. Sometimes young scientists do not make a good impression in the first place they work. I built up their self-confidence and assured them that they could do good work and that the problems we were working on were really in the front line of modern biology.

The emphasis of the program was on genetics which I felt to be the keystone in modern biology. I immediately established a strong biochemistry department which was absolutely essential. I think classical neurobiology was recommended by our research director. However, I felt this was not an area ready for development in 1950. The future showed this to be one of the very important fields 20-30 years later. Consider that this was all planned and carried out in the late 1940s. I went to Oak Ridge in late '46, and by '48, we had a good, structured, and functioning laboratory.

Bureaucratic machinery can interfere as much as it helps. I did a lot of things that directors usually don't do, such as making facilities and equipment available to the individuals, and planning functions that an investigator usually has to do by himself. We helped our young researchers to write their papers, showing them how through our very able editorial office. The papers which came from the Biology Division of the ORNL, some editors have told me, were in beautiful and really perfect shape.

The fundamental work was on bacteria, plants (Maize, Tradescantia), Drosophila, paramecia, mice, and photosynthesis, and on the immunological, biochemical, and biophysical approaches which, at that time, were not very popular, though we had some outstanding biophysicists in our Laboratory (Dick Setlow and others).

It was urgent that we not only work on organisms that were convenient to our approach in biology, but we had to work on a mammal closer to humans, since human work was not possible to conduct in Oak Ridge. We started a mouse genetics study on the recommendations of Herman Muller and Sewall Wright. We were very lucky to get an outstanding geneticist (W.L. Russell) to head it up. It was obvious it would take about 10 years before significant results could come out of the genetics study.

We were also very interested in studying the somatic besides the genetic effects of ionizing radiation. The work was started by Henry Kohn but was soon taken over by Jacob Furth, joined by Arthur Upton. The long-term implications, including carcinogenesis, at acute as well as at low-level exposures were investigated. A good head start in this investigation was gained by taking on the mice surviving the Bikini test in the Pacific. The results were published in a monograph which is often quoted. The work continues under John Storer.

You cannot maintain the morale of a laboratory for 10 years without showing results. Many of the basic studies gave quick results. As a matter of fact, my own work for the first 2-3 years supplied the publications that were necessary to maintain the morale at the Laboratory.

Very extensive studies besides those on mouse genetics concentrated on Drosophila, Paramecium, Neurospora, E. coli, maize, yeast, and a number of other organisms. Whereas the mammalian genetics studies took time to bring to fruition, the Drosophila and microorganism studies gave us relatively quick results and made the Laboratory a mecca for scientists from all over the world to find out about the new approaches developed at the Laboratory.

It was quite remarkable how productive the Laboratory became. All this knowledge was necessary to understand the long-term effects of radiation on man.

The work was successful, as I have been informed by many who have followed the development of ORNL. The years from 1950 to 1966 were the "golden" years of the Biology Division at the ORNL. Over 2,000 publications came out during those years, including papers, abstracts, and books.

Another thing that helped the Laboratory to gain more impetus were the seminars in Gatlinburg which many investigators from abroad attended and which became very well known so that some were translated into Russian.

An effort was made to bring many of the schools in the Southeast into the flow of modern science. There were not many outstanding universities in the South in 1946. Of about 500 colleges in the South, many were weak in basic biological sciences. We thought we ought to do something to help update the curriculum.

We started a traveling lectureship, sending our scientists out to lecture at the local colleges and universities. We paid for their travel and expected the university to pay for their local maintenance; but, if they did not have money, we paid for this, too. This was a very successful program. We found that many of the instructors at these colleges did not have backgrounds in modern biology so we set up a program to bring 10-15 of them to the Biology Division each summer to join in our research projects for 10-12 weeks.

We paid their summer salaries. We soon found that this was not adequate because many young people who were anxious to get into the modern developments didn't have such opportunities. We set up a fellowship for promising, senior undergraduates in the Southeast. This turned out to be an extremely successful program because these students were very anxious to learn about new biology when they did not get the information from their own colleges.

It is most satisfying to work with enthusiastic young people in the laboratory so that they become a part of the developments and of the work that is going on. Again, we got only a little money to pay for this. Later, all of these programs were taken over by the Oak Ridge Institute for Nuclear Studies, a more conservative group that was less willing to take risks as I had. I could not have accomplished all this without the excellent support of the AEC in Washington, D.C.

This program was so successful that at one time we supplied the Rockefeller University with 20-30% of the graduate students from this program. In fact, this was a tremendous source of scientific talent and it could be tapped. Many of these young collegians among undergraduates would love to get involved, and I am afraid we are missing out on a great reservoir of talent by not giving them a chance to become active in research.

In 1961, at the request of the AEC, we organized a series of symposia in Latin America on prominent topics in modern biology, including radiation, genetics, and immunology. We published these proceedings in 15 books, and distributed them free to Latin American scientists. The Ford Foundation, through the International Division of the National Academy of Sciences/National Research Council, supported the Latin American symposia most generously.

This is a very important approach because in Latin America college budgets often do not have the dollars to buy numerous books. Many of these institutions are very isolated and these books help them to get an opening into the developments in modern biology. These symposia were so successful that Latin America is now clearly into the flow of modern biological development.

In the final analysis, the effects of radiation on biological materials are chemical ones. We became interested in the effects of chemicals in our environment. As a matter of fact, this has become the predominant topic of mine for the last 20 years. We have started a number of very interesting efforts to bring people from developing countries from all over the world to training courses that were first set up in Galveston, Texas (Marvin Legator) and often with the help of the United Nations Environmental Programme (UNEP). We gave our visitors the opportunity to get acquainted with specific scientific problems and we took them around to visit different agency laboratories in North Carolina as well as the Food and Drug Administration, the NCI, and Litton Bionetics. We gave them a dinner at the Cosmos Club where prominent scientists spoke to them, including Dr. Noel Brown from UNEP and researchers from NCI. At my office, each visitor was asked to write down what they would like to do with their training. Some of them had very definite ideas and we have helped them to set up their own laboratories.

These laboratories, set up in 11 international locations, have been developed to train people to know more about our chemical environment, especially the use of pesticides and how to protect exposed individuals against the damaging effects of these chemicals. We have had several workshops and symposia on these topics and we have also brought out several books on the subject, authored by researchers from these developing countries. Our Chinese contact is now translating one of the booklets into the Chinese language.

We continue to work cooperatively with our colleagues in developing countries, as a work for peace because this is one way in which we can set up islands of understanding when political propaganda and excitement may obscure the positive activities of the United States' scientists in helping our colleagues in developing countries. We not only help to set up laboratories throughout the world, but we continue to stay in contact with them, sending material such as books, pamphlets, etc., so that they know that we are interested. They then report to us about what they are doing. All of this has been done without the support of the Agency for International Development which, unfortunately, considers our work to be too slow. But, we have no quicker way of training people and showing them how important this area is to solving future problems. We have excellent cooperation in this program with UNEP.

I would like to express my appreciation for the opportunity to develop my ideas; especially, I should mention the people who helped me the most. In Wisconsin, these colleagues were Farrington Daniels, John Warren Williams, C.E. Mendenhall, and B.M. Duggar. In Washington, it has been the National Institutes of Health's Paul Neal, Chester Emmons and Drs. Brackett and Oliphant and quite a few others. In Tennessee, it was Clark Center, the Carbide manager with great respect for basic biology; Clarence Larson, when he was Director of the ORNL; and Alvin Weinberg, in charge of research at ORNL; also, a few members of the staff of the University of Tennessee, Knoxville: Hilton Smith, Arthur Brown, Gordon Carlson, and Alan Solomon.

Of course, special thanks go to the late Shields Warren, the first Chairman of the Biology and Medicine Division of the AEC. He first questioned some of our genetic approaches, but had great respect for scholarship and encouraged me to go ahead. Chuck Dunham, who succeeded him, was also most helpful. I am grateful for the cooperation of Charlie Edington, Bob Rabson, Jim Liverman (when he was in the Washington office), the staff of Associated Universities, Inc. and the Brookhaven National Laboratory, and especially, the many scientists who made the Oak Ridge Laboratory successful.

Since I was not the easiest individual to handle, I should express my thanks to the late Lewis Strauss, a Wall Street banker and member of the AEC, who had sympathy for biological work and who was willing to take a chance for me to take on long-term projects. He was a close friend of President Eisenhower, and of Larson and Gerald Tape who were on the Commission. Mary Bunting was most helpful when we started on the biomedical graduate school and helped me to get young scientists to work in the laboratory. Many thanks should go to Mrs. Henrietta Hollaender who was willing to put up with my endless hours at the Laboratory and working every Saturday and Sunday afternoon, and having at least two parties per week for our staff and visitors. Also, special thanks should go to Stan Carson who, as Associate Director of the Biology Division, was most helpful.

I had a most outstanding administrator in Tennessee, Virginia White, and now have an excellent one in Washington, Claire Wilson. I would especially like to thank the people in the Laboratory who came to us as young PhDs or MDs who were willing to listen to me. We were able to help them to do innovative work and to build up their self-confidence and, thus, their productivity. The best proof that this approach works has been shown by the people who passed through the Laboratory when I was the Director from 1946-66: 17 former or present members of the Biology Division have thus far been elected to the National Academy of Sciences.

What I appreciate the most, in addition to my success in the United States, is my contact with developing countries. There are 2 or 3 ways one can work for peace. One way is to sign petitions to try to influence legislators or to make speeches; but there is another, the way I have practiced in getting our colleagues in developing countries to become involved in the flow of modern science, by showing them that they can do just as well as anybody else. There is no monopoly on intelligence anywhere in the world, it is just the ability to concentrate on the work and to devote time to complicated problems. I am very proud that I have been working so closely with so many people in developing countries, besides our colleagues here in the States. The profound influence of the work on the biological effects of radiation, on the development of molecular biology, genetic engineering, biotechnology, and all the new approaches in biology can best be measured by the extensive quotation of our work and what we have done since leaving Oak Ridge.

It is this combination of profound interest in basic biological mechanisms, the physics and chemistry of life, and the desire for the application of this acquired knowledge to practical problems that made it possible to develop the Biology Division in Oak Ridge.

REFERENCES

1. W.W. Coblentz and H.R. Fulton (1924) Natl. Bur. Standards U.S., Sci. Technol. Papers (495) 19:641-680.
2. F.L. Gates (1929) J. Gen. Physiol. 13:231-248.
3. A. Hollaender and W.D. Claus (1936) J. Gen. Physiol. 19:753-765.
4. A. Hollaender and B.M. Duggar (1936) Irradiation of plant viruses and of microorganisms with monochromatic light. III. Resistance of the virus of typical tobacco mosaic and *Escherichia coli* to radiation from λ3000 to λ2250 A. Proc. Natl. Acad. Sci., USA 22:19-24.
5. Knapp, E., and H. Schreiber (1939) Quantitative analyse der mutationsauslösenden wirkung monochromatischen U-V-lichtes in spermatozoiden von *Sphaerocarpus*. In Proc. 7th Intl. Cong. Genetics, R.C. Punnett, ed. Cambridge University Press, Cambridge, pp. 175-176.
6. Stadler, L.J., and F.M. Uber (1938) Preliminary data on genetic effects of monochromatic ultraviolet radiation in maize. Genetics 23:171.
7. Stadler, L.J., and F.M. Uber (1942) Genetic effects of ultraviolet radiation in maize. IV. Comparison of monochromatic radiation. Genetics 27:84-118.
8. Taylor, Babette, Jesse P. Greenstein, and Alexander Hollaender (1948) Effects of X-radiation on sodium thymus nucleate. Arch. of Biochem. 16(1):19-31.
9 A. Hollaender, K.B. Raper, and R.D. Coghill (1945) The production and characterization of ultraviolet-induced mutations in *Aspergillus terreus*. I. Production of the mutations. Am. J. Botany 32:160-165)
10. Sansome, Eva R., M. Demerec, and Alexander Hollaender (1945) Quantitative irradiation experiments with *Neurospora crassa*. I. Experiments with X rays. Am. J. Bot. 32:218-226.
11. Hollaender, Alexander, Eva R. Sansome, E. Zimmer, and M. Demerec (1945) Quantitative irradiation experiments with *Neurospora crassa*. II. Ultraviolet irradiation. Am. J. Bot. 32:226-235.
12. Jones, Myrna F., and Alexander Hollaender (1944) Effect of long ultraviolet and near visible radiation on the eggs of the nematodes *Enterobius vermicularis* and *Ascaris lumbricoides*. J. Parasitology 30:26-33.
13. Oliphant, John W., and Alexander Hollaender (1946) Homologous serum jaundice. Experimental inactivation of etiologic agent in serum by ultraviolet irradiation. Public Health Reports 61(17):598-602 (Reprint No. 2707).
14. Friedenwald, Jonas, Wilhelm Buschke, Jane Crowell, and Alexander Hollaender (1948) Effects of ultraviolet radiation on the corneal epithelium. II. Exposure to monochromatic radiation. J. Cell. Comp. Physiol. 32(2):161-174.
15. Kaufmann, Berwind P., Alexander Hollaender, and Helen Gay (1946) Modification of the frequency of chromosomal rearrangements induced by X rays in Drosophila. I. Use of near infrared radiation. Genetics 31:349-367.
16. Kaufmann, Berwind P., and Alexander Hollaender (1946) Modification of the frequency of chromosomal rearrangements induced by X rays in Drosophila. II. Use of ultraviolet radiation. Genetics 31:368-376.
17. Hollaender, Alexander (1943) Effect of long ultraviolet and short visible radiation (3500 to 4900 Å) on *Escherichia coli*. J. Bact. 46:531-541.

COMPARISON OF MECHANISMS FOR DNA STRAND BREAK FORMATION BY THE DIRECT AND INDIRECT EFFECT OF RADIATION

Dietrich Schulte-Frohlinde

Max-Planck-Institut fur Strahlenchemie
D-4330 Mulheim a.d. Ruhr, W. Germany

INTRODUCTION

Irradiation of cells may lead to mutations, reproductive cell death and the disappearance of some or all cell activities. These effects, especially reproductive cell death, are believed to be the result of damage to DNA. Two kinds of formation of DNA damage are often distinguished, the so called "direct" and the "indirect" effect of irradiation.[1,2] The direct effect is due to ionization or electronic excitation of the DNA, and the indirect effect is caused by reactive species, in most cases free radicals, which are produced in the vicinity of the DNA. These radicals may be primary radicals produced by energy absorption in water (the solvated electron, the H-atom and the OH radical) or organic radicals produced from organic material other than DNA either by interaction with radiation or by reaction with the primary radicals generated from water.

DOUBLE STRAND BREAK FORMATION AND REPRODUCTIVE CELL DEATH

For E. coli and for yeast cells it has been found that the ratio of double-strand breaks (dsb) in the DNA irradiated in the cell and the number of cell deactivations (reproductive death) is approximately unity.[3,5] From this it follows that at one stage of the events leading to reproductive death the lethal damage is due to double-strand breaks. The so-called "potentially lethal damage" (PLD) consists of these double-strand breaks. The unrepaired dsb leads to cell death but a large number of these dsb's can be repaired.[4,6-9] It is noteworthy that a large number of these dsb's are not produced by radiation-induced free-radical reactions, but are the result of enzymatic (endonuclease) activity at DNA damage sites other than dsb's.[3,10-12] This kind of damage is often named ELS (enzymatic labile sites). The nature of the ELS in DNA consists of changes in the chemical structure of DNA, such as base or sugar loss or alterations, crosslinks within DNA, binding to proteins and addition products from organic and water radicals to the bases. It is not known in detail which of these chemical lesions are the most injurious and which of them are more easily repaired than others. Lethal damage (a dsb) must have its origin in at least two chemically altered sites in the DNA located at opposite strands. The distance between the damage sites of the two strands may have an influence on the rate and success of enzymatic repair. The same consideration applies to the effect of two or more dsb's in the DNA of mammalian cells, where dsb repair is possible. In addition it has been suggested that the accumulation of a larger number of damaged sites within a certain small length of DNA (in the range of a few base pairs, the so-called S 1 nuclease active sites) may be particularly dangerous for the cell.[13,14] For mammalian cells, the success or failure of recombinational repair may be responsible for survival or cell death via the formation of chromosomal aberrations, respectively.[15,16]

A NOVEL SITUATION - A CHALLENGE

On the basis of the results summarized above and from the observation that a much higher dose is necessary in order to achieve the same yield of strand breaks in aqueous solution than in a cell[3], it follows that from a chemical point of view a completely new situation has arisen. It is no longer strand break formation which is the most important radiation-induced lesion in DNA. Now every kind of lesion in DNA is potentially dangerous, mainly because many of the lesions will be converted enzymatically into strand-breaks. If two ssb's are formed in sufficient proximity at opposite strands, or at larger distances if a sufficiently large piece of DNA is cut out by enzymatic action, a dsb will result.

The new situation - that every type of chemical damage in DNA may be potentially dangerous for a cell - raises the question of the chemical nature of all radiation-induced lesions in DNA, and their relative responsiveness to repair.

In order to learn more about the nature of radiation-induced chemical alterations in DNA and of the pathways for their formation, it would be helpful to distinguish the direct from the indirect formation of DNA damage.

CONTRIBUTION OF THE DIRECT AND THE INDIRECT EFFECT TO DNA DEACTIVATION

The question of the contribution of the direct and the indirect effect to DNA deactivation has often been examined in the literature. The results can be discussed as follows: one experimental piece of evidence in support of the indirect effect for ssb formation in the DNA of a cell was based on scavenging OH radicals with high concentrations, e.g., of alcohols (up to moles per liter).[17] The protection effect was found to be roughly a linear function of the rate constants of the reaction of the OH radical with the scavenger. Although the conclusion that OH radicals are responsible seems convincing, there are two difficulties in accepting it. First, a cell containing 10% of an alcohol may have a different repair capacity than a cell without this heavy load. The second difficulty is the result outlined in the present work, i.e. the fact that the direct effect also is quenched by alcohols at high concentrations. This leads to the conclusion that the scavenging experiments with alcohols do not provide conclusive evidence for an indirect effect via OH radicals.

Other experiments with cells use N_2O as a scavenger. However, it has been shown[18] that irradiation of a solution containing N_2O, NaCl, O_2, and phosphate produces compounds which are toxic for bacterial cells. Since in experiments with cells in the presence of N_2O these ingredients can hardly be avoided, it is doubtful whether results obtained with N_2O are convincing. In mammalian cells the addition of N_2O has practically no effect on cell survival.[18] This result does not support the participation of OH radicals in the killing of mammalian cells since N_2O converts solvated electrons into OH radicals. The yield per dose (G value) of the solvated electron is nearly the same as that for OH radicals; therefore the effect of radiation should be doubled in the presence of N_2O if only the OH radical were effective.

Thiols as scavengers are not useful for the determination of the indirect effect because in the presence of oxygen they influence the repair systems[3] and in the absence of oxygen they repair DNA damage, possibly by H donation to DNA free radicals.[19] From the scavenger experiments carried out so far, it would seem that no unequivocal answer concerning a contribution of the direct and indirect effects can be obtained.

Michaels and Hunt[20] have made an attempt to calculate a possible contribution of the indirect effect using known rate constants of OH radical reactions with cell compounds, assuming a homogeneous distribution of the organic material in the cell. The result was that an indirect effect of about 55% due to OH radicals has to be expected. However, in the calculation the presence of thiols in the cells was not taken into account, which would reduce the calculated percentage by 10%. Furthermore, the distribution of the organic material in the cell is not known exactly. In particular, the average amount of protein that is attached to the DNA in addition to the histones permanently associated with the DNA in mammalian cells is not known.

On the other hand, the arguments for a direct effect are equally uncertain. One argument put forward occasionally for an overwhelming contribution of the direct effect is the first order dependence of the logarithm of survival on dose if repair is totally or partly absent. This argument is based on the assumption that a single hit must be responsible for damage on opposite strand, and this may only occur by direct energy absorption in the DNA. However, this argument is not convincing since energy absorption even of 60-Co-gamma rays, occurs in spurs, the average shape of which may be a sphere of ≈ 2 nm diameter with an energy content of 30-120 eV. An average spur may contain 1-4 ion pairs, i.e., 2-8 radicals.[21] The observed first order dependence of DNA damage in cells on dose only allows the conclusion that there is no overlapping of spurs. In other words, no radicals from the spurs of one track contribute to the DNA damage produced by radicals from the spurs of another track (that is from another radiation particle). The damage at the two opposite strands is therefore produced by the various excitations and ionizations from one spur, which will include direct as well as indirect effects, and no determination of their contributions can be made.

Blok and his group,[22] and later Loman and coworkers,[23] have shown that deactivation of biologically active DNA in aqueous solution can be affected by organic radicals. Of particular interest was the result that radicals derived from amino acids can damage DNA. Assuming that this kind of damage can occur in cells and assuming a high concentration of organic material around the DNA in a cell (for the DNA in E.coli cells it has been postulated that 80% of the DNA is covered by proteins),[24] there is a certain probability that a large part of the damage to DNA may occur via secondary organic radicals, in addition to damage by the direct effect. The reaction pathway may also include an interaction of DNA radicals with organic radicals from the surroundings.

A quantitative estimate of the contribution of the direct and the indirect effect is difficult under these circumstances. In addition, it has to be taken into account that DNA contains water molecules which cannot be replaced easily by organic molecules since about 15% of the total associated water[25] is required to maintain the double-helical structure of the DNA. On the other hand, it is not known in which manner this water contributes to the indirect effect. It may well be that ionization of this "structural" water produces OH radicals which react with DNA with high efficiency due to close proximity to the DNA. However, Bakale and Gregg[26] have suggested that the "structural" water (primary layer) and some water attached to DNA in a secondary layer may behave like ice, in which fewer OH radicals are formed but more positive holes and electrons are produced, which diffuse to the DNA and cause damage.

The above discussion shows that the question of the contribution of the direct and the indirect effect to DNA damage is still unresolved. In the following a new appoach will be outlined. The question is whether the formation of strand breaks via the direct effect can be studied in a defined manner and what would be the result of a comparison of the formation of strand breaks via the direct and via the indirect effect. The formation of strand breaks of poly(U) is used as a probe, since ssb's can be measured easily.

MECHANISM OF THE RADICAL-INDUCED STRAND BREAK FORMATION OF POLY(U)

The details of the radical-induced ssb formation have been presented previously. The main features are the following. The H radical adds exclusively at the C=C bond of the uracil moiety, forming mainly the 6-yl radical.[27] The OH radical adds to an extent of more than 80% at the C=C bond of the uracil moiety, leading to 6-yl radicals (80%) and 5-yl radicals (20%) (reaction 1).[28] The rest of the OH radicals abstract H atoms from the various positions of the sugar moiety.

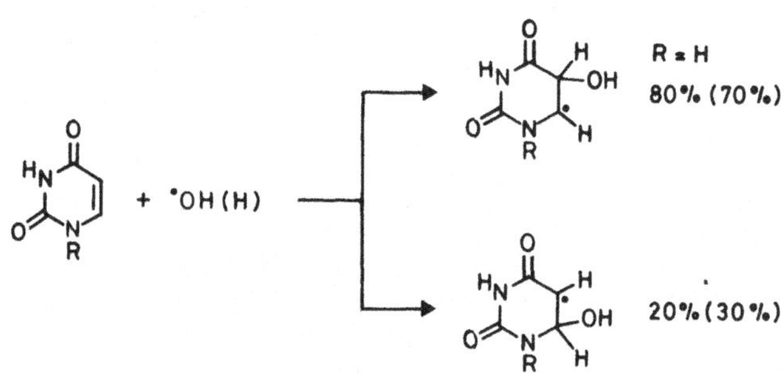

(1)

Since the efficiency of ssb formation is 20% for H atoms,[29] some of the base moiety radicals have to abstract an H atom from the sugar. If they abstract an H atom from position C-2' or C 4' (reaction 2), then ssb formation occurs via the heterolytic splitting of the phosphoric acid ester bond in the β position (reaction 3).[30-34]

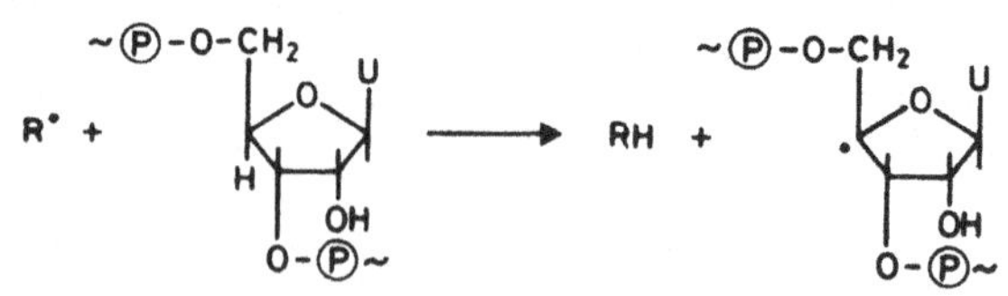

(2)

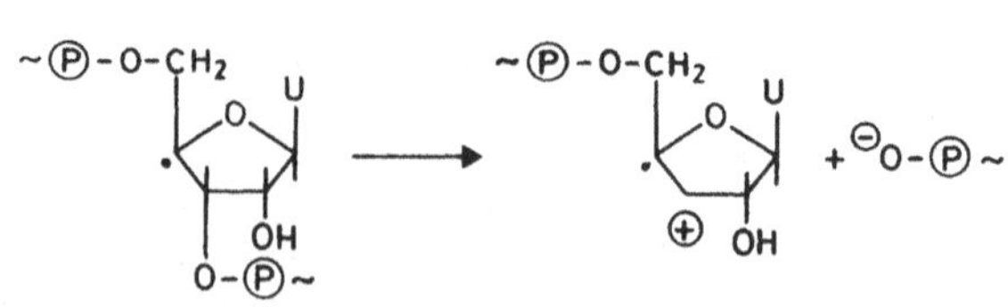

(3)

The OH radical induces ssb's principally by the same mechanism, with an efficiency of 40%.[29] Whereas for the H atom the rate of the H abstraction from the sugar is rate determining,[35] in the case of the OH radical the rate of the heterolytic cleavage of the phosphate bond is rate determining.[33] The kinetic data for OH radical induced ssb formation are presented in Table 1.[36]

Table 1. Kinetic data for ssb formation in poly(U). Pulse radiolysis data were obtained with N_2O-saturated oxygen free solutions.[32] The reacting radical is the OH radical. The photoionization data[36] were obtained with argon-saturated solutions with a pulsed laser, λ = 248 nm. The contribution of the solvated electron to strand break formation is small.

	photoionization	pulse radiolysis
kinetics	first order	first order
k $[s^{-1}]$	1.0 ± 0.2	1.0
E_A [kJ mol^{-1}]	66 ± 6	57 ± 6
A $[s^{-1}]$	5×10^{11}	1×10^{10}
pH dependence	very similar	
k $[M^{-1}s^{-1}]$ with DTT	1.2×10^6	1.7×10^6
k $[M^{-1}s^{-1}]$ with GSH	0.16×10^6	

STRAND BREAK FORMATION INDUCED BY LASER EXCITATION

Poly(U) is a single-stranded polymer. Single-strand break (ssb) formation in this polymer is used as one indicator of the radiation effects. Laser-induced ssb formation in poly(U) occurs biphotonically, which follows from the square root dependence of the ssb yield on laser beam intensity. Since ssb formation is only observed if the solvated electron is generated, it is concluded that the primary step for ssb formation is ionization, i.e., ejection of an electron from the uracil moiety of poly(U) (reaction 4).

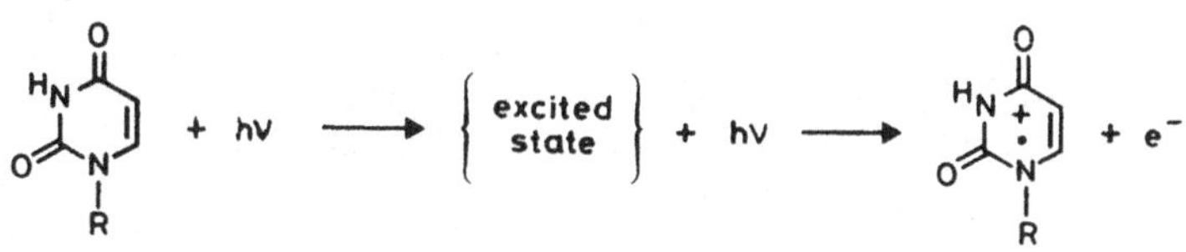

(4)

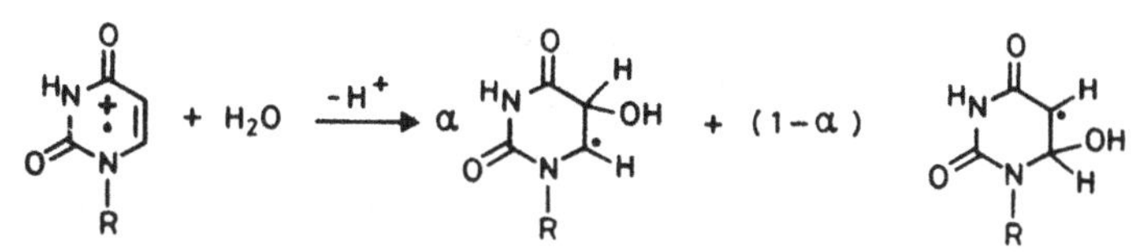

(5)

The kinetic data for the rate determining step of ssb formation induced by laser excitation is summarized in Table 1.

The pH dependence of the rate constant for ssb formation in poly(U) is shown in Figure 1.

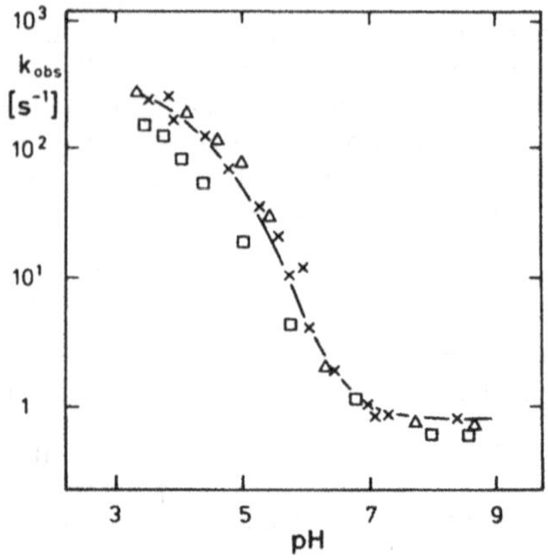

Figure 1. Rate constants for ssb formation of poly(U) measured with pulse radiolysis (OH radicals are the reactive species)[32] and with laser pulses (photoionization) as a function of pH.[36] x: 40 mg poly(U) = 1.6 x 10^{-4} M, laser results; Δ: 12 mg dm^{-3}, and □: 60 mg dm^{-3} poly(U), pulse radiolysis results, oxygen-free.

Surprisingly, the data for five different kinetic parameters are identical with the results obtained for ssb formation with the OH radical. The explanation advanced is that the uracil radical cation of the uracil moiety adds water at the C=C double bond of the uracil moiety, with release of a proton and a radical product distribution similar to that found by addition of the OH radical (reaction 5). This water addition reaction is probably very fast, since proton elimination from the NH-group does not compete. After that stage, ssb formation occurs via the same pathway as that induced by the OH radicals.

This result shows, as proposed earlier,[33] that ssb pathways via the indirect and the direct effect funnel into the same chemical channel. The importance of the difference between the direct and the indirect effect has been overestimated in the past.

Since poly(U) contains only a pyrimidine as a base but no purines, it is of interest whether or not the same pathway for OH-induced or ionization-induced ssb formation can be found for poly(A). Preliminary results show that the yields with poly(A) are much smaller than with poly(U) and that it is difficult to make a comparison, since for poly(A) no mechanism for strand break formation has been published up to now. Probably the comparison of an ionized adenine moiety with that of an adenine moiety which has reacted with an OH radical has to be made on the basis of another method. Nevertheless, it is of interest that ssDNA behaves very similarly to poly(A) under pulsed laser excitation. The mechanism remains to be elucidated.

EFFECT OF SCAVENGERS ON LASER-INDUCED SSB FORMATION IN POLY(U)

The laser-induced formation of ssb's can be inhibited by reaction with a thiol. With dithiothreitol (DTT), the observed reaction rate constant is 1.2 x 10^6 $M^{-1}s^{-1}$, which is the same value as the one obtained for ssb induction with OH radicals (Table 1). The chemical reaction underlying this repair process was identified as a H donation from the RSH to the C-4' radical of poly(U) (reaction 6).[36]

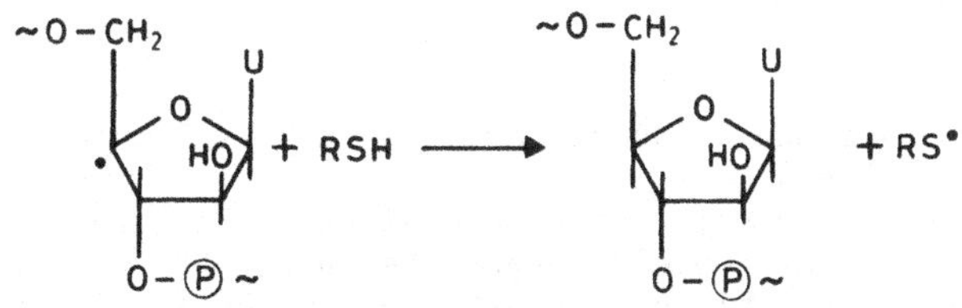

(6)

A decrease in laser-induced ssb formation, however, can also be obtained with high concentrations of alcohols, e.g. ethanol or t-butanol, Figure 2. The half-value-concentration of this decrease ($t_{1/2}$ = 10^{-2} M with isopropanol at pH 6.5) is somewhat smaller than that found for the formation of ssb's in the DNA of cells. In Table 2 a comparison is shown of values obtained for ssb inhibition in the DNA of cells,[17] and in poly(U) in aqueous solutions, by alcohols. Roots and Okada's values[17] were obtained in the presence of air, which was absent in the laser experiments. Results in the presence of oxygen (not shown) indicate an even stronger protection effect of alcohols for ssb formation in poly(U) with laser excitation. The results show that, under direct effect conditions, protection from ssb formation can be obtained by alcohols. The observed protection effect of alcohols for ssb's in the DNA of cells may therefore not be the result of OH scavenging, but could be the result of a quenching of the direct effect. Further experiments are needed to clarify this question.

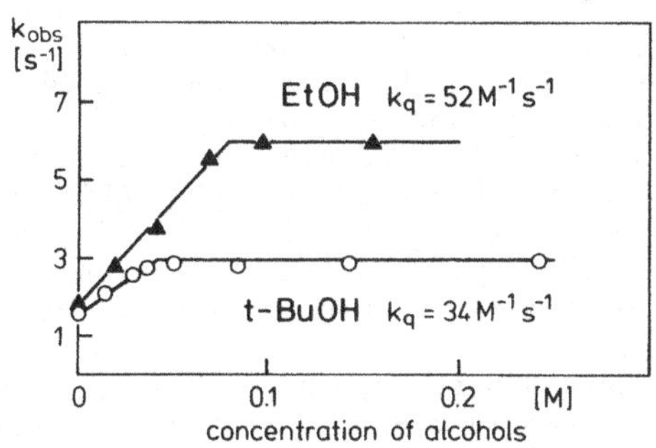

Figure 2. Protection from strand break formation in poly(U) excited by a laser pulse (λ = 248 nm) in the presence of ethanol and tert.-butanol as scavengers in aqueous solution in the absence of oxygen (Opitz and Schulte-Frohlinde, unpublished). The reaction rate constant, k_q, is that for the reaction of the alcohol preventing ssb formation.

Table 2. Protection from ssb formation in DNA and poly(U) by alcohols. $C_{1/2}$, half value concentration for protection; k_q, rate constant for the reaction of an alcohol with a poly(U) intermediate leading to protection against sb formation.[17,36]

	DNA in cells γ-rays	poly(U) in aqueous solution laser excitation	
	$C_{1/2}$	$C_{1/2}$	k_q
Substance	[M]	[M]	$[M^{-1}s^{-1}]$
t-butanol	2.5	0.05	34
ethanol	0.33	0.035	52
2-propanol	-	0.0032	313
	Roots and Okada, 1972	Opitz and Schulte-Frohlinde, 1985	

REFERENCES

1. H. Dertinger and H. Jung, "Molekulare Strahlenbiologie", Springer, Berlin (1969).
2. J. Kiefer, "Biologische Strahlenwirkung", Springer, Berlin (1981).
3. J.W. Hülsewede, Doppelstrangbrüche in der DNA in vivo und vitro nach 60-Co-γ-Bestrahlung, Dissertation Ruhr-Universität Bochum (1985).
4. D. Frankenberg, M. Frankenberg-Schwager, D. Blöcher, and R. Harbich, Evidence for DNA double-strand breaks as the critical lesions in yeast cells irradiated with sparsely or densely ionizing radiation under oxic or anoxic conditions, Radiat. Res. 88:524 (1981).
5. H.S. Kaplan, DNA-strand scission and loss of viability after irradiation of normal and sensitized bacterial cells, Proc. Natl. Acad. Sci. USA 55:1442 (1966).
6. D. Blöcher and W. Pohlit, DNA double strand breaks in Ehrlich ascites tumour cells at low doses of X-rays. II. Can cell death be attributed to double strand breaks?, Int. J. Radiat. Biol. 42:329 (1982).
7. D. Frankenberg, Interpretation of the dose and LET dependence of RBE values for lethal lesions in yeast cells, Radiat. Res. 97:329 (1984).
8. D. Frankenberg, M. Frankenberg-Schwager, and R. Harbich, Split-dose recovery is due to the repair of DNA double-strand breaks, Int. J. Radiat. Biol. 46:541 (1984).
9. H.P. Leenhouts and K.H. Chadwick, The crucial role of DNA double-strand breaks in cellular radiobiological effects, Adv. Radiat. Biol. 7:56 (1978).
10. S.E. Bresler, L.A. Noskin, N.A. Kuzovleva and I. G. Noskina, The nature of the damage to Escherichia coli DNA induced by gamma-irradiation. Int. J. Radiat. Biol. 36:289 (1979).
11. S.E. Bresler, L.A. Noskin, and A.V. Suslov, Induction by gamma irradiation of double-strand breaks of Escherichia coli chromosomes and their role in cell lethality, Biophys. J. 45:749 (1984).
12. G. Ahnström and P.E. Bryant, DNA double-strand breaks generated by the repair of X-ray damage in Chinese hamster cells, Int. J. Radiat. Biol. 41:671 (1982).
13. J. Andrews, H. Martin Bertram, and U. Hagen, S1 nuclease-sensitive sites in yeast DNA: an assay for radiation-induced base damage Int. J. Radiat. Biol. 45:497 (1984).
14. B. Junker, H. Martin-Bertram, and U. Hagen, Distribution of S1 nuclease sensitive sites and double-strand breaks in irradiated -DNA, Int. J. Radiat. Biol. 46:675 (1984).
15. R.P. Virsik and D. Harder, Numerical relationship between cells with radiation-induced chromosome aberrations and cells lethally injured by radiation, Radiat. Environ. Biophys., 18:73 (1980).
16. R.P. Virsik and D. Harder, Analysis of radiation-induced acentric fragments in human G_0 lymphocytes, Radiat. Environ. Biophys., 19:29 (1981).
17. R. Roots and S. Okada, Protection of DNA molecules of cultured mammalian cells from radiation-induced single-strand scissions by various alcohols and SH compounds, Int. J. Radiat. Biol. 21:329 (1972).
18. T. Brustad and E. Wold, Long-lived species in irradiated N_2O-flushed saline phosphate buffer, with toxic effect upon E. coli K-12, Radiat. Res. 66:215 (1976).
19. G.E. Adams, Radiation chemical mechanisms in radiation biology, Adv. Radiat. Chem. 3:125 (1972).
20. H.B. Michaels and J.W. Hunt, A model for radiation damage in cells by direct effect and by indirect effect: a radiation chemistry approach, Radiat. Res., 74:23 (1978).
21. A. Henglein, W. Schnabel, and J. Wendenburg, "Einfuhrüng in die Strahlenchemie", Verlag Chemie, Weinheim (1969).

22. J. de Jong, H. Loman, and J. Blok, Inactivation of biologically-active DNA by radiation-induced phenylalanine radicals, Int. J. Radiat. Biol. 22:11 (1972).
23. F.J. Nabben, H.A. van der Stroom, and H. Loman, Inactivation of biologically active DNA by isopropanol and formate radicals, Int. J. Radiat. Biol. 43:495 (1983).
24. E.D.P. DeRobertis and E.M.F. DeRobertis, "Cell and Molecular Biology", Saunders College, Philadelphia (1980).
25. G. Ebert, "Biopolymere", Steinkopf, Darmstadt, p. 14 (1980).
26. G. Bakale and E.C. Gregg, Conjecture on the role of dry charges in radiosensitization. Br. J. Cancer 37 Suppl. III:24 (1978).
27. S. Das, D.J. Deeble, and C. von Sonntag, Site of H atom attack on uracil and its derivatives in aqueous solution, Z. Naturforsch. 40c:292 (1985).
28. S. Fujita and S. Steenken, Pattern of OH radical addition to uracil and methyl- and carboxyl substituted uracils. Electron transfer of OH adducts with N,N,N',N'-tetramethyl-p-phenylenediamine and tetranitromethane, J. Am. Chem. Soc. 103:2540 (1981).
29. D.G.E. Lemaire, E. Bothe, and D. Schulte-Frohlinde, Yields of radiation-induced main chain scission of poly U in aqueous solution: strand break formation via base radicals, Int. J. Radiat. Biol. 45:351 (1984).
30. M. Dizdaroglu, C. von Sonntag, and D. Schulte-Frohlinde, Strand breaks and sugar release by γ-irradiation of DNA in aqueous solution, J. Am. Chem. Soc. 97:2277 (1975).
31. C. von Sonntag, U. Hagen, A. Schön-Bopp, and D. Schulte-Frohlinde, Radiation-induced strand breaks in DNA: Chemical and enzymatic analysis of end groups and mechanistic aspects, Adv. Radiat. Biol. 9:109 (1981).
32. E. Bothe and D. Schulte-Frohlinde, Release of K^+ and H^+ from poly(U) in aqueous solution upon γ and electron irradiation. Rate of strand break formation in polY U. Z. Naturforsch. 37c:1191 (1982).
33. D. Schulte-Frohlinde, Kinetics and mechanism of polynucleotide and DNA strand break formation, in: "Radiation Protection and Anticarcinogens", O.F. Nygaard and M. G. Simic, eds., Academic Press, New York, p. 53 (1983).
34. C. von Sonntag, Weiss lecture. Carbohydrate radicals: from ethylene glycol to DNA strand breakage. Int. J. Radiat. Biol. 46:507 (1984).
35. E. Bothe and H. Selbach, Rate and rate-determining step of hydrogen-atom-induced strand breakage of poly(U) in aqueous solution under anoxic conditions, Z. Naturforsch. 40c:247 (1985).
36. D. Schulte-Frohlinde, J. Opitz, H. Görner, and E. Bothe, Model studies for the direct effect of high-energy irradiation on DNA. Mechanism of strand break formation induced by photoionization of poly U in aqueous solution. Int. J. Radiat. Biol. in press (1985).

ELECTRON SPIN RESONANCE STUDIES OF THE MECHANISM OF RADIATION DAMAGE TO DNA

Paul M. Cullis and Martyn C.R. Symons

Department of Chemistry
The University
Leicester, LE1 7RH, U.K.

Electron spin resonance spectroscopy has only been used successfully on dry DNA at room temperature or on aqueous DNA at low temperatures. Under these conditions the direct damage mechanism is shown to dominate. The results inicate that electron-loss, which must initially be indiscriminate, rapidly ends up as $G\cdot^{+}$, which is stable up to ca. 210°K. Electrons are trapped at T, giving $T\cdot^{-}$ anions, which are converted into $\cdot$TH in the 130-208°K K range. Above these temperatures, both centres decay without the clear appearance of other intermediate radicals. Arguments are given against the concept that holes and/or electrons are extensively mobile within DNA molecules, and also against the concept that the ionic species studied by ESR spectroscopy recombine to give G and T to a major extent.

More has been learned about the fate of these centres by using a range of additives. In the presence of oxygen, O_2^{-} ions were detected and the primary yield of $T\cdot^{-}$ was reduced. However, both primary centres were lost at relatively low temperatures, with the concomitant formation of $RO_2\cdot$ radicals. The fate of these and the $O_2^{\cdot-}$ anions could not be determined by ESR spectroscopy.

In the presence of hydrogen peroxide or iodoacetamide, electrons were effectively scavenged, giving $\cdot$OH and $H_2\dot{C}ONH_2$ radicals and a reduced yield of $T\cdot^{-}$. These active radicals were rapidly converted into new alkyl-type radicals, thought to be primarily formed by hydrogen atom abstraction. The ESR signals due to these radicals were lost at temperatures below those characteristic of the primary centres. This shows that these may well have been converted into such alkyl radical centres by hydrogen atom transfer despite our inability to detect them.

In contrast, nitroimidazoles gave good ESR signals of their radical anions, with reduction of $T\cdot^{-}$, but no alkyl-type radicals were detected. In this case, strand-break studies showed that these sensitizing drugs actually protect DNA molecules when in contact with DNA. A range of other additives, including transition-metal complexes, have given similar results and it is concluded that both $G\cdot^{+}$ and $T\cdot^{-}$ primary centres give rise to permanent damage to DNA molecules, with a relatively high incidence of double strand breaks.

DNA-histone complexes are remarkable in that, for a given concentration of DNA, the presence of the histone proteins leads to an enhanced initial yield of $T\cdot^{-}$ ($\cdot$TH) with almost no change in the yield of $G\cdot^{+}$. Features assigned to electron-loss centres in the histones were also detected. We conclude that electrons are readily conducted through the proteins and pass onto the DNA, whilst the holes are not mobile. Thus the overall effect is a sensitization of damage. Even when aqueous cell nuclei (calf-thymus) were irradiated, strong signals from ·TH were detected under the usual conditions. This result suggests that electron addition may well be more significant than is often implied.

It is commonly argued that radiation damage to DNA proceeds via damage to water, the most important reagent being the hydroxyl radical. However, DNA molecules are packed as chromosomes very efficiently into cell nuclei, the concentration of water being relatively low. Furthermore, much of this is solvating water rather than being normal 'bulk' water. Thus direct damage, or damage proceeding via radiation effects in neighbouring protein molecules, may well compete effectively with the indirect water-damage mechanism.

Whilst ESR spectroscopy has not previously been used successfully in studies of the water-damage mechanism, it is well suited to a study of the direct mechanism. Dry DNA can be studied at ambient temperatures, but we have favoured a technique used especially by Gregoli and co-workers,[1,2] in which aqueous solutions are cooled to 77°K (or below) and the temperature profile of the reactions of the resulting radicals is studied by ESR spectroscopy.[3] The systems are heterogeneous, the pure ice-phase being quite separate from the hydrated DNA phase. At 77°K the only radicals in irradiated ice are ·OH radicals.[4] These decay, giving H_2O_2 within the ice crystals at ca. 125 K and provided there are no transition metal ion impurities present, there are no DNA reactions induced by damage to these ice crystallites.

There is general agreement that damage to the aqueous DNA phase results very rapidly in the trapping of $G\cdot^+$ and $T\cdot^-$ centres, that is, holes and electrons confined to the guanine and thymine substituents of the DNA molecules.[1-3,5] There is no evidence for extensive mobility of these units, and we suggest that they are trapped in close proximity (Scheme I). Those trapped very close together probably recombine (with light emission), but the majority are thought to be separated in the range of ca. 15-60 Å (Fig. 1).

Electron-loss in the DNA phase must be indiscriminate. We can, however, establish from our spectra that many reasonable hole centres are not trapped. These include H_2O^+ (·OH, see below), $(RO)_2PO_2\cdot$,[6] ROR^+,[7] and other base cations. We conclude that electron-transfer to the G-substituents is rapid from all these centres (Fig. 2).

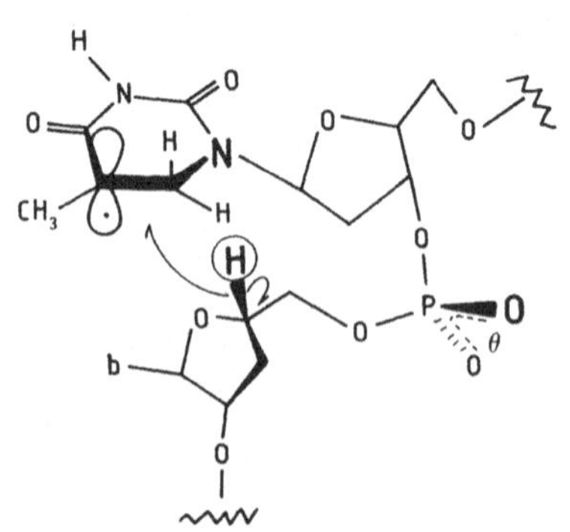

Figure 1. Diagram to indicate the transfer of a hydrogen atom to ·TH centres from a conveniently placed ⩾C-H group.

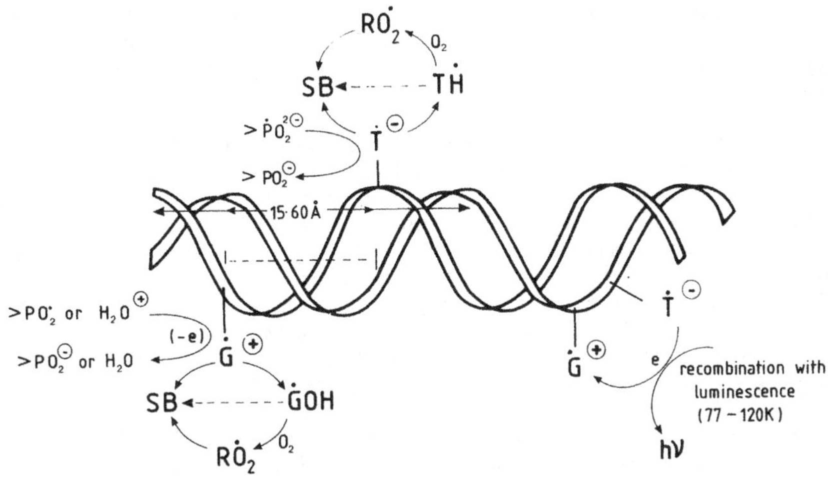

Figure 2. Scheme showing how electron-gain and electron-loss centres of various types become $G^{\cdot +}$ and $T^{\cdot -}$ centres in DNA, in the presence and absence of oxygen.

Similarly, although the ejected electrons could react at a variety of centres, it seems that addition at T is overwhelmingly favoured. In particular, addition to phosphorus, which might have been expected to occur,[8] is not observed. We stress this in view of a recent suggestion that this is a major mode of electron addition to DNA.[9] None of the possible products, indicated in equation 1, are detected by ESR spectroscopy.

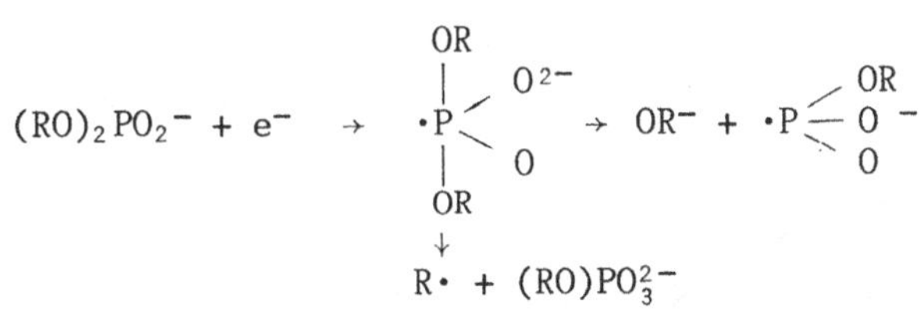

[1]

In the following, we accept the view that the major electron-gain and electron-loss centres initially formed from fully hydrated DNA are $G^{\cdot +}$ and $T^{\cdot -}$. Despite this initial success, ESR spectroscopy has proved to be less useful in probing subsequent reactions of these centres. The only clear reaction is protonation of $T^{\cdot -}$ to give the well-defined octet spectrum characteristic of ·TH (I). This spectrum, which is fortuitously well-defined because of strong hyperfine coupling to the methyl and methylene protons which is almost isotropic, is quite characteristic and a sensitive monitor of electron addition to T. Protonation sets in at ca. 130°K and is complete at ca. 208°K.

(I)

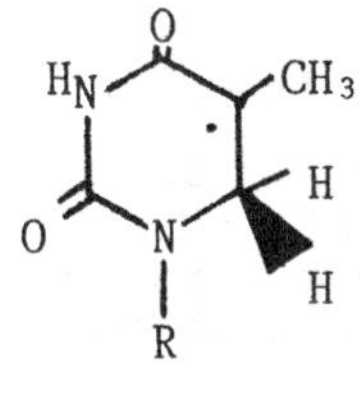

It is probable that G•+ is converted into •GOH by reaction with water. This intermediate has been detected using alkaline conditions, but is not clearly defined under normal conditions. Thus if it is formed, subsequent reactions must be rapid.

It is sometimes suggested that the ultimate fate of these G•+ and T•− centres is that they are lost by recombination on annealing. This may well be true of such centres trapped very close together, in the temperature range 77-120°K. These centres would be present as triplet states and would not contribute to the normal ESR features. Their recombination would account for the weak thermoluminescence in this temperature range,[10] which does not modify the ESR spectral features. However, protonation of T•− is quantitative and electron-loss from •TH, leaving T•+ is most improbable. Our studies of single and double strand-breaks (SSB and DSB) under similar conditions strongly supports this contention and suggests that both centres can lead ultimately to strand-breaking. This implies to us that hydrogen atom transfer from deoxyribose rings to •TH and, say, •GOH, is efficient at low temperatures. We stress that several such C-H bonds are poised permanently in sites close to the radical centres (Fig. 1) and that thermal motions in the temperature range 185-210°K, at which the major radicals decay (Fig. 3), would be sufficient to encourage transfer, possibly with a contribution from tunnelling. Even if this is unfavourable, rapid subsequent reactions such as strand-breaking could tip the equilibrium in favour of reaction.

If this is the case, we need to ask why ESR spectroscopy fails to detect such 'sugar' radicals formed either before or after SSB's. The answer seems to be that the rate of decay of these radicals is faster than the rate of their formation (see below). If this is correct, two factors should be noted. One is that radicals must react together for loss of unpaired electrons. This implies radical mobility in the temperature range 185-210°K. We stress that NMR studies show that water molecules solvating DNA are quite mobile at these temperatures but, nevertheless, in view of the rigidity of the DNA framework, we still have a conceptual problem. We stress that our

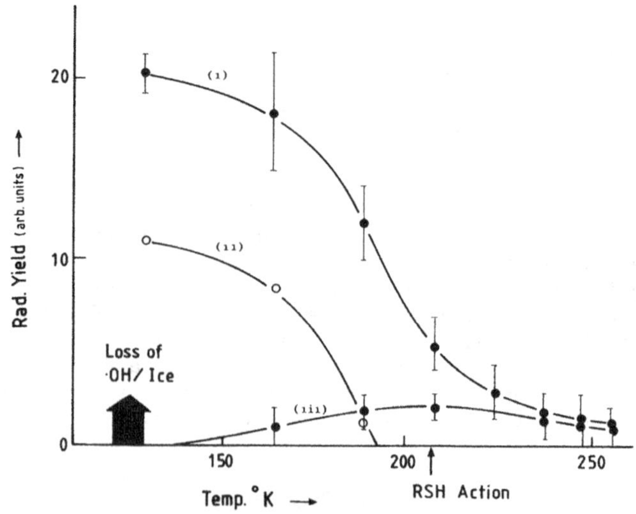

Figure 3. Temperature evolution of various radicals in irradiated DNA. (i) T•− + G•−, (ii) sugar radicals, (iii) •TH. The arrow indicates the temperature at which DNA radicals are lost in the presence of RSH.

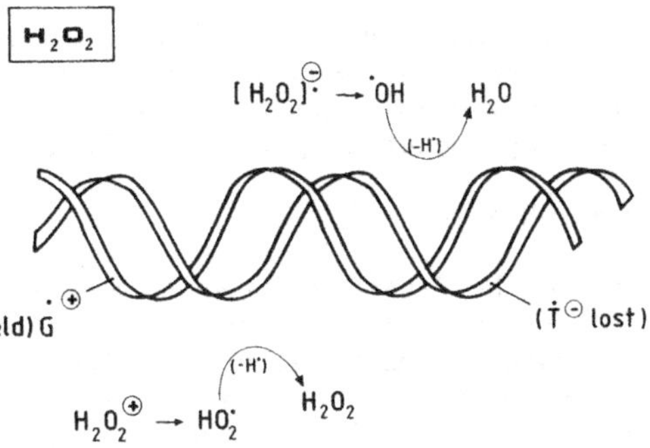

Figure 4. Reaction in the presence of H_2O_2.

strand-break studies support the idea that $G^{\cdot +}$ and $T^{\cdot -}$ centres are trapped quite close together,[11] so the radicals need not migrate far. Why, however, should they migrate towards each other rather than away from each other? Possibly the early losses represent the former mode and the higher temperature losses the latter.

The second important consequence of this mechanism is that, following initial strand-break events radicals are still present. These can continue to react, and such reactions can, at times, lead to further strand-breaks. The consequences of such reactions, which include the formation of DSB's and of small oligonucleotide fragments are currently being studied at Leicester.

EFFECTS OF ADDITIVES

Our aim in studying the effects of additives is threefold: to probe the mechanism of radiation damage discussed above, to modify it by diverting damage from DNA (protection) and to modify it by enhancing DNA damage (sensitization). We start by considering the most important small-molecule additive, oxygen.

Oxygen. In summary, the following ESR results were obtained:[3]

(1) Electrons were diverted from T to oxygen, giving O^{-} ions which were directly detected by ESR spectroscopy.

(2) At ca. 190°K, signals from DNA radicals were lost, with concomitant growth of an anisotropic spectrum characteristic of $RO_2\cdot$ radicals. Unfortunately, it is not yet possible to identify the nature of the R-groups, but the results do suggest efficient conversion of precursor radicals into $RO_2\cdot$. The $RO_2\cdot$ features were lost in the temperature range 200-250°K. This loss presumably occurs by a hydrogen atom transfer process similar to that shown in Figure 1, though different C-H units may be involved. However, we were still unable to detect the resulting carbon-centred radicals and conclude that their rate of loss is again faster than their rates of formation.

The fate of the $O_2^{\cdot -}/HO_2\cdot$ radicals could not be monitored but, under our conditions, attack on DNA is to be expected. Strand-break studies indicated an enhancement ratio of ca. 1.2, so that at least part of the well-studied sensitization of cell damage by oxygen may involve these reactions. We are, of course, well aware that other reactions of oxygen, not involving DNA directly, probably contribute to the overall sensitization.

Hydrogen Peroxide. In the absence of transition-metal ion impurities solutions containing added hydrogen peroxide were stable under our conditions of rapid freezing. The majority of the H O molecules are expected to move into the DNA solvent shell on freezing, in agreement with our results. A marked decrease in the yields of T•⁻/•TH suggested that H O acted as an efficient electron trap, generating trapped •OH radicals [2] (Fig. 4).

$$H_2O_2 + e^- \longrightarrow \cdot OH + OH^- \qquad [2]$$

These reactions cannot be detected directly because of the large signals from 'innocent' •OH radicals in the ice crystallites. However, the appearance of novel ESR features, assignable to carbon-centred (R•) radicals, strongly supports the occurrence of reaction [2], followed, in part, by hydrogen atom abstraction probably primarily from the deoxyribose units. Suggested species are II-IV.

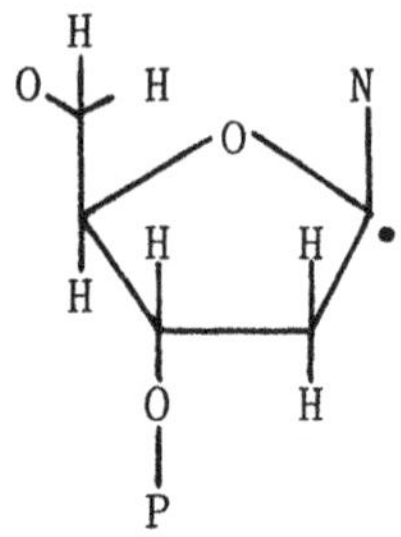

(II)

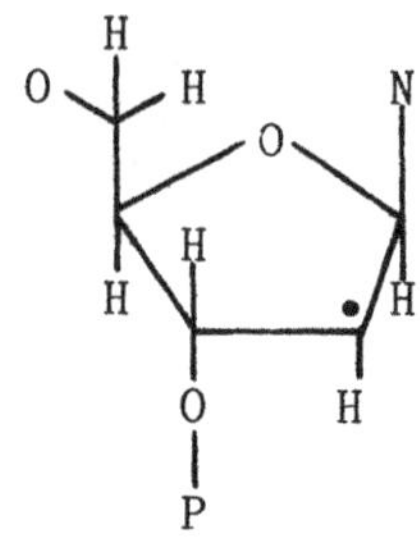

(III)

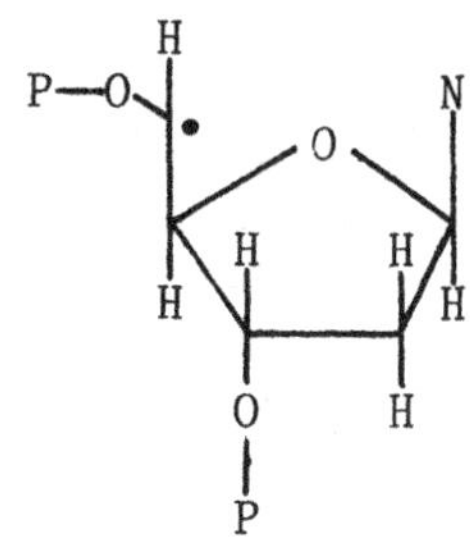

(IV)

An important observation is that these signals were lost at ca. 190°K which is significantly below that for normal loss of the DNA spectra (Fig. 2). This strongly supports our conclusion that the rates of loss of these species is normally faster than their rates of formation and hence that such radicals are reasonable intermediates in the overall mechanism.

Iodoacetamide. This compound also reacts preferentially with electrons to give an active radical product [3], and clear ESR; evidence for this and consequent loss of T•⁻/•TH radicals was obtained.[13] The spectral

$$e^- + I\text{-}CH_2CONH_2 \longrightarrow I^- + H_2\dot{C}CONH_2 \qquad [3]$$

features for $H_2\dot{C}CONH_2$ were lost at ca. 190°K with concomitant growth in features similar to those obtained in the work using H_2O_2, and assigned to carbon-centred radicals. Furthermore, these were lost in the same temperature range. As with the studies using hydrogen peroxide, small increases in the number of strand breaks were detected, suggesting that the radicals, •OH and $H_2\dot{C}CONH_2$, can transmit damage to the deoxyribose units more efficiently than the base-centred radicals (Fig. 5).

Nitroimidazoles. These were selected because of their known ability to sensitize apoxic cells to radiation induced damage. Our results strongly suggest that this sensitizing ability is not related directly to an increase in damage to DNA.[14] Indeed, as expected, ESR spectra characteristic of the drug anions were obtained, with a large diminution in yield of T•⁻/TH. Since nitroimidazole anions are relatively stable, they failed to induce DNA damage so far as we could judge (Fig. 6).

This conclusion was supported by our observation that the numbers of SSB's, and especially DSB's, were reduced. These results have four important corollaries:

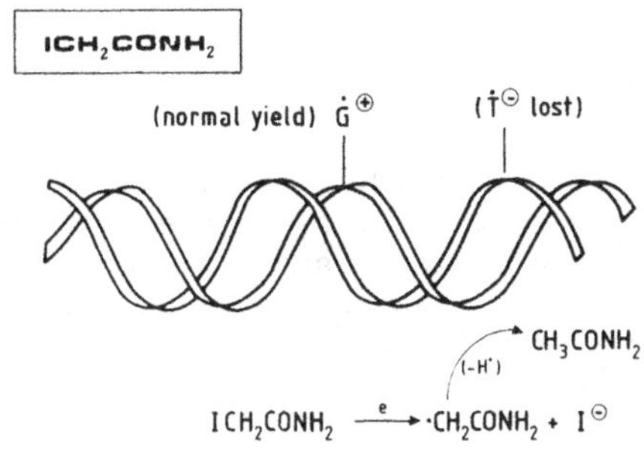

Figure 5. Reaction in the presence of iodoacetamide.

(1) there was no detectable increase in the yield of $G\cdot^+$ centres as had previously been proposed,

(2) when in contact with solvated DNA, these drugs protect rather than sensitize,

(3) the more rapid decrease in DSB's supports our postulate that $G\cdot^+$ and $T\cdot^-$ centres are often so close together that DSB's result, and

(4) both $G\cdot^+$ and $T\cdot^-$ centres can ultimately result in strand breaks.

Thiols. It is generally accepted that thiols can react with DNA centred radicals effectively via hydrogen atom transfer, thereby effecting rapid 'repair'. In particular, glutathione is implicated as playing a major role in such rapid radical repair. A key factor is that RS· radicals are relatively stable and can further stabilise themselves by bonding to a second thiol molecule. We thought it of interest to see to what extent ESR spectroscopy under our low temperature conditions could probe these reactions. Our preliminary results[15] suggest that there is a very rapid loss of all DNA centred radicals at temperatures well below their normal decay range (see Fig. 2). Thus, for example, it is ·TH that is 'repaired' rather than 'sugar' radicals. ESR features thought to be due to $(RSSR)\cdot^-$ radical anions were detected simultaneously, but we were unable to detect RS· radicals directly. The overall mechanism for ·TH is summarised in equations [4]-[6]:

$$\cdot TH + RSH \rightarrow TH_2 + RS\cdot \qquad [4]$$

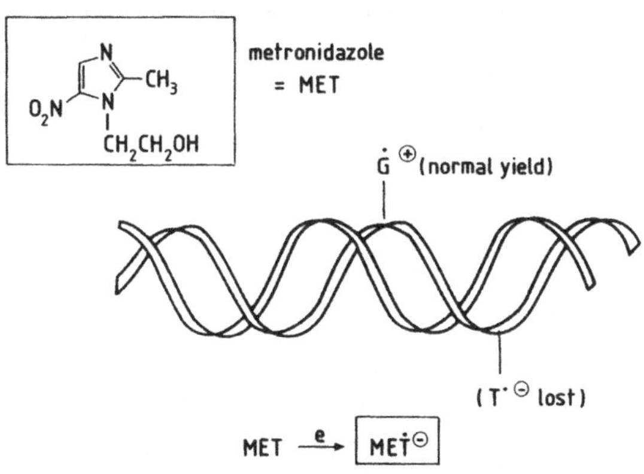

Figure 6. Reaction in the presence of metronidazole.

$$RS\cdot + RSH \rightarrow RS\dot{-}S(H)R \qquad [5]$$

$$RS\dot{-}S(H)R \rightarrow RS\dot{-}SR^- + (H^+) \qquad [6]$$

Presumably RS· radicals are relatively mobile in this temperature region, and RS-S(H)R radicals, which are thought by some to have well defined ESR spectra[15] (which were not detected), are good proton donors under our conditions. The radical anions $RS\text{-}SR^-$ are good electron-donors and may contribute to the overall repair in this capacity. Strand-break studies accord well with these conclusions.

Thus our method of study strongly supports the idea that RSH derivatives act by efficient hydrogen atom, and possibly electron, donation to 'repair' the DNA damage. We stress, however, that in producing species such as TH_2, base defects have been introduced which may, sometimes, have important consequences.

Histones. Nuclear histone complexes have been studied by several workers,[16,17] all of whom detected ·TH formation. Under our conditions, for given concentrations of DNA and identical conditions for samples of pure DNA and its complex, we found a clear enhancement of total radical yield from the DNA phase, as expected. Signals from ·TH grew in at the usual temperature, with a clear enhancement (ca. 30%) for the histone system. However, the central components comprised a superposition of the $G\cdot^+$ signal and a major doublet assigned to the electron-loss centre of the histone protein.[15] Our preliminary conclusions suggest that electrons generated within the proteins are efficiently transmitted through to DNA and are trapped as $T\cdot^-$, whilst the hole centres generated in the protein remain efficiently trapped therein and are not mobile. The net effect is that the DNA is damaged to its normal extent but this is augmented by the gain in electrons from the histone proteins. This represents a clear sensitization (Fig. 7).

Finally, we have begun to extend these studies to whole cell nuclei (calf-thymus). Once again, there is a well-defined yield of ·TH, but the protein signals are now more intense and we have not yet been able to assess the magnitude of the $G\cdot^+$ yield. Both systems are now being studied in the presence of additives.

ACKNOWLEDGEMENTS

We thank the Cancer Research Campaign, the Commission of the European Communities and the Association of International Cancer Research for grants in support of this work and B. W. Wren, P. J. Boon, N. Bartlett, J. McClymont, D. Jones, J. Lea and Professor G. W. Eastland for their extensive contributions.

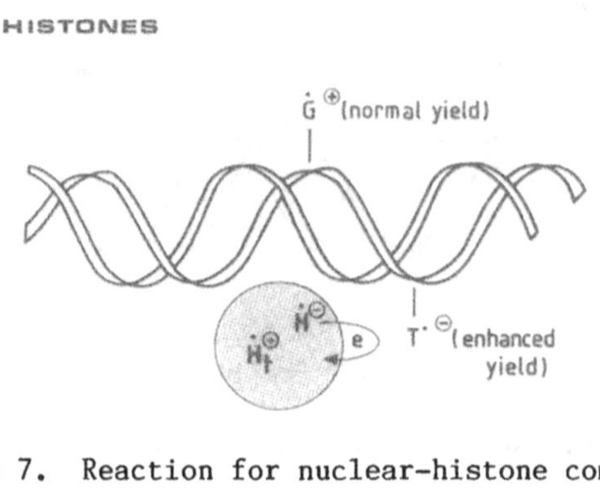

Figure 7. Reaction for nuclear-histone complexes.

REFERENCES

1. S. Gregoli, M. Olast and A. J. Bertinchamps, Radiat. Res., 89:238 (1982).
2. S. Gregoli and A. J. Bertinchamps, Int. J. Radiat. Biol., 21:65 (1972).
3. P. J. Boon, P. M. Cullis, M. C. R. Symons and B. W. Wren, J. Chem. Soc., Perkin Trans. 2, 1393 (1984).
4. J. A. Brivati, M. C. R. Symons, D. J. A. Tinling, H. W. Wardale and D. O. Williams, Trans. Faraday Soc., 2112 (1967).
5. J. Hüttermann, K. Voit, H. Oloff, W. Köhnlein, A. Gräslund and A. Rupprecht, Disc. Faraday Soc., 78:135 (1984).
6. S. Subramanian, M. C. R. Symons and H. W. Wardale, J. Chem. Soc. A, 1239 (1970).
7. M. C. R. Symons and B. W. Wren, J. Chem. Soc., Perkin Trans. 2, 511 (1984).
8. D. J. Nelson and M. C. R. Symons, J. Chem. Soc., Perkin Trans. 2, 286 (1977).
9. J. H. H. Hamerlinck, P. Schipper and H. M. Buck, J. Am. Chem. Soc., 105:385 (1983).
10. A. Gräslund, H. Steen and A. Rupprecht, Int. J. Radiat. Biol., 30:263 (1976).
11. T. Ichikawa and H. Yoshida, J. Phys. Chem., 88:3199 (1984).
12. M. Conder, P. M. Cullis, D. Jones, J. D. McClymont and M. C. R. Symons, J. Chem. Soc., Perkin Trans. 2, submitted for publication.
13. P. M. Cullis, M. C. R. Symons, B. W. Wren and S. Gregoli, J. Chem. Soc., Perkin Trans. 2, submitted for publication.
14. P. J. Boon, P. M. Cullis, M. C. R. Symons and B. W. Wren, J. Chem. Soc., Perkin Trans. 2, 1057 (1985).
15. P. M. Cullis, D. Jones, J. Lea and M. C. R. Symons, unpublished work.
16. M. G. Ormerod, Int. J. Radiat. Biol., 9:295 (1965).
17. M. Kuwabara and G. Yoshi, Biochim. Biophys. Acta, 432:292 (1976).

FREE RADICAL MECHANISMS OF DNA BASE DAMAGE

Michael G. Simic and Slobodan V. Jovanovic
National Bureau of Standards
Gaithersburg, MD 20899, USA

INTRODUCTION

In many respects, DNA is a simple system where the reactions of free radicals are concerned. The reactive components in DNA are the four bases (thymine, T; cytosine, C; adenine, A; and guanine, G) and the sugar moiety (deoxyribose, dR). The sixth component – phosphate – is relatively unreactive towards free radicals.

The purines, Pu (A and G), are more aromatic than the pyrimidines, Py (T and C), which is reflected in their specific free radical reaction mechanisms. The reaction mechanisms within the Pu and Py classes are more similar than between these two classes. Hence, only two types of bases were selected as models,

Thymine

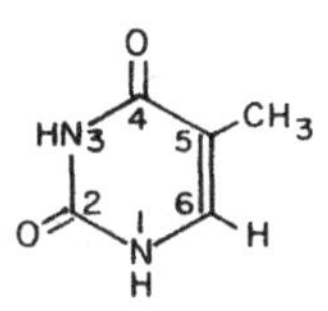

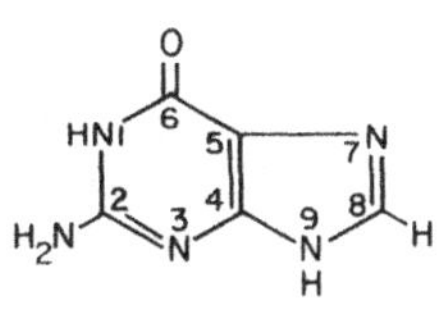

Guanine

Thymidine
or
1-Methylthymine

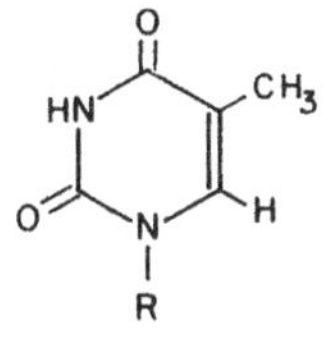

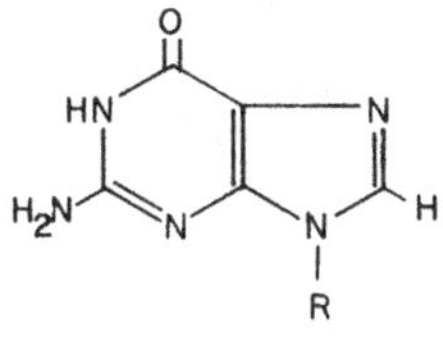

Guanosine
or
9-Methylguanine

Free radicals can oxidize or reduce a base (redox reactions) or add to a base.[1] The abstraction of H atoms is less favored and the main site of occurrence is the methyl group in T.[1] Deoxyribose (dR) is a cyclic aliphatic molecule and, as such, undergoes H atom abstraction only. Reactions of free radicals with dR are particularly significant because the resulting dR• radicals in DNA can lead to either loss of a base (the mechanism not fully understood), a strand break (C-4 position), or internal cyclization with a base attached to it. The details of these interesting reactions can be found elsewhere.[2]

Comprehensive understanding of free radical reactions with DNA is still lacking, in spite of major advances in the last ten years. Therefore, only the general features of model systems can be presented in this short review, and the possible consequences of the overall damage to DNA will be indicated.

GENERATION OF INITIAL RADICALS

The bulk of information on free radical reactions with DNA components comes from kinetic and mechanistic studies in aqueous model systems based on pulse radiolysis and steady-state radiolysis.[3] Ionizing radiations generate the following primary species in aqueous solutions,

$$H_2O \rightsquigarrow e_{aq}^- \ (2.8),\ \cdot OH\ (2.8),\ \cdot H\ (0.6),\ H_2\ (0.4),\ H_2O_2\ (0.8) \qquad (1)$$

The numbers in parentheses indicate the relative yields of these species expressed as G-values (number of molecules/100 eV absorbed).[3]

Because the •OH radical and the H atom are considered to be the species most damaging to DNA, hydrated electrons are usually converted to •OH for mechanistic studies of damage in model systems,

$$e_{aq}^- + N_2O \longrightarrow \cdot OH + OH^- + N_2 \qquad (2)$$

Under these conditions, the major reactive specie is •OH ($G = 5.6$). The •H atom ($G = 0.6$) reacts in a manner similar to •OH. It adds preferentially to the double bonds of bases, and in most cases can be accounted for mechanistically.[3]

PYRIMIDINE RADICALS

Generation of Pyrimidine Radicals

Reaction of •OH radical with thymine leads to formation of three different transients,

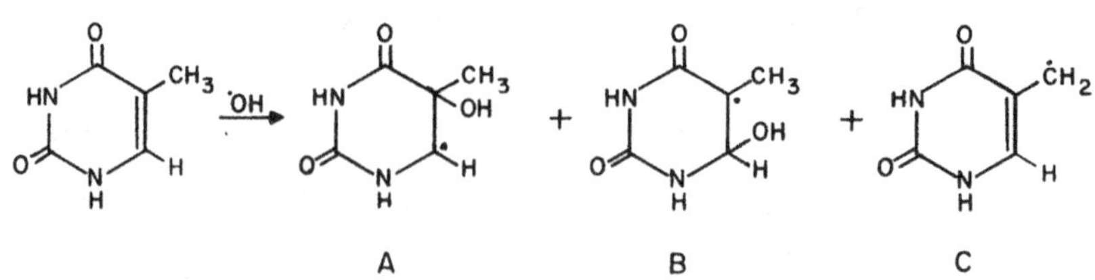

(3)

$k = 6 \times 10^9\ M^{-1}\ s^{-1}$ (ref. 4)

Transients A and B are generated by the addition of •OH to the 5,6 double bond of T.[5] Addition of •OH to position 5 or 6 is nonrandom and depends on the properties and positions of the substituents (Table 1).[6]

Table 1. Yields and redox properties of free radicals from •OH + Py reaction in aqueous solutions[a]

Derivative	Radical	Type	Redox property	% of total P.R.[b]	•OH yield P.A.[c]
Thymine	A	5-OH-6-yl	Reducing	56 (65)[d]	57
	B	6-OH-5-yl	Oxidizing	35 (30)[d]	
	C	5-methyl	Weakly reducing	9	
Uracil	A	5-OH-6-yl	Reducing	80 (82)[d]	77
	B	6-OH-5-yl	Oxidizing	20 (18)[d]	
1-methyluracil	A	5-OH-6-yl	Reducing	65	66
	B	6-OH-5-yl	Oxidizing	20	
	D	1-methyl	Weakly reducing	15	

[a]from reference 6.

[b]As determined from redox yields obtained by pulse radiolysis of aqueous pyrimidine derivatives at pH 7, 20°C.

[c]Glycol yields obtained by HPLC analysis of irradiated aqueous pyrimidines under steady-state conditions at pH 7.

[d]From Ref. 11, 65% pertains to A+C thymine radical yield

Transient C is generated by abstraction of an H atom from the methyl group, and is the least abundant intermediate (~ 9%).[6]

Other C-centered radicals, •R, such as $•CO_2^-$,[7] $•CH_2OH$,[8] and $(CH_3)_2\dot{C}OH$,[8] can also add to the double bond of pyrimidines. The resulting •Py-R radicals are comparable to the •Py-OH radicals. The reaction rates for these radicals, however, are slower $(k(•R + Py) < 10^6\ M^{-1}\ s^{-1})$ than the rates for •OH. The abstraction of an H atom from the methyl group of T by these radicals is energetically unfavorable.

Reactions of pyrimidine radicals with redox agents

Transient A in reaction (3) is a reducing radical,[9] because it is an α amino radical. It can be oxidized by moderately strong oxidants such as nitro derivatives,[10] and by strong oxidants such as p-benzoquinone,[6,9] tetranitromethane,[11] and ferricyanide.[6] Oxidation of transient A leads to formation of a carbocation,[5] which in turn is responsible for the formation of thymine glycol, as shown below,

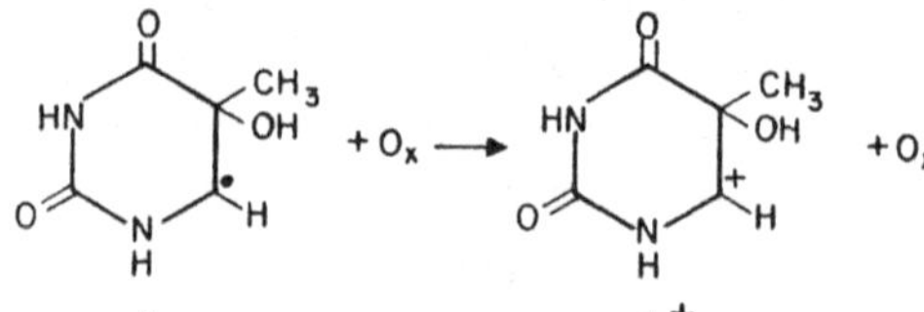

A $\quad$ A^+ (4)

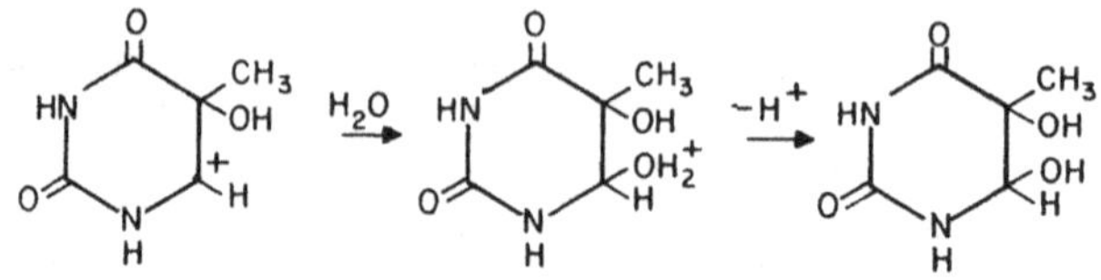

A^+ $\quad$ AOH_2^+ $\quad$ TG (5)

The conclusion that different pyrimidine glycols originate exclusively from type A transients is based on measurements that indicate identical yields of a particular transient A (by pulse radiolysis) and a corresponding glycol (by HPLC) for all uracil derivatives studied (Table 1).

Radical C is also reducing in nature, as indicated by the following resonance,

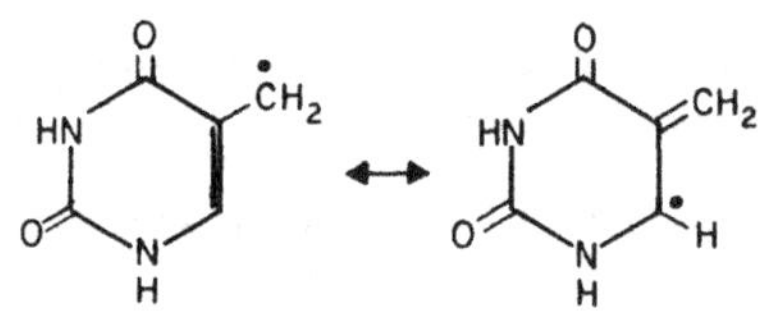

C $\quad$ C′ (6)

Oxidation of radical C is believed to lead to formation of 5'-hydroxymethyluracil, 5-U-CH_2OH,[12]

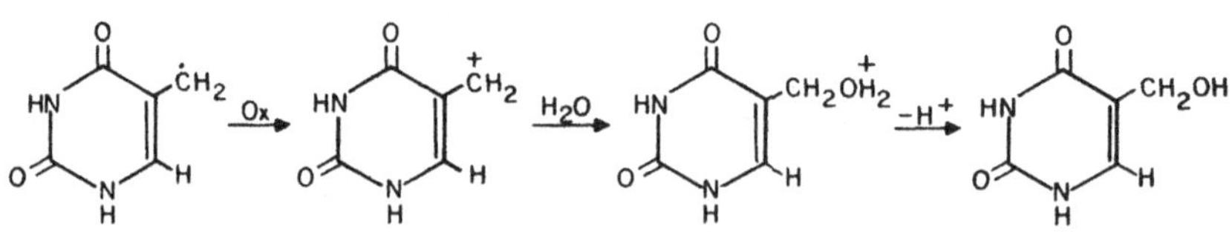

C $\quad$ C^+ $\quad$ COH_2^+ $\quad$ 5-U-CH_2OH (7)

Radical B is an oxidizing radical[11] because its mesomeric form is an oxy radical,

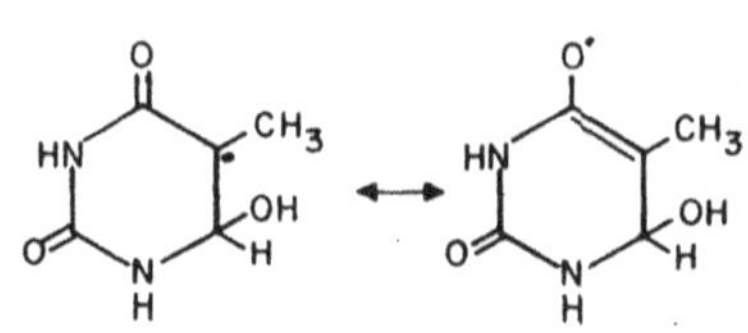

B $\quad$ B′ (8)

Reducing agents, such as ascorbate (AH^-)[6] and N,N,N',N'-tetramethyl-p-phenylenediamine (TMPD),[11] which were used to measure the yield of radical B, reduce B to a hydrate,

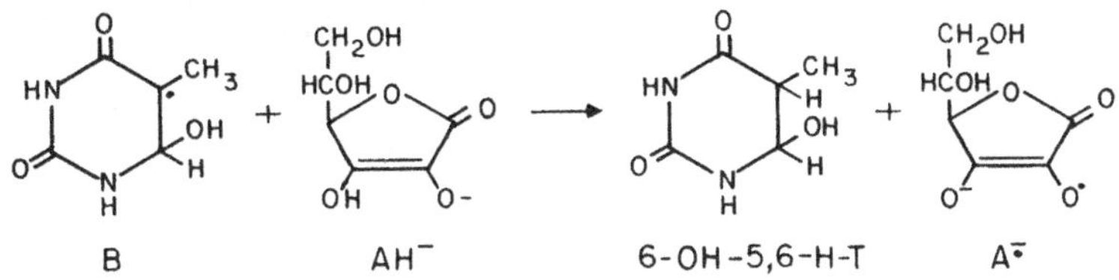

(9)

Reaction of pyrimidine radicals with deoxyribose

The OH adducts of pyrimidines, •Py-OH, appear to be capable of abstracting an H atom from a sugar. This abstraction could be the mechanism by which the •U-OH radical leads to a stand break in poly-U.[13] Similarly, •T-OH radicals in poly-dT were shown to yield thymine hydrate, THOH.[14] The reaction is very slow, as expected, and takes place only within the matrix of an oligomer or a polymer. For example, an isolated •T-OH radical from T does not react with disolved dR. The •T-OH radical within a mononucleotide does not react with its own dR either. This same radical does react, however, in a thymine dinucleotide, T-dR-P-dR-T.[14] Hence, it was concluded that abstraction of H cannot take place from a sugar to which the •Py-OH radical is attached, and that the reaction can occur only with a neighboring sugar.[14]

Reactions of pyrimidine radicals with each other

Pyrimidine radicals are very reactive and disappear rapidly in reactions with each other at close to diffusion controlled rates ($k \sim 10^9\ M^{-1}\ s^{-1}$).[15] If the Py radicals were part of DNA, the rate constant would decrease significantly due to the impaired mobility of macromolecular radicals.

When a reducing radical and an oxidizing radical react with each other, an electron transfer reaction is expected to take place. In the case of radicals A and B generated in reaction (3),

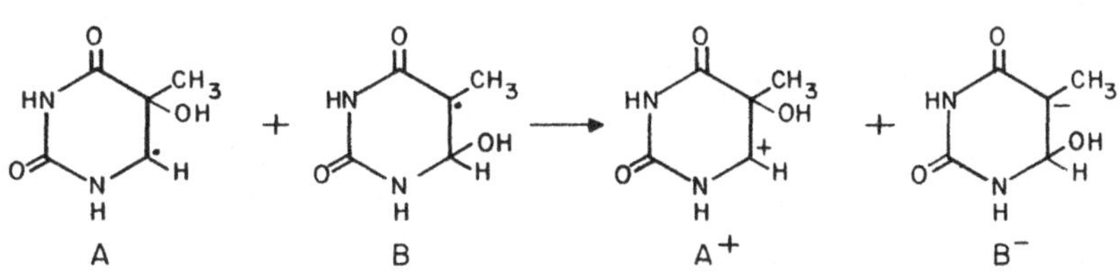

(10)

The thymine carbocation leads to thymine glycol, as shown by reactions (4) and (5). Hence, under these conditions (OH + T reaction only, no additives), the yield of thymine glycol is fairly low.[5,15] The measured G(TG) = 0.6 corresponds only to 11% of the total OH yield.

When two identical radicals react with each other, it appears that only a crosslinked product is formed,

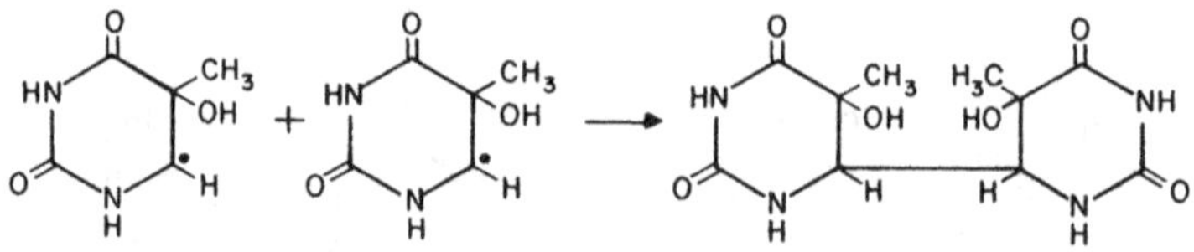

(11)

In many instances the crosslinked OH adducts, such as A-A, may lose one or two water molecules.[16,19]

Reactions of pyrimidine radicals with different radicals

If a reducing Py radical (such as A) reacts with any oxidizing radical, a redox reaction similar to reaction (10) is expected. For example, the yield of TG increases in the presence of phenoxy radicals,[17] which are strong oxidants ($E_7(\phi O\bullet) = 0.85$ V).[18]

$$A + \phi O\bullet \longrightarrow A^+ + \phi O^- \qquad (12)$$

At present, it is not clear whether reaction (12) can also lead to the formation of a T-Tyr crosslink.[19]

When radical A reacts with another radical which is not a strong oxidant, e.g., an OH adduct to the benzene ring of phenylalanine (Phe), crosslinking between these radicals takes place with high probability,[19]

$$A + \bullet\text{Phe-OH} \longrightarrow \text{A-Phe-OH} \qquad (13)$$

The crosslinked product may have a few isomers. Attachment to T may form cis and trans conformers. Attachment to the benzene ring may be through o-, m-, and p-positions. The total yield of all the OH-T-Phe-OH isomers was found[20] to be 33% of all the radicals generated when •T-OH and •Phe-OH radicals were produced in a 1:1 ratio. This type of reaction between a DNA base radical and an amino acid radical[19] may be responsible for the formation of crosslinks between DNA and proteins.[21]

PURINE RADICALS

Formation of purine radicals

Purines are considerably more aromatic than pyrimidines, and give a variety of OH adducts, •Pu-OH.[1] The addition of •OH to C-N double bonds is believed to take place only at C positions.[22] N-positions, being more electro-positive, are expected to be less reactive with an electrophile such as •OH. When 9-methylguanine is used as a model,[23] the resulting OH adducts, 9-Me-Ġ-OH, are shown below,

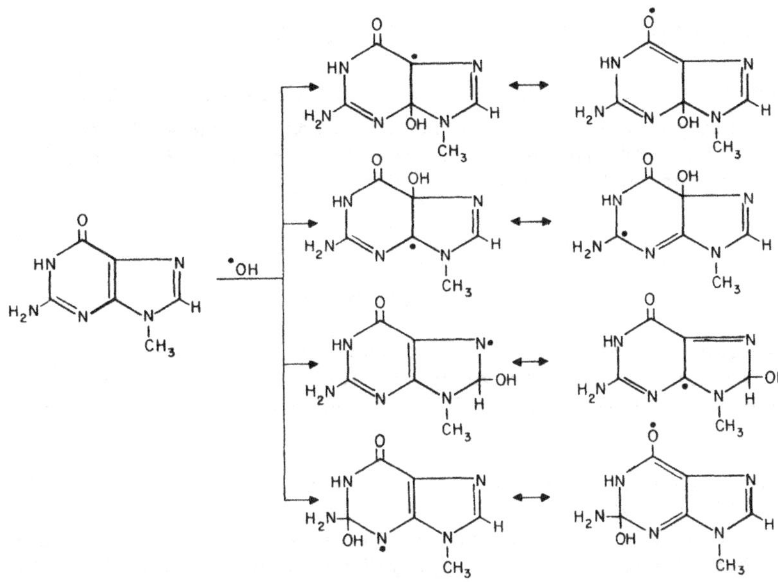

(14a)

(14b)

(14c)

(14d)

The abstraction of H atoms from G, as observed for T, cannot take place. In dR-G, however, substantial abstraction of H atoms from deoxyribose, dR, takes place (~ 20%). Reliable measurements of the percent yield of •OH radical reaction with sugar and the distribution of the resulting radicals on the sugar are not available.

Transients (14a) and (14d) are oxidizing. Transients (14b) and (14c) are reducing radicals since they are α amino radicals.[23] Measurements of the yields of reducing and oxidizing radicals indicate very close to a 1:1 distribution[22,23] (Table 2). This is not surprising if we assume that reactions (14a-14d) have an equal probability. The oxidizing radicals (14a) and (14d) may eliminate OH^-, which would convert them into Me-G$\cdot^+$ radicals. It is also ambiguous at present whether •OH adds to G first or oxidizes G directly,

$$\cdot OH + Me\text{-}G \rightarrow \begin{cases} Me\text{-}\dot{G}\text{-}OH \xrightarrow{-OH^-} Me\text{-}G\cdot^+ \\ Me\text{-}G\cdot^+ + OH^- \end{cases} \quad (15)$$

Purines have lower redox potentials than pyrimidines.[18] Hence, strong oxidants, such as $Br_2^{\overline{\cdot}}$, would be expected to oxidize purines, as suggested previously.[24] In order to eliminate the ambiguity associated with possible addition of •Br to G, Tl(II), which acts only as an oxidant, was used to generate G$\cdot^+$ radicals exclusively,[23]

$$\cdot OH + Tl(I) \rightarrow Tl(II) + OH^- \quad (16)$$

$$Me\text{-}G + Tl(II) \rightarrow MeG\cdot^+ + Tl(I) \quad (17)$$

Tl(II) does not oxidize deoxyribose in dR-G or ribose in R-G. The G$\cdot^+$ radical is a strong acid and in neutral solutions exists in a deprotonated form •G(-H).[23]

$$Me\text{-}G\cdot^+ \rightleftharpoons Me\text{-}\dot{G}(-H) + H^+ \qquad pK_a < 3 \quad (18)$$

Reactions of guanine radicals with redox agents

As shown in Table 2, some of the $Me\text{-}\dot{G}\text{-}OH$ transients are reducing and some are oxidizing radicals. The $Me\text{-}\dot{G}(-H)$ radical is strictly an oxidizing radical.[23] The guanyl radical can be repaired efficiently by an electron donor such as uric acid, UH^{2-},

$$Me\text{-}\dot{G}(-H) + UH^{2-} + H^{+} \longrightarrow Me\text{-}G + UH^{\bullet -}$$
$$k \doteq 1.2 \times 10^{9}\ M^{-1}\ s^{-1} \qquad (19)$$

These measurements indicate 100% formation of the guanine radical by Tl(II) in reaction (17). The guanyl radical, •G(-H), does not abstract H-atom from RSH compounds at pH 7,[23] in contrast to the OH adducts of dGMP, which apparently do.[22]

The distribution of •G-OH isomers has not yet been determined. In one case only, the yield of the OH adduct to the C-8 position of G has been found to be 24% of the reacting •OH radicals.[23] This conclusion is based on measurements of the 8-hydroxy-9-methylguanine yield in the presence of p-benzoquinone, BQ, by HPLC (Table 2),

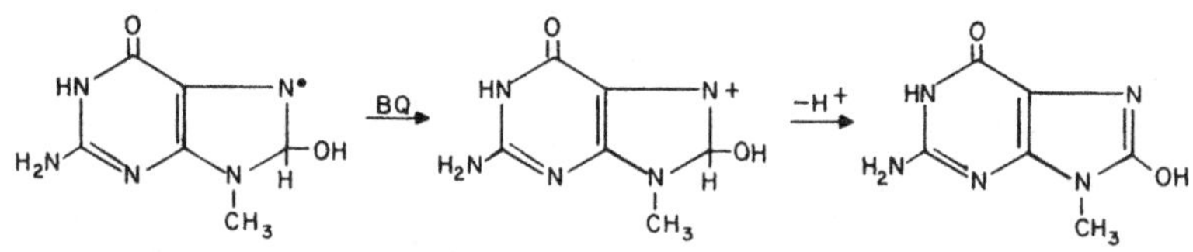

(20)

Table 2. Yields and redox properties of free radicals from •OH + 9-methylguanine raction in aqueous solutions at pH 7[a]

Radical type	Redox property	% of total OH yield	Measured as
8-OH-7-yl	reducing	25	8-OH-9-Me-G[b]
5-OH-4-yl	reducing	24	from difference
total reducing		49	$BQ^{\bullet -}$[c]
•G(-H) or	oxidizing	50	[Promethazine$^{\bullet +}$]
4-OH-5-yl and 2-OH-3-yl			and [TMPD$^{\bullet +}$][c]

[a]From reference (23).

[b]HPLC analysis of stable radiolytic products in N_2O-saturated aqueous solution of 9-methylguanine.

[c]Obtained by pulse radiolysis technique.

Addition of free radicals to purines

Reactions of C-centered free radicals, •R, with the 7,8 double bond of purines have been observed, and the resulting 8-Pu-R products described.[25] These reactions are comparable in principle to the addition of C-centered radicals to pyrimidines.[7,8] For purine nucleosides and nucleotides, however, an intramolecular addition of a sugar radical to the 7,8 double bond may also take place. The feasibility of that internal cyclization was deduced from the observed cyclized 8,5'-cycloadenosine-5'-monophosphate (8,5-cAMP)[26,27] and more recently 8,5'-cyclodeoxyadenosine (8,5'-cdR-A)[28] in irradiated aqueous solutions of AMP and dR-A, respectively. More recently, the analogous product of dR-G, i.e., 8,5'- cyclodeoxyguanosine, was observed in DNA irradiated in aqueous solution.[29] X-ray structural analysis showed that cyclization takes place through the sugar C-5' and the adenine C-8 positions.[30] The existence of two C-5' epimers was also demonstrated.[29,30]

The addition of a ribose or a deoxyribose radical to the C-8 position of either A or G, generates a purine-7-yl radical. As shown above in reaction (20), the Pu-7-yl radical is a reducing (α amino) radical.[23] That radical can be oxidized by the •G(-H) or •A(-H) radical, strong oxidants, or suitable oxidizing radicals. In purine nucleotides only a small fraction (4-6%) of •OH is expected to react to give the C-5' sugar radical. The reaction sequence for dR-G, based on the mechanisms[23] discussed on the previous page is shown below,

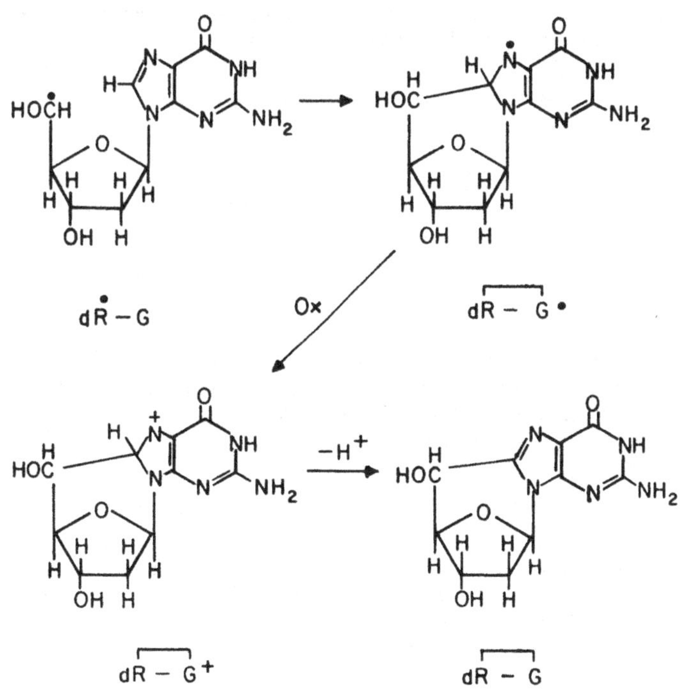

(21)

In the presence of oxygen, reaction (21) will compete with the addition of O_2 to the C-5' sugar radical.

Reactions of purine radicals with each other

Purine radicals are also very reactive and disappear in reactions with each other at close to diffusion controlled rates ($\sim 10^9\ M^{-1}\ s^{-1}$). As mentioned above, the $8\text{-}\dot{G}\text{-}OH$ transient is a reducing radical that can be oxidized to give 8-G-OH products for various guanine derivatives. In the absence of any redox agents, the $8\text{-}\dot{G}\text{-}OH$ radical can be oxidized by $\cdot G_{ox}$ radicals (either •G-OH and/or •G(-H)) which constitute about 50% of the total yield of guanine radicals. Hence, the following reaction is expected to occur,

$$8\text{-}\dot{G}\text{-}OH + \cdot G_{ox} \longrightarrow 8\text{-}G\text{-}OH + G \qquad (22)$$

As in the case of thymine, in the absence of redox additives the yield of 8-G-OH is lower (17% of all •OH radicals) than in the presence of oxidants (25%)[23] because of other "wasteful" radical-radical reactions.

REACTIONS OF PURINE AND PYRIMIDINE RADICALS WITH OXYGEN

Reactions of pyrimidine radicals with oxygen are diffusion controlled ($k \sim 10^9\ M^{-1}\ s^{-1}$). A full account of the chemistry of oxygenated systems has been given elsewhere.[31]

The reactions of purine radicals with oxygen have been studied considerably less. However, an important aspect of these reactions has emerged recently. While the •G-OH intermediates react with oxygen, it has been found that the guanine radical does not,[32]

$$dR\text{-}\dot{G}(\text{-}H) + O_2 \longrightarrow \text{no reaction} \qquad (23)$$

We suggest, therefore, that the lack of reaction (23), which was tested by pulse radiolysis[23] and by oxygen uptake under stationary state conditions[31] ($0.052\ Gy\ s^{-1}$), is due to a highly delocalized unpaired electron within the guanine radical.

ACKNOWLEDGMENT

The many useful editorial suggestions of Ms. Karen A. Taylor and discussions with Drs. George Scholes, Miral Dizdaroglu, and David Bergtold are greatly appreciated. The technical assistance of Mrs. Gloria Wiersma is kindly acknowledged.

REFERENCES

1. "Effects of Ionizing Radiation on DNA," J. Huttermann, W. Kohnlein, R. Teoule, and and A. J. Bertinchamps eds., Springer-Verlag, Berlin, 1978.
2. C. von Sonntag, Adv. Carbohydr. Chem. Biochem. 37: 7 (1980).
3. M. Z. Hoffman, J. Chem. Ed. 58: 83 (1981).
4. Farhataziz and A. B. Ross, "Selected Specific Rates of Reactions of Transients from Water in Aqueous Solution. III. Hydroxl Radical and Perhydroxyl Radical and Their Radical Ions," NSRDS-NBS 59, Wasington, D.C., 1977.
5. G. Scholes, Brit. J. Radiol. 56: 221 (1983).
6. S. V. Jovanovic and M. G. Simic, J. Amer. Chem. Soc., submitted.

7. T. Wada, H. Ide, S. Nishimoto, and T. Kagiya, Int. J. Radiat. Biol. 42: 215 (1982).
8. L. Karam and M. G. Simic, unpublished data.
9. M. Simic and E. Hayon, Biochem. Biophys. Res. Comm. 50: 539 (1973).
10. S. Nishimoto, H. Ide, T. Wada, and T. Kagiya, Int. J. Radiat. Biol. 44: 585 (1983).
11. S. Fujita and S. Steenken, J. Amer. Chem. Soc. 103: 2540 (1981).
12. J. Cadet, p. 189 in ref. 1.
13. D. G. E. Lemaire, E. Bothe, and D. Shulte-Frohlinde, Int. J. Radiat. Biol. 45: 351 (1984).
14. L. Karam, M. Dizdaroglu, and M. Simic, Biochem. Biophys. Res. Comm., submitted.
15. E. Hayon and M. G. Simic, J. Amer. Chem. Soc. 95: 1029 (1973).
16. M. Dizdaroglu and M. G. Simic, Int. J. Radiat. Biol. 46: 241 (1984).
17. M. Dizdaroglu and M. G. Simic, unpublished data.
18. S. V. Jovanovic, A. Harriman, and M. G. Simic, J. Phys. Chem., in press; also S. V. Jovanovic and M. G. Simic, J. Phys. Chem., in press.
19. M. G. Simic and M. Dizdaroglu, Biochemistry 24: 233 (1985).
20. M. Dizdaroglu and M. G. Simic, Int. J. Radiat. Biol. 47: 63 (1985).
21. For more information see review of N. L. Oleinick in this book.
22. P. O'Neil, Radiat. Res. 96: 198 (1983).
23. S. V. Jovanovic and M. G. Simic, unpublished data.
24. R. L. Willson, P. Wardman, and K.-D. Asmus, Nature (London) 252: 323 (1974).
25. H. Steinmans, I. Rosenthal, and E. Elad, J. Amer. Chem. Soc. 91: 17 (1968).
26. K. Keck, Z. Naturforsch. 23b: 1034 (1968).
27. J. A. Raleigh, W. Kremers, and R. Whitehouse, Radiat. Res. 65: 414 (1976).
28. N. Mariaggi, J. Cadet, and R. Teoule, Tetrahedron 32: 2385 (1976).
29. M. Dizdaroglu, Biochemistry, submitted.
30. T. P. Haromy, J. Raleigh, and M. Sundaralingam, Biochemistry 19: 1718 (1980).
31. For more information see review of C. von Sonntag in this book.
32. S. V. Jovanovic, M. AlSheikhly, and M. G. Simic, BBRC.

PEROXYL RADICALS OF NUCLEIC ACIDS AND THEIR COMPONENTS

Clemens von Sonntag

Max-Planck-Institut für Strahlenchemie
Stiftstr. 34 - 36,
D-4330 Mulheim a.d. Ruhr, West Germany

INTRODUCTION

If free radicals add to the bases of nucleic acids or abstract a hydrogen atom from the sugar moiety, mainly carbon-centred radicals are formed. In the presence of oxygen such carbon-centred radicals are converted into the corresponding peroxyl radicals at nearly diffusion-controlled rates ($k \approx 2 \times 10^9$ $M^{-1}s^{-1}$). In the case of the purines, free-radical addition can also lead to radicals which have a high spin density at nitrogen. There are indications that such radicals might not react with oxygen. Our knowledge of the free-radical chemistry of purines, especially in the presence of oxygen, is at present still very limited,[1] so that we must restrict ourselves to pyrimidines and to the sugar moiety. It is not the intention of this article to give a detailed description of pyrimidine[2-5] and sugar[6,7] peroxyl radical chemistry. This has been done recently in the reviews indicated. Rather it will be attempted to make the reader familiar with some of the complexities of peroxy radical reactions and to give some examples of how such reactions are reflected in the degradation of the free components of nucleic acids, and to what extent such results may be extrapolated to the nucleic acids themselves.

THE PEROXYL RADICAL-GENERATING SYSTEM

Our present knowledge is based mostly on experiments in which dilute aqueous solutions of the substrates were saturated with a 4:1 (v/v) mixture of N_2O and O_2 and subjected to ionizing radiation (γ-radiation or high-energy electron pulses). The OH radicals from reactions 1 and 2 react by either adding to double bonds or by abstracting hydrogen atoms (reactions 3 and 4). The resulting carbon centred radicals are converted into the corresponding peroxyl radicals (reaction 5). The H atom is mainly scavenged by O_2 (reaction 6). The $HO_2^{\bullet}$ radical is a fairly strong acid (pH = 4.7), and around neutrality this species is largely associated into $O_2^{\bullet-}$ and H^+. Since the yield of H atoms in this radiolytic system is only 10% of that of the OH radical the yield of primary superoxide ($O^{\bullet}$) is rather low. It will be shown that there are a number of reactions which can yield additional superoxide in the course of the decay of the organic peroxyl radicals.

$$H_2O \rightsquigarrow {}^{\bullet}OH,\ e_{aq}^-,\ H^{\bullet},\ H^+,\ H_2,\ H_2O_2 \qquad (1)$$

$$e_{aq}^- + N_2O \rightarrow {}^{\bullet}OH + N_2 + OH^- \qquad (2)$$

$$^{\bullet}OH + {>}C{=}C{<} \rightarrow HO{-}\overset{|}{\underset{|}{C}}{-}\overset{|}{\underset{|}{C}}{^{\bullet}} \qquad (3)$$

$$\cdot OH + H\text{-}\overset{|}{\underset{|}{C}}\text{-} \rightarrow H_2O + \cdot\overset{|}{\underset{|}{C}}\text{-} \qquad (4)$$

$$\text{-}\overset{|}{\underset{|}{C}}\cdot + O_2 \rightarrow \text{-}\overset{|}{\underset{|}{C}}\text{-}O_2^{\bullet} \qquad (5)$$

$$H\cdot + O_2 \rightarrow HO_2^{\bullet} \quad (\rightleftarrows H^+ + O_2^{\bullet -}) \qquad (6)$$

UNIMOLECULAR DECAY OF ORGANIC PEROXYL RADICALS

Although most of the organic peroxyl radicals decay bimolecularly, those which have an OH group[8] or an >NH group in the α-position can decay unimolecularly by eliminating $HO_2^{\bullet}$. With the OH substituent a spontaneous (reaction 7) and a base-induced reaction (reaction 8) have been observed whereas with >NH substituents the spontaneous $HO_2^{\bullet}$-elimination, if it occurs at all, is too slow to be monitored.

$$HO\text{-}\overset{|}{\underset{|}{C}}\text{-}O_2^{\bullet} \rightarrow HO_2^{\bullet} + \overset{|}{\underset{|}{C}}{=}O \qquad (7)$$

$$HO\text{-}\overset{|}{\underset{|}{C}}\text{-}O_2^{\bullet} + OH^- \xrightarrow[(8)]{-H_2O} {}^-O\text{-}\overset{|}{\underset{|}{C}}\text{-}O_2^{\bullet} \xrightarrow{(9)} O_2^{\bullet -} + \overset{|}{\underset{|}{C}}{=}O$$

In the present context it is noteworthy that the major peroxyl radicals derived from the pyrimidines have an >NH group in the α-position. The OH induced $HO_2^{\bullet}$-elimination in such systems leads to unstable isopyrimidines (e.g. reactions 10 and 12).[9]

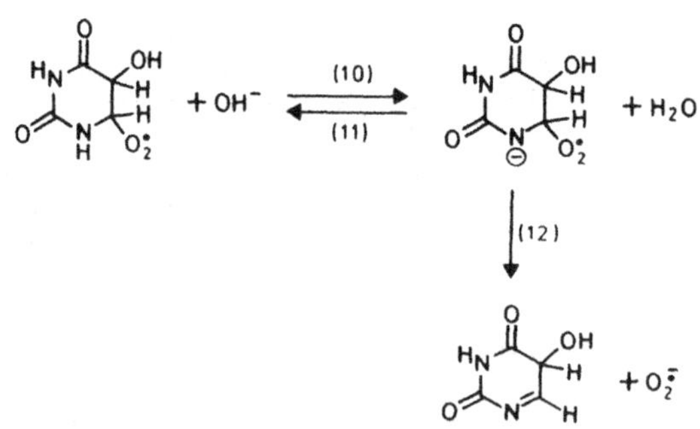

BIMOLECULAR DECAY OF ORGANIC PEROXYL RADICALS

In recent years studies on simple model systems[10] have lead to a picture of the various routes that we followed in the bimolecular decay of organic peroxyl radicals, which is more richly textured than that given a decade ago by Howard in his 1973 review.[11] This could be due to the fact that some of these routes seem to manifest themselves only in aqueous solutions, which in the past had received relatively little attention compared to nonaqueous solutions. Reactions 13 - 23 depict the decay of a secondary (or primary) peroxyl radical. It is obvious from the scheme that tertiary peroxyl radicals can undergo only a restricted number of reactions.

It is now generally accepted that a short-lived tetroxide is an intermediate (reaction 13) which either redissociates into two peroxyl radicals (reaction 14) or decomposes into the products. Three concerted processes are now established: The formation of one molecule of alcohol, carbonyl compound and oxygen (reaction 15, the so-called Russell mechanism); the formation of

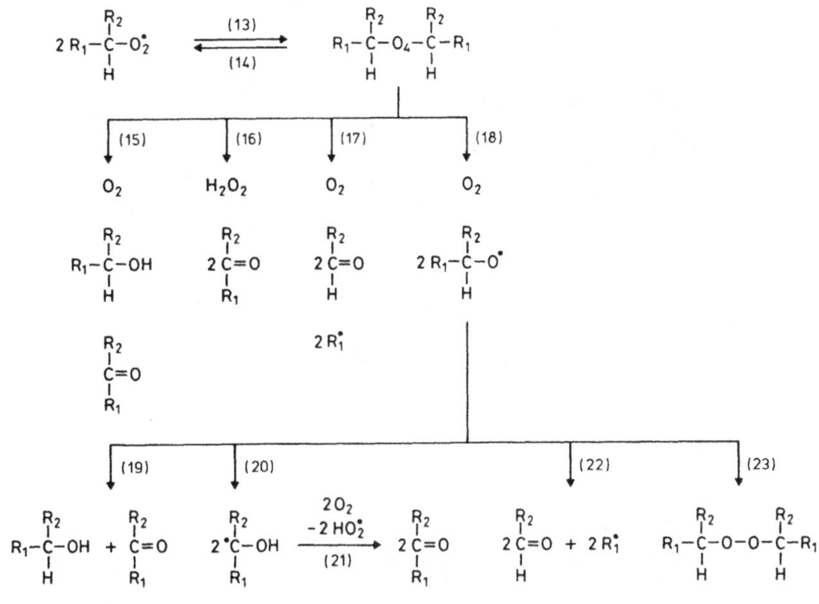

two molecules of carbonyl compound and one molecule of H_2O_2 (reaction 16); and the breakdown into two molecules of a carbonyl compound with a lower carbon content, one molecule of oxygen and two alkyl radicals (reaction 17). Product-wise all these reactions have their free-radical analogues. In reaction 18 one molecule of oxygen and two oxyl radicals are generated. Their disproportionation (reaction 19) leads to the same products as reaction 15. The concerted and the free-radical pathway are not easy to distinguish in this case. However, process 16 and its free-radical analogue (reactions 20 - 21 followed by a disproportionation of the $HO_2^{\bullet}$ ($H^+ + O_2^-$) radicals into H_2O_2 plus O_2 can be distinguished because only the latter yields conducting species ($HO_2^{\bullet} \rightarrow H^+ + O_2^{\bullet-}$) whereas reaction 16 does not. The conducting intermediates can be measured by the pulse radiolysis technique. A further route of intermediate oxyl radicals is the fragmentation (reaction 22) which yields the same products as the concerted pathway 17. If it is possible to create the oxyl radical by a route which does not involve peroxyl radicals, one can distinguish between these two processes (cf. (12)). The combination of two oxyl radicals to give an organic peroxide (reaction 23) is expected to be infrequent (cf. (12)) but it is often difficult to determine such peroxides unequivocally.

On the basis of the knowledge concerning the various routes, it is now possible to account for all of the products identified in the γ-radiolysis of aqueous N_2O/O_2-saturated solutions of uracil.[13] The strong variation induced by the pH of the solution (Table 1) is mainly due to a competition between the OH^--induced $HO_2^{\bullet}$-elimination (reaction 24), followed by the disappearance of the isopyrimidine which is also pH-dependent (reactions 25 and 26) and the bimolecular decay of the peroxyl radicals (reaction 27). The OH group at C(5) facilitates the fragmentation (reaction 30 and/or 33), whereas with two H atoms at this position practically no fragmentation occurs.

THE ROLE OF $HO_2^{\bullet}/O_2^{\bullet-}$

The rate of the bimolecular decay of $HO_2^{\bullet}/O_2^{\bullet-}$ strongly depends on the pH, because $k(HO_2^{\bullet} + O_2^{\bullet-})$ is faster than $k(HO_2^{\bullet} + HO_2^{\bullet})$ and $k(HO_2^{\bullet} + O_2^{\bullet-})$ is negligibly low (for review see ref. 14). Thus even in neutral solutions rather high steady-state concentrations of $O_2^{\bullet-}$ can build up which then could react with an

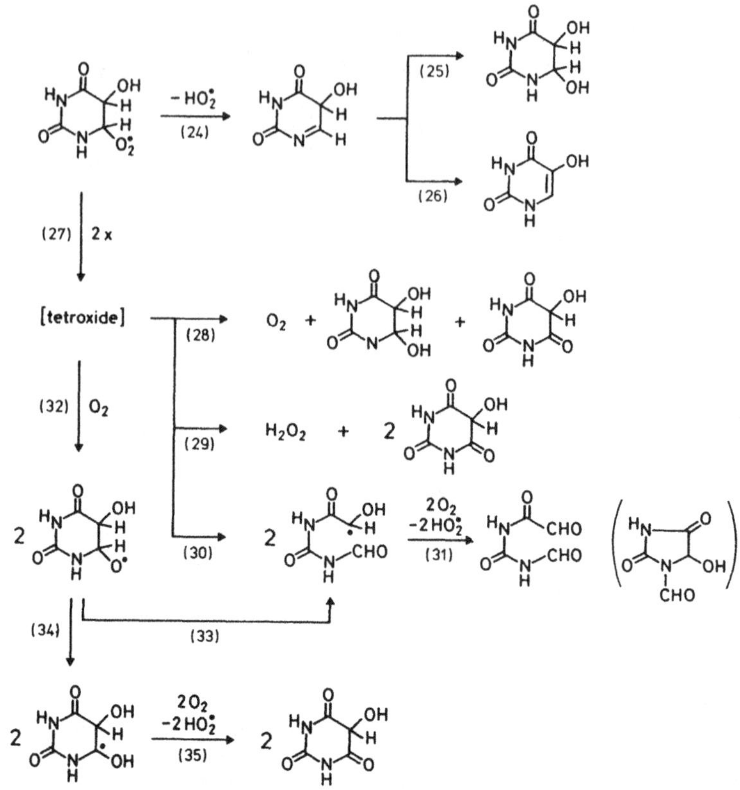

Table 1. γ-Radiolysis of N_2O/O_2-saturated uracil ($2x10^{-4}$ M) solutions under various pH conditions. Products and their G values according to reference 13).

Product	G Value		
	pH 3.0	pH 6.5	pH 10.0
5,6-cis-Dihydroxy-5,6-dihydrouracil	0.6	0.9	1.4
5,6-trans-Dihydroxy-5.6-dihydrouracil	0.5	1.1	1.0
Isobarbituric acid	0	0.2	1.2
1-N-Formyl-5-hydroxy-hydantoin	1.6	1.4	0.2
Dialuric acid	0.9	0.4	0.2
Isodialuric acid	0.1	0.2	0.1
5-Hydroxyhydantoin	0.4	0.4	0.3
Unidentified Product(s)	0.9	0.6	0.9
Hydrogen peroxide	n.d.	3.0	n.d.
Oxygen consumption	n.d.	5.0	n.d.
Uracil consumption	4.9	5.3	5.2

n.d. = not determined

organic peroxyl radical, although the rate constant of reaction 36 is rather small (an estimate of $\underline{k}(36) \leq 10^7\ M^{-1}$ is based on two such cases[15]) compared to the bimolecular decay rate constants of many organic peroxyl radicals which centre around $8 \times 10^8\ M^{-1}s^{-1}$.

$$RO_2^{\cdot} + O_2^{\bar{\cdot}} + H^+ \rightarrow RO_2H + O_2 \qquad (36)$$

Judging from the low bimolecular decay rate constant found for the peroxyl radical derived from dihydrouracil[9] ($k \approx 2 \times 10^7\ M^{-1}\ s^{-1}$), it is conceivable that this constant for thymine peroxyl radicals is about the same or even lower. This could account for the high yield of hydroperoxides formed in aerated solutions of thymine[2] (recall the enhanced yield of $O_2^{\bar{\cdot}}$ due to e_{aq}^--scavenging by O_2 in the absence of N_2O).

Although hydroperoxide formation is the dominant process in the interaction of an organic peroxyl radical with the superoxide radical, there is evidence that in addition to electron transfer, hydrotetroxide formation and breakdown into oxygen and alkoxyl plus hydroxyl radicals may occur (reaction 37).[16]

$$RO_2^{\cdot} + HO_2^{\cdot}/(O_2^{\bar{\cdot}} + H^+) \rightarrow RO\cdot + O_2 + \cdot OH \qquad (37)$$

Thus in the presence of organic peroxyl radicals, the otherwise harmless (i.e. low H-abstractive power) superoxide radical might be converted (albeit with low efficiency) into the reactive hydroxyl radical. As can be seen from the general scheme, the oxyl radical can yield new $HO_2^{\cdot}/O_2^{\bar{\cdot}}$ radicals and the $\cdot OH$ radical new $RO_2^{\cdot}$ radicals. Hence $\cdot OH$ radical formation by this route can occur via a short chain reaction.

PYRIMIDINE RADICAL ANIONS AND OXYGEN

In aerated solutions, if the concentration of the pyrimidine exceeds that of oxygen, the solvated electron is scavenged by the pyrimidine and a pyrimidine radical anion is formed. Pyrimidine radical anions rapidly transfer an electron to O_2 (reaction 38). Alternatively, if an oxygen adduct is an intermediate, its lifetime must be less then one microsecond. The same type of reaction is given by the oxygen protonated species (reaction 41). Thus by the action of oxygen any further reactions of the base radical anion are forestalled so that the integrity of the base is maintained. However, in the presence of buffer protonation at C(6) (reaction 42) is speeded up[17] such that at low oxygen concentrations this protective effect no longer operates, and the pyrimidine is destroyed via the C(5)-peroxyl radical (reaction 43) (D.J. Deeble and C. von Sonntag, unpublished results).

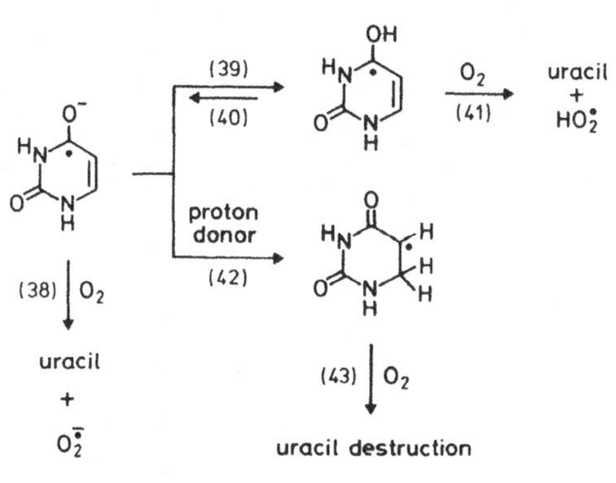

EXTRAPOLATION FROM THE SIMPLE NUCLEOBASES TO THE NUCLEIC ACIDS

All the radiolytic studies on the relatively simple nucleobases have been done in order to learn about the structures which are formed in such reactions and to extrapolate to the nucleic acids. This, however, is only possible if the mechanism of the decay of the base radicals is really understood. Only recently has OH^--induced $HO_2^{\bullet}$-elimination, which also plays a dominant role in neutral solutions at low dose rates of 60-Co γ-irradiation, been recognized and its importance regarding the distribution of the radiolytic products of the pyrimidines been established. It is evident that this reaction, which requires a hydrogen at N(1), cannot proceed in the nucleic acids where the base is linked to the sugar moiety at this position. The nucleosides, or the pyrimidines irradiated at pH 3 where $HO_2^{\bullet}$-elimination is very slow, are therefore better model systems.

The usefulness of the low-molecular-weight model approach is attested to by the work of Teoule and Cadet (cf. (18-22)) which indicates that lesions observed with the nucleobases and the nucleosides can also be found in the nucleic acids. However, as will be shown below, one must be careful not to overstress this analogy because the peroxyl radicals are much longer-lived in nucleic acids due to lower diffusion rates, and they can undergo reactions kinetically of first-order in peroxyl radicals (e.g. reactions with the sugar backbone or reactions with other bases), which are impossible or at least very difficult to investigate with low-molecular-weight material.

THE SUGAR MOIETY

Numerous investigations have been carried out to arrive at an understanding of the peroxyl radicals derived from carbohydrates[6] many of them with the goal to gain some insight into the mechanism of strand breakage of nucleic acids.[7] In such a process either a phosphate linkage at C(3') or at C(5') might be broken. Fragmentation of the carbon skeleton at C(4')-C(5') or C(3')-C(4') and rapid hydrolysis of the former C(1')-C(4') acetal linkage can also lead to strand breakage. One might also consider other lesions which are heat- and alkali-labile (alkali-labile sites) and which do not lead to an immediate strand break.

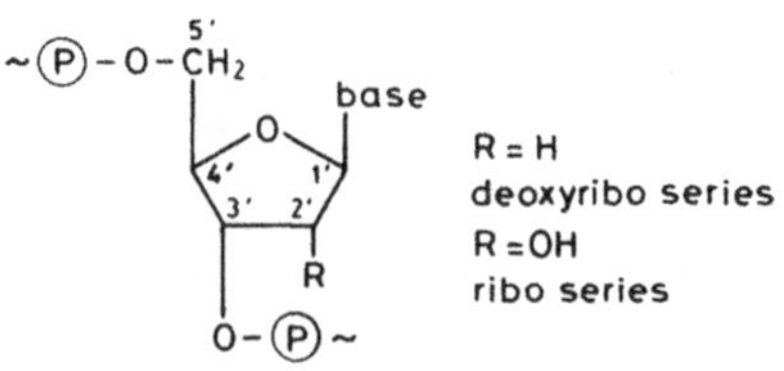

It is obvious from the general scheme of peroxyl radicals (reactions 13-23) that peroxyl radicals at C(3') and C(5') can produce a strand break. Recent investigations on simple model systems[23,24] have shown that the resulting 1-hydroxyalkyl phosphates (A) and acyl phosphates (B) are sufficiently short-lived to group at least the former type of lesion among the immediate strand breaks.

A B

Besides the formation of such intermediates there is efficient fragmentation of C-C bonds in α-position to the phosphate group. A similar fragmentation from the peroxyl radical at C(4') (cf. reactions 17 and/or 22) will preferentially cleave the C-C bond within the ring (C(3')-C(4')), a reaction which has been adduced to account for strand breakage in poly(U)[25] and which may well be a step in the formation of "malondialdehyde" type damage in DNA.[4] Although there is now good evidence for many of the sugar lesions discussed above, further work is needed to sufficiently understand this complex system. One case in point is the question of whether sugar radicals are formed directly by OH attack or whether they have base (and base-peroxyl) radicals as precursors. Concomitant with strand break formation there is nearly always loss of an unaltered base. Thus a study of the process of base release will also yield information on strand breakage.

THE ROLE OF BASE PEROXYL RADICALS AS PRECURSORS OF NUCLEIC ACID STRAND BREAKAGE AND BASE RELEASE

In the formation of 2'-deoxyguanosine-5'-aldehyde from 2'-deoxyguanosine it has been shown that the precursor of this sugar lesion is a base radical, and that OH radicals which abstract a hydrogen atom at C(5') are minor contributors.[26] Such an effect is even more pronounced in the nucleic acid poly(U) where base radicals lead to strand breakage and base release with higher efficiency, not only in the absence of oxygen but also in its presence.[25,27-29] In the latter case there is now increasing evidence that the primary radicals are the base radicals while primary sugar radicals play an insignificant role. In the course of reaction the base peroxyl radicals attack the sugar backbone. The sugar radicals thus formed pick up oxygen, and strand breakage as well as base release results from the subsequent reactions of these sugar-derived peroxyl radicals. The yield of strand breakage and base release is very much higher in poly(U) than in poly(A) or DNA. Thus this nucleic acid demonstrates one possible pathway with high efficiency. Future work must be aimed at a full understanding of the poly(U) system and of the reasons why this type of process occurs with lower efficiency in poly(A) and DNA.

It should be noted that in this report only the indirect effect (OH attack) of radicals on nucleic acid damage has been described, but that similar experiments are in progress (e.g. ref. 30) which allow study of the direct effect, i.e. the direct ionization of the nucleic acid and the subsequent reactions which lead to strand breakage, base release and base damage.

REFERENCES

1. J. Cadet and M. Berger, Radiation-induced decomposition of the purine bases within DNA and related model compounds, Int. J. Radiat. Biol. 47:127 (1985).
2. R. Teoule and J. Cadet, Radiation-induced degradation of the base component in DNA and related substances - final products, in: "Effects of Ionizing Radiation on DNA", A. J. Bertinchamps, J. Hüttermann, W. Köhnlein, and R. Teoule, Springer Verlag, Berlin, p. 171 (1978).
3. G. Scholes, Primary events in the radiolysis of aqueous solutions of nucleic acids and related substances, in: "Effects of Ionizing Radiation on DNA", A. J. Bertinchamps, J. Hüttermann, W. Köhnlein, and R. Teoule, Springer Verlag, Berlin, p. 153 (1978).
4. D. Schulte-Frohlinde and C. von Sonntag, Radiolysis of DNA and model systems in the presence of oxygen, in: "Oxidative Stress", Academic Press, New York (1985).
5. C. von Sonntag and H.-P. Schuchmann, The radiolysis of pyrimidines in aqueous solutions. An updating review, Int. J. Radiat. Biol., in press (1985).

6. C. von Sonntag, Free radical reactions of carbohydrates as studied by radiation techniques, Adv. Carbohydr. Chem. Biochem. 37:7 (1980).
7. C. von Sonntag, U. Hagen, A. Schon-Bopp, and D. Schulte-Frohlinde, Radiation-induced strand breaks in DNA: Chemical and enzymatic analysis of end groups and mechanistic aspects, Adv. Radiat. Biol. 9:109 (1981).
8. E. Bothe, M. N. Schuchmann, D. Schulte-Frohlinde, and C. von Sonntag, $HO_2^{\cdot}$ elimination from α-hydroxyalkylperoxyl radicals in aqueous solution, Photochem. Photobiol. 28:639 (1978).
9. M. I. Al-Sheikhly, A. Hissung, H.-P. Schuchmann, M. N. Schuchmann, A. Garner, and G. Scholes, Radiolysis of dihydrouracil and dihydrothymine in aqueous solutions containing oxygen; first- and second-order reactions of the organic peroxyl radicals; the role of isopyrimidines as intermediates, J. Chem. Soc. Perkin Trans. II:601 (1984).
10. H. Zegota, M. N. Schuchmann, and C. von Sonntag, Cyclopentylperoxyl and cyclohexylperoxyl radicals in aqueous solutions. A study by product analysis and pulse radiolysis, J. Phys. Chem. 88:5589 (1984).
11. J. A. Howard, Homogeneous liquid-phase autoxidations, in "Free Radicals", J. K. Kohi ed., Wiley, New York, Vol. 2, p. 3 (1973).
12. H.-P. Schuchmann and C. von Sonntag, Methylperoxyl radicals: A study of the γ-radiolysis of methane in oxygenated solutions, Z. Naturforsch. 39b:217 (1984).
13. M. N. Schuchmann and C. von Sonntag, The radiolysis of uracil in aqueous solutions. A study by product analysis and pulse radiolysis, J. Chem. Soc. Perkin Trans. II :1525 (1983).
14. B. H. J. Bielski, Reevaluation of the spectral and kinetic properties of $HO_2^{\cdot}$ and of $O_2^{\cdot}$ free radicals, Photochem. Photobiol. 28:645 (1978).
15. M. N. Schuchamnn, H. Zegota, and C. von Sonntag, Acetate peroxyl radicals, $\cdot O_2CH_2CO_2^-$: A study on the γ-radiolysis and pulse radiolysis of acetate in oxygenated aqueous solutions, Z. Naturforsch. 40b:215 (1985).
16. M. N. Schuchmann and C. von Sonntag, Hydroxyl radical-induced oxidation of 2-methyl-2-propanol in oxygenated aqueous solution. A product and pulse radiolysis study, J. Phys. Chem. 83:780 (1979).
17. S. Das, D. J. Deeble, M. N. Schuchmann, and C. von Sonntag, Pulse radiolytic studies on uracil and uracil derivatives. Protonation of their electron adducts at oxygen and carbon, Int. J. Radiat. Biol. 46:7 (1984).
18. R. Teoule, A. Bonicel, C. Bert, J. Cadet, M. Polverelli, and B. Fouque, Radiolyse des acides nucleiques - Hydrolyses acide et enzymatique du fragment thymine modifie et lie a la chaine poly- nucleotidique apres irradiation gamma in vitro de l' ADN de E. coli, C. R. Acad. Sci. Paris 279, Ser.D:1797 (1974).
19. R. Teoule, A. Bonicel, C. Bert, J. Cadet, and M. Polverelli, Identification of radioproducts resulting from the breakage of thymine moiety by gamma-irradiation of E. coli DNA in an aerated aqueous solution, Radiat. Res. 57:46 (1974).
20. M. Polverelli, A. Bonicel, and R. Teoule, Gamma-irradiation of cytosine ^{14}C-2 labelled DNA of Escherichia coli. Identification of the major products resulting from rupture of the N-glycosidic bond, J. Radiat. Res. 17:127 (1976).
21. A. Bonicel and R. Teoule, An indirect method of dosing a major defect: The formamide which appears on the DNA backbone during γ-irradiation of DNA, in: "DNA-Repair and Late Effects", Proc. Symp. IGEGM Vienna, H. Altmann eds., Rotzer-Druck, Eisenstadt, 1976:127 (1977).
22. A. Bonicel and R. Teoule, N-Formamido group formed in the place of thymine fragment under gamma irradiation of DNA. Rapid indirect method of evaluation in both in vitro and in vivo conditions, J. Radiat. Res. 20:166 (1979).

23. M. N. Schuchmann and C. von Sonntag, Radiolysis of di- and tri-methylphosphate in oxygenated aqueous solution: a model system for DNA strand breakage, J. Chem. Soc. Perkin Trans. II :699 (1984).
24. M. N. Schuchmann, H. Zegota, and C. von Sonntag, Reactions of peroxyl radicals derived from alkyl phosphates in aqueous solution - model system for DNA strand breakage, in: "Oxygen Radicals in Chemistry and Biology", W. Bors, M. Saran, and D. Tait, eds., Walter de Gruyter, Berlin, p. 629 (1984).
25. D. Schulte-Frohlinde and E. Bothe, Identifiication of a major pathway of strand break formation in poly U induced by OH radicals in presence of oxygen, Z. Naturforsch. 39c:315 (1984).
26. D. Langfinger and C. von Sonntag, γ-Radiolysis of 2-deoxy-guanosine. The structure of the malondialdehyde-like product, Z. Naturforsch., in press (1985).
27. D.G.E. Lemaire, E. Bothe, and D. Schulte-Frohlinde, Yields of radiation-induced main chain scission of poly U in aqueous solution: strand break formation via base radicals, Int. J. Radiat. Biol. 45:351 (1984).
28. D. J. Deeble and C. von Sonntag, γ-Radiolysis of poly(U) in aqueous solution. The role of primary sugar and base radicals in the release of undamaged uracil, Int. J. Radiat. Biol. 46:247 (1984).
29. E. Bothe and H. Selbach, Rate and rate-determining step of hydrogen-atom-induced strand breakage of poly(U) in aqueous solution under anoxic conditions. Z. Naturforsch. 40c:247 (1985).
30. D. Schulte-Frohlinde, J. Optiz, H. Görner, and E. Bothe, Model studies for the direct effect of high-energy irradiation on DNA. Mechanism of strand break formation induced by photo-ionization of poly U in aqueous solution, Int. J. Radiat. Biol. in press (1985).

RADIATION-INDUCED PEROXIDATION OF DNA AND EFFECTS OF REDOX-ACTIVE AGENTS

Jane E. Frew, Peter Jones, and George Scholes
Radiation and Biophysical Chemistry Laboratory

School of Chemistry, University of Newcastle Upon Tyne
Newcastle Upon Tyne. NE1 7RU
England, U.K.

Damage to DNA initiated by ionizing radiations has been a subject of considerable interest in view of the obvious relevance to primary events in radiobiology. This contribution deals with a particular aspect of this damage, namely peroxidation processes arising in the polynucleotides and in their component pyrimidines when their dilute aqueous solutions are irradiated in the presence of molecular oxygen. Under such experimental conditions (appropriate to fundamental chemical studies of the so-called "indirect effect" of ionizing radiations on living cells) chemical processes in the dissolved substances are a consequence of the initial generation of radical species from the radiation-induced decomposition of water according to the process:

$$H_2O \xrightarrow{\text{radiation}} \cdot OH,\ H\cdot,\ e^-_{aq},\ H_2,\ H_2O_2,\ H_3O^+ \qquad (1)$$

which, in the presence of molecular oxygen is followed by the reactions:

$$H\cdot \ + \ O_2 \longrightarrow HO \qquad (2)$$

$$e^-_{aq} \ + \ O_2 \longrightarrow O_2^- \qquad (3)$$

$$HO_2^{\cdot} \rightleftharpoons O_2^{\cdot -} \ + \ H^+ \qquad (4)$$

The equilibrium constant of reaction (4) is such that under physiological pH conditions the superoxide exists essentially as its anion. Of the two primary radical species from water, OH and O , only the OH (the hydroxyl radical) can initiate chemical reaction with the solutes under consideration. In general terms, peroxidation of organic solutes present in the irradiated solution could then ensue _via_ the formation and reduction of an organic peroxyl radical according to the scheme (RH represents a saturated and R an unsaturated organic molecule):

$$RH \ + \ \cdot OH \longrightarrow R\cdot + H_2O \xrightarrow{O_2} RO_2^{\cdot} \xrightarrow{\text{reduction}} RO_2H \qquad (5)$$

$$R \ + \ \cdot OH \longrightarrow ROH\cdot \xrightarrow{O_2} R(OH)O_2^{\cdot} \xrightarrow{\text{reduction}} R(OH)O_2H \qquad (6)$$

Thus, either hydroperoxide or hydroxyhydroperoxide is produced depending upon whether the initial reaction of the hydroxyl radical is one of dehydrogenation at a saturated carbon-hydrogen bond or of addition to an unsaturated carbon-carbon bond.

With nucleic acids present as solutes in the irradiated solutions the chemical evidence, reviewed briefly below, is that the observed peroxidation is associated with hydroxyhydroperoxides formed on the pyrimidine components of polynucleotides. Such peroxidation of DNA has long interested radiobiologists in view of its possible relevance to the well-known oxygen effect in radiation biology viz. the marked sensitizing effects of oxygen on radiation damage to living cells. In this connection questions arise concerning the post-irradiation fate of the hydroxyhydroperoxides produced in situ on the genetic material. If this were to involve some metal-ion catalysed decomposition (e.g. by transition metal cations) there is the possibility that this could lead to further radical-initiated DNA lesions. With this possibility in mind we considered it of interest, in the first instance, to investigate the effects of some transition metal ions on the stability of hydroxyhydroperoxides produced on irradiation of various pyrimidine compounds and polynucleotides, and some of our results are presented in the second part of this paper. Finally, we present some preliminary results on the effect of bleomycin, a redox-active anticancer drug, on DNA hydroxyhydroperoxide.

Peroxidation of nucleic acid and the role of the pyrimidine components

The peroxidation of both DNA and RNA on X-irradiation of their oxygenated aqueous solutions was first reported in 1956. The conclusion that hydroperoxides (later assigned as hydroxyhydroperoxides) were formed arose from the use of analytical procedures which could differentiate between hydrogen peroxide and total peroxide, i.e. hydrogen peroxide + organic hydroperoxide; for the former, the formation of the colored complex with titanium sulphate reagent was used and for the latter the molybdate catalysed oxidation of I^- to I_3^- was employed. It was found that the hydroperoxide in the DNA solutions underwent post-irradiation decay which could be accounted for by the occurrence of two concurrent first-order processes with reaction half-lives of about 1 hr and 3 hr respectively at 25°C;[1,2] the slower component accounted for about two-thirds of the decay process. Experiments carried out with several nucleosides and nucleotides and the corresponding pyrimidine and purine bases indicated that hydroperoxide formation was associated with oxidation of pyrimidine bases. It was therefore concluded that the two hydroperoxides observed in irradiated DNA solutions were derived from thymine and cytosine. Since, of the various deoxyribonucleotides, only thymidylic acid yielded a relatively stable hydroperoxide it was concluded that the slow decaying component in the DNA solutions was essentially connected with the thymine molecule and, furthermore, that the thymine hydroperoxide remained attached to the main polynucleotide chain since it was completely precipitated with trichloroacetic acid.[3] This assignment was supported by more detailed investigations of the decay kinetics of the DNA hydroperoxide.[4,5] In particular, it was found[5] that the activation energies for the fast and slow decay processes corresponded well with those of the hydroperoxides in irradiated cytidylic acid and thymine solutions, respectively. With regard to the mechanism of decay of the hydroperoxides, it was reported that the total peroxide concentration remained constant during the decay process, suggesting a thermal hydrolytic conversion of hydroperoxide to hydrogen peroxide.[5]

The discovery of the nucleic acid hydroperoxides and their association with the pyrimidine components led naturally to more detailed studies of these products formed from the pyrimidines themselves. In view of the instability of cytosine hydroperoxide in neutral solutions,[3,6] the studies were confined to the thymine hydroperoxide which was sufficiently stable to be isolatable from solution. It was, in fact, suggested[7,8] that these peroxides were hydroxyhydroperoxides formed as a result of OH addition to the C(5)-C(6) pyrimidine bond, reaction with oxygen and subsequent reduction of the pyrimidine peroxyl radical as generally indicated by reaction (6). In the case of thymine this would lead to the structures I and II, which would additionally allow the cis- and trans-isomers.

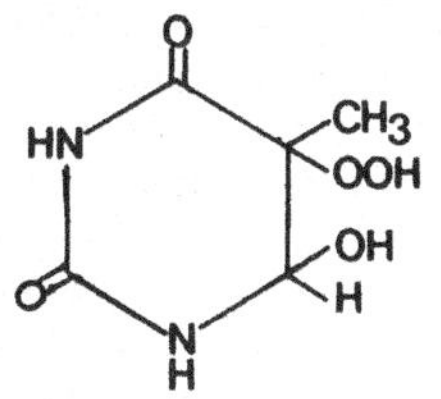

I

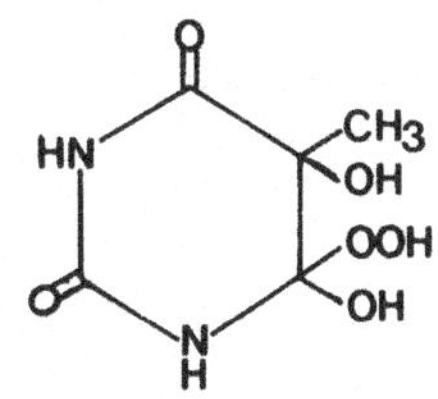

II

The original assignment of the hydroxyhydroperoxide from irradiated thymine solutions as I [7,9] was based on comparisons with the synthesised compounds. However it was later shown (for review see ref. (10)) that the original synthetic methods used led, in fact, to the opposite structural isomers. Indeed, a detailed quantitative study of the products separated from irradiated aerated thymine solutions[11] indicated that both I and II are formed in a ratio of 1:4 and that in both cases the trans:cis ratio is 2. A higher yield of the 6-hydroperoxy compound (II) is to some extent consistent with the recent finding[12] that addition of OH to the C(6) position of thymine is twice as favorable as addition to the C(5) position. Many pyrimidines form hydroperoxides on reaction with •OH radicals in the presence of oxygen and by analogy with thymine are taken to be hydroxyhydroperoxides. It should be noted that the stabilities of both thymine and cytosine hydroxyhydroperoxides are considerable enhanced when formed in the corresponding nucleotides; this allows observation of the decay of the cytosine derivative in irradiated DNA as noted above.

With regard to reduction of the pyrimidine peroxyl radical, it is most likely that this occurs _via_ the superoxide anion,

$$R(OH)O_2^{\bullet} + O_2^{\bullet -} \xrightarrow{(+H^+)} R(OH)(O_2H) + O_2 \qquad (7)$$

a mechanism which suggested itself from studies of the effects of pH on the yields of thymine hydroxyhydroperoxide. It should perhaps be noted that the observed yields of pyrimidine hydroxyhydroperoxides are lower than those expected from the extent of •OH addition to the C(5)-C(6) bonds; in addition to some instability of the pyrimidine hydroxyhydroperoxide, other reactions involving the peroxyl radicals can occur. (These and other details of radical reactions, product formation and thermal decay products have already been reviewed).[10,13]

Such studies with model compounds led to the conclusion that the nucleic acid hydroperoxides are hydroxyhydroperoxides associated with the pyrimidine components. Although no direct evidence is available, it is expected that at least the ratio of the two structural isomers would be the same in the polynucleotide as in the free bases. If, as reported[5], the total peroxide concentration in the DNA solution remains unaltered during the decay process, hydrolysis of hydroxyhydroperoxides to pyrimidine glycols would be a dominant feature. Quantitatively, it has been found that the yield of DNA hydroperoxide in irradiated DNA solutions is strongly concentration dependent.[3,5] It has been found[5] that the yield increases much more with concentration than does the yield of destruction of DNA bases; at the same time the yield of hydrogen peroxide decreases. This was interpreted in terms of an effect of the microviscosity of the DNA solutions in reducing the diffusion of $O_2^{\bar{\cdot}}$; it was considered that this would have a greater effect on the recombination of $O_2^{\bar{\cdot}}$ (producing hydrogen peroxide) than on reaction of $O_2^{\bar{\cdot}}$ with the base peroxyl radicals. That hydroperoxide formation is an important consequence of •OH radical attack on DNA is evidenced by the fact that, in 0.1% DNA solutions, the data[5] would indicate that at least 40% of the OH radicals reacting with the polynucleotides lead to hydroperoxide. It should, of course, be pointed out that, in the polynucleotides, purine as well as pyrimidine bases react with •OH radicals; if hydroperoxides are formed from purine peroxyl radicals they clearly must be very unstable. It is also known that the sugar moeity is attacked by •OH radicals (an early estimate[3] indicated that 20% of the radicals react with the sugar backbone, although more recent evidence may suggest a revision of this value) but, to date, there is no evidence that this results in the formation of stable hydroperoxides.

Effect of Metal Ions

The post-radiolytic decay of the (hydroxy)hydroperoxides of various pyrimidine compounds and of some polynucleotides have now been studied in the absence and presence of various transition metal ions. Solutions, saturated with oxygen, were irradiated with γ-rays with a dose sufficient to generate between $1.7\text{-}3.0 \times 10^{-5}$ mol dm^{-3} of •OH radicals. After removal of hydrogen peroxide with catalase the decay of the hydroperoxide was followed as a function of time, using the iodide method (see reference (14) for details of the methodology). Effects of the cations Co(II), Mn(II), Fe(II), Fe(III) and Cu(II) were studied using concentrations of $1\text{-}6 \times 10^{-6}$ mol dm^{-3} (all as chlorides except Fe(II), which was as sulphate) added to the irradiated solutions after removal of hydrogen peroxide by catalase. It was found that only Fe(II) ions had any significant effects under the conditions used and Fig. 1 shows some of the results obtained with hydroperoxides generated in solutions of thymine, thymidine, thymidylic acid, uracil, uridine, uridylic acid, and the polynucleotides polyuridylic acid, DNA and RNA.

In all cases there is an instantaneous loss of hydroperoxide and, except where all of the hydroperoxide is used up, the time-dependence of the subsequent thermal decay roughly parallels that of the corresponding hydroperoxide in the absence of Fe(II) ions; this is consistent with essentially complete initial consumption of Fe(II) ions. For the pyrimidines and their nucleosides the ratio [hydroperoxide consumed]/[Fe(II) added] is ~1; for DNA the ratio is ~0.5 and for polyuridylic acid ~2. This rather wide variation reflects differences in the mechanisms by which the Fe(II) ions react with the hydroperoxides. An important factor which must be taken into account in this respect is the extent of metal-ion binding to the solutes. The phosphate groups of nucleotides and polynucleotides provide a favorable site for metal-ion binding. Indeed, in separate experiments where the Fe(II) ions were added to the irradiated solutions without removal of hydrogen peroxide by catalase, it was noted that whilst in the irradiated pyrimidine and pyrimidine nucleoside solutions there was preferential reaction of the metal ion with hydrogen peroxide, in the irradiated nucleotide and polynucleotide solutions there was, on the other hand, preferential reaction of the metal ion with the hydroperoxide. The reaction of Fe(II) with the hydroperoxide of polyuridylic acid presents an interesting problem, with results suggesting the occurrence of a chain process. Mechanistic aspects of the metal-ion catalyzed decomposition of these various hydroperoxides appear to be rather complex and are under current investigation. The results clearly demonstrate the instability of polynucleotide hydroperoxides in the presence of Fe(II); the extent to which this leads to further DNA damage, and the detailed chemistry involved, is therefore an area of great interest at the fundamental radiobiological level.

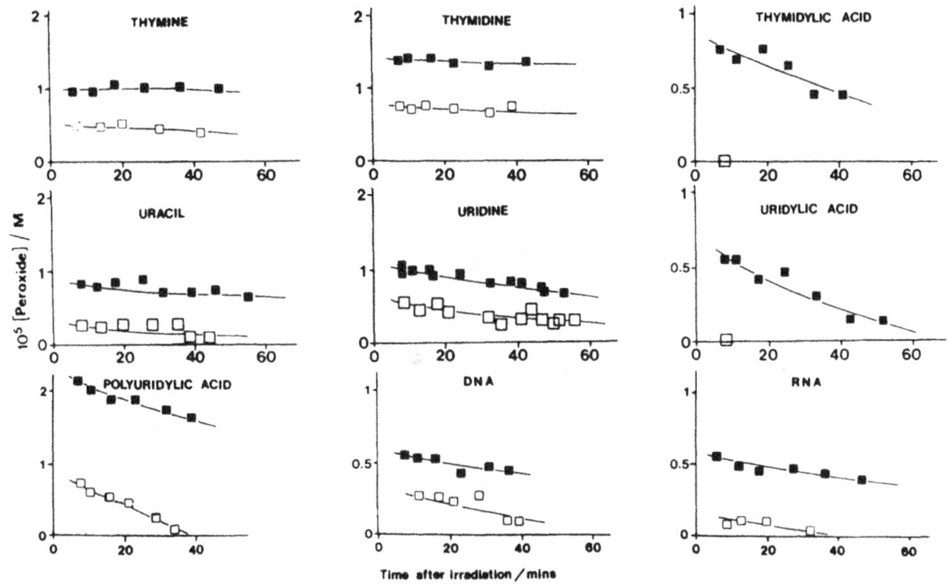

Figure 1. Post-irradiation decay of organic hydroperoxides formed by the γ-radiolysis of aqueous solutions of pyrimidine compounds and polynucleotides. pH = 6.9 (mM phosphate buffer); T 20°C.

Solute concentrations: Polyuridylic acid (4.3×10^{-4} M nucleotide); DNA and RNA, 0.5 g dm^{-3}; other solutions, 3×10^{-4} M.

Dose: Polyuridylic acid, DNA, RNA, 62 Gy; other solutions, 108 Gy.

■, Control
□, In the presence of 6×10^{-6} M $FeSO_4$

Effect of bleomycin

As already discussed above, DNA hydroperoxide under a suitable stimulus could serve to initiate post-radiolytic damage to DNA and hence enhance the oxygen effect in irradiated cells. In addition to certain metal ions, other redox-active compounds can be considered as suitable stimuli, particularly compounds possessing structures which can bring their redox-active centers into intimate contact with the DNA molecule. Bleomycin is such a case and, additionally, provides a good example of how a metal-ion can participate in the mechanism of action of an antitumor agent (see ref. (15) for review).

Bleomycin, chelated with iron, reacts with DNA in the presence of molecular oxygen. The initial stage of the reaction involves reduction of oxygen by the Fe(II)-bleomycin to form a complex ("activated" bleomycin) which then reacts with DNA. Of interest also, in the present context, is that addition to DNA of Fe(III)-bleomycin and either hydrogen peroxide or ethyl hydroperoxide can bring about DNA degradation.[16] In either case, cycling between Fe(II)- and Fe(III)-bleomycin states may occur with the production of •OH and $O_2^{\overline{\cdot}}$ and degradation of the polynucleotide by a radical mechanism. We have therefore carried out some studies of the effect of bleomycin on the stability of the radiation-produced DNA hydroperoxide under such degradative conditions. The results presented in Fig. 2 were obtained using oxygenated DNA solutions

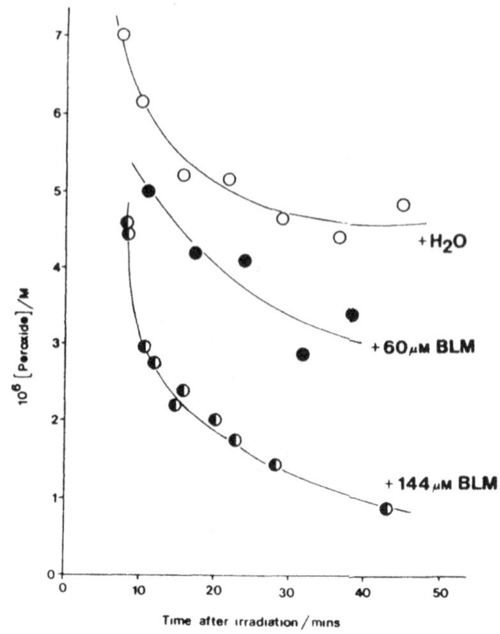

Figure 2. Influence of bleomycin on the decay of DNA hydroperoxide formed on γ-radiolysis of DNA solutions. pH = 6.9 (0.5 mM phosphate buffer); T 20°C; [DNA] = 0.5 g dm^{-3}; dose = 59 Gy. o - control; • - added bleomycin = 6 x 10^{-5} M; ◐ - added bleomycin = 1.4 x 10^{-4} M.

irradiated with γ-rays and subsequently treated with catalase to remove hydrogen peroxide. It can be seen that there is a marked destabilisation of the hydroperoxide which is dependent upon the drug concentration. (N.B. No iron was added to the solution since trace quantities already present are sufficient for activation). In contrast to the DNA hydroperoxide, it was found that there was no effect of comparable amounts of bleomycin on the stability of the radiation-produced thymine hydroperoxide; this would reinforce the view that intercalation of the bleomycin into the DNA and consequent proximity of active sites is an important pre-requisite for reaction with the hydroperoxidic function.

ACKNOWLEDGEMENT

We wish to express our thanks to the North of England Cancer Research Campaign for financial support.

REFERENCES

1. Scholes, G., Weiss, J. and Wheeler, C.M., Formation of hydroperoxides from nucleic acids by irradiation with X-rays in aqueous systems. Nature (London) 178:157 (1956).
2. Daniels, M., Scholes, G., Weiss, J. and Wheeler, C.M., Chemical action of ionizing radiations in solution. Part XVII. Degradation of deoxyribonucleic acid in aqueous solution by irradiation with X-rays (200kV). J. Chem. Soc., 226-234 (1957).
3. Scholes, G., Ward, J.F. and Weiss, J., Mechanism of the radiation-induced degradation of nucleic acids. J. Mol. Biol. 2:379 (1960).

4. Zhizhina, G.P., Zibina, D.L., Kruglyakova, K.E. and Emanuel, N.M., Kinetic characteristics of the decomposition of peroxides in irradiated DNA solutions. Dokl. Acad. Nauk. SSSR 158:935 (1964).
5. Schweibert, M.C. and Daniels, M., ^{60}Co γ-ray induced peroxidation of DNA in aqueous solution. Int. J. Radiat. Phys. Chem. 3:353 (1971).
6. Daniels, M. and Schweibert, M.C., Cytosine hydroperoxide formation in γ-radiolysis of cytosine in aqueous solution. Biochim. Biophys. Acta 134:481 (1967).
7. Scholes, G. and Weiss, J., Oxygen effects and formation of peroxides in aqueous solution. Rad. Res. Suppl. 1:177 (1959).
8. Scholes, G. and Weiss, J., Organic hydroxyhydroperoxides: a class of hydroperoxides formed under the influence of ionizing radiations. Nature (London) 185:305 (1960).
9. Ebert, B. and Monier, R., Structure of thymine hydroperoxide produced by X-irradiation. Nature (London) 184:BA 184 (1959).
10. "Effects of ionizing radiation on DNA", in: Molecular Biology, Biochemistry and Biophysics, Vol. 27, Ed. Huttermann, Kohnlein, W., Teoule, R. and Bertinchamps, A.J. Springer-Verlag, 1978.
11. Teoule, R. and Cadet, J., Radiolysis of thymine in aerated aqueous solution. J. Chem. Soc. Commun., 1269-1270 (1971).
12. Fujita, S. and Steenken, S., Pattern of OH radical addition to uracil and methyl- and carboxyl- substituted uracils: Electron transfer of OH adducts with N,N,N', N'-tetramethyl-phenylenediamine and tetranitromethane. J. Am. Chem. Soc. 103:2540 (1981).
13. Scholes, G, The radiation chemistry of pryimidines, purines and related substances. Photochemistry and Photobiology of Nucleic Acids, Vol. 1 (Ed. S.Y. Wang), 521-577 (1976).
14. Frew, J.E., Jones, P. and Scholes, G., Spectrophotometric determination of hydrogen peroxide and organic hydroperoxides at low concentrations in aqueous solution. Anal. Chim. Acta. 155:139 (1983).
15. Dabrowiak, J.C., Adv. Inorg. Chem. 4:69 (1982).
16. Burger, R.M., Peisach, J. and Horwitx, S.B., Activated bleomycin. A transient complex of drug, iron and oxygen that degrades DNA. J. Biol. Chem. 256:11636 (1981).

THE RADIATION CHEMISTRY OF THE PURINE BASES WITHIN DNA AND RELATED MODEL COMPOUNDS

Jean Cadet, Maurice Berger, and Anthony Shaw

Laboratoires de Chimie
Département de Recherche Fondamentale
Centre d'Etudes Nucleaires de Grenoble
85 X - F38041 Grenoble Cedex

INTRODUCTION

Both the direct and indirect effects of ionizing radiations are believed to contribute to the chemical changes induced in cellular DNA. Relevant information on the possible degradation pathways has been provided by studies using DNA model compounds, the major proportion of which have focused on pyrimidine components[1] and sugar derivatives[2]. With the development of powerful analytical tools such as high performance liquid chromatography (HPLC) and soft ionization mass spectrometry techniques, progress has recently been made in the elucidation of the nature of the radiation-induced chemical modifications of purine bases in DNA and related nucleosides and nucleotides.

This short review details recent aspects of the radiation-induced degradation of adenine and guanine bases in DNA and its model compounds as the result of both direct and indirect effects.

THE INDIRECT EFFECTS

Special emphasis has been placed on the isolation and the identification of the radiation-induced final degradation products of 1 mM 2'-deoxyguanosine in aqueous solutions. Under these experimental conditions the indirect effects predominate through the gamma radiolysis of the water molecules, generating hydroxyl radicals, hydrogen atoms and solvated electrons as the main species. Scavenging experiments have shown that hydroxyl radicals appear to be the main reactive water radicals involved in the degradation pathways. Another feature common to the various irradiated model systems investigated (vide infra) is the notable importance of the restitution reactions involving the purinyl radicals. It is interesting to note that the highest yield of the radiation-induced decomposition of 2'-deoxyguanosine which has been observed in N_2O saturated solution is only 1.5 (Table 1).

Radiolysis of Oxygen-Free Aqueous Solutions

The radiation-induced opening of the imidazole ring of 2'-deoxyguanosine which leads to the formation of the two pyranoid anomers of 2,4-diamino-5-formamidopyrimid-6-one nucleosides or FAPy derivatives is the most important degradation pathway. Initial addition of a hydroxyl radical to the carbon

Table 1. Yields (G Values) of the main radiation-induced[a] degradation products of 2'-deoxyguanosine in oxygen-free aqueous solutions.

Compounds	deaerated	N_2O	t-Butanol[b]
2'-deoxyguanosine	-0.81	-1.50	-0.19
9-(2-deoxy- -D-erythropentopyranosyl)-2,4-diamino-5-formamidopyrimid-6-one	0.08	0.09	0.01
9-(2-deoxy- -D-erythropentopyranosyl)-2,4-diamino-5-formamidopyrimid-6-one	0.26	0.25	0.03
9-(2-deoxy- -D-erythropentopyranosyl)guanine	0.02	0.03	-
9-(2-deoxy- -D-erythropentopyranosyl)guanine	0.01	0.02	-
9-(2-deoxy- -D-erythropentofuranosyl)guanine	0.02	0.02	-
9-(2-deoxy- -L-threopentofuranosyl)guanine	0.02	0.03	-
9-(2-deoxy- -D-erythropento-1,5-dialdo-1,4-furanosyl)guanine	0.07	0.08	-0.01
5',8-cyclo-2',5'-dideoxyguanosine	0.05	0.06	0.01
8-hydroxyl-2'-deoxyguanosine	-	0.24	-
guanine	0.19	0.38	0.03

[a] Determined from the plot of the pecentage of radiation-induced decomposition product against the applied dose for a dose of 400 Gy.

[b] Concentration: 0.5 M

C(8), followed by a hydrogen transfer reaction from a suitable donor which appears to be a hydrogen or electron adduct of a second 2'-deoxyguanosine molecule are the likely initial events in the formation of these FAPy nucleosides.[3] It should be noted that the importance of this base lesion is considerably reduced in DNA when irradiated in deaerated aqueous solution. The FAPy derivative of adenine has been shown to be enzymatically excised from radiation modified polydeoxyadenylic acid by E. coli extracts cotaining an N-glycosylase activity.[4] It has also been proposed that an alternative repair mechanism for this lesion exists which might involve an enzyme catalyzed reclosure for the opened imidazole ring.[5] However, the evidence for the latter seems to be questionable since the DNA substrate used in these experiments showed an unexpectedly high yield of the radiation-induced formation of 4,6-diamino-5-formamidopyrimidine with a very unusual non-linear dependence on the dose.

Formation of 8-hydroxy-2'-deoxyguanosine which is likely to involve the hydroxyl radical adduct to the imidazole carbon C(8) appears to be at best a very inefficient reaction in deaerated aqueous solution (Table 1). Saturation of the irradiated aqueous solution with N_2O leads to a significant formation of this nucleoside. Recently 8-hydroxyadenine has been found to be formed in gamma irradiated aqueous solution of DNA flushed with N O by using an in-line gas chromatography-electron impact mass spectrometry analytical technique.[6]

Another interesting feature of the radiation chemistry of 2'-deoxyguanosine in oxygen-free aqueous solution is the relatively greater importance of the degradation products arising from the fate of osidic radicals in comparison to those of the pyrimidine nucleosides. Rearrangement of the 2-deoxyribose ring leading to the formation of the α furanoid anomer and of the two α and β pyranoid isomers of 2'-deoxyguanosine[3] may be explained by the initial generation of a radical centered at the carbon C(4'). Epimerization has also been shown to occur at position 4' whereas oxidation at carbon C(5') produces the 2'-deoxyguanosine-5'-aldehyde (Fig. 1). It should be noted that the (5'R) and/or (5'S) epimer of 8,5'-cyclo-2'-deoxyguanosine, which is the expected product of the intramolecular aromatic substitution of the purine carbon C(8) by the above 5'-yl radical, as observed for the adenine nucleoside[1] and nucleotide,[7] is not produced. However, a new type of anhydronucleoside with a covalent bond between the carbons C(5') and C(8) but

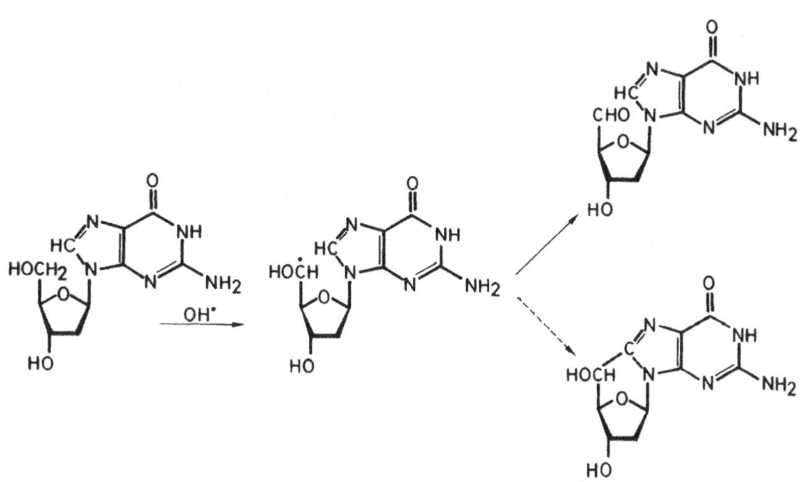

Figure 1. Radiation-induced formation of 2'-deoxyguanosine-5'-aldehyde in oxygen-free aqueous solution.

which has lost the hydroxyl group at the position 5' has been found to be formed through initial hydrogen abstraction at carbon C(4'). It is interesting to note that this 8,5'-cyclo-2',5'-dideoxyguanosine, which is also a degradation product of 2'-deoxyguanosine-5'-monophosphate, represents the first reported radiation-induced event within the sugar moiety of a nucleotide which leads to phosphate release with a concomitant modification of the purine base.[3]

Gamma Irradiation of Aerated Aqueous Solution

The presence of oxygen in the gamma irradiated solution of 2'-deoxyguanosine leads to a pronounced change in the distribution of the final decomposition products. Only 2'-deoxy-D-ribono-1,4-lactone and guanine were found to be formed under air and nitrogen atmospheres. Two major products, which have been isolated as their 3',5'-di-O-acetylated derivatives in order to facilitate their analytical separation by high performance liquid chromatography, have been identified as free 3,5-di-O-acetyl-2-deoxy-D-*erythro*-pentofuranose and an opened imidazole ring nucleoside with an empirical formula $C_3H_6N_4O_2$ for the base moiety. It is noteworthy that these decomposition products have been shown to be generated also through type I photoprocesses in the presence of various photosensitizers such as riboflavin, 3-carbethoxypsoralen and hematoporphyrin,[8] suggesting the occurence of a common intermediate. A reasonable transient reactive species of this photoreaction is a purine radical cation which would result *via* an electron transfer reaction with the photosensitizer in a triplet excited state. The resulting oxidizing radical, which has been shown to be produced in the reaction of 2-deoxyguanosine with Br_2^- or OH•, may be in equilibrium with the neutral radical resulting from interaction with OH^- or H_2O.

It is interesting to note that predominant formation of alkali-labile sites has been observed at the guanine base positions in oligonucleotide chains of DNA when irradiated in aerated aqueous solutions.[10] This might be at least partly related to the generation of DNA apurinic sites by analogy with the specific hydroxyl radical mediated reaction within the purine ring of 2'-deoxyguanosine which gives rise to free 2-deoxyribose (*vide supra*).

The Role of Radiation Modifiers

Steady-state radiolysis of deaerated aqueous solutions of 1 mM 2'-deoxyadenosine or 2'-deoxyguanosine in the presence of 0.5 mM triacetoneamine-N-oxyl (TAN), a stable free radical used as a radiosensitizer, generated a major degradation product which has been identified as the 8-hydroxy substituted derivative of the corresponding nucleoside. A reasonable mechanism for the formation of these oxidized purine nucleosides would involve a recombination reaction of the radiation-induced 8-hydroxy-7,8-dihydropurin-7-yl radical with a molecule of TAN, followed by a Cope rearrangement.[11] A second class of degradation product has been found to result from the binding of TAN to radiation-induced carbon centered radicals at osidic carbons C(4') and C(5'). Epimerization of the former radical gives rise to the α-L-*threo*-pentofuranosyl adduct. A specific radiation-induced binding of TAN to 2'-deoxyadenosine is accounted for by initial intramolecular aromatic substitution of the carbon C(8) by the osidic 5'-yl radical with subsequent formation of a transient anhydronucleoside radical. This latter intermediate is further converted to a stable diamagnetic compound through the binding of a TAN molecule at the carbon C(5').

Oxidizing and reducing radicals have been proposed to be produced as the result of the addition of a hydroxyl radical within the purine ring of 2'-deoxyguanosine or its corresponding 5'-monophosphate ester derivative.[9] It may be inferred from pulse radiolysis experiments that various thiol compounds are able to react with the oxidizing purinyl radical *via* an electron transfer reaction resulting in the restoration of the biomolecule[9] The presence of 0.5 mM cysteine in the gamma irradiated oxygen-free aqueous solution of 1 mM 2'-deoxyguanosine prevents any significant decomposition of the purine ring. This efficient chemical repair contrasts with the increase in the radiation-induced degradation yield of the pyrimidine base of thymidine when irradiated in the presence of cysteine. An almost complete protection of the sugar moiety of 2'-deoxyguanosine is also observed with the exception of the notable increase in the formation of 9-(2-deoxy-L-threo-pentofuranosyl)guanine. This latter compound results from the epimerization of the carbon centered radical at position 4'.

THE DIRECT EFFECTS

To date, studies made on the direct effects of ionizing radiation on DNA and related model compounds have been mostly limited to ESR studies of the radical species produced upon irradiation of the substrate in various solid matrices. The isolation and characterization of the degradation products resulting from the fate of these intermediates encountered serious analytical problems due to the complexity of the radiolysis mixtures and also to the very low yields. Recently important advances have been made in this area, particularly for thymidine and to a lesser extent for 2'-deoxyadenosine. The major decomposition product of the irradiation of 2'-deoxyadenosine in frozen aqueous solution at 196°K has been found to be adenine. The main nucleoside so far isolated resulting from initial formation of a radical within the base moiety is 8-hydroxy-2'-deoxyadenosine. A likely mechanism for its generation would be the addition of OH^- or reaction of H_2O with the adenine radical cation. The formation of an osidic radical centered at carbon C(4') is suggested by the significant formation of the α-furanoid anomer and of the two α and β pyranoid isomers of 2'-deoxyadenosine (Fig. 2), as discussed earlier for the radiation-induced sugar rearrangement of 2'-deoxyguanosine in dilute aqueous solution.[3] Also common to both direct and indirect effects of gamma rays on 2'-deoxyadenosine is the formation of the (5R') epimer of 5',8-cyclo-2'-deoxyadenosine via radical formation at carbon C(5'). The study of the direct effects of ionizing radiation on 2'-deoxyadenosine reveals a predominance of the degradation products resulting from the fate of osidic radicals. This may be accounted for by efficient restitution processes for purinyl radicals.

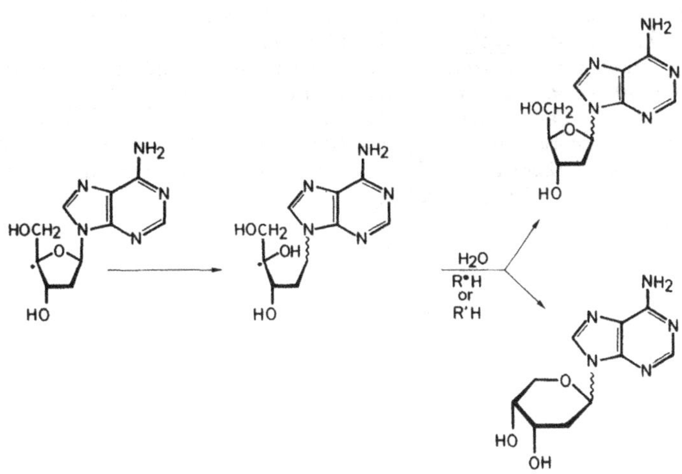

Figure 2. Mechanism of the formation of the α furanoid anomer and of the α and β pyranoid isomers of 2'-deoxyadenosine through the direct effects of gamma rays.

CONCLUSION

Significant progress has been made recently in the determination of the main radiation-induced degradation pathways of the purine DNA nucleosides. These data should facilitate further investigation of the direct and indirect effects of ionizing radiation on more complex substrates such as dinucleoside-monophosphates and DNA itself.

REFERENCES

1. R. Teoule and J. Cadet. Radiation-induced degradation of the base components in DNA and related substances. Final products, in: "Molecular Biology, Biochemistry and Biophysics", J. Hütterman, W. Köhnlein, R. Teoule and A.J. Bertinchamps, eds., Verlag, Berlin, 27: 171 (1978).
2. C. von Sonntag, U. Hagen, A. Schön-Bopp and D. Schulte-Frohlinde. Radiation-induced strand breaks in DNA: chemical and enzymatic analysis of end groups and mechanistic aspects, Adv. Radiat. Biol. 9: 109 (1981).
3. J. Cadet and M. Berger. Radiation-induced decomposition of the purine bases within DNA and related model compounds, Int. J. Radiat. Biol. 47: 127 (1985).
4. L.H. Breimer. Enzymatic excision from γ-irradiated polydeoxyribonucleotides of adenine residues whose imidazole rings have been ruptured, Nucleic Acids Res. 12: 6359 (1984).
5. C.J. Chetsanga and C. Grigorian. In situ enzymatic reclosure of opened imidazole rings of purines in DNA damaged by γ-irradiation, Proc. Natl. Acad. Sci. USA 82: 633 (1985).
6. M. Dizdaroglu. Formation of 8-hydroxyguanine moiety in deoxyribonucleic acid on γ-irradiation in aqueous solution, Biochemistry N. Y. (in press).
7. J.A. Raleigh and A.F. Fuciarelli. Distribution of damage in irradated 5'-AMP: 8,5'-cyclo-AMP, 8-hydroxy-AMP, and adenine release, Radiat. Res. Res. 102: 165 (1985).
8. J. Cadet, C. Decarroz, S.Y. Wang and W.R. Midden. Mechanisms and products of photosensitized degradation of nucleic acids and related model compounds, Israël J. Chem. 23: 420 (1983).
9. P. O'Neill and P.W. Chapman. Potential repair of free radical adducts of dGMP and dG by a series of reductants. A pulse radiolytic study, Int. J. Radiat. Biol. 47: 71 (1985).
10. A.-M. Duplaa and R. Teoule. Sites of gamma radiation-induced DNA strand breaks after alkali treatment, Int. J. Radiat. Biol. 48: 19 (1985).
11. M. Berger, J. Cadet and J. Ulrich. Radiation-induced binding of 2,2,6,6-tetramethyl-1,4-piperidone-N-oxyl to thymidine in oxygen-free aqueous solutions. Isolation and characterization of the adducts, Can. J. Chem. 63: 6 (1985).

BACTERIOPHAGE DNA AS A MODEL FOR CORRELATION OF RADICAL DAMAGE TO DNA AND BIOLOGICAL EFFECTS

Joh. Blok and H. Loman

Biophysics Department
Free University
Amsterdam

INTRODUCTION

The radiolysis of neutral water by γ-rays can be summarized by means of the following symbolical expression:

$$H_2O \rightarrow 0.70\ H_2O_2 + 0.45\ H_2 + 2.7\ \dot{O}H + 2.65\ e^-_{aq} + 0.55\ \dot{H} \qquad (1)$$

The figures in this equation represent the radiation-chemical yields of the products, i.e. the number of molecules or radicals produced per 100 eV of absorbed energy. If oxygen is present in the solution it reacts rapidly with e^-_{aq} and $\cdot H$ yielding a further radical, the anion $O_2^{\cdot -}$.

The purpose of the present paper is to discuss the reactions of the water radiolysis products with DNA and to assess the extent to which the biological functioning of DNA is affected by such reactions. The reactivity of the radiolysis products is most easily studied with single-stranded DNA. On the other hand the replication and transcription of DNA requires the double-stranded form. For these reasons the present discussion is focussed on the radiation effects in the DNA that can be isolated from the bacteriophage ϕX174. This DNA consists of circular single-stranded molecules composed of 5386 nucleotides (Sanger et al.).[1] Under suitable conditions (Guthrie and Sinsheimer;[2] Taketo[3]) it can be transfected to *Escherichia coli* cells. After infection of bacterial cells the first step produces a double-stranded, closed circular DNA molecule that is called RF-DNA (replicative form). A decrease of the biological activity of the single-stranded form after irradiation is therefore either due to damage inhibiting the synthesis of the RF, e.g. strand breaks (see Sinsheimer[4]), or other damage that does not block the synthesis of the first RF molecule but inhibits the further processes required for the production of viable bacteriophage particles.

The double-stranded form of ϕX174 DNA can be isolated from infected bacterial cells and its biological activity can be tested in the same way. For this double-stranded form only radiation damage interferring with the replication and the correct transcription of the RF is biologically relevant. From biochemical experiments it is known that the closed, circular supertwisted form of RF DNA (RFI) does not lose its biological activity after enzymatic introduction of single-strand breaks. The relaxed molecule arising from this treatment is called RFII. If however, double-stranded breaks are introduced in RFI or RFII, the biological activity disappears.

A further advantage of the system is that the biological activity (the number of infective centers or bacteriophage particles produced in a standardized suspension of competent bacterial cells) is strictly proportional to the added DNA-concentration provided the latter is kept below about 0.5 μg/ml. It seems reasonable, therefore, to assume that a fractional decrease of the biological activity by irradiation is proportional to the fraction of molecules whose function is eliminated by inactivating events.

Another type of biologically active DNA that has been used in radiobiological experiments is transforming DNA. This system, which in the past has been studied by DeFilippes and Guild and by Hutchinson, was further explored more recently by Held in a series of papers (see e.g. Held et al.,[5] where further references may be found). The survival curves of this DNA are not exponential and the radiation-induced inactivation does not follow single-event kinetics. Although in a sense this DNA has more resemblance to cellular (bacterial) DNA, it is less suitable than ϕX174-DNA for the establishment of a quantitative relationship between chemical changes and biological effects. It should be stressed however, that statements about relative effectiveness of different radicals or efficiencies of radical reactions pertaining to the ϕX174-DNA system may not always apply to other systems like transforming DNA.

INACTIVATION OF DNA BY WATER FREE RADICALS

Of the irradiation products in water mentioned above, the molecular products H_2, H_2O_2 and the radical anion O_2 do not give a significant contribution to the inactivation of DNA in a pure solution, although a slow reaction is possible, particularly in the presence of contaminating metal ion impurities that may exert a catalytic action. By using suitable radical scavengers, it can be shown that •OH and •H radicals inactivate ϕX174-DNA with roughly equal effectivity (see Blok and Loman[6]).

The survival curves obtained for ϕX174-DNA are exponential over the 4 to 5 decades of inactivation that can be measured. They can be represented by the formula

$$N/N_0 = \exp(-D/D_{37}) \qquad (2)$$

where N/N is the fraction of remaining biological activity after an absorbed dose D. This fraction can be assumed to be equal to the fraction of "surviving" molecules. D_{37} is the dose at which the survival is e^{-1} i.e. 37%. It is the dose at which, on average, one inactivating event has occurred per DNA molecule. The relation between the number of inactivated molecules, G_b, per 100 eV and the value of D_{37} is therefore

$$G_b = (100/rD_{37})N_0 \qquad (3)$$

where r is the absorbed energy (in eV) per unit of mass for one unit of absorbed dose ($r = 6.25 \times 10^{15}$ eV g^{-1} Gy^{-1}) and N is the number of DNA molecules per ml.

In dilute (3.3 μg/ml) solutions of single-stranded DNA, the D_{37}-value is about 0.30 Gy both in O_2 and N_2 in the presence of 0.02 M NaCl, pH about 7, but is reduced to about 0.15 Gy if N_2O is bubbled through the solution (Lafleur et al.[7]). In the latter case hydrated electrons are converted into •OH radicals through the reaction

$$e_{aq}^- + N_2O + H_2O \rightarrow N_2 + OH^- + \cdot OH \quad (4)$$

In the presence of 0.01 M phosphate, pH 7.2, the hydrated electrons are converted into •H radicals by the reaction

$$e_{aq}^- + H_2PO_4 \rightarrow HPO_4^{2-} + \cdot H \quad (5)$$

which has a rate constant of about 5 x 10^6 $M^{-1}s^{-1}$. Under these conditions the D_{37} value is again about 0.30 Gy in O_2 and 0.15 Gy in N_2O, but also about 0.15 Gy in N_2.

These experiments show that the biological yield, G_b, is roughly doubled if e_{aq}^- is replaced either by •OH or •H. In view of the primary radical yields given in expression 1, this means that if there is a biological inactivation by hydrated electrons, it must be very small. In a more sensitive experiment, Nabben et al.[8] eliminated •OH and •H by scavenging these radicals with t-butanol. Moreover, they reduced the scavenging of e_{aq} by protons by working at pH 9.2 (10^{-3} M disodium tetraborate). 10^{-4} M $MgCl_2$ was added in order to reduce the negative charge on DNA, which might reduce the rate constant for e_{aq}^- + DNA. Under these conditions an inactivation was found that must be due to e_{aq}^-. This could be confirmed by measuring the protection afforded by the electron scavenger acetone. This experiment is somewhat less simple than it seems because a correction must be applied for inactivation by acetone radicals formed from acetone + e_{aq}^- . A quantitative evaluation of the data at various t-butanol concentrations up to 0.5 M led to the conclusion that in N_2, the biological effect of the hydrated electron must be about 10 times smaller than that of •OH. This was confirmed by similar experiments with isopropanol (up to 10^{-2} M) as a scavenger of •H and •OH radicals.

There are three reasons for the low contribution of e_{aq}^- to inactivation. Firstly the rate constant of the reaction e_{aq}^- + DNA is about 3 times smaller than that of •OH + DNA. Secondly scavenging of the electrons by impurities in the solution appeared to be about twice as effective as for

Table 1. Effect of gas and buffer conditions on the biological yield double-stranded DNA inactivation

Buffer	Gas	G_{OH}	G_H	G_e	$\frac{G_{OH} + G_H}{G_{OH} \text{ in } O_2}$	$\frac{G_b}{G_b \text{ in } O_2}$
90 mM	O_2	2.7	0	0	1.00	1.0
phosphate	N_2	2.7	3.2	0	2.19	2.4 ± 0.3
pH 7.9	N_2O	5.35	0.55	0	2.19	2.1 ± 0.2
30 mM	O_2	2.7	0	0	1.00	1.0
tetraborate	N_2	2.7	0.55	2.65	1.20	1.2 ± 0.2
pH8+0.2M NaCl	N_2O	5.35	0.55	0	2.19	2.1 ± 0.3
10 mM	O_2	2.7	0	0	1.00	1.0
phosphate	N_2	2.7	3.2	0	2.19	2.1 ± 0.1
pH 7.2	N_2O	5.35	0.55	0	2.19	2.3 ± 0.2

Double-stranded X174-DNA, concentration about 1 μg/ml, was irradiated with γ-rays from ^{60}Co. The D_{37} under oxygen is between 1.5 and 2.0 Gy, depending on the preparation.

•OH radicals. Thirdly the inactivation efficiency (the fraction of the electron-DNA reactions that results in elimination of the biological activity of a DNA molecule) which could roughly be estimated from the experiments (this important factor will be discussed later) was about half that of •OH radical reactions.

Nabben et al.[9] carried out a complete set of experiments to assess the effectiveness of •OH, •H and e^-_{aq} in dilute solutions of doublestranded (RF) DNA. It appeared that the double-stranded form is about 5 times less sensitive than the single-stranded form of ΦX174-DNA, i.e. at the same concentration the D_{37} value is 5 times as large as that of singlestranded DNA under N_2, O_2 or N_2O. Furthermore, •OH and •H radicals are about equally effective and the contribution of hydrated electrons is negligible, like in the case of single-stranded DNA (see Table 1). However, if the possible contribution of electrons was enhanced by applying suitable scavengers, no significant biological effect of hydrated electrons could be detected for double-stranded DNA.

NATURE OF RADIATION DAMAGE

The reaction of water free radicals may bring about a large variety of chemical alterations. If one wants to compare chemical and biological effects at low dose, a complete chemical analysis is not possible, but a number of rather simple techniques are available to discriminate between a few broadly defined categories of damage.

(a) Single-Strand Breaks

If a break is introduced in a single-stranded molecule of ΦX174 DNA, its molecular weight is not reduced because the molecule is circular, but in a sucrose gradient at low salt concentration (0.01 M) it sediments somewhat more slowly due to its increased hydrodynamic radius. Thus for irradiated samples partly overlapping peaks which indicate, respectively, unbroken molecules and molecules with one break, can be observed together with a broad band of more rapidly sedimenting molecules of smaller mass (Lafleur and Loman[10]). At high salt (0.2 M) in sucrose gradients, unbroken molecules and molecules with one break sediment in the same peak. If one assumes a Poisson distribution for the number of breaks per molecule, the fraction of DNA in the peak is $(1 + p)e^{-p}$. The average number of breaks, p, per molecule can then be calculated from this fraction. We refined and standardized our extinction measurements on sucrose gradient fractions to such an extent that the analysis can reliably be carried out with about 1.0 μg/ml of DNA (See Figure 1).

In double-stranded DNA a single-strand break converts the supertwisted RFI into a relaxed RFII molecule which sediments more slowly. In a sucrose gradient both peaks are almost completely separated and their specific biological activity can be measured separately. An even better separation (see Figure 2) is possible by electrophoresis on agarose gels (Van Touw et al.[11]).

By using such techniques one can construct "breakage survival curves" i.e. the fraction o unbroken molecules as a function of dose. Using equation 3 one can then calculate G_{ss}, the number of single-strand breaks produced per 100 eV of absorbed energy, etc.

(b) Double-Strand Breaks

In sucrose gradients or agarose gel patterns of irradiated RFI DNA, a small peak can be seen due to linear DNA molecules arising from a double-

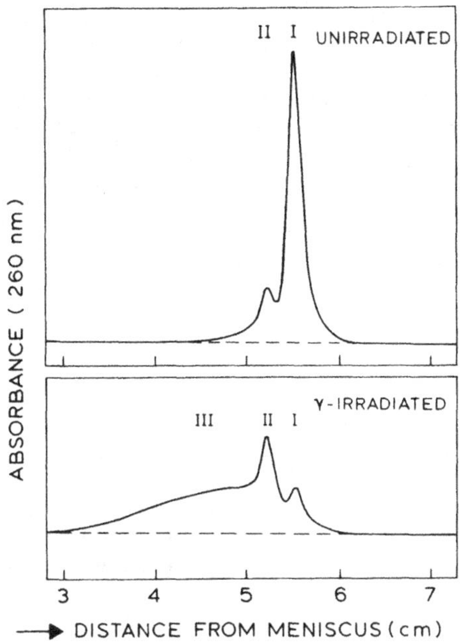

Figure 1. Sucrose sedimentation of unirradiated and irradiated single-stranded ΦX174-DNA at low salt concentration. Unbroken molecules (large peak in unirradiated DNA) sediment faster than molecules with one break. After irradiation a broad band of smaller pieces of DNA appears resulting from two or more breaks in one molecule.

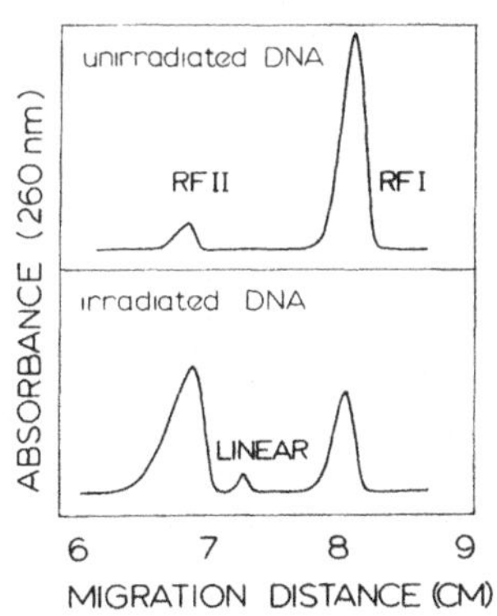

Figure 2. Analysis of double-stranded DNA by electrophoresis on 1% agarose gels (35 volt, 17 hours, 4°C. Amount of DNA about 2 g. After electrophoresis the gel is scanned in a spectrophotometer at 260 nm and the optical density trace is corrected for base line irregularities (monitored at 310 nm) and noise by means of computer procedures (see Van Touw, 1985).[11] Irradiation reduces the amount of RFI, increases the amount of RFII and produces a linear molecule by double-strand breaks (small peak).

strand break in circular molecules. At higher dose double-strand breaks are produced in proportion to the square of the absorbed dose as a consequence of random single-strand breaks due to radical attack on opposite strands of DNA within a certain distance. At low dose however, there is some evidence for a linear production of double-strand breaks (Van der Schans et al.,[12] Blok and Loman[6]) due to a single radical attack. The yield of such linearly produced double-strand breaks is about 100 times lower than the yield of single-strand breaks. Like single-strand breaks in single-stranded bacteriophage. DNA double-strand breaks can apparently not be repaired and are lethal in double-stranded DNA.

In Figure 2 an example is given of the separation of broken and unbroken RF molecules by agarose gel eletrophoresis.

(c) Latent Single-Strand Breaks

If an irradiated sample of single-stranded DNA is brought to a pH of about 12 by adding NaOH, the number of single-strand breaks increases. Apparently, irradiated DNA contains some type of damage that is converted into breaks by alkaline treatment. At room temperature the conversion is completed within 10 minutes. In single-stranded DNA this after-reaction does not reduce the biological activity (see Figure 3). Because breaks are lethal in single-stranded DNA it follows that latent breaks also inactivate a DNA molecule (Lafleur et al.[13]).

In double-stranded DNA latent breaks are also produced. The conversion into breaks is appreciably slower. At present it is not clear whether latent breaks inactivate DNA, because the evidence is conflicting. This point will be further discussed later.

(d) Base Damage

The yield of breaks and latent breaks is not sufficient to account for the biological inactivation of irradiated DNA. This is not surprising

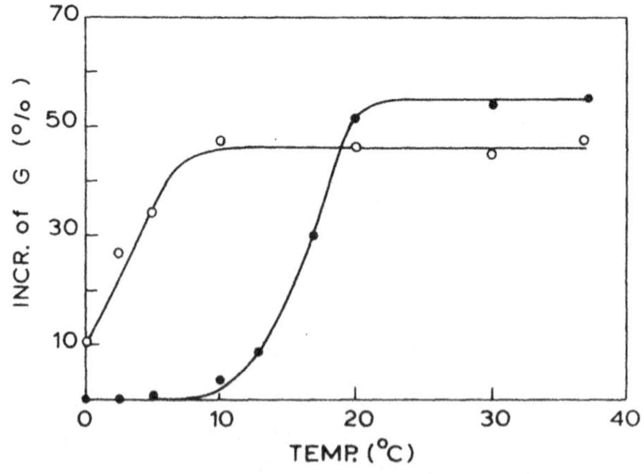

Figure 3. Effect of alkali treatment (10 minutes at pH 12.5) of irradiated (in O_2) single-stranded ϕX174-DNA at various temperatures. Below 10°C the yield of breaks (o) increases without a change in the inactivation yield. The latter increases only (•) at temperatures above 10°C.

because there is ample chemical evidence that a larger part of the radicals reacts with the bases in DNA. One must therefore expect that a large fraction of the DNA molecules are inactivated by a diversity of base alterations. It is, however, quite certain that only a fraction of the base damage causes inactivation.

This is immediately clear from Figure 3. If irradiated single-stranded DNA is exposed to alkali for 10 minutes at about 8°C the biological activity remains unaffected, but a similar treatment at 20°C or higher gives an increase of the inactivation yield of about 30%. This means that a considerable part of the base damage becomes lethal only after a conversion by alkali without further introduction of breaks.

This particular type of latent damage could be tentatively identified. It was shown by Lafleur et al.[14] that the damage introduced in single-stranded DNA by OsO behaves almost exactly in the same way as the latent damage produced by irradiation. The predominant effect of OsO_4 is to saturate the 5,6 double bond of pyrimidines, as was shown by Hariharan et al.[15]. The same authors demonstrated that only about 35% of such reactions causes inactivation. We may therefore conclude that our latent base damage can probably be identified as saturated pyrimidines.

From an analysis of the damage pattern at various radiation doses one can calculate which part of the inactivation is due to each of the above-mentioned categories of damage. A summary of the results is given in Table 2.

INACTIVATION EFFICIENCY OF DNA FREE RADICAL REACTIONS

By means of equation 3 the inactivation yield, G_b, of single-stranded ϕX174 DNA can be calculated from the value of D_{37} (0.3 Gy in O_2 at a DNA concentration of 3.3 μg/ml). The value obtained (G_b = 0.06) is extremely

Table 2. Contribution of various categories of damage to the inactivation of single-strand ϕX174-DNA by water radicals.

	$\cdot OH^1$	$\cdot OH^2$	$\cdot H^3$	$e^{-\,4}_{aq}$
% inactivation by				
breaks	39 ± 8	39 ± 10	10 ± 3	0 ± 1
latent breaks	16 ± 3	17 ± 4	39 ± 3	1 ± 3
other damage	45 ± 6	44 ± 7	51 ± 3	99 ± 3
	100%	100%	100%	100%
% additional inactivation by alkali	30 ± 10	56 ± 10	31 ± 7	5 ± 5

^{1}0.02 M NaCl, N_2O, pH ≃ 7
^{2}0.02 M NaCl, O_2, pH ≃ 7
^{3}0.01 M phosphate + 5 x 10^{-3} M t-butanol, N_2, pH = 7.2
^{4}10^{-3} M borate + 0.5 M t-butanol + 10^{-4} M $MgCl_2$, pH 9.2

low as compared with $G_{OH} = 2.7$. However, before any conclusion can be drawn about the inactivation efficiency, the possibility that the solution is contaminated with radical scavenging impurities has to be examined. If we assume that a scavenger is present at a concentration c_s and that it reacts with ·OH radicals with a rate constant k_s we can write the biological inactivation yield as

$$G_b = \varepsilon G_{OH}\, k_0 N_0/(k_0 N_0 + k_s c_s) \qquad (6)$$

Here k_0 is the rate constant for reaction of ·OH with DNA molecules and ε is the inactivation efficiency. Substitution of this expression in equation 3 gives:

$$D_{37} = (100/r\, G_{OH})(N_0 + k_s c_s/k_0) \qquad (7)$$

By determining the value of D_{37} as a function of the DNA concentration both ε and the impurity-effect can be assessed.

Such an experiment is presented in Figure 4. From the slope one calculates $\varepsilon = 0.24$. The intercept corresponds to a scavenger (impurity) concentration of 8×10^{-5} M if one assumes that its reaction with ·OH radicals has the same rate constant as that of the DNA-nucleotides. Such an impurity concentration is unbelievably high, because the water we use is distilled three times according to standard radiation-chemical procedures and because the DNA radiosensitivity varies less than 10% between different DNA samples. In fact the impurity concentration seems to be of the same order of magnitude as the DNA concentration itself. Most experiments are done with 3.3 μg/ml of DNA, equivalent to 10^{-5} M nucleotide concentration or 1.12×10^{12} DNA molecules per ml. Nevertheless we have no other explanation for the fact that apparently a large fraction of the radicals does not have the opportunity to react with DNA. It can easily be shown by calculation that at the dose rates used a recombination of radicals (e.g. ·OH + ·OH or ·OH + O_2^-) cannot be the cause.

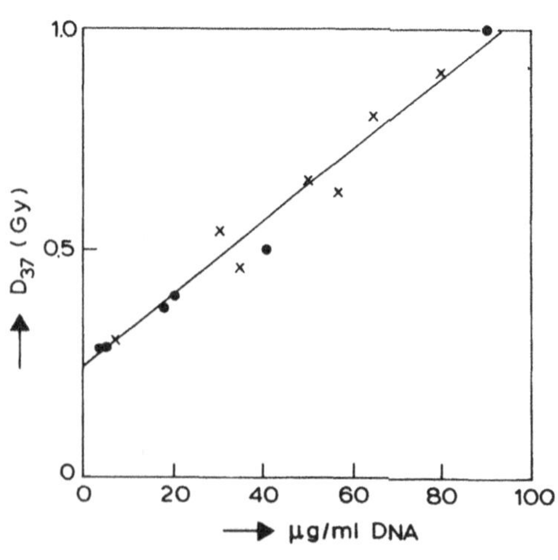

Figure 4. Dependence of the D_{37} value of single-stranded φX174-DNA on DNA concentration in aqueous solutions containing 0.01 M phosphate (x) or 0.02 M NaCl (•). Solutions were saturated with O_2.

For a quantitative understanding of the radiation chemistry of DNA, a further analysis of the rather low inactivation efficiency and the high but reproducible impurity concentration is important. Firstly it should be mentioned that the radiosensitivity of ΦX174 DNA molecules is dependent on their size. With 10^{-2}M phosphate or NaCℓ the D value of a DNA solution of 3.3 μg/ml is 0.25 to 0.30 Gy. If 10^{-4}M MgCl is added, the molecules shrink due to screening of the negative phosphate charges and as a result the D_{37} value becomes 0.40 Gy. This confirms the idea that the DNA is in competition with a certain amount of other solutes, because in a completely pure solution the yield should be independent of the reaction diameter.

Secondly it should be realized that equation 7 is based on the assumption that the DNA concentration is uniform throughout the solution. Instead the DNA is concentrated in random coils with a dimension on the order of 50 nm and (at a concentration of 3.3 μg/ml) these regions of high DNA concentration are separated by an average distance of about 1000 nm. By solving the diffusion equation for a sphere of DNA in a medium in which radicals are produced at random in the presence of small scavenging molecules and comparing the results with the conventional concept of rate constant one can show (Van Rijn et al.[16]) that the molecular rate constant for the reaction of radical with DNA molecules can be written as

$$k_0 = 4\pi RD[1 + R(k_s c_s/D)^{1/2}] \quad (8)$$

where R is the effective radius of the DNA sphere, D the radical diffusion coefficient and $k_s c_s$ the 'scavenging capacity' of the medium.

The mathematical analysis also leads to a rather complex expression for the D_{37} value which appears to vary in a non-linear way with scavenger concentration. Figure 5 shows the results of a series of experiments with various scavengers. By fitting the theory to the experiments using known

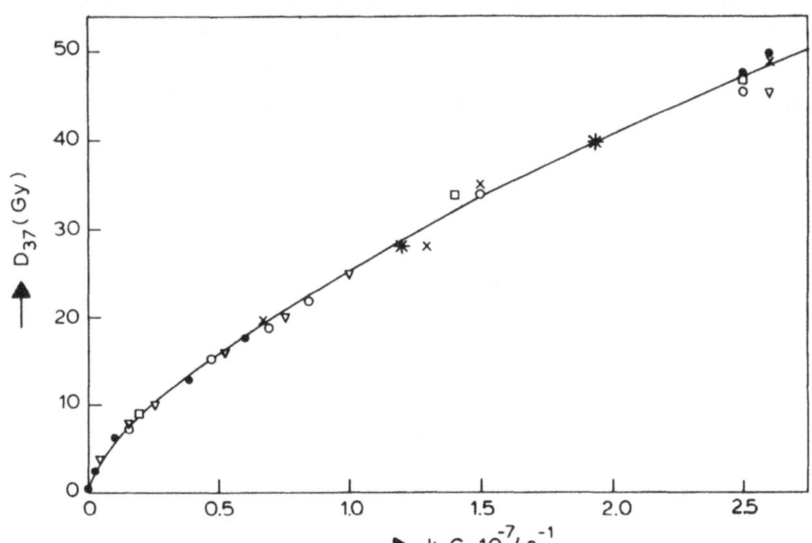

Figure 5. Non-linear dependence of D_{37} of single-stranded ΦX174-DNA on the scavenging capacity (rate constant x concentration) of various scavengers in oxygen-saturated solutions. DNA concentration 3.3 g/ml. ∇ , t-butanol; x, phenol; *, isopropanol; •, nifuroxime; o, misonidazol; □, metronidazol.

data on D and k_s one obtains the values of ε (inactivation efficiency), c_s^0 (remaining scavenger concentration at zero concentration of added scavenger) and R (effective radius of a DNA-sphere in the solution). It turned out that $\varepsilon = 0.14$ to 0.20, $c_s = 2$ to 6×10^{-5} M (assuming the same rate constant as that for t-butanol, viz. 5.2×10^8 M^{-1} s^{-1}) and R = 50 to 70 nm (in the presence of Mg^{++}).

It therefore seems justified to conclude that even in this simple single-stranded DNA only about 20% of the reactions of •OH radicals leads to inactivation. Because roughly half of the inactivation is due to breaks or latent breaks a further conclusion is that only 10% of the OH-radical attack leads to breakage and that the inactivation efficiency of other damage is only about 10%.

Recently the relatively large resistance of single-stranded DNA against inactivation by chemical alterations was confirmed by other experiments in which the DNA was exposed to hydrated electrons, radiation-induced organic radicals and some carcinogenic compounds.

The results are summarized in Table 3. In the case of the formate radical, CO_2^-, the contribution of breaks to inactivation is 23%. For $(CH_3)_2COH$ it is 12%. These radicals do not produce latent breaks. For all the radiation produced radicals the inactivation efficiency of non breakage reactions is therefore about 10%. This low efficiency means that either most of the damage does not prevent the formation of a correct second DNA-strand on the infecting viral strand or that a large part of the damage is removed by biological repair processes. That such a repair is not impossible was shown by experiments of Achey and Wright[17] in which osmium tetroxide treated ϕX174 DNA appeared to be much more active if assayed on spheroplasts in which SOS-repair was induced with a small UV dose.

If RF-DNA is irradiated under the same conditions as the single-stranded form the D_{37}-value is about five times higher. At present this difference cannot yet be analyzed quantitatively in terms of rate constants, impurity effect and inactivation efficiency, but it is quite certain that non-breakage reactions have a very low efficiency. After irradiation of RF-DNA the unbroken RFI molecules can be isolated and tested separately for

Table 3. Inactivation efficiencies of DNA alterations brough about by various reactive chemicals.

Reactant	Inactivation efficiency (%)	Reference
.OH	18 ± 4	Van Rijn et al.[16]
e^-_{aq}	8 ± 1	Nabben et al.[8]
formate radicals	13 ± 2	Nabben et al.[18]
isopropanol radicals	11 ± 1	Nabben et al.[18]
AAF *	100	Lutgerink et al.[19]
AF **	<18	Lutgerink et al., to be published

* AAF = N-acetyl-2-aminofluorene adduct in DNA
** AF = 2-aminofluorene adduct in DNA

their specific biological activity (biological activity of the sample divided by DNA concentration). Lafleur found a decrease of the specific biological activity, but no significant decrease at all if the samples were treated with alkali in order to remove latent breaks. This suggests that latent breaks are 'lethal' in RFI DNA. Recently Van Touw did similar experiments with a sample of DNA with a different history of preparation and found no significant reduction of the specific biological activity even without alkali-treatment. Although the role of latent breaks remains unclear it is evident that the damage remaining after removal of broken molecules has a very low inactivation efficiency. It should be mentioned that the specific biological activity of RFI can only be measured over a limited dose range because at higher doses the amount of RFI becomes too small to be measured.

SECONDARY REACTIONS

When a radiation-induced free radical reacts with a DNA molecule a secondary radical on DNA is formed. Therefore a sequence of further reactions must occur leading to final products in which all electrons are again paired. From pulse radiolysis work it is known that the DNA radicals may have a relatively long life time. This means that there is ample time for substances in the medium, either impurities or material added on purpose (e.g. sensitizers or SH-compounds) to interfere with subsequent radical reactions.

This is what makes the chemistry of free radicals so difficult. Reproducibility can only be achieved by careful purification of all components of the system. Fortunately the D_{37}-values of dilute solutions of ϕX174-DNA are quite reproducible. They show a variation of 5 to 10% only, although we had to conclude that the impurity concentration is stil disappointingly high. The contribution of breaks to inactivation is less reproducible, and it was mentioned already that from some experiments it was concluded that latent breaks inactivate RFI molecules whereas in later samples of DNA they did not seem to affect the biological activity.

Strong evidence for secondary reactions is the finding that the inactivation yield of ϕX174 DNA increases rather abruptly at a concentration (700 to 1000 μg/ml) where the distance between DNA molecules becomes shorter than their radius of gyration. We explained this (see Blok and Loman[6]) by secondary reactions between DNA radicals and undamaged DNA molecules.

In dilute solution one may then expect intramolecular reactions as well and such reactions could change the pattern of damage appreciably and possibly make it more sensitive to interfering reactions with some contaminating material. Recently Lamaire et al. provided convincing evidence that in poly U a rather large part of the strand breaks is due to intramolecular secondary reactions of uracil radicals. If this is also a general phenomenon in irradiated DNA, the damage category of breaks is not independent of that of base damage. This might partly explain why the specific biological activity of RFI DNA is so insensitive to inactivation, for if one removes all molecules with breaks from the total population of damaged molecules, a certain amount of base damage is also eliminated.

IRRADIATION OF INTRACELLULAR DNA

It should be stressed that, although the radiation chemistry of dilute DNA-solutions can provide important information about the nature of radiation-induced free radical reactions occurring also under *in vivo* conditions, such dilute solutions cannot be considered as a model for radiation damage in the living cell, which from a radiation chemical viewpoint is equivalent to a concentrated solution of impurities. Moreover in cells the direct ionization of DNA by fast secondary electrons can give an important contribution to the total damage.

For these reasons Sikkers and Van Touw in our laboratory are working on rather intricate experiments in which E. coli cells are infected with φX174 bacteriophage and irradiated after intracellular accumulation of the double-stranded RF-DNA. After irradiation the damaged viral DNA is separated from the bacterial DNA which is selectively degraded by treatment with restriction enzymes. The viral DNA is then analyzed by electrophoresis on agarose gels. The results obtained so far can be summarized as follows.

(a) The D_{37} of the biological activity is about 1000 times larger than in dilute solution.

(b) The contribution of double-strand breaks, which is at most a few percent in dilute solution, rises to about 20 % for intracellular irradiation.

(c) The specific biological activity of RFI (molecules without breaks) does not decrease significantly in the limited dose range that is accessible to this type of experiment.

(d) Latent breaks can be converted into single-strand breaks by alkali treatment, but not by heat treatment, whereas after irradiation in vitro the conversion can be brough about in both ways. Apparently there exists more than one chemical change that is responsible for this type of damage.

CONCLUSION

It is amazing how much of the radiation-induced damage in DNA is either irrelevant or can be eliminated by repair processes, even in a simple, single-stranded species of DNA. Further research is needed to establish the specific character of the damage that is not eliminated and may lead to mutation or reproductive cell death.

REFERENCES

1. F. Sanger, G.M. Air, B.G. Barell, N.L. Brown, A.R. Coulson, J.C. Fiddes, C.A. Hutchinson III, P.M. Slowcombe and M. Smith, J. Mol. Biol. 125:225 (1978).
2. G.D. Guthrie and R.L. Sinsheimer, Biochim. Biophys. Acta, 72:290 (1963).
3. A. Taketo, J. Ciochem. 72:973 (1972).
4. R.L. Sinsheimer, J. Mol. Biol. 1:37;43 (1959).
5. K.D. Held, H.A. Harrop and B.D. Michael, Int. J. Radiat. Biol. 45:627 (1984).
6. Joh. Blok and H. Loman, Current Topics Rad. Res. 9:165 (1973).
7. M.V.M. Lafleur, H. Loman and Joh. Blok, Int. J. Radiat. Biol. 27:197 (1975).
8. F.J. Nabben, J.P. Karman and H. Loman, Int. J. Radiat. Biol. 42:23 (1982).
9. F.J. Nabben, M.V.M. Lafleur, J.C.M. Sikkers, A.C. Loman, J. Retel and H. Loman, Int. J. Radiat. Biol. 45:379 (1984).
10. M.V.M. Lafleur and H. Loman, Int. J. Radiat. Biol. 41:295 (1982).
11. J.H. van Touw, J.B. Verberne, J. Retel, and H. Loman, Int. J. Radiat. Biol. (1985), in press.
12. G.P. van der Schans, J.F. Bleichrodt and Joh. Blok, Int. J. Radiat.

Biol. 23:133 (1973).
13. M.V.M. Lafleur, M. van Heuvel, H.A. van der Stroom and H. Loman, *Int. J. Radiat. Biol.* 30:223 (1976).
14. V.M. Lafleur, M. van Heuvel, J. Woldhuis and H. Loman, *Int. J. Radiat. Biol.* 33:273 (1978).
15. P.V. Hariharan, P.M. Achey and P.A. Cerutti, *Radiat. Res.* 69:375 (1977).
16. C J.S. van Rijn, G.M. Maier, Joh. Blok, J.B. Verberne and H. Loman, *Int. J. Radiat. Biol.* 47:309 (1985).
17. P.M. Achey and C.F. Wright, *Radiat. Res.* 93:609 (1983).
18. F.J. Nabben, H.A. van der Stroom and H. Loman, *Int. J. Radiat. Biol.* 43:495 (1983).
19. J.T. Lutgerink, J. Retel and H. Loman, *Biochim. Biophys. Acta* 781:81 (1984).
20. M.V.M. Lafleur, J. Woldhuis and H. Loman, *Int. J. Radiat. Biol.* 36:241 (1979).
21. D.G.E. Lemaire, E. Bothe and D. Schulte-Frohlinde, *Int. J. Radiat. Biol.* 45:351 (1984).

ULTRAFAST CHEMICAL REPAIR OF DNA SINGLE AND

DOUBLE STRAND BREAK PRECURSORS IN IRRADIATED V79 CELLS

Barry D. Michael, Susan Davies, and Kathryn D. Held[1]

Cancer Research Campaign Gray Laboratory
Mount Vernon Hospital
Northwood, Middlesex HA6 2RN
England

[1]Department of Radiation Medicine
Massachusetts General Hospital
Boston, MA 02114
U.S.A.

ABSTRACT

The fast kinetics of reactions of free radical precursors of DNA single strand breaks (ssb) and double strand breaks (dsb) have been determined in Chinese hamster V79 cells by fast mixing and irradiation methods using the alkaline unwinding technique to assay breaks. Fast chemical repair of oxygen-dependent ssb and dsb precursors was observed and approached completion within 10-20 ms of irradiation. Treatment of cells with the glutathione synthesis blocking agent, buthionine sulphoximine, showed that approximately half of the chemical repair was attributable to intracellular non-protein thiols. The nature of the residual repair is obscure, but it is apparently not attributable to non-protein thiols. Similar repair rates and thiol dependences were also found for cell kill. With all three endopoints, oxygen competes with and blocks the chemical repair.

INTRODUCTION

Papers in this session address current problems in the study of damage to DNA by fast free radical processes, such as those induced by ionizing radiation. The mechanisms of damage are explored by the measurement of products, their yields, reaction kinetics and reaction rates. The systems used include DNA bases, nucleotides, oligo- and polynucleotides and biologically active DNA. We have adopted a similar mechanistic approach but have sought to measure the kinetics of fast free radical damage to the DNA of intact cells. The object here, as in our earlier studies[1-6], is to compare and link the kinetic information yielded by techniques such as pulse radiolysis from model radiation chemical systems with reaction kinetic data obtained from intact cells and biologically active DNA. This approach has required the development of fast mixing and irradiation techniques which can be used with live cells, retaining their proliferative capacity where necessary. By such methods, it has proved possible to measure the kinetics of radiation-induced intracellular reactions that occur on millisecond timescales. Although these techniques cannot time resolve the fast primary processes such as damage to DNA by direct ionization or indirect attack by water-derived radicals, they do permit observation of the secondary stage in which the initially produced radicals undergo oxidation or reduction to more stable products. The outcome of this secondary stage has a profound influence on the biological effects ultimately produced. Radiosensitization by oxygen and other compounds and radioprotection by thiols occur during this stage. The antagonistic competition between reductive protection by thiols and oxidative radiosensitization has been studied widely by steady-state methods in which the effect of a modifying compound is measured as a function of its concentration. This approach gives valuable information but does not necessarily allow effects on secondary stage chemical modification of damage, as described above, to be distinguished from actions of much slower processes involved in the enzymatic repair of damage[7] or in its eventual expression.

The critical target sites of the damage involved in these secondary stage competing repair and fixation reactions has not been identified, however much evidence points to free radical sites on or associated with nuclear DNA. Cell membranes, particularly membrane-DNA complexes, have also been proposed in this connection[8,9]. Evidence that unrepaired double strand breaks may correlate with lethal events has been reported[10-15]. The object of the studies presented here is an examination of the relationships between the fast radiation-induced secondary stage reactions leading to single and double strand breaks[16] with those that are involved in cell kill[17], and in particular to probe the actions of intracellular thiols and oxygen in these processes. It should be emphasized that although the methodology used necessarily employs ultrafast single-pulse irradiation, essentially the same processes occur during "continuous" radiation exposure which actually consists of a series of virtually instantaneous discrete energy depositions, individually associated with the random passage of ionizing particles through the cell.

METHODS

Cell Culture

V79 379A cells grown in suspension culture were allowed to attach to one side of Thermanox (Lux Plastics) cover slips of 13 mm diameter. Approximately 10^5 cells were applied to each disc and incubated for 18 hours at 37°C in Eagle's MEM (Wellcome Laboratories) supplemented with 10% foetal calf serum (Gibco) and containing 0.1 μCi ^{3}H-thymidine per ml. For experiments with thiol-depleted cells, this incubation was done with 50 μmol dm^{-3} of the glutathione (GSH) synthesis blocking agent buthionine sulphoximine (BSO)[18] present. Before irradiation, cells were cold-chased in medium not containing ^{3}H-thymidine, with or without BSO.

Irradiation and Fast Mixing

Each Thermanox disc was placed in the gas explosion apparatus[1,3] shown in figure 1 and the cells were made hypoxic by flushing the chamber with humidified 95% N_2/5% CO_2 for 2 minutes before irradiation. The cells on each disc were exposed to a single 5 ns pulse of 400 keV electrons from a Febetron 706 electron pulse generator (Hewlett-Packard, McMinnville Division) and to a rapid transition from hypoxic to oxic conditions. Irradiations were done at 23°C. At the surface of the cells exposed to the incoming shot of oxygen its concentration rose from 5 p.p.m. to 10% in 0.1 ms and to 100% in a further 1 ms[3]. By means of electronic time delays, contact with the oxygen shot was preset to occur at a selected instant before, during or after exposure of the cells to the electron pulse.

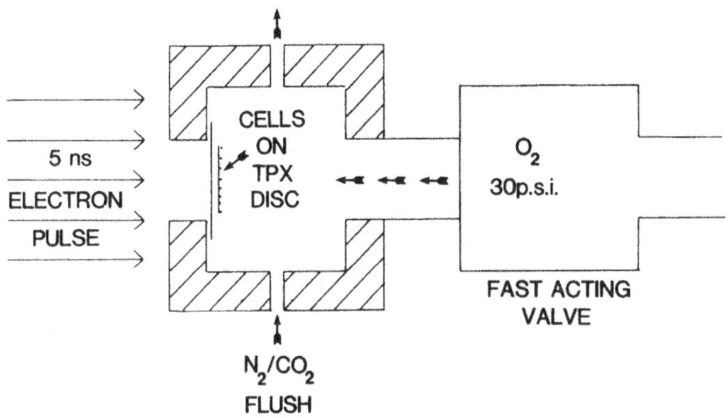

Figure 1. Fast mixing and irradiation apparatus[1,3].

Strand Break Assays

DNA breaks were measured by the alkaline unwinding technique[19-21]. For ssb, discs were placed within 20 s of irradiation into a solution containing 0.05 mol dm^{-3} NaOH and 0.01 mol dm^{-3} Na_2HPO_4 (no NaCl was used) in the dark for 30 minutes at 28°C. Each sample was then neutralized by vigorously adding 0.045 mol dm^{-3} HCl containing phenol red, sonicated for 10 s and SDS added to 0.25%. The solutions were loaded on to columns, maintained at 60°C and containing 150 mg hydroxylapatite (high-resolution grade, Calbiochem), which were then washed through with dilute buffer. Single and double stranded DNA fragments were then eluted with 0.125 and 0.25 mol dm^{-3} phosphate buffer, respectively, and the ^3H activities of each assayed by liquid scintillation counting. For dsb, cells were incubated after irradiation for 2 hours at 37°C to allow ssb to rejoin and the residual breaks assayed as above[22].

RESULTS AND DISCUSSION

Single Strand Breaks

Dose-effect curves for ssb are shown in figure 2. Use of the alkaline unwinding conditions described above gave near-exponential relationships between the double stranded fraction and dose under the various conditions shown, allowing an expression of the type applied by Rydberg[21] to be used.

$$\frac{DS}{SS + DS} = A \exp(-D/D_{37})$$

Here DS and SS are the activities of the double and single stranded eluants, A represents the DS fraction in the unirradiated samples, D is the radiation dose and D_{37} is the dose required to reduce the DS fraction by 1/e. Thus the slopes of the dose-effect lines in figure 2 are proportional to $1/D_{37}$ values which are relative measures of the radiation sensitivities of the DNA to the induction of breaks under the various conditions used. $1/D_{37}$ indicates the yield of breaks per unit dose, although its absolute value depends on the alkaline unwinding conditions employed.

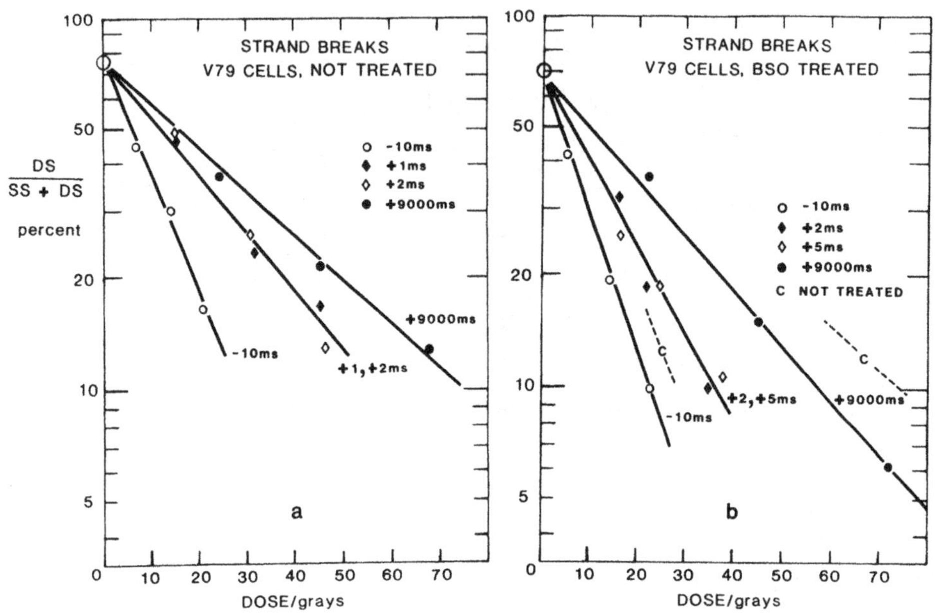

Figure 2. Dose-effect curves for single strand breaks in V79 cells irradiated with the gas explosion technique. Negative times signify O_2 contact before, and positive times after, a single-pulse irradiation[16].

For control cells, not treated with BSO, the data in figure 2(a) show fully oxic (-10 ms) and fully hypoxic (+9000 ms) responses, respectively. With oxygen contact 1 or 2 ms after irradiation (+1, +2 ms) the response is steeper than, although close to, the hypoxic line. Thus there is clearly appreciable post-irradiation sensitization by oxygen. The data in figure 2(b) are for cells treated with 50 mol dm^{-3} of BSO for 20 hours before irradiation. This reduced their non-protein sulphydryl (NPSH) level to below 10% of that of control cells and their GSH content to below 5%. Thiol depletion has two effects: it steepens the oxic (-10 ms) and hypoxic (+9000 ms) responses, also it extends the duration of post-effect sensitization as seen by the data for oxygen contact 2 or 5 ms after irradiation which now lie closer to the oxic than the hypoxic line, contrasting with the +1, +2 ms data in figure 2(a).

The data in figure 2 were from two experiments, one with, the other without BSO treatment. To examine the kinetics in more detail a series of experiments was done, six with and six without BSO treatment. The data from each set were pooled and the time courses obtained are shown in figure 3. Here, values of $1/D_{37}$, as defined above, are shown versus oxygen contact time. As indicated, negative times signify oxygen contact before irradiation and positive times contact after. As in figure 2, thiol depletion increases the yield of breaks throughout and slows down the kinetics of the transition from the fully oxic (-10 ms) to fully hypoxic (+9000 ms) responses. Our interpretation of these kinetics is that about 2/3 of the radiation-induced DNA radicals that can lead to single strand breakage only do so if they react with oxygen. These are oxygen-dependent precursors of strand breaks and, under hypoxia, they decay by a fast chemical repair process which approaches completion at times beyond about 50 ms. When oxygen is suddenly introduced at a high concentration before the decay is complete it reacts rapidly with the radicals that remain unrepaired at that instant and the product leads to strand breakage. Thus the kinetics in figure 3 represent the fast chemical repair of oxygen-dependent precursors of single strand breaks. This interpretation is similar to that made previously for fast kinetic measurements of lethal damage in bacteria[1,3,4] and mammalian cells[2,23] and inactivation of bacterial transforming DNA, probably by strand breakage[5,6].

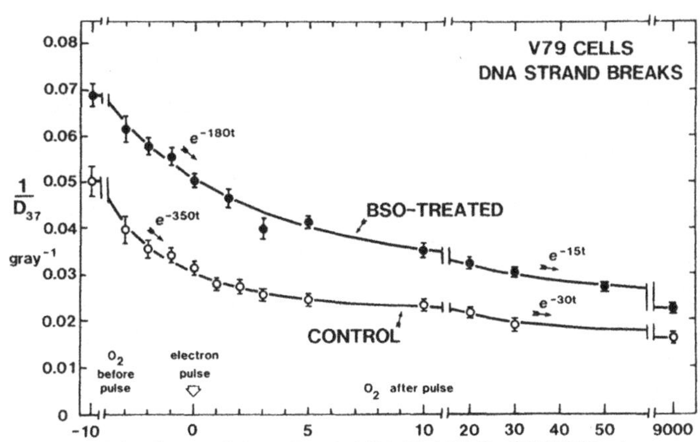

Figure 3. Fast kinetics of chemical repair of O_2-dependent ssb precursors[16].

The requirement for oxygen contact to occur at least several milliseconds before irradiation to give the fully oxic level of sensitivity is explained in terms of the time taken for oxygen to diffuse fully into the cells. The corresponding pre-irradiation contact time requirements with bacteria[3,4] and bacterial transforming DNA[6] are much shorter and this is consistent with the smaller diffusion distances involved.

As in previous studies, we have fitted the decay kinetics with time exponentials and found that there are apparently two stages involved, ~70% of the oxygen-dependent radicals decaying approximately 10-fold faster than the remaining ~30%. Values for the first-order rates are shown in figure 3 and these have been used to generate the lines fitted. As can be seen, thiol depletion slows down the rates of the fast and slow stages of repair by approximately a factor of 2, and this will be discussed later.

Double Strand Breaks

It has been shown that the alkaline unwinding technique can be used to measure the residual double strand breaks that remain after the more rapidly rejoined single strand breaks have been given time to repair[22]. However, it must be pointed out that this finding that residual breaks can be equated with double strand breaks has been brought into question by results obtained using the alkaline elution technique[24]. Rejoining kinetics are shown in figure 4. Approximately 5% of the initial breaks remain after 2 hours' incubation, if it is assumed that the strands on either side of these residual breaks unwind at the same rate as those of the initial, mainly single strand, breaks[22]. We selected a 2-hour incubation before alkaline unwinding as being appropriate to measure residual double strand breaks and found approximately exponential dose-effect relationships (data not shown) similar to those in figure 2, although with about a 20-fold decrease in dose sensitivity.

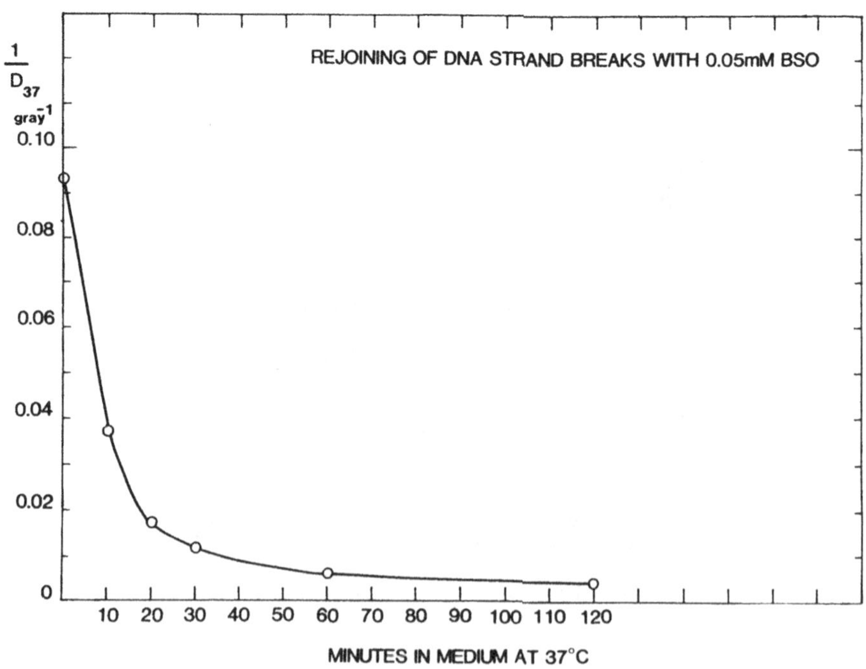

Figure 4. Rejoining of strand breaks in irradiated V79 cells[16].

Fast kinetic experiments, similar to those described above for single strand breaks, were performed but at increased dose levels to offset the reduced yield of breaks. After irradiation, Thermanox discs carrying the cells were placed in medium at 37°C for 2 hours and then transferred to alkali for the unwinding assay. Three experiments were done with BSO treatment and three without and the data of each set were pooled as shown in figure 5. As can be seen by comparing the oxic (-10 ms) and hypoxic (+9000 ms) responses, only about 1/2 of the dsb precursors are sensitized by oxygen as against 2/3 in the case of ssb (figure 3). Thus the oxygen enhancement ratio (OER) is approximately 3 for ssb and 2 for dsb. The yield of dsb in BSO-treated cells is generally greater than that in control cells, as found with ssb (figure 3). Time exponentials were fitted to the data in figure 5 and the first-order rate constants used to generate the curves are shown in the diagram. Unlike the data for ssb, there did not appear to be a significant proportion of oxygen-dependent precursors persisting for longer than 10-20 ms after irradiation (c.f. figure 3) and single time exponentials fitted the data adequately. The first-order rates found to fit for dsb are not significantly different from those for the <u>fast</u> component of ssb.

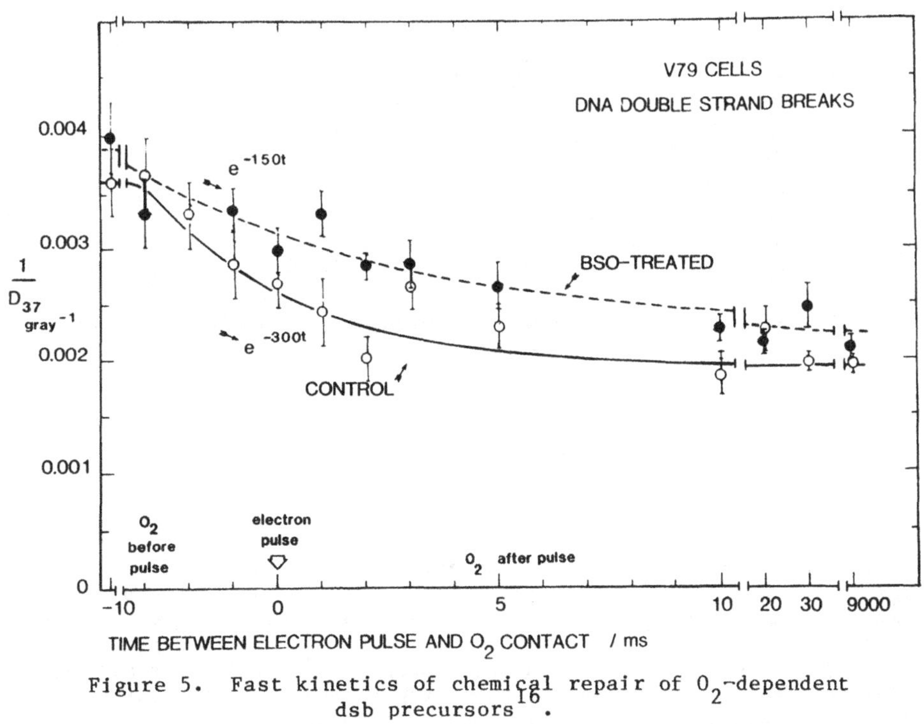

Figure 5. Fast kinetics of chemical repair of O_2-dependent dsb precursors[16].

Cell Survival

In addition to measuring DNA breaks, as described above, cell survival has been scored[17]. Immediately after irradiation, cells were trypsinized off the Thermanox discs, diluted as necessary and plated. After 7 days' incubation, colonies were counted and survival was calculated in the usual way. The plating efficiency was about 65%. Survival curves (data not shown) were shouldered, with an extrapolation number of 2.0 then fell exponentially with dose between survival levels of 10^{-1} and the lower limit measured, 10^{-3}. For the exponential regions, D_o values were obtained from the slopes, representing the dose required to decrease survival by 1/e. The radiosensitivity, defined as $1/D_o$, represents the average yield of lethal events per cell per unit dose in the exponential region. Pooled data from two series of fast kinetics experiments done with and without BSO treatment are shown in figure 6. As found for ssb and dsb, thiol depletion increased the radiosensitivity to cell kill under all conditions of oxygen contact. The decays were fitted with single time exponentials, with first-order rates as indicated in the figure. No significant post-effect sensitization persisted beyond about 10 ms after irradiation.

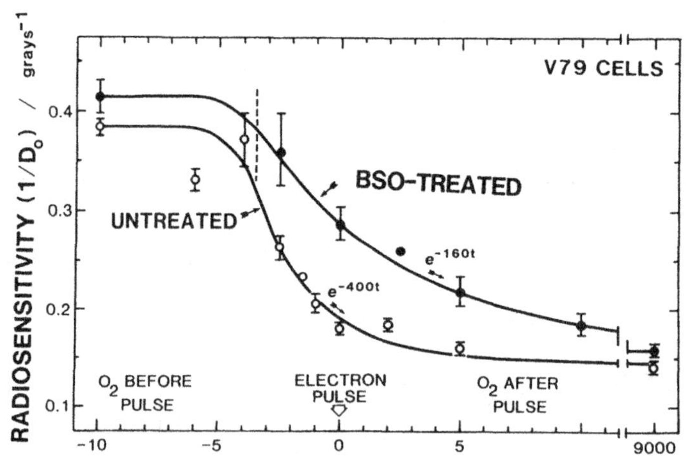

Figure 6. Fast kinetics of chemical repair of O_2-dependent lethal damage[17].

CONCLUSION

The fast chemical repair of radiation-induced damage is an important stage in its progression to biological effect. If chemical repair is incomplete due to competing reactions that block it, for example with oxygen as in the present work, then the damage ultimately expressed is increased. As shown here and in other studies[1-6,23], it is possible to measure the reaction kinetics of these repair processes for various endpoints in mammalian cells. The chemical repair referred to must be discussed in relation to enzymatic repair, not just because the latter occurs on a much slower timescale. Fast chemical repair of damage that would otherwise lead to DNA breaks does not imply fast rejoining of breaks, but rather the ablation of radical sites formed on or near the DNA. Without such repair, these radicals would undergo reactions that ultimately lead to DNA breaks. In enzymatically repair-proficient cells, the majority, though not all, of these breaks would eventually be rejoined. The radical precursors of strand breaks in the present study are those that are sensitive to oxygen. This is either because reaction with oxygen converts them to forms that react further to cause scission of the sugar-phosphate backbone[25], or because it prevents true repair of radical sites formed, for example on the sugar, by loss of a hydrogen atom, which would otherwise lead to strand breakage. Repair of such sites is in keeping with the model for chemical repair by hydrogen atom donation, from thiols, of carbon-centered radicals formed on polymers by irradiation[26], and the validity of this has been demonstrated in model chemical systems[27]. Repair reactions of thiols have been demonstrated in experiments with nucleotides[28-30], polynucleotides[31-33] and DNA itself[6,34,35]. The data presented here demonstrate kinetically that chemical repair of DNA lesions by thiols does occur in mammalian cells. The measured first-order rates of repair are summarized in table 1. The rates shown can be converted to half lives by the expression $t_{1/2}$ = 0.693/(first-order rate), in seconds.

Table 1. Measured first-order rates (s^{-1}) of chemical repair of oxygen-dependent lesions in irradiated V79 cells.

	CONTROL		BSO-TREATED (GSH < 5% NPSH < 10%)	
	fast	slow	fast	slow
SSB	350	30	180	15
DSB	300	—	150	—
LETHAL	400	—	160	—

The main points that emerge from this study are summarized below.

1) The fast chemical repair rates for oxygen-dependent ssb ("fast"), dsb and lethal events are not significantly different from each other. Comparison of the latter two lends further support to the existing evidence that unrepaired dsb are lethal events.

2) However, 30% of ssb precursors ("slow") show a ten-fold longer lifetime than those evident for dsb and for lethal events. This indicates that these "slow" ssb are more efficiently rejoined over 2 hours' incubation than the "fast" ssb. If dsb result from ssb formed in close proximity but on opposite strands, it appears that only the "fast" ssb contribute to this. Conceivably, "fast" ssb are formed in regions of the DNA that are more open and therefore also more accessible to thiols and their chemical repair than are "slow" ssb. "Slow" ssb may be formed in more structured regions where access to thiols is hindered (hence slower chemical repair) but double strand scission is also prevented because the strands are held together until enzymatic repair can take place. A possibility that this suggests is that "slow" ssb are associated with nucleosomal DNA and "fast" ssb (and unrepaired dsb) with linker DNA.

3) Thiol depletion by BSO treatment slows the repair rates by a factor 10
of about 0.5. This is surprising because from simple kinetic considerations at least a 0.1-fold factor would be expected to result from depletion of GSH to <5% and NPSH to <10%. This suggests either that gross cellular measurements of thiol concentrations do not accurately reflect those at the DNA[36], or that an agent, designated below by X and not a non-protein thiol, is responsible for the residual repair. Thus

$$DNA^{\bullet} + O_2 \rightarrow \text{break}$$

$$DNA^{\bullet} + NPSH \rightarrow \text{no break}$$

$$DNA^{\bullet} + X \rightarrow \text{no break}$$

4) The measured intracellular concentration of GSH, which is the major thiol in the control cells used in this work, is approximately 6.5 mmol dm^{-3} [17]. If depletion of GSH to $\leq$5% causes the "fast" repair rate for ssb to fall from 350 to 180 s^{-1} (table 1), then the difference, 170 s^{-1}, indicates a second-order repair rate of 170/0.0065 = 2.6×10^4 dm^3 mol^{-1} s^{-1}, and similarly 2.3×10^3 dm^3 mol^{-1} s^{-1} for the "slow" component. This assumes that the GSH concentration at the DNA does equal the cellular average (see (3) above). These values are substantially lower than the measured rate for repair of bacterial transforming DNA radicals by GSH of $\sim 2 \times 10^5$ dm^3 mol^{-1} s^{-1} [6]. This suggests that conformational factors stearically hinder the reaction of DNA radicals with free thiols in mammalian cells.

ACKNOWLEDGMENTS

The authors wish to acknowledge that this work was supported by the Cancer Research Campaign. Skilled technical assistance from Lisa Karam and Baljinder Gill and processing of the data by Darshan Sehmi are much appreciated, as also are the provision of V79 cells by Michael Woodock and the careful preparation of the script by Dene Hocken and Dorothy Woodman.

REFERENCES

1. B.D. Michael, G.E. Adams, H.B. Hewitt, W.B.G. Jones and M.E. Watts, A posteffect of oxygen in irradiated bacteria: a submillisecond fast mixing study, Radiat. Res. 54: 239 (1973).
2. M.E. Watts, R.L. Maughan and B.D. Michael, Fast kinetics of the oxygen effect in irradiated mammalian cells, Int. J. Radiat. Biol. 33: 195 (1978).
3. B.D. Michael, H.A. Harrop and R.L. Maughan, Fast response methods in the radiation chemistry of lethal damage in intact cells, in: "Radiation Research," S. Okada, M. Imamura, T. Terasima and H. Yamaguchi, eds., Japanese Assoc. for Radiat. Res., Tokyo (1979).
4. B.D. Michael and H.A. Harrop, Timescale and mechanism of radiosensitization and radioprotection at the cellular level, in: "Radiation Sensitizers: Their Use in the Clinical Management of Cancer," L.W. Brady, ed., Masson, New York (1980).
5. B.D. Michael, K.D. Held and H.A. Harrop, Biological aspects of DNA radioprotection, in: "Radioprotectors and Anticarcinogens," O.F. Nygaard and M.G. Simic, eds., Academic Press, New York (1983).

6. K.D. Held, H.A. Harrop and B.D. Michael, Effects of oxygen and sulphydryl-containing compounds on irradiated transforming DNA: III. Reaction rates, Int. J. Radiat. Biol. 45: 627 (1984).
7. J.E. Biaglow, M.E. Varnes, E.P. Clark and E.R. Epp, The role of thiols in cellular response to radiation and drugs, Radiat. Res. 95: 437 (1983).
8. T. Alper, The role of membrane damage in radiation-induced cell death, in: "Membrane Toxicity," M.W. Miller and A.E. Shamoo, eds., Plenum Press, New York and London (1977).
9. T. Alper, "Cellular Radiobiology," pp. 210-224, Cambridge University Press, Cambridge (1979).
10. D.L. Dugle, C.J. Gillespie and J.D. Chapman, DNA strand breaks, repair and survival in X-irradiated mammalian cells, Proc. Natl. Acad. Sci. USA 73: 809 (1976).
11. M.A. Resnick, Similar responses to ionizing radiation of fungal and vertebrate cells and the importance of DNA double-strand breaks, J. Theor. Biol. 71: 339 (1978).
12. D. Frankenberg, M. Frankenberg-Schwager, D. Blocher and R. Harbich, Evidence for DNA double-strand breaks as the critical lesions in yeast cells irradiated with sparsely or densely ionizing radiation under oxic or anoxic conditions, Radiat. Res. 88: 524 (1981).
13. D. Blocher and W. Pohlit, DNA double-strand breaks in Ehrlich ascites tumour cells at low doses of X-rays. II. Can cell death be attributed to double-strand breaks? Int. J. Radiat. Biol. 42: 329 (1982).
14. K. Sakai and S. Okada, Radiation-induced DNA damage and cellular lethality in cultured mammalian cells, Radiat. Res. 98: 479 (1984).
15. I. Radford, The level of induced DNA double-strand breakage correlates with cell killing after X-irradiation, Int. J. Radiat. Biol. 48: 45 (1985).
16. B.D. Michael and S. Davies, in preparation.
17. B.D. Michael, D.C. Shrieve, H.A. Harrop and S. Davies, in preparation.
18. O.W. Griffith and A. Meister, Potent and specific inhibition of glutathione synthesis by buthionine sulphoximine, J. Biol. Chem. 254: 7558 (1979).
19. G. Ahnstrom and K. Erixon, Radiation-induced strand breakage in DNA from mammalian cells: Strand separation in alkaline solution, Int. J. Radiat. Biol. 23: 285 (1973).
20. G. Ahnstrom and K.-A. Edvardsson, Radiation-induced single-strand breaks in DNA determined by rate of alkaline strand separation and hydroxylapatite chromatography: An alternative to velocity sedimentation, Int. J. Radiat. Biol. 26: 493 (1974).
21. B. Rydberg, The rate of strand separation in alkali of DNA of irradiated mammalian cells, Radiat. Res. 61: 274 (1975).
22. P.E. Bryant and D. Blocher, Measurement of the kinetics of DNA double-strand break repair in Ehrlich ascites tumour cells using the unwinding method, Int. J. Radiat. Biol. 38: 335 (1980).
23. R.J. Hodgkiss, N.R. Jones, M.E. Watts and M. Woodcock, Glutathione depletion enhances the lifetime of oxygen-reactive radicals in mammalian cells, Int. J. Radiat. Biol. 46: 673 (1984).
24. G.P. Van der Schans, M.C. Paterson and W.G. Cross, DNA strand breaks and rejoining in cultured human fibroblasts exposed to fast neutrons or gamma rays, Int. J. Radiat. Biol. 44: 75 (1983).
25. D. Schulte-Frohlinde and E. Bothe, Identification of a major pathway of strand break formation in poly U induced by OH radicals in the presence of oxygen, Z. Naturforsch. 39c: 315 (1984).

26. P. Alexander and A. Charlesby, Physico-chemical methods of protection against ionizing radiation, in: "Radiobiology Symposium," Z.M. Bacq and P. Alexander, eds., Butterworths, London (1955).
27. G.E. Adams, R.C. Armstrong, A. Charlesby, B.D. Michael and R.L. Willson, Pulse radiolysis of sulphur compounds. Part 3 - Repair by hydrogen transfer of a macromolecule irradiated in aqueous solution, Trans. Faraday Soc. 65: 732 (1969).
28. J.F. Ward, Deoxynucleotides - models for studying mechanisms of strand breakage in DNA. I. Protection by sulphydryl compounds, Int. J. Radiat. Phys. Chem. 3: 239 (1971).
29. R.L. Willson, P. Wardman and K.-D. Asmus, Interaction of dGMP radical with cysteamine and promethazine as possible model of DNA repair, Nature 252: 323 (1974).
30. P. O'Neill, Pulse radiolytic study of the interaction of thiols and ascorbate with OH adducts of dGMP and dG: Implications for DNA repair processes, Radiat. Res. 96: 198 (1983).
31. E. Bothe and D. Schulte-Frohlinde, Release of K^+ and H^+ from poly U in aqueous solution upon gamma and electron irradiation. Rate of strand breakage formation in poly U, Z. Naturforsch. 37c: 1191 (1982).
32. K. Washino, L. Katsikas and W. Schnabel, Radiation-induced strand breakage in polyriboadenylic acid. A pulse radiolysis study on the protective action of cysteamine under anoxic conditions, Int. J. Radiat. Biol. 46: 747 (1984).
33. D. Schulte-Frohlinde, J. Opitz, H. Gorner and E. Bothe, Model studies for the direct effect of high-energy radiation on DNA. Mechanism of strand break formation induced by laser photoionization of poly U in aqueous solution, Int. J. Radiat. Biol. 48: 397 (1985).
34. J.J. Van Hemmen, W.J.A. Meuling and J. de Jong, Radioprotection of biologically active DNA by cysteamine: A rapid-mix study, Int. J. Radiat. Biol. 25: 455 (1974).
35. M.V. Lafleur, J. Woldhuis and H. Loman, Effects of sulphydryl compounds on radiation damage in biologically active DNA, Int. J. Radiat. Biol. 37: 493 (1980).
36. L. Revesz, The role of endogeneous thiols in intrinsic radio-protection, Int. J. Radiat. Biol. 47: 361 (1985).

ASSESSMENT OF DNA DAMAGE BY FILTER ELUTION ASSAYS

Kurt W. Kohn

Laboratory of Molecular Pharmacology
Division of Cancer Treatment
National Cancer Institute
National Institutes of Health
Bethesda, Maryland 20205

INTRODUCTION

DNA lesions fall into several classes based on the techniques available for sensitive measurement. In mammalian cells, sensitive measurements of the greatest variety of distinguishable DNA lesion classes has been achieved[1] by means of the techniques of alkaline or neutral elution of DNA from filters. The main distinguishable classes of lesions include single-strand breaks (SSB), double-strand breaks (DSB), alkali-labile sites (ALS), interstrand crosslinks (ISC), DNA-protein crosslinks (DPC) and protein-associated strand breaks (PASB). Each of these lesion classes includes a variety of chemical types of breaks or crosslinks at different DNA sites which may differ in their biological consequences; hence generalizations about the biological significance of a given lesion class may not work for all cases.

This paper will review some recent findings from several laboratories which relate to the biological significance of DNA damage measured by filter elution methods. The earlier work was covered in more detail in several previous reviews.[2-5]

BRIEF DESCRIPTION OF THE TECHNIQUE

In brief, the DNA filter elution methods are based upon two main principles. The first is that long DNA strands in a cell lysate, produced by depositing cells on a membrane filter and treating them with an anionic detergent, will be retained on the filter in accordance with DNA strand length; when an eluting solution is pumped slowly through the filter, shorter strands elute sooner on the average. Prior to elution, the lysate on the filter can be treated with proteinase K in order to eliminate the effects of DPC. Usually the eluting solution is alkaline (above pH 12), so that DNA strand pairs are separated and the elution is dependent on the behavior of single strands.

For the measurement of DSB, however, the pH is 10 or less in order to preserve the DNA duplex. SSB and DSB increase the DNA elution rates under these respective conditions. ALS are detected by the increase in DNA elution rate when the pH of the eluting solution is increased, for example from pH

12.1 to 12.6. An increase in elution rate also can be seen with time as ALS are gradually converted to strand breaks by the high pH of the eluting solution.

For the measurement of ISC, chilled cells are x-irradiated prior to lysis, in order to introduce a known frequency of SSB; the frequency of ISC is then gaged on the basis of a reduction in the DNA elution rate.

When multiple lesion types are present simultaneously, for example SSB plus ISC or SSB plus ALS, it is sometimes difficult to obtain unambiguous quantitation for each lesion separately. All of the above assays are best carried out under deproteinizing conditions, so as to eliminate the effects of any DPC which may be present.

The second main principle referred to above is that under appropriate conditions proteins are efficiently adsorbed by the filters, and DNA strands that are covalently linked to proteins are selectively retained. In order to bring out the reduced alkaline elution caused by such DPC, the cells are x-irradiated prior to lysis, as in the case of the ISC assays.

Other variations of the technique can be used to disclose other classes of DNA lesions. For example, the DNA retained on the filter after cell lysis can be treated with a repair endonuclease that cuts DNA strands only at the sites of certain DNA lesions.[6] In another variation, the production of ISC from alkylated DNA is observed during incubation of the DNA on the filter.[7]

The above highly condensed description illustrates the versatility of the filter technique and the potential for further methodologic development.

OVERVIEW OF DNA LESION CLASSES

Single-Strand Breaks (SSB)

The known mechanisms for SSB production in cells are in two categories: free-radical mechanisms and enzyme mechanisms. Free radicals are produced in cells by ionizing radiation and by oxidation-reduction of certain compounds. Hydroxyl radicals or other oxygen radicals can attack DNA bases and may lead to hydrogen abstraction from deoxyribose moieties, causing breakdown of the deoxyribose. The termini of the resulting SSB bear fragments of the disrupted deoxyribose, which must be removed by exonuclease action before the SSB can be repaired.

Hydrogen peroxide (H_2O_2) is a product in some oxidation-reduction reactions in cells and could generate hydroxyl radicals or other oxygen-containing radicals. It was therefore of interest to compare H_2O_2 with X-ray in regard to the production of DNA lesions, mutations and cytotoxicity. Studies in Chinese hamster V79 cells[8] showed that H_2O_2 produces SSB that are repaired at a similar rate to those produced by X-ray.[9] H_2O_2, however, was at least 10-fold less cytotoxic than X-ray, on the basis of doses producing comparable SSB frequencies. H_2O_2 was not detectably mutagenic. Another difference was that H_2O_2 produced no DPC or DSB, lesions that are produced by X-ray. Evidently ionizing radiation produces some types of radical damage in DNA that are not produced by H_2O_2 and that may be the major causes of free radical-induced mutation and cytotoxicity. Thus SSB measurements by themselves may not always indicate the magnitude of biological hazard.

SSB can be generated by repair endonucleases that recognize certain types of DNA damage, for example pyrimidine dimers produced by UV-light. The formation and resealing of these enzymatically induced SSB has been observed in normal human cells and is deficient in the cells of most types of xeroderma pigmentosum.[10] The resealing of such repair intermediate SSB can be slowed by inhibitors of DNA repair synthesis, such as arabinosylcytosine or aphidicolin, and by alkyl-isocyanates, which appear to inhibit the DNA ligase reaction.[11]

Double-Strand Breaks (DSB)

The physical meaning of DSB must be considered briefly. In the filter elution method, a DSB is a lesion that allows a double helical DNA strand to divide into 2 double helical strands under the conditions of hydrodynamic flow through the filter. This could occur due to the presence of two SSB on opposite strands separated by an unknown number of base pairs which however are insufficient to hold the strands together. Thus a DSB in this measurement need not consist of two directly opposed SSB; rather, the opposed SSB may be separated by some distance. It is possible, therefore, that such DSB could be repaired by a mechanism that repairs SSB. This may explain the finding that X-rayinduced DSB and SSB were repaired at similar rates.[9] The fact that the DSB and SSB assays measure different things, however, was shown by the finding that the ratio of DSB to SSB varied for different agents: it was higher for bleomycin than for X-ray, whereas H_2O_2 produced no detectable DSB. X-ray may produce DSB because of the clustering of radicals along an ionization path which would favor multiple radical events in small regions. This effect may be even greater for bleomycin because this drug may act as a radical generator while bound to the DNA.

Alkali-Labile Sites (ALS)

ALS probably arise mainly from base-free sites formed in DNA by the loss of an altered base, either due to spontaneous chemical reactions or, more likely, through the action of glycosylase enzymes. Regardless of how it is formed, a deoxyribose moiety lacking a base can generate an SSB in alkali. The detection of ALS in cells may indicate chemical attack on DNA bases.

Base-free sites are rapidly cleaved in cells by AP-endonuclease, a step which then allows the lesions to be repaired. Hence the level of ALS detected in cells may be viewed as a steady-state balance of formation and repair. According to this picture, it would be possible to enhance the detection of ALS if a specific inhibitor of AP-endonuclease were available.

Not all forms of base damage lead to SSB or ALS. Most notably among these are alkylations of the guanine-O6 position, which are highly mutagenic but would not be detected in these assays. The formation of guanine-O6 alkylations is generally accompanied by alkylation at other sites, especially guanine-N7. However the ratio of alkylation at the two sites varies greatly among compounds, so that a quantitative relation between DNA measurements and the degree of potential hazard may be difficult.

We have recently noted the presence of ALS in DNA strands in which 5-azacytidine was incorporated.[12] This could be due to the chemical instability of 5-azacytidine which could lead to a sequence of reactions analogous to those that yield SSB from base-free sites.

Interstrand Crosslinks (ISC)

The mechanisms of formation and the biological consequences of ISC and DPC have been reviewed recently.[5] In general, ISC result from the actions of compounds that can produce sequentially two covalent reactions with DNA. ISC appear to be highly cytotoxic, and many compounds that produce ISC have anti-tumor activities. Treated cells can remove ISC, usually over a period of many hours, but the mechanism and efficiency of repair are not clear.

We recently studied ISC as well as SSB produced in human cells by the anticancer drug, diazirindinylbenzoquinone.[13-14] The two aziridine groups are potential alkylating groups which may produce ISC; the quinone function may participate in oxidation-reduction reactions in the cell, possibly leading to the production of free radicals and SSB. We found ISC to be the predominant lesions in a human colon carcinoma cell line, whereas SSB were the predominant lesions in a norml and a transformed human embryo cell. The colon carcinoma cell line was by far the most sensitive to killing by the drug, indicating that ISC are more closely related to cytotoxicity than are SSB.

From studies of the action of this compound in isolated cell nuclei;[14] it was found that chemical reduction of the quinone is required for the production of SSB and also enhances the production of ISC. Superoxide dismutase inhibited the production of SSB but not ISC.

DNA-Protein Crosslinks (DPC)

DPC, like ISC, are produced by compounds that can undergo two sequential addition reactions; in this case one covalent reaction links to protein while the second links to DNA. DPC are almost always produced concomittantly by compounds that produce ISC, but appears to have less cytotoxic potential. Some agents, such as heavy metal complexes and formaldehyde, however, produce DPC without ISC; the DPC may be implicated in carcinogenesis or mutagenesis by these agents (for review see reference 5).

It should be noted that DPC may be a highly heterogeneous class of lesions, since the linked proteins (so far unidentified) may be variable, as may be the sites of DNA linkage.

Protein-Associated Strand Breaks (PASB)

PASB are defined as strand breaks that are detected by alkaline elution only under deproteinizing conditions, due to an association with an equal frequency of DPC.[15,16] The criteria of proteinase-dependent SSB and of approximate equivalence between SSB and DPC frequencies must both be met. These lesions are produced by drug-stimulated trapping of complexes between DNA and topoisomerase II.[17,18] The complexes may be similar to intermediates in the topoisomerase reaction, in which an SSB or a DSB is produced with the 5'-termini of each strand break covalently linked to tyrosine groups on the enzyme.

PASB are produced in cells by many, but not all, DNA intercalating agents[15,16,20] and by epipodophilotoxins.[21,22] The trapping of covalent complexes between purified DNA and topoisomerase II by these drugs has been demonstrated.[18,23-25] Although the biological significance of PASB is not clear, it is possible that they may be cytotoxic and may produce chromosome instability and recombinational events.

THE NATURE OF THE DNA ALTERATIONS MEASURED BY ALKALINE ELUTION

Aside from the previously discussed effects of various types of strand breaks and crosslinks, the question has been raised as to whether DNA alkaline elution rate also may be influenced by the state of DNA packing in chromatin.[26,27] This question was raised in connection with some seemingly discordant findings between alkaline elution and other assay methods.

Nicolini et al.[26] reported a discordant result between alkaline elution and an alkaline DNA unwinding assay. Rat liver cell nuclei were lysed in a pH 10 lysis solution (0.2% sodium lauroylsarcosinate, 2 M NaCl, 0.02 M EDTA) and held in this solution at ~20°C for various times before assay. Assays by alkaline elution at pH 12.3 revealed approximately a doubling in elution rate due to holding the lysed nuclei in lysis solution for 48 hr. Assays of alkaline unwinding at pH 12.3, on the other hand, showed a decrease in unwinding rate. Thus alkaline elution assays were suggestive of increasing DNA strand breakage whereas alkaline unwinding assays suggested the opposite. The pardoxical results of the alkaline unwinding assays suggest, as proposed by the authors, that slow conformational changes occur in nuclear DNA when lysed nuclei are held in lysis solution. This is plausible for lysed nuclei suspended in free solution, and there is no easy alternative explanation. The authors then go on to conclude that the increase in elution rate also was due to DNA conformational changes. This conclusion however is not compelling, because it is quite possible that slow chemical DNA strand scission takes place and becomes evident after 24-48 hr in lysis solution. In the alkaline elution assay, the lysed nuclei are partially drawn into the filter pores, and therefore may not undergo further conformational change upon incubation in lysis solution. Hence alkaline elution may in fact be insensitive to conformational factors.

In the same paper, Nicolini et al.[26] presented a physicochemical model of DNA alkaline elution in support of their concept that elution rate depends on DNA strand conformation. The model pictures a DNA strand as having a random-coil conformation encompassing a spherical domain; the probability of elution is pictured as depending on whether the domain of a strand fits over the open area of a filter pore. Differences in DNA packing in chromatin could alter the consequent domain size of DNA strands and would, according to this model, alter the elution kinetics.

This model however is founded on some questionable assumptions. First, the authors assume, on the basis of the commonly observed first-order elution kinetics of DNA from x-irradiated cells,[28] that DNA strands of uniform length would elute according to first-order kinetics. This assumption is inappropriate because x-irradiation produces a random distribution, rather than a uniform distribution of strand lengths. From the observed first-order elution kinetics for a random-length distribution, one can derive mathematically what the elution kinetics for strands of uniform length would be. The result is a step function, which is very different from the first-order kinetics assumed by Nicolini et al. The authors further assume that the single-strands released by alkali have a random-coil conformation (governed by persistance length). This is highly unlikely, because DNA coils will be markedly distorted by the flow of eluting solution into the filter pores.

There is also experimental evidence that strongly argues against Nicolini's model. The model would predict a marked dependence of elution rate on filter pore size. We have found with several types of filters, however, that pore size in the range of 0.45 to 5 μM has little or no effect on the elution kinetics of DNA from x-irradiated cells.

In a recent paper, Nicolini et al.[29] obtained an interesting result in experiments in which cells treated with methylmethanesulfonate (MMS) for 1 hr were lysed on filters and held for various times in pH 10 lysis solution prior to alkaline elution at pH 12.3. The MMS-treated cells showed a much larger increase in DNA elution rate as a function of time at pH 10 than did control cells. Nicolini interprets this result in terms of his model as resulting from unspecified changes in DNA packing in chromatin. A more likely possibility, however, is that some depurination of alkylated bases occurs during the 24-48 hr holding period at pH 10, thereby generating alkali-labile lesions which convert to strand breaks during elution at pH 12.3. The lysis holding procedure may nevertheless provide a sensitive new assay for ALS and DNA alkylation.

Parodi et al.[27] reported a seemingly discordant result between alkaline elution and a sensitive viscometric method. The viscometric method measures the rate of disentanglement of the DNA of detergent-lysed cells in alkali. An increase in the rate of disentanglement was taken to be a measure of DNA strand scission. (An inherent assumption, however, is that the initial degree of entanglement remains unaltered.) The authors measured the effects of several chemicals on nuclear DNA in the liver of treated rats. Some compounds increased both the DNA alkaline elution rate and the viscometric DNA disentanglement rate. Some other compounds, however, increased the alkaline elution rate, but seemed not to affect the disentanglement rate. The authors interpreted the discordant results as suggesting that DNA alkaline elution was responding to an effect other than strand scission, presumably an effect on DNA packing in chromatin. However, the results could also be explained on the basis of the assumption that the initial degree of DNA entanglement affects the viscometric measurement but does not affect the alkaline elution measurement. Another possible origin of discrepancies arises from the fact that, whereas alkaline elution measures all of the DNA equally by mass, the viscometric measurement is most affected by the longest DNA strands in the population.

In summary, the prepoderance of data are consistant with the assumption that alkaline elution (under deproteinizing conditions) measures effective DNA strand length. The reported discrepancies with DNA unwinding and viscometric methods could be due to an effect of the initial DNA conformation on these assays, whereas alkaline elution may be relatively insensitive to the initial DNA conformation.

DNA DAMAGE IN CELLS *IN VITRO*

Hepatocytes

Hepatocytes contain metabolic enzymes that can convert chemicals to active DNA-damaging compounds. Rat hepatocytes can be employed *in vitro* to detect compounds that become genotoxic through metabolic activation in the liver. The DNA damage can be measured either directly in the hepatocytes themselves or in other target cells with which the hepatocytes are admixed. Fresh hepatocytes can be obtained from the liver of rats by means of a collagenase perfusion technique.[30] The dissociated hepatocytes can be placed in bacteriological plates to which the cells do not adhere; the non-adherent cells then can be removed easily from the plates without a possibly damaging treatment with trypsin.[31] Since the hepatocytes do not proliferate, it is not convenient to radioactively label their DNA; therefore the DNA in filter elution fractions is measured fluorometrically.[30,32]

Sinha et al.[31] tested the effects of 91 compounds (64 carcinogens and 27 non-carcinogens) in freshly isolated rat hepatocytes. The cells were exposed at three concentrations (0.03, 0.3 and 3 mM, unless limited by solubility). Cytotoxic effects were assessed by trypan blue staining or by release of the intracellular enzyme, glutomate oxalactate transaminase. Alkaline elution (pH 12.1, 0.035 ml/min) was conducted under fully deproteinizing conditons, including the use of proteinase K, non-adsorptive polycarbonate filters and 0.2% NaDodSO in the eluting solutions. Elution rates were calculated between 3 and 9 hr of elution; by starting the rate calculation at 3 hr, the spurious rapid elution of disintegrated DNA from dead cells is largely eliminated from consideration.

Sinha et al. conclude that the assay accurately identified the genotoxicity of 59/64 (92%) of the carcinogens and of 23/27 (85%) of the probable non-carcinogens. These conclusions, however, may depend on the criteria chosen to define a significant increase in elution rate. I have therefore recalculated the data in a manner fundamentally different from that chosen by the authors, and summarize some of the results in Table 1. The criterion used by Sinha et al. to define a significant increase in elution rate was that there must be at least a 3-fold increase in elution rate compared with the concurrent control. The controls however varied considerably in different experiments, so that the 3-fold criterion in some cases yielded questionable conclusions.

My recalculation instead is based on absolute increase in elution rate, and the criterion for significance is that the increase above the concurrent control must be greater than the elution rate values for 90% of the controls reported. When this criterion indicated a significant increase in elution rate, a measure of potency was obtained by dividing the increase in elution rate by the concentration of the compound (termed "specific elution" in Table 1). Increases in DNA elution rate which occurred only when most of the cells exhibited gross toxicity were excluded (this differs from the analysis of Sinha et al. who included such values).

The recalculated results in Table 1 show significant increases in elution rates and high specific elution values for many known carcinogens of several chemical classes. Among the carcinogenic alkylating agents and nitrosamines, as well as some related compounds which may be metabolised to alkylating agents, all except $CHCl_3$ showed significant increases in alkaline elution rates at one or more of the concentrations tested. The potencies (specific elution) of the active compounds varied over a 60-fold range (excluding 2 extreme cases). Similarly, among the carcinogenic aromatic amines and polycyclic aromatic compounds tested, all showed significant increases in alkaline elution rates with potencies varying over approximately the same range. Much higher potencies were obtained with some natural products: mitomycin, actinomycin D and aflatoxin B1. It remains to be determined how DNA alkaline elution potencies obtained in this way in various cell lines would correlate with carcinogenic potencies.

Table 1. Effects of Various Compounds on DNA Alkaline Elution in Rat Hepatocytes Treated for 3 Hours. (Recalculated from data of Sinha et al.[31])

Compound	Concentration (mM)	Elution Rate (Test-Control)	Specific Elution (Elution Rate/mM)	Carcinogenesis
ALKYLATING AGENTS & HALO COMPOUNDS				
1,2-Br_2-ethane	.03	.457	15.2	+
Propylene oxide	.03	.352	11.7	+
Me_2SO_4	.03	.288	9.6	+
MMS	.03	.179	6.0	+
Lindane	.03	.095	3.2	+
EMS	.03	.081	2.7	+
CCl_4	.03	.081	2.7	+
Azaserine	.03	.071	2.4	+
HN2	.3	.128	.43	+
Aldrin	.3	.078	.26	+
CH_3Cl	3.	.003	<.01	+
NITROSO & RELATED COMPOUNDS				
MNNG	.003	.404	135.	+
MNU	.03	.075	2.5	+
1-nitroso-piperdine	.03	.080	2.7	+

(continued)

Table 1 - Continued

Compound	Concentration (mM)	Elution Rate (Test-Control)	Specific Elution (Elution Rate/mM)	Carcino-genesis
NITROSO & RELATED COMPOUNDS (cont'd)				
Me_2-nitro-samine	.3	.246	.86	+
Ph_2-nitro-samine	3.	.118	.039	+
Piperidine	3.	.000	<.01	-
AROMATIC AMINES				
2-AAF	.003	.059	20.	+
Auramine	.003	.054	18.	+
2-NH_2-anthracene	.03	.081	2.7	+
2-NH_2-naphtha-lene	.3	.136	.45	+
2-NH_2-fluorene	.3	.063	.21	+
1-NH_2-fluorene	3.	.240	.08	?
Benzidine	3.	.203	.08	+
2,4-diamino toluene	3.	.012	<.01	+
Phenacetin	3.	.030	<.01	-
4-NH_2-biphenyl	.3	.057	.19	+
	3.	.238	toxic	
2-NH_2-biphenyl	.3	.007	<.01	?
	3.	.096	toxic	
POLYCYCLIC AROMATIC COMPOUND				
Benzo[a]pyrene	.03	.175	5.8	+
7,12-DMBA	.03	.130	4.3	+
Pyrene	.03	.009	<1.	-
Styrene oxide	.3	.182	.61	+
3-Me-cho-lanthrene	.1	.036	.36	+
Benzo[e]pyrene	.1	.000	<.3	?
Anthracene	.3	.080	.27	±
Phenanthrene	3.	.195	.063	-
Naphthalene	3.	.077	.026 (toxic)	-
Styrene	3.	.064	.021 (toxic)	-
Fluorene	3.	.007	<.01 (toxic)	-
NATURAL PRODUCTS				
Mitomycin C	.0003	.373	1243.	+
Actinomycin D	.0003	.252	840.	+
Aflatoxin B1	.003	.250	83.	+

Abbreviations: MMS, methylmethanesulfonate; EMS, ethylmethanesulfonate; MNU, 1-methyl-1-nitrosourea; MNNG, 1-methyl-1-nitroso-3-nitroguanidine; AAF, acetylaminofluorene; DMBA, dimethylbenza[a]anthracene

(Mitomycin C was unusual in that at higher concentrations the elution rates reverted to normal and even declined to below the controls. This may be due to ISC formation which perhaps becomes predominant at high concentrations.)

These preliminary studies are encouraging, but further confirmatory data are needed at more closely spaced doses. These studies might be extended usefully to human cells admixed with hepatocytes to provide metabolic activation (see below).

The specific elution values might be utilized as indicators of genotoxic potency. As a preliminary exploration of this idea, the results of Sinha et al.[31] in rat hepatocytes were compared with those from alkaline elution assays of the livers of rats treated with some of the same compounds (Table 2). The in vitro/in vivo assay ratios showed an agreement (within a factor of 2) for 4 compounds administered intraperitoneally (AAF, MMS, MNU and EMS); 3 other compounds gave approximately 20-fold lower values (aflatoxin B1, Me_2 nitrosamine, benzidine). Although inconclusive, these results do seem to hold promise if one considers that the in vivo and in vitro experiments were performed in different laboratories, possibly using different strains of rats.

Parodi et al.[33] analysed in vivo data from various sources in an attempt to test correlations among several types of measurements on 21 compounds of various chemical classes. The potencies for increased DNA alkaline elution were found to be correlated with oncogenic potencies and not with mutagenic potencies; the best correlations were with covalent binding and with lethality. Some of the assays compared, however, were performed in different animal species (mice versus rats) using different routes of administration of the compound; oncogenicity in one tissue was sometimes compared with DNA damage in another. It is evident that much more data are needed.

Admixture of Hepatocytes with Other Target Cells

The feasibility of an admixture system was explored by Staiano et al.[34] in a mixture of rat hepatocytes with Salmonella strain TA 98. The mixed cells were treated with acetaminofluorene derivatives; DNA damage was detected in the hepatocytes by alkaline elution while mutagenicity was observed at the histidine locus of Salmonella.

Bradley et al.[30] mixed rat hepatocytes with an equal number of [^{14}C]-thymidine-labeled V79 cells and treated the mixture with Me_2 nitrosamine. The treated V79 cells exhibited increased DNA elution (indicative of DNA strand breaks, or more likely, alkali-labile lesions) only if hepatocytes were present during the exposure to the chemical. The magnitude of the DNA damage, however, was less than half of that produced in the hepatocytes by themselves.

The production of DNA damage in the hepatocytes in large excess over that produced in admixed cells could signal the formation of an active metabolic product in hepatocytes that has a short half-life, compared with the time required to exit one cell and enter another. Such compounds could be specific hepatocarcinogens. It is of course possible to use human target cells admixed with rat hepatocytes, so as to make the test more specific to man.

Pancreatic Cells

Levin et al.[35] reported an interesting study of certain nitrosamine carcinogens in pancreas cells. Cells from hamster or rat pancreas were dissociated by means of a collagenase method. DNA strand breakage was assayed by slow alkaline elution after lysis of cells on polycarbonate filtrers by means of $NaDodSO_4$ with proteinase K included; this is currently the best available

Table 2. Effects of Various Compounds on DNA Elution Rates In Vitro and In Vivo

	Rat Hepatocytes In Vitro[1]	Rat Liver In Vivo[2]	Route	In Vitro / In Vivo
MNNG	135.	0.069	p.o.	1957.
Aflatoxin B1	83.	73.	i.p.	1.1
2-AAF	19.7	0.75	i.p.	26.
MMS	6.0	0.45	i.p.	13.
DMBA	4.3	ns	p.o.	...
MNU	2.5	0.21	i.p.	12.
EMS	1.5	0.20[3]	p.o.	12.5
		0.073	i.p.	21.
Me_2 nitrosamine	0.82	2.3[3]	p.o.	0.36
		1.8	i.p.	0.46
Benzidine	0.068	0.06	i.p.	1.1
1,2-Me_2-hydrazine	0.011	0.18	s.c.	0.06
Urethane	0.003	ns	i.p.	...

[1] Specific elution = (Test - Control)/mM. Hepatocytes exposed to compound for 3 hr. Recalculated from Sinha et al.[31]

[2] DNA fragmentation index = (Test - Control)/(mmol/kg). Recalculated by Parodi et al.,[33] mainly from data of Petzold and Swenberg. ns = not significant.

[3] Brambilla et al.[41]

[4] Abbreviations as in Table 1.

procedure. These authors studied 3 N,N-dialkylnitrosamines in which the alkyl groups were either 2-hydroxy-propyl, 2-oxypropyl or one of each. All 3 produce pancreatic tumors in hamsters but not in rats. Other tumors are produced in both species. The di(oxypropyl) compound is unusual in being highly potent and specific for the production of pancreatic tumors. All 3 compounds probably require metabolic activation to other proximate carcinogens. They also may be metabolically interconvertible, with metabolism tending towards the hydroxypropyl-oxypropyl compound. In their pancreas cell systems, in which the cell DNA was assayed after 1 hr, the authors found that the highest potency for strand breakage occurred with the di-oxypropyl compound in hamster cells. The hydroxypropyl-oxypropyl compound was about 8 times less potent in these cells; in rat cells, this compound was about 15 times less potent than in hamster cells and was the only one of the three compounds that produced detectable DNA damage in the rat cells. The di-hydroxypropyl compound failed to produce detectable DNA damage in hamster as well as rat cells. Evidence was cited that the compound was not taken up by these cells and thus requires metabolic conversion elsewhere; alternatively the pancreatic cells may rapidly convert this compound extracellularly to a product that does not enter cells.

These results were generally consistent with a relationship between DNA damage, as indicated by increased alkaline elution, and the carcinogenicity of these nitrosamines in pancreatic cells. The only quantitative divergence was between the reduced potency of the hydroxypropyl-oxypropyl compound for DNA damage and its high potency for the production of pancreatic tumors in hamsters. This difference may be due to metabolism of the compound, or to a different time course of action.

DNA DAMAGE IN TISSUES OF INTACT ANIMALS

DNA damage in tissues of treated animals has been detected by alkaline elution by several investigators. The DNA in proliferating tissues can be labeled by administration of [^{3}H]thymidine; more commonly however the eluted DNA is assayed fluorometrically. Although there have been several encouraging reports of positive correlations between DNA damage and the carcinogenicity of some compounds, other classes of compounds have failed to show such correlations. Several reasons can be suggested for the instances of poor corrrelation, and it is possible that these problems could be resolved by further studies using modified procedures.

First, the sensitivity of the assays has been limited by relatively high background elution of DNA from control tissues, probably due to DNA damage occurring during tissue homogenization and release of cell nuclei. The background possibly could be reduced by modification of the homogenization and lysis solutions. For example, the DNA in isolated nuclei of cultured cells has been maintained intact in a solution containing EGTA, $MgCl_2$ and Triton X-100.[36,37] For lysis, the published assays of tissues have employed the standard Sarkosyl-2 M NaCl-EDTA lysis solution; our experience with cultured cells, however, suggests that low backgrounds are more consistently obtained using the 2% $NaDodSO_4$-EDTA lysis solution.[1] Proteinase K can be added to eliminate DNA-protein crosslinks that may in some instances obscure the production of DNA strand breaks; proteinase K was usually not employed in the available published work.

Secondly, it may be important to consider the area under the damage-versus-time curve, rather than relying on a measurement of damage at a single time following treatment. For example, Bolognesi et al.[38] find in mice treated with benzidine that 2 hr following treatment, there is more DNA damage in kidney than in liver, whereas after 4 hr the situation is strikingly reversed.

Thirdly, it may be important to consider dose-response relationships carefully, because the doses needed to produce measurable DNA damage are much higher than the chronically administered doses in carcinogenesis assays. This is especially true for compounds that are metabolized to active or inactive products; at high doses the capacity of metabolic enzymes may be exceeded. For example, Parodi et al.[39] found that at high dosage of aniline to rats, the magnitude of DNA damage in liver actually decreased when the dose was increased beyond a certain point.

Some Recent Data

For several classes of compounds, quantitative correlations have been tested between potency for DNA damage and potencies for carcinogenesis or mutagenesis. Potency is defined as magnitude of effect divided by dose of compound administered. In a test of 16 hydrazine derivatives, Parodi et al.[40] found that increased DNA elution in liver and lung of treated mice was correlated with the carcinogenic potencies of these compounds. (The Ames Salmonella-microsome mutagenicity test, however, correlated neither with DNA damage nor with carcinogenicity.)

Brambilla et al.[41] tested 6 nitroso compounds of widely varying potencies, but found little or no correlation among the potencies for DNA damage (SSB), carcinogenesis or mutagenesis. An interesting set of data was obtained for the series: dimethyl, diethyl, di-n-propyl and di-n-butyl nitrosamine. In each successive step of this series, the potency for SSB production in rat liver decreased by approximately a factor of 4. No such consistent pattern was observed for carcinogenic or mutagenic potencies, suggesting that these are quantitatively more complex processes.

In a study of 16 aromatic amines and azo-derivatives, Parodi et al.[42] found no correlation between the potencies for DNA damage and carcinogenicity. Potencies in the Ames mutagenesis test also showed little or no quantitative correlation with carcinogenicity or with DNA damage. The carcinogenicity data, derived from various sources, were very heterogeneous and included different animal species (mice and rats), different routes of administration and different tumor sites.

Skare and Schrotel[43,44] investigated alkaline elution assays as possible short term tests for chemicals that may produce heritable mutations in spermatogonial cells. The production of DNA strand breaks was assayed in rat testicular cells.[43] MMS and MNU exhibited similar potencies for the producton of DNA strand breaks; EMS and ENU also produced strand breaks but were less potent. DNA strand breaks were detected by alkaline elution at subtoxic doses comparable to those which had been reported for the mouse specific locus test. The sensitivity of the alkaline elution assay was comparable to that of the in vivo unscheduled DNA synthesis assay. Excellent reproducibility was reported between individual animals.

Some mutagenic compounds produce DNA interstrand and/or DNA-protein crosslinks without strand breakage detectable by alkaline elution. Skare and Schrotel[44] studied some of these crosslinking agents for their effects on rat testis. Triethylenemelamine produced clear interstrand crosslinking which was greater at 6 hr than at 2 hr following a single dose; about half of the crosslinks observed after 6 hr were still present after 24 hr. Daily dosage for 5 days showed a marked cummulative effect. The rate of removal of interstrand crosslinks therefore may be quite slow, although one also must take into consideration the possibility of prolonged crosslink formation.[45] Mitomycin C exhibited detectable interstrand crosslinking only with the cummulative daily-times-5 administration schedule. Cyclophosphamide exhibited mainly strand breakage with a lesser degree of interstrand crosslinking. A substantial amount of DNA-protein crosslinking was observed in assays in which proteinase K was omitted. The results with triethylenemelamine and cyclophosphamide were consistent with the positive results found by the unscheduled DNA synthesis assay. The latter assay, however, was negative for mitomycin C; one possible reason is that the rate of repair of mitomycin-DNA adducts may be very low.

Tissue-Specified DNA Damage and Carcinogenesis

Significant, but imperfect correlations have been observed between the tendencies of particular carcinogens to produce tumors in certain organs and the production of DNA damage in those organs. In evaluating such correlations, there are at the outset reasons for not expecting perfect correlations. First, organs such as kidney or brain, which are composed of several different cell types, may not give appropriate values for DNA damage when the entire organ is homogenized in performing the assays; a misleading result would be obtained if tumors arise from a minority cell type that sustains more DNA damage than most cells in the organ. Secondly, certain sites of DNA damage (e.g., at or near oncogenes) may contribute to carcinogenicity only if these sites are near, or can be translocated to become near, genes that are active in given differentiated cells; thus the susceptibility of given cell types to carcinogenic DNA damage may depend on where certain oncogenes or proto-oncogenes are located in the chromosomes, and this may depend on the species, or even strain, of animal. Thirdly, the time-course of DNA damage, which may depend on the rate of DNA repair as well as on the rate of pharmacologic clearance of the compound, may differ among tissues and often was not included in the published studies.

Dimethylnitrosamine was found to produce DNA damage (increased alkaline elution rate) in liver more than in other organs of mice,[46] newborn rats[47] and adult rats and hamsters,[48] in agreement with this compound's tendency to produce liver tumors. Diethylnitrosamine also produced DNA damage selectively in the liver and produces liver tumors; however, this compound produces tumors also in rat kidney, although no significant DNA damage was observed in this organ.[48]

Petzold and Swenberg[47] detected DNA damage due to 1,2-dimethylydrazine in liver and duodenum of newborn rats, but not in lung, kidney or thymus; this was consistent with the sites at which tumors are produced in these animals. Brambilla et al.[49] found that this compound produced much more DNA damage in mouse liver than in colonic mucosa, stomach, lung or kidney. This result did not account for the production of colon tumors by this compound in some strains of mice. It may be noted that the compound was negative in the Ames Salmonella-microsome mutagenesis test.

Cavanna et al.[50] reported good agreement between organ sites of DNA damage and tumor production by cycasin, a compound that was negative in the Ames test. DNA damage in rats was extensive in liver, slight in colon and kidney and undetectable in lung. In mice, DNA damage was detected only in liver.

Methylnitrosourea produced extensive DNA damage in kidney, liver, brain and thymus of newborn rats and in mammary gland of pregnant rats.[47] This correlates with tumor formation in brain, thymus and mammary gland, but not with the absence of tumors in liver and kidney. Methylmethanesulfonate produced extensive DNA damage in liver, lung, kidney, brain, thymus, duodenum and stomach, but tumors were reported only in brain. These authors reported analogous partial correlations for aflatoxin, MNNG, 8-propliolactone and streptozotocin. Agreements were obtained between organ sites of DNA damage and tumor production for 1,2-dimethylhydrazine (liver and duodenum), 4-nitroquinoline oxide (lung), and AAF (liver).

Cavanna et al.[51] studied N-diazoacetylglycineamide, a chemical that had been found to induce in newborn and adult mice a high frequency of pulmonary tumors and lymphatic leukemia with short latent periods. Increased DNA alkaline-elution was measurable over a range of doses in liver, lung, kidney, spleen, thymus and bone marrow. The degree of DNA damage was similar in all of these organs.

Brambilla et al.[52] reported significant increases in DNA alkaline elution by procarbazine, an anticancer drug, in liver, lung, bone marrow, kidney, spleen and thymus of mice at doses (about 300 mg/m^2) not far above those used therapeutically in man. The time of maximum DNA damage was determined only in liver and was found to occur 6 hr after an intraperitoneal injection. The magnitude of DNA damage at this time was highest in liver and lung, perhaps slightly lower in bone marrow and kidney and lowest in spleen and thymus. Althought the experiments were conducted in Swiss mice, carcinogenesis data were available only in CD_2F_1 mice which, in response to a single treatment, exhibited a 100% incidence of pulmonary tumors within 25 weeks. Repeated doses produced leukemia and kidney tumors in addition. Despite the DNA damage observed in liver, no liver tumors were reported in the tests cited.

Benzo[a]pyrene and 7,12-dimethylbenz[a]anthracene, although they are highly carcinogenic to skin upon topical application, have little or no carcinogenic effect on liver of adult rats or mice upon systemic administration. Parodi et al.[53] found no significant changes in DNA alkaline elution in the liver of young adult rats and mice injected with either of these compounds. Bolognesi et al.[54] recently studied the effect of benzo[a]pyrene on DNA in fetal CD1 mice. This compound had been reported to induce mainly lung tumors in fetal Swiss mice. Significant increases in DNA alkaline elution were observed both in liver and lung of 15-day old CD1 fetuses 4 hr after either intrafetal or intramaternal injection; 48 hr after intrafetal injection, the DNA damage had disappeared from the liver but was still detectable in lung.

In response to reports of carcinogenic effects produced by aniline, Parodi et al.[39] assayed for DNA damage in various tissues of male rats and mice. (Aniline was negative in Salmonella mutagenesis assays with or without metabolic activation.) Significant increases in DNA alkaline elution were observed in liver and kidney in rats, but not in mice. The carcinogenicity tests cited by the authors revealed a very significant incidence of spleen hemangiosarcomas in rats, but no significant tumor incidence in mice. The DNA damage assays, however, showed no increase in alkaline elution in rat spleen. Thus the DNA assays agreed with the carcinogenicity data in regard to the susceptibililty of rats as opposed to mice, but disagreed strikingly in regard to the spleen as the susceptible rat organ. Perhaps the hemangiosarcomas arise from a minority cell type in normal rat spleen that would exhibit significant DNA damage if it could be assayed in isolation.

APPARENT DNA STRAND SCISSION BY PRESUMABLY INNOCUOUS TREATMENTS

Increases in DNA alkaline elution rate have been reported due to several presumably innocuous agents, including α-tocopheral, ascorbate, visible light and hydrogen peroxide. If alkaline elution assays are to be generally useful in risk assessment one must be able to evaluate such possible spurious cases.

The tests of hydrogen peroxide and ascorbate (which may act by generating hydrogen peroxide) were carried out in cell culture; the intact animal would be better protected than some cultured cells by the presence of larger amounts of catalase and superoxide dismutase. The inclusion of these enzymes in the culture media might eliminate some of these effects; however this question has not yet been studied thoroughly.

Bradley et al.[35] found that illumination of Chinese hamster cells in the cold with light sources emitting only wavelengths above 345 nm produced significant increases in DNA alkaline elution. For example, exposure to 32000 J/m^2 of filtered fluorescent light produced about 0.15 single-strand breaks per million nucleotides (equivalent to an X-ray dose of about 150 rad). Upon warming the cells to 37°C, the breaks were repaired with a half-time of about 20 min. There was no associated cytotoxicity or significant mutagenesis. Thus some types of strand breaks can be accurately repaired without consequent cell damage. Another instance of highly efficient DNA repair is in the case of ultraviolet light, as evidenced by the enormous enhancement of skin lesions in xeroderma pigmentosum patients who have a congential DNA repair defect. Another corollary would be that some individuals may be susceptible to DNA damaging agents that are safe for most of the population.

These considerations come forward in the interpretation of recent results with vitamin E. Russo et al.[56] found that dl-α-tocopherol increased the rate of DNA alkaline elution in rat liver 4 or 24 hr after single injections of doses that may be within the pharmaceutical range. The DNA strand breaks disappeared within 48 hr. The possibility is not excluded that the strand breaks might have arisen during preparation or lysis of the cells due to some alteration in cell membranes or lysosomal enzymes. It seems more likely, however, that oxidation-reductions of α-tocopherol do produce some DNA strand breaks in rat liver. These may or may not produce a significant problem in individuals who have normal DNA repair mechanisms. The possibility must be considered, however, that some individuals have DNA repair deficiencies that could make them more vulnerable to certain substances. These considerations suggest that DNA damage assays in human cells cultured from cancer-prone individuals would be more appropriate than assays of animal tissues.

REFERENCES

1. K. W. Kohn, R. A. G. Ewig, L. C. Erickson, and L. A. Zwelling, Measurement of strand breaks and crosslinks in DNA by alkaline elution, in: "DNA Repair: A Laboratory Manual of Research Techniques," E. Friedberg and P. C. Hanawalt, eds., Marcel Dekker, New York (1981).
2. K. W. Kohn, DNA as a target in cancer chemotherapy: measurements of macromolecular DNA damage produced in mammalian cells by anti-cancer agents and carcinogens, in: "Methods in Cancer Research," Vol. 16, H. Bush and V. DeVita, eds., Academic, New York (1978).
3. K. W. Kohn, DNA damage in mammalian cells, Bio. Science 31:593 (1981).
4. K. W. Kohn, The significance of DNA-damage assays in toxicity and carcinogenicity assessment, Ann. NY Acad. Sci. 407:106 (1983).
5. K. W. Kohn, Biological aspects of DNA damage by crosslinking agents, in: "Molecular Aspects of Anti-cancer Drug Action," S. Neidle and M. J. Waring, eds., MacMillan, London (1983).
6. A. J. Fornace, Measurements of M. luteus endonuclease-sensitive lesions by alkaline elution, Mutat. Res. 94:263 (1982).
7. M. D'Incalci, L. Szmigiero, L. C. Erickson, J. A. Hartley, and K. W. Kohn, A filter incubation method for the determination of potentially crosslinkable sites and labile lesions in DNA in mammalian cells, Analyt. Biochem., in press.
8. M. O. Bradley, L. C. Erickson, Comparison of the effects of hydrogen peroxide and X-ray irradiation on toxicity, mutation, and DNA damage/repair in mammalian cells (V-79), Biochim. Biophys. Acta 654:135 (1981).
9. M. O. Bradley and K. W. Kohn, X-ray induced DNA double strand break production and repair in mammalian cells as measured by neutral filter elution, Nucleic Acids Res. 7:793 (1979).
10. A. J. Fornace, Jr., K. W. Kohn, and H. E. Kann, Jr., DNA single-strand breaks during repair of UV damage in human fibroblasts and abnormalities in xeroderma pigmentosum, Proc. Natl. Acad. Sci. USA 73:39 (1976).
11. H. E. Kann, K. W. Kohn, and J. K. Martin, Inhibition of DNA repair by the 1,3-bis(2-chloroethyl)-1-nitrosourea breakdown product, 2-chloroethyl isocyanate, Cancer Res. 34:398 (1974).
12. M. D'Incalci, J. M. Covey, D. S. Zaharko, and K. W. Kohn, DNA alkali-labile sites induced by 5-aza-2'-deoxycytidine in mouse leukemia L1210 cells, Cancer Res., in press.
13. L. Szmigiero, L. C. Erickson, R. A. G. Ewig, and K. W. Kohn, DNA strand scission and cross-linking by diaziridinylbenzoquinone (diaziquone) in human cells and relation to cell killing, Cancer Res. 44:4447 (1984).
14. L. Szmigiero and K. W. Kohn, Mechanisms of DNA strand breakage and interstrand cross-linking by diziridinylbenzoquinone (diaziquone) in isolated nuclei from human cells, Cancer Res., 44:4453 (1984).
15. W. E. Ross, D. Glaubiger, and K. W. Kohn, Qualitative and quantitative aspects of intercalator-induced DNA strand breaks, Biochim. Biophys. Acta 562:41 (1979).
16. L. A. Zwelling, S. Michaels, L. C. Erickson, R. S. Ungerleider, M. Nichols, and K. W. Kohn, Protein-associated DNA strand breaks in L1210 cells treated with the DNA intercalating agent, 4'-(9-acridinylamino)-methanesulfon-m-anisidide (m-AMSA) and adriamycin, Biochemistry 20:6553 (1981).
17. J. Filipski, J. Yin, and K. W. Kohn, Reconstitution of intercalator-induced DNA scission by an active component from nuclear extracts, Biochim. Biophys. Acta 741:116 (1983).

18. J. Minford, Y. Pommier, J. Filipski, K. W. Kohn, D. Kerrigan, M. Mattern, S. Michaels, R. Schwartz, and L. A. Zwelling, Isolation of intercalator-dependent protein-linked DNA strand cleavage activity from cell nuclei and identification as topoisomerase II, Biochemistry, in press.
19. L. F. Liu, T. C. Rowe, L. Yang, K. M. Tewey, G. L. Chen, Cleavage of DNA by mammalian DNA topoisomerase II, J. Biol. Chem. 258:15365 (1983).
20. L. A. Zwelling, S. Michaels, D. Kerrigan, Y. Pommier, and K. W. Kohn, Protein-concealed DNA strand breaks produced in mouse leukemia L1210 cells by ellipticine and 2-methyl-9-hydroxy-ellipticinium, Biochem. Pharm. 31:3261 (1982).
21. A. Minocha and B. H. Long, Inhibition of the DNA catenation activity of Type II topoisomerase by VP16-213 and VM26, Biochem. Biophys. Acta Commun. 122:165 (1984).
22. B. S. Glisson, S. E. Smallwood, and W. E. Ross, Characterization of VP-16-induced DNA damage in isolated nuclei from L1210 cells, Biochim. Biophys. Acta 783:74 (1984).
23. K. M. Tewey, G. L. Chen, E. M. Nelson, and L. F. Liu, Intercalative antitumor drugs interfere with the breakage-reunion reaction of mammalian DNA topoisomerase II, J. Biol. Chem. 259:9182 (1984).
24. K. M. Tewey, T. C. Rowe, L. Yang, B. D. Halligan, and L. F. Liu, Adriamycin induced DNA damage mediated by mammalian DNA topoisomerase II, Science 226:466 (1984).
25. W. Ross, T. Rowe, B. Glisson, J. Yalowich and L. Liu, Role of topoisomerase II in mediating epipodophyllotoxin-induced DNA cleavage, Cancer Res. 44:5857 (1984).
26. C. Nicolini, A. Belmont, S. Zietz, A. Maura, A. Pino, L. Robiano, and G. Brambilla, Physico-chemical model for DNA alkaline elution: New experimental evidence and differential role of DNA length, chain flexibility and superpacking, J. Theor. Biol. 100:341 (1983).
27. S. Parodi, C. Balbi, M. L. Abelmoschi, M. Pala, P. Russo, and L. Santi, Studies on DNA damage: Discordant responses of rate of DNA elution rate, obtained for several compounds. Possible explanations of the discrepancies, Cell Biophys. 5:285 (1983).
28. K. W. Kohn, L. C. Erickson, R. A. G. Ewig, and C. Friedman, Fractionation of DNA from mammalian cells by alkaline elution, Biochemistry 15:4629 (1976).
29. C. Nicolini, L. Robbiano, A. Pino, A. Maura, R. Finollo, and G. Brambilla, Higher sensitivity for the detection of chemically-induced DNA damage: Role of DNA unfolding in determining alkaline-elution rate, Carcinogenesis 6:385 (1985).
30. M. O. Bradley, G. Dysart, K. Fitzsimmons, P. Harbach, J. Lewin, and G. Wolf, Measurements by filter elution of DNA single- and double-strand breaks in rat hepatocytes: Effects of nitrosamines and gamma-irradiation, Cancer Res. 42:2592 (1982).
31. J. F. Sinha, C. L. Bean, G. R. Dysart, V. I. Taylor, and M. O. Bradley, Evaluation of the alkaline elution/rat hepatocyte assay as a predictor of carcinogenic/mutagenic potential, Mutat. Res. 113:357 (1983).
32. L. C. Erickson, R. Osieka, N. A. Sharkey, and K. W. Kohn, Measurement of DNA damage in unlabeled mammalian cells analyzed by alkaline elution and fluorometric DNA assay. Anal. Biochem. 106:169 (1980).
33. S. Parodi, M. Taningher, P. Boero, and L. Santi, Quantitative correlations amongst alkaline DNA fragmentation, DNA covalent binding, mutagenicity in the AMES test and carcinogenicity, for 21 compounds, Mutat. Res. 93:1 (1982).
34. N. Staiano, L. C. Erickson, and S. S. Thorgeirsson, Bacterial mutagenesis and host cell DNA damage by chemical carcinogens in the Salmonella/hepatocyte system, Biochem. Biophys. Res. Commun. 94:837 (1980).
35. J. Levin, B. Phillips, Z. M. Iqbal, DNA damage in isolated hamster and rat pancreas cells by pancreatic carcinogens, Chem. Biol. Interact. 48:59 (1984).

36. J. Filipski and K. W. Kohn, Ellipticine-induced protein-associated DNA breaks in isolated L1210 nuclei, Biochim. Biophys. Acta 698:280 (1982).
37. Y. Pommier, K. W. Kohn, R. E. Schwartz, and L. A. Zwelling, Formation and rejoining of deoxyribonucleic acid double-strand breaks induced in isolated cell nuclei by antineoplastic intercalating agents, Biochemistry 23:3194 (1984).
38. C. Bolognesi, C. F. Cesarone, and L. Santi, Evaluation of DNA damage by alkaline elution technique after in vivo treatment with aromatic amines, Carcinogenesis 2:265 (1981).
39. S. Parodi, M. Pala, P. Russo, A. Zunino, C. Balbi, A. Albini, F. Valerio, M. R. Cimbrle, and L. Santi, DNA damage in liver, kidney, bone marrow, and spleen of rats and mice treated with commercial and purified aniline as determined by alkaline elution assay and sister chromatid exchange induction, Cancer Res. 42:2277 (1982).
40. S. Parodi, S. DeFlora, M. Cavanna, A. Pino, L. Robbiano, C. Bennicelli, and G. Brambilla, DNA-damaging activity in vivo and bacterial mutagenicity of sixteen hydrazine derivatives as related quantitatively to their carcinogenicity, Cancer Res. 41:1469 (1981).
41. G. Brambilla, M. Cavanna, A. Pino, and L. Robbiano, Quantitative correlation among DNA damaging potency of six N-nitroso compounds and their potency in inducing tumor growth and bacterial mutations, Carcinogensis 2:425 (1981).
42. S. Parodi, M. Taningher, P. Russo, M. Pala, M. Tamaro, C. Monti-Bragandin, DNA-damaging activity in vivo and bacterial mutagenicity of sixteen aromatic amines and azo-derivatives, as related quantitatively to their carcinogenicity, Carcinogensis 2:1317 (1981).
43. J. A. Skare and K. R. Schrotel, Alkaline elution of rat testicular DNA: Detection of DNA strand breaks after in vivo treatment with chemical mutagens, Mutat. Res. 130:283 (1984).
44. J. A. Skare and K. R. Schrotel, Alkaline elution of rat testicular DNA: Detection of DNA cross-links after in vivo treatment with chemical mutagens, Mutat. Res. 130:295 (1984).
45. K. Micetich, L. A. Zwelling, and K. W. Kohn, Quenching of DNA-platinum-(II) monoadducts as a possible mechanism of resistance to cis-platinum(II)diamminedichloride in L1210 cells. Cancer Res. 43:3609 (1983).
46. S. Parodi, M. Taningher, L. Santi, M. Cavanna, L. Sciaba, A. Maura, and G. Brambilla, A practical procedure for testing DNA damage in vivo, proposed for a pre-screening of chemical carcinogens, Mutat. Res. 54:39 (1978).
47. G. L. Petzold and J. A. Swenberg, Detection of DNA damage induced in vivo following exposure of rats to carcinogens, Cancer Res. 38:1589 (1978).
48. A. Barbin, J. C. Bereziat, and H. Bartsch, Evaluation of DNA damage by the alkaline elution technique in liver, kidney and lungs of rats and hamsters treated with N-nitrosodialkylamines, Carcinogenesis 4:541 (1983).
49. G. Brambilla, M. Cavanna, S. Parodi, L. Sciaba, A. Pino, and L. Robbiano, DNA damage in liver, colon, stomach, lung and kidney of BALB/C mice treated with 1,2-dimethyldrazine, Int. J. Cancer 22:174 (1978).
50. M. Cavanna, S. Parodi, M. Taningher, C. Bolognesi, L. Sciaba, and G. Brambilla, DNA fragmentation in some organs of rats and mice treated with cycasin, Br. J. Cancer 39:383 (1979).
51. M. Cavanna, S. Parodi, L. Robbiano, A. Pino, L. Sciaba, and G. Brambilla, Alkaline elution assay as a potentially useful method for assessing DNA damage induced in vivo by diazoalkanes, GANN 71:251 (1980).
52. G. Brambilla, M. Cavanna, A. Pino, and L. Robbiano, DNA damage and repair in mouse tissues following procarbazine administration, Pharmacol. Res. Commun. 13:213 (1981).
53. S. Parodi, M. Taningher, M. Pala, and L. Santi, Alkaline DNA fragmentation in vivo: Borderline or negative results obtained respectively with 7,12-dimethylbenz[a]anthracene and benzo[a]pyrene, Tumori 67:87 (1981).

54. C. Bolognesi, L. Rossi, O. Barbieri, and L. Santi, Benzo[a]pyrene-induced DNA damage in mouse fetal tissues, Carcinogenesis, in press.
55. M. O. Bradley, L. C. Erickson, and K. W. Kohn, Non-enzymatic DNA strand breaks induced in mammalian cells by fluorescent light, Biochim. Biophys. Acta 520: 11 (1978).
56. P. Russo, M. Pala, S. Parodi, C. Ghiara, N. Ferrari, and G. Vidali, Effects of vitamin E on liver DNA, Cancer Lett. 25:163 (1984).

DNA MODIFICATION BY SULFUR MUSTARDS AND NITROSOUREAS AND REPAIR OF THESE LESIONS

David B. Ludlum[1] and Bruno Papirmeister[2]

[1]Division of Oncology, Department of Medicine, Albany Medical College, Albany, NY 12208 USA; [2]U.S. Army Medical Research Institute of Chemical Defense, Aberdeen Proving Ground, MD 21010 USA

INTRODUCTION

In this chapter, the nature and significance of DNA modifications caused by a particular mutagenic agent, chloroethyl ethyl sulfide (CEES), will be compared with those produced by a therapeutic agent, chloroethyl cyclohexyl nitrosourea (CCNU). This comparison illustrates the differences in the kind of biological response which can arise from DNA modification by different agents and the role of DNA repair in determining this response. In particular, the ability of tumor cells to become resistant to therapeutic agents has some important implications for the ability of cells in general to tolerate environmental mutagens.

DNA modification by the synthetic products, CEES and CCNU, should be viewed in the context of DNA modification caused by naturally-occurring compounds which react with DNA. For example, cycasin, a compound derived from plants, and the important metabolite, S-adenosylmethionine, both methylate DNA.[1,2] Not surprisingly, a variety of repair mechanisms has evolved which serve to maintain the integrity of DNA in the presence of such naturally-occurring DNA modifiers. The ability of these enzymes to repair other, more varied DNA lesions is currently under active investigation as described below.

Just as many different kinds of agents can modify DNA, the range of modifications which can be produced is also very broad. Indeed, no simple classification of DNA modifications is yet available, although some general types are widely recognized--for example, simple substitutions by alkyl groups, more bulky substitutions by larger adducts, and modifications which involve the introduction of crosslinks into DNA.

Between them, CEES and CCNU cause most of these different kinds of DNA modifications. These differences apparently call, in turn, upon a variety of different mechanisms for repairing this damage.

DNA MODIFICATION BY CEES AND CCNU

Differences in the way in which DNA is modified by CEES and CCNU have their origin in the different chemical structures of these two agents and in

Figure 1. Chloroethyl ethyl sulfide and its reactive intermediate.

the different reactive intermediates which they generate. These reactive intermediates are contrasted in Figures 1 and 2. As shown in Figure 1, CEES cyclizes in polar solvents to generate an intermediate sulphonium compound which reacts with the electron-rich atoms in DNA. The nature of this sulphonium ion dictates the kind of DNA substitution which is produced; as described below, the ring nitrogens of DNA are the primary sites of attack.

CEES is of interest because it is a one-armed analog of the highly toxic compound, bis-chloroethyl sulfide, or sulfur mustard. However, its reactive intermediate also has a structural similarity to the important metabolic intermediate, S-adenosylmethionine, another sulphonium compound. Furthermore, the reactive intermediate generated from CEES is similar to the alkylating species generated from the reaction of dibromoethane and glutathione.[3] Thus, CEES is probably a good model for some fairly ubiquitous compounds which react with DNA.

As shown in Figure 2, the antitumor agent, CCNU, generates a variety of reactive species, each one somewhat different from the reactive species generated from CEES.[4] Chloroethyl diazonium hydroxide, generated by decompostion along route I, is probably the most important of these intermediates in explaining the cytotoxic action of the haloethylnitrosoureas since it can transfer a chloroethyl group to DNA which undergoes a subsequent crosslinking reaction.[5] However, the other intermediates generated by decomposition along routes II and III also cause a great deal of DNA modification, substituting DNA bases with hydroxyethyl groups. Mechanisms also have been proposed for DNA crosslinking reactions which would involve these intermediates.[6]

Quite apart from differences in their ability to cause DNA-DNA crosslinking, CCNU and the other haloethylnitrosoureas differ from CEES in the sites of initial substitution, at least quantitatively. The more polar sulphonium ion from CEES reacts primarily with the N-7 position of guanine and with the N-3 position of adenine to produce the two derivatives shown at the top of Figure 3.[7] Together, these two derivatives account for most of the DNA modification by CEES, although other derivatives have been found when

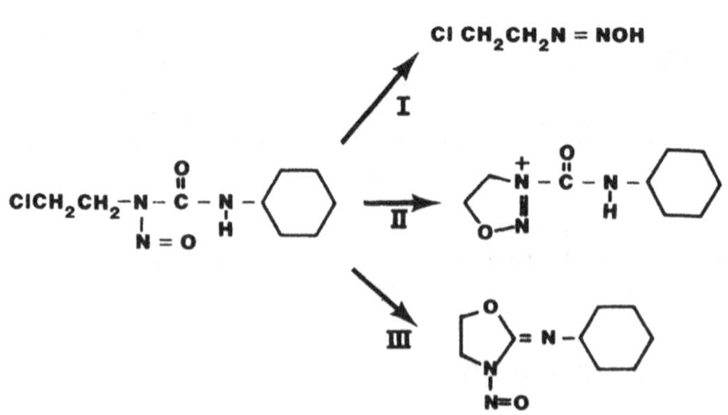

Figure 2. Chloroethyl cyclohexyl nitrosourea and its reactive intermediates.

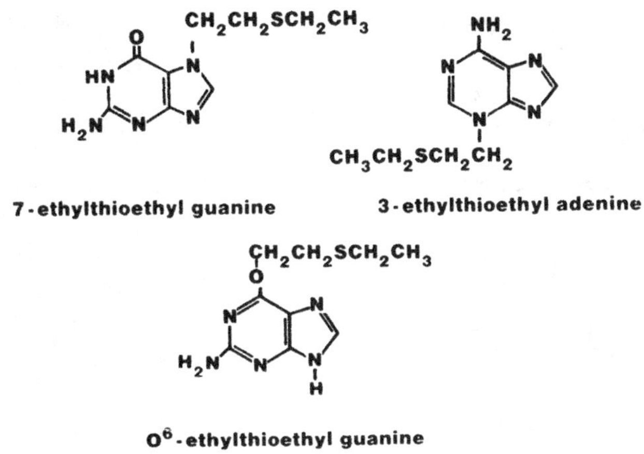

Figure 3. Some DNA modifications caused by chloroethyl ethyl sulfide.

reactions are carried out at the monomer level.[8] Much smaller amounts of O^6-ethylthioethylguanine are formed also, but these could still have considerable biological significance.[9] No evidence for phosphate alkylation has been obtained and, in general, the tendency for CEES to alkylate ring nitrogens is very much greater than its tendency to alkylate oxygen atoms.

The range of DNA modifications produced by CCNU is much broader. In part, this reflects the fact that CCNU can generate more than one reactive intermediate. However, these intermediates also have a much greater tendency to alkylate oxygens, including phosphate groups in DNA. As described by Singer and Kusmierek for ethylnitrosourea,[10] all of the DNA N's and O's probably can be substituted by the haloethylnitrosoureas.

Some of the CCNU-induced DNA modifications which have been identified are shown in Figure 4; others are published elsewhere.[6] The derivatives shown here include examples of substitution by hydroxyethyl and haloethyl groups at the O^6 and N-7 positions of guanine, as well as two crosslinked bases.

As mentioned above, the transfer of a haloethyl group can lead to DNA crosslinking. Thus, 7-haloethylguanine can react with a neighboring guanine to produce 1,2-bis (7-guanyl)-ethane, while a guanine substituted with a haloethyl group in the O^6 position can rearrange and react with a neighboring cytosine to produce 1-(3-cytosinyl),2-(1-guanyl)-ethane. The first crosslinked structure probably represents intrastrand crosslinking and the second, interstrand crosslinking.

Not shown in this figure, but probably of importance because it can lead to DNA-strand breakage, is alkylation of the sugar-phosphate backbone. A comparison of Figure 3 and Figure 4 shows that CEES and CCNU produce different kinds of DNA modifications that are evidently repaired with varying efficiency by various DNA repair mechanisms.

SIGNIFICANCE OF DNA MODIFICATIONS

Sulfur mustards and haloethylnitrosoureas are cytotoxic and carcinogenic to animals and probably to humans as well. These biological responses appear to be related to DNA modification, but not all modifications are of equal importance. The importance of a particular modification is determined by whether or not it conveys misinformation in replicative processes, whether it is chemically stable, and whether effective mechanisms exist for its repair.

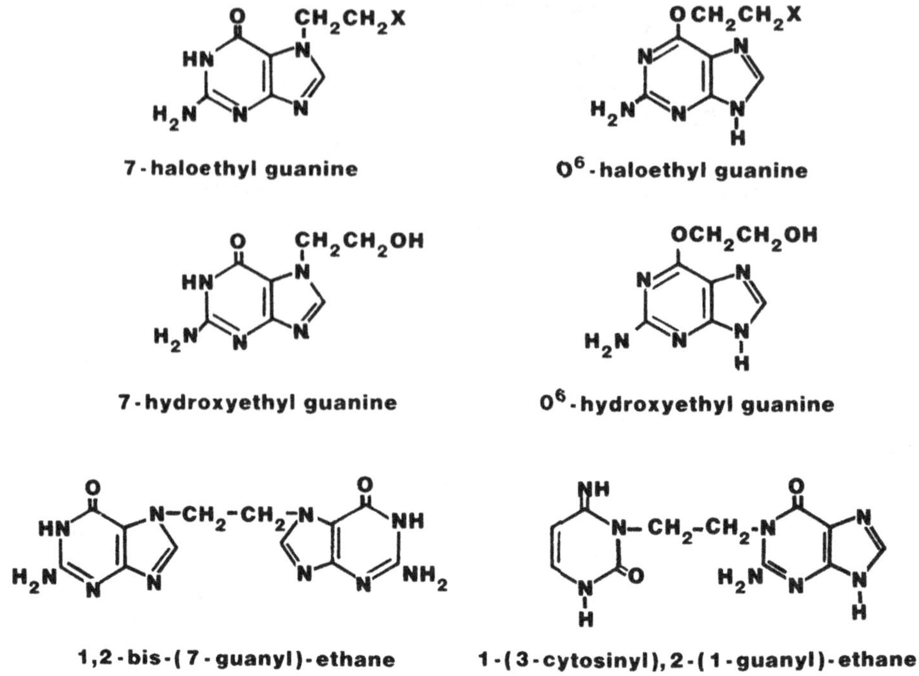

Figure 4. Some DNA modifications caused by the haloethylnitrosoureas.

Numerous attempts have been made to assign relative importance to different DNA modifications so that repair of the most important ones can be investigated first.

Of the CEES-induced modifications shown in Figure 3, both 7-ethylthioethylguanine and 3-ethylthioethyladenine are released spontaneously from DNA to leave apurinic sites. Since such apurinic sites are formed to some extent in nature, one would anticipate that small amounts of these lesions would be repaired efficiently and might not be particularly significant.

O^6 alkylation of guanine, as represented by formation of O^6-ethylthioethylguanine, is apparently more significant. Not only have such modifications been shown to convey misinformation, but the formation of O^6-alkylguanines seems to be associated with increased tumor incidence in animals fed alkylating carcinogens.

Both N-7 and O^6-substituted guanines are found among the modifications produced by the haloethylnitrosoureas, and similar comments would apply to them. However, as mentioned above, the reactive intermediates formed from CCNU produce a much wider spectrum of DNA modifications than can be formed by CEES. Several of these modifications, including phosphate alkylation, are probably quite important.

Most significantly, the bifunctional nature of the haloethylnitrosoureas permits the formation of crosslinked nucleosides. Numerous correlations have been found between cytotoxicity and the amount of DNA-DNA interstrand crosslinking, but intrastrand crosslinks may be of considerable biological significance as well. Thus, crosslinked structures like those shown at the bottom of Figure 4 are probably especially important.

Table 1. Proposed repair mechanisms for CEES-induced DNA modifications.

Repair Process	Modified Base Probably Recognized	Reference
Excision Repair		
uvrabc excision nuclease	O^6-ethylthioethylguanine (?) and/or helical distortions	14
DNA glycosylases	3-ethylthioethyladenine	7
AP endonucleases	Depurinated sites	16
RecA-dependent repair	Not yet identified	14

CELLULAR RESPONSE TO CEES- AND CCNU-INDUCED DNA MODIFICATIONS

The most compelling evidence that cells can adapt to a wide variety of DNA lesions is provided by observations that repair-deficient cells are much more sensitive to the cytotoxic and mutagenic action of DNA-modifying agents than competent cells. This is true for both the monofunctional and difunctional sulfur mustards and haloethylnitrosoureas, as well as for the simpler methylating agents.[12–15] Additional studies are needed to determine what mechanisms compensate for each kind of DNA lesion, but it would appear that many kinds of DNA damage can be repaired to some extent.

Tables 1 and 2 list repair mechanisms which are important in repairing CEES-induced and CCNU-induced lesions, respectively. These tables are drawn

Table 2. Proposed repair mechanisms for haloethylnitrosourea induced DNA modifications.

Repair Process	Modified Base Probably Recognized	Reference
Excision Repair		
uvrabc excision nuclease	Crosslinks and/or helical distortions	15
DNA glycosylases	7-substituted guanines; ring-opened 7-substituted guanines; 3-substituted adenines	
AP endonucleases	Depurinated sites	
Transferase repair	O^6-alkylguanines	17,18
RecA-dependent repair	Phosphotriesters and/or other unidentified lesions	15

Table 3. DNA modification by mutagens and therapeutic agents.

	Mutagen (CEES)	Therapeutic Agent (CCNU)
DNA Modification	Toxic	Therapeutic
DNA Repair	Prevents Damage	Interferes with Treatment

primarily from studies of E. coli mutants deficient in various repair pathways which show that loss of the indicated repair pathway is associated with decreased survival, increased numbers of mutants, or both.

Ultimately, by using purified repair factors and defined substrates, it will be possible to establish the efficiency with which each mechanism repairs each lesion. Data are beginning to accumulate, however, which indicate that the cellular repair mechanisms deal more efficiently with methyl groups than with larger substituents. For example, guanine substituted in the 6 position with bulkier groups appears to be repaired less readily than is O^6-methylguanine.[18,19]

Nevertheless, it does appear that cells possess adaptive processes which can deal with a wide variety of lesions. Evidence for this is obtained from a study of chemotherapeutic agents like CCNU. Here, antitumor activity is associated with DNA damage and excessive repair can cause resistance to treatment. The contrasting role of DNA repair as it relates to cancer treatment and prevention is shown in Table 3.

Tumors often become resistant to treatment with DNA-modifying agents and this resistance frequently seems to be associated with the development of repair mechanisms. To the extent that these processes are true repair processes and restore the original genetic information, they would serve to protect normal cells from environmental mutagens and carcinogens and would normally be of benefit. It seems probable that further study of cellular repair processes will provide both increased protection from environmental agents and more effective chemotherapy as well.

ACKNOWLEDGEMENTS

This work was supported by grant #CA32171 from the National Institutes of Health and by Army Contract #DAMD17-82-C-2203 from the USAMRDC. We gratefully acknowledge the editorial assistance of Suzanne Wissel.

REFERENCES

1. E. C. Miller, Mechanisms of chemical carcinogenesis, Cancer 47:1055 (1981).
2. B. Rydberg and T. Lindahl, Nonenzymatic methylation of DNA by the intracellular methyl group donor S-adenosyl-L-methionine is a potentially mutagenic reaction, Europ. Molec. Biol. Organizational Journal 1:211 (1982).
3. N. Ozawa and F. P. Guengerich, Evidence for formation of an S-[2-(N^7-guanyl)ethyl]glutathione adduct in glutathione-mediated binding of the

carcinogen 1,2-dibromoethane to DNA, Proc. Natl. Acad. Sci. USA 80:5266 (1983).
4. J. W. Lown and S. M. S. Chauhan, Mechanism of action of (2-halo-ethyl)-nitrosoureas on DNA. Isolation and reactions of postulated 2-(alkyl-imino)-3-nitrosooxazolidine intermediates in the decomposition of 1,3-bis(2-chloroethyl)-3-(4'-trans methylcyclohexyl)-1-nitrosourea, J. Med. Chem. 24:270 (1981).
5. D. B. Ludlum, B. S. Kramer, J. Wang, and C. Fenselau, Reaction of 1,3-Bis(2-Chloroethyl)-1-Nitrosourea with Synthetic Polynucleotides, Biochemistry 14:5480 (1975).
6. D. B. Ludlum and W. P.Tong, DNA modification by the nitrosoureas: Chemical nature and cellular repair, in: Cancer Chemotherapy, Volume II, F. M. Muggia, ed., Martinus Nijhoff, Boston (1985).
7. B. Papirmeister, J. K. Dorsey, C. L. Davison, and C. L. Gross, Sensitization of DNA to endonuclease by adenine alkylation and its biological significance, Fed. Proc. 29:726 (1970).
8. G. H. Sack, Jr., C. Fenselau, M.-N. N. Kan, L. S. Kan, G. W. Wood, and P.-Y. Lau, Characterization of the products of alkylation of 2'-deoxy-adenosine and 2'-deoxyguanosine by chloroethyl ethyl sulfide, J. Org. Chem. 43:3932 (1978).
9. B. Ludlum and B. Papirmeister, unpublished data.
10. B. Singer and J. T. Kusmierek, Chemical Mutagenesis, Ann. Rev. Biochem. 52:655 (1982).
11. J. W. Lown and L. W. McLaughlin, Nitrosourea-induced DNA single-strand breaks, Biochem. Pharmacol. 28:1631 (1979).
12. M. Fox and D. Scott, The genetic toxicology of nitrogen and sulphur mustard, Mutat. Res. 75:131 (1980).
13. P. D. Lawley and P. Brookes, Molecular mechanism of the cytotoxic action of difunctional alkylating agents and of resistance to this action, Nature 206:480 (1965).
14. R. M. Gilbert, S. Rowland, C. L. Davison, and B. Papirmeister, Involvement of separate pathways in the repair of mutational and lethal lesions induced by a monofunctional sulfur mustard, Mutat. Res. 28:257 (1975).
15. B. M. Kacinski, W. D. Rupp, and D. B. Ludlum, Repair of haloethylnitrosourea-induced DNA damage in mutant and adapted bacteria, submitted manuscript.
16. W. G. Verly, Prereplicative error-free DNA repair, Biochem. Pharmacol. 29:977 (1980).
17. P. Robins, H. L. Harris, I. Goldsmith, and T. Lindahl, Crosslinking of DNA induced by chloroethylnitrosourea is prevented by O^6-methylguanine-DNA transferase, Nucl. Acid Res. Mol. Biol. 11:7743 (1983).
18. T. P. Brent, Suppression of cross-link formation in chloroethylnitrosourea-treated DNA by an activity in extracts of human leukemic lymphoblasts, Cancer Res. 44:1887 (1984).
19. P. Schendel, B. V. Edington, J. G. McCarthy, and M. L. Todd, Repair of alkylation damage in E. coli. in: Cellular Responses to DNA Damage, E. C. Friedberg and B. A. Bridges, eds., Alan R. Liss, Inc., New York (1983).

INTRAGENOMIC LOCALIZATION OF CARCINOGEN-DNA DAMAGE

G.N. Wogan and G. McMahon

Department of Applied Biological Sciences
Massachusetts Institute of Technology
Cambridge, MA 02139

Covalent modification of DNA is thought to be a critical step in the induction of mutations as well as in the transformation of cells. Adducts formed with DNA by a large number of chemical carcinogens have been chemically characterized, and a great deal of current research is directed at the elucidation of molecular mechanisms through which adduct formation leads to specific changes in gene function. Attempts to construct quantitative relationships between levels of adduction and measurable effects such as mutation or transformation frequency, have been constrained by the necessary assumption that adducts are randomly distributed along the genome. However, several lines of evidence have been produced which indicate that adduct distribution is non-random, and a number of factors associated with localized binding of carcinogens to DNA have thus far been identified.

At the simplest level of organization, base composition of DNA sequences can cause preferential binding of those carcinogens which form adducts only at specific nucleophilic sites. For example compounds which form only quanine adducts would obviously be disproportionately concentrated in G-C rich regions of DNA. Additional factors operative at higher levels of DNA organization have also been found to be important determinants of adduct localization in non-random fashion. Specific nucleotide sequence in the region immediately surrounding target nucleotides has been shown to affect reactivity and binding levels for several carcinogens. DNA conformation (e.g.. B- vs Z-form) also strongly influences the binding of several carcinogens in *in vitro* models. The impacts of chromatin structure on carcinogen-DNA binding are particularly evident from *in vivo* studies. Fractionation of chromatin by partial digestion with DNAse I or by microsomal nuclease has revealed a general tendency for carcinogen localization in regions of active transcription compared to bulk DNA, and preferential binding to internucleosomal over nucleosomal core DNA. Recently, preferential binding has also been demonstrated in specific sequences, such as repetitive elements and multi-copy genes. The nature of these lines of evidence will be summarized briefly in the following sections; some aspects have been more extensively reviewed by Osborne[1] and Kriek et al.[2]

BASE COMPOSITION

Adduct profiles formed by the covalent interactions of electrophilic carcinogen derivatives with DNA have now been characterized for a large number of specific carcinogenic chemicals. For a given carcinogen, characteristic profiles of adducts are formed in specific experimental systems. It is therefore predictable that carcinogen binding would be higher in those regions where the target nucleotide (e.g., guanine) is more concentrated, as in G-C rich regions. For instance, it has been shown that N-acetoxy-acetylaminoflourene (N-acetoxy-AAF) shows a 2-fold greater reactivity with calf thymus satellite DNA when compared to its reactivity with bulk DNA.[3] In addition, several equilibrium binding studies using synthetic polynucleotides have suggested that the composition of target residues within the DNA substantially affects the level of carcinogen binding.[4]

NUCLEOTIDE SEQUENCE

However, similar experiments have suggested that the neighboring DNA sequence proximal to a given target nucleotide also can have a profound effect on defining the relative probability of attachment at a particular nucleotide locus. As summarized in Table 1, a number of carcinogens have shown non-random reactivity with DNA of defined sequence. Generally, the methods used for detection of sites of putative carcinogen-DNA binding utilized <u>in vitro</u> reaction of pre-activated carcinogen with a DNA molecule of defined sequence. Detection of sites of carcinogen binding usually relied upon strand-scission of the DNA at the site of putative carcinogen-DNA binding, followed by mapping of the sites of strand-scission by gel electrophoresis relative to a unique site. For example, the primary N^7-Guanine-Aflatoxin B_1 (N^7-Gua-AFB_1) adduct, when treated with alkali, readily depurinates to form a single-strand DNA break. Using this method, Misra et al.[5] have determined that Aflatoxin B_1 (AFB_1) reacts with single-strand DNA in a random fashion while its reaction with double-strand DNA shows a 6 to 10-fold difference in guanine reactivity, depending on the neighboring nucleotide sequences. In addition, it has been determined that the profile of AFB_1 reactivity is similar for both microsome- and chloroperoxybenzoic acid-activated (CPB-activated) AFB_1.[5,6] While photoactivation of AFB_1 leads to non-random DNA binding, it is unknown how the spectrum of DNA adducts generated by this method compare to those adducts generated by microsome or chemical activation of AFB_1. Similarly. analysis of both UV-induced[7] and alkali-induced[8] DNA cleavage sites after reaction of DNA with benzo(a)pyrene diol-epoxide (BPDE) implicate a strong site specificity of BPDE with the DNA molecule which is not simply related to the relative guanine density. Using an assay in which the exonucleolytic activity of T4 DNA polymerase is blocked in the vicinity of an adducted site, Fuchs[9] has determined that N-acetoxy-AAF shows up to a 40-fold difference in reactivity with guanine, depending on its position in the sequence. Likewise, analysis of piperidine-induced sites of DNA cleavage suggest up to an 8-fold difference in guanine reactivity after adduction of DNA with N-OAc-Glu-P-1 (2-acetoxyamino-6-methyldipyrido[1,2-a:3',2'-d]imidazole).[10]

One basic assumption which underlies these studies is that the major adduct from a particular carcinogen with the DNA molecule readily forms a strand break under the experimental conditions used in the assay. At low levels of adduction, the relative intensity of a particular band on the autoradiogram would represent the probability of binding for a particular carcinogen at a given nucleotide-specific locus. For instance, the binding of AFB_1 to guanine residues at the N^7 position is the primary adduct form and easily depurinates when treated with alkali. However, other carcinogens are known to form several stable adduct forms for which differences in the relative rates of depurination would complicate quantitative interpretation of the sites of putative carcinogen-DNA binding.

DNA CONFORMATION

In reactions between carcinogens and DNA, both the accessibility and electrostatic potential of target atoms can play important roles in determining binding levels. These relationships have been clearly shown in studies of the

Table 1. DNA Sequence Specificity and Carcinogen Binding

Carcinogen	DNA	Assay	Conclusion	Ref.
AFB	ϕX174	CPB-, microsome-, and photoactivated AFB_1 binding followed by alkali-induced DNA cleavage	Binding to SS DNA random, DS DNA non-random	(5)
	ϕX174		6 to 10-fold difference in relative quanine reactivity	(6)
	E. coli lac gene	Alkali-induced cleavage after treatment of the DNA with microsome-activated AFB	Non-random binding	(41)
	pBR322	Alkali-induced cleavage	Non-random guanine reactivity	(42)
BPDE	chicken globin	UV-induced cleavage of BPDE-adducted DNA	Site-specific BPDE binding is not related to guanine density in a simple fashion	(7)
	E. coli lac gene	Alkali-induced cleavage of BPDE-adducted DNA	Non-random nucleotide reactivity	(8)
N-acetoxy-AAF	pBR322	Blockage of T4 DNA polymerase exonuclease activity in the vicinity of the adduct	Up to a 40-fold difference in reactivity of guanine residues depending on position	(9)
N-OAc-Glu-P-1	pBR322	Piperidine-induced DNA cleavage after reaction of the DNA with N-OAc-Glu-P-1	Up to 8-fold difference in reactivity of guanine residues depending on position	(10)

binding of N-OH-AAF to poly(dG-dC)·poly(dG-dC) in both B and Z conformations.[11] The carcinogen was shown to bind strongly to B-form DNA but not significantly to Z form, characteristics primarily attributed to accessibility of the C-8 atom of guanine, the preferred site of linking. Reaction of B-form poly(dG-dC)·poly(dG-dC) with AAF facilitates the transition to the Z conformation.[12,13] Similarly, N^7-methylating agents favor the B to Z conversion.[14] These latter findings are in contrast to the effects of AFB_1 binding on the B to Z transition. In this case, the formation of AFB_1-DNA adducts (which occurs at the N^7 position of guanine) with poly(dG-dC)·poly(dG-dC) inhibits the salt-induced conversion of B to Z-form DNA.[15]

While the functional role of Z-form DNA in vivo has not been fully established, the implication of these findings is that carcinogens could affect the stabilization or destabilization of secondary DNA conformations. This would further imply that the binding of carcinogens might affect DNA structures known to be proximal to actively transcribing genes. In addition, it has been shown that these DNA conformations are also S1 nuclease-sensitive and may be stabilized by the supercoiling of the DNA molecule. Such secondary DNA conformations appear to be intrinsic to the DNA sequence since intra-strand hydrogen bonding imparts enough single-stranded character to the region to allow these sites to be accessible to S1 nuclease attack. However, a study which compared the reactivity and distribution of adducts after reaction of BPDE with supercoiled and linear X174 DNA revealed no important differences between the two.[16]

CHROMATIN STRUCTURE

While many aspects of DNA conformation are intrinsic to the nucleotide sequence, experimental study of carcinogen interactions with chromatin provide further evidence concerning other aspects of chromatin structure which determine whether carcinogens are preferentially associated with specific domains. A number of investigators have studied this question by treatment of cells with particular carcinogens followed by fractionation of the chromatin by partial digestion of isolated nuclei with nucleases. At one level of organization, it has been shown that some carcinogens preferentially bind to internucleosomal (linker) DNA when compared to nucleosomal (core) DNA. Specifically, such a preference has been shown for methylating agents such as N-methyl-N-nitrosourea(MNU) in both HeLa cells[17] and rat hepatocytes.[18] Treatment of rainbow trout with AFB_1 has shown up to a 5-fold preference of binding of AFB for linker DNA compared to core DNA of liver chromatin.[19] Several other studies using 4-nitroquinoline-1-oxide (4NQO)[20] and BPDE[21-24] have similarly shown preferences for adduct formation in the linker region of chromatin. Using SV40, Seidman et al.[25] also have shown that the binding of BPDE to SV40 in vivo shows a 3-fold preference for linker versus core DNA.

The accessibility of chromatin to digestion by DNAse I has allowed categorization of chromatin into transcriptionally active (euchromatin) and inactive (heterochromatin) domains. It has been shown that carcinogen-DNA adducts were preferentially solubilized by nuclease digestion in rat liver DNA isolated from animals exposed to AAF,[26] N-OH-AAF,[29] dimethyl-N-nitrosamine (DMN),[27] MNU,[18] and BPDE.[22] These results suggest that the adducts are preferentially localized in the euchromatin fraction. In addition, benzo(a)pyrene-treated bronchial epithelial cells and fibroblasts showed preferential binding to DNAse I-sensitive regions of the chromatin.[28] Recently, it has been shown that AAF-DNA adducts in rat liver parenchymal cell nuclei were more DNAse I-sensitive than adducts formed in the nuclei of non-parenchymal cells.[26] In contrast to the above studies, N-OH-AAF[29,30] and dimethyl sulfate (DMS)[18] did not lead to enrichment of adducts in nuclease-sensitive chromatin regions.

DNA REPLICATION AND AMPLIFICATION

Several lines of evidence suggest that replicating DNA exists in a more open and nuclease-accessible chromatin conformation than non-replicating DNA. In addition, in vitro studies using carcinogen-adducted DNA as a template for DNA polymerization have suggested that the covalent binding of bulky adducts may lead to direct effects on replicon function. Taken together, these observations would suggest that carcinogen binding should occur preferentially in replicating DNA compared to non-replicating DNA. Preferential adduct formation in repair-synthesized DNA has been reported in several studies using AAF,[31] MNU,[18,32,33] DMN,[34] and 4NQO,[20] which would support that suggestion. However, these studies are difficult to interpret since repair-synthesized DNA is operatively defined on the basis of incorporation of radiolabeled precursors. Considering this, Cordeiro-Stone et al.[35] have shown that replication fork-associated DNA isolated from MNU-treated C3H 10T1/2 cells showed 2.6 to 5-fold more methylation than randomly methylated DNA. In addition, it has been shown recently that BPDE-treated human fibroblasts exhibit 3-fold preferential binding of BPDE to linker DNA versus core DNA for DNA analyzed from cells in the S phase of the cell cycle.[36]

These studies would suggest that replicating DNA may be a preferential target for carcinogen modification. This hypothesis is pertinent to recent findings which show that amplification of specific genes can be stimulated by the treatment of cells with carcinogens and DNA damaging agents. Specifically, it has been shown that UV light and N-acetoxy-AAF facilitate development of methotrexate-(MTX-)resistance when DNA synthesis is allowed to occur prior to exposure of the cells to MTX.[37] Importantly, a large percentage of MTX-resistant colonies showed an increase in copy number of the dihydrofolate reductase gene. In addition, Lavi[38] has shown that amplification of SV40-specific DNA in SV40-transformed Chinese hamster embryo cells occured when the cells were treated with a variety of direct-acting and indirect-acting carcinogens.

The suggestion that carcinogen modification of the DNA molecule may lead to specific gene amplification is especially intriguing in light of the finding that the copy number of some proto-oncogenes is increased in tumor-derived DNA when compared to DNA derived from normal tissues. Since the carcinogenicity of many chemicals in adult animals requires relatively long-term exposure, one might speculate that selective amplification of specific DNA domains may serve to generate genetic variability in a cell population that would otherwise be slow to fix genetic changes by cell proliferation. Selection of the transformed cell phenotype could occur in a situation where the carcinogen-induced change in copy number of a particular DNA sequence leads to a steady-state level of gene products which provides the cell with a distinct growth advantage. In this regard, proto-oncogenes are good candidates for carcinogen-induced DNA amplification since many proto-oncogene products are known to be involved in growth regulation and cell proliferation. The accumulation of these dominant genetic changes may be envisioned as part of the multi-stage nature of the malignant process.

TRANSCRIPTION

As noted above, transcriptionally-active chromatin indirectly has been shown to bind carcinogens preferentially in vivo. In general, direct measurement of carcinogen-DNA adducts within actively-transcribing genes is hampered by the low copy number of most genes in the eukaryotic cell. This has recently been accomplished, however, by the isolation and analysis of the DNA of the ribosomal genes from the liver of rats treated with AFB_1.[39] The DNA sequences coding for the 45S precursor to 18S and 28S rRNA are present in approximately 300 copies per cell in rat hepatocytes, making it feasible to isolate sufficient DNA for direct adduct analysis. In rats treated with a single dose of AFB_1, rRNA was found to contain 4 to 5 times more AFB_1-N^7-Gua adducts than unfractionated nuclear DNA. This difference could not be accounted for by G-C enrichment and was therefore attributed to greater accessibility due to transcriptional activity. In addition, in vitro modification of rDNA by AFB_1 showed only a 40-70% enrichment in adducts when compared to adduction of nuclear DNA.

These studies illustrate ways in which the various levels of chromatin organization can influence significantly the identity and localization of carcinogen-mediated covalent modifications of the DNA molecule in living cells. As the sensitivity for detection of carcinogen-DNA adducts improves, so will the technical feasibility for the quantitation of particular carcinogens within known DNA sequences. Recently, Gupta[40] has been able to examine the relative distribution of N-OH-AAF-DNA adduct forms in repetitive DNA sequences isolated from rat hepatocytes, which showed preferential binding of carcinogen. The non-random nature of the binding of carcinogen in this family of DNA sequences is not consistent with the hypothesis that such repetitive elements are located in inert heterochromatin. However, both of these studies indicate promise for further studies in which it may be possible to measure directly the formation of carcinogen-DNA adducts in specific genes which are being transcribed actively.

REFERENCES

1. M.R. Osborne, DNA interactions of reactive intermediates derived from carcinogens, in: "Chemical Carcinogens," C. E. Searle. ed., American Chemical Society, Washington, D.C. (1984).
2. E. Kriek, L. Engelse, E. Scherer and J.G. Westra, Formation of DNA modifications by chemical carcinogens - identification, localization and quantification, Biochim. Biophys. Acta., 738:181 (1983).
3. W.B. Melchior and F.A. Beland, Preferential reaction of the carcinogen N-acetoxy-2-acetylaminofluroene with satellite DNA, Chem.-Biol. Inter. 49: 177 (1984).
4. W.T. Briscoe and L.E. Cotter, The effects of neighboring bases on N-methyl-N-nitrosourea alkylation of DNA, Chem.-Biol. Inter. 52:103 (1984).
5. R.P. Misra, K.F. Muench and M.Z. Humayun, Covalent and noncovalent interactions of aflatoxin with defined deoxyribonucleic acid sequences. Biochem. 22:3351 (1983).
6. K.F. Muench, R.P. Misra and M.Z. Humayun, Sequence specificity in aflatoxin B -DNA interactions, Proc. Natl. Acad. Sci. USA 80:6 (1983).
7. T.C. Boles and M.E. Hogan, Site-specific carcinogen binding to DNA, Proc. Natl. Acad. Sci. USA 81:5623 (1984).
8. W.A. Haseltine, K.M. Lo and A.D. D'Andrea, Preferred sites of strand scission in DNA modified by anti-diol epoxide of benzo[a]pyrene. Science 209:929 (1980).
9. R.P. Fuchs, DNA binding spectrum of the carcinogen N-acetoxy-N-2-acetylaminofluorene significantly differs from the mutation spectrum, J. Mol. Biol. 177:173 (1983).
10. Y. Hashimoto and K. Shudo, Sequence selective modification of DNA with muta-carcinogenic 2-amino-6-methyldipyrido[1,2-a:3',2'd] imidazole, Biochem. Biophys. Res. Commun. 116:1100 (1983).
11. P. Rio and M. Levy, M., Preferential binding of the chemical carcinogen N-hydroxy-2-aminofluorene to B-DNA as compared to Z-DNA, Nucl. Acid Res. 11:4947 (1983).
12. R.M. Santella, D. Grunberger, I.B. Weinstein and A. Rich, Induction of the Z conformation in poly(dG-dC)•poly(dG-dC) by binding of N-2-acetylamino fluorene to guanine residues, Proc. Natl. Acad. Sci. USA 78:1451 (1981).
13. E. Sage and M. Levy, Conformation of poly(dG-dC)•poly(dG-dC) modified by the carcinogens of N-acetoxy-N-acetyl-2-aminofluorene and N-hydroxy-N-2-aminofluorene, Proc. Natl. Acad. Sci. USA 77:4597 (1980).
14. A. Moller, A. Nordheim, S.R. Nichols and A. Rich, Proc. Natl. Acad. Sci. USA 78:4777.
15. A. Nordheim, W.M. Hao, G.N. Wogan and A. Rich, Salt-induced conversion of B-DNA to Z-DNA inhibited by aflatoxin B , Science 219:1434 (1983).
16. M.C. MacLeod and M.S. Tang, Interactions of benzo[a]pyrene diolepoxide with linear and supercoiled DNA, Cancer Res. 45:51 (1985).
17. S. Sudhakar, K.D. Tew, P.S. Schein, P.V. Wooley and M.E. Smulsen, Nitrosourea interaction with chromatin and effect on poly(adenosine diphosphate ribose) polymerase activity. Cancer Res. 39:1411 (1979).
18. E.M.L. Berkowitz and H. Silk, Methylation of chromosomal DNA by two alkylating agents differing in carcinogenic potential. Canc. Lett. 12:311-321 (1981).
19. G.S. Bailey, J.E. Nixon, J.D. Hendricks, R.O. Sinnhumber and K.E. Van Holde, Carcinogen aflatoxin B is located preferentially in internucleosomal deoxyribonucleic acid following exposure in vivo in rainbow trout, Biochem. 19:5836 (1980).
20. K. Nose, Distribution of DNA bound carcinogen 4-nitroquinoline 1-oxide and of repair-synthesized DNA in chromatin of WI-38 cells, Canc. Lett. 14:205 (1981).

21. A. Kootstra, T.J. Slaga and D.E. Olins, Chem.-Biol. Inter. 28:225 (1979).
22. C.L. Jahn and G.W. Litman, Accessibility of deoxyribonucleic acid in chromatin to the covalent binding of the chemical carcinogen benzo(a)-pyrene, Biochem. 18:1442 (1979).
23. F. Feldman, J. Remsen, T.V. Wang and P. Cerutti, Formation and excision of covalent deoxyribonucleic acid adducts of benzo(a)pyrene 4,5-expoxide and benzo(a)pyrenediol epoxide I in human lung cells A549. Biochem. 19:1095 (1980).
24. P.L. Jack and P. Brookes, The distribution of benzo(a)pyrene DNA adducts in mammalian chromatin. Nucl. Acid Res. 9:5533 (1981).
25. M. Seidman, J. Slor and M. Bustin, The binding of a carcinogen to the nucleosomal and non-nucleosomal regions of the simian virus 40 chromosome in vivo, J. Biol. Chem. 258:5215 (1983).
26. B.L. Baranyi-Furlong and J.I. Goodman, Non-random interaction of N-2-acetylaminofluorene and its N-hydroxy metabolite with DNA of hepatic parenchymal and non-parenchymal cell nuclei following in vivo administration of carcinogen, Chem.-Biol. Inter. 48:15 (1984).
27. R. Ramanathan, S. Rajalakshmi and D.S.R. Sarma, Nonandom nature of in vivo methylation by dimethylnitrosamine and the subsequent removal of methylated products from rat liver chromatin DNA. Canc. Res. 36:2073 (1976).
28. J.E. Arrand and A.M. Murray, Benzpyrene groups bind preferentially to the DNA of active chromatin in human lung cells. Nucl. Acid Res. 10:1547 (1982).
29. G. Metzger, F.X. Wilhem and M.L. Wilhem, Non-random binding of a chemical carcinogen to the DNA in chromatin. Biochem. Biophys. Res. Commun. 75:703 (1977).
30. R. Ramanathan, S. Rajalakshmi and D.S.R. Sarma, Non-random nature of in vivo interaction of -N-hydroxy-2-acetylaminofluorene and its subsequent removal from rat liver chromatin-DNA. Chem.-Biol. Inter. 14:375 (1977).
31. T.D. Tlsty and M.W. Lieberman, The distribution of DNA repair synthesis in chromatin and its rearrangement following damage with N-acetyoxy-2-acetylaminofluorene. Nucl. Acid Res. 5:3261 (1978).
32. W.J. Bodell and M.R. Banerjee, The influence of chromatin structure on the distribution of DNA repair synthesis studied by nuclease digestion. Nucl. Acid Res. 6:359 (1979).
33. R. Cox, Canc. Res. 37:2675 (1977).
34. R. Ramanathan, S. Rajalakshmi, D.S.R. Sarma and E. Farber, Non random nature of in vivo interactions of H-N-hydroxy-2-acetylaminofluorene and its subsequent removal from rat liver chromatin DNA, Canc. Res. 36:2073 (1976).
35. M. Codeio-Stone, M.D. Topal and D.G. Kaufman, DNA in proximity in the site of replication is preferentially alkylated in S phase 10T1/2 cells treated with N-methyl-N-nitrosourea, Carcinog. 3:1119 (1982).
36. P. Kurian, A.M. Jeffrey and G.E. Milo, Preferential binding of benzo-[a]pyrene diol epoxide to the linker DNA of human foreskin fibroblasts in S phase in the presence of benzamide, Proc. Natl. Acad. Sci. USA 82:2769 (1985).
37. T.D. Tlsty, P.C. Brown and R.T. Schimke, UV radiation facilitates methotrexate resistance and amplification of the dihydrofolate reductase gene in cultured 3T6 mouse cells. Mol. Cell. Biol. 4:1050 (1984).
38. S. Lavi, Carcinogen-mediated amplification of specific DNA sequences, J. Cell. Biochem. 18:664 (1982).
39. T.R. Irvin and G.N. Wogan, Quantitation of aflatoxin B adduction within the ribosomal RNA gene sequences of rat liver DNA, Proc. Natl. Acad. Sci. USA 81:664 (1984).
40. R.C. Gupta, Nonrandom binding of the carcinogen N-hydroxy-2-acetylaminofluorene to repetitive sequences of rat liver DNA in vivo, Proc. Natl. Acad. Sci. USA 81:6943 (1984).

41. A.D. D'Andrea and W.A. Haseltine, Modification of DNA by aflatoxin B creates alkali-labile lesions in DNA at positions of guanine and adenine. Proc. Natl. Acad. Sci. USA 75:4120 (1978).
42. T. Suzuki, J. Kuwahari and Y. Sugiura, Nucleotide sequence cleavage of guanine-modified DNA with aflatoxin B , dimethyl sulfate, and mitomycin C by bleomycin and deoxyribonuclease I, Biochem. Biophys. Res. Commun. 117:916 (1983).

IONIZING RADIATION INDUCED DNA DAMAGE: IDENTITIES AND DNA REPAIR

J. F. Ward

University of California at San Diego, M010
La Jolla, CA 92093

DAMAGE PRODUCTION MECHANISMS

When cells are irradiated with low LET radiation the total dose of energy deposited per unit mass can be measured with great accuracy. In contrast, treatments with other agents can have problems of agent activation, diffusion or delivery to the target. With ionizing radiation the major problem is in the assignment of biological significance to any particular lesion. Model studies have indicated that a myriad of lesions (chemically distinct) can be produced.[1]

The mechanisms by which ionizing radiation energy is deposited in matter presents an additional variation in the nature of the damage. Figure 1

Figure 1. Diagrammatic representation of the relative sizes of the DNA double strand and the average energy deposition event from low LET ionizing radiation.

indicates in cartoon fashion the DNA double strand on the same scale as the average energy deposition event from low LET (X or gamma) radiation. The energy associated with this average energy deposition event is 60 eV.[2] This energy is sufficient to produce 3-4 radical pairs.[3] If this volume of energy deposition overlaps the DNA structure, then, either by ionizations of the DNA or by reactions of other reactive species (mainly OH radicals from water closely associated with the macromolecule), locally multiply damaged sites (LMDS) can be produced. Of course, if the energy deposition event does not totally overlap the DNA structure (a more probable occurence), singly damaged sites can be formed.

The ratio of the amounts of significant damage produced by OH radicals or via direct ionization is generally considered to be 2 (for a summary see Ref. 4). It should be mentioned that the OH radical reactions involved are not necessarily the same as those of diffusing OH radicals that occur in dilute aqueous solution,[5] but are modulated to some extent by their site of formation so they are not governed necessarily by dilute solution diffusion constants. Some recent evidence[6] indicates that the specific mechanism of damage production may not be important; damage, in the form of DNA double strand breaks, produced either by direct or indirect ionization is repaired equally rapidly intracellularly to damage produced preponderantly by direct ionization of the DNA.

Additional recent evidence indicates that singly damaged sites produced in cellular DNA by OH radicals from hydrogen peroxide are insignificant in killing cells.[7] Presumably the presence of intact template is sufficient for accurate, efficient repair.

IDENTITIES OF SINGLY DAMAGED LESIONS

Some recent calculations[4] indicate that the ratio of base damage to sugar damage in cellular DNA is greater than 2. A broad range of base damage identities has been established from model system studies; for thymine alone 21 damaged products have been identified,[8] and the range may be equally great for the other bases. It is not likely that the variety of damage will be significantly less in DNA irradiated in the cell. It would be natural to assume that such damage could be removed by classical excision repair processes. Exceptions to this generalization do exist, notably the 8, 5' cycloadenosine studied by Raleigh and his colleagues.[9] In this product, the 8 position of the base moiety is connected to the 5' position of the deoxyribose, presenting problems to both glycosylase mechanisms and exonuclease mechanisms of damage removal.

Radiation damage to the deoxyribose moieties needs to be considered carefully. This damage is the major source of the observed yields of DNA strand breaks. Additional contributions to the single strand break yield, as measured under alkaline conditions, arise from damage which results in loss of a base, leaving an AP site (alkali labile), or from the production of a base damage product whose glycosylic bond to deoxyribose is easily hydrolysed by alkali. The strand breaks formed by sugar damage by definition contain chemically altered sites which must be removed prior to insertion of the correct deoxynucleotide; the undamaged base cannot be retained during the necessary removal of the damaged sugar prior to resynthesis. There is no feasible mechanism for the radiation induced production of a DNA strand break which can be repaired by simple ligation. Even under the conditions of a model system that totally favors the formation and detection of simple dissociation of the phosphate ester bond, the yield of such damage is minimal;[10] strongly competing reactions for the hydrated electrons that cause this dissociation almost totally proscribe the formation of directly ligatable breaks. It is clear, therefore, that all individually damaged sites produced by ionizing radiation require multiple enzymatic processes for their repair, probably analogous to the classical processes described for other DNA lesions.

DNA DAMAGE MEASUREMENTS

If enzymatic repair processes are to be defined for ionizing radiation, then it would be useful to describe the types of damage involved as well as their relationship to the types of damage being assayed. The most commonly measured damage sites, single strand breaks (ssb) and double strand breaks (dsb), are questionable in this regard. As mentioned previously, ssb are generally measured under alkaline conditions (alkaline sucrose gradients, alkaline filter elution and alkaline unwinding). During the alkali treatment, damage sites (base damage, AP sites, etc.) are recruited to the ssb form to become unknown fractions of the total measured. In addition, there are a variety of chemical termini left even at true ssb.[11] Thus the assay procedures used for ssb measurement do not specifically measure a single class of damaged sites.

The duration and pH of the alkali treatment could alter the ratios of the various contributing lesions--model systems have shown the effects of these parameters on the yields of breaks.[12] With this in mind, it is unexpected that 90% of the lesions of the type measured by alkaline filter elution are repaired at the same rate, with a half-life of 3-4 mins.[13]

The question of double strand break (dsb) measurement is highly significant. It has been suggested[4,14] that these arise because of damage in opposite strands from a single energy deposition event. Therefore such lesions, which we have termed locally multiply damaged sites (LMDS), will be composed of combinations of the types of lesions classified as singly damaged sites. Of course all the sites described above will not be recruited, since dsb's are generally measured under "neutral" conditions, although the pH of 9.5 generally used for netural filter elution may recruit other damage. However, it is clear that LMDS will not consist solely of frank dsbs.

Three complications arise:

(1) Ionizing radiation produces alterations in deoxyribose which do not cause breaks immediately, but which on incubation (in the absence of enzymes) hydrolyse to form breaks over a time period of 1-2 hrs, depending on the temperature and the pH. These lesions could be involved in the production of some dsb's found in cellular DNA.[4] Since they do not cause a break immediately, but will if no repair occurs, if the damage on the opposite strand is repaired prior to the breakage of the labile bond, so that the correct template is available, then no dsb would result. We have previously suggested this phenomenon, arguing that release of DNA from cells and measurement of the damage therein might give an exaggeration of the dsb yield. Permitting the cells to "repair" would remove the lesions described above and therefore the measured yield of dsb would be less. Note no repair of actual frank dsb need be invoked here.

(2) The LMDS contain base damage sites also; an extrapolation from the single lesion case indicates that the ratios of double breaks to breaks plus base damage to double base damage would be 1/2.7/7.3, respectively. Consequently, base damage repair mechanisms achieve significance, particularly in the presence of damaged sites on the opposite strand, and the question of hierarchies of repair processes arises.

(3) The distribution and multiplicity of individual damage sites in a LMDS must be considered. It is clear that sites directly juxtaposed can lead

to loss of genetic information, which could lead to deletion or insertion of incorrect bases. If the sites are offset by more bases, then the patch size and the direction of insertion in the repair process becomes crucial. The average size of a low LET event is approximately 17 Å in radius, encompassing a maximum of about 10 base pairs with an equivalent maximum separation of individually damaged sites. Of course larger energy deposition events occur, in particular with high LET radiation. The consequences resulting are: greater possible separation of damage sites and greater numbers of damaged sites per energy deposition event.

Unfolding the complexities of the repair mechanisms of such complex lesions is a major challenge.

ACKNOWLEDGEMENTS

The work in the author's laboratory was supported by PHS Grants CA26279 and CA28640, awarded by the National Cancer Institute, DHHS.

REFERENCES

1. J. F. Ward, Adv. Radiat. Biol. 5:182 (1975).
2. A. Mozumder and J. L. Magee, Radiat. Res. 28:203 (1966).
3. J. L. Magee and A. Chatterjee, J. Phys. Chem. 20:2219 (1978).
4. J. F. Ward, Radiat. Res. (accepted for publication, 1985).
5. 9th Annual Meeting of Microdosimetry -- General Discussion.
6. J. W. Evans and J. F. Ward, 33rd Annual Meeting of the Radiat. Res. Soc. Abstract Eh12 (1985).
7. J. F. Ward, W. F. Blakely, and E. I. Joner, Radiat. Res. (Accepted publication, 1985).
8. J. Cadet and R. Teoule, Chem. Commun. 20:1269 (1971).
9. J. A. Raleigh, W. Kremers and R. Whitehouse, Radiat. Res. 65:414 (1976).
10. C. von Sonntag, Int. J. Radiat. Biol. 46:507 (1984).
11. J. F. Ward and I. Kuo, Int. J. Radiat. Biol. 23:543 (1973).
12. W. F. Blakely, J. F. Ward and E. I. Joner, Radiat. Res. (Submitted).
13. J. F. Ward, Radiat. Res. 86:185 (1981).
14. C. M. L. Kerr, M. K. Webster and F. Williams, J. Phys. Chem. 76:2848 (1972).

EFFECTS OF LET_∞ ON THE FATE OF DNA DAMAGE INDUCED IN RABBIT SENSORY CELLS *IN SITU*: FUNDAMENTAL ASPECTS

J.T. Lett, D.S. Bertold*, and P.C. Keng**

Dept. Radiology & Radiation Biology, Colorado State Univ. Ft. Collins, CO 80523; *Center for Radiation Res., Natl. Bureau of Standards, Gaithersburg, MD 20899; **Cancer Center Univer. of Rochester Medical Center, Rochester, NY 14627

INTRODUCTION AND RATIONALE

For half a century, radiobiological theory was dominated by the notion that cellular radiosensitivity to ionizing radiations could be explained solely on the basis of the physics of energy deposition[1,2]. The process was extended even to explanations for the variation in relative biological effectiveness (RBE) with increased linear energy transfer (LET_∞) associated with densely-ionizing, particulate radiations[3,4]. In more recent years, examination of the radiation responses of the repair-deficient S/S variant of the L5178Y murine leukemic lymphoblast has demonstrated that radiobiological theories which do not incorporate metabolic modification of radiochemical damage are untenable[2].

Cellular Responses to Ionizing Radiations and Radiobiological Theory

Many mammalian cells exhibit a biphasic, cell cycle response to X-(γ) photons. Peaks of radioresistance occur usually when such cells are irradiated with X (γ) photons while in progression through the G_1- and late S and/or early G_2 phases of the cycle. The repair deficiency of the L5178Y S/S variant with respect to its parent line stems from the absence, or inadequacy, of the repair mechanisms that operate when normally radioresistant mammalian cells are irradiated in G_1[2], and may be related to a deletion from the Y chromosome[5]. Henceforth, this repair mechanism will be called the G_1 repair mechanism, although there is no assurance that it actually operates only during G_1. When the S/S variant was exposed in G_1 to selected portions of beams of such densely-ionizing, particulate radiations as the heavy ions ^{20}Ne, ^{28}Si, ^{40}Ar and ^{56}Fe, for which the LET_∞'s were circa 35, 65, 90 and 220 keV/μm, respectively, the RBE was ≤ 1[6,7]. It seems likely, therefore, that the progressive increase in RBE with LET_∞ to the peak exhibited by normally radioresistant cells around 100-200 keV/μm[8,9] can be attributed in large measure to reduction in the efficacy of the G_1 repair mechanism[6,7].

Extended studies with the S/S variant, which were begun in the early 1960's[10,11,12], have demonstrated the need for a theory of cellular radiosensitivity that incorporates metabolic modification of radiation damage, and hence involves the laws of thermodynamics. An approach to the formulation of such a theory, which will accommodate genetic control of repair mechanisms and thus marry radiobiological theory with modern molecular biology, has been suggested[2], and it has as its basis the postulate that radiochemical damage in cellular DNA is modified by competing sets of enzymatic reactions that either repair or fix the radiochemical damage[2,13]. Prior to the mathematical formulation of the theory, however, there is a necessity to examine other concepts that may be basic to it, and a consideration of long-term experiments involved in that examination is the principal objective of this article.

Once the occurrence of cellular, and then DNA, repair mechanisms had been established, it became popular to propose mechanisms whereby cell death resulted from the inability of repair mechanisms to complete their allotted tasks. Usually this inability was thought to have its origins in certain types of irreparable DNA damage, so-called lethal or non-repairable lesions, or errors made by otherwise competent repair processes, and so forth. Yet despite numerous attempts at mathematical modelling, the relationship between cellular lethality and lethal, non-repairable or unrepaired lesions has not been quantified with a biological endpoint, and the boundary conditions of the resulting equations have not been established. For example, the idea that a radiation-induced (unrepaired) double strand break or crosslink in DNA can be lethal to a mammalian cell has been in vogue, primarily via multiple reincarnations, for at least a quarter of a century, but is without successful demonstration to date. Although, as we shall see later, the primary reasons for such failure are experimental, the net result is that the gap between radiation physics and biochemistry has not been bridged in the manner necessary for the evaluation of promising theories of cellular radiosensitivity.

Possible Modulation of DNA Repair by Endogeneous Cellular Processes

What is not clear from the various models and theories that have been proposed is why unrepaired damage cannot result from the timing of endogeneous cellular functions not necessarily related to the repair mechanisms themselves. Requirements such as the need to engage in DNA replication or to undergo mitosis could terminate, or interfere, with repair processes during the parts of the cycle in question. These possibilities are difficult to explore directly, so we chose to approach them from the opposite standpoint, namely: how effective are the repair mechanisms in cells that undergo neither DNA replication nor mitosis during their lifetimes? In other words, what happens to fully differentiated, non-dividing cells exposed to ionizing radiations *in situ*?

Evidence that DNA replication plays other than an inadvertant role in the recovery of cells from damage by ionizing radiations is circumstantial at best, and there is direct evidence to the contrary. For example, the X-ray survival curves of the L5178Y S/S variant are identical in the G_1 phase of the cell cycle and during the first half of S [2]. If normal human fibroblasts are maintained in plateau (G_0) phase culture *in vitro*, and repair is allowed to proceed to its maximum, the number of aberrations induced by X-photons and expressed in precondensed *chromatids*[14] is the same as the number of *chromosome* aberrations observed if the cells are allowed, subsequently, to proceed to the next metaphase[15]. Selective irradiation of the DNA replication fork does not seem to influence cellular radiosensitivity[16]. Observations of this nature carry the implication that DNA repair (and certainly G_1 repair) and DNA replication are not necessarily coupled in the processes of cellular recovery.

An experimental difficulty with cell populations in plateau phase *in vitro* is that a certain percentage of the cells (<5%) do cycle. We chose to examine cells that are locked permanently in G_0 *in situ*, namely the photoreceptor cells (rods) of the retina of the New Zealand white (NZW) rabbit (*Oryctolagus cuniculus*)[17,18,19], and since those cells were examined throughout the subsequent lifespans of the irradiated animals, the experiments had practical, as well as theoretical, objectives.

CHROMOSOMAL ARCHITECTURE AND THE EXTENT TO WHICH REPAIR CAN BE MEASURED

On average, a mammalian chromosome contains about 10^{11} daltons of DNA. Very special difficulties presented by a structure (molecule) of that size preclude the use of more usual experimental procedures for its quantification and necessitate employment of such physicochemical techniques as gradient sedimentation[20] and viscoelastometry[21]. From experiments utilizing those techniques, evidence has accumulated that mammalian DNA is present in chromatin in a highly condensed form, composed of supercoiled loops and even higher forms of architecture. In irradiated mammalian cells, the basic chemical and physicochemical forms of DNA damage, viz., changes to bases and sugars, single and double strand breaks, crosslinks and so forth, can also result in changes in the chromosomal architecture or superstructure of DNA. Repair of damaged DNA in irradiated cells, therefore, cannot be considered as complete until the chromosomal architecture of DNA has been restored.

Radiation Damage and the Superstructure of DNA

Increasing doses of ionizing radiation induce a distinct sequence of changes in the superstructure of chromosomal DNA in mammalian cells. If those changes are measured under neutral conditions by the sedimentation of nucleoids[22] or viscoelastometric analysis of retardation times[23,24], a discontinuity in behavior occurs at a dose of *circa* 1000 rads (10 Gy)- or perhaps better, in the region 500-1000 rads, see Figure 2 in reference 24. After radiation doses below the critical value (range), nonlinear molecules are present and "...serve to caution against the indiscriminate interpretation of changes in neutral sucrose gradient profiles in terms of molecular weight changes alone"[23]. From earlier studies, in which DNA damage was induced with large doses of X (γ) photons and analyzed by alkaline sucrose gradient sedimentation[25,26], neutral sucrose gradient sedimentation[27,28], and the amount of DNA bound to other cellular components[29], as well as the use of α-particles of limited range[28], it was concluded that DNA exists *in vivo* as (supercoiled) loops of size $1\text{-}2\times10^{9}$ daltons that are probably attached to the nuclear matrix at radiosensitive sites. Analogous conclusions about DNA structure were reached from experiments with alkaline hydrolysis[25,30]. Subsidiary arrays (and attachment points) of molecules that are approximately the sizes of replicons probably exist within the loops[25,27,28,31].

Originally, the sizes of the DNA loops were estimated by the use of large doses of X(γ) photons. If breaks are introduced into a distribution of polymer molecules by radiation, the changes in molecular size are related to the dose, D, by the equation:

$$N\left(\frac{1}{M_{irr}} - \frac{1}{M_o}\right) = kD$$

where M_{irr} and M_0 are the number average molecular weights of the irradiated and unirradiated polymer distributions, respectively, N is Avogadro's number and k is a constant related to the efficiency of strand breakage. Should $1/M_{irr}$ be recorded against D, therefore, extrapolation to D=0 will yield the molecular weight of the DNA in the unirradiated cell. When such analyses were performed by alkaline

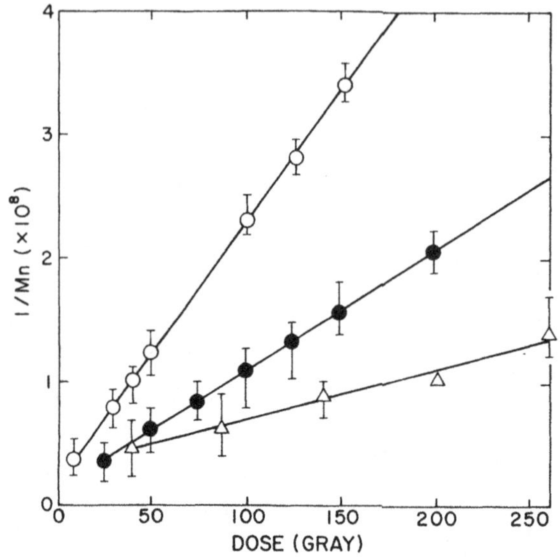

Fig. 1. Changes in the reciprocals of number average molecular weights (Mn) with dose provide the efficiencies of total strand breakage for 300 kVp X-photons, o, 365 MeV/u ^{20}Ne ions,●, or 530 MeV/u ^{40}Ar ions,▲. The lines are best fits by linear least squares regression analysis and the error bars represent standard errors of the means.

sucrose sedimentation of DNA from mammalian cells that had received large doses of X(γ) photons, M_o (for single-stranded DNA) was found to be ~5x10^{8} daltons[25,26], which suggested that in the mammalian cell the DNA is composed of units (loops)[25,31] of size circa 10^{9} daltons joined at sites of specific radiation action (radiosensitive sites). As is shown in Figure 1, similar results were obtained when the photoreceptor cells in isolated retinas were irradiated at 0° C with large doses of the heavy ions ^{20}Ne and ^{40}Ar; it is clear that the efficiency of total strand breakage (∝ 1/k) decreased with LET_∞.

The significance of the 1000 rad dose of X(γ) photons to the DNA structure in mammalian cells also is clear. Measurements of the efficiencies with which X(γ) photons produce breaks in single strands of DNA generally have yielded the value 50±10 eV[25,26], which is close to the energy deposited per primary ionization. After a dose of 1000 rads, each (single-stranded) molecule of size ~5x10^{8} daltons will be broken once on average[25,26]. At doses in the range 500-1000 rads, therefore, the chances are high that each strand in a DNA duplex (loop) of size circa 10^{9} daltons will be broken at least once and the configuration of the supercoiled loops will be completely relaxed. Of course, the damage will be more extensive if specific sites (regions) of radiation action are also disrupted.

Strand Breakage Induced in the DNA of Photoreceptor Cells by Heavy Ions

Under conditions where DNA molecules are present as single strands during sedimentation, the equation on p. Ms3 can be used to calculate the efficiencies of total strand breakage (TSB) in DNA, see Table 1. The values for double strand breakage (DSB) were made from a survey of the data in the literature, and the efficiencies of single strand breakage (SSB) were computed from those figures and the experimental values for TSB.

Table 1. Efficiencies of double, single and total strand breakage in the retinal photoreceptor DNA of the NZW rabbit as a function of LET_{∞}

LET_{∞} (keV/μm)	Efficiency of Strand Breakage (eV/break)		
	TSB	DSB	SSB
1	50	1100	55
35	110	940	140
90	240	790	620

A question to be asked at this point is, why were double strand breaks not measured directly by using neutral sucrose gradients? The reasons are simple. First, the efficiency of double strand breakage for X(γ) photons, for example, is twenty times less than that for single strand breakage. Secondly, the caveat propounded by Chase & Shafer, see p. Ms , is a serious one. Thirdly, the sedimentation of double stranded DNA, whether it is truly in the form of linear molecules or not, is beset with severe difficulties[20]. Hence the technique of sedimentation through neutral sucrose gradients has neither the sensitivity nor the reliability needed for analysis of the repair of strand breaks in the dose range of radiobiological significance for mammalian systems.

Ten to twenty years ago, there were few alternatives to experiments involving large doses of radiation from which the majority of mammalian cells would not survive[12]. One could work with a radioresistant bacterium such as *Micrococcus* (now *Deinococcus*) *radiodurans*, of course, and indeed, the repair of double strand breaks was demonstrated first with that organism[32], but it hardly seems justifiable nowadays to pursue studies of mammalian DNA repair mechanisms only in dying cells. So, the very sensitive technique of sedimentation through reoriented alkaline sucrose gradients was used in these studies on the grounds that complete rejoining of total strand breaks must signify complete rejoining of all the single- and all the double strand breaks in DNA. Radiations of different LET_{∞} were used deliberately in order to provide a wide range for the ratio of double strand breaks to single strand breaks in DNA because the former could not be measured directly.

Limiting Sensitivity of Reorienting Gradient Zonal Ultracentrifugation

All the DNA studies that have been described here are consistent with the idea that chromosomal DNA is arranged in supercoiled loops, composed themselves perhaps of rosettes of replicons, all of which are attached to the nuclear matrix. Mammalian DNA, therefore, is subject to two general forms of damage at the level of chromosomal structure: relatively specific damage in the region of the attachment sites and approximately random damage to the DNA between them.

When cells are exposed to low doses of ionizing radiations, rapidly sedimenting species appear in the alkaline sedimentation profile[20,25], in the manner indicated by the examples in Figure 2. This behavior is entirely consistent with the changes observed by viscoelastometry and it represents intermediate levels of destruction to the superstructure of mammalian DNA, which sediments at an apparent sedimentation coefficient, S_A, of ~600S[19] under the lytic conditions employed in our experiments.

To date, sedimentation through reoriented alkaline sucrose gradients is the most sensitive method available for detecting damage to

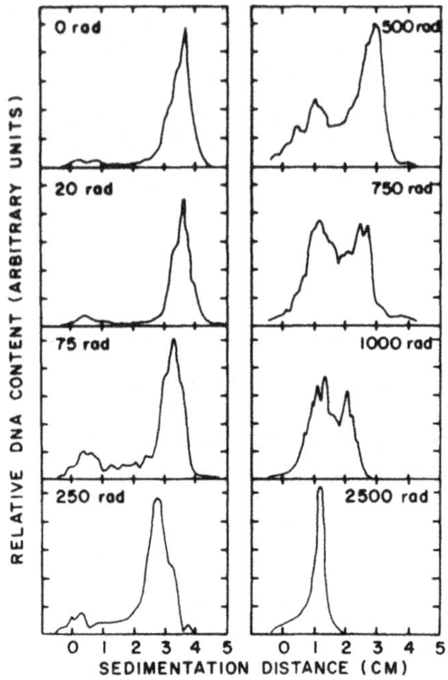

Fig. 2. Representative sedimentation profiles obtained from photoreceptor DNA after retinas had been irradiated with 365 MeV/u ^{20}Ne ions at $0^{o}C$ in vitro. For sedimentation conditions see reference 19.

the DNA superstructure in nondividing cells. Changes in the sedimentation pattern are produced by 5-10 rads of X(γ) photons[19]. For dividing cells, i.e., situations where the DNA can be prelabelled with radioactive precursors, the limiting sensitivity is 1-2.5 rads of X(γ) photons. Such levels of sensitivity also determine the extent to which DNA repair can be followed.

PHOTORECEPTOR CELLS OF THE RABBIT RETINA AS A MODEL SYSTEM

The extents to which the rabbit retinal photoreceptor cells fulfilled the need for an experimental model system are as follows. The full complement of photoreceptor cells is all of one type, rods, which are fully differentiated at the birth of the animal, or shortly thereafter, and dead cells are never replaced in situ. Retinas can be excised by simple surgical procedures[18] and stored in liquid N_2 without damage to the photoreceptor DNA[19], and their photoreceptor complements, ~$6x10^7$ rods, can be measured by simple means[7]. Such cells remain in G_0 throughout the life span of the animal, except perhaps in very old age[33], and do not progress perceptibly into the S phase after irradiation[34]. Finally, the size of the Beckman Ti-15 titanium zonal rotor is such that it can be used to examine all the DNA in the entire photoreceptor complement of an individual retina in one sedimentation profile. The retina of the second eye serves for confirmation, etc., in both irradiated and unirradiated situations.

PHOTORECEPTOR DNA IN THE RETINA OF THE AGING NZW RABBIT

Radiation therapists long have known that nondividing tissues are relatively resistant to ionizing radiations, a phenomenon identified as

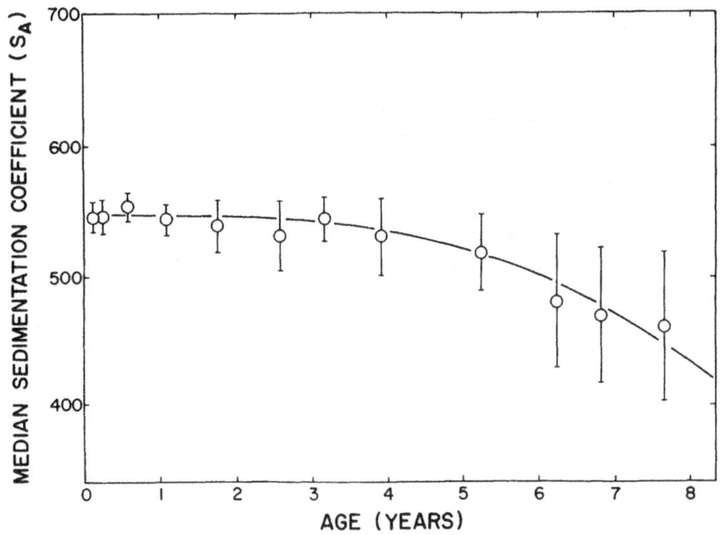

Fig. 3. The median (apparent) sedimentation coefficient, S_A, of photoreceptor DNA from the NZW rabbit retina as a function of animal age. The error bars represent the range of values obtained for a given age group.

the law of Bergonie and Tribondeau. The likelihood was, therefore, that the photoreceptors in the rabbit retina would withstand considerable doses of localized irradiation and that damage would be manifested as radiation sequelae late in the lifespan of the species[33]. Such considerations necessitated investigation of naturally occurring changes in the DNA of retinal photoreceptor cells throughout the lifetimes of the animals, because those measurements would provide the baseline against which the changes wrought by radiation could be compared[6,7,33].

A preliminary synopsis of the results from the lifespan study[7,33], which took about ten years to obtain, is shown in Figure 3. The median sedimentation coefficient of the DNA from the whole complement of photoreceptor cells in the retina decreases with age, and the largest changes are comparable to the initial effects produced by several hundred rads of X(γ) photons in young (6-8 wk-old) animals. As the age of the animal increases, the (degradative) changes in DNA structure begin before loss in photoreceptor cells can be detected (data to be presented elsewhere) and eventually they occur in all the cells. Analysis of the sedimentation profiles shows that the changes in DNA structure cannot be explained by extensive DNA degradation in some photoreceptors while the DNA in others remains intact[33].

RESTORATION OF DNA DAMAGE IN IRRADIATED PHOTORECEPTOR CELLS IN SITU

In order to provide the maximum period for the examination of the formation and fate of DNA damage in the retinal photoreceptor cells of the NZW rabbit, the eyes of young animals were irradiated with localized doses of ^{60}Co γ photons, ^{20}Ne ions and ^{40}Ar ions [19,33]. Following even comparatively large doses of radiation, damage caused to the photoreceptor DNA was repaired by the photoreceptor cells, regardless of LET_∞, e.g. see Figure 4, and the photoreceptor complement of the retina was not altered. Even though the photoreceptor cells survived large doses of radiation, at least initially (Figure 5), the processes that restored the DNA damage took several weeks to complete their tasks (Figure 4).

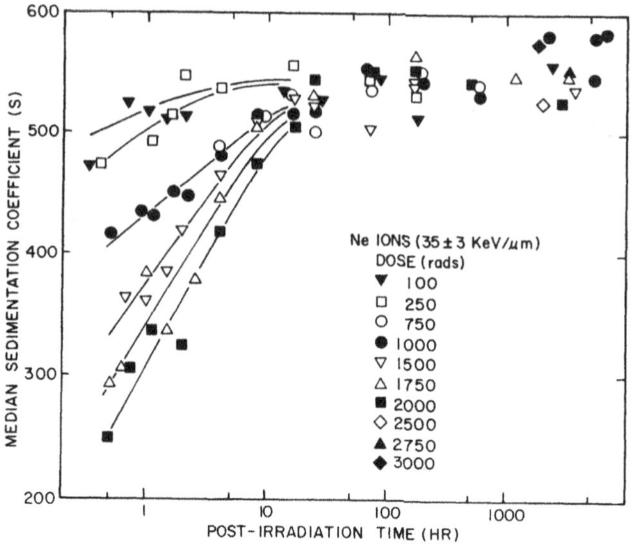

Fig. 4. Changes in median (apparent) sedimentation coefficient, S_A, with postirradiation time for the DNA of retinal photoreceptor cells irradiated with 365 MeV/u ^{20}Ne ions _in situ_.

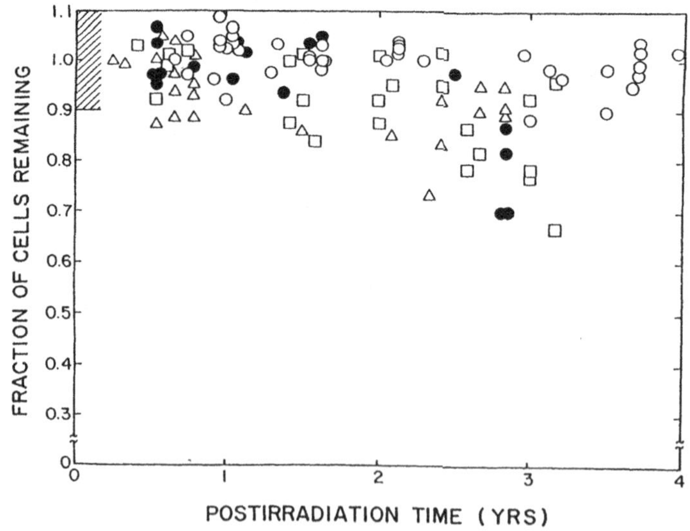

Fig. 5. Complement of photoreceptor cells present in the retina of the NZW rabbit after local irradiation of optic and proximate tissues at 6-8 wk of age with ^{60}Co γ-photons, ^{20}Ne ions and ^{40}Ar ions. Radiation: ^{60}Co γ-photons, ○, 10 Gy; ●, 20 Gy; ^{20}Ne ions, □, 10 Gy; ^{40}Ar ions, △, 5 Gy. The shaded area represents the range of values about the mean for unirradiated cells. Taken with permission from reference 7.

Reconstitution of chromosomal DNA architecture seemed to proceed in a progressive manner and DNA species of ever-increasing sedimentation coefficients appeared in the sedimentation profiles. Except at the very highest doses employed, the sedimentation profiles from irradiated animals ultimately became indistinguishable from those of control animals of similar ages. Given the sensitivity of the technique of reorienting gradient zonal ultracentrifugation used in these studies, therefore, retinal photoreceptor cells that had received a dose of a thousand or more rads of ionizing radiation, seemingly restored the damage in their DNA superstuctures to less than that caused by 5-10 rads of X(γ) photons, and all the photoreceptor cells survived. Moreover, since the LET_∞ of radiations employed ranged from 0.3-95 keV/μm, the increased propensity for producing double strand breaks in the cellular DNA did not seem to affect the outcome.

FATE OF IRRADIATED PHOTORECEPTOR DNA IN SITU

Despite the efficacy with which the repair processes in retinal photoreceptor cells disposed of DNA damage, the question of whether repair was complete or correct cannot be answered yet. As irradiated animals progressed through the lifespan, the "repaired" DNA in the photoreceptor cells began to exhibit symptoms of damage akin to those seen in unirradiated animals, but the damage appeared earlier the larger the radiation dose employed. This behavior has been adumbrated previously [6,33], (Figures 6,7), and will be presented in detail elsewhere. The appearance of damage in the erstwhile repaired DNA preceded the loss of photoreceptor cells from the retina, in a fashion similar to that found with natural aging, but there can be no more than the possibility at this juncture that the two processes are similar. Analyses of the degradation processes carry also the suggestion of RBE's for ^{20}Ne and ^{40}Ar ions comparable to those found for cellular lethality in proliferating tissues. It must be borne in mind, of course, that

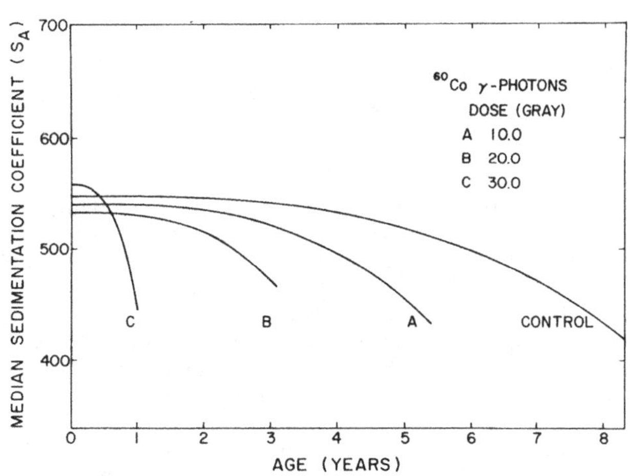

Fig. 6. Changes in the median apparent sedimentation coefficient, S_A, of the DNA in the photoreceptor cells of the retina of the NZW rabbit, as a function of age or post-irradiation time after "complete" repair in situ, following single, acute doses of ^{60}Co γ-photons delivered locally to optic and proximate tissues at age circa 6 wks. Dose: A, 10 Gy; B, 20 Gy, C, 30 Gy. The curves are best fits by regression analysis to data that will be published in detail elsewhere. Taken with permission from reference 6.

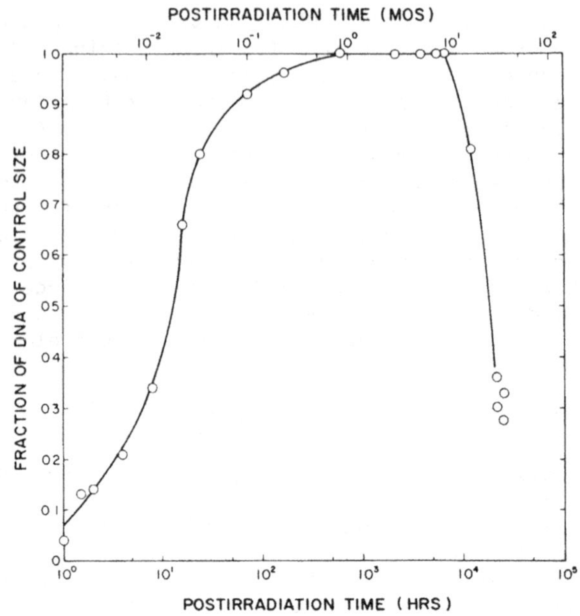

Fig. 7. Changes in the composition of the DNA of retinal photoreceptor cells from the NZW rabbit after local irradiation of optic and proximate tissues with 10 Gy of ^{20}Ne ions at 6-8 wks of age. Taken with permission from reference 7.

degeneration of photoreceptor DNA could be initiated, or mediated, by the radiation responses of adjacent proliferating tissues <u>in situ</u>.

CONCLUSIONS

1. Following doses of ionizing radiation from which they survive *in situ*, at least initially, nondividing photoreceptor cells of the rabbit retina seem able to repair DNA damage and restore DNA superstructure, although they may take as long as several weeks so to do.
2. Since, to the best of current knowledge, retinal photoreceptor cells remain locked in the G_0 phase of the cell cycle throughout their lifetimes, the restoration of DNA superstructure (architecture) from various kinds of DNA damage does not require the participation of DNA replication.
3. Whether the repair of DNA damage and the restitution of DNA superstructure can be completed in the G_0 phase of the cell cycle is still unknown not only because of the limiting sensitivity of the analytical techniques, but also because the "repaired" DNA in irradiated cells deteriorates earlier than the DNA in the unirradiated counterparts.
4. The implicit assumption that DNA replication plays an intrinsic role in DNA repair processes requires critical examination.
5. Proliferating cells may die from radiation damage because they engage in other cellular functions before otherwise competent repair mechanisms have time sufficient to finish their tasks.

6. Fully differentiated, nondividing cells in the central nervous systems of individuals exposed to densely ionizing radiations, and especially astronauts exposed to heavy ions in outer space, are likely to exhibit late radiation sequelae that are dependent on LET_∞.

ACKNOWLEDGEMENTS

This research was supported by the National Aeronautics and Space Administration with grant #NSG 9045, by the Department of Health and Human Services with grant #AG00005 from the National Institute on Aging and grant #CA 10714 from the National Cancer Institute, and by the Department of Energy with contract #DOE-AC03-76SF00098 to the Lawrence Berkeley Laboratory, University of California.

REFERENCES

1. D. E. Lea, Actions of Radiations on Living Cells, Cambridge University Press, London (1946).
2. H. Nagasawa, A. B. Cox and J. T. Lett, The radiation responses of synchronous L5178Y S/S cells and their significance for radiobiological theory, Proc. Roy. Soc. Lond. B 211:25 (1980).
3. D. E. Lea, R. B. Haines and C. A. Coulson, Bactericidal action of radioactive radiations, Proc. Roy. Soc. Lond. B 120:47 (1936).
4. A. M. Kellerer and H. H. Rossi, The theory of dual radiation action, Curr. Top. Radiat. Res. Q 8:85 (1972).
5. M. H. Rutledge and J. T. Lett, The L5178Y S/S murine leukemic lymphoblast: a radiosensitive, malignant cell of stable karyotype, Radiat. Res. 81:282 (1980).
6. J. T. Lett, D. S. Bergtold, A. B. Cox, P. C. Keng and R. Okayasu, Further evidence that the survival of irradiated mammalian cells is controlled by temporal processes, Br. J. Cancer 49, Suppl. VI:163 (1984).
7. J. T. Lett, A. B. Cox and D. S. Bergtold, Cellular and tissue responses to heavy ions: basic considerations, Radiat. Environ. Biophys. (in press, 1985).
8. G. W. Barendson, Responses of cultured cells, tumours and normal tissues to radiations of different linear energy transfer, Curr. Top. Radiat. Res. Q. 4:293 (1968).
9. E. A. Blakely, F. Q. H. Ngo, S. B. Curtis and C. A. Tobias, Heavy-ion radiobiology: cellular studies, in: "Advances in Radiation Biology," J. T. Lett, ed., Academic Press, New York (1984).
10. J. Z. Beer, J. T. Lett and P. Alexander, Influence of temperature and medium on the X-ray sensitivities of leukaemia cells in vitro, Nature 199:193 (1963).
11. J. T. Lett, G. Parkins, P. Alexander and M. G. Ormerod, Mechanisms of sensitization to X-rays of mammalian cells by 5-bromodeoxyuridine, Nature 203:593 (1964).
12. J. T. Lett, I. Caldwell, C. J. Dean and P. Alexander, Rejoining of X-ray induced strand breaks in the DNA of leukaemia cells, Nature 214:790 (1967).
13. P. Alexander, J. T. Lett and C. J. Dean, The role of post-irradiation repair processes in chemical protection and sensitization, Progr. Biochem. Pharmacol. 1:22 (1965).
14. M. N. Cornforth and J. S. Bedford, On the nature of a defect in cells from individuals with ataxia-telangiectasia, Science 227:1589 (1985).
15. M. N. Cornforth, Rejoining of X-ray induced chromosome breaks in human cells and its relationship to cellular repair, Doctoral Dissertation, Colorado State University (1985).

16. K. G. Hofer and R. L. Warters, Cell lethality after selective irradiation of the DNA replication fork, Radiat. Environ. Biophys. 24:161 (1985).
17. K. T. Wheeler, E. L. Pautler and J. T. Lett, γ-irradiation of photoreceptor cells in vivo: DNA repair and retinal function, Expt. Cell Res. 74:281 (1972).
18. K. T. Wheeler, R. E. Sheridan, E. L. Pautler and J. T. Lett, In vivo restitution of the DNA structure in gamma irradiated rabbit retinas, Radiat. Res. 53:414 (1973).
19. P. C. Keng, D. S. Bergtold and J. T. Lett, Effects of heavy ions on rabbit tissues: analysis of low levels of DNA damage in retinal photoreceptor cells, Int. J. Radiat. Biol. 43:219 (1983).
20. J. T. Lett, Measurement of single-strand breaks by sedimentation in alkaline sucrose gradients, in: "DNA Repair: a laboratory manual of research procedures, vol. 1, part B," E. C. Friedberg and P. C. Hanawalt, eds., Marcel Dekker, Inc., New York (1981).
21. C. S. Lange, J. Ostashevsky, E. Perlmutter, J. DeLeon and G. Grossman, The use of viscoelastometry to determine survival curves for intact genomes, Radiat. Res. 100:1 (1984).
22. P. R. Cook and I. A. Brazell, Supercoils in human DNA, J. Cell Sci. 19:261 (1975).
23. E. S. Chase and R. H. Shafer, Viscoelastic behavior of mammalian DNA, Biophys. J. 28:93 (1979).
24. K. L. W. Wun and R. H. Shafer, Structural changes in mammalian cell DNA induced by low-dose X-ray damage and subsequent post-irradiation incubation in the presence and absence of caffeine, Radiat. Res. 90:310 (1982).
25. J. T. Lett, E. S. Klucis and C. Sun, On the size of the DNA in the mammalian chromosome: structural subunits, Biophys. J. 10:277 (1970).
26. J. T. Lett and C. Sun, The production of strand breaks in mammalian DNA by X-rays: at different stages in the cell cycle, Radiat. Res. 44:771 (1970).
27. A. Cole, F. Shonka, P. Corry and W. G. Cooper, CHO cell repair of single-strand and double-strand DNA breaks by γ- and α-radiations, in: "Molecular Mechanisms for Repair of DNA, part B," P. C. Hanawalt and R. B. Setlow, eds., Plenum Press, New York (1978).
28. R. Datta, A. Cole and S. Robinson, Use of track-end alpha particles from ^{241}Am to study radiosensitive sites in CHO cells, Radiat. Res. 65:139 (1976).
29. K. Regel, K. Gunther and G. Kampf, Some conclusions as to the in vivo organization of DNA in eukaryotic cells following from sedimentation studies after irradiation, Studia Biophys. 76:11 (1979).
30. E. S. Klucis and J. T. Lett, Zonal centrifugation of mammalian DNA, Analyt. Biochem. 35:480 (1970).
31. J. T. Lett, Linkers in mammalian chromosomal DNA, in: "Aging Carcinogenesis and Radiation Biology," K. C. Smith, ed., Plenum Press, New York (1976).
32. C. J. Dean, P. Feldschreiber and J. T. Lett, Repair of X-ray damage to the DNA in Micrococcus radiodurans, Nature 209:49 (1966).
33. D. S. Bergtold and J. T. Lett, Alterations in chromosomal DNA and aging: an overview, in: "Molecular Biology of Aging: Gene Stability and Gene Expression," R. S. Sohal, L. Birnbaum and R. G. Cutler, eds., Raven Press, New York (in press, 1985).
34. K. T. Wheeler and J. T. Lett, On the possibility that DNA repair is related to age in non-dividing cells, Proc. Natl. Acad. Sci. USA, 71:1862 (1974).

RADIATION DAMAGE TO DNA IN VARIOUS ANIMAL TISSUES: A COMPARISON OF YIELDS AND REPAIR *IN VIVO* AND *IN VITRO*

Raymond E. Meyn, W. Timothy Jenkins and David Murray

University of Texas System Cancer Center
6723 Bertner Ave., Houston, TX 77030

In general, DNA is assumed to be the critical target in mammalian cells exposed to ionizing radiation. However, the spectrum of lesions induced, their subsequent repair, and the interrelationships between these damage and repair mechanisms to cellular endpoints such as cell killing, mutation, and carcinogenesis remain important topics for research. Of the various types of lesions known to be produced by radiation, DNA strand breaks have been the subject of intensive study, primarily owing to the availability of sensitive methods for their detection in cultured cells irradiated *in vitro*. Although the results from such work have contributed significantly to our understanding of the effects of radiation on mammalian cells, the models developed from these studies may not be strictly applicable to the situation *in vivo*. The evidence that has accumulated over the last few years suggests that there may be significant differences between the *in vitro* and *in vivo* situations that influence both the formation and repair of radiation-induced DNA lesions. Such factors include the degree of oxygenation,[1] intracellular levels of reduced glutathione (GSH),[2] and degree of cellular differentiation.[3-7]

In order to examine in greater detail how these factors may influence the formation and repair of DNA damage *in vivo*, we measured the efficiency of production and the kinetics of rejoining of stand breaks in various tissues of mice that had been exposed to whole-body doses of X rays. DNA damage was measured using the alkaline elution technique[8] modified for use with cells from animal tissues *in vivo* using a microfluorometric DNA assay, which allows these measurements to be made in nonproliferating cells (i.e., cells unable to incorporate radioactive nucleosides).

Methods

The methodology used in these investigations will only be briefly outlined since the details have been described previously.[9,10] C_3H mice from a specific-pathogen-free breeding colony were used at 12-16 weeks of age. They were restrained in a Perspex chamber and whole-body irradiated using a General Electric Maximar 250 KVP X ray machine (dose rate, 3.53 Gy min^{-1}). The dose-response relationships shown in this paper have been adjusted for the difference in dose received by the different tissues. The animals were sacrificed at various times following irradiation, and their tissues were immediately removed and immersed in ice-cold Puck's saline A containing 5 mM EDTA. A total of about 3 min was required for irradiation and dissection under conditions where repair of DNA damage could occur. To induce acute tissue hypoxia, the mice were either killed by cervical dislocation 15 min prior to irradiation or were asphyxiated with nitrogen gas. Suspensions were prepared from each tissue, and DNA damage was measured using the alkaline elution technique[8] modified for use with cells from animal tissues. DNA was assayed using the fluorescent dye Hoechst 33258. This method generates elution profiles identical to those obtained with radioactivity assays using [^{14}C]-dThd-labeled CHO cells. DNA strand breaks were quantitated from such DNA elution profiles by calculating a relative strand scission factor (SSF), defined as the absolute log of the ratio of the percentage of DNA retained on the filter for the irradiated sample to that for an unirradiated, control sample, in each case after an eluted volume of 17.5 ml. An SSF of 0 indicates no strand break formation.

Intertissue Variations in Sensitivity to X Ray-induced DNA Strand Breakage In Vivo and in Vitro

The efficiency of DNA strand-break production in different mouse tissues irradiated *in vivo* was compared. Figure 1 shows the dose responses for spleen and jejunum, which relate the relative degree of DNA strand breakage to the radiation dose. Similar dose responses were generated for other tissues. Cell suspensions also were prepared from various tissues from unirradiated mice, and then irradiated with graded doses of X rays on ice *in vitro*; these results for spleen and jejunum are also shown in Fig. 1. The slopes of the resulting dose-responses, calculated by linear regression analysis, are included in Table 1, and provide an indication of the relative radiosensitivity of the tissues based on the efficiency of strand-break production. Liver, brain, and kidney were analyzed after a single dose of 4 Gy *in vitro*, and the calculated slopes of the extrapolated SSFs are listed in Table 1 for comparison.

The efficiency of DNA strand-break formation among tissues irradiated *in vivo* varies by a factor of more than two, each tissue having its own characteristic response. This confirms the observations of Ono and Okada,[11,12] who showed that the efficiency of DNA strand breakage was considerably less in mouse testicular cells than in either liver or thymus cells irradiated *in vivo*. They attributed this finding to a hypoxic condition of testicular cells. In contrast to the effects of irradiation *in vivo*, cells from normal tissues irradiated *in vitro* had similar sensitivities (Table 1). Furthermore, each tissue was 2- to 5-fold more sensitive when irradiated *in vitro* than when irradiated *in vivo*.

Table 1. Slopes of *in vivo* and *in vitro* dose-response curves[a,b]

Tissue	In vivo	In vitro	Ratio
Bone marrow	0.68±0.05[c]	1.20±0.11	1.76
Spleen	0.60±0.04	1.24±0.09	2.07
Brain	0.53±0.05	1.26[d]	2.38
Kidney	0.43±0.06	1.28[d]	2.98
Liver	0.35±0.05	1.35[d]	3.86
Jejunum	0.30±0.03	1.34±0.21	4.47

a: Adapted from reference 9;
b: Units of slope are SSF per 10 Gy;
c: 95% confidence interval;
d: values were extrapolated from data obtained using a single dose of 4Gy.

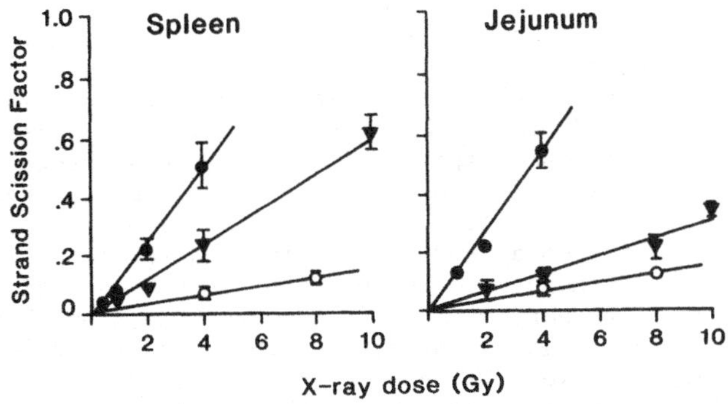

Figure 1. Dose responses for DNA strand-break production in spleen and jejunal cells immediately after irradiation with various doses of x rays. Cells were irradiated under 3 different conditions: (▼) - in vivo; (●) - in vitro on ice; (O) - after N -asphyxiation (actually hypoxic).

There are two likely explanations for this behavior. These effects could be the result of intertissue differences in strand-break repair occurring during the 3 minutes required for irradiation and tissue preparation in vivo, an effect that is eliminated when the cells are irradiated in vitro on ice. Such an effect has, in fact, been suggested in a study by Ryabchenko et al.,[13] although their doses and irradiation times were considerably greater than those in the present study. However, it is difficult to explain the present data on the basis of repair during irradiation since, as will be shown later in this paper, these tissues have a rather uniform initial rate of repair, with a half-time in the range of 8-10 min.

Alternately based on the assumption that O_2 would be the major factor modifying the efficiency of X-ray-induced strand breaks,[1,14] and because the sensitivities of the tissues with respect to strand-break induction appeared to be fairly uniform after irradiation on ice in air (Table 1), O_2 levels in normal mouse tissues in vivo may be considerably lower than those experienced by cells irradiated in air, and these levels may differ from tissue to tissue. In fact, the O_2 levels would have to be so low that the cells would be considered "radiobiologically hypoxic". This is not a new concept: Hendry[15] suggested that the radiosensitization of tissues in animals breathing O_2 instead of air could be explained if either the critical normal cells were at lower oxygen tensions than venous blood, or if the "K value" of normal tissues was inordinately high. Although the first of these situations would seem rather surprising, some tissue O_2 levels may be lower than previously thought.[16] Moreover, cellular respiration may further reduce the O_2 concentration at the site of DNA in the nucleus. On the other hand, the K values for cells irradiated in vivo may well be higher than those for cells in vitro. In a recent study of the radioprotective effects of WR-2721 on mouse skin in vivo, Denekamp et al.[17] concluded that "many normal tissues may already be much closer to the transition between oxic sensitivity and hypoxic resistance than is generally appreciated." Our data are consistent with this hypothesis.

The role of O_2 in modifying the DNA lesions produced in mammalian cells exposed to X rays is rather poorly understood. In addition to the well characterized quantitative modification of DNA strand breaks as a function of O_2 tension, the degree of oxygenation may also have a qualitative effect on lesion production. This suggestion arises from in vitro studies showing that the degree of modification of DNA damage and cell killing by oxygen is not equivalent. Koch and Painter[1] using alkaline sucrose gradients, demonstrated that the oxygen enhancement ratio (OER) for single-strand breaks is 4, while the OER for cell killing is only 3. They suggested that this difference might be due to subtle changes in the spectrum of lesions produced under the different conditions. Using the experimental approach described earlier, we determined dose responses for initial DNA strand breaks in tissues irradiated in N_2-asphyxiated animals (i.e., under complete hypoxia). The results for spleen and jejunal cells are shown in Fig. 1, together with the results for cells from air-breathing mice and for cells irradiated in vitro. The slopes of these dose responses are shown in Table 2, along with data for two mouse fibrosarcoma tumors, FSa and NFSa, that have been included because they contain a significant proportion of hypoxic cells in vivo. By inducing acute hypoxia, we were able to further reduce the frequency of strand breaks in each tissue relative to that observed in air-breathing animals; this also had the effect of eliminating the intertissue variation observed after in vivo irradiation.

A pseudo-OER of 8-9 was calculated for each tissue by comparing the efficiency of strand-break production for cells irradiated in vitro with that for cells from mice irradiated after N_2-asphyxiation. Such a high OER is difficult to explain on the basis of changes in efficiency of strand-break production alone; however, Fornace and Little[18] have shown that X rays can also induce DNA-protein cross-links, especially under hypoxic conditions. Using CHO cells irradiated under aerated and hypoxic conditions, we also have observed a high OER (7.8) for strand breaks, as determined by alkaline elution (Meyn et al., unpublished data). However, when the OER for strand breaks was reevaluated under conditions designed to eliminate the influence of DNA-bound proteins by incorporating the enzyme proteinase-K (PK) into the alkaline elution protocol, the OER decreased to approximately 4, presumably because of the detection of additional breaks in the hypoxic cells that must have been masked by DNA-protein adducts. If such DNA-protein adducts are preferentially produced in cells irradiated at low oxygen levels, then similar lesions may be produced in animal tissues irradiated in vivo, if these are indeed radiobiologically hypoxic.

To examine this possibility, tissues were irradiated in situ, either in air-breathing mice or in N_2-asphyxiated animals, or cell suspensions derived from these same tissues were irradiated on ice in vitro. Initial DNA strand breaks were again assayed using alkaline elution, but in these experiments the elution kinetics were measured both with and without PK in order to reveal any DNA-protein adducts. The slopes of the resulting dose responses are shown in Table 2 for the NFSa tumor. When PK digestion was included,

Table 2. Pseudo-OERs for normal and neoplastic mouse tissues.

	Slope (SSF per 10 Gy)			
Tissue	In vitro	In vivo	N_2-asphyxiated	pseudo-OER
Spleen	1.24	0.60	0.15	8.32
Jejunum	1.34	0.30	0.15	8.93
FSa	1.64	0.37	0.19	8.53
NFSa (-PK)	1.51	0.32	0.17	8.70
NFSa (+PK)	1.64(1.09)[a]	0.42(1.33)	0.29(1.67)	5.65

a: numbers in parentheses refer to the fractional increase in SSF observed after PK digestion.

there was relatively little effect on the DNA elution rate of cells irradiated in vitro. However, for cells from tissues irradiated in vivo the elution rate was increased by PK digestion, indicating a significant level of DNA-protein adducts. This effect was even more marked for the DNA from cells from irradiated, acutely hypoxic animals. Qualitatively similar results were seen in the other normal tissues, which supports the hypothesis that tissues in air-breathing animals may be radiobiologically hypoxic, since DNA-protein adducts were detected in tissues irradiated in vivo. Thus, in confirmation of the idea originally proposed by Koch and Painter,[1] irradiation under different conditions of oxygenation appears to result in both qualitative and quantitative changes in DNA lesions. The possible contribution of these different lesions to biological endpoints (cell killing, carcinogenesis, and mutagenesis) is presently unknown.

The Efficiency of DNA Strand-Break Repair in Normal Mouse Tissues Irradiated in Vivo with X rays

The rate of repair of DNA damage may be an even more important factor in determining the overall effect of radiation on a tissue than is the initial yield of lesions induced by the treatment. Cultured mammalian cells repair the majority of strand breaks within 2 hr after irradiation (e.g., references 19 and 20). An accumulating body of data (e.g., reference 21) suggests that there is a correlation between lethality and the level of residual unrepaired strand breaks. The level of such unrepaired breaks may be determined by a competition between repair and fixation of precursor lesions.[22] Consequently, the rate at which such lesions are repaired may represent an important means of modifying lethality, and may be influenced by factors such as oxygenation,[23] thiols,[2] cell cycle stage[21] and cell differentiation.[3-7]

Intertissue differences in repair characteristics also have been demonstrated in vivo. Wang and Wheeler[24] observed that cerebellar neurons and intracerebral 9L tumor cells in the same rat exhibited different repair kinetics. While the amount of initial damage was the same in each cell type, the tumor cells repaired more rapidly, although the two tissues ultimately repaired to the same extent. Wheeler and Wierowski[7] subsequently showed that the repair kinetics for both the 9L tumor cells and the neurons were biphasic, and suggested that this represents the removal of lesions from regions of the genome that vary in their accessibility to repair enzymes rather than the repair of two different types of lesions. Furthermore, they suggested that for the differentiated neurons, a fraction of the genome may be completely inaccessible to repair enzymes.

The repair of DNA damage in vivo is therefore complex. We have characterized and compared the kinetics of strand-break repair in selected tissues from mice irradiated in vivo with 10 Gy of X rays.[10] These results are shown in Figure 2 (the data represent actual measured SSF values). The data for each tissue, expressed as the percentage of remaining strand breaks versus time after irradiation, are summarized in Table 3.

Of the tissue studied, liver repaired the fastest within the first hour. The kinetics of repair for the other tissues were similar; each tissue showed an appreciable level (7-16)% of unrepaired lesions after 2 hr. After 12 hr there was significant residual damage only in testes and spleen cells. Residual damage was also assayed at 24 hr in some tissues. With the exception of spleen cells, all of the profiles were indistinguishable from unirradiated controls at that time. Spleen cells, however, actually displayed more breaks at these later times, which suggests possible degradation of DNA in the irradiated cells. Some contribution from enzymatically induced strand breaks (either by repair or degradation) cannot be ruled out for any of the results shown in Table 3.

Table 3. Percentage of strand breaks remaining as a function of time after irradiation of mouse tissues in vivo (taken from reference 10).

Tissue	Time After Irradiation (min) 5	10	20	30	60	120	360	720
Liver	65±20	30±9	29±10	19±7	9±7	7±6	7±7	5±5
Spleen	82±14	47±10	39±9	32±8	23±7	8±5	12±4	16±6
Kidney	55±12	48±13	35±11	25±9	14±8	8±7	5±7	2±5
Testes	79±13	56±8	35±4	22±4	19±5	13±9	13±5	15±8
Jejunum	86±15	51±10	43±12	33±12	20±6	7±6	6±5	5±7
Brain	86±16	53±9	49±11	39±9	19±8	11±5	4±3	1±6
Bone-marrow	66±11	46±8	34±5	28±6	23±6	16±4	7±5	6±6

In contrast to the biphasic kinetics reported for cultured cells in vitro (e.g. references 18,19,22) and some tissues in vivo,[7] the repair kinetics for these tissues were multiphasic. This may be related to the heterogeneity of both the cell types and microenvironments within the tissue population. We stress that these data characterize the overall cell population of the tissue, and not specifically that of any subpopulations that may play a critical role in the cytotoxic, mutagenic or carcinogenic response, and which may have different repair characteristics from those representative of the overall tissue.

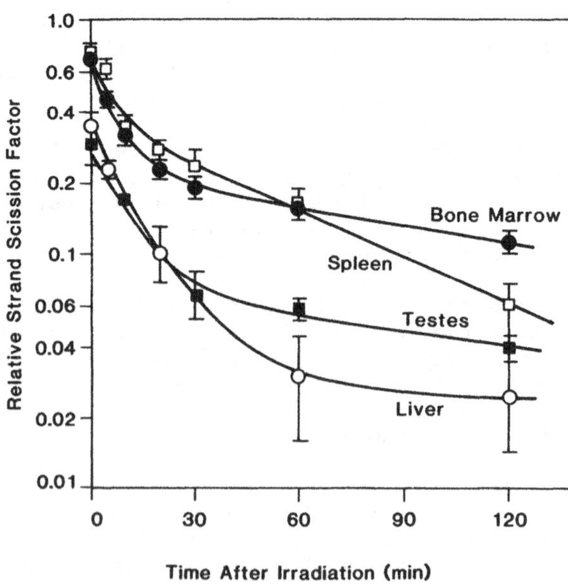

Figure 2. Repair kinetics of DNA strand breaks in mouse tissues in vivo after whole-body irradiation with 10 Gy of X rays. Data are shown for the following selected normal tissues: spleen (□); bone marrow (●); liver (o); and testes (■). The "repair time" refers to the period after the completion of irradiation and removal of tissues. (Adapted from reference 10).

For mammalian cells in vitro, hypoxia inhibits strand-break repair.[23] However, the degree of variation in the repair rate among normal tissues is rather small, despite the apparent intertissue differences in oxygenation suggested by the data in the previous section. Thiols such as GSH also have been implicated as potential repair modifiers.[2] There is indeed a significant intertissue variability in tissue GSH levels in these mice, with liver and testes containing the highest levels.[10] However, there is no obvious correlation between the repair efficiency and the overall GSH concentration in vivo.

CONCLUSIONS

The results reviewed here add weight to the idea that cultured mammalian cells irradiated in equilibrium with air may not appropriately mimic the response of cells comprising normal tissues irradiated in vivo. The major differences appear to involve at least two parameters: the degree of oxygenation and the state of differentiation. The degree of oxygenation maymhave more effect on the initial yield of lesions produced (both in type and number), whereas the state of differentiation (or proliferative versus nonproliferative state) may influence the repair rates. Thus, these factors could have important consequences with respect to radiation-induced cell killing, carcinogenesis, and mutation because of the envisioned role of DNA damage and repair mechanisms in these processes.

ACKNOWLEDGEMENT

This work was supported by PHS grant CA26312 from the National Cancer Institute, DHHS.

REFERENCES

1. C.J. Koch and R.B. Painter, The effect of extreme hypoxia on the repair of DNA single-strand breaks in mammalian cells. Radiat. Res. 64:256 (1975).
2. L. Revesz and M. Edgren, Glutathione-dependent yield and repair of single-strand DNA breaks in irradiated cells. Br. J. Cancer 49(VI):55 (1984).
3. P. Karran and M.G. Ormerod, Is the ability to repair damage to DNA related to the proliferative capacity of a cell? The rejoining of X-ray-produced strand breaks. Biochim. Biophys. Acta 299:54 (1973).
4. C. Kidson, DNA repair in differentiation. In DNA Repair Mechanisms, (P.C. Hanawalt, E.C. Friedberg, and C.F. Fox, Eds.), New York, Academic Press. pp. 761-768 (1978).
5. J.T. Lett, P.C. Keng and C. Sun, Rejoining of DNA strand breaks in non dividing cells irradiated in situ. In DNA Repair Mechanisms, (P.C. Hanawalt, E.C. Frieberg and C.F. Fox, Eds.), New York, Academic Press. pp. 481-484 (1978).
6. H.J. Rahmsdorf, H. Ponta, M. Bachle, U. Mallick, K-F. Weibezahn, and P. Herrlick, Differentiated cells from BALB/c mice differ in their radiosensitivity. Exp. Cell. Res. 136:111 (1981).
7. K.T. Wheeler and J.V. Wierowski, DNA accessibility: A determinant of mammmalian cell differentiation? Radiat. Res. 93:312 (1983).
8. K.W. Kohn, DNA as a target in cancer chemotherapy, Measurement of macromolecular DNA damage produced in mammalian cells by anticancer agents and carcinogens. Methods Cancer Res. 16:291 (1979).
9. R.E. Meyn and W.T. Jenkins, Variation in normal and tumor tissue sensitivity of mice to ionizing radiation-induced DNA strand-breaks in vivo. Cancer Res. 43:5668 (1983).

10. D. Murray, W.T. Jenking and R.E. Meyn, The efficiency of DNA strand-break repair in two fibrosarcoma tumors and in normal tissues of mice irradiated in vivo with X-rays. Radiat. Res. 100:171 (1984).
11. T. Ono and S. Okada, Radiation-induced DNA scissions and their rejoining in testicular cells of mouse. Mutat. Res. 36:213 (1976).
12. T. Ono and S. Okada, Radiation-induced DNA single-strand scission and its rejoining in spermatogonia and spermatozoa of mouse. Mutat. Res. 43:25 (1977).
13. N.I. Ryabchenko, B.P. Ivannik and S.Ya. Proskuryakov, Viscosimetric analysis of the occurrence and repair of DNA single-strand breaks in irradiated animal tissues. Int. J. Radiat. Biol. 41:389 (1982).
14. J.D. Chapman, D.L. Dugle, A.P. Reuvers, B.E. Mecker and J. Borsa, Studies on the radiosensitizing effect of oxygen in Chinese hamster cells. Int. J. Radiat. Biol. 26:383 (1974).
15. J.H. Hendry, Quantitation of the radiotherapeutic importance of naturally-hypoxic tissues from collated experiments with rodents using single doses. Int. J. Radiat. Oncol. Biol. Phys. 5:971 (1979).
16. H.G. Bohlen, Intestinal tissue PO_2 and microvascular responses during glucose exposure. Am. J. Physiol. 238:164 (1980).
17. J. Denekamp, B.D. Michael, A. Rojas and F.A. Stewart, Thiol radioprotection in vivo: The critical role of tissue oxygen concentration. Br. J. Radiol. 54:1112 (1981).
18. A.J. Fornace and J.B. Little, DNA cross-linking induced by X-rays and chemical agents. Biochim. Biophys. Acta 477:343 (1977).
19. A. Cole, R.E. Meyn, R. Chen, P.M. Corry and W. Hittelman, Mechanisms of cell injury. In Radiation Biology in Cancer Research (R.E. Meyn and H.R. Withers, Eds.), New York, Raven Press, pp. 3358 (1980).
20. J.T. Lett, I. Caldwell, C.J. Dean and P. Alexander, Rejoining of X-ray induced breaks in the DNA of leukemia cells. Nature 214:790 (1967).
21. K. Sakai and S. Okada, Radiation-induced DNA damage and cellular lethality in cultured mammalian cells. Radiat. Res. 98:479 (1984).
22. J.F. Ward, Some biological consequences of the spatial distribution of ionizing radiation-produced free radicals. Radiat. Res. 86:185 (1981).
23. C.J. Koch, The effect of oxygen on the repair of radiation damage by cells and tissues. Adv. Radiat. Biol. 8:273 (1979).
24. T.S. Wang and K.T. Wheeler, Repair of X-ray-induced DNA damage in rat cerebellar neurons and brain tumor cells. Radiat. Res. 73:464 (1978).
25. D.L. Dugle, C.J. Gillespie and J.D. Chapman, DNA strand breaks, repair, and survival in X-irradiated mammalian cells. Proc. Natl. Acad. Sci. USA 73:809 (1976).

DIFFERENTIAL REGULATION OF BASE AND NUCLEOTIDE EXCISION REPAIR IN MAMMALIAN CELLS

Klaus Erixon

Department of Radiation Biology
University of Stockholm
S-10691 Stockholm, Sweden

INTRODUCTION

During excision repair in mammalian cells the first event which has been possible to measure is the formation of a single-strand break (SSB). This is produced when a DNA lesion is recognized and incised by a lesion-specific endonuclease. In this study, repair mechanisms have been deduced through the use of a sensitive and convenient technique to measure SSB in cellular DNA, named Alkaline DNA Unwinding (ADU).[1] The level of SSB has been followed in human fibroblasts of normal xeroderma pigmentosum (xp) origin after exposure to 8 different genotoxic agents. These agents fall into two distinct groups. The first group is represented by UV light (254 nm) and UV-mimetic agents that will introduce "bulky" DNA adducts, for which xp-cells have a decreased repair capacity and survival.[2] The agents selected to represent this class are N-Acetoxy-N-Acetylaminofluorene (NA-AAF), 7-Bromomethyl-Benzanthracene (7-BrMeBA) and 4-Nitroquinoline-N-Oxide (4-NQNO). The second group contains simple alkylating agents that will introduce methyl and ethyl-groups on DNA for which xp-cells have more or less normal repair and survival.[2]

In analogy with the specificity of the isolated uvrABC endonuclease complex from *E. coli*,[3,4] lesions produced by the first group are believed to be recognized by direct acting endonucleases by virtue of the local distortion of the DNA helix. Lesions formed by the second group of agents are recognized by a different set of enzymes, glycosylases, that cleave off modified bases, e.g. 3-methyladenine (3-MeAd) and 7-methylguanine (7-MeGu), the major adducts after treatment with dimethylsulphate (DMS) or methyl methansulphonate (MMS). The resulting apurinic sites (AP-sites) are then cut by AP-endonucleases, giving rise to SSBs.[5] This repair pathway, sometimes called base excision repair or short patch repair, has been claimed to produce primarily short repair patches, 1-3 inserted deoxynucleotides, although a significant fraction of longer patches have also been observed.[6] Repair of UV-mimetic lesions by nucleotide excision results in patches 25-100 deoxynucleotides long, hence the name long patch repair. The reason for the difference in patch size is not known.

REPAIR EVENTS AFTER EXPOSURE TO UV-MIMETIC AGENTS

In order to follow the level of SSB present during repair of UV-induced damage, normal human fibroblasts were radioactively labeled and grown to con-

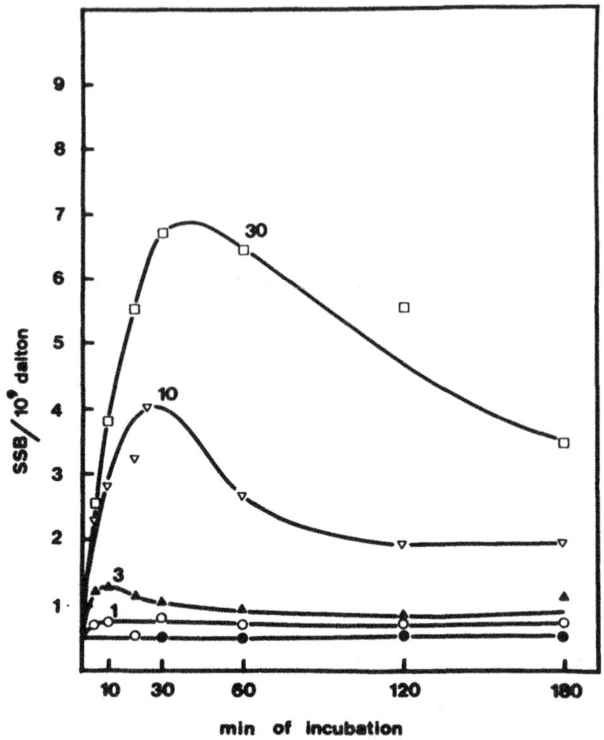

Figure 1. Normal human fibroblasts (NHF) were labeled by ^{3}H-TdR and grown to confluency, irradiated with different fluences of UV-light at 0°C and incubated in growth medium at 37°C. SSB were determined by ADU at different times of incubation (□ 30 J/m², ▽ 10 J/m², ▲ 3 J/m², o 1 J/m², ● control).

fluency, irradiated at 0°C and incubated in growth media. ADU was performed at various times post irradiation and the data in Figure 1 were obtained. Strand-breaks appear in a dose dependent fashion at 37°C. No SSB are formed at 0°C. They are an expression of ongoing incision, excision, polymerization and ligation at sites under repair. The time course of these transient SSB is the result of consecutive reactions:

$$L \xrightarrow{k1} SB \xrightarrow{k2} R \qquad (1)$$

By endonucleolytic action, lesions (L) are converted to strand breaks (SB) with a reaction rate k1. These strand breaks are then processed to give rise to repaired and ligated sites (R) with a reaction rate k2. The number and time course of observed strand breaks at any time is a function of the rate of incision and the lifetime of these breaks. Only a small fraction of the lesions induced is found in the intermediate stages between incision and ligation at any time. At a dose of 10 J/m^2 a maximum of 4 SSB/10^9 dalton are observed, even though 300 dimers/10^9 dalton are actually present. Thus the processes of recognition and incision are rate-limiting in deoxynucleotide excision repair.

These incision-events may be trapped by the addition of inhibitors that block DNA polymerase alpha, like aphidicolin or cytosinarabinoside (araC),[7,8] which thus will decrease k2 in equation 1 to a negligible value. Because the sealing activities now seem fully inhibited, we may study endonucleolytic activity separately, Figure 2.

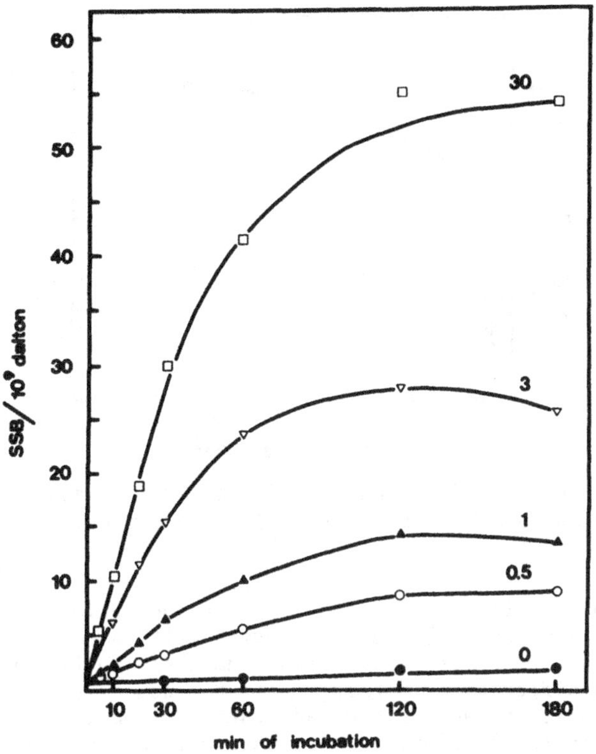

Figure 2. Number of strand breaks as a function of time, after different fluences of UV, as indicated in the figure. The cultures were incubated with 20 μM araC + 2 mM hydroxyurea (HU) 30 min before and also after irradiation (symbols represent fluences as in Figure 1).

Exposure to the other three UV-mimetic agents of the group cause a very similar response, i.e only a small increase in SSB until araC or aphidicolin is added, when the level may increase up to 50 times during a 2 hour period and saturation is reached (data not shown).

Essentially no increase in the level of SSB was obtained in xp-A cells, either in the absence or presence of araC, indicating the lack of incision in these cells. It thus seems that the repair events triggered by UV and UV-mimetic agents are quite similar. A common step for recognition and/or incision of these different DNA adducts is thus suggested.

The initial rate of incision and the maximal yield of SSB was measured as a function of UV-dose, Figure 3. The rate of incision starts to saturate at about 3 J/m^3 and reaches a maximal rate of about 1 nick per minute per 10^9 dalton. The final level of strand breaks which is reached after 2 hours will saturate at the same dose with a maximum of about 50 SSB/10^9 dalton, i.e. on the average the nicks are 30 kb apart. A corresponding behavior was observed for the other UV-mimetic agents.

Nucleotide Excision Repair is Regulated by Factors that Prepare Chromatin for Recognition and Incision

As concluded earlier, the rate of incision and/or recognition is rate-limiting in repair of UV and UV-mimetic damage. Such lesions are removed with different rates in different parts of the genome. This has been clearly demonstrated in a series of studies by Hanawalt and collaborators,[9,10] who in

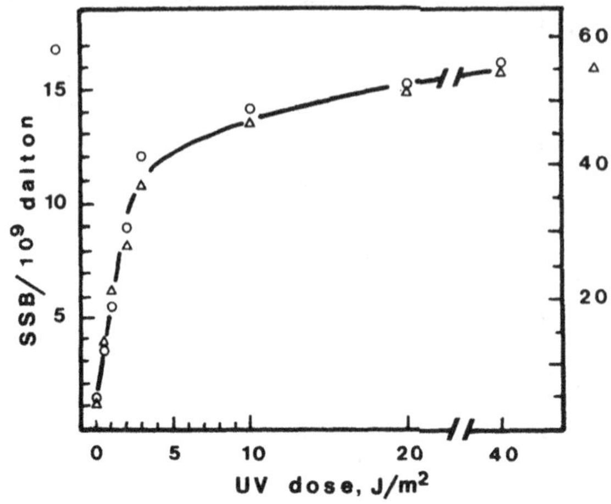

Figure 3. Number of strand breaks as a function of UV dose in the presence of HU and araC after 15 min (o) and 120 min (Δ) of incubation.

addition showed a preferential repair of a transcriptionally active sequence in CHO cells.[11] Kantor and Ritter[12] observed multiphase kinetics in the removal of pyrimidine dimers from normal human and xp fibroblasts. We have also observed a differential repair within the human genome and correlated this repair capacity to accessible and transcriptionally active chromatin domains (paper in preparation). These data suggest that the rate of recognition of lesions in genomic domains with different accessibilities determines the rate of repair. The rate of incision is then a measure of the rate with which lesions are recognized, i.e. become available to the endonucleases. Thus, if this is the case, we actually measure the rate of recognition respective to the total number of lesions in Figure 3. Why then are only a maximum of 50 lesions recognized?

A Domain Model for Nucleotide Excision Repair

In order to explain the maximal level of strand breaks reached and the identical shape of the dose-response curves in Figure 3, a domain model was proposed.[8] According to this model the identical dose dependence for the initial rate of incision and for the final yield of accumulated SSB may be explained by two concepts: 1) a limited number of potentially available domains and 2) a limit in the number of incisions that simultaneously may be tolerated in each domain. The incision rate is thus determined by the chromatin structure of certain units, domains, that may or may not expose the substrate, the lesions, to the repair enzymes. These domains may be equivalent to the supercoiled DNA loops present in chromatin.[8]

If only one endonucleolytic event is possible within such a region at any given time, the maximum number of strand breaks in the presence of araC (55 per 10^9 dalton), represents the total number of such regions; an increased number of damage sites within a domain would not lead to more than one nick in that region. The initial, linear part of the dose-response curve represents the dose range within which the number of domains containing damage increases. At higher doses where the probability of having more than one site of damage per domain becomes significant, the fraction of damaged sites which are nicked decreases. Saturation occurs when all domains have at least one lesion. The maximal rate of incision, 1 SSB/10^9 dalton/min, Figure 3, then represents the rate at which these potentially accessible domains become available to the repair enzymes. Half of the genome is scanned by this process in about 30 minutes and almost all potentially accessible domains identified in 120 min (Figure 2). The maximal incision rate, 1 SSB per min per 10^9 dalton, corresponds to a scanning of about 3000 domains or 2% of the genome per minute in the human fibroblasts.

One possible reason for a restriction to only one nick per domain may be that incision is limited to topologically intact chromatin domains, equivalent to supercoiled DNA loops. There may be a requirement for a torsional strain in these loops for incision to take place. Once a nick is created in such a loop, it will relax and the strain will be lost, thus preventing further incision. The average distance between these domains, calculated from the data, is about 30 kb and is in the same range as has been estimated for the size of the DNA loops bound to the nuclear matrix.[13] There are several possibilities for the induction of torsional strain in these loops. Topoisomerases, both of type I and II, have been found in mammalian cells. The topoisomerase II inhibitor novobiocin inhibits UV-dependent incision,[8,14] suggesting the involvment of this enzyme in making the lesions available to the repair endonucleases. Novobiocin has been reported to reduce the degree of negative supercoiling in mammalian cells,[15] and the bacterial analog to topoisomerase II, gyrase, is able to induce negative supercoiling in DNA.[16] To conclude, the rate of incision and repair may be governed by processes that keep chromatin domains available and supercoiled with a torsional strain. A schematic representation of the model is depicted in Figure 4.

REPAIR EVENTS AFTER EXPOSURE TO ALKYLATING AGENTS

In contrast to the UV-mimetic agents, alkylating agents give rise initially to a significant level of SSB which rapidly decreases with time after treatment, both in normal and in xp-cells, Figure 5. The addition of araC or aphidicolin causes an increase in this level in normal cells, and to a lesser extent in xp-A cells. Thus polymerase α seems to participate in the repair of alkylated bases, even in xp cells. However, alkylating agents and UV-mimetic agents trigger quite different repair responses, as judged from the kinetics for SSB in the absence of inhibitors, compare Figures 1 and 5. In view of the proposed domain model for deoxynucleotide excision repair, we should not consider differences in incision enzymes only. The possible influence of the structure of the substrate, i.e. the chromatin, may also be fundamental for base excision.

The level of SSB observed must again be the product of the frequency of incision and the life time of these strand breaks, equation 1. According to the domain model, the rate of incision at UV-mimetic lesions is governed by a relatively slow, temperatuur sensitive process, that scans through the potentially available domains at a rate of about 2% per minute. The endonucleases thus do not "see" the lesions until some kind of change in the chromatin conformation has occured. Once incision has occured, the SSB has a life time of several minutes.[17] As already discussed, there is also a restriction in the number of incisions tolerated.

The life time of the intermediate SSB during base excision repair, on the other hand, may be quite short. This explains why no increase in the level of SSB is observed after treatment, only a decrease (Figure 5). In order to explore this possibility, cells were treated with the more reactive methylating agent, DMS, at ice-temperature. Even at 0°C a certain level of SSB is obtained, which however stays reasonably constant after removal of the drug. As the temperature is quickly raised to 20 or 37°C , additional incision and ligation is observed, Figure 6. Obviously these reactions are fast and their relative rates depend differently on temperature.

An efficient incision and excision of lesions and a short lifetime of the intermediate SSB may explain the initial high level of strand breaks for

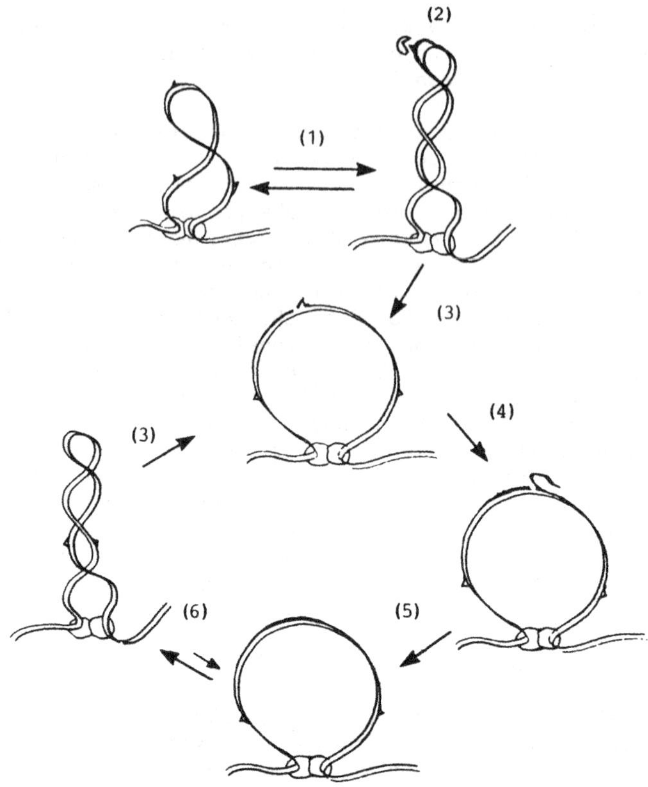

Figure 4. Schematic representation of a domain model for nucleotide excision repair. As torsional strain is introduced into a DNA loop (1), local denaturation will take place at sites of damage (2). This modified structure will be recognized by the endonuclease for deoxynucleotide excision repair (2) and introduce a nick, which will relax the loop (3). Without the torsion no further incision will take place until polymerization (4), excision and ligation (5) have taken place. Only after this stage may negative supercoils be introduces again, and the tension restored (6), such that incision and a new cycle of repair can take place.

alkylating agents, as well as our inability to detect an actual increase of SSB. In the domain model for repair of UV-mimetic lesions, araC-susceptible sites reflect availability for incision followed by synthesis by polymerase α. The formation of SSB in the presence of araC is, however, a slow process also during alkylation-induced repair (Figure 5). Thus, other processes must occur to explain the fast incision observed at room temperature, and possibly even at 0°C, Figure 6.

Poly-ADP-ribosylation

A chromatin associated process, poly-ADP-ribosylation (pADPR) of nuclear proteins, has been reported to be triggered by DNA strand breaks and to be inhibited by, e.g., 3-aminobenzamide (3-ABA).[18] When this agent is added to cell cultures previously exposed to MMS or DMS, a rapid increase in strand breaks occurs. Upon removal of 3-ABA there is a fast decrease back to the original level of strand breaks, Figure 7.

These rapid, reversible changes in the level of strand breaks indicate a possible involvement of, and a rapid turnover of pADPR, in accordance with

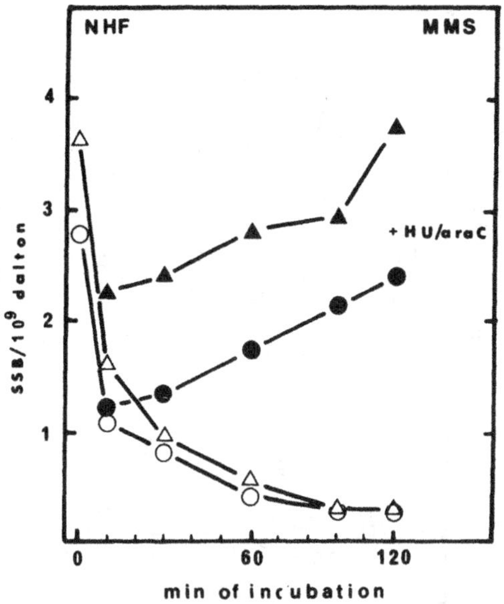

Figure 5. Normal human fibroblasts (NHF) were exposed for 10 min at 20°C to 4 or 8 mM MMS. After washing, the cells were incubated in growth medium at 37°C with or without 2 mM HU, 20 μM araC.

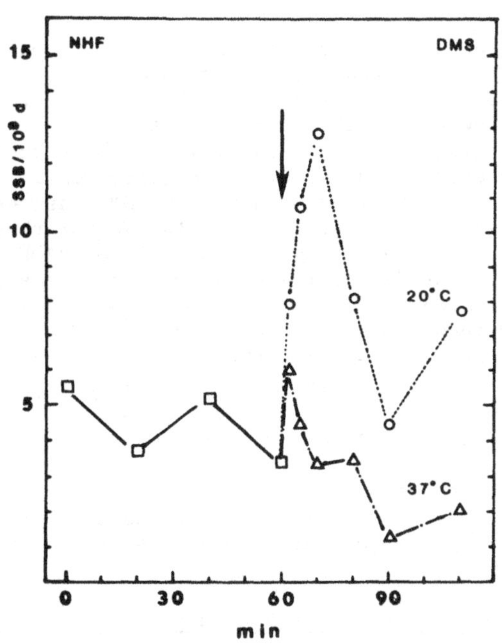

Figure 6. NHF were exposed to 2 mM DMS at 0°C for 10 min. After washing, the cells were incubated in growth medium at 0°C for 60 min. The temperature was then raised to 20 or 37°C.

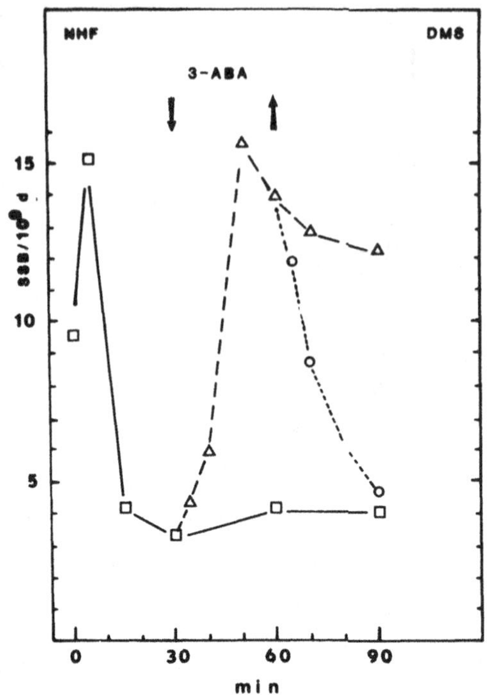

Figure 7. NHF were exposed to 2 mM DMS at 0°C for 10 min. After washing, the cells were transfered to 20°C (zero time), (□). At 30 min 3-ABA was added to 20 mM, (Δ). At 50 min, some of these samples received fresh medium without 3-ABA, (●).

other published data. The effect of the drug may then be caused by a) a temporary increase in the frequency of incision or b) a sudden decrease in an activity that is responsible for closing the breaks, e.g. polymerization or ligation. The following hypothesis is consistent with the data: pADPR occurs on chromosomal proteins in close proximity to SSB produced by glycosylases plus AP-endonucleases.[20] These enzymes are quite small, with MW's around 20-30 kdalton and may have greater access to DNA lesions in compact chromatin than DNA-polymerases and ligases. Access to DNA for these latter enzymes, responsible for closing the breaks, is gained upon pADPR of the local proteins covering the breaks. 3-ABA interferes with this accessibility process so that ligation is delayed while incision continues. A similar mechanism of action has already been proposed, and is supported by several observations: 1) Electron microscopy studies show that pADPR results in opening of the tightly coiled nucleosome structure.[21] 2) _In vitro_ experiments demonstrate that ligase is able to seal nicks produced by AP-endonuclease on pure DNA[22] but is not able to seal all strand breaks produced by DNase on SV40 minichromosomes.[23] The suggestion here is: to avoid premature closure of nicks induced by AP-endonucleases and abortion of repair, the chromatin structure may prevent access of the ligase until excision has taken place. Thus, chromatin modification by poly-ADP-ribosylation may be essential for completion of base excision repair.

Distribution of Long and Short Patch Repair in the Genome

The effects of 3-ABA and araC, both separately and combined, were compared after exposue of cells to MMS or UV. The effect of the inhibitors was additive in cells exposed to MMS, whereas 3-ABA failed to increase the number of strand breaks after exposure to UV-mimetic agents, either in the absence or presence of araC (data not shown). The effect of 3-ABA thus seems specific for agents that trigger short patch repair while the effect of araC is common to the repair of both types of lesions. The additive response of 3-ABA and araC may indicate that both short and long patch repair operate in parallel after alkylation, as has been suggested by Snyder and Regan.[24]

Cells with different repair capacities for UV-mimetic damage were treated with MMS and the responses to the inhibitors were compared. AraC gave the most pronounced increase of strand breaks in normal human fibroblasts with the highest capacity to remove dimers, while 3-ABA gave the strongest response in xp-A and V79 Chinese Hamster cells with a low capacity for UV-induced repair (not shown). These results may indicate that alkylation-induced damage can be processed by either pathway of repair, whichever is available. Where long patch repair is deficient, more damage is left to be handled by short patch repair.

Strand breaks accumulate also in MMS-treated xp cells on exposure to araC. Both in normal and xp cells, the effect of araC may be explained by the inhibition of polymerase α. Recognition and incision are performed by glycosylases plus AP-endonucleases but the particular chromatin structure determines which DNA polymerase will subsequently be used. Thus, polymerase α may operate in the same "accessible" chromatin both for UV- and MMS-induced repair. UV-lesions are not incised in xp cells, but alkylated bases are, so that excision and polymerization can proceed from these nicks. Once incision has taken place, both polymerase α and β may carry out repair synthesis, where polymerase α presumably gives rise to long repair patches in "open" chromatin. Repair replication in condensed chromatin may then be performed by polymerase β. An additive response of araC and dideoxythymidine, inhibitors of polymerase α and β, respectively, indicates that both polymerases are involved in the repair of alkylation-induced damage (not shown).

A crucial element in the domain model for deoxynucleotide excision repair is the restraint in the number of incisions simultaneously tolerated. These restraints do not seem to exist for incision at alkylated lesions, either in the absence or presence of araC (data not shown). In addition, MMS induced repair is not inhibited by novobiocin[25] contrary to UV-induced repair.[8,14] The data support the suggestion that initiation of base excision is not restricted in the same way as deoxynucleotide excision. It should be noted that base excision gives rise to both short and long patches of repair, depending on the structure of chromatin where the repair occurs. It is possible that a small fraction of true UV-mimetic repair takes place after alkylation, as suggested by the somewhat lower response to araC in xp-A cells. The properties of base and deoxynucleotide excision repair are compared and summarized in Table 1.

In conclusion, differences in the structure of chromatin may determine where long and short patch repair take place. Repair of UV-mimetic lesions is initiated in potentially accessible domains by a particular recognition mechanism which "prepares" these domains for incision, Figure 8. This is the mechanism xp cells lack. Repair replication in these domains is performed by polymerase α, which gives rise to long repair patches. Repair of alkylated bases is initiated in both open and more condensed chromatin. Incision events in open chromatin may be processed by polymerase α directly. Repair replication after incision in condensed chromatin requires the opening of the structure by pADPR to give access to polymerase β and ligase to complete the repair.

Table 1. Characteristics of Base and Nucleotide Excision Repair

	Deoxynucleotide Excision	Base Excision
1.	deficient in xp cells	proficient in xp cells
2.	long repair patches	short and long patches
3.	performed by polymerase α	performed by polymerases α and β
4.	not effected by 3-ABA	3-ABA will increase SSB
5.	inhibited by Novobiocin	not effected by Novobiocin
6.	lifetime of SB ~ 3 min	shorter lifetime of SB
7.	preferential repair of "open" domains	long patches in "open" domains, short patches in condensed domains
8.	rate of incision governed by a "scanning" factor, which xp cells lack	incision in condensed chromatin requires pADPR in order to complete repair

ACKNOWLEDGEMENT

This work was supported by the Swedish Natural Science Research Council. For excellent technical assistant I want to thank Miss Eva Jakobson and Miss Riitta Partanen.

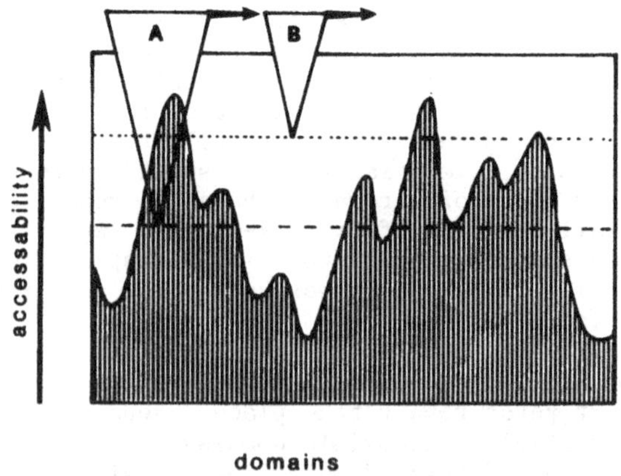

Figure 8. A schematic representation of the concept of preferential repair in potentially accessible domains. Individual domains are represented by bars of varying length, representing the degree of accessibility and hypothetical distribution within the genome. These domains are made available to repair endonucleases by a scanning mechanism, which prepares these domains for incision. This scanning mechanism can be more or less efficient, depending on cell type, thus reaching to different degrees of "compactness", A versus B.

REFERENCES

1. G. Ahnstrom and K. Erixon, Measurement of strand breaks by alkaline denaturation and hydroxyapatite chromatography, in "DNA Repair. A Laboratory Manual of Research Procedures, 1 B", Eds. E.C.Friedberg and P.C.Hanawalt (Marcel Dekker, N.Y.) (1981).
2. E.C. Friedberg, U. K. Ehmann, and J. I. Williams, Human diseases associated with defective DNA repair, Adv. Radiat. Biol., 8:85 (1979).
3. E. Seeberg, Purification and properties of the uvrA protein from E.coli, Proc. Natl. Acad. Sci., 79:988 (1982).
4. B. M. Kacinski and W. D. Rupp, Interaction of the UVRABC endonuclease in vivo and in vitro with DNA damage produced by antineoplastic anthracyclines, Cancer Res., 44:3489 (1984).
5. T. Lindahl, DNA repair enzymes, Ann. Rev. Biochem., 51:61 (1982).
6. A. Francis, R. O. Snyder, W. C. Dunn and J. D. Regan, Classification of chemical agents as to their ability to induce long- or short-patch repair in human cells, Mutat. Res., 83:156 (1981).
7. K. Erixon and G. Ahnstrom, Single-strand breaks in DNA during repair of UV-induced damage in normal human and Xeroderma pigmentosum cells as determined by alkaline DNA unwinding and hydroxylapatite chromatography. Effects of hydroxyurea, 5-fluorodeoxyuridine and 1-β-D-arabinofuranosylcytosine on the kinetics of repair, Mutat. Res., 59:257 (1979).
8. K. Erixon, DNA excision repair in mammalian cells. A model, based on a quantitative analysis of the kinetics of incision and sealing of single-strand breaks after UV-irradiation of human fibroblasts. Thesis, University of Stockholm (1980).
9. S. A. Leadon, M. E. Zolan and P. C. Hanawalt, Restricted repair of aflatoxin B1 induced damage in alpha-DNA of monkey cells, Nucl. Acid Res., 11:16:5675 (1983).
10. J. Mansbridge and P. C. Hanawalt, Domain-limited repair of DNA in UV-irradiated fibroblasts from xeroderma pigmentosum group C. In "Cellular Responses to DNA Damage", UCLA symp.,vol 11, 195-207, Eds. E.C. Friedberg and B.A. Bridges (1983).
11. V. A. Bohr, C. A. Smith, D. S. Okumoto and P. C. Hanawalt, DNA repair in an active gene: removal of pyrimidine dimers from the DHFR gene of CHO cells is much more efficient than in the genome overall, Cell, 40:359 (1985).
12. G. J. Kantor and C. Ritter, The rate of removal of sunlight-induced pyrimidine dimers from the DNA of cultured human diploid fibroblasts, Photochem. Photobiol. 40:137 (1984).
13. J. R. Paulson and U. K. Laemmli, The structure of histone-depleted metaphase chromosomes, Cell, 12:817 (1977).
14. A. R. Collins and R. T. Johnson, Novobiocin: inhibitor of the repair of UV-induced but not X-ray induced damage in mammalian cells, Nucl. Acid Res., 7:1311 (1979).
15. M. R. Matern and R. B. Painter, Dependence of mammalian DNA replication on DNA supercoiling. II. Effects of novobiocin on DNA synthesis in Chinese hamster ovary cells, Biochem. Biophys. Acta, 563:306 (1979).
16. N. R. Cozzarelli, DNA gyrase and the supercoiling of DNA, Science, 207:953 (1980).
17. K. Erixon and G. Ahnstrom, Kinetics of UV-induced excision repair in human fibroblasts. In "DNA Repair Mechanisms". Eds., P.C. Hanawalt and E.C. Friedberg, (Academic Press,N.Y.) (1978).
18. N. A. Berger, Poly(ADP-ribose) in the cellular response to DNA damage, Radiat. Res., 101:4 (1985).
19. K. Wielckens, E. George, T. Pless and H. Hilz, Stimulation of poly(ADP-ribosyl)ation during Ehrlich ascites tumor cell "starvation" and suppression of concomitant DNA fragmentation by benzamide, J. Biol. Chem., 58:4098 (1983).

20. N. Malik, M. Miwa, T. Sugimura, P. Thraves and M. E. Smulson, Immunoaffinity fractionation of the poly ADP-ribosylated domains of chromatin, Proc. Natl. Acad. Sci. USA, 80:2554 (1983).
21. G. G. Poirier, G. DeMurcia, J. Jongstra-Bilen, C. Niedergang and P. Mandel, Poly(ADP-ribosyl)ation of polynucleosomes causes relaxation of chromatin structure, Proc. Natl. Acad. Sci. USA, 79:3423 (1982).
22. C. Goffin and W. G. Verly, Repair of depurinated DNA with enzymes from rat liver chromatin, Biochem. J., 220:133 (1984).
23. D. H. Evans and S. Linn, Excision repair of pyrimidine dimers from Simian virus 40 minichromosomes in vitro, J. Biol. Chem., 259:10252 (1984).
24. R. D. Snyder and J. D. Regan, DNA repair in normal human and xeroderma pigmentosum group A fibroblasts following treatment with various methanesulfonates the demonstration of a long-patch (UV-like) repair component, Carcinogenesis, 3:7 (1982).
25. R. D. Snyder, B. Van Houten and J. D. Regan, Studies on the inhibition of repair of ultraviolet and methyl methansulfonate-induced damage in the DNA of human fibroblasts by novobiocin, Nucl. Acids Res., 10:6207 (1982).

IMMEDIATE AND REPAIR INDUCED DNA DOUBLE STRAND BREAKS IN MAMMALIAN CELLS

Peter E. Bryant

Department of Anatomy and Experimental Pathology
University of St. Andrews
St. Andrews KY16 9TS, Fife, U.K.

INTRODUCTION

Although ionizing radiation is a non-specific damaging agent, it is clear that the biological responses of cells to radiation are principally attributable to damage induced in the DNA. The extent to which this primary damage is expressed however may depend on both physiological and organizational factors within a cell.

Ionizing radiation is known to cause a number of different types of lesion in the DNA of cells, and evidence suggests that the double-strand break (dsb) is potentially the most dangerous of these lesions, since if not correctly repaired, it can lead to extensive loss or rearrangement of genetic material with serious consequences for the cell. Ionizing radiation damage to DNA in cells falls into two main categories; one in which the covalently linked backbone of the DNA is directly broken, i.e. at the time of radiation exposure (frank breaks), and the second type, including a wide variety of lesions in the structures of bases or sugar moieties, which leads via the intermediacy of certain enzymes, to strand breaks. This second type of lesion has been detected in two basic ways; the first in which specific changes have been studied chemically, e.g. in the case of thymine lesions of the 5,6-dihydroxydihydro-thymine type[1] and various damages to the deoxyribose sugar moeity,[2] and the second in which DNA lesions are converted to strand breaks by the use of endonuclease preparations, e.g. those detected by treatment of free DNA from irradiated cells with extracts of the bacterium Micrococcus luteus[3,4] and so-called "bulky lesions" detected by treatment of DNA with S1 nuclease.[5]

The evidence will be reviewed that not only frank dsbs arise in the DNA of cells exposed to ionizing radiation, but also double-stranded base or sugar damage which, as outlined above, may lead to additional dsb during post-irradiation incubation, i.e. during the repair of such lesions. Then briefly the data pertaining to the kinetics and mechanisms of repair of dsb and to their possible biological consequences are reviewed.

INDUCTION OF FRANK DOUBLE-STRAND BREAKS IN DNA OF CELLS

Dsb's are known to be formed as a linear function of absorbed radiation dose with a frequency under euoxic conditions of about 0.1 - 0.2 dsb/10^9

daltons/ Gy.[6-9] The linear relationship between radiation induced dsb and dose indicates that these breaks arise as a result of single ionizing events, which are probably clusters of ionizations having energy values of between 100 eV and 1 keV.[10-13] Dsb thus arise by a so-called "one-hit" process, although they may be induced by a cluster of several simultaneously produced ionizations.

REPAIR INDUCED DOUBLE-STRAND BREAKS

If the above mentioned pattern of energy deposition within the tracks of ionizing particles leads to the induction of frank dsb, then it might be expected that double-stranded lesions also will occur that are comprised of either a single-strand break (ssb) in one strand opposite a base or sugar damage in the other, or even two base or sugar damages opposite one another (see Figure 1).

The type of lesion denoted as an endonuclease sensitive site (es) in Figure 1 may include a wide spectrum of lesion types. Such lesions might occur with a similar frequency to that for frank dsb, since single-stranded endonuclease sensitive sites are known to occur, in DNA from irradiated cells treated with *M. luteus* extract, at a frequency relative to ssb of es/ssb = 0.5 [14], and in supercoiled DNA (phage), at frequencies of 1.7 and 1.8 using endonuclease preparations from yeast and *E. coli*, respectively.[15] The frequencies of such latent lesions in the DNA of cells is, however, an open question, since work by Bresler et al.[16] has shown that inhibition of the nuclease activity of *E. coli*, using a high concentration of EDTA applied immediately after radiation exposure, yielded only 15% of the ssb observed in the absence of this substance. On the basis of this result the authors claim that 85% of ssb arise by endonuclease cleavage of "enzyme labile" sites during irradiation.

It is possible that if such es sites occur opposite one another, as indicated in Figure 1, then such lesions would be equally as damaging to a cell as a frank dsb, assuming the cell incises the DNA at the es sites so as to create a dsb. It might be possible, for example, that a cell first repairs the ssb (a in Fig. 1), at the cost of a point mutation, and then subsequently cleaves the DNA at the es. In the same way, one es of a pair

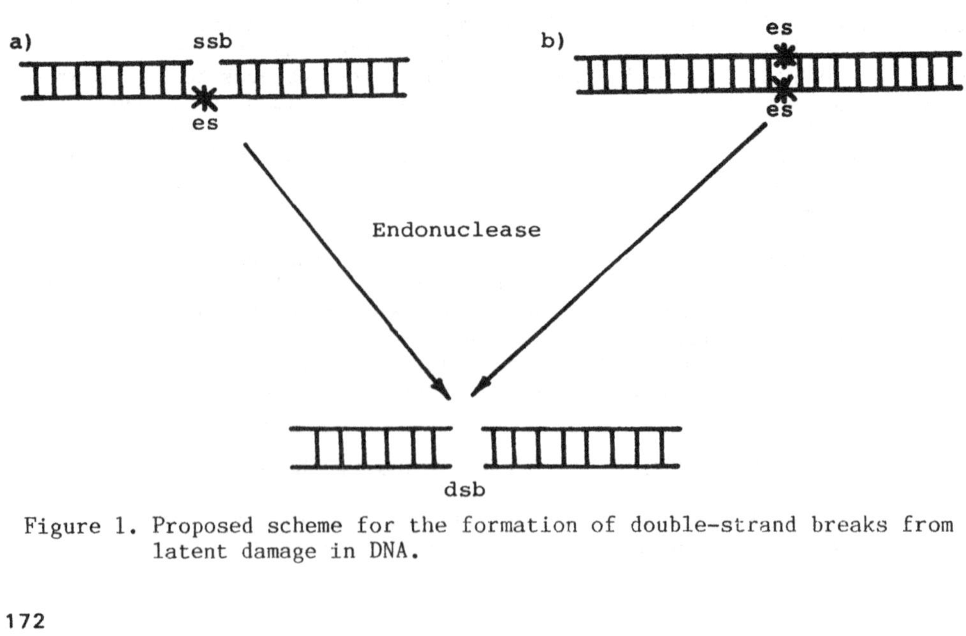

Figure 1. Proposed scheme for the formation of double-strand breaks from latent damage in DNA.

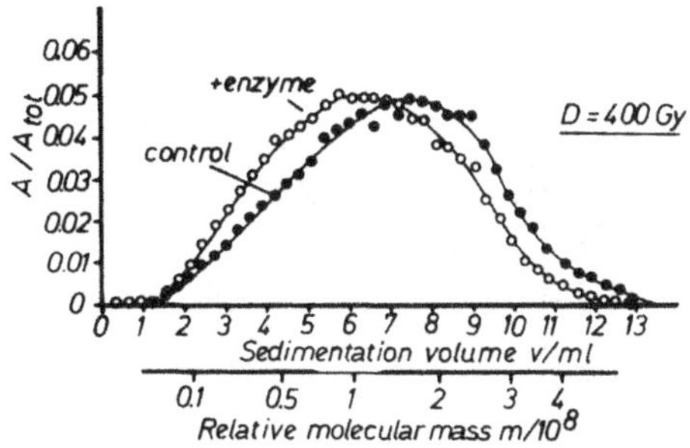

Figure 2. Neutral velocity gradient profiles of DNA from irradiated cells treated (O) with or (●) without M.luteus endonuclease.[21]

(b in Figure 1) might be repaired before the other is cut,[17,18] in an anologous way to that suggested for repair of interstrand crosslinks. It is known, however, that in the case of 8-methoxypsoralen photo-induced cross-links in yeast, cells induce dsb as well as ssb during repair.[19] There is some evidence from work with bacteria that additional dsb are induced enzymatically after X-irradiation.[20] It was shown by these authors that the polymerase mutant of E. coli K12 pol A1 yielded 2.9 times the number of dsb as the wild type (WT) strain after irradiation at room temperature. If the mutant strain was irradiated at 3°C, there was no difference in the numbers of dsb between mutant and WT, indicating that the dsb in the polymerase deficient strain occur as a result of endonuclease incision of the DNA. These authors concluded that the additional dsb are formed as a result of overlapping excision regions, or an excision region and a ssb. Apparently they did not consider the possibility described above, namely that the enzymatically induced dsb may occur from two base or sugar damages (es) occuring opposite one another or an es opposite an ssb. Some years ago we investigated the presence of double-stranded base or sugar lesions in mammalian cells,[21] and whether cells yield additional dsb during the repair of such lesions.

ASSAY OF DOUBLE-STRANDED ENDONUCLEASE SENSITIVE SITES IN MAMMALIAN CELLS

Chinese hamster cells (V79) were treated after various doses of X-rays with an "incubation solution" containing a low concentration of the detergent Brij 58, to permeabilize cells, and an M.luteus endonuclease preparation. Briefly, 50 µl of incubation solution: pooled phosphocellulose fractions prepared according to Riazuddin and Grossman,[22] transferred on Sephadex G25 to a solution of 4 g/l Brij 58, 5 mmol/l phosphate buffer pH 7.5, 1 mmol/l 2-mercaptoethanol, 1 mmol/l EDTA , was added to 50 µl of a suspension of irradiated cells (2×10^6 cells/ml) in phosphate-buffered saline, and incubated for 35 min at 37°C. After incubation the cells were applied to neutral sucrose gradients and centrifuged according to Lehmann and Stevens.[7] Typical results of gradient profiles are presented in Figure 2. When the x-ray exposed cells were treated with the M. luteus extract, the DNA profile shifted to a lower average molecular mass, indicating the presence of additional dsb due to the enzyme treatment.

The frequencies of these additional dsb were examined in further experiments as a function of absorbed dose. These are presented in Figure 3.

The data presented in Figure 3 shows firstly that frank dsb (open circles) are induced as a linear function of dose with a frequency similar to that in other cell lines, e.g. in mouse Ehrlich ascites tumour cells.[9] Secondly, the data show that in endonuclease treated cells the frequency of

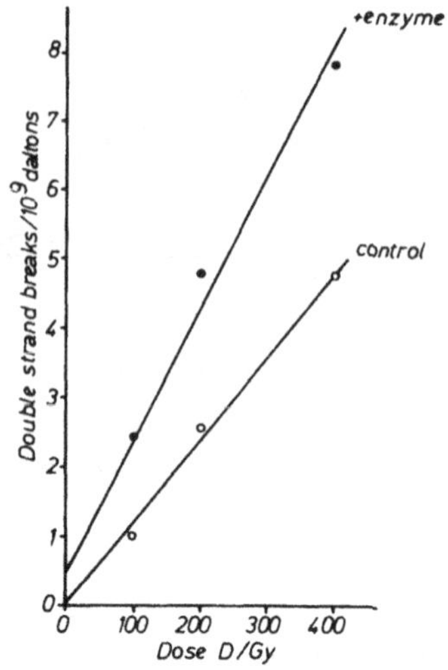

Figure 3. Double strand breaks as a function of dose in DNA treated (●) with or (O) without M.luteus enclnuclease.[21]

dsb increased by a factor of 1.7 (closed circles). These additional dsb were also induced as a linear function of radiation dose indicating that double es or es-ssb sites are formed by the passage of single ionizing particles. The line formed by the closed circles extrapolates back at zero dose to near the origin, indicating a very small contribution fron non-specific endonuclease activity in the preparation.

INDUCTION OF DOUBLE-STRAND BREAKS DURING REPAIR

In view of the data in Figures 2 and 3 showing that double-stranded base or sugar damage in the DNA was induced by X-irradiation, the question arose as to whether the incision of such sites by the cell would be detectable during the repair incubation after irradiation. We therefore carried out experiments in which cells were incubated after irradiation in growth medium with or without cytosine arabinoside present (Figure 4). It was argued that ara C (which is known to inhibit the repair of dsb) might lead to an accumulation of dsb which would not occur in medium alone. The results, however, showed that both with or without araC present dsb accumulated after 500 Gy of X-rays. Dsb were measured using neutral sucrose gradient sedimentation. In the presence of araC however, more dsb accumulated than in its absence, indicating that a simultaneous incision and repair of double-strand damaged sites is taking place in these cells. These data appear to be confirmed by those of Dugle et al,[23] who used the same cell line. Their data (also discussed by Ward[12]) indicate an increase in dsb over the first two hours of incubation after irradiation.

What evidence is there from other systems for the presence of double-stranded sites damaged in irradiated DNA or cells? One example of this is provided from work with an in vitro system; Ward[12] showed that when free DNA is irradiated in the presence of a high scavenger concentration (1 mol/1 t-butyl alcohol) to simulate the radical scavenging property of the cellular environment, in addition to directly induced dsb, a gradual increase in dsb occured due to hydrolysis of sugar damaged sites. The increase was rapid over the first 20 min to a value of about 1.3 after which dsb increased more slowly over a period of some hours. This work indicates the presence in

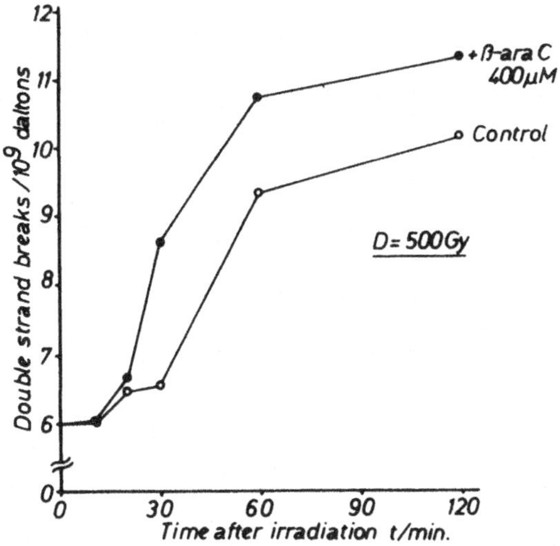

Figure 4. Double strand breaks as a function of time after irradiation in cells treated (●) with or (O) without cytosine arabinoside.[21]

irradiated DNA of double-stranded damaged sites which might, in cells, be enzymatically converted to dsb.

Treatment of DNA, isolated from irradiated mouse L5178Y cells and from mouse brain tissue, with S1 nuclease (from Aspergillus oryzae) has also been shown to lead to an increase in the number of measured dsb by factors of 2.85 and 2.3, respectively.[24,25] This effect was also observed in yeast,[5] where twice the number of S1 sensitive sites were observed, as compared with the number of frank dsb. The latter authors postulated that the S1 nuclease cleaves the DNA at sites of multiple base damages, which have given rise to a localized denatured region or "bulky lesion".

A close examination of the dsb repair kinetics of other cell lines, however, shows no strong indication of the type of incision indicated by the data presented in Figure 4.[26] It is not clear whether cell lines such as Ehrlich ascites tumour cells repair double-stranded es sites by a mechanism involving sequential repair of single-strand damage, so that no additional dsb occur, or whether the procedures adopted by these authors (detergent lysis and pronase digestion at 50°C overnight) could have already led to the incision or hydrolysis of these sites to dsb. An additional possibility is that the rate of incision of such sites in Ehrlich ascites cells is small compared with that of the repair, so that dsb do not accumulate as in V79 cells. Thus clearly there are a number of open questions related to the frequencies and kinetics of repair of dsb in mammalian cells.

KINETICS OF DSB REPAIR

The repair of dsb in mammalian cells and in yeast appears to be approximately first order with a half time of some 2-4h depending on conditions.[7,8,26,27] The kinetics of repair of dsb in yeast is not affected by radiation doses of up to 2400 Gy[28] and in mammalian cells up to at least 500 Gy.[7] The repair of dsb was shown to be independent of the cell cycle phase in mammalian cells.[9] Dsb repair is slow when compared with that for ssb, which in mammalian cells has a half time of approximately 5 min followed by a slower component that may be distorted by the involvement of incision of base damages.[29] The repair of dsb corresponds closely with the repair of chromosome breaks, as monitored in human cells by visual observation of prematurely condensed chromosomes, where repair was found to be initially first order with a half time of repair of 2h. The latter measurements were made at doses of 3.3 and 6 Gy.[30]

Much faster kinetics of dsb repair were measured using the neutral elution technique of Bradley and Kohn.[31] Half-times of the order of 10 min[32-34] were interpreted as rejoining of a class of dsb by ligation.[33] The rapid repair found with this method compared to those of neutral gradient sedimentation and premature chromosome condensation, suggests that it measures some other type of damage in addition to dsb. One problem with this method may be the use of a high pH value (9.6), which has been shown to lead to the formation of additional dsb.[35] These could be lesions in which two ssb occur some nucleotides apart and would normally be repaired by the ssb repair mechanism. However this may not be the whole explanation since recent data from our laboratory show that at an elution pH of 7, the repair measured by the elution method is rapid, and appears complete at about 1h after irradiation (Bryant, unpublished results).

POSSIBLE MECHANISMS OF DSB REPAIR

The mechanism of dsb repair is not known; however evidence from work with bacteria and yeast indicates that the mechanism involves some kind of recombinational event between homologous chromosomes. Data show, for example, that the rec A^- strain of E. coli K12 is unable to repair dsb, as is the WT strain K12 when grown on a medium yielding an average of only 1.3 genome copies per cell.[36] When the WT cells were grown in a rich medium, where 4-5 genome copies were present, dsb repair occured. Similarly in yeast the rad 52 mutant is both dsb repair and recombinationally deficient.[37,38] Also haploid yeast are unable to repair dsb in the G1 phase, i.e. where only one genome copy is present, whereas they are able to repair dsb if irradiated in the G2 phase after duplication of the genome.[39,40] It is also interesting in this respect to note that integration of linearized (but not circular) plasmids into yeast requires the presence of the rad 52 gene.[41] Resnick[42] proposed a recombinational model for the repair of dsb in which either a temporary or permanent exchange of homologous genetic information takes place. Evidence for such hybrid intermediates in G2 irradiated mammalian cells was provided by Resnick and Moore[43] using unifiliar substitution of thymidine with bromodeoxyuridine and CsCl sedimentation of isolated DNA. Fonck et al.[44] showed a two component formation of "heavy-heavy" hybrid intermediates, using a similar technique in a mouse tumor cell line, the cells of which were principally in Gn phase, and showed that the hybrids dissociated again after some hours. They also showed that the hybrids formed in the presence of adenine arabinoside, which is known to inhibit the repair of dsb by blocking DNA synthesis. It was thus concluded that hybrid formation is an earlier step involving pairing of homologous strands of DNA. These data thus support the model proposed by Resnick.

BIOLOGICAL CONSEQUENCES OF DSB

A dsb left unrepaired may have dire consequences for a cell, and experiments have shown that the rad 52 mutant of yeast (dsb repair negative) is extremely radiosensitive, such that 1-2 dsb constitute a lethal event.[37,45] In mammalian cells, and probably also in WT yeast where most dsb appear to be repaired, misrepair may also play an important role in cell killing whereby illegitimately joined dsb leads to "exchange" type chromosomal aberrations, which may be lethal for the cell as a result of loss of large fragments of

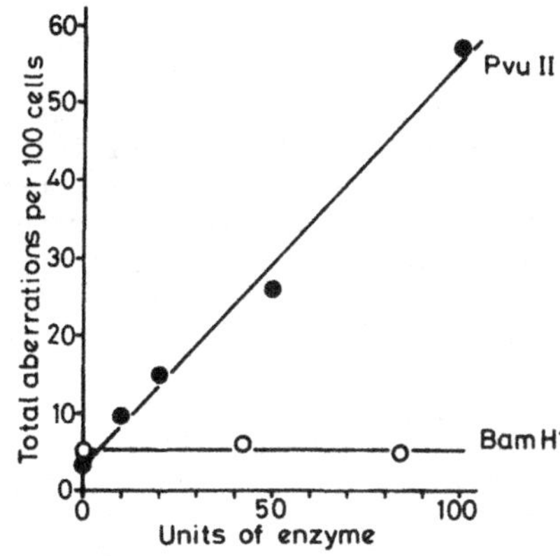

Figure 5. Chromosomal abberations in Chinese hamster cells (V79) as a function of restriction enzyme concentration (units per flask).[47]

chromosomes or due to the formation of anaphase "bridges" during mitosis.[46] This hypothesis is supported by recent evidence (Figure 5) in which "blunt-ended" dsb, induced in permeabilized mammalian cells by treatment with the restriction endonuclease Pvu II, led to chromosomal aberration formation, whereas treatment with Bam H1 (inducing cohesive-ended dsb with a base overlap of 4) did not increase the aberration frequency above the background level.[47]

It should be noted that restriction endonuclease induced dsb are formed in free DNA with 3'OH and 5'PO4 ends, and as such could in theory, in a cell, be repaired by a ligation mechanism.[47] However, the mechanism of repair of this type of lesion in chromatin of a mammalian cell is not yet known.

In view of the close correlation between cell killing and the occurence of chromosome fragments which has been established in a wide variety of studies,[48] it has been suggested that after X-ray exposure, cells undergo a "chromosomal" mode of cell death.[46] In order to test this hypothesis, experiments were carried out[49] in which restriction endonucleases inducing blunt and cohesive ended dsb were tested for their ability to kill cells. In support of this notion, it was shown that blunt ended dsb generated in the DNA of Chinese hamster cells by Pvu, II led to a dose-dependent loss of clonogenic ability, while cohesive-ended dsb generated by Bam H1 did not affect cell survival.

It seems plausible therefore that in mammalian cells, loss of clonogenic ability after irradiation is attributable to the non-repair or misrepair of a small fraction of dsb (1-2%) which lead to the induction of deletion and exchange type chromosomal aberrations, respectively. The consequent loss or rearrangement of genetic material then leads to death of a cell. The non-lethal loss or rearrangement of genetic material resulting from non-repair or misrepair of dsb may also have serious consequences for a cell, leading for example to chromosomal mutations[50] or to oncogenic transformation.[51]

CONCLUSIONS

It seems logical to postulate that dsb arising both directly, i.e. at the time of irradiation, as well as indirectly via repair processes as described here, will be potentially equally damaging for a cell in terms of the potential to induce chromosomal aberrations. However, in some cell systems the repair of double es or es-ssb sites may run concurrently with the incision so that these lesions do not remain open for long: hence the lack of accumulation of dsb during repair. The rate of incision will thus determine both the accumulation and the probability of exchanges (leading to chromosomal aberrations) between these and other frank dsb. Rapid incision leading to a large additional pool of dsb appears to be the case in Chinese hamster V79 cells, as described here. Some evidence also exists for the conversion of base damage, via dsb, into deletion type chromatid aberrations which accumulate in irradiated G2 human cells treated with ara C.[52]

A small fraction of dsb, probably arising both at the time of irradiation as well as (to various extents in different cell lines) enzymatically during repair of base or sugar damage, appears to be either left unrepaired, yielding deletion type chromosomal aberrations, or is misrepaired, yielding exchange aberrations. The induction of these aberrations appears to be of central importance in the biological effects of ionizing radiation such as mutations, oncogenic transformation, and cell death.

REFERENCES

1. P. A. Cerutti, Effects of ionizing radiation on mammalian cells, Naturwissenschaften, 61:51 (1974).
2. C. Von Sonntag, U. Hagen, A. Schön-Bopp, and D. Schulte-Frohlinde, Radiation induced strand breaks DNA; Chemical and enzymatic analysis of end-groups and mechanistic aspects, in "Advances in Radiation Biology", 9, J. T. Lett and H. Adler eds., Academic Press, pp. 109-142 (1981).
3. M. C. Paterson, B. P. Smith, P. H. M. Lohman, A. K. Anderson and L. Fishman, Defective excision repair of gamma ray damage DNA in human (ataxia telangiectasia) fibroblasts, Nature, 260:444 (1976).
4. A. Schön-Bopp, G. Schafer and U. Hagen, Endonuclease activities in extracts of M. luteus that act on gamma irradiated DNA, Int. J. Radiat. Biol., 31:227 (1977).
5. J. Andrews, H. Martin-Bertram and U. Hagen, S1 nuclease sites in yeast DNA: an assay for radiation induced base damage, Int. J. Radiat. Biol., 45:497 (1984).
6. A. R. Lehman and M. Ormerod, Double strand breaks in the DNA of a mammalian cell after x-irradiation, Biochem. Biophys. Acta, 217:268 (1970).
7. A. R. Lehman and S. Stevens, The production and repair of double strand breaks in cells from normal humans and from patients with ataxia telangiectasia, Biochem. Biophys. Acta, 474:49 (1977).
8. M. Frankenberg-Schwager, D. Frankenberg, D. Blocher and C. Adamczyk, The influence of oxygen on the survival and yield of DNA double strand breaks in irradiated yeast, Int. J. Radiat. Biol., 36:261 (1979).
9. D. Blocher, DNA double strand breaks in Ehrlich ascites tumor cells at low doses of x-rays. I. Determination of induced breaks by centrifugation at reduced speed, Int. J. Radiat. Biol., 42:317 (1982).
10. H. G. Paretzke, An appraisal of the relative importance for radiobiology of slow electrons, in: 5th Symp. on Microdosimetry, H. G. Ebert and B. G. R. Smith, eds. (CEC) pp. 41 (1976).
11. D. Frankenberg, M. Frankenberg-Schwager and R. Harbich, Evidence for DNA double strand breaks as the critical lesion in yeast cells irradiated with sparsely ionizing radiations under oxic or anoxic conditions, Radiat. Res., 88:524 (1981).
12. J. Ward, Some biochemical consequences of the spatial distribution of ionizing radiation produced free radicals, Symp on Radical Processes in Radiobiology and Carcinogenesis, Radiat. Res., 86:185 (1981).
13. D. Frankenberg, Interpretation of the dose and LET dependence of RBE

values for lethal lesions in yeast cells, Rad. Res., 97:329 (1984).
14. M. C. Paterson, Ataxia telangiectasia: A model inherited disease linking deficient DNA repair with radiosensitivity and cancer proneness, in: DNA Repair Mechanisms, P. C. Hanawalt, E. C. Friedberg and C. F. Fox, eds., Academic Press, pp. 637 (1976).
15. S. S. Wallace, H. L. Katcher and P. R. Armel, Measurement of x-ray damage in supercoiled phage by means of enzyme probes, in: DNA Repair: A Laboratory Manual of Research Procedures, 1, Part A, E.C. Friedberg and P. C. Hanawalt eds., (Marcel Dekker, N.Y. and Basel) pp. 113 (1981).
16. S. E. Bresler, L. A. Noskin, N. A. Kuzovleva and I. G. Noskina, The nature of the damage to E. coli DNA induced by gamma irradiation, Int. J. Radiat. Biol., 36:289 (1979).
17. K. W. Kohn, N. H. Steigbigel and C. L. Spears, Crosslinking and repair of DNA in sensitive and resistant strains of E. coli treated with nitrogen mustard, Proc. Natl. Acad. Sci. USA, 53:1154 (1965).
18. R. S. Cole, Repair of DNA containing interstrand crosslinks in E. coli, Proc. Natl. Acad. Sci. USA, 70:1064 (1973).
19. N. Magana-Schwencke, J.A.P. Henriques, R. Chanet and E. Moustacchi, The fate of 8-methoxypsoralen photoinduced crosslinks in nuclear and mitochondrial yeast DNA: Comparison of WT and repair deficient strains, Proc. Natl. Acad. Sci. USA 79:1722 (1982).
20. T. Bonura, C. D. Town, K. C. Smith and H. S. Kaplan, Enzymatic induction of DNA double strand breaks in gamma irradiated E. coli K12, Proc. Natl. Acad. Sci. USA., 72:4265 (1975).
21. G. Ahnstrom and P. E. Bryant, DNA double-strand breaks generated by the repair of x-ray damage in Chinese hamster cells, Int. J. Radiat. Biol., 41:671 (1982).
22. S. Riazuddin and L. Grossman, Micrococcus luteus correndonucleases I. Resolution and purification of two endonucleases specific for DNA containing pyrimidine dimers, J. Biol. Chem., 252:6280 (1977).
23. D. L. Dugle, C. J. Gillespie and J. D. Chapman, DNA strand break repair and survival in x-irradiated mammalian cells, Proc. Natl. Acad. Sci.-USA, 73:809 (1976).
24. K Yoshizawa, I. Furuno, T. Yada and H. Matsudaira, Induction and repair of strand breaks and 3'hydroxy-terminals in the DNA of mouse brain following gamma irradiation, Biochem. Biophys. Acta., 521:144 (1978).
25. I. Furuno, T. Yada, H. Matsudaira and T. Maruyama, Induction and repair of DNA strand breaks in cultured mammalian cells following fast neutron irradiation, Int. J. Radiat. Biol., 36:639 (1979).
26. D. Blocher and W. Pohlit, DNA double strand breaks in Ehrlich ascites tumor cells at low doses of x-rays. II. Can cell death be attributed to double strand breaks?, Int. J. Radiat. Biol., 42:329 (1982)
27. P. E. Bryant and D. Blocher, Measurement of the kinetics of DNA double strand break repair in Ehrlich ascites tumor cells using the unwinding method, Int. J. Radiat. Biol., 38:335 (1980).
28. M. Frankenberg-Schwager, D. Frankenberg, D. Blocher and C. Adamczyk, Repair of DNA double strand breaks in irradiated yeast cells under non growth conditions, Radiat. Res., 82:498 (1980).
29. P. E. Bryant, R. Warring and G. Anstrom, DNA repair kinetics after low doses of x-rays: A comparison of results obtained by the unwinding and nucleoid sedimentation methods, Mutat. Res., 131:15 (1984).
30. M. N. Cornforth and J. S. Bedford, X-ray induced breakage and rejoining of human interphase chromosomes, Science, 222:1141 (1983).
31. M. O. Bradley and K. W. Kohn, X-ray induced DNA double strand break production and repair in mammalian cells as measured by neutral filter elution, Nucl. Acids Res., 7:793 (1979).
32. G. P. Van der Schans, M. C. Paterson and W. G. Cross, DNA strand break and rejoining in cultured human fibroblasts exposed to fast neutrons or gamma rays, Int. J. Radiat. Biol., 44:75 (1983).
33. K. F. Weibezahn and T. Coquerella, Radiation induced double strand

breaks are joined by ligation and recombination processes, Nucleic Acids Res., 9:3139 (1981).
34. H. J. Rahmsdorf, H. Ponta, M. Bachle, U. Mallich, K. F. Weibezhan and P. Herrlich, Differentiated cells from BALB/c mice differ in their radiosensitivity, Exptl. Cell Res., 136:111 (1981).
35. M. J. Tilby, P. S. Loverock and E. M. Fielding, An effect of elevated post-irradiation pH on the yield of double-strand breaks in DNA from irradiated bacterial cells, Radiat. Res., 98:284 (1984).
36. F. Krasin and F. Hutchinson, Repair of double strand breaks in E. coli which requires the rec A function and the presence of a duplicate genome, J. Mol. Biol., 116:81 (1977).
37. K.S.Y. Ho, Induction of DNA double strand breaks by x-rays in a radiosensitive strain of the yeast S. cerevisiae, Mutat. Res., 30:327 (1975).
38. J. C. Game and R. K. Mortimer, A genetic study of x-ray sensitive mutants in yeast, Mutation Res., 24:281 (1974).
39. E. Chlebowicz and W. J. Jachymczyk, Repair of MMS induced DNA double strand breaks in haploid cells of S. cerevisiae requires the presence of a duplicate genome, Mol. Gen. Genet., 167:279 (1983).
40. G. Brunborg, M. A. Resnick and D. H. Williamson, Cell cycle specific repair of DNA double strand breaks in S. cerevisiae, Radiat. Res. 82:547 (1980).
41. T. L. Orr-Weaver, J. W. Szostak and R. J. Rothstein, Yeast transformation: A model system for the study of recombination, Proc. Natl. - Acad. Sci., 78:6345 (1981).
42. M. A. Resnick, The repair of double strand breaks in DNA: A model involving recombination, J. Theor. Biol., 59:97 (1976).
43. M. A. Resnick and P. Moore, Molecular recombination and the repair of DNA double strand breaks in CHO cells, Nucl. Acids Res., 6:3145 (1979).
44. K. Fonck, P. Barthel and P. E. Bryant, Kinetics of recombinational hybrid formation in x-irradiated mammalian cells: A possible first step in the repair of DNA double strand breaks, Mutation Res., DNA Repair Reports, 132:113 (1984).
45. M. A. Resnick and P. Martin, The repair of double strand breaks in the nuclear DNA of S. cerevisiae and its genetic control, Mol. Gen. Genet., 143:119 (1976).
46. P. E. Bryant, Effects of ara A and fresh medium on chromosome damage and DNA double strand break repair in x-irradiated stationary cells, Brit. J. Cancer, 45, Suppl. VI:61 (1984).
47. P. E. Bryant, Enzymatic restriction of mammalian cell DNA using Pvu II and Bam H1: Evidence for the double strand break origin of chromosomal aberrations, Int. J. Radiat. Biol., 46:57 (1984).
48. G. P. Joshi, W. J. Nelson, S. H. Revell and C. A. Shaw, X-ray induced chromosome damage in live mammalian cells and improved measurement of its effects on their colony forming ability, Int. J. Radiat. Biol., 41:161 (1982).
49. P. E. Bryant, Enzymatic restriction of mammalian cell DNA: Evidence for double strand breaks as potentially lethal lesions, Int. J. Radiat. Biol., 48:55 (1985).
50. J. Thacker, The chromosomes of a V79 Chinese hamster cell line and a mutant subline lacking HPRT activity, Cytogenet. Cell Genet., 29:16 (1981).
51. G. Klein and E. Klein, Oncogene activation and tumor progression, Carcinogenesis, 5:429 (1984).
52. R. J. Preston, The effect of cytosine arabinoside on the frequency of x-ray induced chromosome aberrations in normal human leukocytes, Mutat. Res., 69:71 (1980).

DNA-PROTEIN CROSS-LINKS: NEW INSIGHTS INTO THEIR FORMATION AND REPAIR IN IRRADIATED MAMMALIAN CELLS

Nancy L. Oleinick, Song-mao Chiu, Libby R. Friedman, Liang-yan Xue, and Narayani Ramakrishnan

Division of Biochemical Oncology, Department of Radiology
Case Western Reserve University, Cleveland, OH 44106

INTRODUCTION

The production of strong binding between DNA and protein by radiations and chemicals has been known for many years. DNA-protein cross-links (DPCs) were first recognized as a distinct lesion in ultraviolet light (UV)-irradiated bacteria by Smith [1] and by Alexander and Moroson.[2] The importance of DPCs for cellular lethality was clearly demonstrated in *E. coli*.[3] Smith has reviewed various aspects of this work on several occasions.[4-6]

In contrast to the abundance of information concerning the biochemistry and radiobiology of UV-induced DPCs in bacteria, relatively little is known about DPCs in mammalian cells, and the data are particularly sparse with regard to DPCs formed in cells exposed to ionizing radiation. Earlier work on this subject was reviewed by Yamamoto.[7] However, the application of newer technologies for the investigation of both DNA and protein has permitted the investigation of the biochemistry of radiation-induced DPCs at a level not previously possible. The purpose of this review is to summarize the present state of knowledge of DPCs formed in mammalian cells and in mammalian cell chromatin in response to x- or γ-radiation. Some reference will be made to correlative studies with UV; however, a review of DPCs induced by both UV and visible light has been written by Peak for this volume. Ultimately, a thorough understanding of DPCs will depend upon a precise definition of the chemistry of the interaction of DNA nucleotides with the amino acids of nuclear proteins. Since isolation of the two linked moieties has not yet been achieved, information concerning the chemistry of cross-linking is derived from various model systems and the use of highly sophisticated methodology for the separation and positive identification of the various possible base-amino acid heterodimers. This exciting area of investigation is reviewed by Dizdaroglu and Simic in this volume.

MEASUREMENT OF DNA-PROTEIN CROSS-LINKS

DPCs can be detected in cells by a variety of techniques. Most depend upon the separation of DNA from protein and the observation that with increasing dose of radiation or chemical cross-linking agent, more of one component is found in association with the other. Thus, organic extraction, equilibrium density gradient sedimentation in CsCl, gel filtration on large pore Sepharose 4B,[8] and the binding of protein but not DNA to nitrocellulose filters[9,10] have all been employed. None of these is highly sensitive; to produce measurable levels of DPCs by any of the above procedures, doses of ionizing radiation in excess of 10 Gy or fluences of UV in excess of 20 J/m^2 are required, which precludes the measurement of DPCs in mammalian cells at doses leaving a large fraction of survivors. Within the useful dose range, the amount of DNA cross-linked to protein is an approximately linear function of the total dose.[9,10] Examples of dose-response data obtained either by the nitrocellulose filter technique or by phenol extraction are shown in figure 1. It should be noted that all of the above techniques measure the amount of DNA associated with protein and not the actual number of covalent cross-links. Both the average size of the cross-linked DNA fragments and the amount of DNA bound to protein must be known to calculate the number of DPCs (see below).

Another technique for the detection of the presence of DPCs is alkaline elution.[11] As an alkaline solution is slowly percolated through lysed cells on a filter, fragmented DNA is eluted at a rate dependent upon the size of the fragments. If the DNA is cross-linked to protein, the elution of part of the DNA is retarded. Prior digestion with a protease permits the free elution of the DNA, demonstrating that an attachment to protein is responsible for the reduced rate of elution. Alkaline elution has been used to demonstrate the dose-response and to indicate a partial

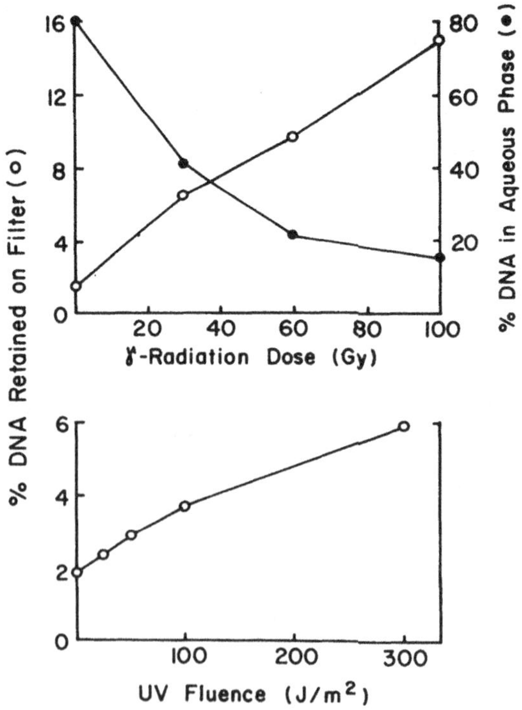

Fig. 1. Dose response for the formation of DPCs in Chinese hamster V79 cells. Exponentially growing monolayer cultures, prelabeled with ^3H-thymidine (0.1 μCi/ml), were exposed to ^{60}Co γ-radiation (upper panel) or to UV-radiation (lower panel) and assayed for the presence of DPCs either by nitrocellulose filter binding (O) or by phenol extraction (●). Some of these data are from ref. 10.

repair of DPCs.[12] However, the quantitation of DPCs immediately after irradiation is complicated by the simultaneous presence of DNA strand breaks. Fornace and Little[12] showed that by allowing the bulk of the strand breaks to be rejoined during a two hour incubation of the cells postirradiation, the residual cross-links were readily detected for doses of 50-600 Gy. In addition, alkaline elution does not allow the isolation of the DPCs for further molecular analysis. We have found that nitrocellulose filter binding is particularly useful in this regard.

THE FORMATION OF DNA-PROTEIN CROSS-LINKS

DPCs can be formed by ionizing radiation-induced hydroxyl radicals. Mee and Adelstein[13] have conducted radiobiological studies on chromatin prepared from Chinese hamster V79 lung fibroblasts. They assayed the presence of DPCs by sedimentation of the chromatin in alkaline sucrose gradients, by extraction with sodium dodecyl sulfate (SDS) followed by precipitation of the detergent and proteins with KCl,[1] and by the binding of protein and associated DNA to nitrocellulose filters.[8] The best correlation of level of DPCs with radical yield under various scavenging conditions was for hydroxyl radicals. Mee and Adelstein[13] also compared the level of DPCs for irradiation of intact cells suspended in Ca^{++}-free balanced salt solution. While a dose-dependent increase in DPCs was detected after 10-50 Gy in isolated chromatin, 1000 Gy was used to produce a similar level of DPCs in intact cells. Nonetheless, hydroxyl radicals appeared to be responsible for the DPCs. In intact V79 cells, we found that the production of DPCs was severely depressed when irradiation was in the presence of dimethylsulfoxide (DMSO), a scavenger of hydroxyl radicals. Hydrogen donation from dithiothreitol (DTT) could further reduce the yield of DPCs.

Hypoxic conditions enhance the yield of ionizing radiation-induced DPCs. Both Fornace and Little[12] and Meyn and Jenkins[14] have reported that the yield of DPCs, as detected by alkaline elution, is greater in mammalian cells irradiated under hypoxic than under aerated conditions. A similar result was recently reported by Mee and Adelstein[15] in irradiated isolated chromatin. It would appear that some of the initial DNA lesions react with oxygen to produce strand breaks faster than with proteins to produce DPCs. Interestingly, the opposite result has been observed in mammalian cells irradiated with visible light.[16] While DPC formation after irradiation with UV light is the result of the absorption of light by DNA bases, in the case of light in the visible range, the action spectrum suggests that the mechanism involves the absorption of light by a cellular porphyrin, which leads to damage through the production of activated oxygen species.

The DNA-protein cross-links are covalent. Until the linked nucleotide and amino acid moieties are isolated and identified, the types of chemical bonds involved in ionizing radiation-induced DPCs can not be known. However, the linkages survive treatment with high salt concentrations, detergents, alkali, urea, or guanidine hydrochloride.[9] Cress and Bowden[9] showed that only 20% of the linkages were labile to hydroxylamine, an agent which disrupts P-N types of bonds. Treatment with a protease before any of the assays for DPCs universally abolishes these lesions (e.g., [10,12]). This provides the best evidence that the lesions being measured are covalent adducts of protein to DNA.

When isolated chromatin is γ-irradiated, DPCs form with histones of the nucleosomal core and with non-histone proteins. Mee and Adelstein[17] analyzed the proteins tightly bound to chromatin after irradiation of chromatin preparations with 50 Gy of γ-radiation. They found by gel electrophoresis that histones H2a, H2b, H3, and H4, but not H1 had become bound to DNA. Although many other non-histone proteins are associated with DNA in the cell, the preparations used may have disrupted those associations, precluding the observation of cross-links to other types of proteins. Using antisera to dehistonized HeLa cell chromatin, Olinski et al.[18] found that the immunological reactivity of the chromatin was significantly increased by 100-1000 Gy of γ-radiation. DNA became covalently cross-linked to several non-histone proteins, some of which reacted with antiserum prepared against a 0.35 M NaCl extract of HeLa nuclei and some with antiserum to the residual 0.35 M NaCl washed nuclei.[19] The non-histone proteins were not identified, although it was suggested that some may be chromosomal scaffold proteins.[18] It may be concluded that radiation can cross-link a variety of proteins to DNA depending on the structure of the model DNA-protein complexes at the time of irradiation. Which model is related to the complexes found in intact cells can only be determined by irradiation of cells during normal growth.

When intact cells are irradiated, various possibilities exist for nonrandom selection of DNA regions and proteins to form DPCs. Our approach to the characterization of DPCs has been to irradiate intact cell monolayers, so that normal DNA-protein associations are preserved, then to isolate and identify classes of DNA bound to protein and, conversely, classes of protein bound to DNA.[20] A similar approach has been employed by Cress[21] to identify proteins involved in DPCs. Proteins could be histones, nonhistone chromosomal proteins, regulatory proteins, proteins which modify the superhelicity of DNA, such as topoisomerases, or proteins of the nuclear matrix, to which DNA is normally bound during replication and transcription. The DNA which becomes covalently linked to protein could be randomly distributed or might be preferentially found among specific sequences, conformational classes, or nuclear regions.

Untreated cells contain 1-2% of their DNA tightly associated to a subset of proteins of the nuclear matrix. Before proceeding to study the proteins linked to DNA by irradiation, it was first necessary to identify the classes of nuclear proteins bound to DNA in the normal untreated cells. To this end, our laboratory has purified the DPCs relatively free of non-cross-linked DNA fragments and proteins by centrifugation and filtration on nitrocellulose (Fig. 2a). These have been compared with DNA-associated proteins which band with DNA on CsCl (Fig. 2c) and with proteins of the nuclear matrix (Fig. 2e) by electrophoresis on polyacrylamide gels. The major protein band from filter-bound DPCs has a relative molecular mass of 43 kd (Fig. 2a). Numerous other proteins are complexed with DNA but are not necessarily covalently bound to DNA (Fig. 2c). Most of these peptides either do not bind to or are not recovered from the nitrocellulose filters. Similar profiles of DNA-bound proteins have been observed by Cress.[21] The apparent sizes of the proteins in our profiles of DNA-associated and nuclear matrix proteins appear to be nearly identical to those published by Razin et al.[22] We conclude that the major (43 kd) protein of DPCs is one of a subset of DNA-binding proteins of the nuclear matrix.

Irradiation of intact mammalian cells links DNA to the same subset of proteins of the nuclear matrix. As a result of ionizing radiation, there is an increase in the amount of DNA cross-linked to protein.[9,10] When V79 cells were exposed to 10-100 Gy of γ-radiation and the bound proteins were isolated and electrophoresed as above, the same major protein bands were observed (Fig. 2b,d), suggesting that the majority of radiation-induced DPCs are to specific proteins of the nuclear matrix. Cress has isolated the proteins apparent in the DPC complex from HeLa cells immediately after irradiation with 5 Gy by comigration of the proteins with genomic DNA using CsCl density gradients (similarly to Fig. 2c,d). Analysis of the proteins by polyacrylamide gel electrophoresis in SDS revealed

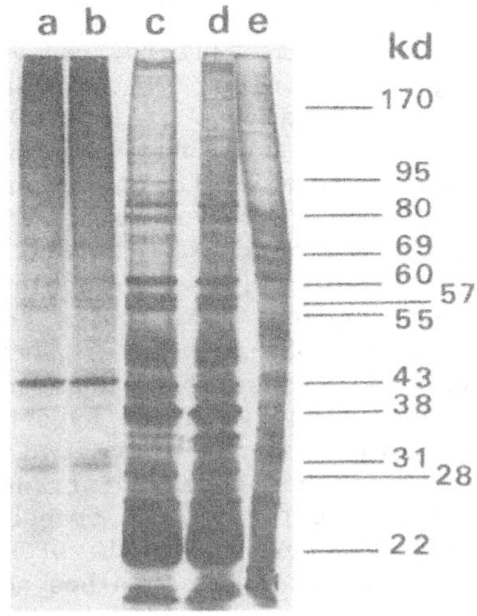

Fig. 2. Proteins of DNA-protein complexes of V79 cells. Cultures were given 0 or 100 Gy then lysed in 1 M $NaClO_4$, 1% Sarkosyl. The DNA-protein complexes were collected by centrifugation at 100,000xg for 18 hrs. The pellet, containing >98% of the DNA and 6-7% of cellular protein, was resuspended in the same solution. For a and b, the preparation was filtered through nitrocellulose,[10] and the filter-bound material (1-2% of the DNA, 40-50% of input protein) was recovered in 1% SDS at 60° for 2 hrs. For c and d, the DNA and protein were banded in CsCl gradients, and the DNA peak was collected. For e, nuclei were isolated from unirradiated cells, treated with 2 M NaCl to remove histones, then digested with EcoRI. The nuclear matrix and DNA fragments still bound to it were separated from released DNA by centrifugation. After digestion of all samples with DNase I, the proteins were dissolved and subjected to electrophoresis in 7.5% polyacrylamide gels containing SDS and observed by silver staining.

a: Proteins of filter-bound DPCs, 0 Gy.
b: Same as a, 100 Gy.
c: DNA-associated proteins from CsCl gradient, 0 Gy.
d: Same as b, 100 Gy.
e: Proteins of the nuclear matrix, 0 Gy.

major apparent molecular weight classes of 170, 95, 68, 55, 45, and 29 kd. Thus, the DNA-binding proteins of human and hamster cells appear to be of similar size. The peptides corresponding to 68 and 170 kd do not immunoreact with polyclonal antibodies specific for lamin A and C and for topoisomerase II, respectively (A. Cress, unpublished observation). Thus, radiation cross-links DNA to a subset of nuclear matrix proteins normally found in association with DNA.

The low level of DPCs found in unirradiated cells preferentially involves DNA regions containing actively transcribing sequences. We have analyzed the DNA which is bound to protein in control, unirradiated cells by collecting DPCs on nitrocellulose filters, recovering and deproteinizing the DNA, and hybridizing it to various DNA and RNA probes. In each case, the ratio of probe hybridized to amount of input DNA has been compared to the ratio of probe hybridized to total nuclear DNA isolated from the same population of cells in the same experiment. Thus, we can calculate the abundance or frequency of the probed sequences in any fraction of DNA relative to the frequency in the bulk DNA. Using ^{125}I-poly(A+)RNA or ^{125}I-rRNA as probes of DNA from V79 cells, or ^{32}P-labeled cloned Drosophila hsp-70 gene as a probe of the integrated gene copies transfected into mouse L-hs4 cells,[23] we find that the DNA-protein complexes are between 2 and 15 times enriched in the actively transcribing sequences hybridizing to these probes. Our results are consistent with the hypothesis that at least part of the DNA which is bound to protein in untreated, exponentially growing cells consists of sequences in the process of transcription on the nuclear matrix (e.g., [24]). Since replication, like transcription, is reported to occur while DNA is matrix-bound (e.g., [25]), the DPCs from control cells may also be enriched in replicating DNA, but this has not yet been demonstrated in our preparations.

Exposure to ionizing radiation but not to UV-radiation produces DPCs which, like control DPCs, are enriched in actively transcribing DNA. The additional DNA cross-linked by γ-radiation has been analyzed in the same manner as for control DPCs. This produces a probe/DNA ratio which is consistently greater for protein-bound DNA than for bulk DNA. In this case, the contribution of the control DPCs must be subtracted from the total DPCs resulting from irradiation. In order to estimate the relative frequency of active sequences in the newly cross-linked DNA, we described the irradiated condition by a simple mathematical model and observed whether or not the measured probe/DNA ratio agreed with that calculated for the model condition. For the model, we assumed (a) that the total DPC content after irradiation was the arithmetic sum of the DPCs in unirradiated cells and the DPCs formed by irradiation and (b) that the DNA cross-linked by radiation was not enriched for any particular sequence, i.e., it had the same frequency of probed sequence as the bulk DNA. Statement (b) holds if DNA damage is truly random. This calculation is described in greater detail in ref. 20. Analyzing the DNA of DPCs by this method, it was found that γ-radiation resulted in DPCs which were enriched in actively transcribing sequences, but UV-radiation yielded DPCs which had the same or lower frequency of active probed sequences as did the bulk DNA.[20]

The apparently random UV-induced formation of DPCs in the DNA is consistent with the observations that thymine dimer production by UV is random.[26,27] In contrast, hydroxyl radical attack resulting from ionizing radiation may occur with greater frequency in regions of the nucleus containing stretches of chromatin open and accessible to other diffusible agents, such as polymerases and nucleases. We demonstrated earlier [28,29] that regions of DNA containing active sequences are also more susceptible to ionizing radiation-induced production of single-strand breaks.

Gamma-radiation results in the cross-linking of an increased number of DNA fragments to protein. The filter-binding assay measures the percentage of input DNA which becomes trapped on the nitrocellulose filter due to protein and not the number of cross-links formed. Two interpretations exist for the radiation-induced increase in bound DNA: an increase in number of DNA fragments which are cross-linked to protein or an increase in size of the bound DNA fragments with no change in the number of covalent attachment sites. The latter situation could result if another radiation-induced lesion caused fragmentation of the DNA at a site further away from the cross-link than the normal fragmentation site. To address this problem, we irradiated ^{3}H-thymidine-labeled V79 cells with 0-100 Gy of γ-radiation and collected the DPCs from nitrocellulose filters. After deproteinization, the DNA was electrophoresed on agarose gels under non-denaturing conditions, the radioactivity of the gel slices was determined, and the number average molecular weight (M_n) of each DNA sample was calculated. In five experiments, 2.40% $\pm$ 0.04% (n=5) of the DNA from unirradiated cells became bound to nitrocellulose, and M_n was 1.22 $\pm$ 0.02 x 10^7 d (m $\pm$ S.E.; n=19) and did not vary with radiation dose (Ramakrishnan, Chiu, and Oleinick, unpublished). From these data and assuming 5 pg of DNA per V79 cell, it was possible to calculate that on the average, there are 6085 $\pm$ 98 (n=5) DPCs per normal unirradiated V79 cell in exponential growth and that 157 $\pm$ 5 (n=14) additional DPCs are produced for each Gy of γ-radiation.

THE REPAIR OF DNA-PROTEIN CROSS-LINKS

DNA is removed from protein after gamma-irradiation with biphasic kinetics. When irradiated L-hs4 or V79 cell monolayers are incubated at 37° postirradiation, the level of DPCs decreases such that 50% remain after 1 hour and <10% after 4 hours (Fig. 3a; cf. 10). In contrast, no repair of DPCs could be observed if the cells were removed from the monolayer and incubated in suspension. Furthermore, V79 cells irradiated during metaphase yield the same initial level of DPCs as do interphase cells, but DPCs produced in colcemid-arrested metaphase cells are not repaired even after removal of the colcemid and completion of cell division.[30] The chromosome scaffold proteins to which the metaphase DNA becomes bound appears to be identical on SDS-PAGE to the nuclear matrix proteins which cross-link to interphase DNA. Metaphase cells do not have an intact nuclear matrix. Since in these cells single-strand breaks appear to rejoin,[29] although at a slightly reduced rate compared to interphase cells, the data suggest that an intact nuclear matrix may be important for repair of DPCs but not for repair of strand breaks.

During repair of DPCs, a dynamic interaction between DNA and the nuclear matrix suggests that at least some of the repair occurs on the matrix. The majority of γ-radiation-induced DPCs are removed during the first few hours after irradiation of Chinese hamster V79 or mouse L-hs4 cells (Fig. 3a). In order to study the fate of the active DNA sequences cross-linked to protein, at various times up to 4 hours postirradiation, the residual DPCs have been isolated from the repairing cells. The relative frequency of actively transcribing genes in the DNA from those DPCs was determined by hybridization to one or more probes as described above.

An example is shown in Fig. 3 for mouse L-hs4 cells. These cells contain about 20 integrated copies of the Drosophila hsp-70 heat shock gene [23]. In cells maintained at 37°, there is minimal transcription of hsp-70, and DPCs in the bulk chromatin are repaired rapidly (panel a). In heat-shocked cells (46° for 6 min followed by incubation at 37° for 1.5 hours to permit activation of heat shock genes), repair of DPCs in bulk chromatin is less efficient (panel b); however, active hsp-70 genes which are cross-linked to protein are removed from DPCs within the first hour after 60 Gy γ-radiation (panel f). This is demonstrated by the change in the calculated enrichment of the DPCs isolated at various times postirradiation. The Enrichment Factors (panels e and f) are calculated from the ratio of ^{32}P-labeled probe/^{3}H-DNA (panels c and d) as described above and in ref. 20. Immediately after 60 Gy, DPCs are enriched in hsp-70 sequences (Enrichment Factor >1); one hour later, the residual DPCs have become depleted in active hsp-70 sequences (panel f). Thus, hsp-70 sequences are removed from DPCs faster than is bulk DNA. At later times, the DPCs regain the original pre-irradiation complement of hsp-70 sequences. A kinetic pattern nearly identical to the one shown in panels d and f has been observed employing other probes of actively transcribing DNA.[20] A

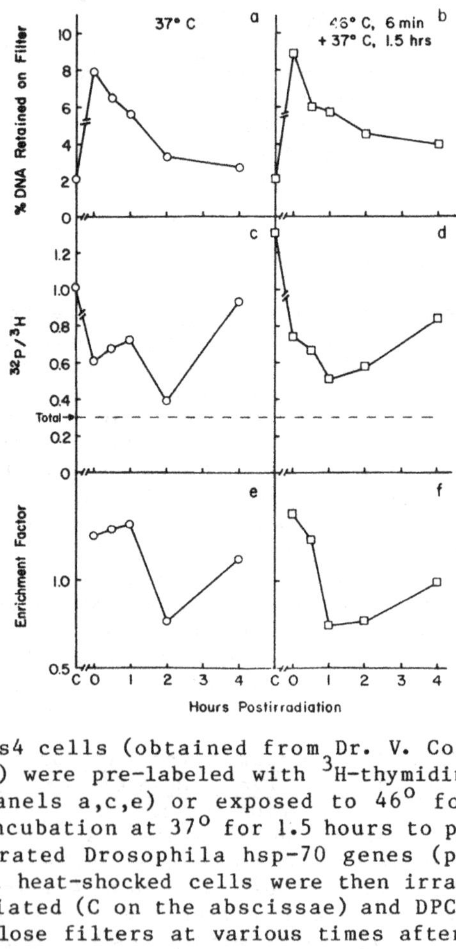

Fig. 3: Mouse L-hs4 cells (obtained from Dr. V. Corces, Johns Hopkins University) were pre-labeled with ^{3}H-thymidine, then either kept at 37° (panels a,c,e) or exposed to 46° for 6 minutes before further incubation at 37° for 1.5 hours to permit activation of the integrated Drosophila hsp-70 genes (panels b,d,f). Both normal and heat-shocked cells were then irradiated with 60 Gy or not irradiated (C on the abscissae) and DPCs were collected on nitrocellulose filters at various times after irradiation (panels a and b). The DPCs were recovered from replicate filters and deproteinized, and various amounts of the DNA were loaded onto additional filters for hybridization with cloned, nick translated hsp-70. The ratio of amount of ^{32}P-hsp-70 hybridized to amount of ^{3}H-DNA is recorded in panels c and d. The dashed line indicates the ^{32}P/^{3}H ratio of total cellular DNA. From these data for irradiated and control cells, the enrichment of the DPCs for hsp-70 sequences was calculated as described in the text (panels e and f). An Enrichment Factor of 1 indicates the same frequency of hsp-70 in the experimental DNA sample as in the bulk DNA.

striking contrast to these kinetics is seen in the fate of the hsp-70 sequences in non-heat-shocked cells (panels c and e). Following 60 Gy γ-radiation, there is a delay for the first hour before the DPCs become depleted in hsp-70 sequences. In this case, the removal of bulk DNA from DPCs is faster than the removal of hsp-70 sequences from DPCs. The data are consistent with an active involvement of the nuclear matrix binding proteins during the repair of DNA damage and the restoration of normal DNA-protein associations on the nuclear matrix following the completion of repair.

THE IMPORTANCE OF DNA-PROTEIN CROSS-LINKS FOR ASSESSMENT OF RADIATION DAMAGE AND THE BIOCHEMISTRY OF REPAIR

DPCs are composed of both normal and abnormal associations of DNA primarily with proteins of the nuclear matrix. Although the number of fragments of DNA linked to protein increases after irradiation, the normal and radiation-induced DPCs share two characteristics, the identity of the proteins involved and the enrichment of the DNA for actively transcribing sequences. There are several implications of these central features of DPCs. First, DPC formation in irradiated cells appears to occur in a non-random fashion preferentially to regions of DNA containing active sequences at or near the nuclear matrix. DNA single-strand breaks were likewise observed to be produced preferentially in active regions of the genome [28]. We speculate that a special environment provided by the nuclear matrix makes it an ideal region for DPC formation. Perhaps it allows accessibility of hydroxyl radicals to the DNA, because of the more open conformation of active DNA or a more highly hydrated milieu as compared to regions of the nucleus containing tightly condensed chromatin.

The second implication is that the nature of the interaction of DNA with the binding proteins of the matrix may be conducive to covalent bond formation upon free radical attack. Nothing is known about the specific interactions of DNA and matrix proteins. However, other types of DNA-protein complexes have been studied in great detail. In the nucleosome (model 1), DNA wraps around a histone octamer in a manner which is not specific for the sequence of DNA. In contrast, numerous sequence-specific proteins bind to DNA in the process of regulation of gene transcription. These proteins, such as the catabolite activator protein.[30] interact directly with functional groups of DNA bases which are in the major groove of the double helix (model 2). A third model of a DNA-protein complex is the covalent attachment of topoisomerase to one end of a broken single strand of DNA; this is an intermediate in the strand-passing reaction of the enzyme. Can any of these known complexes serve as a model for the interaction of DNA with binding proteins of the nuclear matrix? We suggest that the nucleosome is not a good model, because of the absence of sequence specificity. There are data which argue for specific classes of sequences which serve as attachment sites of DNA loops to the matrix.[32] Models 2 and 3 are both potentially useful at present. Model 3, a protein-associated DNA strand break, is consistent with the occurrence of single-strand breaks in matrix-associated DNA,[33] and model 2 provides close interactions between DNA bases and amino acids which could be converted to covalent bonds by irradiation.

Another implication of the data is that there are two categories of interaction of DNA with the nuclear matrix, strong and weak. The strong interactions are those registered as DPCs in unirradiated cells. The weak ones are converted to strong interactions, DPCs, by irradiation. Further, it is possible that the strongly attached DNA sequences are those undergoing transcription at the time of observation, while the weak ones are potentially active in transcription but not being traversed by RNA polymerase at the time. Further experiments will be required to test this hypothesis.

The final implication of these data is that there appears to be a hierarchy for repair. Actively transcribing sequences, which are essential for cell survival, seem to be repaired in preference to inactive sequences and to bulk DNA, most of which is inactive. A similar preferential repair of transcriptionally active DNA has been reported for CHO cells treated with UV light.[34] Thus, damage in active DNA may not be the most deleterious to the cell, because it is rapidly repaired. Damage in bulk DNA, most of which is never transcribed by the cell, may be tolerated rather than repaired or may lead to abnormal chromosome segregation at mitosis. Damage in potentially active sequences may be the most severe, if the need for transcription preceeds repair.

ACKNOWLEDGMENT

Research in the authors' laboratory is supported by Research Grant CA-15378 from the National Cancer Institute, DHHS.

REFERENCES

1. K. C. Smith, Dose dependent decrease in extractability of DNA from bacteria following irradiation with ultraviolet light or with visible light plus dye, Biochem. Biophys. Res. Commun. 8:157 (1962).
2. P. Alexander and H. L. Moroson, Cross-linking of deoxyribonucleic acid to protein following ultra-violet irradiation of different cells, Nature 194: 882 (1962).
3. K. C. Smith, B. Hodgkins, and M. E. O'Leary, The biological importance of ultraviolet light induced DNA-protein crosslinks in Escherichia coli 15 TAU, Biochim. Biophys. Acta 114:1 (1966).
4. K. C. Smith, The biological importance of U.V.-induced DNA-protein cross-linking in vivo and its probable chemical mechanism, Photochem. Photobiol. 7:651 (1968).
5. K. C. Smith, The radiation-induced addition of proteins and other molecules to nucleic acids, in: "Photochemistry and Photobiology of Nucleic Acids, Vol. II," S. Y. Wang, ed., p. 187, Academic Press, New York (1976).
6. K. C. Smith, Radiation-induced cross-linking of DNA and protein in bacteria, in: "Aging, Carcinogenesis, and Radiation Biology," K. C. Smith, ed., p. 67, Plenum, New York (1976).
7. O. Yamamoto, Ionizing radiation-induced DNA-protein cross-linking, in: "Aging, Carcinogenesis, and Radiation Biology," K. C. Smith, ed., p. 165, Plenum, New York (1976).
8. G. F. Strniste and S. C. Rall, Induction of stable protein-deoxyribonucleic acid adducts in Chinese hamster cell chromatin by ultraviolet light, Biochemistry 15:1712 (1976).
9. A. E. Cress and G. T. Bowden, Covalent DNA-protein crosslinking occurs after hyperthermia and radiation, Radiat. Res. 95:610 (1983).
10. S. M. Chiu, N. M. Sokany, L. R. Friedman, and N. L. Oleinick, Differential processing of ultraviolet or ionizing radiation-induced DNA-protein cross-links in Chinese hamster cells, Int. J. Radiat. Biol. 46:681 (1984).
11. K. W. Kohn, R. A. G. Ewig, L. C. Erickson, and L. A. Zwelling, Measurement of strand breaks and cross-links by alkaline elution, in: "DNA Repair: A Laboratory Manual of Research Procedures, Vol. 1, part B," E. C. Friedberg and P. C. Hanawalt, eds., p. 379,

Marcel Dekker, Inc., New York (1981).
12. A. J. Fornace, Jr., and J. B. Little, DNA crosslinking induced by X-rays and chemical agents, Biochim. Biophys. Acta 477:343 (1977).
13. L. K. Mee and S. J. Adelstein, Radiolysis of chromatin extracted from cultured mammalian cells: Formation of DNA-protein cross links, Int. J. Radiat. Biol. 36:359 (1979).
14. R. E. Meyn and W. T. Jenkins, Modification of radiation-induced DNA lesions by oxygen, Radiat. Res. Abstracts p. 83 (1984).
15. L. K. Mee and S. J. Adelstein, DNA-protein crosslinks in gamma-irradiated chromatin, presented at Intl. Conference on Mechanisms of DNA Damage and Repair, Gaithersberg, MD, June (1985).
16. J. G. Peak, M. J. Peak, R. S. Sikorski and C. A. Jones, Induction of DNA-protein crosslinks in human cells by ultraviolet and visible radiations: Action spectrum, Photochem. Photobiol. 41:295 (1985).
17. L. K. Mee and S. J. Adelstein, Predominance of core histones in formation of DNA-protein crosslinks in γ-irradiated chromatin, Proc. Natl. Acad. Sci. USA 78:2194 (1981).
18. R. Olinski, R. C. Briggs, L. S. Hnilica, J. Stein, and G. Stein, Gamma-radiation-induced crosslinking of cell-specific chromosomal nonhistone protein-DNA complexes in HeLa chromatin, Radiat. Res. 86:102 (1981).
19. Z. M. Banjar, L. S. Hnilica, R. C. Briggs, J. Stein, and G. Stein, Crosslinking of chromosomal proteins to DNA in HeLa cells by UV, gamma radiation and some antitumor drugs, Biochem. Biophys. Res. Commun. 114:767 (1983).
20. S. M. Chiu, L. R. Friedman, N. M. Sokany, and N. L. Oleinick, A role for the nuclear matrix in repair of radiation-induced DNA-protein cross-links, Radiat. Res. (1985, submitted).
21. A. Cress, Nuclear matrix proteins are covalently linked to DNA after ionizing radiation, Radiat. Res. Abstracts, p. 94 (1985).
22. S. V. Razin, V. V. Chernokhvostov, A. V. Roodyn, I. B. Zbarsky, and G. P. Georgiev, Proteins tightly bound to DNA in the regions of DNA attachment to the skeletal structures of interphase nuclei and metaphase chromosomes, Cell 27:65 (1981).
23. V. Corces, A. Pellicer, R. Axel, and M. Meselson, Integration, transcription, and control of a Drosophila heat shock gene in mouse cells, Proc. Natl. Acad. Sci. USA 78:7038 (1981).
24. S. I. Robinson, B. D. Nelkin, and B. Vogelstein, The ovalbumin gene is associated with the nuclear matrix of chicken oviduct cells, Cell 28:99 (1982).
25. R. Berezney and D. S. Coffey, Nuclear protein matrix: association with newly synthesized DNA, Science 189:291 (1975).
26. J. I. Williams and E. C. Friedberg, Deoxyribonucleic acid excision repair in chromatin after ultraviolet irradiation of human fibroblasts in culture, Biochemistry 18:3965 (1979).
27. H. J. Niggli and P. A. Cerutti, Nucleosomal distribution of thymine photodimers following far- and near ultraviolet irradiation, Biochem. Biophys. Res. Commun. 105:1215 (1982).
28. S. M. Chiu, N. L. Oleinick, L. R. Friedman, and P. J. Stambrook, Hypersensitivity of DNA in transcriptionally active chromatin to ionizing radiation, Biochim. Biophys. Acta 699:15 (1982).
29. N. L. Oleinick, S. M. Chiu, and L. R. Friedman, Gamma-radiation as a probe of chromatin structure: damage to and repair of active chromatin in the metaphase chromosome, Radiat. Res. 98:629 (1984).
30. S. M. Chiu, L. R. Friedman, L. Y. Xue, and N. L. Oleinick, DNA-protein cross-links in metaphase chromatin, Radiat. Res. Abstracts p. 75. (1985).
31. I. T. Weber and T. A. Steitz, Model of specific complex between catabolite gene activator protein and B-DNA suggested by electrostatic complementarity, Proc. Natl. Acad. Sci. USA 81:3973 (1984).

32. J. Mirkovitch, M.-E. Mirault, and U. K. Laemmli, Organization of the higher-order chromatin loop: Specific DNA attachment sites on nuclear scaffold, Cell 39:223 (1984).
33. H. Probst and R. Herzog, DNA regions associated with the nuclear matrix of Ehrlich ascites cells expose single-stranded sites after deproteinization, Eur. J. Biochem. 146:167 (1985)
34. V. A. Bohr, C. A. Smith, D. S. Okumoto, and P. C. Hanawalt, DNA repair in an active gene: Removal of pyrimidine dimers from the DHFR gene of CHO cells is much more efficient than in the genome overall, Cell 40:359 (1985).

DNA-TO-PROTEIN CROSSLINKS AND BACKBONE BREAKS CAUSED BY FAR- AND NEAR-ULTRAVIOLET, AND VISIBLE LIGHT RADIATIONS IN MAMMALIAN CELLS

Meyrick J. Peak and Jennifer G. Peak

Photobiology Group, Division of Biological and
Medical Research, Argonne National Laboratory
Argonne, Illinois 60439

INTRODUCTION

The cytotoxic, mutagenic and carcinogenic potential of germicidal far-UV* radiation has been known for decades. This radiation is not a major component of the solar radiations that reach the surface of the earth, unlike the abundant near-UV and visible radiations present in the solar spectrum. Although it has been recognized for decades that the mixture of mid-UV, near-UV, and visible radiation (wavelengths longer than 290 nm) that comprise sunlight is cytotoxic, mutagenic, and carcinogenic,[1] surprisingly little attention has been paid to the specific DNA damages that may be caused in cells by these particular wavelengths of nonionizing radiation--even though these radiations are environmentally abundant, penetrate deeply through skin,[2] and are increasingly used for cosmetic purposes (natural and artificial solar exposure of human skin).

Far-ultraviolet light (normally the readily available low-pressure mercury line at 254 nm) induces pyrimidine photoproducts [mainly dimers of the cyclobutane type and adducts of the (6-4) configuration] as a result of chemical matching between tautomeric bases and the energy of the photons. Many studies have shown good correlations in the far-UV range in the spectral efficiencies for the induction of DNA pyrimidine photoproduct formation and the killing and mutation of cells.[3,4] The use of such action spectra for elucidation of the molecular events that cause genetic effects in cells was recently reviewed,[5] and these correlations constitute strong evidence that pyrimdine photoproducts caused by the far-UV are important lesions in lethality and mutagenesis.

Until recently, it has been assumed that the same or a similar situation holds for wavelengths longer than 320 nm, but at very much reduced spectral efficiencies due to the rapid decline in the absorption of DNA in this wavelength region. Several studies, recently reviewed by Kantor,[6] have shown that both broad-band and monochromatic near-UV and visible light radiations are lethal and cause mutations in a variety of mammalian and bacterial cells. Two general conclusions may be made as a result of these studies [especially those of Smith and Paterson[7] and Wells and Han,[8] who used monochromatic radiation, thus avoiding problems of interpretation inherent in results obtained

* Far-UV, wavelengths shorter than 320 nm; near-UV, wavelengths longer than 320 nm; visible light, wavelengths longer than about 420 nm.

from the use of broad-band sources]. First, DNA is the critical target for mutagenesis, and at least a portion of the lethal events are caused by near-UV radiation. Second, more lethal and mutagenic events occur than can be accounted for by the formation of pyrimidine dimer photoproducts evidence that critical lesions other than these DNA photoproducts are involved, especially at the longer wavelengths. These conclusions are supported by the growing evidence (see below) that near-UV and far-UV exert their biological effects via totally different mechanisms, i.e., photodynamic (involving nonDNA intermediates) and direct (involving DNA alone), respectively. The most important characteristic difference between the responses to near- and far-UV is that the effects of the former are mediated via reactive intermediates, most probably reactive oxygen species generated by nonDNA sensitizing molecules that become excited by the radiation.[9] The events occurring as a result of exposure of cells to mid-UV radiation (radiation at about 310 nm) are mixed far- and near-UV responses, as discussed by Peak et al.[4]

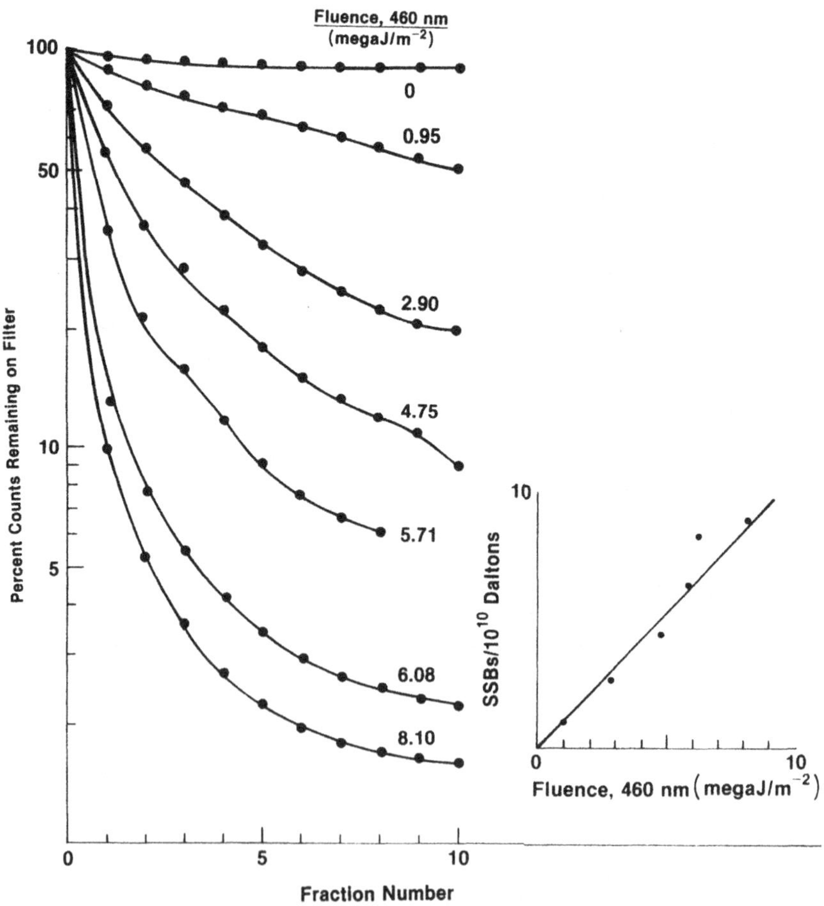

Figure 1. Alkaline elution profiles of DNA and dose response of SSBs (inset) induced in the DNA of human P3 cells[10] irradiated with 460 nm blue light. Preparation and irradiation of the cells and alkaline elution methodology[11,12] were as described by Peak et al.[13] and Peak and Peak.[9] Cells were irradiated at + 0.5°C and lysed immediately after exposure to the radiation for DNA analysis.

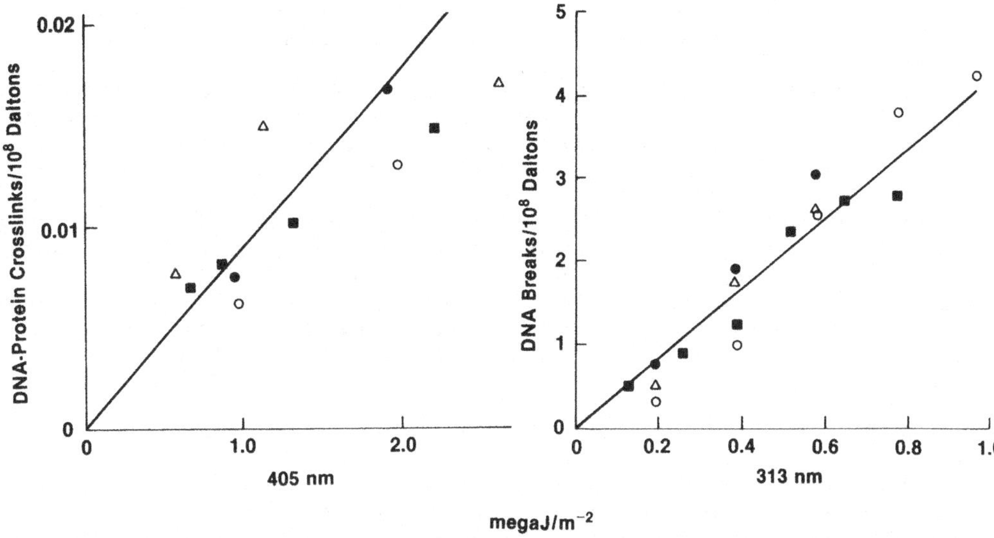

Figure 2. Dose responses for the induction of DNA-to-protein crosslinks and SSBs in the DNA of human P3 cells by 405-nm and 313-nm radiation, respectively. Methods were as described in the legend to Figure 1. Both panels are reproduced with permission from Photochemistry and Photobiology.

We have initiated an investigation into the induction of DNA lesions other than pyrimidine photoproducts in human cells by monochromatic near-UV and visible-light radiations, and have included far-UV radiations for comparison. Several recent studies into the formation of DNA backbone breaks (SSBs*) and DNA-to-protein crosslinks have been published[7,13-19] (Peak and Peak, unpublished), and the formation of DNA-to-protein crosslinks by UV radiation was recently reviewed by Shetlar.[20] Here we summarize the information regarding the mechanisms of induction of these lesions by UV and visible radiations and their possible biological significance.

INITIAL DNA LESIONS CAUSED BY NEAR-UV AND VISIBLE RADIATION: FORMATION AND BREAKAGE OF COVALENT BONDS

DNA backbone breaks and also DNA-to-protein crosslinks are induced in human cells by far-UV, mid-UV, near-UV, and visible radiations[13,19] (Peak and Peak, unpublished) (Figures 1 and 2). For instance, Figure 1 shows alkaline elution profiles of DNA from cells that had been exposed to narrow-band light at 460 nm (blue light) after treatment with proteinase K. The progressively enhanced elution of the DNA from the filters with increased fluence of blue light indicates an increase breakage of the DNA. The inset of Figure 1 is the fluence response curve for SSB formation derived from the alkaline elution profiles determined from the initial rates of elution (elution rate constant, Kohn et al.[12]). The longest wavelength at which we have detected SSBs in these cells is 490 nm (blue light). Also using alkaline-elution techniques, we have assayed DNA-to-protein crosslinks (Figure 2), and the

* For the purpose of this chapter, SSBs refer to both alkali-labile lesions and true frank breaks. Peak and Peak[21] presented evidence that a majority (80%) of the SSBs induced by 356-nm radiation in bacterial-cell DNA are true frank breaks; no determinations are available of the proportion of frank breaks to alkali-labile sites caused by near-UV and visible radiations in mammalian cells.

longest wavelength for which we have obtained positive results is 434 nm. No evidence for either DNA damage was observed with green light at 545 nm.

Repeat determinations of the yields of these DNA changes, as shown in Figure 2, permit accurate determinations of the relative yields per quantum induced by the different wavelengths used in this study, and the action spectra derived from these calculations are shown by Figure 3, superimposed upon the Setlow[3] spectrum of DNA. Both lesions are induced most efficiently in DNA by far-UV radiation. They are also induced by near-UV and blue radiations, but at an efficiency that is reduced by several orders of magnitude. At wavelengths of radiation shorter than about 300 nm, the data points for both DNA damages fall reasonably close to the DNA spectrum, whereas at longer wavelengths, the DNA damage spectra diverge increasingly above the DNA spectrum. This finding is taken as evidence for the participation of nonDNA intermediates. The spectra for the induction of SSBs and DNA-to-protein crosslinks are essentially the same for wavelengths shorter than 334 nm, whereas at longer near-UV and visible wavlengths of radiation, they are completely different. For SSBs, there is a near-UV shoulder extending into the blue region, and a minor secondary peak with a maximum near 460 nm, also qualitatively observed by Rosenstein and Ducore,[22] whereas for DNA-to-protein crosslinks, there is a secondary peak in the near-UV region with a maximum for the yield of this product near 405 nm. Production of crosslinks at this wavelength is some four orders of magnitude less efficient than at 254 nm. It has not yet been determined if there is a minor peak for the induction of DNA-protein crosslinks at 460 nm.

MECHANISMS OF INDUCTION OF DNA DAMAGES CAUSED BY NEAR-UV AND VISIBLE RADIATIONS

Table 1 indicates no differences in the yield of DNA photoproducts when human cells were irradiated by far-UV radiation (290 nm) under anoxic or aerobic conditions or in a deuterated or hydrated water environment--evidence that the formation of these photoproducts does not require oxygen. This finding is consistent with the hypothesis of direct DNA damages caused by this wavelength region, which is also supported by the correlation between the spectra at shorter wavelengths (Figure 3). The same treatments during irradiation by near-UV radiations alter the rates of induction of these damages. Anoxic irradiation reduced the rate of induction of crosslinks by 405 nm radiation by 80%, indicating that oxygen is a required molecule for the induction of most of this damage by this radiation. The enhancement of the induction of both lesions by 365 and 405 nm radiation in the presence of D_2O is evidence that reactive oxygen species, such as singlet oxygen, superoxide anion, and other species are generated in the cell by these radiations and cause the DNA damages. D_2O slows the decay of singlet oxygen[23] and possibly superoxide anion[24] by more than an order of magnitude, thus enhancing the probability of its arrival in the vicinity of DNA before decay. Further investigations of the role of reactive oxygen species induced by near-UV indicate that singlet oxygen may play a role in the inactivation of transforming DNA[25] since diazabicyclo[2.2.2]octane protects the DNA from damage caused by near-UV radiation. Other attempts to identify reactive oxygen species involved in DNA SSB formation caused by near-UV radiations (Peak et al., submitted) indicate that the issue is complex and that several pathways of reactions involving peroxide and hydroxyl radical, as well as singlet oxygen and superoxide radical, may be important. A role for peroxide in visible-light-induced SSBs was also supported by Roza et al.,[26] who showed that catalase inhibits the induction of SSBs in human fibroblasts. The chemical basis for the induction of DNA-to-protein crosslinks by reactive oxygen species generated by near-UV and visible radiations is not known, but the observation that ionizing radiation induces crosslinks between thymine and phenylalanine[27] may be relevant. Much more research is needed in this new area of investigation.

It is also evident that most of the SSBs and crosslinks caused by near-UV radiation require the participation of nonDNA molecules other than oxygen, since neither DNA[28] nor oxygen absorb photons of this energy. Rosenstein et al.[29] and Rosenstein and Ducore[30] postulated that the minor peak for SSB induction at 460 nm may be due to bilirubin photosensitization and demonstrated a large fluence enhancement factor for bilirubin-sensitized SSBs (Table 1). The secondary peak at 405 nm that we demonstrated for crosslinking similarly may be due to porphyrin sensitization, and Blazek and Hariharan[31] showed that added hematoporphyrin derivative enhances the formation of crosslinks in Chinese hamster ovary cells by broad-band visible and near-UV radiation. We have confirmed this enhancement in P3 cells using monochromatic 405-nm radiation and protoporphyrin IX (Table 1). Also, Peak et al.[32] showed that several other naturally occuring metabolites, such as reduced nicotinamide coenzymes and riboflavin, have the potential to damage cells irradiated with near-UV radiation. Cunningham et al.[35] (and Cunningham, Krinsky, Giovanazzi, and Peak, unpublished) demonstrated that irradiated NADH, NADPH, riboflavin, and the nucleosides 2-thiouracil and 4-thiouridine all produce superoxide anion when irradiated by far- and near-UV radiations and also by solar UV radiation. In our hands, superoxide-anion-initiated reactions effectively broke the DNA of Chinese hamster ovary cells but did not cause DNA-to-protein crosslinks (unpublished observations); thus, the mechanism for these two lesions must involve different pathways. It is obvious that DNA damages resulting from exposures of cells to near-UV occur as the complex results of many different possible sensitization pathways.

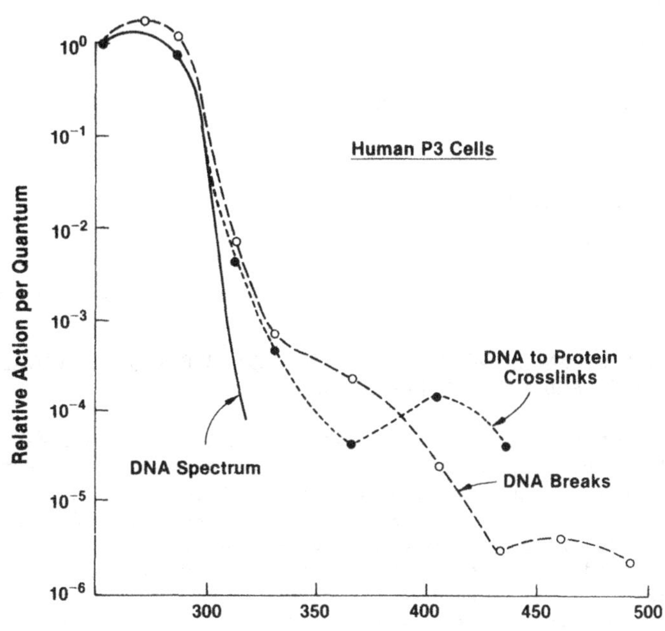

Figure 3. Action spectra for the induction of SSBs (open symbols) and DNA-to-protein crosslinks (closed symbols), compared with the spectrum of DNA, redrawn from Setlow.[3] The doses used were corrected for cellular scattering by the method of Morowitz[33] and the rates of induction of the DNA damages, obtained from replicated data such as shown in Figure 2, were corrected for quantum efficiency before the action spectra were plotted. The spectra are redrawn from Peak et al.[13] and Peak and Peak[9] to superimpose upon each other.

Table 1. Fluence Enhancement Factors (FEF) for Modifiers of Radiation-Induced Crosslink and SSB Formation in Human Cells. The table shows FEFs for cells irradiated in D_2O vs. H_2O; anoxically vs. equilibrated with air; and plus or minus bilirubin (100 μg/mL) or protoporphyrin XI (10^{-5} M).

Treatment	Wavelength	FEF Crosslinks	FEF SSBs
D_2O	290	1.0[b]	1.0[c]
	365	ND	0.5[c]
	405	0.5[b]	0.2[c]
Anoxia	290	1.0[b]	ND
	405	5.2[b]	ND
Bilirubin	460	ND	.04[d]
Protoporphyrin IX	405	0.1[e]	ND
Catalase	sunlamp	ND	0.2[f]

[a]Ratio of the fluence that induces n SSBs or crosslinks in the presence of the modifier to the fluence that induces n SSBs or crosslinks in the absence of the modifier.

[b]Human P3 cells.[19]

[c]Human P3 cells, Peak and Peak, submitted.

[d]Human fibroblasts.[30]

[e]P3 cells, unpublished data.

[f]Human lymphocytes.[34]

BIOLOGICAL SIGNIFICANCES OF THE FORMATION AND BREAKAGE OF COVALENT BONDS IN DNA BY NEAR-UV AND VISIBLE RADIATIONS

Our analyses of spectral responses for DNA damages caused by far-UV, near-UV, and visible light radiations have been valuable in that they enhance our understanding of the chemical mechanisms involved. The near congruence of the spectra for far-UV damages and the spectrum of DNA (Figure 3) is good evidence that the mechanism is the same for the induction of breaks, crosslinks, and pyrimidine dimers.[4] This mechanism involves the direct absorption of photons by DNA as the primary chromophore resulting in transfer of photon energy to the DNA, which results in the formation of several types of that DNA damage. For near-UV, the different spectra imply that at least several nonDNA sensitizer molecules act as primary chromophores but that DNA damage eventually results. With our recent understanding that near-UV and visible radiations produce a variety of chemically potent reactive oxygen species within the cell, we recognize the possibility for many types of DNA damage. If we ignore the presence of mitochondrial DNA (less than 1%) in the cell and assume that SSBs and DNA-to-protein crosslinks are random single events along the genome, it becomes possible to compute the number of events per cell genome per lethal event caused by the different energies used, Table 2. The biological roles of the different DNA damages caused by far-UV radiations were discussed previously.[13] In the near-UV and visible region, many more breaks and crosslinks are formed per lethal event than by far-UV; for instance, at 365 nm, a dead cell has sustained 200 crosslinks and 12,000 SSBs, and at 405 nm, 1,000 and 6,000, respectively. Either the cells can repair many such damages, or else they can tolerate them. About 20 times more SSBs per lethal event are caused by 365-nm radiation than by x-rays, strong evidence that these breaks are effectively repaired. Experimental evidence for the repair of visible-light-induced breaks was described by Bradley et al.,[36] Holmberg et al.,[34] and Roza et al.[26] Therefore, it is likely that SSBs are not a serious event with regard to cellular lethality, a conclusion reached by Ward et al.[37] for the SSBs induced by x-rays. The role of crosslinks and their repair in lethal events is less clear. Fornace and Kohn[38] and Chiu et al.[39] have shown that far-UV- and x-ray-induced crosslinks, respectively, are removed; however the crosslinks produced by light sensitizations are not necessarily the same as those produced by far-UV or x-rays, and the question remains open. The lack of any correlation at all between the action spectra for SSBs, or crosslinks, and lethality and mutagenesis in the same cells (Jones, Huberman, Cunningham, and Peak, unpublished) is evidence that another lesion or lesions are involved in these events. The multitude of chemical events that can be caused in cellular metabolites by the reactive species generated by these long wavelengths of radiation (see, for example, the recent papers by Dizdaroglu and Simic,[40,41,27] Gajewski et al.;[42] Karam et al.[43]) means that death is attributable to the total spectrum of changed chemicals delivered by a lethal dose, only some of which are DNA changes leading to SSBs and crosslinks.

Table 2. Induction of DNA Damages in Human Cells by X-rays, Far-UV, Near-UV, and Visible Light Radiations (reproduced with permission from Photochemistry and Photobiology.

Radiation	F_{37}[a]	DNA lesions/genome/lethal event		
		Dimers[b]	Crosslinks[c,d]	Breaks[d]
254 nm	14	727,000	48	260
290 nm	31		26	600
313 nm	3,650	954,000e	58	460
334 nm	510,000		950	6,000
365 nm	1,710,000	61,020	210	11,800
405 nm	3,590,000		990	5,700
434 nm	> 4,200,000		> 450	> 2,200
x-rays	200f			570f

[a] $J/m2$

[b] Smith and Paterson[7] with human fibroblasts.

[c] DNA-to-protein crosslinks.[9]

[d] Human P3 cells, this study.

[e] 310 nm monochromator setting, mostly 313 nm radiation.

[f] Hill, Chang-Liu, Peak, and Peak (unpublished).

ACKNOWLEDGMENT

This investigation is supported in part by PHS Grants 85456 and 85461 awarded by the National Cancer Institute, DHHS (M.J.P.), and in part by the U.S. Department of Energy under contract No. W-31-109-ENG-38.

REFERENCES

1. Blum, H.F., "Carcinogenesis by Ultraviolet Light," Princeton University Press, Princeton, N.J. (1959).
2. Bruls, W.A.G., Slaper, H., van der Leun, J.C., and Berrens, L., Transmission of human epidermis and stratum corneum as a function of thickness in the ultraviolet and visible wavelengths., Photochem. Photobiol., 40:485 (1984).
3. Setlow, R.B., The wavelengths in sunlight effective in producing skin cancer: A theoretical analysis. Proc. Natl. Acad. Sci. U.S.A., 71:3363 (1974).
4. Peak, M.J., J.G. Peak, M.P. Moehring and R.B. Webb, Ultraviolet action spectra for DNA dimer formation, lethality, and mutagenesis in E.coli with emphasis on the UVB region. Photochem. Photobiol., 40:613 (1984).
5. Peak, M.J. and J.G. Peak, Use of action spectra for identifying molecular targets and mechanisms of action of solar ultraviolet light. Physiol. Plant., 58:367 (1983).
6. Kantor, G.J., Effects of sunlight on mammalian cells. Photochem. Photobiol. 41: 741 (1985).
7. Smith, P.J. and M.C. Paterson, Lethality and the induction and repair of DNA damage in far, mid or near UV irradiated human fibroblasts: Comparison of effects in normal, xeroderma pigmentosum and Bloom's syndrome cells. Photochem. Photobiol., 36:333 (1982).
8. Wells, R.L. and A. Han, Action spectrum for killing and mutation of Chinese hamster ovary cells exposed to mid- and near-ultraviolet monochromatic radiation. Mutation Res., 129:251 (1984).
9. Peak, M.J. and J. G. Peak, Molecular Photobiology of UVA, in: "Monograph on UVA", W. Gange and F. Urbach, eds., Plenum Publishing Corporation, New York (in press).
10. Huberman, E., C.K. McKeown, C.A. Jones, D.R. Hoffman and S. Murao, Induction of mutations by chemical agents at the hypoxanthine guanine phosphoribosyl transferase locus in human epithelial cells. Mutat. Res., 130:127 (1984).
11. Kohn, K.W., C.A. Freidman, R.A.G. Ewig and Z.M. Igbal, DNA chain growth during replication of asynchronous L1210 cells. Alkaline elution of large DNA segments from cells lysed on filters. Biochem., 13:4134 (1974).
12. Kohn, K.W., R.A.G. Ewig, L.G. Erickson and L.A. Zwellig, in: "DNA Repair 1. A Laboratory Manual of Research Procedures," E.C. Freidberg and P.C. Hanawalt, eds., Marcel Dekker, New York, pp. 379-401 (1981).
13. Peak, J.G., M.J. Peak, R.A. Sikorski and C.A. Jones, Induction of DNA to protein crosslinks in human cells by ultraviolet and visible radiations: Action spectrum. Photochem. Photobiol., 41:295 (1984).
14. Elkind, M.M. and A. Han, DNA strand lesions due to "sunlight" and UV light; A comparison of their induction in Chinese hamster and human cells, and their fate in Chinese hamster cells. Photochem. Photobiol., 27:717 (1978).
15. Gantt, R., G.M. Jones, E.V. Stephens, A.E. Baech and K.K. Sanford, Visible light-induced DNA crosslinks in cultured mouse and human cells, Biochim. Biophys. Acta, 565:231 (1979).
16. Bradley, M.D., I.C. Hsu and C. C. Harris, Relationships between sister chromatid exchange and mutagenicity, toxicity and DNA damage. Nature, 289:189 (1979).

17. Anderson, E., Y. Nakashima and W. Konigsberg, Photo-induced crosslinkage of gene-5 protein and bacteriophage fd DNA. Nucleic Acids Res., 2:361 (1976).
18. Han, A., M.J.Peak and J.G. Peak, Induction of DNA-protein cross-linking in Chinese hamster ovary cells by monochromatic 365 and 405 nm ultraviolet light. Photochem. Photobiol., 39:343 (1984).
19. Peak, M.J., J.G. Peak and C.A. Jones, Different (direct and indirect) mechanisms for the induction of DNA-protein crosslinks in human cells by far- and near-ultraviolet radiations (290 and 405 nm), Photochem. Photobiol., (in press).
20. Shetlar, M.D., Cross-linking of protein to nucleic acid by ultraviolet light, in: "Photochem. Photobiol. Reviews," K.C. Smith, ed., Academic Press, New York, Vol. 5, 107-197 (1970).
21. Peak, M.J. and J.G. Peak, Single-strand breaks induced in Bacillus subtilis DNA by ultraviolet light: Action spectrum and properties. Photochem. Photobiol., 35:675 (1982).
22. Rosenstein, B.S. and J.M. Ducore, Induction of DNA strand breaks in normal human fibroblasts exposed to monochromatic ultraviolet and visible wavelengths in the 260-546 nm range. Photochem Photobiol., 38:51 (1983).
23. Merkel, P.B., R. Nilsson and D.R. Kearns, Deuterium effects on singlet oxygen lifetimes in solutions. A new test of singlet oxygen reactions. J. Am. Chem. Soc., 94:1030 (1972).
24. Bielski, B.H.J. and E. Saito, Deuterium isotope effect on the decay kinetics of perhydroxyl radical. J. Phys. Chem., 75:2263 (1971).
25. J.G. Peak, C.S. Foote and M.J. Peak, Protection by DABCO against inactivation of transforming DNA by near-ultraviolet light: Action spectrum and implications for involvement of singlet oxygen. Photochem. Photobiol. 34:45 (1981).
26. Roza, L., G.P. van der Schans and P.H.M. Lohman, The induction and repair of DNA damage and its influence on cell death in primary human fibroblasts exposed to UV-A or UV-C radiation. Mutat. Res., 146:89 (1985).
27. Dizdaruglu, M. and M.G. Simic, Radiation-induced crosslinks between thymine and phenylalanine. Int. J. Radiat. Biol., 47:63 (1985).
28. Sutherland, J.C. and K.P. Griffin, Absorption spectrum of DNA for wavelengths longer than 320 nm. Radiat. Res., 86:399 (1981).
29. Rosenstein, B.S., J.M. Ducore and S.W. Cummings, The mechanism of bilirubin photosensitized DNA strand breakage in human cells exposed to phototherapy light. Mutat. Res., 112:397 (1983).
30. Rosenstein, B.S. and J.M. Ducore, Enhancement by bilirubin of DNA damage induced in human cells exposed to phototherapy light. Pediat. Res., 18:3 (1984).
31. Blazek, E.R. and P.V. Hariharan, Alkaline elution studies of hematoporphyrin-derivative photosensitized DNA damage and repair in Chinese hamster ovary cells. Photochem. Photobiol., 40:5 (1984).
32. Peak, J.G., M.J. Peak and M. MacCoss, DNA breakage caused by 334-nm ultraviolet light is enhanced by naturally occurring nucleic acid components and nucleotide coenzymes. Photochem. Photobiol., 39:713 (1984).
33. Morowitz, H.J., Absorption effects in volume irradiation of micro organisms. Science, 111:229 (1950).
34. Holmberg, M., Zs. Almassy, M. Langerberg and B. Niejahr, The repair of strand breaks in human lymphocytes exposed to near UV-radiation (UVA) and far UV-radiation (UVC). Photochem. Photobiol., 41:437 (1985).
35. Cunningham, M.J., J.S. Johnson, S.M. Giovanazzi and M.J. Peak, Photosensitized production of superoxide anion by monochromatic (290-405 nm) ultraviolet irradiation of NADH and NADPH coenzymes. Photochem. Photobiol., (in press).

36. Bradley, M.O., L.C. Erickson and K.W. Kohn, Non-enzymatic strand breaks induced in mammalian cells by fluorescent light. Biochim. Biophys. Acta, 520:11 (1978).
37. Ward, J.F., W.F. Blakely and E.I. Jones, Mammalian cells are not killed by DNA single strand breaks caused by hydroxyl radicals from hydrogen peroxide. Radiat. Res., (in press).
38. Fornace, A.J. and K.W. Kohn, DNA-protein crosslinking by ultraviolet radiation in normal human and xeroderma pigmentosum fibroblasts. Biochim. Biophys. Acta, 435:95 (1976).
39. Chiu, S-M., N.M. Sokany, L.R. Friedman and N.L. Oleinick, Differential processing of ultraviolet or ionizing radiation-induced DNA-protein cross-links in Chinese hamster cells. Int. J. Radiat. Biol., 46:681 (1984).
40. Dizdaroglu, M. and M.G. Simic, Radiation-induced crosslinks of cytosine. Radiat. Res., 100:41 (1984).
41. Dizdaroglu, M. and M.G. Simic, Radiation-induced formation of thymine crosslinks. Int. J. Radiat. Biol., 46:241 (1984).
42. Gajewski, E., M. Dizdaroglu, H.C. Krutzsch and M.G. Simic, OH radical-induced crosslinks of methionine peptides. Int. J. Radiat. Biol., 46:47 (1984).
43. Karam, L.R., M. Dizdaroglu, and M.G. Simic, OH radical-induced products of tyrosine peptides. Int. J. Radiat. Biol., 46:715 (1984).

SEQUENCE SPECIFICITY AND BIOLOGICAL CONSEQUENCES OF DRUGS THAT BIND COVALENTLY IN THE MINOR GROOVE OF DNA

Laurence H. Hurley and Donald R. Needham-VanDevanter

Drug Dynamics Institute, College of Pharmacy
University of Texas at Austin
Austin, TX 78712

INTRODUCTION

A variety of synthetic and naturally occurring DNA ligands have been shown to bind partially or totally within the minor groove of B-form DNA. The distamycins, netropsin, stilbamidine, and SN18071 all associate non-covalently with the minor groove[1,2] while the intercalators adriamycin and actinomycin have substituents which reside in the minor groove and stabilize their respective complexes. The pyrrolo(1,4)benzodiazepine [P(1,4)B] anti-tumor antibiotics, anthramycin, tomaymycin, and sibiromycin (see Figure 1) bind covalently in the minor groove to N^2 of guanine.[3,4] CC-1065, another antitumor antibiotic, binds covalently to N3 of adenine in the minor groove of DNA.[5,6]

DNA ligands which bind within the minor groove of DNA exhibit varying degrees of sequence selectivity. Factors which contribute to nucleotide sequence recognition by minor groove ligands have been extensively investigated.[2,7] Electrostatic interactions, ligand and DNA dehydration energies, hydrophobic interactions and steric factors all play significant roles in sequence selectivity in the minor groove. Interestingly, ligand recognition of nucleotide sequence in the minor groove does not involve significant hydrogen bonding. This is in sharp contrast to cellular enzyme and protein recognition of nucleotide sequence, which is achieved in the major groove via specific hydrogen bond formatlon between individual bases and the ligand. The ability to "read" nucleotide sequence via hydrogen bonding allows precise binding of proteins to specific DNA sequences. Minor groove ligands examined to date exhibit a much lower sequence specifity, generally binding to a subset of possible sequences, rather than a single sequence.

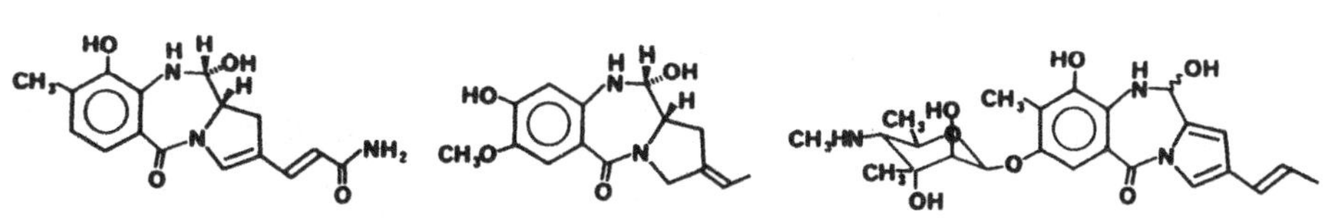

Anthramycin Tomaymycin Sibiromyin

Figure 1. Structures of representatives of the P(1,4)B antibiotic group.

COVALENT BINDING OF CC-1065 AND THE P(1,4)Bs TO DNA

CC-1065 and the P(1,4)Bs will only form covalent adducts with double-stranded DNA and appear to be specifc for B form DNA. While CC-1065 has an absolute requirement for the AT base pairs (bps) the P(1,4)Bs require GC bps. Upon acidic or thermal denaturation of the P(1,4)B-DNA adducts, the drugs are released in their active form from DNA to leave an unmodified template.[8] In contrast, when the CC-1065-DNA adduct is heat denatured a modified CC-1065 molecule is released, leaving a single strand break and an apurinic acid site on DNA.[6] Consequently, two different strategies were required to elucidate the structures of the CC-1065 and P(1,4)B-DNA adducts.

Structure of the P(1,4)B-DNA Adduct

Due to the labile nature of the covalent linkage between the P(1,4)Bs and DNA, our initially proposed structure of the DNA adduct was based upon indirect evidence.[3,4] This structure was subsequently confirmed by ^{1}H and ^{13}C-NMR studies on a self-complementary 6mer[d(ATGCAT)$_2$] and calf thymus DNA adducts, respectively. Based upon ^{1}H-NMR studies and steric considerations we believe the imine (see Figure 2) is the ultimate DNA reactive species.[10]

Structure of the CC-1065-DNA Adduct

Thermal treatment of CC-1065-DNA adducts leads to release of a butanol soluble product which has a similar chromophore to CC-1065. Subsequent purification of this by reverse phase HPLC gave sufficient material for ^{1}H and ^{13}C-NMR, which identified the compound structure as the CC-1065-N3-adenine adduct (see Figure 3). Since the absolute configuration of CC-1065 was not revealed by X-ray crystallography, the configuration of C-4a in the adduct initially could not be assigned (see later).

STEREOCHEMISTRY AT THE COVALENT LINKAGE SITE AND ORIENTATION OF P(1,4)Bs AND CC-1065 IN THE MINOR GROOVE OF DNA

CC-1065, and to lesser extent the P(1,4)Bs, possess a pronounced right handed twist along the long axis of the drug molecules. In CC-1065 this results from out of plane amide bonds between subunits A and B and B and C, while in the P(1,4)B the stereochemistry at the C-11a position (S) provides the drug molecules with a right handed twist. In some cases (CC-1065 and tomaymycin), modeling studies showed that it was possible to accommodate the drug molecules in either orientation in the minor groove of DNA, while still being able to form the required covalent linkage to DNA. For CC-1065, the stereochemistry at 4a (see Figure 3) is reversed for each orientation of the

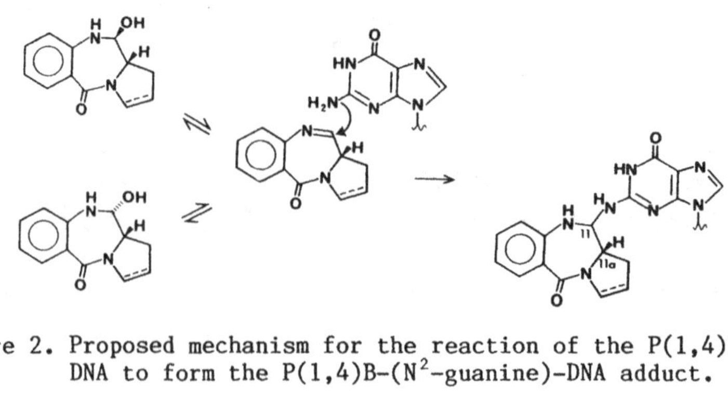

Figure 2. Proposed mechanism for the reaction of the P(1,4)Bs with DNA to form the P(1,4)B-(N^2-guanine)-DNA adduct.

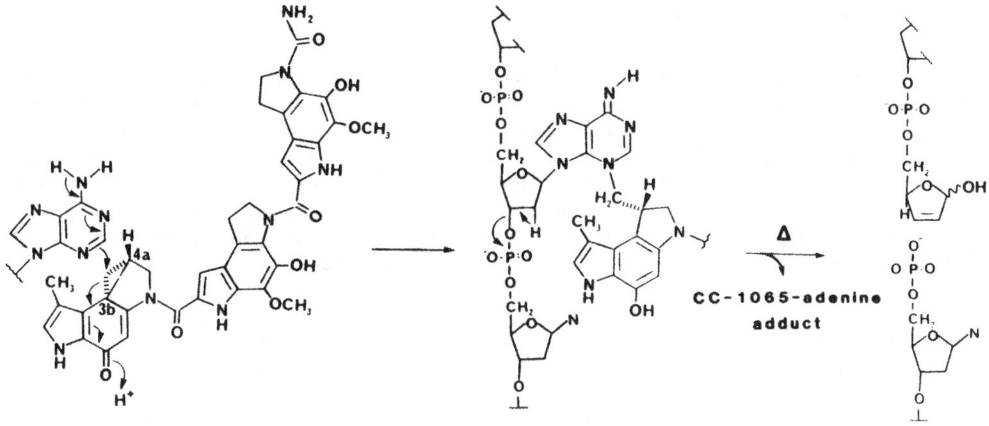

Figure 3. Reaction of CC-1065 with DNA to form the CC-1065-(N3-adenine)-DNA adduct and the consequences of thermal treatment of CC-1065-DNA adducts.

drug in the minor groove, while for tomaymycin the stereochemistry at C-11 (see Figure 2) has to be reversed. Since the absolute configuration of the cyclopropane ring of CC-1065 was not revealed by X ray crystallography and the imine of the P(1,4)B (see Figure 2) is believed to be the DNA reactive species, it was not possible at this stage to predict the orientation of these drugs in the minor groove of DNA.

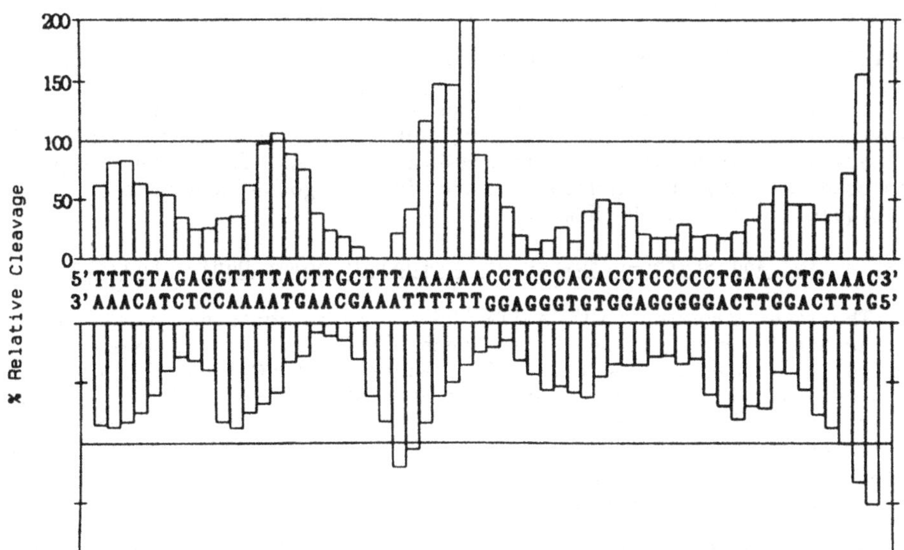

Figure 4. Histograms of MPE•Fe(II) footprints of anthramycin on the opposite strands of a 242 bp fragment isolated from SV40 DNA.[14] The fragment was either 5' or 3' ^{32}P single-end labelled and preincubated with anthramycin (50 μM) for 4 hrs at 37°C.

DNA SEQUENCE SPECIFICITY OF THE P(1,4)Bs AND CC-1065

Minor groove ligand DNA binding sites can be determined by combining Maxam-Gilbert DNA sequencing of ^{32}P-end labelled restriction enzyme fragments with degradation of identical restriction fragments in the presence or absence of the ligand.[11] The footprinting method uses nonspecific DNA degradation by DNase I or MPE•Fe(II), which leaves footprints at drug binding sites upon electrophoresis and autoradiography.[12] Alternatively, the binding sites of ligands which can be induced to generate their own DNA strand breaks after binding can be identified directly by electrophoresis and autoradiography.[6,13] The footprinting technology using MPE•Fe(II) has been used to determine the sequence preference for the P(1,4)Bs,[14] while the direct method in which thermal treatment of CC-1065-DNA adducts leads to single strand breaks in DNA has been used to determine the sequence specificity of CC-1065.[5,6]

The P(1,4)Bs have a 2-3 bp sequence specificity which includes the covalently modified guanine residue. The most preferred binding sequence for the P(1,4)Bs is 5'PuGPu, while 5'PyGPy is the least preferred.[14] Footprinting analysis by MPE•Fe(II) reveals a minimum of a 3-4 bp footprint size for the P(1,4)B with the expected offset (2-3 bp) on opposite strands[15] which is illustrated for anthramycin (see Figure 4). There is an extremely large enhancement of MPE•Fe(II) cleavage between drug binding sites in AT rich regions probably resulting from a drug induced conformation change in DNA.[14]

A thermally induced β-elimination on the 3' side of the covalently modified adenine (see Figure 3) gives rise to a single strand break which has been used to locate CC-1065 binding sites on DNA.[6] Analysis of over 1000 bases of SV40 and T7 DNA to which CC-1065 was bound reveals that CC-1065 binds with greatest affinity to 5'PuNTTA and 5'AAAAA sequences.[5] Furthermore, since the DNA sequence specificity lies exclusively to the 5' side of the covalently modified adenine, this establishes the orientation of the drug molecule in the minor groove and consequently the stereochemistry at C-4a of C-1065 must be S.

CONSTRUCTION OF DEFINED OLIGOMER ADDUCTS OF CC-1065 AND THE P(1,4)Bs

The sequence specificity experiments described in the previous section allowed us to predict short (5-14 bps) defined DNA sequences to which the P(1,4)Bs and CC-1065 will bind in unambiguous ways. These duplexes have been used to study the structure of the drug-oligoduplex adducts by sequencing techniques and 1D and 2D-^{1}H-NMR.

Construction And Stereo-Drawings of CC-1065-Oligomer Duplex Adducts

The 14mer sequence (5'CGGAGTTAGGGGCG), which contains the CC-1065 adenine binding site within the sequence 5'TTA, and its complement were chemically synthesized and the duplex reacted with CC-1065 to give an adduct which maintained the B form.[16] Drug mediated strand breakage was used to confirm the CC-1065 binding site (see Figure 5). Recently an 8mer self complementary duplex has been shown by ^{1}H-NMR to bind CC-1065 in the orientation predicted by sequence specificity studies.[17] Stereo drawings of the CC-1065-DNA adduct are shown in Figure 6.

Construction and Stereo-Drawings of the Anthramycin-Oligomer Duplex Adduct

Although the P(1,4)Bs have a sequence preference for 5'PuGPu,[4] anthramycin has been demonstrated to form a defined adduct with the self-complementary 6mer 5'd(ATGCAT)$_2$,[9,18] Contrary to the case where tomaymycin may be able to bind in two orientations on DNA,[10] CPK modelinq studies predict anthramycin binding in only one orientation on DNA,[4] in which the stereochemistry at the

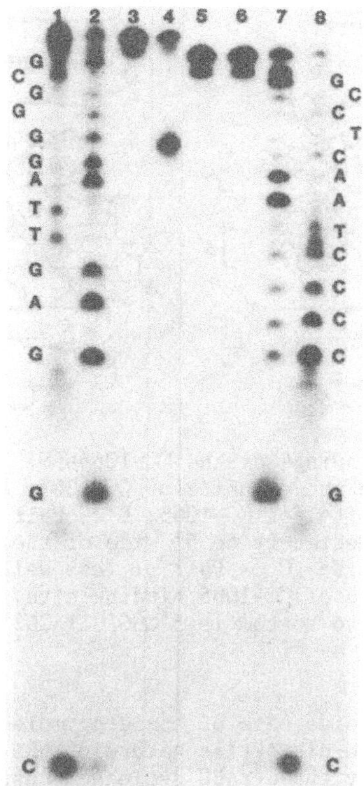

Figure 5. Autoradiogram of the 14 base pair synthetic DNA duplex 5'-CGGAGTTAGGGGCG showing CC-1065 binding and thermal
GCCTCAATCCCCGC-5'
strand scission at a unique site.[16] In lanes 1-4 the duplex has a unique 5'-(^{32}P) end-label on the covalently modified (upper) strand. In lanes 5-8, the end-label is on the complementary (lower) strand. Lanes 1,8: pyrimidine specific cleavage. Lanes 2,7: purine specific cleavage. Lanes 3,6: unmodified duplex heated to 90°C for 30 mins. Lanes 4,5: duplex modified with CC-1065, followed by 30 min heating at 90°C.

linkage site is 11R and the aromatic ring of anthramycin lies to the 3' side of the covalently modified guanine. This has been confirmed by ^{1}H-NMR studies using 1D-difference NOE measurements.[18] Computer graphics modeling studies have been used to prepare stereo-drawings of the anthramycin 5'd-$(ATGCAT)_2$ (see Figure 7).

BIOLOGICAL CONSEQUENCES OF A NON-DISTORTIVE COVALENT MODIFICATION IN THE MINOR GROOVE OF DNA

The P(1,4)Bs and CC-1065 are amongst the most potent antitumor agents to be discovered. Based upon our present knowledge of the structures of these DNA adducts, it would seem reasonable to predict that the repair of these lesions might represent a difficult problem for cells. Both classes of antibiotics form covalent adducts with DNA that are non-distortive and helix stabilizing, and the attachment site is between the floor of the minor

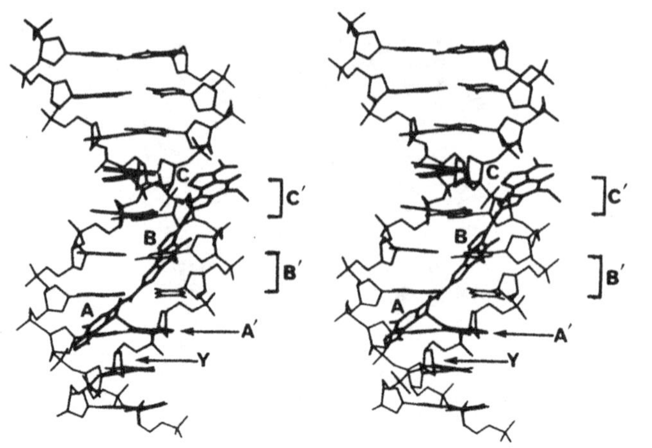

Figure 6. Stereo-drawings of the CC-1065-DNA adduct. Key: A, B, and C are the subunits of CC-1065; A'- Adenine covalent binding site for CC-1065; B' - Pair of highly conserved bases immediately to 5' side of the adenine covalent binding site; C' - Pair of less well conserved bases at the 5' end of CC-1065 binding site. The base sequence from top to bottom is 5'CGGAGTTAGG3'.

groove of DNA and the inside edge of the drug molecules. Somewhat surprisingly, in view of the non-distortive nature of the adduct, anthramycin lesions are recognized by the uvrABC nuclease (unpublished results) and repair proficient human fibroblast cells remove anthramycin lesions more efficiently than XP-cell lines in complemention groups A and D.[19] Significantly, excision dependent single and double strand breaks are produced in human fibroblast cell lines. Whether the uvrABC nuclease complex is able to discern faithfully which strand of DNA is covalently modified is an interesting but as yet unanswered question. Little is known about the repair of CC-1065-DNA lesions except that in CC-1065 treated cells (normal and XP) NAD levels fall, most likely as a consequence of poly(ADP)ribosylation, and do not recover even after a period of 24 hrs (unpublished results).

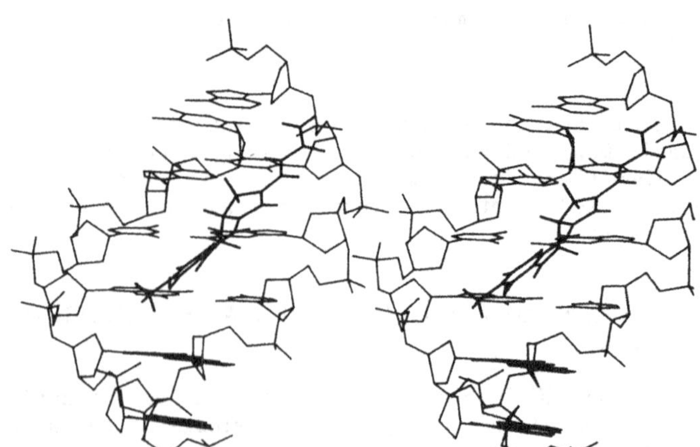

Figure 7. Stereo-drawings of the anthramycin-5'd(ATGCAT)$_2$.

ACKNOWLEDGEMENTS

This work was supported by grants from the Department of Health Education and Welfare (CA-30349; CA-31232; CA-35318) and the Welch Foundation to LHH. Some of this work was the result of collaborative research programs with colleagues at the Upjohn (CC-1065) and Smith, Kline and French Companies [P(1,4)Bs], and the authors gratefully acknowledge their contributions.

REFERENCES

1. C. Zimmer, Molecular Aspects of the Specific Interaction of Nonintercalatinq Oligopeptide Antibiotics with DNA, Comments Mol. Cell. Biophys., 1:399 (1983).
2. B. Pullman, Electrostatics and Specificity in Nucleic Acid Reactions, in: Specificity in Biological Interactions, C. Chagas and B. Pullman, eds., The Vatican Press. (1984).
3. L.H. Hurley and R.L. Petrusek. Proposed Structure of the Anthramycin DNA Adduct, Nature, 282:529 (1980).
4. R.L. Petrusek, G.L. Anderson, T.F. Garner, Q.L. Fannin, D.J. Kaplan. S.G. Zimmer, and L.H. Hurley, Pyrrolo(1,4)Benzodiazepine Antiobitics, Proposed Structures and Characteristics of the in vitro DNA Adducts of Anthramycin, Tomaymycin, Sibiromycin and Neothramycins A and B, Biochemistry, 20:1111 (1981).
5. L.H. Hurley, V.L. Reynolds, D.H. Swenson, T. Scahill, Reaction of the Antitumor Antibiotic CC-1065 with DNA: Structure of a DNA Adduct with DNA Sequence Specificity, Science, 226:843 (1984).
6. V.L. Reynolds, I.J. Molineux, D.J. Kaplan, D.H. Swenson and L.H. Hurley, DNA Sequence Specificity of the Antitumor Antibiotic CC-1065, Biochemistry, in press (1985).
7. K. Zakrzewska, R. Lavery, and B. Pullman, The Solvation Contribution to the Binding Energy of DNA with Non-Intercalating Antibiotics, Nucleic Acids Research, 12:6559 (1984).
8. L.H. Hurley, C. Gairola and M.J. Zmijewski, Studies on the In Vitro Reactivity of the Pyrrolo(1,4)Benzodiazepine Antibiotics Towards DNA, Using Specifically Radiolabelled Molecules, Biochem. Biophys. Acta, 475:5215 (1977).
9. D.E. Graves, C. Pattaroni, C. Balakrishnan, J.M. Ostrander. L.H. Hurley and T.R. Krugh, The Reaction of Anthramycin with DNA: Proton and Carbon Nuclear Magnetic Resonance Studies on the Structure of the Anthramycin-DNA Adduct, J. Biol. Chem. 259:8202 (1984).
10. M. Barkley, S. Cheatham, D.E. Thurston and L.H. Hurley, Pyrrolo(1,4)-Benzodiazepine Antitumor Antibiotics, Evidence for Two Forms of Tomaymycin Bound to DNA, Submitted for publication.
11. J.C. Dabrowiak, Sequence Specificity of Drug DNA Interactions, Life Sciences, 32:2915 (1983).
12. M.W. Van Dyke and P.B. Dervan, Methidium-EDTA-FE(II) and DNase I Footprinting Report Different Small Molecule Binding Site Sizes on DNA, Nucleic Acids Res. 11:5555 (1983).
13. J.S. Taylor, P.G. Schultz and P.B. Dervan, DNA Affinity Cleaving, Sequence Specific Cleavage of DNA by Distamycin-EDTA-Fe(II) and EDTA-Distamycin-Fe(II), Tetrahedron, 40:457 (1984).
14. R.P. Hertzberg, S.M. Hecht, V.L. Reynolds, I.J. Molineux and L.H. Hurley, DNA Sequence Specificity of the Pyrrolo(1,4)Benzodiazepine Antitumor Antibiotics. MPE-Fe(II) Footprinting Analysis of DNA Binding Sites for Anthramycin and Related Drugs, Submitted for publication.
15. M.W. Van Dyke and P.B. Dervan, Footprinting with MPE-Fe(II). Complimentary-Strand Analysis of Distamycin and Actinomycin-Binding Sites on Heterogeneous DNA, Cold Spring Harbor Symposium on Quantitative Biology, 47:347 (1983).

16. D.R. Needham-VanDevanter, L.H. Hurley, V.L. Reynolds, N.Y. Theriault, W.C. Krueger and W. Wierenga, Characterization of an Adduct Between CC-1065 and a Defined Oligodeoxynucleotide Duplex, Nucleic Acids Research, 12:6159 (1984).
17. T.A. Scahill, unpublished results.
18. D.E. Graves, M.P. Stone and T.R. Krugh, Structure of the Anthramycin-d(ATGCAT), Adduct from One and Two Dimensional Proton NMR Experimentation in Solution, Biochemistry, in press (1985).
19. R.L. Petrusek, E.L. Uhlenhopp, N. Duteau and L.H. Hurley, Reaction of Anthramycin with DNA. Biological Consequences of DNA Damage in Normal and Xeroderma Pigmentosum Cell Lines, J. Biol. Chem. 257:6207 (1982).

MOLECULAR RECOGNITION OF DNA BINDING AGENTS: HIGH-FIELD ^1H AND ^{31}P ONE- AND TWO-DIMENSIONAL NMR STUDIES ON THE 1:1 INTERCALATION COMPLEXES OF MITOXANTRONE WITH SELECTED OLIGODEOXYRIBONUCLEOTIDES

[1]Albin Otter, [1]Christopher C. Hanstock, [1]George Kotovych, [2]Bernard Rayner, [2]Jacques J. Vasseur, [2]Jean-Louis Imbach, and [1]J. William Lown*

[1]Department of Chemistry, University of Alberta, Edmonton, Alberta, T6G 2G2, Canada; [2]Laboratoire de Chimie Bio-organique, Universite des Sciences et Techniques u Languedoc, Montpellier 34060, France

INTRODUCTION

The mechanisms whereby peptides, control proteins and xenobiotics recognize nucleic acids are fundamental to many important processes in biology and chemotherapy and appear to underlie their characteristic properties in vivo.[1-4] This applies to the recognition between enzymes or regulatory proteins and complementary binding sites,[2-3] to DNA-histone interactions,[3,4] and in many cases of current clinical interest in anticancer and antiviral chemotherapy.[5-7] The binding of such antineoplastic agents as doxorubicin, actinomycin, cis-platinum, mitomycin C, and echinomycin to cell target DNA invokes precise and highly specific interactions in molecular recognition.[5,6] There is increasing knowledge from molecular biology of gene expression and control, DNA structure and topology and the recognition of sequences of unusual susceptibility to xenobiotics.[8] This new information increases the prospects for rational anticancer drug design provided one can determine the structural factors controlling molecular recognition and binding.

The techniques that have been applied to drug complexation in the past include electron microscopy and X-ray diffraction[9] for studying solid state behavior. NMR is potentially one of the more valuable techniques to provide detailed information on the structure of DNA complexes in solution. Until recently however progress has been slow owing to the complexity of ^1H-NMR spectra of oligodeoxyribonucleotides.[10] However the advent of cryospectrometers and the development of 2D-techniques[11] permitted applications including COSY,[12] SECSY,[13] HH-INADEQUATE,[17,18] J-resolved,[14,15] and NOESY[13,16] which have been valuable in more detailed analysis of oligonucleoides.

The analysis of the ^1H-NMR spectra of drug:DNA complexes poses additional problems both in the complexity of the resulting spectra and, more particularly, in the line broadening resulting from equilibrium binding. Further progress thus depends both on the careful selection of significant agents which bind sufficiently strongly to give well resolved spectra and on the development of new NMR techniques to permit reliable assignments.

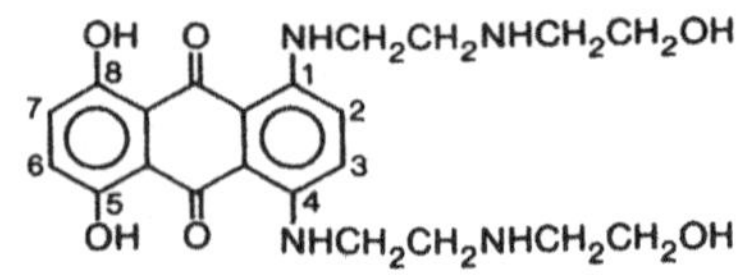

Figure 1. Structure and numbering of mitoxantrone.

Application of One- and Two-Dimensional H-NMR Methods to the Analysis of Drug: Oligonucleotide Complexes Illustrated by the Example of Mitoxantrone

The synthetic agent mitoxantrone (Fig. 1) and its congeners show considerable promise as anticancer agents[19, 20] and as clinical alternatives to the apparently more cardiotoxic doxorubicin.[21, 22] Biochemical evidence suggests that, in common with the anthracyclines,[5, 6, 8] nucleic acids are among the principal cell targets of these new agents and that they cause inter alia marked changes in chromatin structure.[23, 25] While there are strong indications that the interaction of mitoxantrone with cellular DNA contributes significantly to its cytotoxic action, the exact nature of the DNA interactions is, at present, unclear. Therefore we report high-field ^{1}H-NMR analysis of the stoichiometric 1:1 complex of mitoxantrone with $[d(CpGpCpG)]_2$ and the development of ^{31}P-NMR techniques for the assignment of the oligomers $[d(ApGpCpT)]_2$ and $[d(ApCpGpT)]_2$ and their 1:1 complexes with this agent.

Complete and unambiguous ^{1}H-NMR assignments for the duplex oligomer $[d(CpGpCpG)]_2$ were achieved, employing 2D-COSY,[12] 1D-NOE differences and the 2D-double quantum technique 2D-HH-INADEQUATE.[17, 18] Measurement of the resonances in the 1:1 complex was completed by observation of the spectra after cumulative additions of the drug at 1:5, 1:2 and 1:1 drug to DNA ratios. Upfield shifts for C(1)5 (0.22 ppm) and C(3)5 (0.19 ppm) and line broadening of G(2)8 and C(3)5 suggest that they lie close to the binding

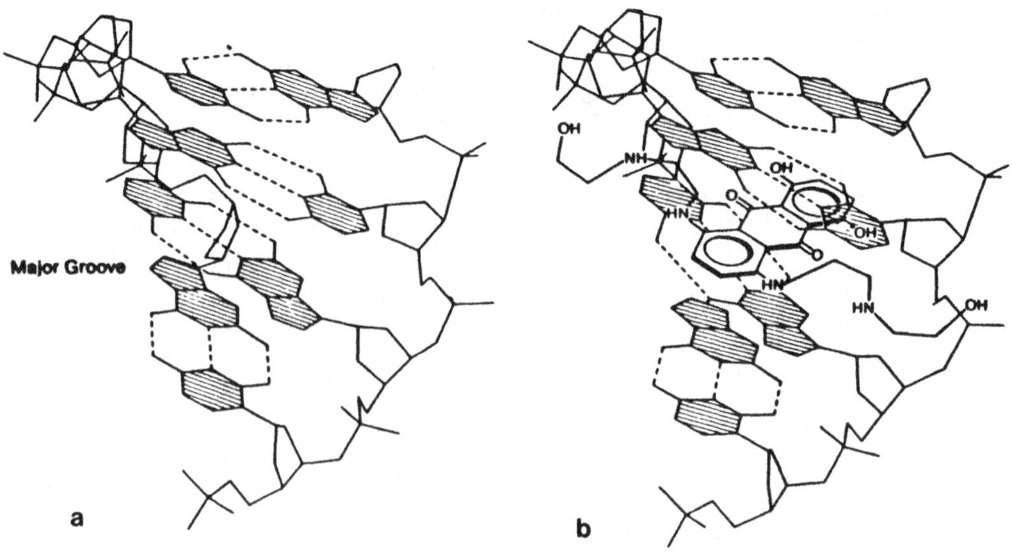

Figure 2. (a) Computer generated depiction of the carbon skeleton of $[d(CpGpCpG)]_2$.
(b) 1:1 mitoxantrone complex with $[d(CpGpCpG)]_2$ allowing a 17° unwinding angle[31] at the intercalation site and showing binding via the major groove.

site. This indicates that since the former protons lie closer to the center of a base pair than C6 or C8 that they are experiencing to a greater extent a diamagnetic shift due to the ring current of the intercalated chromophore of mitoxantrone. The ^{13}C-NMR signals of the 1:1 complex show no doubling of individual signals compared with the tetramer alone, in accord with formation of a complex which retains elements of symmetry.

The results of the sugar proton assignments are also in accord with an intercalated chromophore,[27] rather than one externally bound on the duplex,[28,29] since the 3', 5'1 and 5'2 protons form links to the sugar phosphate backbone on the outside edges of the miniature double helix. The aromatic protons of the chromophore and the methylene group directly attached to the aromatic system of the drug show the largest shifts to higher fields on formation of the complex. This shows that the major influence on these groups is due to the ring currents of the DNA bases. This influence diminishes along the side-chain to the terminal OH group. Saturation of the aromatic protons of the drug in the complex shows no NOE's for the 6,7 protons with the base protons. This result is in accord with an alignment of the chromophore as depicted in Fig. 2b [similar to the intercalation complex of doxorubicin[30]] with protrusion of the periphery of the chromophore beyond the area of the adjacent base pairs. The terminal $-CH_2OH$ groups show weak NOE's with the C(1)6 and C(3)6 resonances. In addition the CH_2NHCH_2 methylenes show weak NOE's with the G(4)8 proton. All three of these positions lie in the major groove.[3,4] Therefore the data suggest that the central NH group and the terminal-CH_2OH groups of the side chains bind to the neighboring bases from the intercalation site and in the major groove and also to the central phosphate groups on the backbone (see Fig. 2).

The binding constant for mitoxantrone with calf thymus DNA in 0.1M NaCl, pH 7.0, is ca. 6×10^6 M^{-1} and the rate constant for SDS driven dissociation of the drug from this complex under the same conditions and 20°C is 1.3 sec^{-1}, so that these data indicating firm binding are in accord with the properties of the complex. The unwinding angle of mitoxantrone, determined independently by viscosity measurements and by a topoisomerase assay, is 17°.[31] Therefore this value was allowed for in arriving at the structure for the intercalation complex shown in Fig. 2b.

Problems Posed in the Assignment of ^1H-NMR of Drug: Oligonucleotide Complexes and the Application of ^{31}P-^1H Chemical Shift Correlation Spectroscopy

A more challenging problem is to deduce the structure of complexes of drugs such as mitoxantrone to longer oligonucleotide sequences. As we have seen, owing to the severe overlap of many of the proton resonances, the application of commonly used decoupling experiments for structure elucidation is of limited utility.

Many of these problems can be overcome by the combined application of 2D-NMR techniques such as COSY,[12] SECSY,[13] NOESY [13,16] and HH-INADEQUATE.[17,18] While the latter is useful in assigning the resonances within one nucleotide,[32] we now wished to examine the utility of ^{31}P-^1H-shift correlation spectroscopy[33-36] for the unambiguous assignment of the nucleotides relative to each other by using the ^{31}P-^1H couplings in the sugar phosphate backbone of the oligodeoxyribonucleotides.

The assignment of the ^{31}P-resonances becomes quite straight forward and provides the investigator with another parameter for the study of the effect of drug intercalation. This is of particular importance in cases where other techniques (e.g. NOE) fail due to extensive line broadening upon binding of the drug. Our present studies on $[d(ApCpGpT)]_2$[32] and $[d(ApGpCpT)]_2$[32] and their 1:1 complexes with mitoxantrone show the application of one and two-dimensional ^{31}P techniques in this field of research.

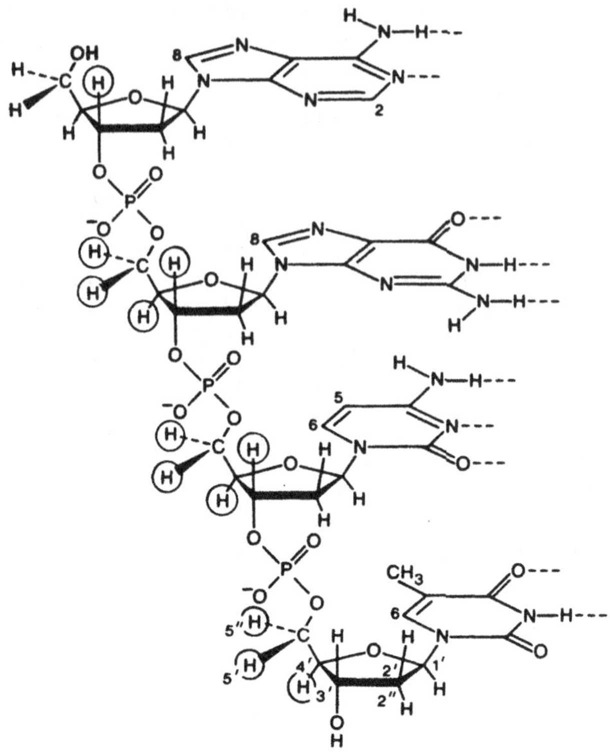

Figure 3. One strand of the duplex $[d(ApGpCpT)]_2$ showing numbering; the protons encircled show cross peaks with the phosphorus nuclei.

Only the protons bound by scalar couplings to the phosphate groups give rise to cross-peaks in the 2D- $^{31}P-^{1}H$ spectrum. Therefore for oligodeoxyribonucleotides the 3' and 5' protons (^{3}J couplings) and the 4' protons (^{4}J couplings) are anticipated to produce cross-peaks in the experiment.

The applied pulse sequence eliminates $^{31}P-^{1}H$ couplings but homonuclear $^{1}H-^{1}H$ couplings are retained as can be seen in the ^{1}H projection of Figure 4.

The three well separated ^{31}P signals show cross-peaks with the 3', 4' and 5' protons of the directly bonded sugar moieties. Careful analysis of the cross sections through each of the ^{31}P resonances (Fig. 5) allows unambiguous assignments of all the directly bound sugar protons.

The (3'p5') linkage for each of the phosphorus atoms can be established, and together with the previously published[32] results obtained by HH-INADEQUATE techniques,[17,18] thus gives the sequence of nucleotides as well as the assignment of the three ^{31}P resonances. The problem of severe overlap of signals, which is encountered in one-dimensional H-spectra is completely removed in the cross sections.

The $^{31}P-^{1}H$ shift correlation experiments of two other tetramers - $[d(ApCpGpT)]_2$ and $[d(CpGpCpG)]_2$ were recorded and analyzed in the same way and the results are summarized in Table 1. The narrow absorption range of the ^{31}P signals is noteworthy, considering the wide range of ^{31}P resonances (>500 ppm).

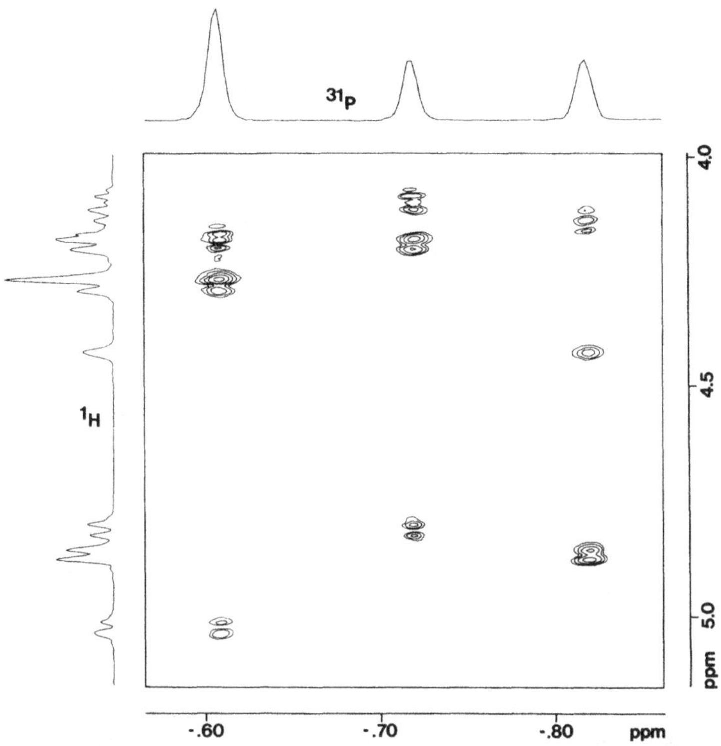

Figure 4. 2D-^{31}P-^{1}H-shift correlation contour plot of $[d(ApGpCpT)]_2$ showing protections on both axes. A 160 x 256 data set was collected (96 scans for each t_1-value) with a total acquisition time of about 15 hours at a concentration of 20 mM (solvent: D_2O buffered at pH 7.0 containing 40 mM potassium phosphate and 20 mM sodium chloride, temperature: 295°K). The sweep widths in t_1 and t_2 were 600 Hz and 100 Hz respectively. Chemical shifts are reported relative to TSP for the ^{1}H-domain and to external 10% phosphoric acid in D_2O for the ^{31}P-domain.

Additional difficulties arise in the application of ^{31}P-^{1}H-2D-techniques to the study of complexation of anticancer drugs, including mitoxantrone, to oligonucleotides. The transfer of magnetization from the protons to the ^{31}P nuclei is severely hampered owing to extensive line broadening (very short T_2's) upon intercalation of the drug. Raising the temperature of the sample,

Table 1. ^{31}P Chemical shifts of 10 mM oligonucleotide-tetramers at 295°K, except $[d(ApCpGpT)]_2$ at 300°K[a]

[d(C_pGpCpG)]	C1pG2: -0.69	G2pC3: -0.54	C3pG4: -0.69
[d(ApCpGpT)]	ApC: -0.70	CpG: -0.52	GpT: -0.71
[d(ApGpCpT)]	ApG: -0.82	GpC: -0.61	CpT: -0.72

[a] Relative to external 10% H_3PO_4.

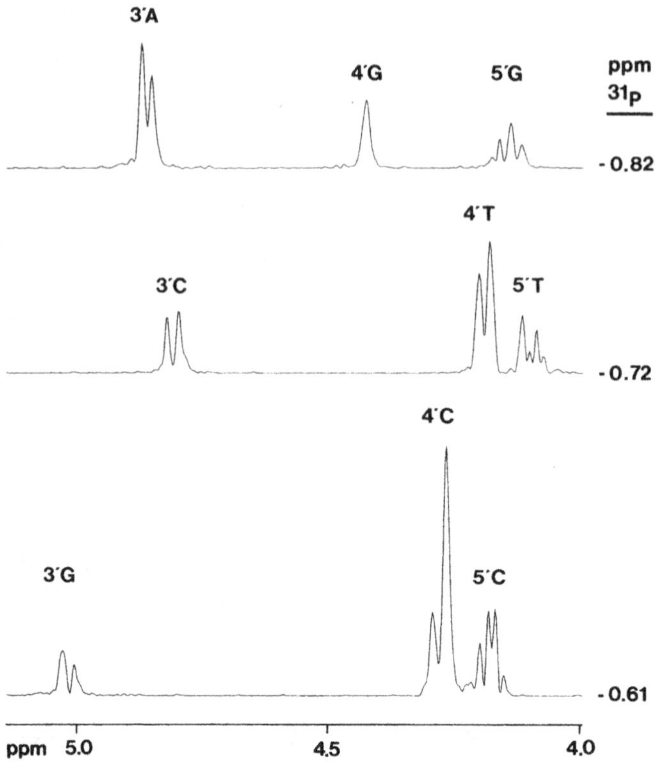

Figure 5. Cross sections through each of the three ^{31}P-resonances of the 2D-^{31}P-^{1}H-shift correlation contour plot of $[d(ApGpCpT)]_2$ shown in Figure 4. With 96 scans for each t_1 value a 160 x 256 data set was collected within a total acquisition time of about 15 hours at a concentration of 20 mM (solvent: D_2O buffered at pH 7.0 containing 40 mM potassium phosphate and 20 mM sodium chloride, temperature: 295°K). The sweep widths were 60 Hz in t_1 and 100 Hz in t_2. Chemical shifts in the ^{1}H-domain are relative to TSP whereas in the ^{31}P-domain they are relative to external 10% phosphoric acid in D_2O.

a technique used to overcome such problems as stacking, is inappropriate since the complexes are known to dissociate at 40°C. However, now that the individual ^{31}P -signals have been assigned, it is possible to observe the effects of intercalation of the chromophore by comparing the one-dimensional ^{31}P-spectra, as illustrated in Fig. 6 for $[d(ApCpGpT)]_2$. The signal at lowest field (top trace) identified as CpG experiences substantial upfield shift (0.36 ppm) whereas the other two resonances are shifted downfield by- $\lesssim$ 0.1 ppm. Furthermore the shifted signal undergoes considerable broadening (ca. 4 times wider at half-height). This suggests a C,G preference with the chromophore intercalating in the central C,G site, in accord with the previous binding studies and NMR experiments on the complex with $[d(CpGpCpG)]_2$. This conclusion would not have been possible from ^{1}H-NMR experiments alone and illustrates the potential of the ^{31}P-^{1}H correlation technique.

With these powerful and adaptable NMR methods in hand, the stage is set for a more detailed analysis of the molecular recognition and binding characteristics of clinically useful anticancer drugs.

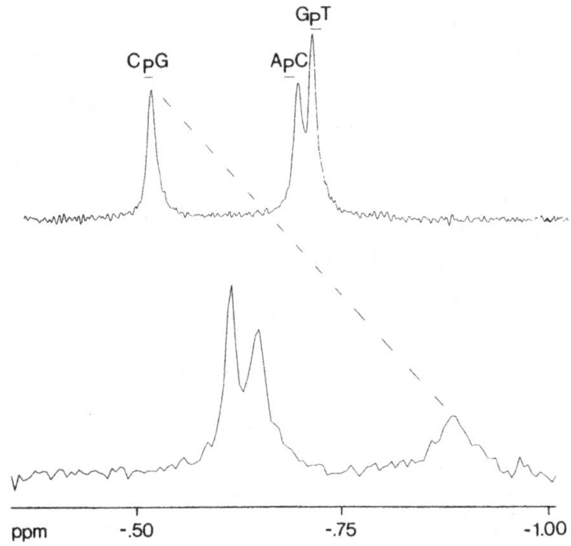

Figure 6. H-Broadband decoupled ^{31}P-spectra of $[d(ApCpGpT)]_2$ at 300°K. Top: without drug, bottom: 1:1 complex with mitoxantrone. Both spectra were recorded at ca. 10 mM concentration in D_2O buffered at pH 7 (40 mM potassium phosphate, 20 mM sodium chloride) using a sweep width of 200 Hz. 120 scans for the top and 256 scans for the lower spectrum were recorded and the FID's were resolution enhanced by means of a Gaussian multiplication. Chemical shifts are relative to external 10% phosphoric acid in D_2O.

ACKNOWLEDGEMENTS

This investigation was supported by grants (to JWL) from the National Cancer Institute of Canada, (to JWL and GK) from the Natural Sciences and Engineering Council of Canada, by a contract (JWL) with the National Foundation for Cancer Research, and (to J-L.I.) from the Association pour la Recherche sur le Cancer (France). AO and CCH thank the Alberta Heritage Foundation for Medical Research for the award of fellowships.

REFERENCES AND FOOTNOTES

1. Abbreviations: NMR nuclear magnetic resonance; COSY, correlation spectroscopy; NOESY, nuclear Overhauser effect correlation spectroscopy; TSP, 3(trimethylsilyl)-1-propanesulfonic acid, sodium salt hydrate. The symbols used to represent oligodeoxyribonucleotides and their protected derivatives follow the recommendations of the IUPAC-IUC Commission on Biochemical Nomenclature Eur. J. Biochem. 9:131 (1983).
2. M. W. Caruthers, Accts. Chem. Res., 13:155 (1980).
3. S. H. Kim, J. L. Sussman, G. M. Church, "Structure and Conformation of Nucleic Acids and Protein-Nucleic Acid Interactions" M. Sunderalingam and S. T. Rao, Eds. University Park Press, Baltimore, (1975), p.571.
4. G. V. Garsky, V. G. Tumanyan, A. S. Zasedatelev, A. L. Srruke, S. L. Grokhousky and B. P. Gottliph, "Nucleic Acid-Protein Recognition", H. J. Vogel, Ed., Academic Press, New York, (1977), p.189.
5. Molecular Aspects of Anti-Cancer Drug Action, S. Neidle and M. J. Waring, Eds., Macmillan, London (1983).

6. E. F. Gale, E. Cundliffe, P. D. Reynolds, M. Richmond and M. J. Waring, The Molecular Basis of Antibiotic Action, second edition, Wiley, New York, (1981).
7. T. J. Bardos and T. L. Kalman, (Eds) New Approaches to the Design of Antineoplastic Agents, Elsevier Biomedical, New York (1982).
8. R. W. Ruddon, "Cancer Biology" Oxford University Press, Oxford (1981).
9. M. Wu, N. Davidson, Nucleic Acids Res., 5:2825 (1978).
10. H. M. Sobell, "Nucleic Acid Geometry and Dynamics", Ed. R. H. Sarma, Pergamon Press, New York, (1980), p.389.
11. R. Benn and H. Gunther, Modern Pulse Methods in High-Resolution NMR Spectroscopy, Angew. Chem. Int. Ed. Engl. 22:350 (1983).
12. D. Frechet, D. M. Cheng, L. S. Kam and P.O.P. Ts'O, Biochemistry 22:5194, (1983).
13. J. Frigon, J. M. Wright, W. Leupin, W. A. Denny and D. R. Kearns, J. Amer. Chem. Soc. 104:5540 (1982).
14. J. Doornbos, C. T. J. Wreesmann, J. H. van Boom and C. Altona, Eur. J. Biochem. 13:571 (1983).
15. P. P. Lankhorst, G. Wilhe, J. H. van Boom, C. Altona and C.A.G. Haasnoot, Nucl. Acids Res. 11:2839 (1983).
16. R. M. Scheek, N. Russo, R. Boelens, R. Kapstein, and J. H. van Boom, J. Amer. Chem. Soc. 105:2914 (1983).
17. A. Bax, R. Freeman and T. A. Frenkiel, J. Amer. Chem. Soc. 103:2101 (1981).
18. T. H. Mareci, and R. Freeman, J. Magn. Res. 51:531 (1983).
19. K. C. Murdock, R. G. ChP. F. Fabio, R. B. Angier, R. E. Wallace, F. E. Durr and R. V. Citarella, J. Med. Chem. 22:1024 (1979).
20. D. S. Alberts, K. S. Griffith, G. E. Goodman, T. S. Herman and E. Murray, Cancer Chemother. Pharmacol. 5:11 (1980).
21. B. Smith, Br. Heart J. 31:607 (1969).
22. G. Bonnadona and S. Monfardini, Lancet 1:833, (1969).
23. F. Traganos, D. P. Evenson, L. Staiano-Coico, Z. Darzynkiewicz and M. R. Melamed, Cancer Res. 40:671 (1980).
24. H. Waldes and M. S. Center, Biochem. Pharmacol. 31:1057 (1982).
25. G. T. Bowden, D. Garcia, Y. M. Peng and D. S. Alberts, Cancer Res. 42:2660 (1982).
26. R. V. Citarella, R. E. Wallace, K. C. Murdock, R. B. Angier, F. E. Durr, and M. Forbes, Cancer Res. 42:440 (1982).
27. J. W. Lown, C. C. Hanstock, R. D. Bradley and D. G. Scraba, Molec. Pharmacol. 25:178 (1984).
28. J. Kapuscinski, Z. Darzynkiewicz, F. Traganos and M. R. Melamed, Biochem. Pharamcol. 30:231 (1981).
29. W. O. Foye, O. Vajragupta, and S. K. Sengupta, J. Pharm. Sci. 7:253 (1982).
30. G. J. Quigley, A. Wang, G. Ughetto, G. van der Marel, J. van Boom and A. Rich, Proc. Natl. Acad. Sci. USA, 77:6453 (1980).
31. J. W. Lown, A. R. Morgan, S-F. Yen, Y-H. Wang and W. D. Wilson, Biochemistry, in press (1985).
32. C. C. Hanstock and J. W. Lown, J. Magn. Res. 58:167 (1984).
33. G. Bodenhausen and R. Freeman, J. Magn. Res. 25:471 (1977).
34. A. Pardi, R. Walker, H. Rapoport, G. Wider and K. Wuthrich, J. Am. Chem. Soc. 105:1652 (1983).
35. A. Bax and G. A. Morris, J. Magn. Res. 42:501 (1981).
36. D.G. Gorenstein, K. Lai, and D.O. Shah, Biochemistry, 23:6717 (1984).
37. J. W. Lown and C. C. Hansrock, J. Biomol. Str. Dyn. in press (1985).

DNA BINDING AND BIOLOGICAL PROPERTIES OF BIS AND TRIS INTERACTING MOLECULES

Jean-Bernard Le Pecq[1] and Bernard P. Roques[2]

[1]Institut Gustave Roussy, rue Camille Demoulins, Villejuif, France
[2]Université René Descartes, 4, av. de l' Observatoire, Paris, France

INTRODUCTION

Many biological processes are controlled by reversible bimolecular associations which are characterized by a high selectivity and a high binding constant. Such interactions are involved in the recognition of antigens by antibodies, of hormones by receptors and of nucleic acid sequences by proteins. Binding constants of the order of 10^8 to 10^{14} M^{-1} are often characteristic of such interactions. Many small organic molecules able to reversibly bind to nucleic acids have been discovered.[1] Some of them are endowed with interesting biological properties. A few anti-tumor agents, such as actinomycin D doxorubicin, are DNA intercalating molecules.[1] Many DNA binding molecules have also been found to be mutagenic. However, the binding constant for DNA of these molecules, which varies from 10^4 to 10^6 M^{-1}, is much smaller than that of the natural DNA binding proteins[2] which is generally larger than 10^{10} M^{-1}. Therefore, we thought it might be of interest to design simple molecules capable of binding to DNA with an affinity constant higher than 10^{10} M^{-1}. Such molecules might be able to compete in vivo with the DNA binding proteins which regulate genetic expression, and might be endowed with interesting biological properties. Animal cells might respond to agents able to form high affinity complexes with DNA in an unpredictable way. For instance, one can wonder whether the modifications of the DNA structure resulting from such complexes would be recognized by the cellular DNA repair machinery. Reversible DNA complexes of long life time might be difficult to distinguish from covalent DNA adducts. In this paper, we will describe the strategy which we used to design molecules able to bind to double stranded nucleic acids with high affinity. This strategy takes advantage of the characteristics of the DNA intercalation process and of the properties of oligomeric ligands. The properties of bis- and tris-intercalating molecules will be discussed. Because several antitumor agents were discovered among DNA bisintercalating molecules, the studies dealing with the mechanism of action of these compounds in relation to their DNA binding properties will also be reviewed.

DESIGN OF SIMPLE MOLECULES WITH HIGH DNA AFFINITY

Let us compare the interactions of a monomeric and a dimeric ligand, as schematized in Figure 1, where K_{dim} and K_{mon} are the binding constants associated with the dimeric and monomeric system, respectively. If $(\Delta G_{dim})_{max}$ is

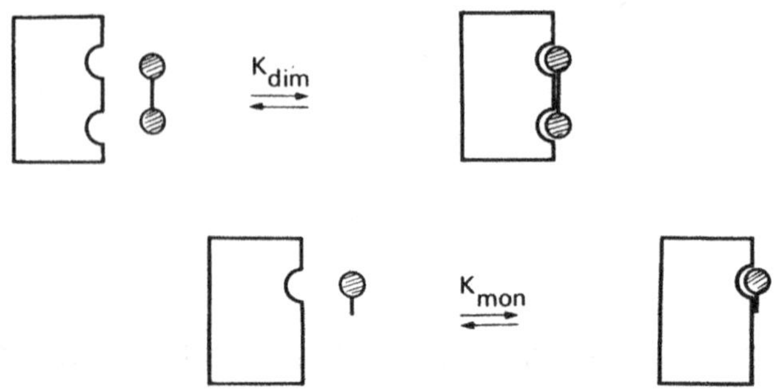

Figure 1.

the maximum possible free energy of interaction of the dimeric system, and ΔG_{mon} is the corresponding free energy of the monomeric system, then equation 1 applies.

$$(\Delta G_{dim})_{max} = 2\,\Delta G_{mon} - RT\ln X \qquad (1)$$

The last term, RTlnX, (with X being the molar concentration of water in the solution), arises because the entropies of mixing do not cancel out in the addition of the subunit free energies. Therefore:

$$RT\ln(K_{dim})_{max} = 2RT\ln K_{mon} - RT\ln X \qquad (2)$$

or $$(K_{dim})_{max} = 55.6\ K_{mon} \qquad (3)$$

Chipman and Sharon[3] studied the interaction of monosaccharides and disaccharides with lysozyme and found that this relationship could be applied. Provided that the amount of free energy lost in the dimeric system remains low, because of entropy variation, the binding affinity associated with the dimeric system will be much larger than the one associated with the monomeric system. Intercalating molecules can bind to DNA every two base pairs. Therefore, the distance along the DNA axis between two adjacent potential binding sites is 3 x 3.4 Å. In principle, a dimeric molecule made up of two intercalating subunits linked by a chain at least 10.2 Å long might bisintercalate in DNA. Consequently, such dimeric molecules are expected to have much greater DNA affinity than the corresponding monomers. Because potential intercalation binding sites are almost evenly spaced along the DNA axis, trisintercalating agents can be imagined which are made up of three intercalating units linked by a chain able to position the intercalating units properly. Trisintercalating molecules are expected to have a DNA affinity still higher than that of bisintercalating molecules. More generally, oligointercalating molecules can be imagined. The DNA binding of these compounds will increase with the degree of polymerization. In the last few years, such bis and trisintercalating molecules were prepared in our laboratories and it was demonstrated that these molecules are indeed able to bind to DNA with a very high binding constant.

DNA BINDING PROPERTIES OF BIS AND TRISINTERCALATORS

In order to compare the DNA binding affinity of mono, bis and trisintercalating molecules, an homologous series of chloromethoxyacridine derivatives, made up of one, two or three intercalating units linked by polyamine chains, were prepared (Figure 2).[4,5] Because the binding constants for the dimeric and trimeric molecules are very high, they could not be measured directly by equilibrium methods. Nevertheless, apparent binding constants could be

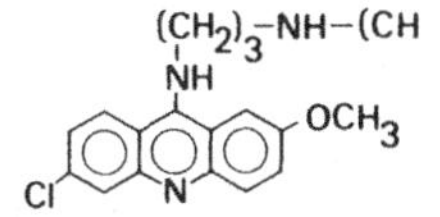

AcMo

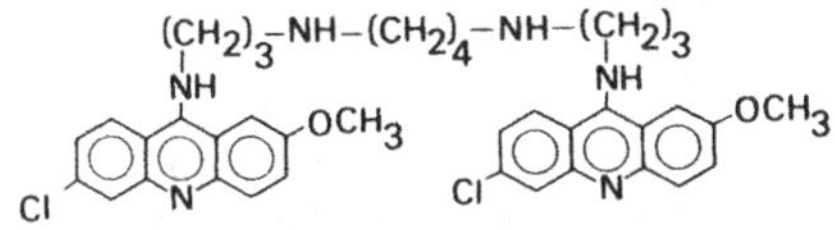

AcDi

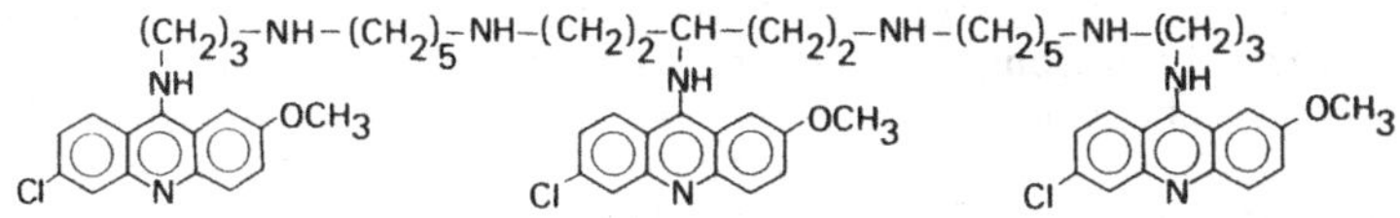

AcTri

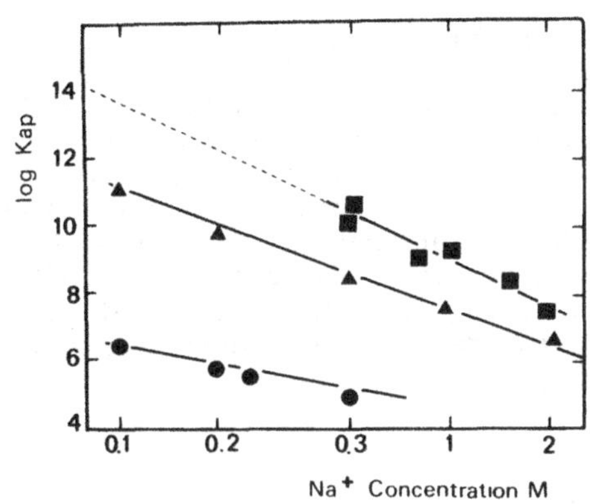

Figure 2. Comparison of binding constants as a function of Na^+ concentration for an acridine monomer (●-●), dimer (▲-▲) and entrimer (■-■). The formulae are shown at the left.

deduced from kinetic experiments. The on-rate constants for binding to poly [d(A-T))] could be measured by stopped flow using fluorescence detection. Off-rate constants could be measured by displacing the dyes bound to poly [d(A-T)] by an excess of GC rich DNA. This displacement is accompanied by a large decrease of dye fluorescence, because these dyes are fluorescent only when intercalated in between two adjacent AT base pairs. At high salt concentration, the true equilibrium constants and the apparent constants derived from kinetic measurements could be measured. They were found to be identical. Therefore, these apparent binding constants are probably close to the true equilibrium constants. The apparent binding constants for poly [d(A-T)] as a function of salt concentration for the mono, bis and trisintercalating acridine derivatives are shown in Figure 2. The binding of the trisintercalating derivative was found to be so high that determinations could only be performed in high salt conditions. Nevertheless, because the logarithm of the binding constant varies linearly with the logarithm of the salt concentration, extrapolated values could be obtained for the salt concentrations corresponding to the biological environment. For instance, in 0.1 M NaCl, the DNA binding constant for the mono, bis and trisintercalating derivatives could be estimated to be 3×10^6, 2×10^{11} and 1×10^{14} M^{-1} respectively. The binding constant increment between the mono and bisintercalators is larger than that between the bis and trisintercalators. This probably results from the greater constraints put on the DNA structure by the trisintercalating molecule.

The mode of DNA binding for bis and trisintercalating compounds can be complex, as schematized in Figure 3. Many different kinds of bisintercalating molecules, and a few tris-intercalating molecules, have been prepared and their mode of DNA binding studied.[1,4-24] It has been shown that, depending on the size of the intercalated aromatic ring and the linking chain length, molecules can bisintercalate and leave one or two base pairs in between the intercalated rings. Multiple modes of binding also arise because the aromatic rings of the molecules can either intercalate or stay on the outside of the DNA helix. For a given molecule, the equilibrium between these different modes of binding is strongly dependent on the DNA torsional constraint, the DNA sequence and the number of molecules bound per base pair. For different molecules the binding behavior was found to be strongly dependent on the length, structure and rigidity of the linking chain as well as on the nature of the intercalating subunit. In addition, the conformation of the free molecules in solution strongly affects DNA binding.[15,25] Strong self stacking which leads to a closed conformation decreases DNA affinity. Intermolecular associations are often observed even at low concentrations. In order to prevent self stacking, we prepared dimeric molecules with rigid chains. Among such molecules, we discovered several compounds, such as Ditercalinium (Figure 4), which elicited potent antitumor activities.[16,26] In this paper, we would like to concentrate on the most recent developments dealing with the study of these molecules: the interaction of the bisintercalating molecules with B and Z DNA; the conformational dynamics of the DNA complexes; and the NMR study of the conformational properties of oligointercalators and of the structure of an oligonucleotide bisintercalator complex.

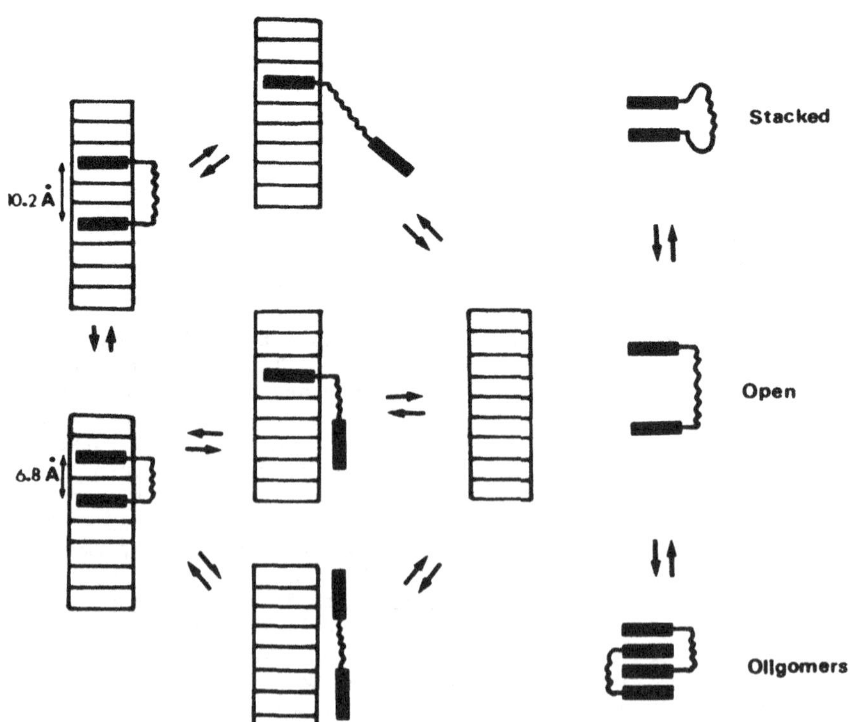

Figure 3.

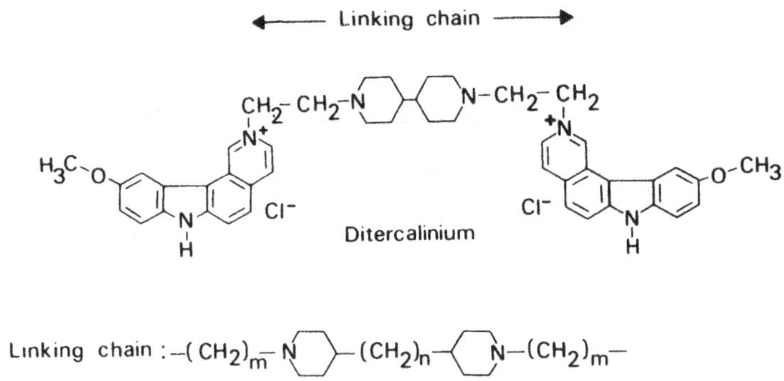

	m	n
Ditercalinium	2	0
I	2	1
II	2	2
III	2	3
IV	3	0

Figure 4.

Bisintercalator Binding to B and Z DNA

The monointercalating drug ethidium bromide binds better to B DNA than to Z DNA. The addition of ethidium to Z DNA cooperatively induces the Z to B transition.[27,28] Several other monointercalating molecules, such as daunorubicin, behave similarly.[29,30] Shafer et al.[31] showed that Z DNA can bind ethidium and the bis(methidium) spermine dimer according to an intercalation process. Recently, we compared the binding of ethidium and ethidium dimer to poly[(dGm^{5}dC)] in the Z or B form.[28] In agreement with Shafer et al.[31], we observed that ethidium and ethidium dimer can intercalate in Z DNA. The binding affinity of ethidium is more than ten times greater for B DNA than for Z DNA. The binding of ethidium to poly[(dGm^{5}dC)], in buffers where the Z form is stable, is highly cooperative. The Z to B transition occurs when only a small number of dye molecules are bound per DNA base pair. The binding affinities of ethidium dimer for the Z or B form of poly[(dGm^{5}dC)] are almost identical. The addition of ethidium dimer to poly[(dGm^{5}dC)] in conditions where the Z form is stable does not induce the Z to B transition, as seen by circular dichroism. In agreement with these observations, it was observed that the addition of ethidium to poly[(dGm^{5}dC)] in the Z form inhibits the precipitation of the polymer by anti-Z DNA antibodies. This inhibition results from the Z to B transition caused by ethidium. On the contrary, the binding of ethidium dimer to the Z form of the polymer does not inhibit the DNA-antibody reaction, because in that case the Z to B transition does not occur.[32] In order to determine whether the ability to bind to Z DNA was a specific property of bisintercalators, the effect on the Z to B transition of various bisintercalators, whose formulae are shown in Figure 4, was studied (J. Ramstein et al., unpublished results from our laboratories). The Z to B transition of poly [(dGm^{5}dC)] was followed by circular dichroism, as described.[28] Ditercalinium, which has the shortest chain, was found to be able to induce the Z to B transition when the number of molecules bound per nucleotide was still small ($\simeq$ 0.02). On the contrary, compounds I, II, and III shown in Figure 4 were unable to induce the Z to B transition when they were bound to the Z form of poly[(dGm^{5}dC)]. Ditercalinium and compound I have antitumor activities whereas compounds II, III and IV have not. This result suggests that the antitumor activity is not related to the ability of the molecules to interfere with the Z to B equilibrium. This also suggests that bisintercalating molecules able to recognize preferentially either B or Z DNA can be obtained. Other studies are needed to determine the bisintercalator properties which are responsible for the Z or B preference.

Conformational Dynamics of DNA-Bisintercalator Complexes

The fluorescence of ethidium and ethidium dimer is strongly enhanced when the protons of their amino groups are excluded with deuterium. The rate of exchange can be followed simply by stopped-flow using fluorescence detection. When ethidium dimer binds to DNA, the rate of hydrogen-deuterium (H-D) exchange is slowed down considerably.[33] It was shown that the rate of H-D exchange is limited by the dynamics of the structural fluctuation of the complex. The model used to account for the H-D exchange of DNA protons[34] could be used also to account for the H-D exchange of DNA-bound ethidium dimer amino protons. In this model, exchange can only take place from an open state according to the reaction:

$$\text{closed} \rightleftarrows \text{open} \rightarrow \text{exchange}$$

The effects of catalysts on H-D exchange could be explained by such a model. In addition, it was observed that the values of the rate constants for the exchange of DNA protons and DNA-bound ethidium dimer amino protons were very similar. This observation suggested that the mechanisms which control the H-D exchange in the two cases might be correlated. Fluctuations of the structure of the ethidium dimer-DNA complex might result from the dynamic fluctuations of the DNA structure itself. In order to study this problem, the ethidium dimer H-D exchange was measured when the dye was bound to nucleic acids of differrent structures.[28] Because the rate of proton exchange for Z DNA is much slower than that for B DNA,[35] the H-D exchange of ethidium dimer bound to the Z or B form of poly[(dGm^{5}dC)] was measured. It was observed that, indeed, the rate of H-D exchange for the dye is smaller with Z DNA than with B DNA.[28] The possibility of a correlation between the conformational dynamics of the DNA and of one of its bound ligands is worthy of consideration. Such conformational dynamics also might interfere with the recognition of such DNA-complexes by repair enzymes and be important in determining the biological properties of bisintercalating molecules.

^{1}H NMR Study of the Conformational Properties of DNA-Intercalating Oligomers

H NMR is one of the most powerful methods to study the self association of planar aromatic dyes and their interaction with self-complementary di- or oligonucleotide phosphate in aqueous solution.[36] The changes in the chemical shifts of the drug and/or nucleic acid protons induced by the ring current shifts of the aromatic rings are currently used: 1) to monitor formation of the complex; 2) to determine the binding mode (i.e. intercalation) of the drug; 3) to propose a geometry for the complex.

One of the most important factors governing both the ability of a dimer or a trimer to bis- or tris-intercalate into DNA and the stability of the complexes formed is the conformational behaviour of the oligomer in aqueous solution. Due to their planar aromatic ring and hydrophobic character, DNA intercalating moieties (acridine, ethidium, pyridocarbazoles...) form stacked aggregates in aqueous solution, with self-association constants in the range of 1^- M to 5.000 M^{-1}.[25,37-38] In the case of dimeric or trimeric compounds, aromatic rings linked by flexible chains stack one above another, leading to intramolecularly stacked forms which are strongly stabilized in physiological conditions.[10,38] This feature is very unfavourable for DNA bis-intercalation since this process requires the opening of the folded form. Moreover, in the stacked conformation the proximity of the charged intercalating rings decreases the pKa of the nitrogen of the linking chain, therefore removing the electrostatic interaction of the spacer with the negatively charged DNA ribose-phosphate backbone.[10-38] A similar feature, i.e. intramolecular stacking of the three acridine moieties, also occurs in an acridine trimer.[5]

These unfavorable conformational features were overcome through the synthesis of bis-intercalators (i.e. ditercalinium) bearing rigid linking chains, such as N,N'-ethyl bis-piperidine.[16] As expected, this kind of rigid chain inhibits intramolecular stacking interactions, as clearly shown (at low dimer concentration) by the similarity between the ^{1}H NMR spectra of the monomer and the derived dimer (Figure 5 a and e), whereas the aromatic protons of the flexible dimer were strongly upfield shifted by the shielding current of the superimposed ring (not shown). Moreover, in the rigid dimer, distinct signals were observed for axial and equatorial protons of the bis-piperidine moiety, demonstrating the rigidity of the bis-(ethyl-piperidinyl) bridge under physiological conditions. In contrast, in the corresponding monomer the chemical shifts are averaged by the rapid equilibrium between different conformations of the piperidine ring.[39]

Bis-Intercalation at the Level of the DNA Duplex with Self-Complementary Oligonucleotides: Structural Analysis by NMR Spectroscopy

Definite evidence of intercalation can be obtained by NMR and X-ray studies using self-complementary oligonucleotides or polynucleotides. Thus, a comparative ^{1}H NMR study of the interactions between poly(dA-dT), ethidium bromide and ethidium dimer under various conditions (ionic strength and temperature) demonstrated the bis-intercalation of ethidium dimer (EtDi) into poly(dA-dT). Indeed, at low temperature and high salt concentrations, EtDi

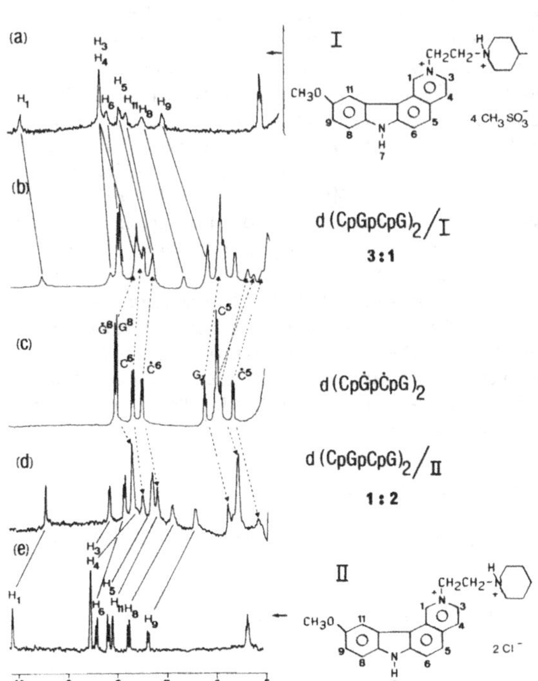

Figure 5. ^{1}H NMR spectra (a) free ditercalinium (I) (e) free monomer (II), (c) self paired complementary oligonucleotide d[CpGpCpG] , (b) oligonucleotide complex with I, (d) oligonucleotide complex with II (reprinted from ref. 42).

induces upfield shifts in some protons of poly(dA-dT) which are twice as large as those for the monomer when both the drugs are at the same DNA to drug ratio. Moreover, the dimer protects about six base pairs against thermal denaturation, while ethidium bromide protects only three. This first demonstration of DNA bis-intercalation at the molecular level was confirmed by ^{19}F NMR studies on fluorinated acridine dimers.[41] In accordance with previous experiments, the DNA binding mode of these dimers changes from bis to mono-intercalation during the helix to coil transition.[40-41] Among pyridocarbazole dimers exibiting high DNA affinity, only those belonging to the series of 7H-pyrido [4,3-c]-carbazole dimers (i.e. ditercalinium) elicit strong antitumor activity.[16,26] Such differences could be due, at least partly, to the structure of their DNA intercalation complexes. Therefore, as a first step, we studied the complex between ditercalinium and its related monomer with the self-complementary tetra-nucleotide d[CpGpCpG] (Figure 5).[42]

As shown in Figure 5 (c,d,e), addition of monomer II to d[CpGpCpG] leads to large upfield shifts of the aromatic protons of both the pyridocarbazole and the nucleotide bases, with maximum effects for a d(CpGpCpG) /monomer II ratio of 1:2. At this ratio, the melting temperature increases from 37°C in the free tetranucleotide to 54°C in the complex. Although the exchange between the free helix and the complex remains fast on the NMR time scale, the imino protons of the bases and the indole NH of the drug are shielded and their exchange rate is strongly decreased. Addition of the rigid dimer I to d[CpGpCpG] up to a 1/3 ratio leads to a single set of signals for the pyridocarbazole protons, while tetranucleotide protons are split into two sets (Figure 5b). One series of signals has chemical shifts identical to those of uncomplexed tetranucleotide (Figure 5 c), whereas the smaller upfield signals assigned by saturation transfer correspond to protons of the minihelix complex in slow exchange with the free oligonucleotide. The high field shifts induced on all aromatic resonances in the 3/1,d[CpGpCpG] /dimer I, and in the 1/2,d[CpGpCpG] /monomer II complexes are identical, indicating that pyridocarbazole rings are in similar environments in both complexes. Isoshielding curves of CG base pairs establishes that the 0.8-1.1 ppm upfield shifts measured on the ring protons of dimer I or monomer II in the complexes are consistent with intercalation of the pyridocarbazole moieties between adjacent paired nucleobases. Intermolecular NOE's were observed from sugar $H_{2'-2'}$ protons to protons (H5, H8, OCH_3) of the convex face of the pyridocarbazole ring, suggesting intercalation of the aromatic moieties with the quaternizing chain located in the major groove. If confirmed, this orientation should be different from that established by crystallographic analysis of the natural dimer triostin with the self-complementary hexadeoxynucleotide CGTACG. In this case the two aromatic rings of triostin are intercalated and the peptide bridge is located in the minor groove.[43] It remains to be demonstrated that biologically inactive pyridocarbazole dimers lead to different structures with oligonucleotides and that these putative changes could be related to the loss of potency.

Biological Properties of Bisintercalators

Two different types of bisintercalating molecules have been found to have antitumor activities:

a) The naturally occuring quinoxaline antibiotics, such as echinomycin and luzopeptine.[44]

b) The synthetic molecules derived from acridines,[45] anthracyclines[46-47] or 7H pyridocarbazole, such as ditercalinium.[16,26]

Since among monointercalating agents several antitumor drugs of clinical importance have been discovered, it is interesting to ask the following questions: Are the mechanisms of action of mono- and bisintercalating agents related? Is the high DNA binding affinity of the latter derivatives responsible for their original properties?

The antibiotics related to echinomycin bind to DNA with an affinity which is weaker than that of synthetic bisintercalators.[48] Their mechanisms of action, although not yet studied in detail, seem to be similar to that of actinomycin D, i.e. transcription is preferentially inhibited. In this case, bisintercalators seem to behave like monointercalators.

The mechanism of the antitumor action of ditercalinium and related derivatives has been studied recently.[49-50] It appears to be quite different from that of other antitumor agents. Surprisingly, ditercalinium induces a delayed cytotoxicity. When animal cells are incubated _in vitro_ for a short period with this drug, and later incubated in a fresh medium, they are able to continue their growth for 5 to 6 generations. After drug exposure, cells cloned in solid medium form abortive small colonies made up of 50 to 100 cells. Almost all physical or chemical toxic agents which alter DNA structure induce cell arrest in the G2 phase of the cell cycle. In this respect as well, ditercalinium behaves differently. Cells become arrested almost randomly in the different phases of the cell cycle. The cytological modifications which accompany cell death are also quite different from those observed with other cytotoxic drugs. Interestingly, echinomycin does not induce such a delayed toxicity.

Because analysis of the cytotoxic action of ditercalinium on animal cells appears difficult, we studied the effects of this drug on bacterial cells. An _E. coli_ mutant strain sensitive to ditercalinium and related compounds was isolated.[51] This strain is only slightly more sensitive to monointercalating agents than the wild type strain. The mutation responsible for the sensitivity to ditercalinium was characterized recently (B. Lambert _et al._, unpublished results from our laboratories) and found to be identical to _polA_. The induction of the SOS system in this strain by ditercalinium was also measured. It was found that SOS induction occured at extremely low concentration ($\simeq 1$ ng/ml). The drug was also cytotoxic in this strain, whereas monointercalating agents elicit only cytostatic action. Previous studies have shown that in the wild type _E. coli_ strain, the DNA is not accessible to intercalating drugs.[52] It was recently proposed that this was related to the association of the bacterial DNA with proteins in a chromatin-like structure rather than being due to the impermeability of the membrane. The fact that the _polA_ mutation, which is not known to alter any membrane function, induces ditercalinium sensitivity further underlines the importance of _in vivo_ DNA structure in determining sensitivity to DNA reactive drugs. It is thought that one of the functions of polymerase I in _E. coli_ is to fill the gaps that appear in DNA at the replication fork, since replication is discontinuous. Therefore, it is possible that in _polA_- strains the number of DNA gaps in growing cells is much higher than in wild type cells. These gaps might be more accessible than bulk DNA for recognition by some bisintercalating agents. Therefore, we would like to propose the following hypothesis. Some drugs might be unable to bind _in vivo_ to the bulk DNA which is in the regular B structure and well protected by its association with proteins. These drugs might recognize transient alteration of DNA structure, which could appear in low number during replication, transription, recombination or any other event involving a change of the DNA protein associations, and/or DNA structural alteration. It would therefore be of interest to study not only the interaction of such drugs with regular double stranded B DNA, but also with modified DNA structures, for instance DNA with mispaired bases or gaps. The discovery that echinomycin induces a very unusual change in DNA conformation[43] is extremely interesting in this respect.

ACKNOWLEDGEMENTS

This work was supported through grants from the CNRS (UA 498, LA 147), INSERM (U 266, U 140), Université Paris V and Paris VI, Association pour la Recherche sur le Cancer and Fondation pour la Recherche Médicale. The authors are grateful to Dr. S. Brown for helpful discussions.

REFERENCES

1. M.J. Waring, DNA modification and cancer, Ann. Rev. Biochem. 50:154 (1981).
2. T.M. Jovin, Recognition mechanisms of DNA-specific enzymes, Ann. Rev. Biochem. 45:889 (1976).
3. D.M. Chipman and N. Sharon, Mechanism of lysozyme action, Science, 165:454 (1969).
4. B. Gaugain, J. Markovits, J.H. Le Pecq and B.P. Roques, DNA poly-intercalation: Comparison of DNA binding properties of an acridine dimer and trimer, FEBS Lett. 169:123 (1984).
5. Ph. Laugaa, J. Markovits, A. Delbarre, J.B. Le Pecq and B.P. Roques, DNA tris-intercalation: First acridine trimer with DNA affinity in the range of DNA regulatory proteins-kinetics studies, Biochemistry, in press (1985).
6. J.B. Le Pecq, M. Le Bret, J. Barbet and B.P. Roques, DNA polyintercalating drugs: DNA binding of diacridine derivatives, Proc. Natl. Acad. Sci. USA 72:2915 (1975).
7. B.F. Cain, B.C. Baguley and W.A. Denny, Potential antitumor agents. 28. Deoxyribonucleic acid polyintercalating agents, J. Med. Chem. 21:658 (1978).
8. T.K. Chen, R. Fico and E.S. Canellakis, Diacridines, bifunctional intercalators. Chemistry and antitumor activity, J. Med. Chem. 21:868 (1978).
9. P.B. Dervan and M.M. Becker, Molecular recognition of DNA by small molecules. Synthesis of bis(methidium)spermine, a DNA polyintercalating molecule, J. Am. Chem. Soc. 100:1968 (1978).
10. B. Gaugain, J. Barbet, R. Oberlin, B.P. Roques and J.B. Le Pecq, DNA bifunctional intercalators. 1. Synthesis and conformational properties of an ethidium homodimer and of an acridine ethidium hetero-dimer, Biochemistry 17:5071 (1978).
11. B. Gaugain, J. Barbet, N. Capelle, B.P. Roques and J.B. Le Pecq, DNA bifunctional intercalators. 2. Fluorescence properties and DNA binding interaction of an ethidium homodimer and an acridine ethidium heterodimer, Biochemistry 17:5078 (1978).
12. K.F. Kuhlmann, N.J. Charbeneau and C.W. Mosher, Synthesis, DNA-binding and biological activity of double intercalating analog of ethidium bromide, Nucleic Acids Res. 5:2629 (1978).
13. J.W. Lown, B.C. Gunn, R.Y. Chang, K.C. Mjumdar and J.S. Lee, Bis-intercalative binding to DNA of novel bis(10-methyl)acridinium chlorides and its dependence on chain length of linker, Can. J. Biochem. 56:1006 (1978).
14. M. Becker and P.B. Dervan, Molecular recognition of nucleic acid by small molecules. Binding affinity and structural specificity of bis-(methidium)spermine, J. Am. Chem. Soc. 101:3664 (1979).
15. N. Capelle, J. Barbet, P. Dessen, S. Blanquet, B.P. Roques and J.B. Le Pecq, Deoxyribonucleic acid bifunctional intercalators: kinetic investigations of the binding of several acridine dimers to deoxyribonucleic acid, Biochemistry, 15:3354 (1979).
16. D. Pelaprat, A. Delbarre, I. Le Guen, B.P. Roques and J.B. Le Pecq, DNA intercalating compounds as potential antitumor agents. 2. Preparation and properties of 7H-pyridocarbazole dimers, J. Med. Chem. 23:1336 (1980).

17. R.G.McR. Wright, L.P.G. Wakelin, A. Fieldes, R.M. Acheson and M.J. Waring, Effects of ring substituents and linker chain on the bifunctional intercalation of diacridines into deoxyribonucleic acid, Biochemistry 19:5825 (1980).
18. K.F. Kuhlman and C.W. Mosher, New dimeric analogues of ethidium: Synthesis, interaction with DNA and antitumor activity, J. Med. Chem. 24:1333 (1981).
19. H.D. King, W.D. Wilson and E.J. Gabbay, Interactions of some novel amide-linked bis(acridines) with deoxyribonucleic acid, Biochemistry 21:4982 (1982).
20. W.D. Wilson, R.A. Keel, R.L. Jones and C.W. Mosher, Viscosimetric analysis of the interaction of bisphenamthridinium compounds with closed circular supercoiled and linear DNA, Nucl. Acids Res. 10:4093 (1982).
21. G.J. Atwell, W. Leupin, S.J. Twigden and W.A. Denny, Triacridine derivative: First DNA tris-intercalating ligand, J. Am. Chem. Soc. 105:2913 (1983).
22. J.B. Hansen, E. Langvad, F. Frandsen and O. Buchardt, 9-Acridinyl and 2-Methoxy-6-Chloro-9-Acridinyl derivatives of aliphatic Di-, Tri-, and Tetraamines. Chemistry, cytostatic activity and Schistosomicidal activity, J. Med. Chem. 26:1510 (1983).
23. W.A. Denny, G.J. Atwell, G.A. Wilhmott and L.P.G. Wakelin, Interaction of paired homologous series of diacridines and triacridines with deoxyriboucleic acid, Biophys. Chem. in press (1985).
24. N. Helbecque, J.L. Bernier and J.P. Henichart, Design of a new DNA-polyintercalating drug, a bisacridinyl peptidic analogue of Triostin A, Biochem. J. 225:829 (1985).
25. J. Barbet, B.P. Roques, S. Combrisson and J.B. Le Pecq, DNA poly-intercalating drugs. Proton magnetic resonance studies of a new acridine dimer. Conformations and interactions with mono- dinucleotides, Biochemistry, 15:2642 (1976).
26. B.P. Roques, D. Pelaprat, I. Le Guen, G. Porcher, Ch. Gosse and J.B. Le Pecq, DNA bifunctional intercalators. Antileukemic activities of new pyridocarbazole dimers, Biochem. Pharmacol. 28:1811 (1979).
27. F. Pohl, T.M. Jovin, W. Baehr and J.J. Holbrook, Ethidium bromide as a cooperative effector of a DNA structure, Proc. Natl. Acad. Sci. USA, 69:3805 (1972).
28. J. Markovits, J. Ramstein, B.P. Roques and J.B. Le Pecq, Effect of B-Z transition and nucleic acid structure on the conformational dynamics of bound ethidium dimer by hydrogen deuterium exchange kinetics, Nucleic Acids Res. in press (1985).
29. C.W. Chen, R.H. Knop and J.S. Cohen, Adriamycin inhibits the B to Z transition of Poly(dGm^{5}dC).Poly(dGm^{5}dC), Biochemistry 22:5468 (1983).
30. P.A. Mirau and D.R. Kearns, The effect of intercalating drugs on the kinetics of the B to Z transition of poly(dG-dC), Nucleic Acids Res. 11:1931 (1983).
31. R.H. Shafer, S.C. Brown, A. Delbarre and D. Wade, Binding of ethidium and bis(methidium)spermine to Z DNA by intercalation, Nucleic Acids Res. 12:4679 (1984).
32. C. Lagravère, B. Malfoy, M. Leng and J. Laval, Ring-opened alkylated guanine is not repaired in Z-DNA, Nature, 310:798 (1984).
33. J. Markovits, J. Ramstein, B.P. Roques and J.B. Le Pecq, Dynamic structure of DNA complexes. Fluorometric measurement of hydrogen-deuterium exchange kinetics of DNA-bound ethidium dimer and acridine-ethidium dimer, Biochemistry, 22:3231 (1983).
34. S.W. Englander and N.R. Kallenbach, Hydrogen exchange and structural dynamics of proteins and nucleic acids, Quarterly Rev. Biophys. 16:521 (1984).
35. J. Ramstein and M. Leng, Salt dependent dynamic structure of poly(dG-dC).poly(dG-dC), Nature 288:413 (1980).
36. T.R. Krugh and M.E. Nuss, Nuclear magnetic resonance studies of drug-nucleic acid complexes, in: "Biological Applications of Magnetic

Resonance", R.G. Schulman ed., p.113-178, Academic Press New York (1979).

37. G. Thomas and B.P. Roques, Proton magnetic resonance studies of ethidium bromide and its sodium borohydride reduced derivative, FEBS Lett. 26:169 (1972).
38. A. Delbarre, B.P. Roques, J.B. Le Pecq, J.Y. Lallemand and N. Dat-Xuong, PMR study of the self-association of DNA intercalating ellipticine derivatives in aqueous solution, Biophys. Chem. 4:275 (1976).
39. A. Delbarre, B. Gaugain, J. Markovits, A. Vilar, J.B. Le Pecq and B.P. Roques, H NMR study of the nature of bonding interactions involved in complexes between nucleic acids and intercalating compounds, in: "Proceeding of Jerusalem Symposium in Quantum Chemistry and Biochemistry Intermolecular Forces", B. Pullman ed., D. Reidel Publ., vol. 14 p.273-283 (1981).
40. A. Delbarre, J.B. Le Pecq and B.P. Roques, ^{1}H NMR study of an ethidium dimer poly(dA-dT) complex: Evidence of a transition between bis and monointercalation, Nucleic Acids Res. 11:4467 (1983).
41. A. Delbarre, R.H. Shafer and T.L. James, ^{19}F-NMR study of DNA-bisintercalation, Biopolymers 22:2497 (1983).
42. A. Delbarre, M. Delepierre, J. lgolen, J.B Le Pecq and B.P. Roques, ^{1}H NMR study of the structure of a pyridocarbazole dimer-d(CpGpCpG) complex, Biochimie. in press (1985).
43. A.H.J. Wang, G. Ughetto, G.J. Quigley, T. Hakoshima, G.A. Van der Marel, J.H. Von Boom and A. Rich, The molecular structure of a DNA-Triostin A complex, Science, 225:1115 (1984).
44. W.C. Rose, J.E. Schurig, J.B. Huftalen and W.T. Bradner, Experimental antitumor activity and toxicity of a new chemotherapeutic agent, BBM 928 A, Cancer Res. 43:1504 (1983).
45. W.A. Denny, B.C. Baguley, B.F. Cain and M.J. Waring, Antitumour acridines, in: "Molecular Aspects of Anticancer Drug Action", S. Neidle and M.J. Waring eds., Verlag Chemie Publ., p.1-34 (1983).
46. K.M. Dawson, Activity of SC 33428, a novel bishydrazone-bridged derivative of 4-demethoxydaunorubicin, against experimental tumors in mice, Cancer Res. 43:2880 (1983).
47. J.R. Brown, New natural, semisynthetic and synthetic anthracycline drugs, in: "Molecular Aspects of Anticancer Drug Action", S.Neidle and M.J. Waring eds., Verlag Chemie publ., p.57-92 (1983).
48. M.J. Waring and K.R. Fox, Molecular aspects of the interaction between quinoxaline antibiotics and nucleic acids, in: "Molecular Aspects of Anticancer Drug Action", S. Neidle and M.J. Waring eds., Verlag Chemie Publ. p.127-156 (1983).
49. J.P. Bendirdjian, C. Delaporte, B.P. Roques and A. Jaoquemin-Sablon, Effects of 7H-pyridocarbazole mono- and bifunctional DNA-intercalators on Chinese hamster lung cells in vitro, Biochem. Pharmacol. 33:3681 (1984).
50. C. Esnault, B.P. Roques, A. Jacquemin-Sablon and J.B Le Pecq, Effects of new antitumor bifunctional intercalators derived from 7H-pyridocarbazole on sensitive and resistant L1210 cells, Cancer Res. 44:4335 (1984).
51. B. Lambert and J.B. Le Pecq, Isolement et Caractérisation de souches d'Escherichia Coli sensibles à des antibiotiques et toxiques hydrophiles et/ou chargés, C. R. Acad. Sci. Paris, 294:447 (1982).
52. B. Lambert and J.B. Le Pecq, Effect of mutation, electric membrane potential and metabolic inhibitors on the accessibility of nucleic acids to ethidium bromide in Escherichia coli cells, Biochemistry, 23:166 (1984).
53. A. Kornberg, "DNA Replication", W.H. Freeman and Co., San Franscisco (1980).

NOVEL TYPES OF DNA-SUGAR DAMAGE IN NEOCARZINOSTATIN CYTOTOXICITY AND MUTAGENESIS

Irving H. Goldberg

Department of Pharmacology
Harvard Medical School
Boston, MA 02115, U.S.A.

Although a number of antitumor antibiotics interact with DNA[1] to form physical complexes, or covalent adducts with the bases, relatively few damage DNA by specifically interacting with the deoxyribose moiety. Neocarzinostatin (NCS), a member of a family of macromolecular ($M_r > 10,000$) antibiotics obtained from culture filtrates of Streptomyces, is such an agent (for pertinent literature on NCS prior to 1980, see Ref. 2, and between 1980 and 1984, see Ref. 3). Many of the biochemical and cellular effects of NCS resemble those of ionizing radiation. Most, possibly all, of the DNA lesions caused by NCS appear to result from the direct attack of an activated form of the drug on the deoxyribose of DNA. This is to be contrasted with ionizing radiation or the antibiotic bleomycin, that damage DNA deoxyribose through the intervention of a reduced form of oxygen. This paper will describe current information on the nature of the interaction between the active component of NCS and DNA, on the mechanism of the ensuing deoxyribose damage, and on some of the biological consequences of these actions.

NCS NONPROTEIN CHROMOPHORE AND APOPROTEIN

All the biological activity of the antibiotic resides in a methanol-extractable nonprotein chromophore ($M_r = 659$) that is tightly ($K_D \sim 10^{-10}$ M) and specifically bound to an apoprotein ($M_r = 10,700$). The latter (apo-NCS) acts as a carrier and protects the labile chromophore (NCS-Chrom) against degradation. The X-ray crystallographic structure of the related antibiotic, actinoxanthin, at 2.5 Å resolution shows that the protein molecule is kidney-shaped and has a well-defined cavity into which the nonprotein chromophore is proposed to fit[4]. Since the three-dimensional structure of NCS and actinoxanthin are very similar, it seems reasonable to expect a similar binding mode in both cases.

INTERACTION WITH DNA

NCS-Chrom has a strong affinity for DNA ($K_D = 0.25 \times 10^{-6}$ M). Binding was shown by: 1) quenching by DNA of the 440 nm fluorescence and shifting of the emission peak to 420 nm, 2) protection by DNA against spontaneous loss of activity in aqueous solution, and 3) inhibition by DNA of the spontaneous generation of 490 nm fluorescence due to degradation of NCS-Chrom to a highly fluorescent product. Chromophore has a preference for DNA high in T and A residues, in agreement with earlier studies showing that DNA's rich in these nucleotides were better targets for attack by NCS. Spectroscopic titrations suggest at least two types of chromophore binding sites on DNA: the tight binding one is saturated at r_b = 0.125 chromophore molecules/nucleotide (4 base pairs), and the other at r_b = 0.25 (2 base pairs) (D. Dasgupta and I.H. Goldberg, submitted). Only the former site is found in synthetic polynucleotides such as poly(dA-dT)·poly(dA-dT). Since there is no physical evidence for the interaction of native NCS or apo-NCS with DNA, it appears that dissociation of the chromophore from the protein and association with DNA are steps in the degradation of DNA by NCS. Thiol-treated (and inactivated) NCS-Chrom has a much lower affinity for apo-NCS than does untreated chromophore, but both forms of NCS-Chrom bind equally well to DNA.

NCS-Chrom unwinds supercoiled pMB9 DNA, suggesting an intercalation mechanism of binding. In addition, NCS-Chrom shares many of the other characteristcs of classical intercalators in its interaction with DNA. Thus, viscosity studies indicate that the DNA helix unwinding induced by the chromophore is 21°, and electric dichroism measurements show that each bound chromophore molecule lengthens DNA by 3.3 Å (Fig. 1). Further, absorbance transitions of the chromophore at 315-385 nm (range over which the naphthoate moiety absorbs) are oriented approximately parallel to the DNA bases, as expected for an intercalated aromatic ring such as the naphthoate subunit of NCS-Chrom (Fig. 2). Also, supercoiled DNA is nearly twice as effective as relaxed DNA in protecting the chromophore from degradation. Equilibrium and stopped-flow kinetic studies (D. Dasgupta and I.H. Goldberg, unpublished data) on the reversible interaction between NCS-Chrom and nucleic acids show that: 1) NCS-Chrom binds in the minor groove of B-DNA, as revealed by the lack of effect of bulky moieties in the major groove of DNA and the inhibition of binding by minor groove specific agents, such as the antibiotics netropsin and distamycin; 2) the binding mode is dependent on the base sequence (for poly(dA-dT) it is intercalative, whereas for poly(dG-dC) binding is external); and 3) rapid external binding of NCS-Chrom to DNA (presumably electrostatic) is followed by a slower intercalative binding.

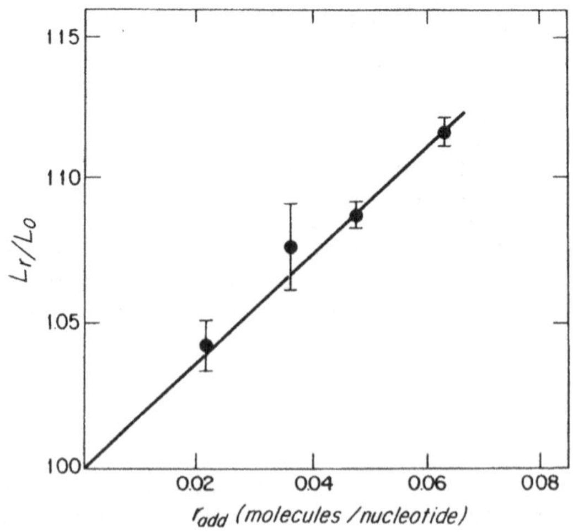

Figure 1. Change in DNA length induced by NCS-Chrom binding. DNA length was determined by measurement of electric dichroism.

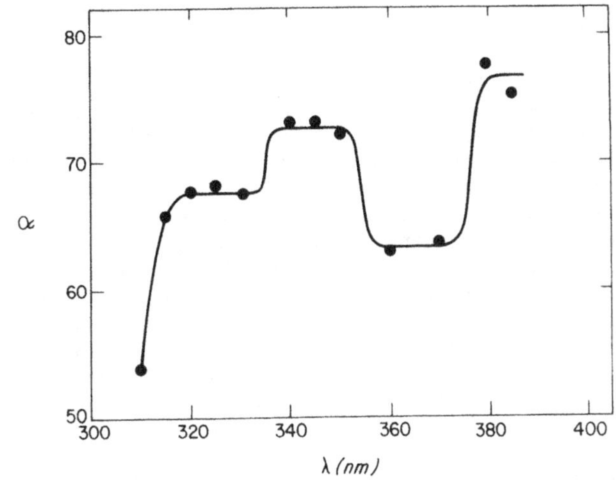

Figure 2. Electric dichroism spectrum of DNA-NCS-Chrom complex plotted as apparent angle α between chromophore transition moments and DNA helix axis.

CHEMISTRY OF NCS-CHROM

NCS-Chrom has been purified by high-pressure liquid chromatography, and a structure has been proposed for its major form[5-7] (Fig. 3). It consists of three subunits: a 5-methyl-7-methoxy naphthoate, a 2,6-dideoxy-2-methylamino-galactose moiety and an interconnecting highly unsaturated C_{12}-subunit bearing a cyclic carbonate moiety and an epoxide. The C_{12}-subunit consists of a novel, highly strained bicyclo(7,3,0)dodecadiyne system. Mercaptan treatment in methanol results in the addition of one molecule of mercaptan, as well as reduction of the C_{12}-substructure, consuming a total of three sulfhydryls. Spectroscopic and other data suggest that the naphthoate moiety of NCS-Chrom intercalates DNA and that the positively charged amino-sugar interacts electrostatically with the negatively charged oxygen of the DNA sugar-phosphate backbone. These types of binding may serve to anchor the NCS-Chrom to DNA so as to place the C_{12} connecting piece of the chromophore in juxtaposition with the DNA deoxyribose. The C_{12}-subunit undergoes rearrangement upon thiol-induced reduction to an activated species, possibly a free radical form, that damages the DNA. ESR studies show that free radical signals are generated during the thiol reaction with NCS, but their relationship to "activated" drug is not yet clear.

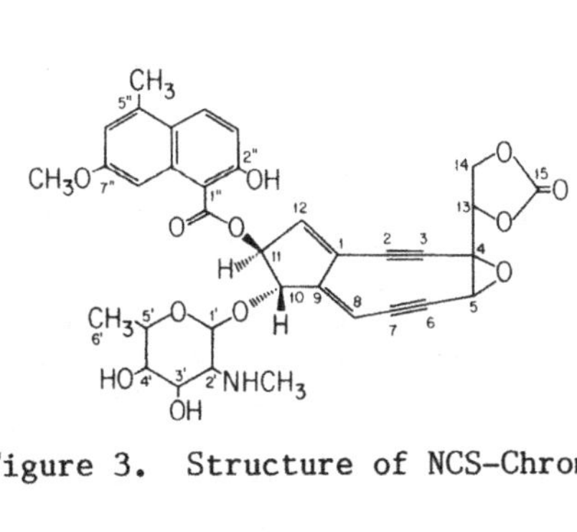

Figure 3. Structure of NCS-Chrom.

TYPES OF DNA DAMAGE

Low doses of NCS selectively inhibit DNA replicon initiation in mammalian cells, while having little or no effect on chain elongation. NCS or NCS-Chrom treatment of mammalian cells induces single-strand breaks in the DNA, that can be repaired to a considerable extent.[2,3,8,9] Experiments using the nucleoid sedimentation technique show a quantitative correlation between NCS-induced inhibition of replicon initiation and generation of sufficient strand breakage to relax domains of supercoiling in the DNA (target size of about 10^9 daltons) of mammalian cells. Double-strand breaks appear to result from the coincidence of single-strand breaks on opposite strands within a few base pairs of one another. DNA damage can be reproduced in a system devoid of cells or cellular components, provided that a reducing agent, such as a thiol, DNA and drug are simultaneously present. Not only does NCS degrade naked DNA, it also solubilizes chromatin from mammalian cell nuclei by attacking primarily the linker DNA between nucleosomes.

At least four types of lesions are formed in DNA by NCS-Chrom in the presence of sulfhydryl cofactors. The predominant lesion is a strand break (T>A>>C>G), resulting from selective, oxygen-dependent oxidation of deoxyribose to a 5'-aldehyde. The next most frequent lesion is release of free base (T>A>>C>G), presumably associated with some as yet undetermined form of sugar oxidation. In addition, at least two types of adducts between DNA and NCS-Chrom have been detected. In the presence of 2-mercaptoethanol and oxygen, a labile adduct containing an oxidized deoxyribose is formed at a very low yield. In the presence of dithiothreitol, a much more stable adduct species is produced at a somewhat higher yield.[10] Under anaerobic conditions, the yield of this type of adduct is further increased (while strand breakage and base release are suppressed), so that stable adducts become the predominant DNA lesion. This stable adduct species, and probably the labile adduct species as well, involves covalent linkage of the chromophore to deoxyribose in DNA.

In Vitro Strand Breakage, Base Release, and Deoxyribose Damage

The identification of T and A residues as targets in double-stranded DNA in the scission reaction was first revealed by experiments in which various synthetic and natural DNA's of different base composition were tested for their ability to protect against the degradation of a radioactive DNA by NCS. Subsequent studies using the DNA sequencing technique of Maxam and Gilbert where $^{32}PO_4$ is attached to the 5'-end of the DNA, also showed that NCS cleaves double-stranded DNA restriction fragments almost exclusively at T and A residues. Overall, T residues are attacked much more frequently than are A residues, although there is variability in the attack rate for both at different locations in the DNA molecule, presumably because of tertiary structural constraints. While all T residues are sites of scission, not all A residues are cleavage sites, but there appears to be no clear-cut nucleotide sequence specificity in determining cleavage frequency. These studies also confirm earlier studies that show there is a phosphate moiety at the 3'-end of the break.

Studies with DNA's labeled with different forms of radioactive thymidine indicated that DNA strand breakage is accompanied by damage to the deoxyribose moiety. These findings were confirmed by DNA sequencing experiments, using DNA restriction fragments with $^{32}PO_4$ at the 3'-end, that showed that the 5'-end of the strand-break possessed a modified nucleoside, not an exposed phosphate group (Fig. 4). In these studies the radioactive band at the point of a break moved about two nucleotides more slowly on the electrophoresis gel than the marker produced by chemical treatment in the Maxam-Gilbert procedure, which has a 5'-phosphate end (compare lanes 4 and 5 with T + C marker lane).

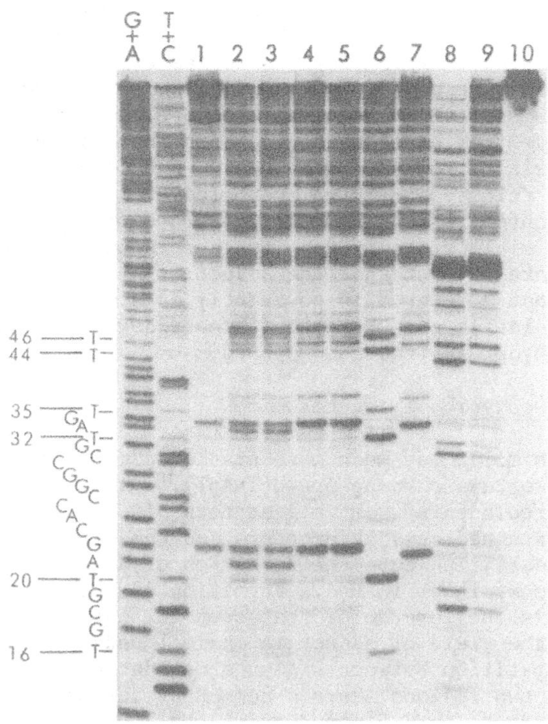

Figure 4. Cleavage of 3'-^{32}P end-labeled DNA restriction fragment by NCS-Chrom.

The modified nucleoside could be removed by alkali so as to generate a phosphate-ended fragment (lane 6). The modified nucleoside in thymidine-labeled DNA was liberated by nuclease digestion and was characterized as thymidine 5'-aldehyde (see 3 in Scheme I). Since nucleoside 5'-aldehyde would be expected to release its base (and sugar) via β-elimination reactions upon alkaline treatment, the finding that over 80% of base-release is alkali-dependent indicates that most of the breaks result from oxidation at the C-5' of deoxyribose in DNA. Similarly, DNA sequencing experiments show that strand breaks (gaps) with phosphate moieties at both ends account for less than 20% of the spontaneous breakage. In addition to spontaneous breaks there are alkali-labile breaks,[3,11] the exact chemistry of which is yet to be elucidated (see below).

Labile NCS-Chrom-Deoxyribose Adducts

In addition to causing DNA strand breaks with nucleoside 5'-aldehyde and 5'-phosphate termini, NCS-Chrom forms novel covalent adducts with DNA. The predominant adduct recovered from nuclease digests of NCS-Chrom-treated poly(dA-dT)·poly(dA-dT) in the presence of 2-mercaptoethanol and dioxygen is a compound with the structure chromophore-d(TpApT), in which the attached chromophore rendered both phosphodiester bonds refractory to endonuclease S_1. This adduct fragment was completely hydrolyzed at pH 12, releasing adenine, 3'-dTMP, and 5'-dTMP. At pH 7 the adduct fragment slowly released chromophore and 3'-dTMP with parallel kinetics, leaving a modified d(ApT), which was cleaved by snake venom phosphodiesterase to yield 5'-dTMP and a modified deoxyadenosine. These hydrolyses patterns are unlike those of any previously characterized base or phosphotriester DNA adduct, but indicate the presence of an altered deoxyadenosine sugar with chemical properties similar to those of thymidine 5'-aldehyde. The modified deoxyadenosine was identified as deoxyadenosine 5'-aldehyde by a series of chemical derivatizations. The formation of adducts containing a modified deoxyribose suggests that deoxyribose is the site of covalent chromophore attachment. Support for such a possibility comes from experiments in which this compound was subjected to mild acid hydrolysis to release free adenine and to snake venom exonuclease (pH 6.5) treatment to release 5'-TMP so as to leave, in both cases, adducts of slightly altered chromatographic mobility. These results eliminate adenine and 5'-dTMP as possible sites of covalent chromophore attachment. In addition, electrophoresis data suggest that the adduct is not a phosphotriester. These results suggest that NCS-Chrom is covalently attached to the C-5' of deoxyadenosine in some labile structure that breaks down at neutral pH to deoxyadenosine 5'-aldehyde and free chromophore.

Stable NCS-Chrom-Deoxyribose Adducts

Recently, a similar but much more stable adduct species has been isolated, again with the structure chromophore-d(TpApT).[10] Acid hydrolysis of this adduct released free adenine, as well as both 3'-dTMP and 5'-dTMP, leaving a compound which contained only chromophore and the deoxyadenosine sugar moiety. These results clearly indicate the formation of a covalent linkage between chromophore and deoxyribose in DNA. Formation of this adduct shows an oxygen dependence which is the inverse of that seen for nucleoside 5'-aldehyde formation, i.e., the yield of adduct is maximal under anaerobic conditions. This apparent competition between DNA sugar oxidation and adduct formation suggests that the two lesions share a common precursor. A chromophore-induced carbon-centered radical on C-5' of deoxyribose, that could react by addition with the same chromophore molecule to form the stable covalent adduct or with O_2 to form the 5'-aldehyde (see below), would be the most logical candidate for such a precursor.

The resistance of both types of chromophore-oligonucleotide adducts to further nuclease digestion suggests that the adducts block exonucleolytic digestion of the DNA and that the sequence specificity of adduct formation might then be determined by mapping adducts as exonuclease termination sites in defined-sequence DNA. A 5'-end-labeled DNA restriction fragment was treated with NCS-Chrom under anoxia in the presence of dithiothreitol, conditions known to maximize formation of chromophore-deoxyribose adducts. Under conditions where unmodified DNA was digested to completion, chromophore-treated DNA was highly resistant to digestion by exonuclease III plus the 3' → 5' exonucleolytic activity of T4 DNA polymerase, and partially resistant to digestion by exonuclease III plus snake venom exonuclease. Since the exonucleolytic activity of T4 polymerase does not have the pause sites of exonuclease III and degrades single-stranded DNA (unlike exonuclease III), its use alone or in combination with exonuclease III provides a much "cleaner" picture of the sites of obstruction due to adduct. The electrophoretic mobilities of the products of exonucleolytic digestion suggested that: 1) digestion by exonuclease III or T4 polymerase terminated one nucleotide before the nucleotide containing the adduct; 2) the remaining nucleotide directly adjacent to the adduct (3' side) could be removed by snake venom phosphodiesterase, but at a slow rate; 3) the covalently linked chromophore decreased the electrophoretic mobilities of the digestion products by the equivalent of approximately 3 nucleotides; and 4) adducts formed under anaerobic conditions occurred at the same nucleotide positions as the strand breaks formed under aerobic conditions (primarily at T, and to a lesser extent, A residues). The close similarity in sequence specificity of adducts and strand breaks suggests that a common form of nascent DNA damage may be a precursor to both lesions. Again, a chromophore-induced free radical on C-5' of deoxyribose, subject to competitive fixation by addition reactions with either oxygen or chromophore, is the most likely candidate for such a precursor.

Adducts on the sugar in DNA have not been described before. Most adducts of nucleic acids are on the base or phosphate. It is clear that in the case of NCS a chemical reaction on C-5' in DNA is necessary before a suitable site is generated for adduct formation. Any agent capable of inducing carbon-centered deoxyribose radicals could in principle also induce deoxyribose adducts as a result of addition reactions between these radicals and other molecules. Formation of N-ethylmaleimide adducts on deoxyribose in DNA in the presence of ionizing radiation has been proposed to occur by such a mechanism.[13] Bulky adducts on DNA sugars obviously could not be removed by repair glycosylases, and it seems unlikely, though not impossible, that the DNA would be recognized by polymerases as a normal substrate. Furthermore, the phosphodiester bonds of adduct-containing DNA sugars appear to be relatively resistant to a variety of nucleases.

KINETICS AND STOICHIOMETRY OF DIOXYGEN AND THIOL CONSUMPTION

DNA strand breakage, base release and the formation of nucleoside 5'-aldehyde by NCS require O_2. Kinetic studies have indicated that, in the presence of DNA, O_2 and sulfhydryls, the chromophore undergoes a single oxidation-reduction cycle, which efficently damages DNA and leaves an inactive product devoid of further redox activity. The process can be resolved into three steps. The first step, which is accompanied by marked changes in chromophore spectral properties, but which does not require O_2, is a reaction between the chromophore and a single sulfhydryl, probably an addition reaction. Formation of this activated chromophore species is rapidly followed (within a few seconds) by the uptake of one mole of O_2 per mole of chromophore. O_2 uptake is then followed by uptake of an additional sulfhydryl group. Since consumption of the second sulfhydryl does not occur in the absence of O_2, it may represent a one-electron reduction of a reaction intermediate which includes some form of reduced oxygen (see below). The oxygen that ends up in the 5'-aldehyde at the strand break is derived entirely from O_2 and not H_2O, as shown by O^{18} incorporation experiments.

Nitroaromatic radiation sensitizers, such as misonidazole, nitrofurazone and metronidazole, substitute for O_2 in NCS-induced DNA strand breaking (Fig. 5).[14] Compounds with higher one-electron reduction potentials are more effective in stimulating the anaerobic reaction. It is of interest, however, that while the DNA base attack site specificity for NCS remains the same as in the aerated reaction, the DNA damage products are different. Spontaneous base release increases and is associated with spontaneous (and alkali-labile) breaks with phosphates at both the 3'- and 5'-ends (i.e., gaps) and no nucleoside 5'-aldehyde.

MECHANISM OF DNA DAMAGE

Thiol-activated NCS-Chrom abstracts tritium from the 5', but not from the 1' or 2', position of deoxyribose in DNA and incorporates it into a stable, non-exchangeable form.[15,16] The abstracted tritium remains covalently associated with the chromophore. Drug activation and the consequent hydrogen abstraction reaction, presumably generating a carbon-centered radical at C-5', do not require molecular oxygen but have a dose-dependent relation with thiol.[15] Under aerobic conditions, where base release and DNA strand breaks with nucleoside 5'-aldehyde at the 5'-ends are produced, hydrogen abstraction from C-5' parallels these parameters of DNA damage. Based on these and earlier mentioned data, it is possible to formulate a reaction scheme in

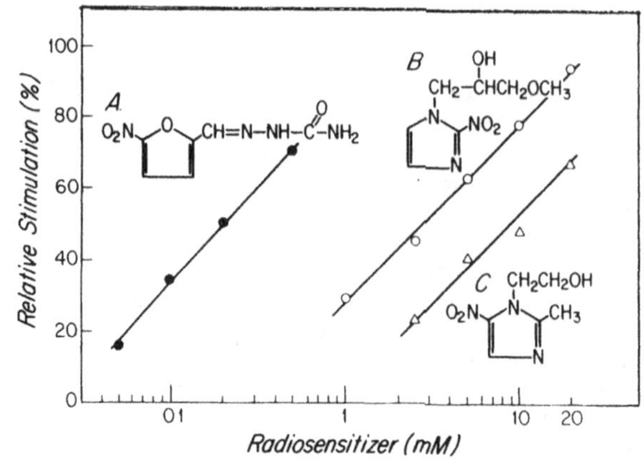

Figure 5. Comparison of the ability of different nitroaromatic radiation sensitizers to substitute for oxygen in the stimulation of NCS-dependent DNA damage in nitrogen. (A) Nitrofurazone, E^1_7/mv = -257; (B) Misonidazole, E^1_7/mv = -389; and (C) Metronidazole, E^1_7/mv = -486.

which a carbon-centered radical at C-5' is an intermediate in the formation of the various DNA damage products found under both aerobic and anaerobic conditions (Scheme I).[16]

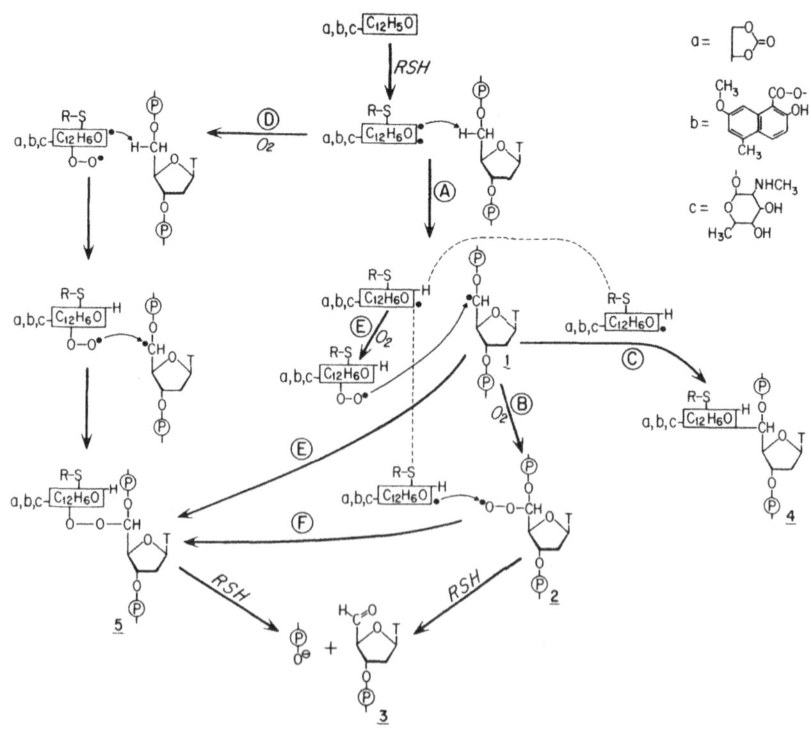

Scheme I

The first step involves the generation of an active species of the drug (shown as a free radical) upon thiol addition to the highly unsaturated C_{12} subunit of the chromophore. The activated drug abstracts a hydrogen from the 5'-carbon of deoxyribose (pathway A) to generate a carbon-centered radical (1) on the DNA, the fate of which depends on whether or not O_2 is available. Addition of O_2 to the 5'-carbon-centered radical (pathway B) gives rise to a peroxyl radical (2),[17] which degrades in the presence of thiol (via formation of a hemiacetal) to produce DNA strand breaks with 5'-terminal nucleoside aldehyde (3). The involvement of thiol in the degradation of a peroxyl structure to an aldehyde is compatible with the experiments in which at least a second molecule of thiol is consumed in the reaction beyond the one used in the initial activation of the drug. In the absence of O_2 the carbon-centered radical on the deoxyribose (1) may instead react with the chromophore containing the abstracted hydrogen (pathway C) to form a stable covalent chromophore-DNA adduct (4), possibly by addition of the C-5' radical across an unsaturated bond in the chromophore, or to be repaired by abstraction of a hydrogen from thiol. This analysis, involving a common DNA intermediate, is consistent with studies[10] showing that there is a competition between stable chromophore-DNA adduct formation in anoxia and DNA strand breakage and base release in O_2.

Although drug activation and hydrogen atom abstraction from the 5'-position of deoxyribose in DNA can proceed in the absence of O_2, this does not rule out the possibility that in the presence of O_2 the activated chromophore can add O_2 to form a peroxyl derivative of the activated chromophore (pathway D). The peroxyl radical on the drug cannot itself be responsible for hydrogen atom abstraction, since the latter is incorporated into the chromophore in a non-exchangeable form. Nevertheless, a radical elsewhere in the peroxyl form of the drug could abstract the hydrogen atom and then react (possibly as part of a concerted reaction) with the carbon-centered radical on the DNA by way of its peroxyl radical to form a labile adduct, possibly with the structure shown in (5), as an intermediate in strand break formation. Again, this peroxy adduct may be cleaved by thiol to form 3. Consistent with such a second role for thiol in the formation of the 5'-aldehyde are experiments showing that 2-mercaptoethanol accelerates the decomposition of the labile adduct (L.F. Povirk and I.H. Goldberg, unpublished results) and that high thiol concentrations increase both the absolute yield of 5'-aldehyde and the ratio of aldehyde to released free base (L.S. Kappen and I.H. Goldberg, unpublished results).

It is also possible that the activated chromophore adds O_2 after it abstracts the hydrogen atom from DNA (pathway E) and then reacts with (1) to form the labile adduct, (5); or it may instead react with the peroxyl form of the deoxyribose (2) to form (5) via pathway F. Since the chromophore is anchored on the DNA by intercalation and electrostatic interactions, it would not be surprising if it participated in such reactions after the initial hydrogen abstraction. Reaction pathway E, involving a peroxyl form of the drug and a labile chromophore-DNA adduct intermediate, best fits data showing that one molecule of O_2 is consumed for each molecule of chromophore, whereas DNA damage is somewhat less than stoichiometric. Furthermore, pathway E is more attractive than D, since it does not require postulating a different hydrogen-abstracting form of the chromophore under aerobic and anaerobic conditions. Finally, it should be noted that it is possible to combine pathways A and E in a concerted mechanism involving a ternary complex of activated chromophore, O_2 and C-5' of deoxyribose.

The proposed scheme focuses on the 5'-carbon of deoxyribose as the predominant site of attack. It is consistent with the findings that over 80% of the DNA strand breaks (major lesion) have nucleoside 5'-aldehyde at their 5'-termini and that C-5' oxidation also occurs in the formation of the labile chromophore-DNA adducts. The absence of any ^3H abstraction from C-1' and C-2' of thymidine in DNA eliminates these positions as targets. It should be noted, however, that C-1' appears to be the primary site of hydrogen abstraction when radiolytically-activated NCS reacts with DNA.[18] Furthermore, since the mechanism of spontaneous base release and the chemistry of the alkali-labile lesions are unknown, it is not possible at this stage to invoke attack at C-5' in the generation of these lesions.

When misonidazole substitutes for O_2 in the NCS-DNA reaction, the attack site specificity is the same as in O_2, but in the presence of misonidazole base release and gap formation (with 3'- and 5'-phosphoryl termini) predominate. Despite the difference in final DNA products, abstraction of 5'-hydrogen by thiol-activated drug to form a carbon-centered radical on C-5' of deoxyribose appears to be a common initial step in both the O_2- and misonidazole-dependent reactions; in fact, there appears to be a competition between the misonidazole-dependent and the oxygen-dependent pathways. It is likely that after the induction of the same nascent lesion on the DNA, participation of misonidazole in the subsequent steps, by a mechanism different from that of O_2, leads to different products.

It should be emphasized, as mentioned earlier, that while the two antitumor antibiotics, NCS and bleomycin, cause hydrogen atom abstraction from deoxyribose in DNA (from the 5'-position with NCS and from the 4'-position with bleomycin), the mechanisms are fundamentally different. NCS is activated to a species that abstracts the hydrogen atom itself, whereas iron in bleomycin activates oxygen to do the abstraction.

BASIS OF MUTAGENICITY

NCS is an SOS-dependent mutagen that induces base substitutions and, to a lesser extent, frameshifts. NCS and a number of other "oxidative mutagens" have been found to be mutagenic for a new Salmonella tester strain, TA 102, that contains A-T base pairs at the site of mutation in a multicopy plasmid,[19] consistent with the *in vitro* base specificity studies. On the other hand, studies of NCS-induced mutations in the *lacI* gene revealed that all types of monitorable base substitutions were induced, with some preference for G:C to A:T transitions (A:T to G:C transitions are not monitorable in this system).[20] The spectrum was dominated by a G:C to A:T transition hotspot at the ochre 21 locus. Several less prominent hotspots were also present, while no mutations at all were detected at nearly half the monitorable sites. Since the predominent lesions induced in DNA by NCS or NCS-Chrom are strand breaks or covalent NCS-Chrom-deoxyribose adducts occurring mainly at T and A residues, it appears that many of the mutations, particularly those occurring at G:C base pairs, result from other, minor lesions, which are formed with a different base specificity.

Since NCS releases free bases from DNA, it seemed likely that AP sites with an intact sugar-phosphate backbone are formed. In fact, AP endonuclease-sensitive sites have been detected in NCS-treated supercoiled DNA and DNA restriction fragments. When defined-sequence DNA from the *lacI* region of plasmid pMC1 was treated with NCS-Chrom in the presence of glutathione, alkali-dependent strand breaks, occurring at certain cytosine residues, were detected in addition to the direct strand breaks at T and A residues[21] (Fig. 6). Although these sites were substrates for endonuclease IV, a five-fold greater concentration of enzyme was required for their cleavage than was required for cleavage of apurinic sites in depurinated DNA. These sites were also at least five-fold less sensitive to *E. coli* endonuclease VI (exonuclease III) and at least ten-fold less sensitive to *E. coli* endonuclease III. The correlation between alkali-dependent cleavage at C residues and release of

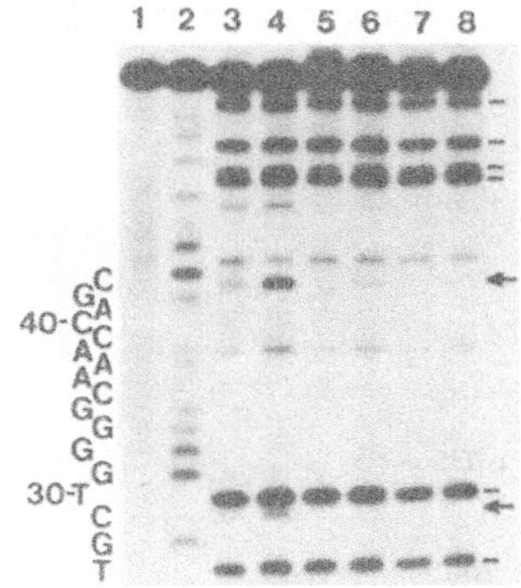

Figure 6. Formation of apyrimidinic sites at C residues.[21] 5'-end-labeled lacI DNA fragment was either depurinated (lanes 1 and 2), or treated with NCS-Chrom in presence of glutathione (lanes 3 and 4), dithiothreitol (lanes 5 and 6) or 2-mercaptoethanol (lanes 7 and 8). Samples in even lanes were heated in alkali. Arrows show C residues at which alkali-dependent cleavage occurs.

free cytosine base, as well as the susceptibility of the lesions to endonuclease IV of E. coli, strongly suggest that they are AP sites. It seems reasonable to propose that the NCS-induced sites are resistant because the sugar residue has a chemical structure different from that of apurinic sites generated by acid treatment. In fact, when NCS-induced AP sites at C residues in 5'-end-labeled DNA are cleaved by heating at pH 8, the resulting fragments comigrate with the 3' phosphate-ended marker; however, fragments corresponding to heat-induced cleavage of authentic AP sites migrate one nucleotide more slowly, suggesting that the sugar remains attached to the 3'-end. Thus, the 5' sugar-phosphate linkage is apparently more labile in the NCS-induced site than in the authentic AP site.

When chromophore-induced apyrimidinic sites were quantitated as alkali-dependent breaks at eleven specific sites in the lacI gene, a correlation (coefficient of 0.94) was found between occurrence of these lesions[21] and the reported frequencies of G:C to A:T transitions at the same sites[20] (Fig. 7). All occurrences of the trinucleotide sequence AGC, including the ochre 21 mutational hotspot, were particularly prominent sites. The selective formation of endonuclease-resistant apyrimidinic sites at specific C residues may explain the high frequency of G:C to A:T transitions in the mutational spectrum of NCS, especially since it has been shown that there is a selective incorporation of A residues opposite AP sites.[22]

The observed occurrence of adducts almost exclusively at T and A residues[12] does not suggest a role for adducts in generating GC:AT transitions. One hypothesis which cannot be excluded at present, however, is that adducts on T or A residues may result in mutations at adjacent G:C base pairs. In fact, the T residues directly preceding the G residues associated with two G:C to A:T transition hotspots (ochre 21 and amber 19) were found to be prominent sites of adduct formation. Since, due to the nature of the genetic code, all monitorable G:C to A:T transitions in the lacI system occur at G residues preceded by a T residue[23], the NCS spectrum in lacI provides little evidence either for or against such a mechanism. Since chromophore adducts involve modification of DNA sugars, rather than bases, it is possible that normal base-pairing can still occur during replication, and that little or

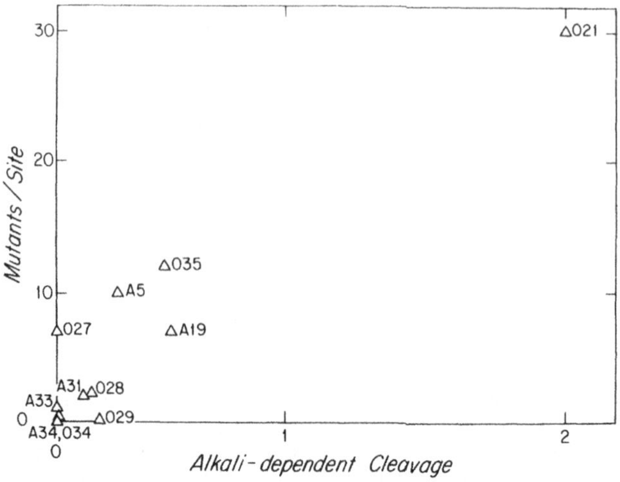

Figure 7. Occurrence of G:C to A:T transitions in vivo vs. alkali-dependent cleavage in vitro at sites in the lacI gene.[21] Values represent relative probability of alkali-dependent cleavage at each site. Data on mutation frequency are those of Foster and Eisenstadt.[20] A = amber sites, O = ochre sites.

no base substitution mutagenesis occurs at adduct sites. On the other hand, NCS-Chrom is an intercalator, and other covalently bound intercalators, such as the acridine mustards,[24] are potent frameshift mutagens. Thus, adducts may be involved in NCS-induced frameshift mutagenesis, rather than base substitution mutagenesis. The sequence specificity of NCS-induced frameshifts is unknown.

As an alternative approach for studying the biological consequences of various NCS-induced lesions, we have also been examining the effects of NCS-Chrom on intact lambda phage (L.F. Povirk and I.H. Goldberg, unpublished data). Preliminary results suggest that chromophore treatment is about two-fold less lethal and five-fold less mutagenic under anaerobic conditions. These results suggest that adducts are less lethal and less mutagenic than oxygen-dependent lesions, and are probably repaired in E. coli.

REFERENCES

1. I.H. Goldberg and P.A. Friedman, Antibiotics and Nucleic Acids. Ann. Rev. Biochem. 40:775 (1971).
2. I.H. Goldberg, T. Hatayama, L.S. Kappen, M.A. Napier, and L.F. Povirk, Protein antibiotics as DNA damaging agents. in: "Molecular Actions and Targets for Cancer Chemotherapeutic Agents," A.C. Sartorelli, J.R. Bertino and J.S. Lazo, eds., Second Annual Bristol-Myers Symposium in Cancer Research, Academic Press, New York, p. 163 (1981).
3. I.H. Goldberg, L.F. Povirk, L.S. Kappen, and D.-H. Chin, Mechanism of genotoxicity by a radiomimetic antitumor antibiotic. in: "Cellular Regulation and Malignant Growth", S. Ebashi, ed., Japan Sci. Soc. Press, Tokyo/Springer-Verlag, Berlin, p. 483 (1985).
4. V.Z. Pletnev, A.P. Kuzin, S.D. Trakhanov, and P.V. Kostetsky, Three-dimensional structure of actinoxanthin. IV. A 2.5Å resolution. Biopolymers, 21:287 (1982).

5. O.D. Hensens, R.S. Dewey, J.M. Liesch, M.A. Napier, R.A. Reamer, J.L. Smith, G. Albers-Schonberg, and I.H. Goldberg, Neocarzinostatin chromophore: Presence of a highly strained ether ring and its reaction with mercaptan and sodium borohydride. Biochem. Biophys. Res. Commun. 113:538 (1983).
6. M. Shibuya, K. Toyooka, and S. Kubota, Synthesis of the naphthalene-carboxylic acid derivative obtained from neocarzinostatin (NCS): a structure revision. Tetradedron Lett. 25:1171 (1984).
7. K. Edo, M. Mizugaki, Y. Koide, H. Seto, K. Furikata, N. Otake, and N. Ishida, The structure of neocarzinostatin chromophore possessing a novel bicyclo[7,3,0]dodecadiyne system. Tetradedron Lett. 26:331 (1985).
8. W.N. Hittelman and M. Pollard, Induction and repair of DNA and chromosome damage by neocarzinostatin in quiescent normal human fibroblasts. Cancer Res. 42:4584 (1982).
9. W.-L. Kuo, R.E. Meyn, and C.W. Haidle, Neocarzinostatin-mediated DNA damage and repair in wild-type and repair-deficient chinese hamster ovary cells. Cancer Res. 44:1748 (1984).
10. L.F. Povirk and I.H. Goldberg, Competition between anaerobic covalent linkage of neocarzinostatin chromophore to deoxyribose in DNA and oxygen-dependent strand breakage and base release. Biochemistry, 23:6304 (1984).
11. E. Boye, W. Köhnlein, and K. Skarstad, Characterization of intracellular DNA strand breaks induced by neocarzinostatin in Escherichia coli cells. Nucleic Acids Res. 12:8281 (1984).
12. L.F. Povirk and I.H. Goldberg, Detection of neocarzinostatin chromophore-deoxyribose adducts as exonuclease-resistant sites in defined-sequence DNA. Biochemistry, in press (1985).
13. J.F. Ward, Molecular mechanisms of radiation-induced damage to nucleic acids. Adv. Radiat. Biol. 5:181 (1975).
14. L.S. Kappen and I.H. Goldberg, Nitroaromatic radiation sensitizers substitute for oxygen in neocarzinostatin-induced DNA damage. Proc. Natl. Acad. Sci. USA 81:3312.(1984).
15. R.L. Charnas and I.H. Goldberg, Neocarzinostatin abstracts a hydrogen during formation of nucleotide 5'-aldehyde on DNA. Biochem. Biophys. Res. Commun. 122:642 (1984).
16. L.S. Kappen and I.H. Goldberg, Activation of neocarzinostatin chromophore and formation of nascent DNA damage do not require molecular oxygen. Nucleic Acids Res. 13:1637 (1985).
17. C. von Sonntag, U. Hagen, A. Schön-Bopp, and D. Schulte-Frohlinde, Radiation-induced strand breaks in DNA: Chemical and enzymatic analysis of end groups and mechanistic aspects. Adv. Radiat. Bol. 9:109 (1981).
18. V. Favaudon, R.L. Charnas, and I.H. Goldberg, Polydeoxyadenylic-deoxythymidylic acid damage by radiolytically activated neocarzinostatin. Biochemistry 24:250 (1985).
19. D.E. Levin, M. Hollstein, M.F. Christman, E.A. Schwiers, and B.N. Ames, A new Salmonella tester strain (TA102) with A·T base pairs at the site of mutation detects oxidative mutagens. Proc. Natl. Acad. Sci. USA 79:7445 (1982).
20. P.L. Foster and E. Eisenstadt, Distribution and specificity of mutations induced by neocarzinostatin in the lac I gene of Escherichia coli. J. Bacteriol. 153:379 (1983).
21. L.F. Povirk and I.H. Goldberg, Endonuclease-resistant apyrimidinic sites formed by neocarzinostatin at cytosine residues in DNA: Evidence for a possible role in mutagenesis. Proc. Natl. Acad. Sci. USA, 82:3182 (1985).
22. L.A. Loeb, Apurinic sites as mutagenic intermediates. Cell 40:483 (1985).

23. C. Coulondre and J.H. Miller, Genetic studies of the lac repressor. III. Additional correlation of mutational sites with specific amino acid residues. J. Mol. Biol. 117:525 (1977).
24. J.H. Creech, R.K. Preston R.M. Peck, A.P. O'Connell, and B.N. Ames. Antitumor and mutagenic properties of a variety of heterocyclic nitrogen and sulfur mustards. J. Med. Chem. 15:739 (1972).

DNA-DRUG BINDING AND CONTROL OF GENETIC INFORMATION

Richard E. Dickerson

Molecular Biology Institute
University of California, Los Angeles
Los Angeles, California 90024

INTRODUCTION

It is generally accepted that the coding problem in molecular genetics has been solved--that is, that we understand the manner in which a particular amino acid sequence is coded within the progression of bases along DNA, and how the information present in the DNA is transcribed and used to assemble the polypeptide chain. Many experimental details remain to be cleared up, but the fundamental concept of triplet codons and a genetic code is well established.

The central problem in molecular genetics and molecular biology today appears to be one of control of information, rather than coding. Every cell in a living organism contains the same complement of genetic information. How is this information controlled so that one cell line leads to muscle cells, another to liver cells, and others to all of the great number of differentiated cell types in a mature organism? This question is important enough, and broad enough, to motivate half or more of the research going on in molecular biology today, from development and differentation, to genetic engineering and the study of oncogenes. We are involved with one aspect of the question of control--regulation of the flow of genetic information from DNA by the binding of small molecules such as drugs, or macromolecules such as repressors, to the DNA.

Our immediate goals are to use DNA oligonucleotide synthesis and x-ray crystal structure analysis to study the molecular basis for the interaction of groove-binding and intercalating antitumor antibiotics with double-helical DNA, and then to use this information to design more efficient DNA-binding agents, including ones capable of recognizing particular sequences of ten or more base pairs. This work is a logical extension of our recent x-ray structure analysis of the complex of netropsin, a minor groove binding drug that recognizes and binds to regions of A:T base pairs, with a B-DNA dodecamer of sequence: CGCGAATTCGCG.[1-3] Groove-binding drugs probably are more promising candidates for the design of sequence-specific DNA-binding agents than are intercalators, because the information about base sequence resides in the hydrogen bond donor and acceptor atoms at the edges of the bases on the floor of the major and minor grooves. From these studies should come new analogues of the present drugs, such as netropsin, and possibly other sequence-specific groove-binding drugs, with entirely new backbones, that can be targeted toward particular base sequences. Such newly developed drugs potentially

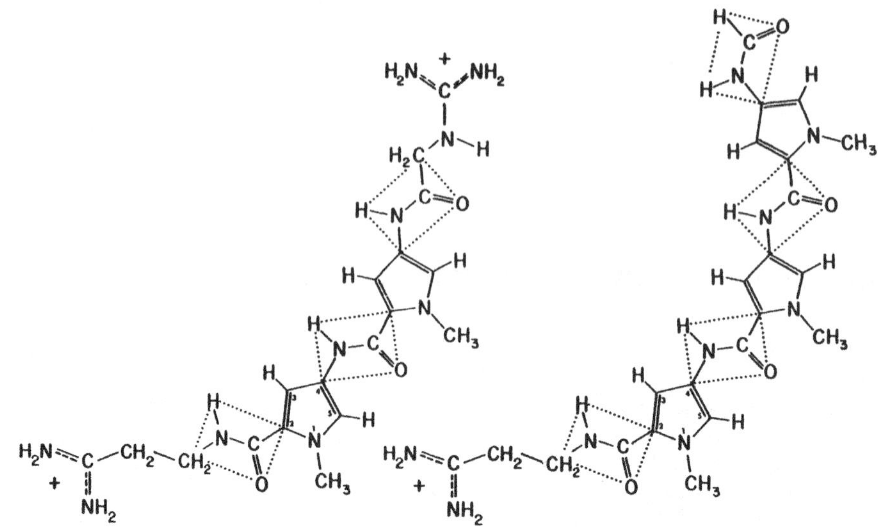

Figure 1. Netropsin (left) and distamycin (right). Both may be regarded as polymers of repeating pyrrole-amide units. Netropsin has two cationic ends; distamycin has only one positive charge. These are the naturally-occurring species. These and synthetic analogues with greater or fewer pyrrole-amide units sometimes are numbered according to the number of pyrroles in the molecule. The natural compounds then would be designated as Nt-2 and Dst-3, respectively.

could be directed toward invader cells in preference to the host cells of a patient, or toward neoplastic cells rather than normal cells.

Many of the more effective antineoplastic agents in cancer chemotherapy exert their effect by binding to double helical DNA, thereby interfering both with replication of new DNA and transcription to messenger RNA.[4-6] Actinomycin D, adriamycin, daunomycin and bleomycin are examples of this type of drug. Netropsin and the related distamycin (Figure 1), although too toxic for routine clinical use, also are DNA-binding drug molecules. Drugs such as actinomycin and daunomycin are intercalators--that is, they have a flat multi-ring group that inserts between two adjacent base pairs of the helix, along with a non-planar residue that nests into one groove or the other. They act essentially like molecular "grit", jamming the teeth of the DNA zipper.[7,8] Intercalation has become nearly the classical mode of binding of drug molecules to DNA, and much attention has focused on other simple intercalators that have no pharmacological use, such as proflavin, other acridines and phenanthridines.

More recently it has been realized that a second important category of DNA-binding drug molecules acts, not by intercalating between base pairs, but by filling one of the grooves of the double helix.[9-10] Netropsin and distamycin are of this class,[11,12] and it begins to appear that the DNA-cleaving bleomycin may be a groove-binder rather than an intercalator, as frequently has been assumed.[13-15]

Relatively few DNA-binding antitumor drugs show clear evidence of base sequence specificity--the ability to recognize and bind to one base sequence in preference to another, as do repressors and other DNA control proteins. Actinomycin D intercalates best between bases of sequence GC,[16] and bleomycin prefers GC or GT.[17] Netropsin and distamycin are unusually specific, exhibiting strong preference for regions of four or more A:T base pairs, in double-helical

B-DNA only. Dervan has shown that the minimal binding site is directly related to the length of the drug molecule.[18] In the nomenclature of Figure 1, Nt-x or Dst-x, with x pyrroles and x + 1 amides, require x + 2 consecutive A:T base pairs as their binding sites.

The various natural and synthetic netropsins and distamycins are a class of antitumor antibiotics with the ability to bind selectively to various sized regions of consecutive A:T base pairs. We are attempting to create "lexitropsins", or DNA sequence-reading netropsin analogues, that will bind selectively to any desired base sequence in the size region of six to twelve base pairs. This is a collaborative project with Professor J. William Lown of the University of Alberta in Edmonton, Canada, in which Lown synthesizes new drug analogues, we make the requisite DNA sequences, and we examine the complexes by x-ray crystallography while he uses nmr. If we succeed in our drug design efforts, then we should be able to target the drug molecules specifically toward certain sequences of DNA at will, ignoring the vast bulk of DNA.

This ability to vector a drug molecule toward a particular DNA region should have great clinical utility. One of the central problems of cancer chemotherapy is that it represents little more than carefully controlled poisoning of the patient. In an ideal world one would want the drug to go only to neoplastic cells and to avoid the normal cells of the patient. But we have no dependable criteria by which neoplastic and normal cells can be differentiated. Drugs that act by binding to DNA, interfering with transcription and replication, cause damage to any rapidly growing and dividing cell line. This includes tumor cells, it is true, but it also includes many normal cell lines such as in the epithelial lining of the stomach, bone marrow, and hair follicles. Unintended damage to these cells produces the undesirable side effects of chemotherapy: nausea and vomiting, anemia, and hair loss. If one could synthesize a drug molecule that targeted neoplastic cells but tended not to attack these normal cells, the side effects might be ameliorated.

It would not be necessary to produce a "magic bullet" that bound to tumor cells with 100% efficiency and totally avoided normal cells. It might be sufficient merely to bind to a control region of an oncogene and reduce its expression down to normal levels. Some oncogenes are the result of point mutations in a normal developmental gene, and a drug molecule could be designed to bind at the region of the mutation, turning off the oncogene but not the normal protooncogene. If a lexitropsin could be designed that would recognize one particular six-base-pair sequence, then this same sequence would be expected to occur at random one time out of $4^6 = 4096$ in normal DNA. One would continue to hit normal DNA, but at a frequency only one-four-thousandth that for the case of the usual, nonspecific anticancer drugs. This alone would be of profound benefit in cutting down side effects.

These are only some of the more obvious applications of "lexitropsins". If we can design a dependable set of sequence-seeking, DNA-binding drug molecules, then they might equally well serve as carriers, being covalently coupled to other molecules whose effects on the DNA are of interest. They might be coupled to EDTA:Fe, as in the work of Dervan and colleagues, and used for sequence-selective DNA cleavage within the cell. This offers the prospect of "micro-surgery" *in vivo* to correct certain types of genetic mistakes. Many other possibilities will suggest themselves, if we can design a good DNA sequence vector.

We suspect that drugs and repressors utilize two different channels of information in reading base sequence in double-helical DNA: the minor groove for drugs such as netropsin, and the major groove for control proteins. The key "semantophore", or information-bearing element, in the minor groove is the

$-NH_2$ group of guanine, or its absence in adenine. This group is critical in the stabilization of the B double helix by the spine of hydration,[19-21] by cations such as Cs^+,[22] or by drug molecules.[1-3] In the major groove the key recognition elements are hydrogen bond acceptors such as N7 of purines and carbonyl O6 of G or O4 of T, and hydrogen bond donors $-NH_2$ at N6 of A or N4 of C. These groups form patterns of acceptors (a) and donors (d) across the base pair edges: a-d-a for an A:T base pair, or a-a-d for G:C. It has been suggested that amino acid side chains from repressors extend down to the floor of the major groove and recognize these patterns.[23]

The spine of hydration and stabilization of the B helix

One of the most striking features of the x-ray crystal structure analysis of CGCGAATTCGCG was a regular, zig-zag spine of water molecules running down the center of the minor groove in the AATT center, and fraying away in the two CGCG ends.[19,21] This is shown in Figure 2a. The first shell atoms of this spine of hydration form hydrogen-bonded bridges connecting adenine N3 and thymine O2 atoms, on bases in opposite strands and adjacent base pairs, that are brought closer together by virtue of the helix rotation. These first shell waters then are given local near-tetrahedral environments by bridging second-shell water molecules, to form the entire zig-zag spine. As outlined in detail,[19,21] we believe that this spine of hydration stabilizes B-DNA relative to the A form and also relative to the opening up that is required for replication and transcription. The absence of a N2 amine group in A:T base pairs favors formation of a spine of hydration, whereas the presence of the N2 amine on guanine in G:C pairs disrupts the spine and destabilizes the B form. Minor groove stabilization of the B double helix in A:T-rich polymers also can be produced by ions such as Cs^+, as reported by Bartenev et al.[22] In such cases, the cesium ions occupy precisely the same locations as the first hydration shell of the spine of hydration, bridging A and T.

Sequence-specific minor groove binding of netropsin

Mary Kopka and co-workers have carried out the x-ray crystal structure analysis of a complex of netropsin with CGCGAATTCGCG.[1-3] This binding is illustrated in Figure 2b. It was known from the outset, from earlier work of other groups, that both netropsin and distamycin (Figure 1) exhibit a remarkable specificity for A:T binding sites. They show little or no binding affinity for single-stranded DNA or RNA, or for double-stranded RNA or DNA/RNA hybrids,[24-26] suggesting that they do not bind to an A helix. Neither do they bind to left-handed Z-DNA; in fact, binding of netropsin to DNA favors A-to-B and Z-to-B helix transitions.[27-28] Natural netropsin attaches only to clusters of four or more A:T or I:C base pairs, never to G:C,[29,30,46] although distamycin is slightly more tolerant of isolated G:C base pairs.[18,31,32] Alternating ATAT regions bind netropsin less well than continuous runs of A or T on the same strand.[33-34] Chemical protection studies[35,46-47] and nmr experiments[36] indicate that netropsin does not intercalate between base pairs, but binds within the minor groove of the intact double helix, using hydrogen bonds between netropsin amide NH and the exposed adenine N3 and thymine O2 on the floor of the minor groove. As has been mentioned earlier, the size of the minimal binding site increases with the length of the netropsin or distamycin molecule.[18]

Figure 2 shows that netropsin binds within the minor groove of B-DNA by displacing the spine of hydration. Each of the three amide NH groups occupies the site formerly filled by a first-shell water molecule. This "super-spine" now is held in place not by hydrogen bonding between individual water molecules, but by covalent bonds along the backbone of the drug molecule. The drug molecule is twisted slightly so that each of its two pyrrole rings fits perpendicularly into the minor groove crevice. The minor groove is so narrow

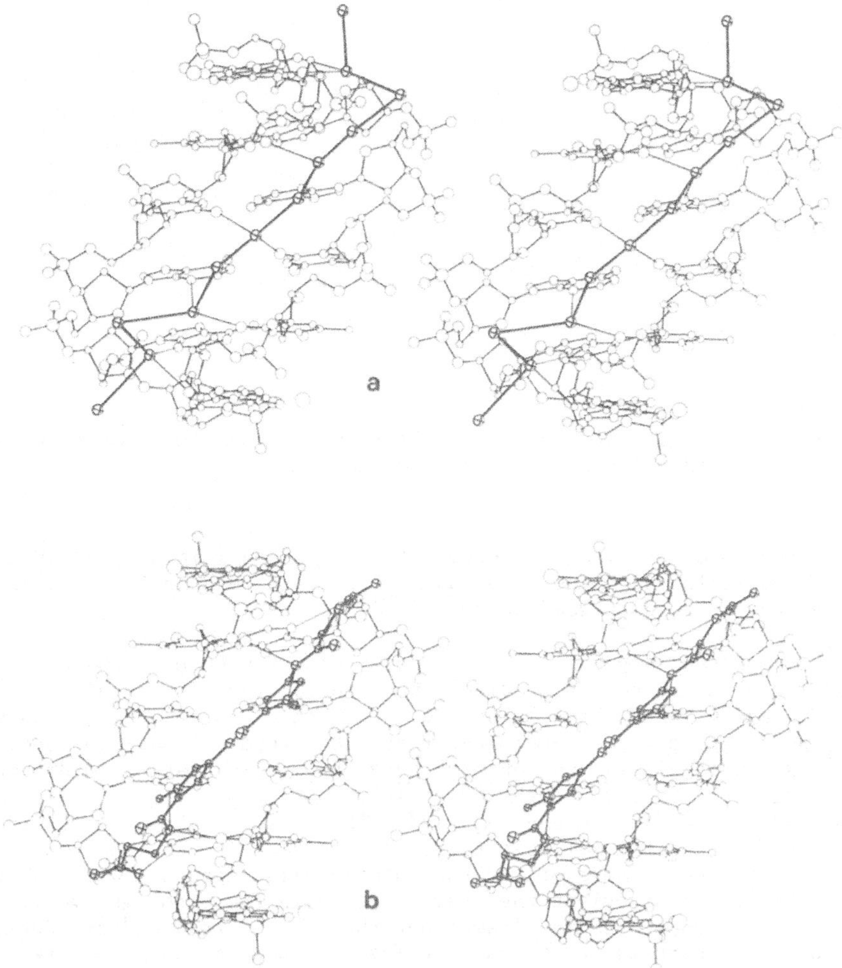

Figure 2. Comparison of (a) minor groove spine of hydration and (b) netropsin, within the minor groove. Only the central six base pairs of CGCGA-ATTCGCG are shown for clarity. Open circles and open bonds are DNA; crossed circles and dark bonds are water spine or netropsin.

in A:T base pair regions (opening of 3.5Å) that there is barely room for insertion of a flat organic ring such as a pyrrole. In fact, insertion of the drug molecule actually pushes open the minor groove somewhat, and bends back the axis of the helix by 8°. Hence, in a concerted fitting process, the DNA deforms the drug molecule slightly, and the drug deforms the DNA by a similar small amount. No gross overall unwinding or elongation is observed, however, contrary to what had been reported from some earlier solution studies. What had been interpreted as elongation probably is to be ascribed to a stiffening of the double helix by combination with the groove-filling drug molecule.

Details of the bonding between netropsin and CGCGAATTCGCG are shown in Figure 3. Amide atom N_4 makes one normal hydrogen bond to the O2 of thymine 20, and a longer bond to the N3 of adenine 6. The central amide N_6 makes two long hydrogen bonds to O2's of thymines 19 and 7, and amide N_8 bridges adenine 18 and thymine 8. The amidinium tail of the drug makes a normal hydrogen

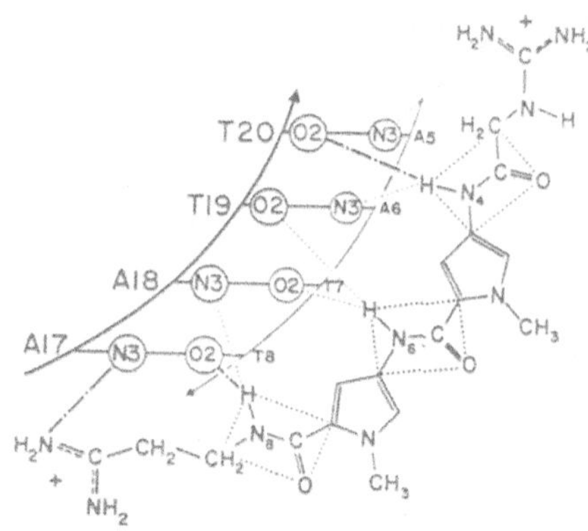

Figure 3. Interaction between netropsin and thymine 02 or adenine N3 atoms of floor of the minor groove.

bond to the N3 of adenine 17, whereas the upper guanidinium end makes a looser association with the N3 of adenine 5. The reasons for these long hydrogen bonds are discussed in detail.[1-3]

The important aspect of the netropsin-DNA structure is that the x-ray analysis enabled us to see precisely why the binding site required at least four base pairs, with these being A:T rather than G:C, and suggested ways of making modified netropsin analogues that would read G:C base pairs at selected positions. Careful examination of Figure 4b reveals that the hydrogen bonding amide NH's are not close to the C2 positions of adenines, where a guanine amine group would be recognized and rejected. The closest drug atoms to the adenine C2 on the four central A:T base pairs are pyrrole ring CH, and $-CH_2-$ methylene groups at the far corners of the two outside amide groups. The hydrogen bonding amide groups orient the drug properly in the minor groove of the DNA and provide a proper reading frame; but the actual "reading" of adenine vs. guanine is carried out by van der Waals nonbonded contacts between methylenes and ring CH on the drug, and adenine base edges on the DNA. This binding and reading is schematized in Figure 4a.

Distances between adenine C2 ring atoms and the close approach atoms on netropsin are exactly what would be expected for van der Waals packing: 3.5 Å. If one constructs a hypothetical $-NH_2$ group on the C2 atom, as in guanine, then the separation between this nitrogen N2 and the netropsin atoms is around 2.8 Å. This is too short for a nonbonded contact but is exactly right for a hydrogen bond. This observation set us to thinking about how one might design a modified netropsin that would accept a guanine -NH . The result is the lexitropsin, or sequence-reading netropsin analogue, depicted in Figure 4b. If the pyrrole CH is replaced by N: as in imidazole, or by O: as in furan, then enough room is produced to allow the fitting in of a -NH on the base. Furthermore, since both N: and O: would be proton acceptors, a new hydrogen bond could be formed between DNA and drug, strengthening the bond. Substitution of imidazole or furan for one pyrrole in netropsin should not only permit, but actually favor, binding to a G:C base pair at that particular locus.

Prof. J. William Lown of the University of Alberta arrived at identical conclusions on the basis of an examination of space-filling CPK models, and acted upon his hunch by synthesizing a series of eighteen netropsin variants or lexitropsins. These include those with systematic exchange of pyrrole for imidazole or triazole at the first and second ring positions or both, others in which the central C=O is replaced by $-SO_2-$, and still others in which the

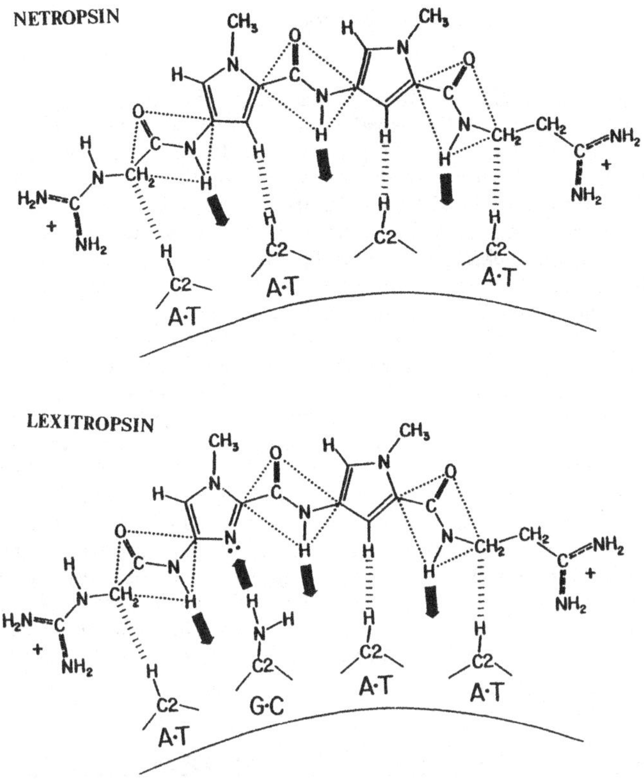

Figure 4. DNA-drug binding and recognition in (a) netropsin, and (b) a hypothetical "lexitropsin" capable of reading and accepting a G:C base pair at the second position. Black arrows are hydrogen bonds, pointing from donor to acceptor. Bands of short parallel lines are nonbonding van der Waals packing contacts.

guanidinium end of netropsin is altered in length and/or charge. Several of these have been sent to outside investigators for screening as possible anti-tumor agents (J.W. Lown, private communication). Some have shown initial possibilities against Herpes Simplex, and against some types of leukemia. Although the parent netropsin is too toxic for clinical use, these initial screenings lead us to hope that suitable modified lexitropsins may be valuable in cancer chemotherapy.

A collaboration has been set up with Prof. Lown, by which he will supply us with lexitropsins, and we will synthesize DNA for his purposes and ours. He will look at the complexes by nmr, while we crystallize them and work out their structure in the solid state. As a first step in this direction, we have synthesized a new DNA dodecamer of sequence: CGTGTTAACACG. This contains a normal netropsin site, TTAA, in the middle, and a site specifically tailored to the lexitropsin shown in Figure 4b, TGTT, toward one end. If the alteration to make a lexitropsin merely tolerates G:C base pairs, then the lexitropsin probably will bind in the old familiar TTAA site, as representing the least obstruction. However, if it now prefers a G:C base pair at the second position, as Figure 4b suggests, then the drug will bind at one of the two off center sites, TGTT or its symmetrical partner at the other end, AACA. Hence the dodecamer that we have synthesized has been chosen not only to bind the lexitropsin, but to discriminate between an active vs. permissive role for the intruding $-NH_2$.

At the same time that we are attempting the x-ray structure analyses of these complexes, Lown will be examining them in solution by nmr and measuring binding constants. What he learns will help us to decide which are the most profitable further x-ray analyses to pursue. Our success with these and parallel structures should enable us to decide whether it is worthwhile to synthesize longer lexitropsins, with more detailed patterns of imidazole vs. pyrrole and longer DNA sequences to match them. It may be that imidazole is too basic, and that furan should be used instead, with O in place of N. It also may be that the misregistration between pyrrole-amide backbones and the floor of the minor groove of B-DNA will induce us to make intermittent modifications in the pyrrole-amide backbone, or even to go over entirely to a new backbone type. An attractive possibility might be "tethered blocks", in which a series of short netropsin-like molecules, each recognizing one of a succession of 6 base pair DNA regions, would be connected by flexible links to allow for adjustment to the DNA helix geometry. Ultimately we would like to look at the complex of a drug with DNA which involves the reading of a dozen or more successive base pairs in a pre-determined mix of A:T and G:C base pairs. At that stage we will feel confident that we understand the principles of drug binding and how to design drugs for general sequences. We shall have to proceed slowly, and let the results at each stage shape our next steps. But we have the DNA synthesis capability and x-ray structure analysis experience, and Lown at Alberta has the drug synthesis capability and nmr expertise so that between the two groups, we should expect to make good progress.

Structure analysis of the complex of Hoechst 33258 with CGCGAATTCGCG

Hoechst 33258 is a fluorescent dye that binds tightly to B-DNA, making it a widely-used DNA stain.[37] This very tightness of binding makes it a potent carcinogen. It is of interest in the context of our lexitropsin studies because it may furnish an alternative to the pyrrole-amide as a backbone for a DNA-reading drug. The structure of Hoechst 33258 is shown in Figure 5, where it is contrasted with netropsin and with the bithiazole DNA-binding moiety of bleomycin A_2.

Graduate student Philip Pjura has co-crystallized a complex of Hoechst 33258 with CGCGAATTCGCG, observed good intensity changes, and collected a three-dimensional x-ray data set from the complex. He finds that the differences between data from the complex and DNA alone are almost exactly those observed by Kopka with the netropsin complex.[1-3] He has commenced refinement of the DNA structure against the data for the complex, and shortly will calculate a difference map to locate the drug molecule. Refinement to completion should follow without difficulty, giving us yet a second complex of DNA with a drug occupying the minor groove (if our predictions are right).

Computer modeling of bleomycin binding to DNA

Bleomycin is a widely used antitumor drug that binds to DNA and then cleaves the sugar-phosphate backbone. It favors cleavage at GC or GT steps, cutting at the pyrimidine C3' position and releasing a free base propionaldehyde.[38,39] The drug conveniently can be regarded as being made up of three parts: a DNA-binding bithiazole (Figure 5), a connecting peptide chain, and a cluster of iron-chelating groups.[40,41] These latter provide five of the six octahedral ligands for the iron, the sixth site remaining open for the binding of O_2 as the first step toward DNA chain cleavage. Available spectral data in the past have been interpreted as favoring an intercalative mode of binding of the bithiazole to DNA.[42,43] The great similarity in size, shape, charged ends, and hydrogen-bonding propensity between the bithiazole and Hoechst and

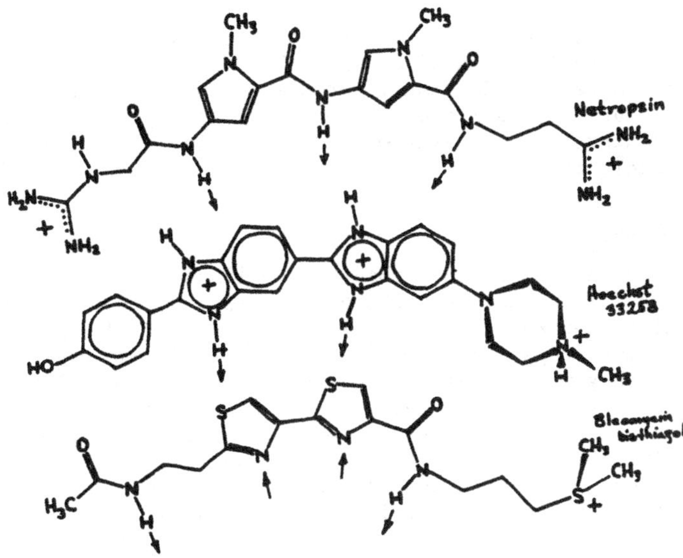

Figure 5. Comparison of skeletons of netropsin, Hoechst 33258, and the DNA-binding bithiazole portion of bleomycin A_2. Arrows indicate potential hydrogen bond donors or acceptors. Note the common crescent shape of the three molecules.

netropsin have led us to consider whether the bithiazole might not be a minor groove binder instead. Graduate students David Goodsell, Chun Yoon and Gil Prive have used energy-refinement and minimization procedures to propose a groove-binding structure for a DNA-bithiazole complex, as depicted in Figure 6.

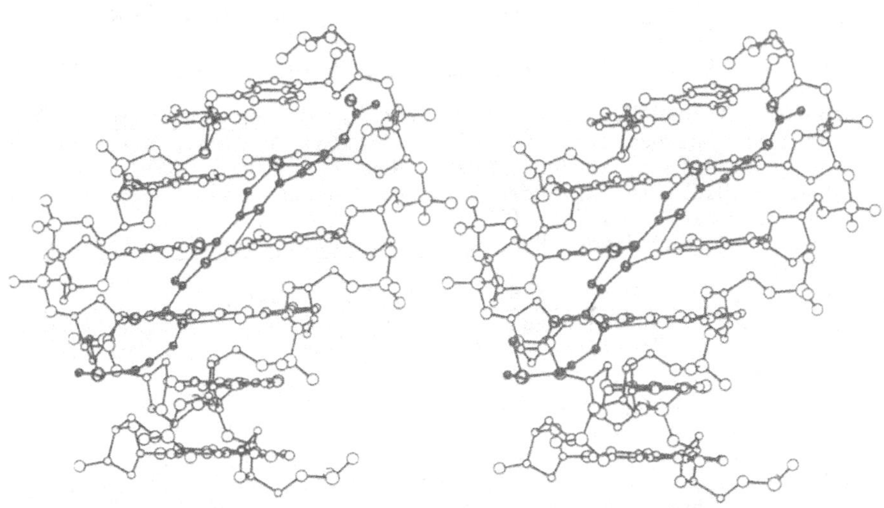

Figure 6. Stereo pair computer drawing of the hypothetical complex of the bithiazole group from bleomycin with an alternating GC double helix of B-DNA. Thin lines represent hydrogen bonds. The iron-chelating continuation of an entire bleomycin molecule would extend down at the lower left.

This model is stereochemically satisfactory, explains the observed GT or GC base specificity in terms of hydrogen bonding, and satisfies the chemical data at least as well as does an intercalative model. They have described this proposed structure and submitted it for publication, as a preliminary to the x-ray crystal structure analysis of a real complex of DNA with either bleomycin or the bithiazole moiety.

Laboratory Goals

This paper has illustrated some of the lines of research currently under way in our laboratory and the goals toward which we are working. The ultimate goal is the knowledge that will enable us to design better DNA-binding antitumor drugs, capable of being targeted more specifically toward regions where they will be effective against neoplastic cells, and away from regions where they will cause undesirable side effects. Our working principle has been that it is desirable to be able to localize a DNA-binding drug by binding it to a pre-determined base sequence. The observation of how natural antitumor drugs bind should aid us in the preparation of synthetic analogues with the binding specificity that we need.

REFERENCES

1. Kopka, M.L., Yoon, C., Goodsell, D., Pjura, P. and Dickerson, R.E. The Molecular Origin of DNA-Drug Specificity to Netropsin and Distamycin. Proc. Natl. Acad. Sci. USA., 82:1376 (1985).
2. Kopka, M.L., Pjura, P., Yoon, C., Goodsell, D. and Dickerson, R.E. The Binding of Netropsin to Double-Helical B-DNA of Sequence C-G-C-G-A-A-T-T-^{Br}C-G-C-G: Single Crystal X-Ray Structure Analysis. In "Structure and Motion: Membranes, Nucleic Acids and Proteins", (E. Clementi and R. Sarma, Eds.), Adenine Press, New York pp. 461-483 (1985).
3. Kopka, M.L., Yoon, C., Goodsell, D., Pjura, P. and Dickerson, R.E. The Binding of an Antitumor Drug to DNA: Netropsin and C-G-C-G-A-A-T-T-^{Br}C-G-C-G. J. Mol. Biol., 183:553 (1985).
4. AMA Drug Evaluations, 5th Ed. (1985) American Medical Association, Chicago, IL.
5. Dorr, R.T. and Fritz, W.L. Cancer Chemotherapy Handbook, Elsevier, New York (1980).
6. Crooke, S.T. and Prestayko, A.W. (Eds.). Cancer and Chemotherapy, Vol. I: Introduction; Vol. II: Clinical Oncology; Vol. III: Antineoplastic Agents, Academic Press, New York (1980,1981).
7. Schwartz, H.S. In "Advances in Cancer Chemotherapy" (A Roswosky Ed.), Marcel Dekker, New York (1979).
8. Wilson, W.D. and Jones, R.L. Adv. Pharmacology and Chemotherapy. 18:177 (1981).
9. Krey, A.K. Prog. Mol. Subcell. Biol. 7:43 (1980).
10. Braithwaite, A.W. and Baguley, B.C. Biochemistry, 19:1101 (1980).
11. Hahn, F.E. In "Antibiotics" (J.W. Corcoran and F.E. Hahn, Eds.) Vol. III, pp. 79-100, Springer Verlag, New York (1975).
12. Zimmer, C. Comments Mol. Cell. Biophys., 1:399 (1983).
13. Crooke, S.T. In Ref. 6, Vol. III, pp. 221-233 (1981).
14. Crooke, S.T. In Ref. 6, Vol. III, pp. 343-350 (1981).
15. Glickson, J.D., Pillai, R.P., Riordan, J.M., Sakai, T.T., Lenkinski, R.E., Krishna, N.R. and Elgavish, G.A. In "Biomolecular Stereodynamics (R.H. Sarma, Ed.). Adenine Press, New York, Vol. II, pp. 409-436 (1981).
16. Sobell, H.M. and Jain, S.C. J. Mol. Biol., 68:21 (1972).
17. Sugiura, Y., Suzuki, T., Otsuka, M., Kobayshi, S., Ohno, M., Takita, T. and Umezawa, H. J. Biol. Chem., 258:1328 (1983).
18. Schultz, P.G. and Dervan, P.B. J. Biomol. Str. Dyn., 1:1133 (1984).

19. Drew, H.R. and Dickerson, R.E. *J. Mol. Biol.*, 151:535 (1981).
20. Dickerson, R.E., Drew, H.R., Conner, B.N., Kopka, M.L. and Pjura, P.E. Helix Geometry and Hydration in A-DNA, B-DNA and Z-DNA. *Cold Spr. Har. Symp. Quant. Biol.* 47:13 (1982).
21. Kopka, M.L., Fratini, A.V., Drew, H.R. and Dickerson, R.E. Ordered Water Structure Around a B-DNA Dodecamer. *J. Mol. Biol.*, 163:129 (1983).
22. Bartenev, V.N., Golovamov, E.I., Kapitonova, K.A., Mokulskii, M.A., Volkova, L.I., and Skurabovskii, I.Y. *J. Mol. Biol.*, 169:217 (1983).
23. Seeman, N.C., Rosenberg, J.M. and Rich, A. *Proc. Natl. Acad. Sci. USA*, 73:804 (1976).
24. Wartell, R.M., Larson, J.E. and Wells, R.E. *J. Biol. Chem.*, 250:2698 (1975).
25. Zimmer, C. *Prog. Nucl. Acid Res. Mol. Biol.*, 15:285 (1975).
26. Zimmer, C. Reinert, K.E., Luck, G., Wahnert, U., Lober, G. and Thrum, H. *J. Mol. Biol.*, 58:329 (1971).
27. Ivanov, V.I., Minchenkova, L.A., Minyat, E.E., Frank-Kamentskii, M.D. and Schyolkina, A.K. *J. Mol. Biol.*, 87:817 (1974).
28. Zimmer, C., Marck, C. and Guschlbauer, W. *FEBS Lett.*, 154: 6 (1983).
29. Reinert, K.E. and Thrum, H. *Studia Biophysica*, 24/25:319(1970).
30. Van Dyke, M.W., Hertzberg, R.P. and Dervan, P.B. *Proc. Natl. Acad. Sci. USA* 79:5470 (1982).
31. Luck, G., Zimmer, C., Reinert, K.E. and Arcamone, F. *Nucl. Acids Res.*, 4:2655 (1977).
32. Zimmer, C., Luck, G., Birch-Hirschfeld, E., Weiss, R., Arcamone, F. and Guschlbauer, W. *Biochim. Biophys. Acta*, 741:15 (1983).
33. Gursky, G.V., Tumanyan, V.G., Zasadatelev, A.S., Zhuze, A.L., Grokhovsky, S.L. and Gottikh, B.P. *In* "Nucleic AcidProtein Recognition" (H.J. Vogel, Ed.), Academic Press, New York, pp. 189-217 (1977).
34. Zimmer, C., Marck, C., Schneider, C. and Guschlbauer, W. *Nucl. Acids Res*, 6:2831 (1979).
35. Kolchinskii, A.M., Mirazabekov, A.D., Zasedatelev, A.S., Gurskii, G.V., Grokhovskii, S.L., Zhuze, A.L. and Gottikh, B.P. *Mol. Biol.* (1975).
36. Patel, D.J. *Proc. Natl. Acad. Sci. USA*, 78:7281 (1982).
37. Hilwig, I. and Gropp, A. *Expt. Cell Res.*, 75:122(1972).
38. Sugiura, Y., Suzuki, T., Otsuka, M., Kobayashi, S., Ohno, M., Takia, T. and Umezawa, H. *J. Biol. Chem.*, 258:1328 (1983).
39. Giloni, L., Takeshita, M., Johnson, F., Iden, C. and Grollman, A.P. *J. Biol. Chem.*, 256:8608 (1981).
40. Pillai R.P., Krishna, N.R., Sakai, T.T. and Glickson, J.D. *Biochem. Biophys. Res. Comm.*, 7:270 (1980).
41. Glickson, J.D., Pillai, R.P., Riordan, J.M., Sakai, T.T., Lankinski, R.E., Krishna, G.A. and Elgavish, G.A. *In* "Biomolecular Sterodynamics, (Adenine Press, New York), pp. 409-436 (1981).
42. Glickson, J.D., Sakai, T.T., Geckle, J.M. and Chen, D.M. *In* "Peptides: Structure and Biological Function" (1979).
43. Sakai, T.T., Riordan, T.E., Booth, J.D. and Glickson, J. *J. Med. Chem.*, 24:279 (1981).
44. McKay, D.B., Steitz, T.A., Weber, I.T., West, S.C. and Howard-Flanders, P. *J. Biol. Chem.*, 256:6662 (1980).

COMPUTER-AIDED DESIGN OF NEW DNA INTERCALATORS

Stephen Neidle

Cancer Research Campaign Biomolecular Structure Unit
Institute of Cancer Research
Clifton Avenue, Sutton, Surrey SM2 5PX, U.K.

INTRODUCTION

The overwhelming majority of the approximately thirty currently clinically-useful anti-neoplastic agents are believed to interfere with aspects of DNA function and/or metabolism.[1-4] Those compounds that directly interact with DNA constitute an increasingly important group of such drugs, exemplified by cis-platinum, amsacrine, bleomycin and doxorubicin, as well as the more traditional alkylating agents. Interference with processes of DNA transcription, replication, RNA synthesis and topoisomerase action, as well as mutagenicity, have been variously noted for these agents; the precise mechanisms whereby they have a differential effect on tumour cells remain unclear, although their differential effect on rapidly-proliferating cells is an important factor. In the absence of such evidence, approaches to the development of new DNA-reactive anti-cancer drugs remain semi-empirical, although their clinical need is of very considerable importance on account of both the narrow spectrum of activity and narrow therapeutic index of current drugs.

Doxorubicin has proven activity against a wider range of human tumours than any other drug, yet its applicability is often inhibited by severe dose-related cumulative cardio-toxicity. The origins of this have not been unequivocally elucidated, but the weight of evidence points to the involvement of lipid peroxidation mechanisms.

There has been intensive effort to circumvent this problem, whilst maintaining or even improving the range of tumour conditions that are treatable. These have resulted on the one hand in new anthracycline analogues such as 4'-epidoxorubicin, and on the other in anthraquinones that have radically simpler chemistry. Mitoxanthrone is the first such compound to have progressed through clinical trial into usage, and it appears to have promise in the treatment of breast cancer.[5]

The development of new DNA-interactive compounds is currently underway in a number of laboratories; this has been given added impetus by the realization that DNA-binding is a necessary and critical (though in itself insufficient) factor that may correlate with inhibition of cell growth in culture and even anti-tumour activity *in vivo*. This paper describes an approach to the rational design of such compounds that attempts to optimise DNA-intercalative factors.

FEATURES OF INTERCALATION

The analysis of drug-DNA interactions at the molecular level has employed many of the physico-chemical tools for examining drug-receptor complexes in general. However, DNA is not a homogeneous macromolecule with a unique type of receptor site; there are sixteen sequence-distinct dinucleoside intercalation sites. Measurements of binding parameters with natural-sequence DNA produces averaged results that tend to obscure more subtle sequence-dependent effects, such as differences in affinities or dissociation constants between different sequences. The development of footprinting techniques[6] holds out some promise for the quantitative study of these differences, which may be of importance in the future design of drugs for binding to particular DNA sequences within a gene. In general, the structurally simple intercalators, such as daunomycin or the acridines, appear to show relatively small sequence preferences, in contrast to the more complex antibiotics actinomycin and echinomycin.

The classic studies of Lerman and Waring showed that the intercalation process necessitates the possession of a planar aromatic chromophore by the drug molecule. The fundamental feature of this non-covalent, equilibrium binding is the insertion of a drug molecule between two adjacent base pairs, thus unwinding and extending the DNA molecule to a length dependent on the number of binding sites. Intercalation is an energetically favourable process, and thus relates to the dynamic base-pair breathing properties of DNA itself.

A detailed picture of the molecular geometry of intercalation has been obtained from high-resolution X-ray crystallographic studies of a variety of drugs complexed with short lengths of oligonucleotide.[7,8] The majority are of Watson-Crick base-paired self-complementary DNA (or RNA) with drug chromophore sandwiched at a separation of 3.4A from each base-pair (Fig. 1). The complex between the bis-intercalator echinomycin and a hexanucleotide has a more unusual arrangement, with non-standard base-pairing that may relate to mechanisms of mutagenesis (a characteristic feature of all intercalators).[9] None of these X-ray analyses has provided structural information on drug binding sites that are embedded within a DNA sequence; all have drug bound at the end of a sequence, and thus in principle may be susceptible to end-effects. It is perhaps significant that all of these complexes have a pyrimidine-3',5'-purine sequence at the binding site, which has been shown to have a ~5 kcal $mole^{-1}$ advantage over other sequences.[10]

Detailed examination of the molecular geometries of these complexes has shown that they are describable by a single consensus set of conformations for the nucleotide residues comprising the binding site.[11,12] Features of base....base movement and bending modes differ in detail, and appear to be dependent on the nature of the drug, its position in the binding site and its direction of entry.

This consensus intercalation geometry provides an opportunistic starting-point for the molecular design of new intercalating agents.

METHODOLOGY

A combination of interactive computer graphics and quasi real-time energy calculations have been employed in our laboratory as basic molecular design tools.[13] This has enabled simulations to be made of new molecules being intercalated. In contrast to X-ray crystallography, which provides mainly static information, this approach allows the dynamics of binding to be examined, alternative modes to be studied, and estimations of the relative strengths of interactions to be made.

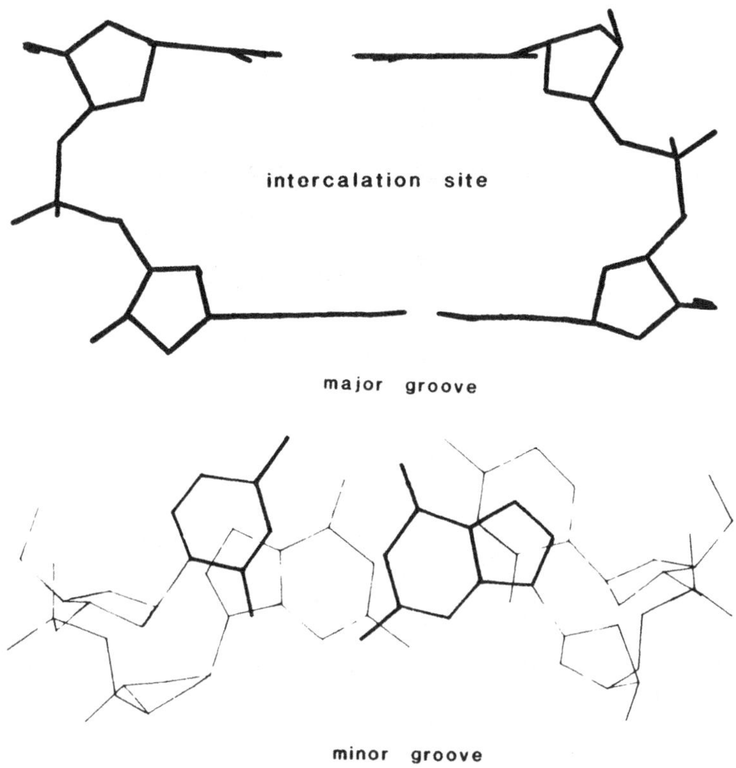

Figure 1. The molecular geometry of intercalated dinucleosides.

Interaction energies are calculated as the sum of non-bonded and electrostatic terms:

$$E_{TOT} = \frac{A}{r^6} - \frac{B}{r^{12}} + \frac{q_1 q_2}{er}$$

for atoms of charges q_1 and q_2 at distance r apart. The non-bonded constants A and B for each atom type are chosen so as to control atom-atom contact distances. Partial charges q are calculated by quantum-mechanical methods. Their inclusion is important as nucleic acids themselves, and sometimes the interacting drugs, are charged species and thus contribute significantly to the total energy of interaction.

It is perhaps insufficiently realised that such calculations need to be interpreted with caution in view of the many approximations inherent in them. In particular, it has to be borne in mind that the intercalation model used is restricted to the immediate vicinity of the binding site; effects on neighbouring residues are ignored. Since entropic and solvent factors are not considered, the calculated energies are not absolute but relative ones, and thus cannot be directly used to derive thermodynamic or kinetic parameters.

INTERCALATOR GEOMETRY AND INTERACTION

It is a common misconception that the planar part of a DNA-intercalating molecule necessarily requires a formal positive charge. This factor is

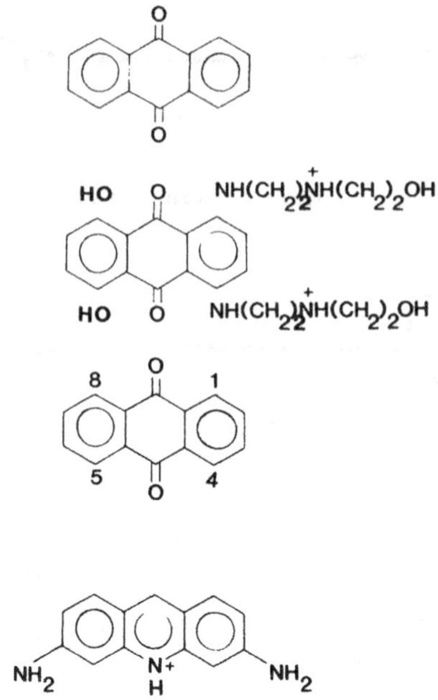

Figure 2. (a) Anthraquinone (b) Mitoxanthrone (c) Substitution positions in anthraquinone. (d) Proflavine

present in the classic intercalators proflavine and ethidium, but not in the more complex peptide-containing actinomycins and echinomycin. We examine the charge factor in this section, together with other steric aspects. Chromophore planarity is an important feature for true intercalation to take place.

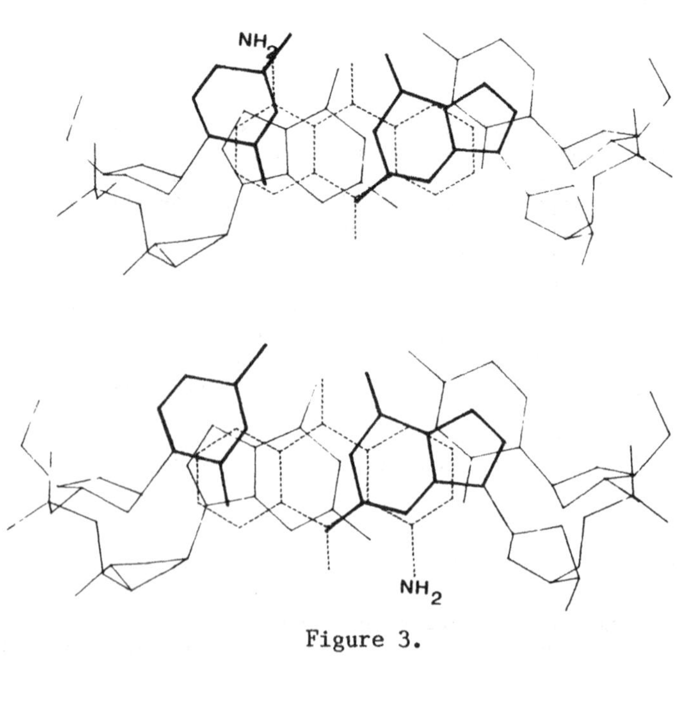

Figure 3.

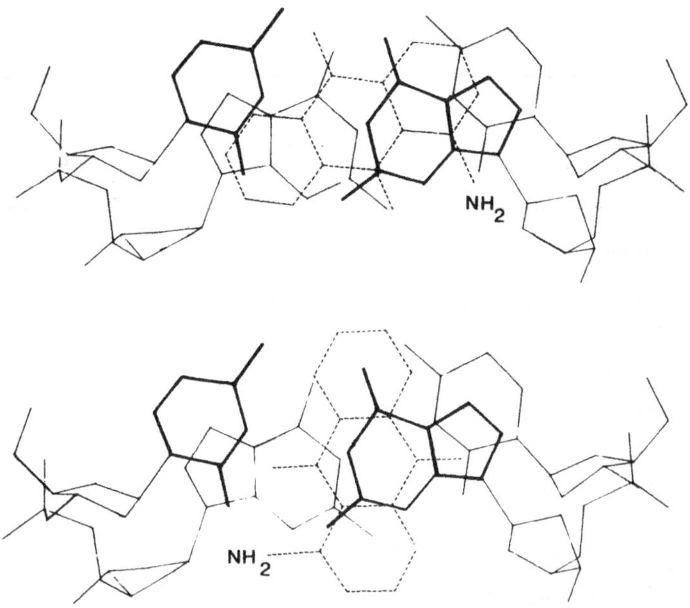

Figure 4.

In the case of the formally uncharged anthraquinone chromophore (AQ), two derivatives have been examined. One is the 1-amine (Fig. 2), and the other is the 1,4-diamine-5,8-dihydroxyl, which is the substitution pattern in mitoxanthrone (which has an alkylamino alkyl hydroxyl side-chain attached to each amine group). Crystallographic data have shown that these anthraquinones have intramolecular hydrogen bonding involving quinone oxygen atoms as donors and hydroxyl/amine groups as donors. This increases the effective planar area of the molecule; the interaction with base-pairs is an attractive one (often termed 'stacking' forces), and relates to chromophore surface area. Modelling studies have shown that these anthraquinones can interact with the intercalation site in a variety of orientations (Table 1). Several are illustrated in Figs. 3 and 4.

The simpler anthraquinone has more possible modes of interaction on account of its lack of molecular symmetry. The electrostatic contributions to the total stabilisations are slight, reflecting the small partial charges on the anthraquinone atoms. The dominance of stacking interactions is shown by the significantly lower stabilities of perpendicular modes. It is perhaps more surprising that the dihydroxyl-diamino anthraquinone molecule does not interact more strongly; the superior anti-tumour properties of several of its derivatives compared to[14] non-hydroxylated ones are thus not directly related to DNA-binding features.

Results for proflavine are given in Table 2. These indicate that the addition of a positive charge to a chromophore itself results in a dramatic improvement in the calculated total interaction energy. The electrostatic component is now as important as that from stacking interactions; indeed when the latter are reduced, as in the perpendicular case, electrostatic interactions dominate the situation. The crystal structure of the proflavine-dCpG complex shows the minimum-energy, parallel major groove geometry, as calculated.[15]

Table 1. Low-Energy Intercalated Anthraquinone Structures

	$E_{non-bonded}$	$E_{electro-static}$	E_{Total}
(a) 1-amino-AQ			
(i) parallel orientation, amine in major groove	-61.3	-0.9	-62.2
(ii) parallel orientation, amine in minor groove	-61.7	-1.1	-62.8
(iii) perpendicular orientation, amine in major groove	-49.2	3.0	-46.2
(iv) perpendicular orientation, amine in minor groove	-45.6	2.0	-43.6
(v) diagonal mode, with amine into major groove	-52.6	0.8	-51.8
(vi) diagonal mode, with amine into minor groove	-50.5	-0.1	-50.6
(b) 1,4-diamino-5, 8-dihydroxyl-AQ			
(i) parallel orientation	-65.7	-0.6	-66.3
(ii) diagonal orientation	-63.0	-1.4	-64.4
(iii) perpendicular mode with amines in minor groove	-57.0	0.9	-56.1
(iv) perpendicular mode with amines in major groove	-57.3	3.1	-54.2

Energies are in kcal mole^{-1}

These results are for chromophores only; when significant side-chains or other substituents are attached, the situation becomes more complicated.[16-17] The effects of protonated alkyl-aminoalkyl side-chains attached to an anthraquinone chromophore have been systematically examined. These studies have shown that the side-chains themselves can be major contributors to the total stabilisation energy, by virtue of generalised attractive van der Waals forces. The possession of a basic nitrogen atom at or near the terminus of the side-chain provides an electro-static component that amounts to 1/3 of the total energy. Further formal positive charges, for example on the chromophore itself, would certainly lead to a stronger total interaction. However, it would also lead to problems for transporting such molecules through cell membranes to target DNA.

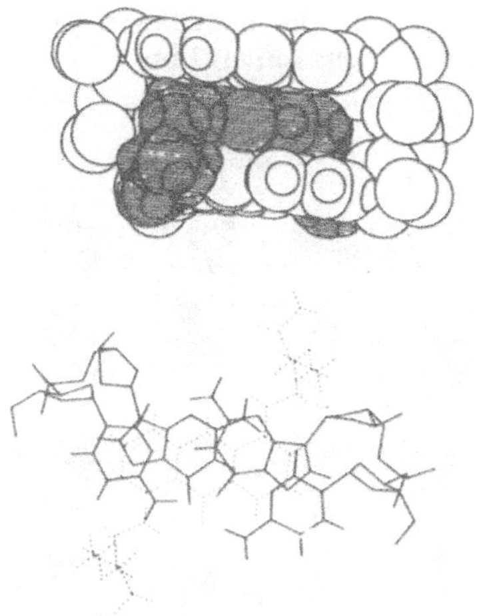

Figure 5.

melted (i.e. A-T rich) regions of DNA. Thus, dissociation would be predicted to be much slower than for the other compounds in the series, a prediction borne out by subsquent experiment.

CONCLUSIONS

Several general features of intercalating molecules are now apparent from rational design considerations:

(i) The size and the planarity of the chromophore are important for optimal fit into the receptor site.

(ii) Its charge distribution, and possession of formal electron-deficiency can significantly enhance interactions with base pairs.

(iii) The nature, charge, length and substitution pattern of side-chains are constitutive factors in the enhancement of overall DNA binding ability. Flexibility in side-chain conformation is a desirable feature, since this can allow a protonated amino group to be n close, energetically favourable contact with regions of purine and pyrimidine bases. Changes in side-chains that alter this critical [chromophore-NH-$(CH_2)_2\overset{+}{N}$] feature are generally deleterious to binding energies (and to biological activity).

(iv) The possession of two such side-chains can more than counterbalance a less than optimal intercalating chromophore.

(v) The design of intercalators that have simultaneously both major and minor groove binding implies that DNA damage induced by such intercalators will be relatively long-lasting; this may confer chemotherapeutic advantage in the inhibition of DNA replication.

Computerised molecular modelling is capable of predicting the ranking order of DNA-interaction parameters, as well as kinetic ones of perhaps more significant anti-tumour interest. Whether or not the methodology will lead directly to new anti-cancer drugs, rather than merely new DNA-binding agents, will in large measure depend on the successful incorporation of pharmacokinetic, metabolic and other considerations.

REFERENCES

1. M.J.Waring, Ann. Rev. Biochem., 50:159 (1981).
2. E.F.Gale, E. Cundliffe,P.E. Reynolds, M.H. Richmond and M.J. Waring, The Molecular Basis of Antibiotic Action, 2nd ed., Wiley, London, 1981.
3. S. Neidle and M.J. Waring, (eds.), Molecular Aspects of Anti-Cancer Drug Action, Macmillan Press, London, 1983.
4. W.D. Wilson and R.L. Jones, Adv. Pharmacol. Chemother., 18:177 (1981).
5. R.C. Stuart-Harris, T. Bozek, N.A. Pavlidis and I.E. Smith, Cancer Chemother. Pharmacol., 12:1 (1984).
6. M.M. Van Dyke and P.B. Dervan, Science, 225:1122 (1984).
7. S. Neidle and H.M. Berman, Prog. Biophys. Mol. Biol., 41:43 (1983).
8. A. Aggarwal, S.A. Islam, R. Kuroda and S. Neidle, Biopolymers, 23:1025 (1984).
9. A.H.-J. Wang, G. Ughetto, G.J. Quigley, T. Hakoshima, G. van der Marel, J.H. Van Boom and A. Rich, Science, 225:1115 (1984).
10. S. Broyde and B. Hingerty, Biopolymers, 18:2905 (1979).
11. H.M. Berman, S. Neidle and R.K. Stodola, Proc. Natl. Acad. Sci. USA. 75:828 (1978).
12. H.M. Berman and S. Neidle, in Nucleic Acid Geometry and Dynamics, (Sarma, R.H., ed.), Pergamon Press, New York, 1980, pp. 367-382.
13. S. Neidle, Biochem. Soc. Trans., 12:1008 (1984).
14. K. Murdock, R.G. Child, P.F. Fabio, R.B. Angier, R.E. Wallace, F.E. Durr and R.V. Citarella, J. Med. Chem, 22:1024 (1979).
15. S.A. Islam and S. Neidle, Acta Crystallogr., B40:424 (1984).
16. S.A. Islam, S. Neidle, B.M. Gandecha and J.R. Brown, Biochem. Pharmacol., 32:2801 (1983).
17. S.A. Islam, S. Neidle, B.M. Gandecha, M. Partridge, L.H. Patterson and J.R. Brown, J. Med. Chem., in press (1985).
18. B.M. Gandecha, J.R. Brown and M.R. Crampton, Biochem. Pharmacol., 34:733 (1985).

DIFFERENTIAL EXPRESSION OF SOS GENES IN AN E. COLI MUTANT PRODUCING UNSTABLE LEXA PROTEIN ENHANCES EXCISION REPAIR BUT INHIBITS MUTAGENESIS

Kenneth R. Peterson[1], Ann K. Ganesan[2], and David W. Mount[1]

[1]Department of Molecular and Cellular Biology, Biosciences West, University of Arizona, Tucson, AZ 85721, and
[2]Department of Biological Sciences, Stanford University Stanford, CA 94305

ABSTRACT

The lexA41 mutant of E. coli is a UV-resistant derivative of another mutant, lexA3, which produces a repressor that is not cleaved following inducing treatments. lexA41 carried an additional mutation which changed amino acid 132 in the LexA protein from Ala to Thr. The resultant protein was unstable and was degraded both before and after an inducing treatment. This instability was greater at 42° than at 30°. The protein was more stable in Lon⁻ mutants at both temperatures. lac operon fusions to most of the genes in the SOS regulon were used to show that the various damage-inducible genes were derepressed to different extents. uvrA, B, and D were almost fully derepressed. Consistant with this finding, the rate of removal of T4 endonuclease V-sensitive sites was more rapid in the UV-irradiated lexA41 mutant than in normal cells, suggesting a more active excision repair system.

We propose that the instability of the LexA41 protein reduces the intracellular concentration of repressor to a level that allows a high level of excision repair. The additional observation that SOS mutagenesis was only weakly induced in a lexA41 uvrA⁻ mutant implies that the mutant protein partially represses one or more genes whose products promote SOS mutagenesis.

INTRODUCTION

The SOS response of Escherichia coli, a complex physiological response, is displayed following treatments which damage DNA or inhibit DNA replication.[1,2] Two significant activities include enhanced capacity for DNA repair resulting from derepression of the recA, uvrA, uvrB and uvrD genes and increased mutagenesis due to derepression of recA, umuC and umuD.[3,4]

These and other physiological changes are the consequence of the derepression of at least seventeen unlinked operons negatively regulated by LexA repressor. Following treatments that induce the SOS response, a signal molecule, presumably a DNA metabolic product(s), interacts with RecA protein converting it to an activated form. Activated RecA protein facilitates the proteolytic cleavage of LexA repressor, which results in derepression of the regulon. The cell then enters a new physiological state during which time DNA repair processes are augmented. Once damage has been repaired or bypassed in some way, the level of signal molecule(s) drops. RecA protein ceases to be activated, LexA repressor accumulates, represses its target genes, and the cell returns to a normal state. Detailed reviews of the SOS regulatory circuit have appeared elsewhere.[1,2]

Among mutants isolated in the lexA structural gene are those of the lexA(Ind⁻) class, which synthesize repressors that are not cleaved [5,6] (Peterson and Mount, unpublished experiments). Unable to induce the SOS response, they are sensitive to DNA-damaging treatments. One particular mutant, lexA3, is deficient in induction of all the known SOS functions, and produces a repressor resistant to cleavage due to a mutation that alters the cleavage site in the protein.[7] The lexA41 (formerly tsl-1) mutant, a lexA(Ts) allele, is a UV-resistant, temperature-sensitive derivative of a lexA3 strain.[8] Cells carrying this mutation exhibit a so-called "split-phenotype", a phenomenum in which only a subset of the SOS responses can be detected physiologically following inducing treatments.[9] Specifically, lexA41 strains are resistant to DNA damage, but show temperature sensitive growth and abnormal response to mutagenesis by UV light.

We have recently sought to explain the "split-phenotype" associated with lexA(Ts) mutants by analyzing the structural and functional interactions of LexA41 repressor with target genes and activated RecA protein. These studies will be published in detail elsewhere and are summarized below.

RESULTS AND DISCUSSION

1. Cloning of lexA41

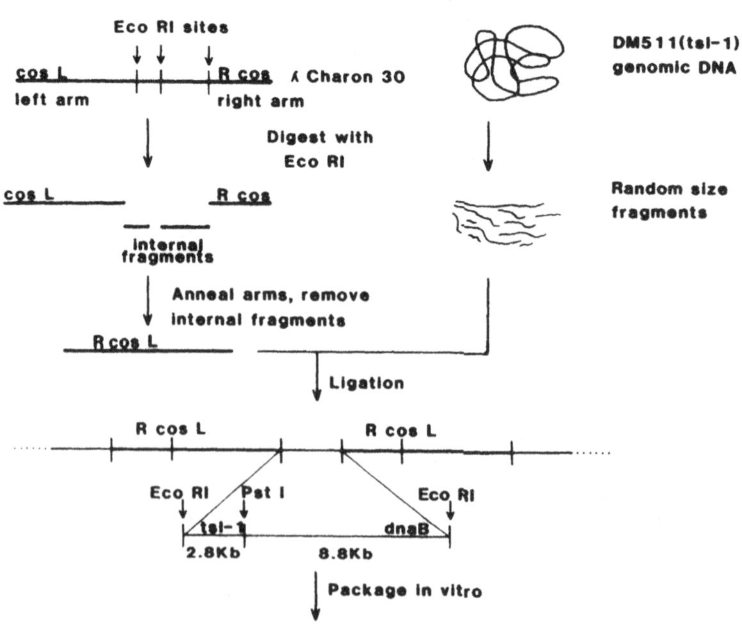

Figure 1. Construction of E. coli gene library containing lexA41 (tsl-1).

Our objective in cloning the lexA41 allele was to determine the DNA sequence change(s) in the mutant in order to derive the sequence of the mutant protein. With this information we might be able to speculate on why the altered sites on the repressor result in a protein that has varied affinities for different operator targets as well as conferring thermolability on the protein.

Since lexA41 is recessive to wild-type and most mutant lexA alleles, identifying the cloned gene genetically presented problems. However, because lexA41 was on the same 12 kilobase Eco RI fragment as dnaB, screening for clones bearing the later gene could be easily performed with the assumption that lexA41 would also be present.[10] Figures 1 and 2 summarize the cloning of this fragment and the identification of $dnaB^+$ clones, respectively. A smaller fragment containing lexA41 was subsequently subcloned into a plasmid vector. The presence of the mutant was confirmed by the ability of the plasmid to repress a SOSop::lac operon fusion in a

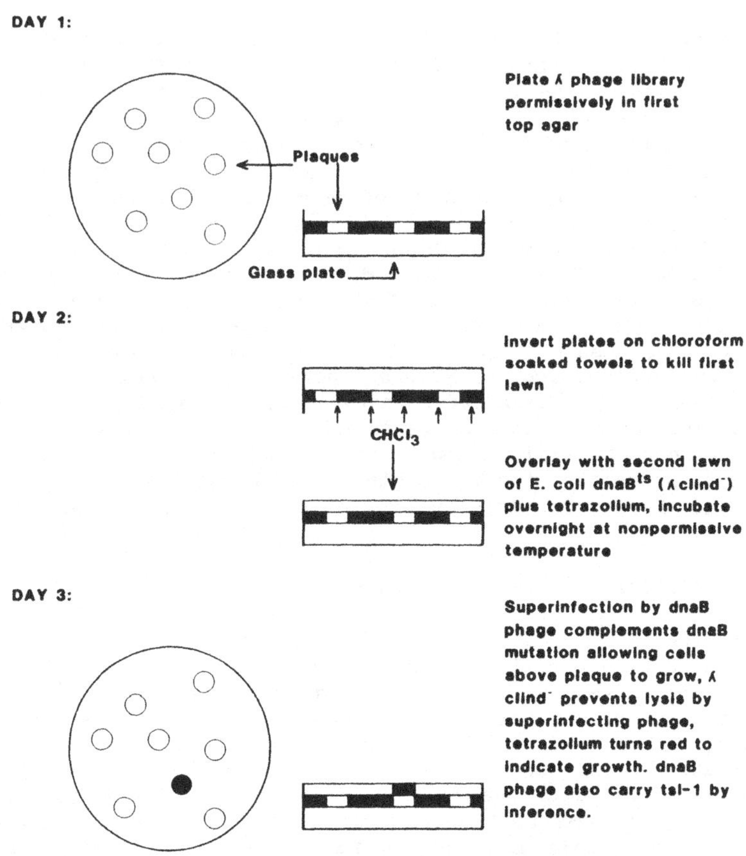

Figure 2. Screen for phage clones bearing lexA41 (tsl-1) and dnaB.

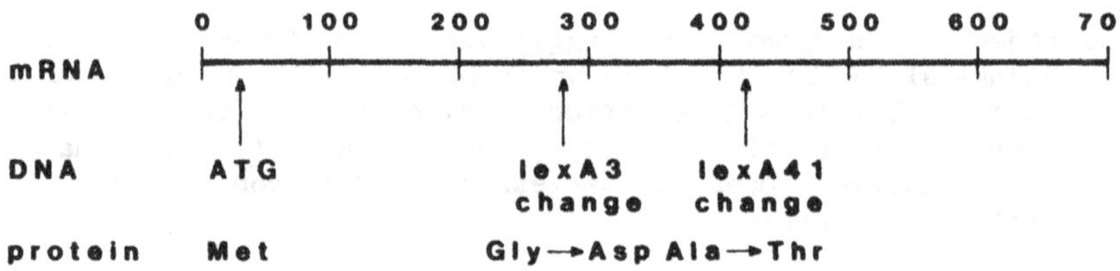

Figure 3. Mutations in the lexA41 gene.

lexA null mutant background at 32° but not at 42°. Smaller restriction fragments were further subcloned into a M13 vector where they were sequenced by the dideoxy chain-termination method.[11] In addition to the original lexA3 mutation which changes amino acid 85 from Gly to Asp, a second base change was found located downstream in the end of the gene that codes for C-terminal half of the repressor. This mutation changed amino acid residue 132 from Ala to Thr (Fig. 3). This region, in phage repressors, determines interaction of monomers to form dimers, the form that binds tightly to operators

2. Instability of LexA41 repressor

The stability of the mutant LexA41 protein was examined by pulse labeling the protein, immunoprecipitating it from crude cell extracts and detecting the amount present by polyacrylamide gel electrophoresis. Wild-type LexA protein is stable in normally growing cells at 32° or 42° with a half-life of about 60 minutes.[12] LexA41 repressor had a half-life of approximately 5 minutes at 32° and 1.5 minutes at 42° (data not shown). Mutant LexA41 protein was unstable and degraded in a similar, recA-independent, manner both before and after an inducing treatment. The stability of the protein was markedly increased in lon⁻ mutants; LexA41 protein had a half-life greater than 20 minutes at either temperature. Apparently, some abnormal feature of the protein, resulting from it being a double mutant, makes it a substrate for the lon-directed proteolytic system of E. coli. Lon protease has recently been shown to be a heat shock protein and at 42° one would expect to have an increased intracellular concentration of this protein.[13] The greater instability of LexA41 protein at 42° may be a consequence of increased protease that can remove repressor faster than at 32°. An important implication of these findings is that the degradation of the LexA41 protein by the Lon protease decreases the concentration of the repressor and thereby derepresses many of the genes in the SOS regulon (data not shown).

3. Excision repair in lexA41 mutants

One of the phenotypes expressed in lexA41 strains is UV-resistance. We hypothesized that this might be due, in part, to increased expression of the excision repair genes uvrA, uvrB, uvrC and uvrD. LexA41 repressor may have a lower affinity for the operators of these genes and they are consitutively expressed at higher basal levels. Other genes may still bind repressor adequately and are not expressed above wild-type basal levels. Alternately, following UV-irradiation, the rate of repressor cleavage is such that weaker bound operators are preferentially derepressed while those that bind repressor tighter remain partially or fully repressed. Either mechanism would explain the "split-phenotype" of UV-resistance, but decreased SOS mutagenesis, associated with lexA(Ts) strains.

Our data support the former model. Analysis of labeled LexA41 protein immunoprecipitates following UV-irradiation has demonstrated that this repressor was not cleaved via the RecA protein-catalyzed route[12] (data not shown). This concurs well with the fact that lexA41 retained the lexA3 mutation which results in a non-cleavable protein. LexA41 repressor was less stable then LexA^{+} protein in vivo at permissive temperature, suggesting that there may be a lower steady-state concentration of protein sufficient to bind only operators for which the altered repressor has greatest affinity (see above). The extent of derepression of several of the excision repair genes was measured in lexA41 strains using lac operon fusions to the operator-promoter regions of these genes. Figure 4 shows that basal level expression of uvrA and uvrB was increased in lexA41 mutants at permissive temperature approximately 2.5 and 2.0-fold, respectively. These genes were almost fully derepressed when compared to wild-type backgrounds following inducing treatments (data not shown). uvrD showed a similar pattern (data not shown).

Consistant with increased expression of uvrA and uvrB in a lexA41 mutant was the finding that the rate of removal of T4 endonuclease V-sensitive sites (ESS) was more rapid in the lexA41 recA(Def) mutant than in normal cells or lexA^{+} recA(Def) cells (Table 1). T4 endonuclease V was used to detect pyrimidine dimers in the DNA of irradiated cells by virtue of its ability to nick DNA at sites containing dimers. This suggests a more active excision repair system which may occur because the instability of LexA41 protein reduces the intracellular concentration of repressor to a level that allows a high degree of derepression of the excision repair genes.

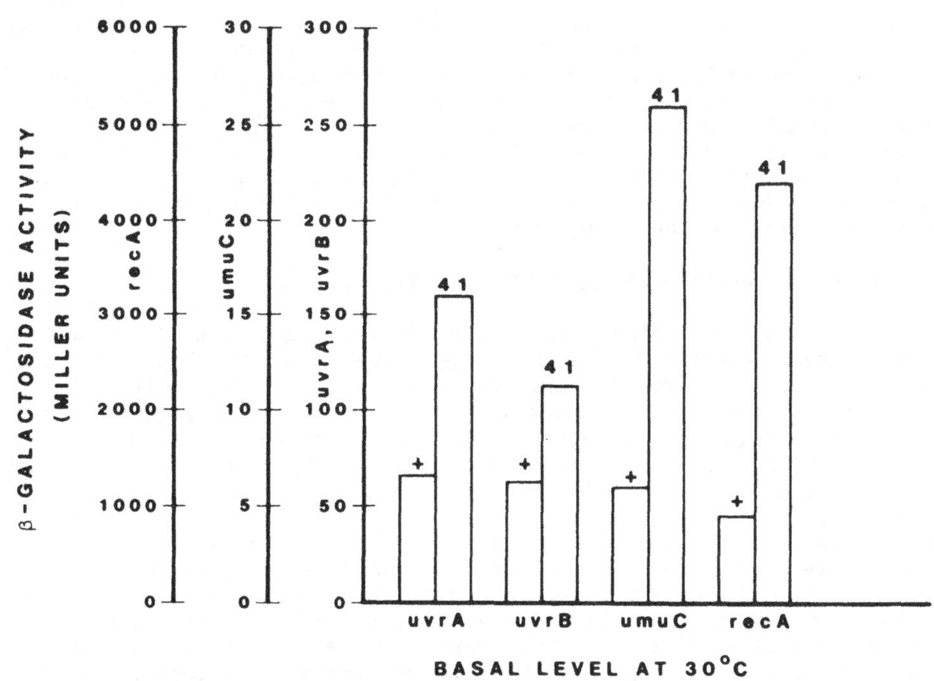

Figure 4. β-galactosidase activity of lexA^{+} and lexA41 SOSop::lac fusions at 30°.

Table 1. Effect of lexA41 on rate of removal of ESS (endonuclease-sensitive sites) in UV-irradiated cells.

UV dose (J/m^2)	min. after UV	% ESS remaining: AG201 recA(Def)	% ESS remaining: AG202 recA(Def) lexA41
2	0	100	100
	5	ND	39*
	10	75	17
	20	43	<1
	40	<1*	ND
40	0	100	100
	10	94	38
	20	80	29
	30	55	14
	45	55	2

* The values are from one experiment, the rest are from two experiments.

4. Inducible Mutagenesis in lexA41 strains

SOS mutagenesis has been shown to be considerably lower in lexA(Ts) strains.[14,15] The products of the umuC, umuD, and recA genes are necessary for inducible mutagenesis.[2] RecA protein is necessary first in its regulatory capacity for LexA repressor proteolysis, thereby derepressing umuC and umuD, and second it plays a direct role in the mutagenic process(es).[4,16] We have examined the expression of recA and umuC using lac operon fusions to these genes in lexA41 hosts. Figure 4 shows that the basal level of umuC was increased about 4-fold and that recA expression was increased approximately 5-fold. In lexA(Def) strains, which lack functional repressor, umuC and recA expression were increased about 29 fold and 11 fold, respectively (data not shown). Unlike uvrA and uvrB, which were close to fully derepressed in lexA41 cells, umuC and recA were only 14% and 45% fully expressed. We suggest that either umuD, C or recA or both are limiting lexA41 mutants and that this is sufficient to account for the lower level of mutation induction. Alternately, an as yet unidentified gene(s) in the SOS regulon that is required for inducible mutagenesis remains repressed by the lexA41 mutant.

5. Model for "split-phenotype" and temperature sensitivity

Based on these findings, we propose the following model to account for the phenotypes associated with the lexA41 mutant. At 30°, LexA41 protein is degraded by Lon protease such that the intracellular concentration of repressor is low enough to allow almost full derepression of target genes with operator sequences for which LexA41 repressor has low affinity, partial derepression of genes for which repressor has an intermediate affinity and no derepression of genes that LexA41 protein has great affinity for. uvrA, uvrB and uvrD fall into the first class, umuD, C and recA fall into the second class and sulA falls into the third class. Phenotypically then, lexA41 cells are UV-resistant because they constitutively express excision repair functions, they are mutagenesis-deficient because recA and umuD, C are not expressed at adequate levels and the mutant does not lethally filament because sulA remains fully repressed. At 42°, LexA41 protein is more unstable because of greater turnover by increased concentration of Lon protease, which is a heat shock protein. With less repressor available to bind operator DNA, more SOS genes are derepressed to a greater extent, including sulA. This results in production of the SulA protein, a cell division inhibitor, and lexA41 strains lethally filament. The lon gene product has been shown to be responsible for the degradation of the SulA division inhibitor, hence the derepression of sulA is sufficient to allow production of enough protein to overwhelm the degradative process

ACKNOWLEDGEMENTS

This work was supported by grants CM24496 fr the National Institutes of Health and PCM-S20460 the National Science Foundation.

REFERENCES

1. C. C. Walker, Mutagenesis and inducible responses to deoxyribonucleic acid damage in Escherichia coli, Microbiol. Rev. 48:60 (1984).
2. J. W. Little and D. W. Mount, The SOS regulatory system of Escherichia coli, Cell 29:11 (1982).
3. S. J. Elledge and G. C. Walker, Proteins required for ultraviolet light and chemical mutagenesis: identification of the products of the umuC locus of Escherichia coli, J. Mol. Biol. 164:175 (1983).
4. E. M. Witkin and T. Kogoma, Involvement of the activated form of RecA protein in SOS mutagenesis and stable DNA replication in E. coli, Proc. Natl. Acad. Sci. USA 81:7139 (1984).
5. D. W. Mount, K. B. Low, and S. Edmiston, Dominant mutations (lex) in Escherichia coli K-12 which affect radiation sensitivity and frequency of ultraviolet light-induced mutations, J. Bacteriol. 112:286 (1972).
6. J. W. Little, S. H. Edmiston, L. Z. Pacelli, and D. W. Mount, Cleavage of the Escherichia coli lexA protein by the recA protease, Proc. Natl. Acad. Sci. USA 77:3225 (1980).
7. B. E. Markham, J. W. Little, and D. W. Mount, Nucleotide sequence of the lexA gene of Escherichia coli K-12, Nucleic Acids Res. 9:4149 (1981).
8. D. W. Mount, A. C. Walker and C. Kosel, Suppression of lex mutations affecting deoxyribonucleic acid repair in Escherichia coli K-12 by closely linked thermosensitive mutations, J. Bacteriol. 116:950 (1973).
9. E. M. Witkin, Ultraviolet mutagenesis and inducible DNA repair in Escherichia coli, Bacteriol. Rev. 40:869 (1976).
10. T. Miki, Y. Ebina, F. Kishi, and A. Nakazawa, Organization of the lexA gene of Escherichia coli and nucleotide sequence of the regulatory region, Nucleic Acids Res. 9:529 (1981).
11. F. Sanger, A. R. Coulson, B. G. Barrell, A. J. H. Smith, and B. A. Roe, J. Mol. Biol. 143:161 (1980).
12. J. W. Little, The SOS regulatory system: control of its state by the level of recA protease, J. Mol. Biol. 167:791 (1983).
13. T. A. Phillips, R. A. Van Bogelen, and F. C. Neidhardt, lon gene product of Escherichia coli is a heat-shock protein, J. Bacteriol. 159:283 (1984).
14. D. W. Mount and C. Kosel, Ultraviolet light-induced mutation in UV-resistant, thermosensitive derivatives of $lexA^-$ strains of Escherichia coli K-12, Mol. Gen. Genet. 136:95 (1975).
15. D. W. Mount, C. Kosel, and A. Walker, Inducible error-free DNA repair in tsl recA mutants of E. coli, Mol. Gen. Genet. 146:37 (1976).
16. D. Ennis, B. Fisher, S. Edmiston, and D. W. Mount, Dual role for E. coli RecA protein in SOS mutagenesis, Proc. Natl. Acad. Sci. USA (in press).
17. S. Mizusawa and S. Gottesman, Protein degradation in Escherichia coli: the lon gene controls the stability of SulA protein, Proc. Natl. Acad. Sci. USA 80:358 (1983).

MUTAGENIC DNA REPAIR IN BACTERIA:

THE ROLE OF UmuDC AND MucAB

Lorraine Marsh and Graham C. Walker

Department of Biology
Massachusetts Institute of Technology
Cambridge, MA

INTRODUCTION

Bacterial systems offer several advantages for studying mutagenic processes. The ease with which mutagenesis can be assayed has been harnessed in the well-established Ames test of mutagenicity. Likewise, the powerful genetics available for analysis of the gene product involved in mutagenesis have made Escherichia coli the best understood organism with respect to mutagenic processes.

Mutagenesis by some alkylating agents and base analogs in Escherichia coli appears to involve a simple mispairing during replication. For example 2-aminopurine is incorporated into DNA as adenine but mispairs frequently.[1] However, mutagenesis by UV light and many chemcials in E. coli is not a passive event, but requires the active participation of a number of proteins including RecA and the UmuD and UmuC proteins.[2] Mutations of either umuD or umuC completely prevent this mutagenesis and slightly increase the sensitivity of strains to DNA-damaging agents.[3,4] This umuDC-dependent mutagenic repair process has been termed "error-prone repair" or "SOS processing".[2]

The bulk of the mutagens which are also carcinogens require Umu functions in order to be mutagenic. In general lesions produced by Umu-dependent mutagens seem to block DNA replication in E. coli. These include UV light induced lesions, apurinic/apyrimidinic (AP) sites[5] and a variety of base modifications by alkylating agents[2,6] and agents which form bulky adducts.[7] Some of these latter may be mutagenic through an AP site intermediate.[5] This chapter will focus exclusively on Umu-dependent mutagenic repair.

The mucA and mucB genes, carried on the plasmid pKM101 together complement a defect in either of the umu genes. The plasmid pKM101 (which is included in the Ames tester strains) derives from a clinically isolated plasmid R46[8,9] and can transfer to a variety of gram negative bacteria. Many of these bacteria are naturally unmutable by UV light but become mutable and more UV resistant when carrying pKM101 or R46.[10] Other broad host range plasmids conferring UV resistance/mutability are also known and it seems likely that muc-like genes are quite common. This suggests that mutability is a dispensable trait in bacteria dependent largely on the presence or absence of Umu or Muc-like proteins. Two reports have recently been presented suggesting that, after introduction into eukaryotic cell, the mucAB genes can affect mutability or UV resistance, but since the expression of the MucA and MucB proteins was not assayed in these systems the results await confirmation[11,12] Muc and umu are at least superficially interchangeable for the mutagens: No cases are known in which only muc or only umu will allow mutagenesis though differences in mutagenic spectrum are observed as will be discussed later.

The umuDC and mucAB operons have been cloned and sequenced [13,14,15,16]. Both operons are similarly arranged (Fig. 1) with coding for a 16 kilodalton (Kd) protein followed by coding for a 45 Kd protein. Over all there is about 50% homology between the derived amino acid sequence for the Muc and the Umu proteins suggesting that they may have arisen from a common ancestor.[15]

Both umuDC and mucAB are SOS-inducible[17,18] and are tightly under the control of the lexA repressor in E. coli. LexA binding sites are found at the beginning of each operon. Many studies of mutagenesis in E. coli are complicated by the fact that the same lesions (or co-produced lesions) which are premutagenic are also required for the induction of umuDC.

Though umuD and umuC are regulated by the SOS system in E. coli there is no reason per se why they could not function constitutively. Though constitutive overexpression of umuDC makes E. coli cold-sensitive for DNA replication[19] moderate constitutive expression (i.e. in a lexA(Def) strain) is tolerated and overexpression of mucAB does not kill E. coli (L. Marsh and G.C. Walker, unpublished observation). Therefore, while mutagenic repair is inducible in wild type E. coli, it may not be inducible in all bacteria or, extrapolating, in eukaryotic systems.[10]

MODEL FOR SOS PROCESSING

Though the details of the mechanism of mutagenic DNA repair (SOS processing) are not known the model shown in Fig. 2 is probably essentially correct. This model suggests that when DNA polymerase III holoenzyme, the normal replicative enzyme of E. coli is unable to synthesize past a lesion, the UmuD and UmuC gene products along with activated RecA protein somehow allow this blocked synthesis to continue in an abnormal fashion creating a trans-lesion complementary strand of DNA. Normal DNA synthesis then continues or the new DNA is ligated to DNA previously synthesized.

The most easily interpretable evidence for this model comes from studies of single-stranded DNA phages for which other forms of DNA repair will not function.[20] UmuC mutations block both mutagenesis and repair of UV damaged M13[21] (L. Marsh and G.C. Walker, unpublished observations). The unique feature of this repair is that, unlike other DNA repair pathways SOS-processing does not make use of the inherent informational redundancy of double stranded DNA. So why does a cell with double-stranded DNA use such a system? It has been proposed that SOS processing acts on the chromosome

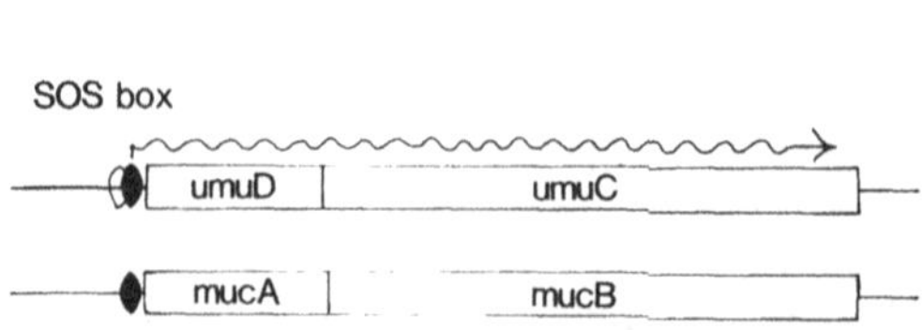

Figure 1. Schematic organization of the umuDC and mucAB operons.

primarily when lesions occur close to one another on both strands and are not repaired before replication. Under these circumstances the cell apparently cannot use error-free recombinational or excision pathways to repair its damage and has no choice but to attempt trans-lesion synthesis.[22,23]

An interesting corollary of this model is that the product of SOS processing is still lesion-containing DNA. However, this DNA is double-stranded and can in turn be repaired by excision or recombinational repair systems.

Recent experiments show that UV mutagenesis can occur in the absence of umuC if UV-induced lesions are photoreversed after an initial induction period. This suggests that the Umu products are perhaps not needed for trans-lesion synthesis itself but for some later DNA elongation step.[24] It is possible however that in Umu^+ strains a modified polymerase contributes to trans-lesion synthesis as well as elongation.

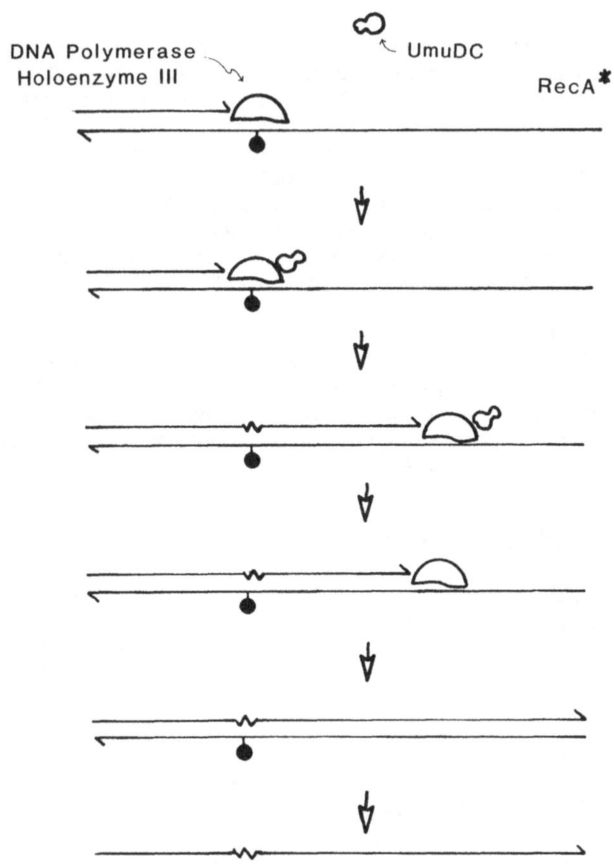

Figure 2. Outline of postulated steps in SOS processing of lesion-containing DNA. See text for details.

SPECIFIC INTERACTIONS INVOLVED IN SOS PROCESSING

Though the biochemical interactions involved in SOS-processing are not known, we are beginning to make progress in understanding these interactions through genetic and physiological studies. These studies suggest interactions of the UmuDC proteins with one another, RecA protein, and the cellular DNA replication apparatus.

The phenotype of strains with mutations in either umuD or umuC is identical, i.e. a failure to carry out SOS-processing. Though the intact mucAB operon complements mutations in either umuD or umuC, mucA alone will not complement a defect in umuD and vice versa. That is, each 45Kd protein requires its cognate 16Kd protein for activity. This suggests that interaction between the 16Kd and 45Kd proteins is required for function and that the umu and muc have evolved far enough apart that umu and muc proteins will not work with one another.[15]

Overproduction of UmuDC proteins in a lexA(Def) genetic background makes strains cold sensitive for growth. This cold sensitivity is due to a rapid and reversible inhibition if DNA replication at the nonpermissive temperature.[19] This ability of mumDC to interfere with DNA synthesis, though not a normal function, suggests an interaction of the Umu proteins with DNA replication machinery or a DNA substrate. Perhaps this interaction is related to the modifications in replicative machinery necessary to allow bypass of lesions.

RecA protein serves at least two roles in mutagenesis. The first is to promote the cleavage of LexA repressor and induce the SOS system. However, even if the lexA gene is genetically inactivated so that SOS functions are expressed constitutively, RecA protein is still required for mutagenesis.[25,26] An analysis of the sequence of umuD and mucA shows that extensive homology exists between the predicted amino acid sequence and the amino acid sequence of the carboxyl-terminal domain of LexA protein. The carboxyl-terminal region of LexA protein contains RecA binding sites, the specific site of protein cleavage, and possibly dimerization sites, but not DNA binding sequences. This raises the possibility that MucA and UmuD protein might be cleaved and activated by RecA. However, to date we have been unable to detect cleavage or autodigestion of UmuD protein in vitro (L. Dodson and G.C. Walker, unpublished observation) and a specific mutation at the putative protein cleavage site of MucA does not destroy activity (L. Marsh and G.C. Walker, unpublished observation). Perhaps an interaction of UmuD or MucA protein with RecA protein is required for mutagenesis even if they are not cleaved like LexA repressor.

CONTRIBUTIONS TO MUTAGENIC SPECTRUM

The mutagenic spectrum observed in E. coli after DNA damage has several components. First there is the spectrum of damage to DNA. Most DNA-damaging agents have specific base preferences and often there is a longer range sequence specificity. Many DNA-damaging agents induce more than one type of lesion and this adds to the spectrum complexity.

Secondly, there is bias in error-free repair. Lesions which are efficiently repaired in an error-free manner are obviously no longer substrates for an error-prone process. One lacI gene "hot spot" for UV-induced mutagenesis which exists in a strain defective in excision repair is not found in its excision proficient counterpart.[27]

Lastly, there is a spectrum of preference for SOS processing. Some lesions appear to be better substrates for SOS-processing than others. For example, survival after treatment with angelicin and near UV is much more dependent on the UmuDC^{+} phenotype than survival after UV-treatment (28). Presumably Angelicin-induced lesions are well-handled by SOS-processing while (some?) UV-induced lesions are not.

Do umuDc and mucAB play any role in determining the specificity of mutagenesis, that is in the choice of targeted mutagenesis sites or the mis-inserted bases? The clearest evidence that this is so comes from a comparison of mutagenesis induced by umuDC as opposed to mucAB. Since muc has diverged greatly from umu during evolution but still serves a similar function muc can be treated as a special mutant of umu. Though the general spectrum of UV-induced lacI mutations is similar with muc and with umu one striking exception occurs. The relative frequency of a UV-induced C T transition at one location in lacI is 3.5 times higher in a strain without pKM101 than with (27). Thus umu creates a "hot spot" for mutagenesis at this site. Unfortunately pKM101 carries many functions besides muc and one could argue that some other repair function is affecting the spectrum.

We have recently created a specific mutation (muc*) in cloned mucAB which changes the gly at codon 16 of mucA to a glu (L. Marsh and G.C. Walker, in preparation). This mutant muc has several interesting properties one of which is that it alters the mutagenic spectrum. Reversion of the argE3 or his-4 ochre mutations of E. coli provides a crude but useful system for looking at mutagenic spectrum changes since "true" revertants at the site of the original mutation are easily distinguished from induced ochre suppressor mutations.[29] Figure 3 presents bar graphs of the ratio of true (gene specific) revertants to total revertants in excision repair proficient and deficient strains. In these experiments the only difference between the two strains is a single mutation in the cloned muc genes and overall reversion frequencies are similar, Clearly this mutation can change the specificity of at least this simple test system independently of the

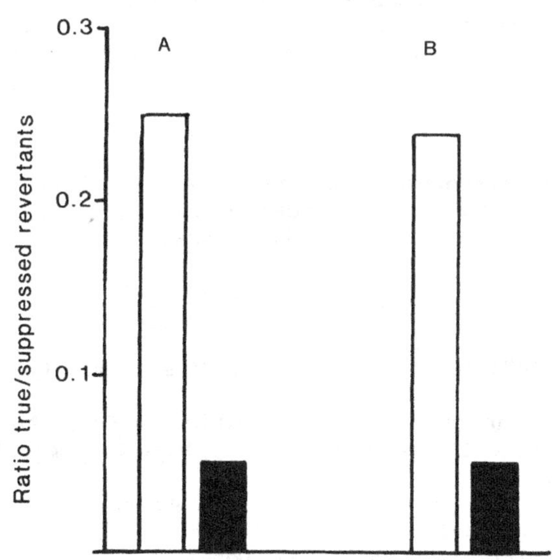

Figure 3. Relative frequency of true revertants and suppressor mutations in umuC strains carrying muc plasmids. Open bars, wild type mucAB plasmid; filled bars, mucAB* (see text). A TK610 host [uvrA]; B GW2100 host [Uvr^{+}].

excision repair status of the strain that it is in. It is interesting to note that in both of the cases mentioned potential hairpin structures can form at one of the two sites being compared: in lacI around internal palidromes within the gene and in the arg^{oc} reversion system at the tRNA genes. Glickman has suggested that such potential hairpins might be hotspots for mutagenesis.[27] In any case muc and umu may contribute to specificity of mutagenesis at least at some sites. This suggests an interaction with the lesion or the DNA around it. At other sites the spectrum of recovered mutants may be most determined by the spectrum of damage or the relative efficiency of repair.

RELEVANCE TO HIGHER ORGANISMS

It is clear that in bacteria bulky and non-informational lesions in DNA require the active participation of UmuDc-like or MucAB-like proteins in order to cause mutations. Similar systems may operate in higher cells. The yeast REV genes bear at least a formal resemblance to the umu locus. Mutations in these genes make yeast strains unmutable by UV-light and other DNA damaging agents without greatly affecting UV-resistance.[30] It will be of interest to see whether these genes show any homology to umu or muc genes. Though mammalian DNA polymerases are able to synthesize past UV-induced lesions *in vitro* these enzyme preparations may lack accessory proteins found *in vivo*. The fidelity of *E. coli* DNA polymerase III increases dramatically when it is in the holoenzyme form, complexed with other subunits of the complete enzyme.[31]

Though the *E. coli* umu genes are inducible by DNA damage, inducibility is not an essential part of their function. It may be that in some organisms or tissues similar functions are expressed constitutively as in *E. coli* lexA(Def) strains.

Since muc and umu genes introduce an element of sequence specificity into the spectrum of mutagenesis independent of the initial spectrum of DNA damage a umu-like system in higher cells might be responsible for some "hot spots" of mutagenesis which do not correspond to "hot spots" of DNA damage.

ACKNOWLEDGEMENTS

This work was supported by training grant ES07020 from the National Institutes of Health and was performed in the laboratory of G. C. Walker, whose encouragement is gratefully acknowledged.

REFERENCES

1. A. Korberg, "DNA Replication," W. H. Freeman and Co., San Francisco (1980).
2. G. C. Walker, Mutagenesis and inducible responses to deoxyribonucleic acid damage in *Escherichia coli*, *Microbiological Rev.* 48:60-93 (1984).
3. T. Kato and Y. Shinoura, Isolation and characterization of mutants of *Escherichia coli* deficient in induction of mutations by ultraviolet light, *Mol. Gen. Genet.* 156:121-131 (1977).
4. G. Steinborn, Uvm mutants of *Escherichia coli* K-12 deficient in UV mutagenesis. I. Isolation of umv mutants and their phenotypical characterization in DNA repair and mutagenesis, *Mol. Gen. Genet.* 165:87-93 (1978).
5. L. A. Loeb, Apurinic sites as mutagenic intermediates, *Cell* 40:483-484 (1985).

6. P. L. Foster and E. Eisenstadt, Induction of transversion mutations in Escherichia coli by N-methly-N'-nitro-N-nitrosoguanidine is SOS dependent, J. Bacteriol. in press (1985).
7. J. H. Miller, Carcinogens induce targeted mutations in Escherichia coli, Cell 31:5-7 (1982).
8. K. E. Mortelmans, B. A. D. Stocker, Segregation of the mutator property of plasmid R46 from its ultraviolet-protecting property, Mol. Gen. Genet. 167:317-328 (1979).
9. P. J. Langer and G. C. Walker, Restriction endonuclease cleavage map of pKM101: Relationship to parental plasmid R46, Mol. Gen. Genet. 182:268-272 (1981).
10. G. C. Walker, L. Marsh and L. A. Dodson, Genetic analyses of DNA repair:Inference and Extrapolation, Ann. Rev. Genet. (in press).
11. A. A. Potter, E. R. Nestmann, and V. N. Iyer, Introduction of the plasmid pKM101-associated muc genes into Saccharomyces cerevisiae, Mut. Res. 131:197-204 (1984).
12. R. Elli, L. Marcucci, R. Bosi, C. Oppi, P. A. Battaglia and F. Gigliani, The pR UV$^+$ plasmid, transfected into mammalian cells, enhances their UV survival, Nucleic A. Res. 11:3679-3683 (1983).
13. S. J. Elledge and G. C. Walker, Proteins required for ultraviolet light and chemical mutagenesis:Identification of the products of the umuC locus of Escherichia coli, J. Mol. Biol. 164:175-192 (1983).
14. K. L. Perry and G. C. Walker, Identification of plasmid (pKM101)-coded proteins involved in mutagenesis and UV resistance, Nature (London) 300:278-281 (1982).
15. K. L. Perry, S. J. Elledge, B. B. Mitchell, L. Marsh and G. C. Walker, The umuDC and mucAB operons whose products are required for UV and chemical mutagenesis:UmuD, MucA, and LexA proteins share homology, Proc. Natl. Acad. Sci. (in press) (1985).
16. Y. Kitagawa, E. Akaboshi, K. Matubara, H. Shinagawa, T. Horii, T. Kato, Proc. Natl. Acad. USA, in press (1985).
17. A. C. Bagg, C. J. Kenyon, and G. C. Walker, Inducibility of a gene product required for UV and chemical mutagenesis in Escherichia coli. Proc. Natl. Acad. Sci. USA 78:5749-5753 (1981).
18. S. J. Elledge and G. C. Walker, The muc genes of pKM101 are inducible by DNA-damage. J. Bacteriol. 155:1306-1311 (1983).
19. L. Marsh and G. C. Walker, Cold sensitivity induced by overproduction of UmuDC in Escherichia coli, J. Bacteriol. 162:155-161 (1985).
20. P. Caillet-Fauquet, M. Defais, and M. Radman, Molecular mechanisms of induced mutagenesis. Replication in vivo of bacteriophage ØX174 single-stranded, ultraviolet light-irradiated DNA in intact and irradiated host cells, J. Mol. Biol. 117:95-112 (1977).
21. M. Defais, Role of the E. coli umuC gene product in the repair of single-stranded DNA phage, Mol. Gen. Genet. 192:509-511 (1983).
22. M. Radman, SOS repair hypothesis:Phenomenology of an inducible DNA repair which is accompanied by mutagenesis, p. 355-367, in: "Molecular Mechanisms for Repair of DNA, Part A," P. Hanawalt and R. B. Setlow, ed., Plenum Press, New York, (1975).
23. T. V. Wang and K. C. Smith, Role of the umuC gene in post-replication repair in UV-irradiated Escherichia coli K-12 uvrB, Mut. Res. 145:107-112 (1985).
24. B. A. Bridges, R. Woodgate, Mutagenic repair in Escherichia coli. X. The umuC gene product may be required for replication past pyrimidine dimers but not for the coding error in UV-mutagenesis, Mol. Gen. Genet. 196:364-366 (1984).
25. M. Blanco, G. Herrera, P. Collado, J. Rebollo, and L. M. Botella, Influence of RecA protein on induced mutagenesis, Biochemie 64:633-636 (1982).
26. D. G. Ennis, B. Fisher, S. Edmiston, and D. W. Mount, Dual role for E. coli RecA protein in SOS mutagenesis. In press.
27. B. Glickman, Mutational specificity of UV light in E. coli: Influence

of excision repair and the mutator plasmid pKM101, _in_: "Induced Mutagenesis", C. W. Lawrence, ed., Plenum Press, New York (1983).

28. S. S. Miller and E. Eisenstadt, Enhanced sensitivity of UmuC⁻ _Escherichia coli_ to photodynamic inactivation by angelicin (isopsoralen), _J. Bacter._ in press (1985).

29. T. Kato, Y. Shinoura, A. Templin, and A. J. Clark, Analysis of ultraviolet light-induced suppressor mutations in the strain of _Escherichia coli_ K-12 AB1157: An implication for molecular mechanisms of UV mutagenesis, _Mol. Gen. Genet._ 180:283-291 (1980).

30. C. W. Lawrence, Mutagenesis in _Saccharomyces cerevisiae_, _Adv. Genet._ 21:173-254 (1982).

31. R. H. Scheuermann and H. Echols, A separate editing exonuclease for DNA replication: the subunit of _Escherichia coli_ DNA polymerase III holoenzyme, _Proc. Natl. Acad. Sci._ 81:7747-7751 (1984).

THE REPAIR OF PYRIMIDINE DIMERS VIA A DNA-GLYCOSYLASE MECHANISM

Robert H. Grafstrom

Department of Microbiology
Jefferson Medical College
Thomas Jefferson University
Philadelphia, PA

INTRODUCTION

The pyrimidine dimer is the major photoproduct produced in DNA after UV irradiation. Ever since the repair of UV damage was first described, the pathway for the excision repair of these adducts was presumed to consist of four steps: incision near the damage site; excision of the damaged nucleotides; replacement of the missing nucleotides by DNA polymerase and finally ligation. Although the understanding of excision repair has increased in recent years [1] the requirement for these four basic steps has remained for nucleotide excision repair. The discovery of DNA glycosylases which are very specific for a particular damaged base in DNA[2] led to a two step pathway for the incision process. The glycosylic bond between the base and the sugar is broken resulting in an apurinic or apyrimidinic (AP) site in the DNA leaving the phosphodiester backbone intact. These AP sites are then cleaved by the damage specific AP endonucleases and the AP site is removed as in nucleotide excision repair.

In experiments that focused on the incision step in the repair of pyrimidine dimers, two enzymes have been studied extensively as model systems: the "UV endonucleases" isolated either from M. luteus or from bacteriophage T4 infected E. coli. The T4 "UV endonuclease" is encoded by the denV gene and maps at 64kb on the T4 physical map.[3,4] Both of these enzymes have very similar properties and are very different from the Uvr ABC repair enzyme complex which is responsible for the repair of pyrimidine dimers and other bulky adducts in E. coli. The M. luteus and T4 "UV endonucleases" are specific for pyrimidine dimers and do not repair other bulky adducts. These enzymes are low molecular weight, single polypeptide chains that are easy to isolate and relatively stable. As a result of these properties, these enzymes have been used analytically in studies of DNA repair in other cell systems to detect the removal of pyrimidine dimers from DNA. The T4 "UV endonuclease" has also been demonstrated to complement both mammalian and bacterial cell lines deficient in the repair of pyrimidine dimers.[5,6] Thus, these "UV endonucleases" were excellent candidates to study the requirements of the incision mechanism using the purified enzyme and DNA substrates of defined sequence.[7] As a result of these experiments, the incision step of these simple repair endonucleases was shown to consist to two discrete steps that were similar to that used by DNA glycosylases for base excision repair.[8,9,10,11] This mechanism is

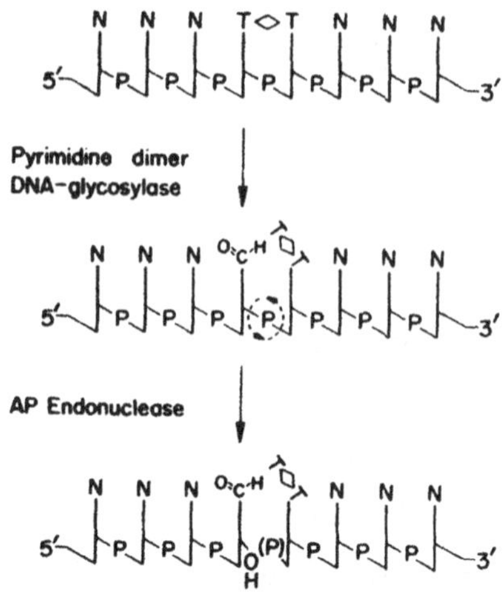

Figure 1. A two step model for the incision at sites of pyrimidine dimers by the "UV endonucleases" of T4 denV gene and M. luteus.

outlined in Figure 1. The first step in the incision pathway is the cleavage of the N-glycosylic bond of the 5' pyrimidine of the dimer resulting in a AP site and a mixed thymine-thymidylate dimer still attached to the DNA through the N-glycosylic bond of the 3' pyrimidine of the dimer. The second step is the cleavage 3' to the AP site by an associated AP endonuclease resulting in DNA damage on both sides of the final incision site. The experiments with both the T4 and M. luteus "UV endonucleases" that led to the elucidation of this mechanism are discussed in more detail below.

RESULTS

The two step incision mechanism for the repair of pyrimidine dimers outlined in Figure 1 provides several, testable predictions that are true of the "UV endonucleases" as well as other DNA glycosylases.

Site of Incision

The first conclusion from the application of DNA sequencing analysis to the study of the mechanism of the "UV endonuclease" incision step was totally unexpected: the M. luteus "UV endonuclease" cut the DNA between the pyrimidine of the dimer[7,8,11] and not near or next to the site as suggested by earlier experiments.[12] This result was based entirely on the DNA sequencing data in which the fragments were one nucleotide longer than expected. The 5' pyrimidine of the dimer was shown to be the target of the N-glycosylase by the sequencing data. This result was later confirmed in a more classical approach in which 5'-glycosylase cleavage was distinguished from 3'-glycosylase cleavage based on the resulting end products. In the former case, after treatment with yeast photolyase to release the 5' pyrimidine of the dimer (see below) and alkaline hydrolysis to remove the AP site, all of the 5' nucleotides at the site of incision will be pyrimidines. In the case of 3'-glycosylase cleavage the 5' nucleotides will be a random mixture of all four deoxynucleotides.[13]

Unlike the other DNA glycosylases which release the free base from the DNA backbone, the pyrimidine dimer-DNA glycosylases leave the pyrimidine dimer still attached to the DNA through the N-glycosylic bond of the 3' pyrimidine of the dimer. This structure was confirmed as an intermediate after pyrimidine dimer-DNA glycosylase action by the release of free thymine from the DNA. The release of ^{3}H-thymine was dependent on treatment either with yeast photolyase or by photolysis which is a second dose of UV irradiation.[8,9,10,13] Using this latter procedure to assay exclusively for the pyrimidine dimer-DNA glycosylase activity, this enzyme was shown to prefer T-T dimers over C-T dimers and double stranded DNA over single stranded DNA.[13]

As a result of cleavage of the N-glycosylic bond an apyrimidinic site is formed in the DNA. The existence of an apyrimidinic site in the DNA was inferred from the DNA sequencing data[7,8,11] which demonstrated a shift in the electrophoretic mobilities of the DNA fragments cut with the M. luteus "UV endonuclease". This shift in the electrophoretic mobility was dependent on alkaline hydrolysis of the sample before analysis on the sequencing gel. Later these apyrimidinic sites were confirmed as intermediates in the incision reaction by the stimulation of DNA nicking activity of the pyrimidine dimer-DNA glycosylase either by purified placental AP endonuclease or by increased sensitivity to alkaline hydrolysis [8,13,14,15]. The reduction of these apyrimidinic sites using sodium borohydride made these lesions impervious to subsequent alkaline hydrolysis further suggesting that they were indeed AP sites, since the alkaline hydrolysis reaction proceeds by a β-elimination mechanism.[16]

Associated AP Endonuclease

An interesting feature of these pyrimidine dimer-DNA glycosylases that set them apart from the other DNA glycosylases was that both appeared to have an associated AP endonuclease activity.[8,13,15,17,18] The data suggested that these small molecular weight proteins had two catalytic sites: one for glycosylic cleavage of the dimer and a second for the cleavage of the AP site. The well defined genetic system of T4, in which mutants in the denV gene are available, allowed for the testing of this hypothesis. Using crude extracts of T4 infected E. coli it was shown that various denV mutants lacked the glycosylase activity.[14,15] Furthermore, temperature sensitive denV mutants had a temperature sensitive DNA glycosylase.[14] However, these mutant extracts still possessed near normal levels of the AP endonuclease activity. These studies were complicated by the numerous (at least three) AP endonucleases found in E. coli. After partial purification, it was convincingly demonstrated that the purified mutant extracts lacked both the DNA glycosylase and the associated AP endonuclease activities.[17,18] Suppression of one of the amber mutants, T4, uvs-13, partially restored the glycosylase activity but not the associated AP endonuclease activity suggesting that there were two different domains of enzymatic activity within the single polypeptide chain.[19] These two enzymatic activities also showed different salt and pH optima and different heat inactivation profiles.[18,19] Others have reported the same heat inactivation profiles for the pyrimidine dimer-DNA glycosylase and the associated AP endonuclease activities for both the T4[17] and M. luteus[13] enzymes. UV sensitive mutants of M. luteus also lack the pyrimidine dimer-DNA glycosylase activity but experiments designed to test the effect of these mutations on the associated AP endonuclease activity were also complicated by the other AP endonucleases found in M. luteus.[20] In competition experiments using the T4 "UV endonuclease" and UV irradiated DNA, AP DNA was shown to compete for the pyrimidine dimer-DNA glycosylase activity while native DNA did not. Both AP DNA and native DNA could compete for cleavage by the associated AP endonuclease. These results independently suggest that the glycosylase and AP endonuclease activities occur within two different domains on the same polypeptide chain.[15]

Based on the original DNA sequencing data[7,8,11] the AP endonuclease activity was shown to cleave exclusively on the 3' side of the AP site, leaving DNA damage on both sides of the nick. These results were later substantiated using apyrimidinic DNA to identify the site of cleavage.[13] Other investigators[19,21] have shown that cleavage of either UV irradiated DNA or AP DNA by the pyrimidine dimer-DNA glycosylase and its associated AP endonuclease produced a DNA that had a 3' terminus that was not a substrate for DNA polymerase I. These results were interpreted to mean that the DNA contained a 3' terminal AP site since treatment of this DNA with a Class II AP endonuclease restored DNA polymerase activity. The Class II AP endonucleases cut 5' to the AP site and hence can remove 3' terminal AP sites leaving DNA with a 3' terminal hydroxy group. The priming activity could also be restored by the combined action of the alkali (to remove the 3' AP site by β-elimination) and alkaline phosphatase (to remove the resulting 3' PO_4 group). An interesting aspect of this work was the demonstration that the 3' -> 5' exonuclease activity of Pol I was inefficient in the removal of these 3' terminal AP sites. These findings suggest a reason for the existence of multiple AP endonucleases in cells. However, a human placental AP endonuclease that had both Class I and Class II AP endonucleolytic activities was unable to remove 3' terminal AP sites.[22]

Finally, the AP endonuclease associated with the M. luteus pyrimidine dimer-DNA glycosylase has been shown to prefer apyrimidinic sites resulting from the cleavage of UV irradiated DNA by the pyrimidine dimer-DNA glycosylase over other apurinic or apyrimidinic sites. Other AP endonucleases do not show this preference for dimer apyrimidinic DNA.[13]

In Vivo Activity

In order to test whether the two step incision mechanism was operating in vivo, Radany and Friedberg[23] UV irradiated uvr mutants of E. coli that had been labeled with ^{3}H-thymidine. The cells were then infected with bacteriophage T4 and after a brief incubation, the acid soluble extracts were prepared. If the oligonucleotides produced as a consequence of DNA repair contained mixed thymine-thymidylate dimers as had been proposed by the in vitro data presented above, then exposing these oligonucleotides to a second dose of UV irradiation would break these dimers releasing free ^{3}H-thymine. ^{3}H-thymine was released only from cells infected with wild type T4 phage and not with T4 denV mutants or with mock infected cells. The release of ^{3}H-thymine was also dependent upon the second dose of UV irradiation. Similar results were found in UV irradiated wild type M. luteus but not in UV sensitive mutants. In neither of these experiments, however, was the fate of the subsequent AP site determined.

Cloning of the denV Gene

The cloning of the denV gene was first reported by Loyd and Hanawalt,[24] however, their constructs were apparently very unstable in vivo and have not been further characterized. More recently, two independent groups[3,4] have reported the sequence of the denV gene using a combined genetic and DNA sequencing approach in which regions surrounding the denV gene were sequenced. The gene was eventually identified through the sequencing of the denV mutants. The gene maps at the 64kb position on the T4 physical map. Using this information, the gene was cloned into a multicopy number plasmid[25] The cloned denV gene partially restores UV resistance to uvr A,B, and

C mutants of E. coli as well as recA and recBC mutants. The reason for this latter result remains unclear.

SUMMARY

The "UV endonuclease" isolated either from M. luteus or bacteriophage T4 infected E. coli (the denV gene product) consists of two enzymatic activities on a single polypeptide chain: a pyrimidine dimer-DNA glycosylase and an AP endonuclease. The repair of pyrimidine dimers by this enzyme is initiated by the cleavage of the N-glycosylic bond of the 5' pyrimidine of the dimer that leaves the cyclobutane dimer still attached to the DNA through the N-glycosylic bond of the 3' pyrimidine of the dimer. This reaction results in the formation of an apyrimidinic site in the DNA. The second step in this repair pathway is the endonucleolytic cleavage of the DNA 3' to the AP site by the associated AP endonuclease. As a result, the nicked DNA contains DNA damage on both sides of the incision site: an apyrimidinic moiety on the 3' end and a thymine-thymidylate dimer on the 5' end. The enzymes prefer double stranded DNA over single stranded DNA, and thymine over cytosine at the 5' position of the dimer. The AP endonuclease activity prefers the AP site created by the pyrimidine dimer-DNA glycosylase on UV irradiated DNA over either apurinic or apyrimidinic DNA. This repair mechanism appears to be operative in vivo since DNA intermediates containing thymine-thymidylate dimer sites have been detected in UV irradiated T4 infected E. coli and in UV irradiated M. luteus. The cloned denV gene partially complements the UV repair deficient uvr A,B,C strains of E. coli.

REFERENCES

1. L. Grossman, Enzymes involved in the repair of damaged DNA, Arch. Biochem. Biophys. 211:511 (1981).
2. T. Lindahl, DNA repair enzymes, Ann. Rev. Biochem. 51:61 (1982).
3. K. Valerie, E. E. Henderson, and J. K. deRiel, Identification, physical map location and sequence of the denV gene from bacteriophage T4 Nuc. Acids Res. 12:8085 (1984).
4. E. H. Radany, L. Naumovski, J. D. Love, K. A. Gutekunst, D. H. Hall, and E. C. Friedberg, Physical mapping and complete nucleotide sequence of the denV gene of bacteriophage T4, J. Virol. 52:846 (1984).
5. K. Shimizu and M. Sekiguchi, Introduction of an active enzyme into permeable cells of E. coli, Mol. Gen. Genet. 168:37 (1979).
6. K. Tanaka, M. Sekiguchi, and Y. Okada, Restoration of UV-induced unscheduled DNA synthesis of xeroderma pigmentossum cells by the concomitant treatment with bacteriophage T4 endonuclease V and HVJ (Sendai Virus), Proc. Natl. Acad. Sci. USA 72:4071 (1975).
7. L. Grossman, S. Riazuddin, W.A. Hasetine, and C. Lindan, Nucleotide excision repair of damaged DNA, Cold Spring Harbor Symp. Quant. Biol. 43:947 (1979).
8. W. A. Haseltine, L. K. Gordon, C. P. Lindan, R. H. Grafstrom, N. L. Shaper, and L. Grossman, Cleavage of pyrimidine dimers in specific DNA sequences by a pyrimidine dimer-DNA glycosylase of M. luteus Nature 285:634 (1980).
9. E. H. Radany and E. C. Friedberg, A pyrimidine dimer-DNA glycosylase activity associated with the v gene product of bacteriophage T4, Nature 286:182 (1980).
10. B. Demple and S. Linn, DNA N-glycosylases and UV repair, Nature 287:203 (1980).
11. L. K. Gordon and W. A. Haseltine, Comparison of the cleavage of pyrimidine dimers by the bacteriophage T4 and Micrococcus luteus UV-specific endonucleases, J. Biol. Chem. 255:12047 (1980).

12. S. Riazuddin and L. Grossman, Micrococcus luteus correndonucleases. II. Mechanism of action of two endonucleases specific for DNA containing pyrimidine dimers, J. Biol. Chem. 252:6287 (1977).
13. R. H. Grafstrom, L. Park, and L. Grossman, Enzymatic repair of pyrimidine dimer-containing DNA, J. Biol. Chem. 257:13465 (1982).
14. P. A. Seawell, C. A. Smith, and A. K. Ganesan, denV gene of bacteriophage T4 determines a DNA glycosylase specific for pyrimidine dimers in DNA, J. Virol. 35:790 (1980).
15. S. McMillan, H. J. Edenberg, E. H. Radany, R. C. Friedberg, and E. C. Friedberg, denV gene of bacteriophage T4 codes for both pyrimidine dimer-DNA glycosylase and apyrimidinic endonuclease activities, J. Virol. 40:211 (1981).
16. L. Grossman and R. H. Grafstrom, AP sites and AP endonucleases, Biochimie 64:577 (9182).
17. H. R. Warner, L. M. Christensen, and M. L. Persson, Evidence that the UV endonuclease activity induced by bacteriophage T4 contains both pyrimidine dimer-DNA glycosylase and apyrimidinic/apurinic endonuclease activities in the enzyme molecule, J. Virol. 40:205 (1981).
18. Y. Nakabeppu and M. Sekiguchi, Physical association of pyrimidine dimer-DNA glycosylase and apurinic/apyrimidinic endonuclease essential for repair of UV-damaged DNA, Proc. Natl. Acad. Sci. USA 78:2742 (1981).
19. Y. Nakabeppu, K. Yamashita, and M. Sekiguchi, Purification and characterization of normal and mutant forms of T4 endonuclease V, J. Biol. Chem. 257:2556 (1982).
20. J. Pierre and J. Laval, M. luteus endonucleases for apurinic/apyrimidinic sites in DNA. I. Purification and general properties, Biochem. 19:5018 (1980).
21. H. R. Warner, B. F. Demple, W. A. Deutsch, C. M. Kane, and S. Linn, Apurinic/apyrimidinic endonucleases in repair of pyrimidine dimers and other lesions in DNA, Proc. Natl. Acad. Sci. USA 77:4602 (1980).
22. R. H. Grafstrom, N. L. Shaper, and L. Grossman, Human placental apurinic/apyrimidinic endonuclease. Mechanism of action, J. Biol. Chem. 257:13459 (1982).
23. E. H. Radany and E. C. Friedberg, Demonstration of pyrimidine dimer-DNA glycosylase activity in vivo: bacteriophage T4-infected E. coli as a model system, J. Virol. 41:88 (1982).
24. R. S. Lloyd and P. C. Hanawalt, Expression of the denV gene of bacteriophage T4 cloned in E. coli, Proc. Natl. Acad. Sci. USA 78:2796 (1981).
25. K. Valerie, E. E. Henderson, and J. K. deRiel, Expression of a cloned denV gene of phage T4 in E. coli, Proc. Natl. Acad. Sci. USA in press.

THE INVOLVEMENT OF AN E. COLI MULTIPROTEIN COMPLEX IN THE COMPLETE REPAIR OF UV-DAMAGED DNA

Lawrence Grossman, Paul R. Caron, and Euk Y. Oh

The Johns Hopkins University, Department of Biochemistry
615 North Wolfe Street
Baltimore, Maryland 21205

With the plethora of new chemicals and agents being introduced into our environment, it is unlikely that most organisms will have had sufficient time to evolve new enzyme systems to accomodate to such rapid evolutionary pressures. It can be anticipated, therefore, that the most accomodating of repair systems would be those with broad specificity such that a wide variety of DNA adducts with differing primary chemical structures can be recognized to initiate repair processes. Such an enzyme system has been identified in the relatively simple model system from Escherichia coli referred to as the uvr repair system which is responsible for the repair of DNA containing ultraviolet light induced pyrimidine dimers, cross-linked pyrimidines or purines, alkylated purines in which the alkyl groups are ethyl or larger, polycyclic aromatic hydrocarbon adducts such as benzo[α]pyrene-guanine adducts and acetylaminofluorene-guanine adducts. In all likelihood what is recognized by the uvr system are the distortions arising as a consequence of this disparate chemistry. Because certain DNA repair deficient human cell lines show a similar spectrum of sensitivity as E. coli uvr^{s} lines it is likely that there are many similarities between the bacterial and mammalian repair systems. Because the patients with such diseases are predisposed towards skin cancer it is imperative that we characterize this mechanism in a genetically and technically managable bacterial model system.

Just as the E. coli system was chosen for its laboratory advantages ultraviolet light induced pyrimidine dimers were chosen as the model DNA 'substrate' because of the ease with which they can be manipulated, predicted and followed. Furthermore, they are one of the few ultimate carcinogens whose absolute structure is known; furthermore, the major form of cancer is skin cancer which is induced by this specific perturbation to DNA.

The removal of ultraviolet light-induced pyrimidine cyclobutane dimers from damaged DNA in bacteria is catalyzed by two essentially different mechanisms. As was discussed by the previous speaker, Dr. Robert Grafstrom, in Micrococcus luteus [1-3] and T4 phage-infected E. coli [4,5] the 5'-pyrimidine N-glycosylic linkage of the pyrimidine dimer is initially acted upon by a dimer specific endonuclease. This 17 Kd protein has two activities in which a pyrimidine dimer DNA glycosylase activity initially hydrolyzes the N-glycosyl linkage of the 5'-pyrimidine moiety of the dimer generating an apyrimidinic site juxtaposed to a thymine:thymidylate dimer

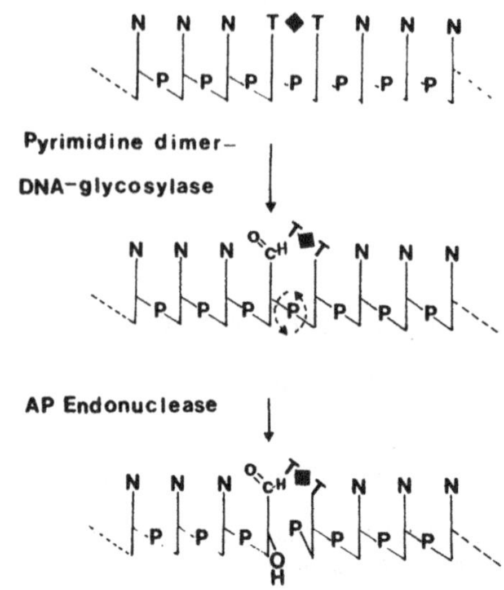

Figure 1. The Pyrimidine Dimer DNA Glycosylase Mechanism in M. luteus

(Figure I) A 3'-apyrimidinic endonucleolytic activity associated with the same enzyme generates a 3'-hydroxylated apyrimidinic terminus and a 5'-phosphorylated thymine:thymidylate dimer terminus.[3] Although these seemingly different activities are associated with the same polypeptide a common modified Michael addition reaction can be invoked to explain both activities (preliminary results).

The incision reaction in uninfected E. coli, which requires the uvrA, uvrB and the uvrC gene products,[6-8] results in two endonucleolytic breaks in which one break occurs seven nucleotides 5' to a dimer and a second site of hydrolysis is three to four nucleotides 3' to the same pyrimidine dimer[9-11] [Figure 2]. Prelimary evidence suggests a sequential mechanism in which the 5' break is made first. Although the size of the incised fragment in the DNA suggests that incision may be coordinated with excision reactions neither fragment release under nondenaturing conditions nor turnover of the uvrABC proteins could be demonstrated.[9-11] The implication of these findings is that the in vitro conditions using purified uvrA, uvrB and uvrC proteins lack important components limiting the repair process to uncoupled repair reactions. The uvrABC-incised DNA complex is, therefore, amenable to an examination of those other proteins that may facilitate the excision reaction.

Although the uvrA, uvrB and uvrC genes are required for incision of damaged DNA in vivo additional gene products have been implicated in the excision-repair process. Those which have been implicated include the uvrD[12,13], polA[14,15], polC[15], recA[12,14], recB[12,14], lexA[16] and lig[17] genes. The last four genes control steps beyond the excision step or are the regulatory genes influencing the expression of the uvr system. The gene product of uvrD has been identified by a number of laboratories as helicase II[18-22]. DNA polymerase I as the gene product of polA participates in excision repair by virtue of its coordinated polymerase:5' to 3' exonuclease functions[23] as well as its strong affinity for nicked sites on damaged DNA.

The generation of the predicted excision product, as a consequence of the initial bimodal incision reaction catalyzed by the uvrABC complex is released from the uvrABC reaction only following denaturation of the

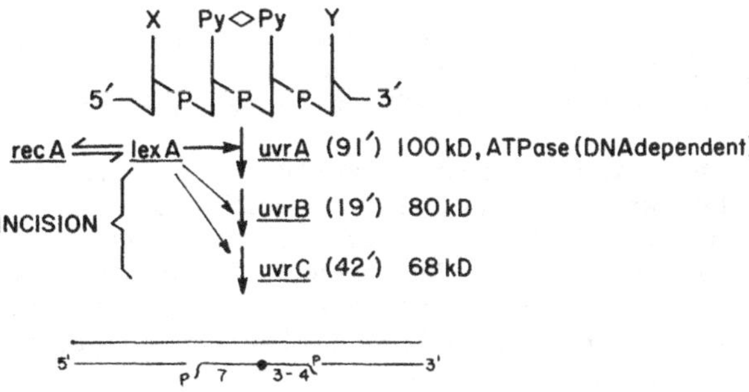

SPECIFICITY INCLUDES:

1. UV INDUCED CYCLOBUTANE DIMERS
2. PSORALEN AND MITOMYCIN INDUCED CROSS LINKS
3. O^6-ETHYLGUANINE ALKYLATION ADDUCTS
4. BENZ(A)PYRENE-GUANINE ADDUCTS
5. AAF-GUANINE ADDUCTS

MUTANT PHENOTYPE OF $\underline{uvrA}^-$, $\underline{B}^-$ OR $\underline{C}^-$

- DAMAGE SENSITIVITY
- INCREASED RATES OF MUTATION IN SURVIVORS
- DECREASED HOST CELL REACTIVATION (HCR) OF DAMAGED PHAGE
- DAMAGE SENSITIVITY OF MULTIPLE $\underline{uvr}^-$ IS NOT ADDITIVE

Figure 2. The *E. coli* uvrABC Endonucleolytic Scheme

protein-damaged DNA complex or through the action of chelators. The 12-13 nucleotide long DNA product predicted by analysis of incision sites on DNA of defined sequences[9-11] can be identified on a 20% polyacrylamide gel using generally labeled DNA as a substrate. This fragment is only generated when the uvrA, uvrB and uvrC proteins are present and when the DNA used as a substrate has been damaged by ultraviolet irradiation. When the DNA is undamaged or when one of the protein components is withheld from the reaction mixtures, the fragment is not generated. This fragment appears polydisperse rather than as two sharp bands because it is comprised of a population of many different sequences which in this gel system have significant sequence dependent mobility differences. This fragment can also be end-labeled using polynucleotide kinase only after alkaline phosphatase treatment and thus must have a phosphorylated 5'-terminus. A free 3' hydroxyl on the parental DNA strand is inferred because it provides a priming site for DNA polymerase I (see below)-similar results were obtained by Sancar and Rupp.[9]

Identification of the fragment as it is released under non-denaturing conditions is limited to the use of those techniques in which fragments of defined sizes are resolvable under extremely benign conditions such that fragment release when facilitated by other protein factors can be followed.[24] The entire reaction was passed through a DEAE-Sephacel column equilibrated and resolved. It was observed that only the simultaneous addition of both uvrD protein and DNA polymerase I resulted in release of an oligonucleotide peak, the extent of which was identical to the fragment released by EDTA treatment. Neither uvrD alone or polymerase I alone stimulated release of radioactivity after uvrABC incision of damaged DNA. The background peak is seen with the DNA substrate alone and is not the result of UV irradiation or uvrABC action. The net amount of radioactivity associated with an oligomeric species which is released by EDTA and by the combined action of uvrD and polymerase I corresponds to the removal of 12-13 base oligomer at approximately 25% of the damaged sites.

The incorporation of deoxynucleoside triphosphates into an acid precipitable DNA when catalyzed by DNA polymerase I is dependent on prior incision of the UV-irradiated DNA by the uvrABC endonuclease and on the addition of helicase II. In the absence of incision due either to the lack of a damaged DNA substrate or the lack of one of the components of the uvrABC endonuclease there is only a minimal stimulation of DNA polymerase I nucleotide incorporation by uvrD into endogenous nicks in the DNA (see Figure 3).

The ordered sequence of excision repair reactions terminate with the restoration of the intergrity of the repaired strands by ligation reactions following incision, excision and resynthesis. It is to be expected, therefore, that the requirements for ligation should be internally consistent with the roles of helicase II-DNA polymerase I mediated excision-resynthesis reactions. In order to determine which protein factors were necessary to complete the repair reaction and restore the DNA to a covalently closed molecule a nicking-reclosing assay was used. It was observed that the addition of helicase II, DNA polymerase I, and DNA ligase leads to almost complete resealing of the nicked DNA. Furthermore, T4 DNA ligase and E. coli DNA ligase were found to be completely interchangeable in the ligation assay whereas DNA polymerase I could not be substituted for by either the Klenow fragment, DNA polymerase III holoenzyme or T4 DNA polymerase. The incorporation of nucleotides into the DNA as a result of the repair process was determined by quantitating the total number of nucleotides incorporated and the number of sites repaired. This leads to approximately 12 nucleotides per site in good agreement with the size of the gapped DNA which is expected to result after release of the DNA fragment containing the damaged site.

The nicking of UV irradiated DNA was found to be directly related to the amount of the uvrC protein added in the endonuclease assay under those circumstances in which excess amounts of the proteins uvrA and uvrB were included with the DNA substrate. The further addition of uvrC protein after the initial reaction had reached completion led to a rapid increase in the nicking of damaged DNA. This catalytic enhancement was not observed in those reactions augmented with either uvrA or uvrB proteins or under prolonged incubation conditions. Addition of the uvrD protein led to an effect similar to the further addition of uvrC protein, suggesting that the uvrC protein is now turning over. Addition of helicase II in control reactions in which only the uvrA and uvrB proteins were preincubated with the DNA did not lead to an increase in incision suggesting that the helicase II preparation contained no uvrC-like activity. Furthermore, those reactions including uvrD protein proceed with linear kinetics.

The current perception of excision repair mechanisms in E. coli and in lower eukaryotes[27,28] must take into account the large number of genes participating in the incision reaction of DNA containing 'bulky adducts' such as pyrimidine dimers. The initial studies by Seeberg and his colleagues[6,9] gave evidence from enzyme complementation studies that three protiens, in the presence of ATP, are required to effectively incise UV damaged DNA. It was implicit in these and subsequent in vitro studies that the E. coli uvr system represented an order of molecular complexity not observed for other repair systems. It is now apparent that the uvrA protein in the presence of ATP finds a damaged site on DNA[7,10] as well as provides a site for the binding of the uvrB protein.[29] Catalysis is, however, dependent on the subsequent binding of the uvrC protein to the uvrAB-DNA complex.[10,11]

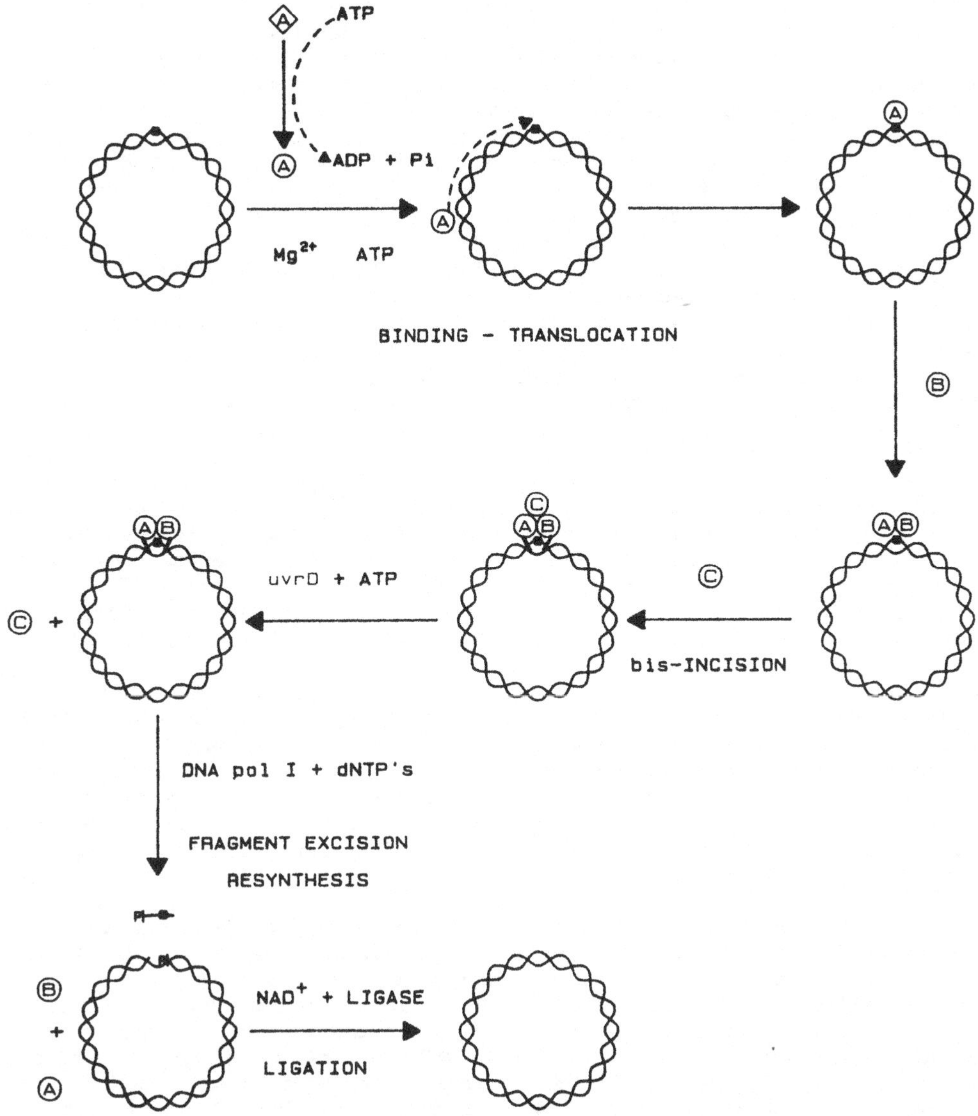

Figure 3. The **REPAIROSOME** Model of Excision Repair in E. coli

From protein-DNA binding studies the stability of this complex is persistent following catalysis and, unlike a typical enzyme system, protein turnover is not observed[10,11] suggesting that there is a partial repair reaction in the presence of limiting protein species. The release of the potentially excisable 12-13 nucleotide oligomer product of the incision reaction can be demonstrated by either the presence of protein denaturing conditions, chelating conditions, or by the action of helicase II and DNA polymerase I. When the presence of a protein complex at the damaged site was examined by diagnostic use of E. coli exonuclease III sensitivity using linear damaged DNA substrates it was indicative of protection rendered by the post-incision uvrABC complex. This protection was partially relieved by helicase II and completely relieved by the combination of helicase II and DNA polymerase I in the presence of deoxynucleoside triphosphates, suggesting that the combined action of helicase II and DNA polymerase I facilitated the release of the uvrABC complex. Restoration of the covalently closed double stranded DNA molecule requires the action of polynucleotide ligase. Furthermore, using limited amounts of the uvrC gene product in the incision reaction, it is possible to demonstrate that the addition of helicase II leads to turnover of the uvrC protein. A similar conclusion was reached from in vivo studies in which Ben-Ishai and Sharon found that in uvrD mutants incision is saturated at significantly lower fluences than in wild type strains.[30]

This evidence, and the data reported for these in vitro leads to the suggestion that in the absence of DNA polymerase I the uvrD protein is capable of interacting with the uvrABC protein complex to carry out a partial reaction in which an intermediate in the excision repair mechanism can accumulate. The uvrD gene product is believed to act via protein-protein interactions with the uvrABC complex rather than simply binding at one of the nicks in the DNA created by the incision step because other studies have shown that helicase II requires a 12 base single strand region of DNA to bind.[25] DNA polymerase I is also incapable of initiating nucleotide incorporation into this nicked DNA in the absence of helicase II presumably because the nicks are protected by the uvrABC protein complex. The physical association between helicase II and DNA polymerase I has been observed as well as the inhibition of polymerization by helicase II (unpublished results). Furthermore, these two proteins copurify during the early stages of purification (Hoffman-Berling, personal communication).

This evidence for a series of multiprotein interactions leads to the proposition of a **repairosome** complex (Figure 3) by analogy to other protein complexes such as the primosome involved in DNA replication.[26] In the absence of protein architectural studies the suggestion of the repairosome complex is, however, an implied one just as the primosome or replisome models are. Becuase these in vitro systems are amenable to complementation experiments it is anticipated that structure-function relationships can be more firmly established.

In addition to the potential role of such a complex facilitating coordinated reactions, the persistence of the uvrABC complex binding to the DNA after incision may also be important in vivo in reducing the vulnerability of nicks and single stranded gaps to various nucleases within the cell, thereby protecting the genome until the rest of the repair machinery arrives. It is not believed that the uvrA, uvrB, or uvrC proteins are covalently attached to the DNA after incision because the fragment can be released by treatment with EDTA (this work) or by phenol extraction.[9] UvrD mutants exhibit normal levels of DNA incision after UV irradiation but are defective in the excision step of DNA repair and in postincision rejoining of repaired regions.[31,32] Neither uvrD nor polA mutants are as sensitive to ultraviolet light as the individual uvr mutants implying alternative pathways for the repair of ultraviolet damaged DNA which can bypass the requirement for helicase II or DNA polymerase I. In addition a number of other genes have been identified which are implicated in the excision repair pathway but whose in vitro role in the uvrABC reaction have yet to be studied. The ability of other DNA polymerases to carry out excision repair in polA mutants may require additional protein factors besides uvrA, uvrB, uvrC and uvrD. The use of purified proteins makes the analysis of the excision reaction by reconstitution amenable to further study. For example, exonuclease VII of E. coli is capable of excising pyrimidine dimers from UV irradiated DNA preincised with the T4 or M. luteus dimer-specific DNA glycosylase-AP endonuclease.[33] That such an enzyme can participate in 'back up' excision mechanisms in vivo is suggested from the increase excision deficiency of polAex xseA mutants[33] when compared to single mutants in either of these genes. Whereas incision is dependent on each of the uvrABC genes the excision process has the potential for the action of a number of different genes.

SUMMARY

The bimodal nature of the E. coli uvrABC catalyzed incision reaction of UV irradiated DNA leads to potential excision of a 12-13 base long damaged fragment. However, the oligonucleotide fragment containing the UV-induced pyrimidine dimer is not released under non-denaturing in vitro reaction conditions. The uvrABC proteins, also, are stably bound to the incised DNA and do not turn over following the incision event. In this communication it is shown that damaged fragment release from the parental uvrABC incised DNA is dependent on either chelating conditions or upon the simultaneous addition of the uvrD gene product (helicase II) and the polA gene product (DNA polymerase I) when catalyzing concommitant polymerization of deoxynucleoside triphosphate substrates. The product of this multiprotein catalyzed series of reactions serves as a substrate for polynucleotide ligase which results in the restoration of the integrity of the strands of DNA. The addition of the uvrD protein to the incised DNA-uvrABC complex also results in turnover of only the uvrC protein. It is suggested that the repair processes of incision, excision, resynthesis and ligation are coordinately catalyzed by a protective complex of proteins in a 'repairosome' type of configuration.

ACKNOWLEDGEMENTS

This work was supported by grants from the U.S. Department of Energy (EY-76-S-02-2814) and the National Institutes of Health (GM-22846, GM-31110 and ES-CA-03130).

REFERENCES

1. Grossman, L., Riazuddin, S., Haseltine, W.A. and Lindan, C.P. (1980) Cold Spring Harbor Symp. Quant. Biol., 43:947-955.
2. Haseltine, W.A., Gordon, L.K., Lindan, C.P., Grafstrom, R.H., Shaper, N.L. and Grossman, L. (1980) Nature (London), 285:634-641.
3. Grafstrom, R.H., Park, L. and Grossman, L. (1982) J. Biol. Chem. 257:13465-13474.
4. Radany, E. and Friedberg, E.C. (1980) Nature (London), 286:182-185.
5. Demple, B. and Linn, S. (1980) Nature (London), 287:203-208.
6. Seeberg, E. (1978) Proc. Natl. Acad. Sci. USA, 75:2569-2573.
7. Seeberg, E. (1981) Prog. Nucleic Acids Res. Mol. Biol., 26:217-226.
8. Seeberg, E. and Steinum, A.L. (1982) Proc. Natl. Acad. Sci. USA, 79:988-992.
9. Sancar, A. and Rupp, W.D. (1983) Cell, 33: 249-260.
10. Yeung, A.T., Mattes, W.B., Oh, E.Y. and Grossman, L. (1983) Proc. Natl. Acad. Sci. USA, 80:6157-6161.
11. Yeung, A.T., Mattes, W.B., Oh, E.Y. and Grossman, L. (1983) in Cellular Responses to DNA Damage (eds. Errol C. Friedberg and Bryn A. Bridges), pp. 77-86 Alan R. Liss, Inc., N.Y., N.Y.
12. Youngs, D.A., Van der Schuren, E. and Smith, K.C. (1974) J. Bacteriol., 117:717-725.
13. Youngs, D.A. and Smith, K.C. (1976) J. Bacteriol. 125:102-110.
14. Tait, R.C., Harris, A.L. and Smith, K.C. (1974) Proc. Natl. Acad. Sci. USA, 71:675-679.
15. Smith, K.C., Youngs, D.A., Van der Schuren, E., Carlson, K.M. and Sargenti, N.J. (1978) in DNA Repair Mechanisms (eds. Hanawalt, P.C., Friedberg, E.C. and Fox, F.C.), pp 247-250.

16. Little, J.W., and Mount, D.W. (1982) Cell, 29:11-22.
17. Youngs, D.A. and Smith, K.C. (1977) Mol. Gen. Genet., 152:37-41.
18. Oeda, K., Horiuchi, T. and Sekiguchi, M. (1982) Nature (London), 298:98-100.
19. Hickson, I.D., Arthur, H.M., Bramhill, D., and Emmerson, P.T. (1983) Mol. Gen. Genet., 190:265-270.
20. Taucher-Scholz, G. and Hoffman-Berling, H. (1983) Eur. J. Biochem., 137:573-580.
21. Kushner, S.R., Maples, V.F., Easton, A., Farrance, I. and Peramachi, P. (1983) in Cellular Responses to DNA Damage (eds. Errol C. Friedberg and Bryn A. Bridges), pp 153-159 Alan R. Liss, Inc., N.Y., N.Y.
22. Kamura, K. and Sekiguchi, M. (1984) J. Biol. Chem., 259:1560-1565.
23. Lehman, I.R. and Chien, J.R. (1973) J. Biol. Chem., 248:7717-7723.
24. Caron, P.R., Kushner, S.R. and Grossman, L. (1985) Proc. Natl. Acad. Sci. USA, in press.
25. Kuhn, B., Abdel-Monem, M., Krell, H., and Hoffman-Berling, H. (1979) J. Biol. Chem., 254:11343-11350.
26. Arai, K. and Kornberg, A. (1981) Proc. Natl. Acad. Sci. USA, 78:69-73.
27. Reynolds, R.J. and Friedberg, E.C. (1981) J. Bacteriol., 146:692-704.
28. Wilcox, D.R. and Prakash, L. (1981) J. Bacteriol., 148:618-623.
29. Kacinski, B.M. and Rupp, W.D. (1981) Nature 294:400-401.
30. Ben-Ishai, R. and Sharon, R. (1981) in Chromosome Damage and Repair (eds. Erling Seeberg and Kjell Kleppe), pp 147-151, NATO Advanced Study Institute Series A-Life Sciences.
31. Kuemmerle, N.B. and Masker, W.E. (1980) J. Bacteriol., 142:535-546.
32. Kuemmerle, N.B., Ley, R.D., and Masker, W.E. (1982) Mut. Research, 94:285-297.
33. Chase, J.W., Masker, W.E., and Murphy, J.B. (1979) J. Bacteriol., 137:234-242.

PYRIMIDINE DIMERS INDUCED ALTERATION OF DNA TERTIARY

STRUCTURE EVIDENCE FOR UNWINDING AND SHORTENING

[1]Antonia M. Pedrini, Silvia Tornaletti, [2]Paola Menichini, and Angelo Abbondandolo

[1]Instituto di Genetica Biochimica ed Evoluzionistica C.N.R. Pavia Italy, and [2]Istituto Scientifico Tumori, Genova, Italy

INTRODUCTION

Excision-repair of DNA is an enzymatically mediated process by which modified bases are removed from cellular DNA together with adjacent normal nucleotides. In Escherichia coli the endonucleolitic activity that initiates repair of pyrimidine dimers and bulky adducts resulting from the reaction of DNA bases with carcinogens such as B(a)P, N-acetyl-2-aminofluorene, 7-bromomethylbenz(a)anthracene, 4-nitroquinoline 1-oxide, cis-DDP and crosslinking agents, consists of the complex of the uvr ABC genes products. It has been proposed that the versatility of this incision complex is probably based on its ability to recognize a structural distortion of the DNA helix common to this broad class of damage rather than the chemical nature of the adducts themselves [1]

The highly purified uvr ABC endonuclease complex has been recently shown to act bifunctionally hydrolyzing one phosphodiester bond on each site of the damage. In the case of pyrimidine dimers it acts on the same strand containing the dimer; 7 nucleotides 5'- and 3 to 4 nucleotides 3' - to the dimer. Because of its unique specificity it has been suggested that such a complex recognizes and acts at the extremes of the structural distortion imposed by these disparate chemical and photochemical structures [1]

Based primarily on the enhanced susceptibility of damaged DNA to denaturation by heat or formaldehyde[2,3,4] and to digestion by S1 nuclease,[5] several authors have suggested that the conformational changes induced by this class of damage consists of the partial denaturation of duplex DNA at the site of damage. In the case of pyrimidine dimers the denatured area should consist of about 4.3 bp.[3,4] However circular discroism and viscosity measurement of sample irradiated at low fluences have evidenced conformational variations that cannot be identified as partial denaturation.[6]

These contradictory results prompted us to reexamine the nature of the distortion caused by pyrimidine dimers by taking advantage of the unique properties of circular closed DNA molecules.

The properties of circular closed DNA molecules have in fact introduced new and more sensitive ways of detecting modification of DNA structure.

METHODS

Use of supercoiled DNA for probing damage induced structural changes to DNA

In the Watson-Crick structure of DNA, the two antiparallel strand are coiled around each other. A direct consequences of this intertwining is that if a double-stranded DNA is in the form of a ring with no discontinuity in the backbond bonds of either strand, the complementary single-stranded rings are linked. The parameter that describes quantitatively the linking of the pair of single-stranded ring in such a DNA is the linking number: Lk. Roughly speaking Lk is the number of times one strand goes around the other in the duplex ring and as long as the two strands are continuous, Lk is a topological invariant. If Lk deviates from the value that it will assume when the molecule is in its most stable configuration, namely in its relaxed state, the axis of the duplex molecule winds in space. When viewed with an electron microscope the molecules appear twisted and the term supertwisted or supercoiled has been used to described such molecules.[7]

The reciprocal relationship between twisting (T) of the strands of the double helix and supertwisting of the helix axis (W) may be expressed in a quantitative form through the relationship:

$$Lk=T+W$$

While the linking number is an integer and it is an invariant so long as none of the backbone bonds are broken, the twist (T) and the writhing number (W) are continuously variable quantities. From this relationship it follows that the degree of supercoiling is tightly coupled to the helical structure of DNA. Thus changes in the helix geometry can be easily detected by monitoring changes in supercoiling. Since the hydrodynamic properties of a closed circular DNA are very sensitive to changes in its superhelicity, very small changes in the helical geometry can be followed as changes in the degree of supercoiling. The introduction of the agarose gel electrophoresis technique for the resolution of molecules with different degree of supercoiling has therefore greatly facilitated the detection of small changes in helix geometry.

When an homogeneous population of circular closed molecules are subjected to cycles of nicking and resealing with DNA topoisomerases, a population of topological isomers are obtained. By choosing the appropriate relaxation and agarose gel electrophoresis conditions, "partially relaxed" topoisomers of different linking number can be resolved as a series of bands. The spacing between a pair of adjacent topoisomers represents the difference in mobility when the linking difference is changed by 1.

When the formation of damage causes a change in helix configuration such alteration will result in a change in the electrophoretic mobility of the "partially relaxed" topoisomers. The topoisomer bands will shift continuously as a function of the damage load and of the degree of supercoiling. This "band shift method" is a very informative technique since i) from the direction of the shift is possible to establish whether the damage overwinds or unwinds the DNA molecule, ii) from the amount of damage required to shift the topoisomers one superhelical turn, it is possible to calculate the angular desplacement (unwinding angle) between adjacent bases caused by the damage, iii) from the degree of damage required to relax each topoisomers one superhelical turn, is possible to establish a possible effect of the degree of supercoiling on the damage induced structural alteration of DNA.

The unwinding angle obtained by monitoring the damage induced changes in the electrophoretic mobility of "partially relaxed" DNA sample depends on the assumption that the major effect of damage on DNA migration is DNA unwinding. But damage could cause additional effects to DNA structure

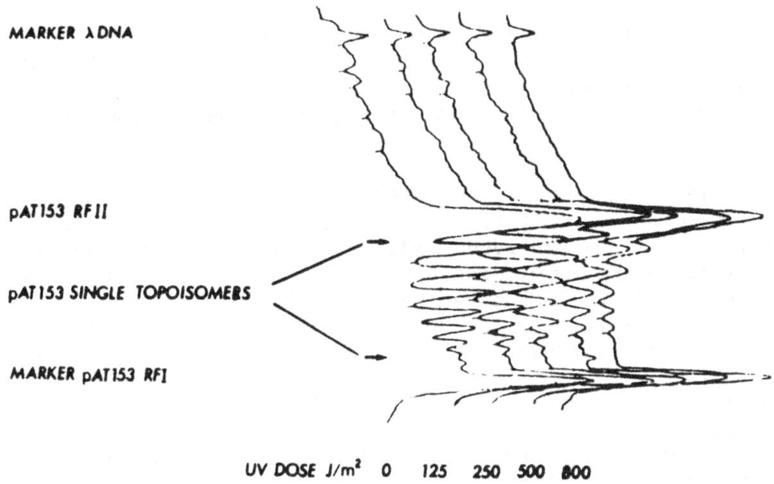

Figure 1. Reduction in electrophoretic mobility of individual pAT153 topoisomers as a function of UV dose. Microdensitometric scanning of negatives of agarose gel photographs.

besides unwinding. In order to predict whether these additional effects might become important, it is useful to run control experiments with nicked circular DNA which is supposed to respond most nearly like partially relaxed DNA to effects other than unwinding. In those instance where indications are available that the observed mobility changes are not only due to unwinding but also to other effects such as changes in effective DNA length, the unwinding angle can be obtained by applying the "plus/minus" method developed by Wang[8] for the calculation of the helical repeat of DNA in solution.

This method consists of analyzing the damage induced mobility shift of "partially relaxed" DNA under two different sets of conditions. In one set, the electrophoresis conditions are so selected that the DNAs are negatively supercoiled, in the other set, DNAs are positively supercoiled. Averaging the magnitudes of the shifts in these two sets of condition one can correct for mobility changes associated with damage formation that are not caused directly by unwinding.

RESULTS

Conformational changes of ultraviolet-irradiated DNA

The use of "partially relaxed" DNA has been applied to the study of the effect on DNA structure of two bulky adducts such as those produced by reaction of DNA with psoralene and benzo (a) pyrene diolepoxide.[10] It has been shown that both adducts cause the unwinding of the helix of 28° and 31° respectively.

We have applied the "band shift" method to the study of UV induced structural alteration of DNA structure. We observed that the irradiation of "partially relaxed" topoisomers of plasmid pAT153 (3657 bp) at UV doses that induced the formation from 1 up to 40 dimers per molecule, caused a continuous reduction in the electrophoretic mobility of all topoisomerase bands as shown in Figure 1. The mobility shift caused by photodamage was rendered quantitative by calculating the relative mobility (Rf) of each band with respect to two nonirradiated marker DNAs: -DNA was used as origin and pAT153-RFI as front. The amount of unwinding was determined by plotting the number of superhelical turns as a function of the UV dose. By assuming that this mobility change was primarily due to the unwinding caused by pyrimidine dimers, the major photoproduct at 254 nm, an unwinding angle of 14.3° per dimer was determined.[11]

The estimate of a 14.3° unwinding of the sugar-phosphate backbone caused by one pyrimidine dimer is indeed subject not only to the uncertainties that the observed mobility changes could be due to structural alterations other than unwinding but also to the possibilities that other types of photoproducts influence the duplex twist number. The former possibility is substantiated by the observation that also linear duplex[11] or circular nicked molecules show a slight but measurable increase in their electrophoretic mobility upon UV irradiation. Since the electrophoretic mobility of linear or nicked circular DNAs' has been shown to be proportional to their molecular weight, the increase in electrophoretic mobility suggests that photodamage produces a more compact structure by reducing the effective DNA length.

In order to establish whether the unwinding angle of 14.3° per pyrimidine dimer calculated by using negatively supercoiled topoisomers is correct or has to be modified, we applied the "plus/minus" method outlined "methods" paragraph. Subsequently we performed the same experimental protocol after photoreactivation of all dimers with Streptomyces griseus photoreactivating enzyme.

In Figure 2 is shown the effect of increasing UV light on the electrophoretic mobility of negatively (left) and positively (right) shift and a downward shift of the negatively and positively supercoiled DNAs shift and a downward shift of the negatively and positively supercoiled DNAs respectively. But when we have calculated the UV dose necessary to unwind the helix the superhelical turn in the two sets of conditions, a marked difference has been observed: a dose which causes the formation of 25 dimers per molecule is required for the negatively supercoiled DNA, while for the positively supercoiled topoisomers a dose producing 42 dimers is needed. The discrepancy between these two values indicates that the observed mobility shift of single irradiated topoisomers is not solely caused by unwinding but also by some other effects on DNA structure such as changes in DNA persistence length. Averaging the magnitude of these two sets of shifts, we have cancelled out these effects and recalculate the unwinding angle which is reduced from 14.3° to 10.7° (360°:33.5).

This first correction is followed by a second one after the photoreactivation experiment. In fact, as already indicated by preliminary observations,[12] selective removal of all pyrimidine dimers by treatment with photoreactivating enzyme does not restore the original electrophoretic mobility of both positvely and negatively supercoiled topoisomers. The electrophoretic mobility of completely photoreactivated samples is close but not exactly the same as that of the unirradiated topoisomers; 20% of the original topoisomer shift has indeed been maintained and represents the overall contribution of photodamage other than pyrimidine dimers. The unwinding angle has been therefore further recalculated to be 8.6°. Surprisingly this very small value is not significantly different from the value obtained by Camerman and Camerman[13] by X-ray diffraction of thymine dimers in frozen solution.

The use of circular closed DNA molecules irradiated at low UV doses has allowed us to see that the structural alterations caused by dimers consist of the unwinding of the duplex and the production of a more compact DNA structure. These results might be interpreted as due to the disruption of the hydrogen bonds at the damaged sites which will cause localized

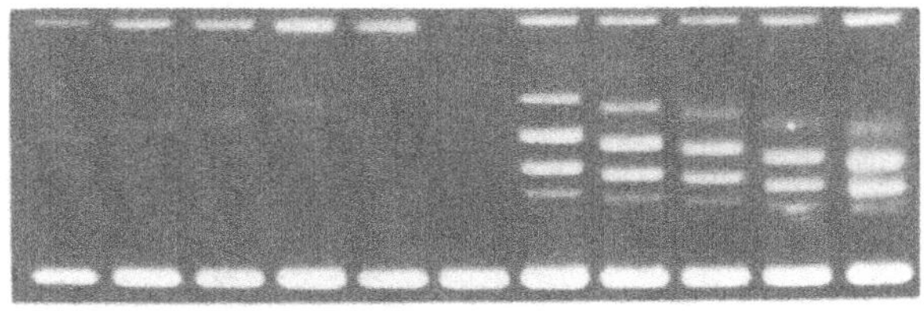

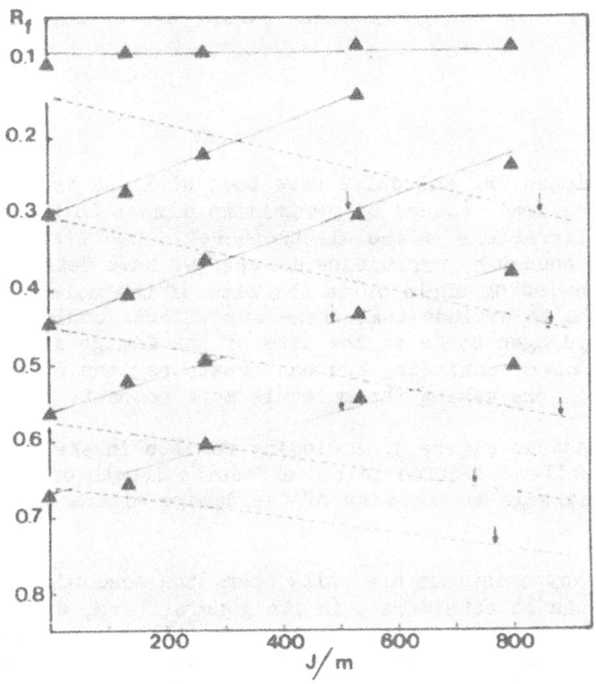

Figure 2. (top) Effect of increasing UV dose on the electrophoretic mobility of negatively (left) and positively (right) supercoiled topoisomers. (bottom) Regression lines obtained by calculating the relative mobility (Rf) of each topoisomer band as a function of the UV dose. The arrows indicate the dose necessary to unwind each topoisomer one superhelical turn. (———) negatively supercoiled topoisomers, (------) positively supercoiled topoisomers.

unwinding. Furthermore, the interruption of base pairing would produce single-stranded regions that would collapse and reduce the effective DNA length. This possibility can be considered quite unlikely for two reasons. First of all, the value of the unwinding angle we have calculated is much smaller than the unwinding angle of about 36° (360°:10.4 bp) caused by the disruption of the hydrogen bond of a single base pair. Second, we have demonstrated that the UV induced increase in S1 sensitivity, that has been considered for a long time as an indication of the single-stranded character of pyrimidine dimer containing DNA, is not abolished by complete removal of dimers with photoreactivating enzyme. The often reported S1 sensitivity of UV irradiated DNA is therefore due to photoproducts other than dimers which have greater destabilizing effects.

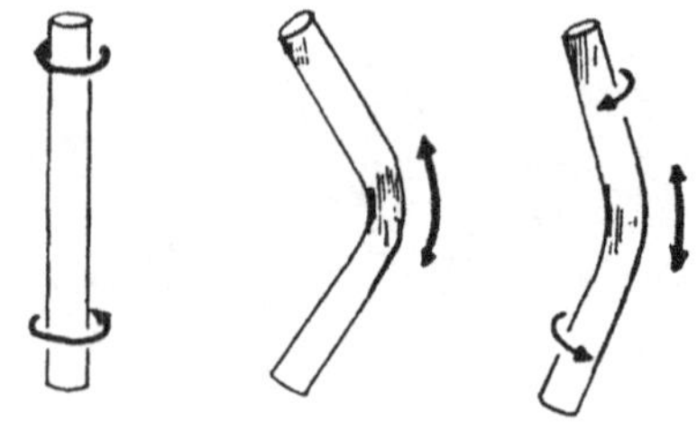

Figure 3. Schematic drawing of the effect of pyrimidine dimers on the two geometrical parameters of supercoiled DNA.

CONCLUSION

Circular closed DNA molecules have been utilized for the study of the structural alterations induced by pyrimidine dimers to DNA in solution. By analyzing the alterations in the electrophoretic mobility on agarose gel of these molecules caused by pyrimidine dimers, we have detected a modification either in the unwinding angle or in the size of the molecules. Futhermore we have been able to exclude that these two effects could be due to the disruption of hydrogen bonds at the site of the damage although these alterations can cause unwinding and can create regions of flexibility that easily would collapse making the molecule more compact.

As visualized in Figure 3, unwinding results in the variation of the twist number, while reduction in the effective length of DNA resulting by bending the helix axis at the site of the damage alters the writhing number.

Although many questions are still open, the concomitant unwinding and bending of DNA can be considered, in its general term, a valid model for the description of the structural alteration created by pyrimidine dimers to DNA. Such a model is also substantiated by recent model building sites carried on by several authors. This conclusion has been reached by incorporating into a model of B-DNA the conformation obtained by theoretical calculation of a deoxymonophosphate pyrimidine dimer. The structural distortion emerging from this type of approach, predicts an overall helix deformation substantially similar to the one we have proposed in Figure 3, namely unwinding and bending of the helix axis at the site of the damage.[14,15]

A similar picture, emerging either from experimental data or theoretical calculations, can be proposed for the structural alterations induced by other bulky adducts all repaired by the uvr ABC incision system [8,10,11]. We propose, therefore, that the distortion resulting from the concomitant unwinding and bending of the helix at the damaged site is the signal that stimulates the action of the E. coli incision complex.

Since it is reasonable to expect that the extent of this alteration, will be different for each kind of adduct, one may predict that, if the complex acts at the extreme of the distortion, the size of the fragment excised will vary from one kind of damage to another.

ACKNOWLEDGEMENT

We thank Dr. L. Grossman for his interest and encouragement. We thank also Miss D. Travarne for expert assistance in the preparation of the manuscript. This work was supported by a grant from the Progetto Finalizzato del Consiglio Nazionale delle Richerche "Medicina Preventive e Riabilitativa: Rischio Tossicologico: Genotossicita".

REFERENCES

1. A.T. Yeung, W.B. Mattes, E.Y. Oh and L. Grossman, Enzymatic properties of the purified Escherichia coli uvr ABC complex. In: "Cellular responsed to DNA damage" E.C. Friedberg and B.A. Bridges (eds). Alan R. Liss Inc., New York (1983).
2. M. Kahn, The effect of thymine dimers on DNA: DNA hybridization. Biopolymers 13:669 (1974).
3. F.N. Hayes, D.L. Williams, R.L. Ratliff, A.J. Varghese and C.S. Rupert, Effect of a single thymine photodimer on the oligodeoxythymidilate polydeoxyadenylate interaction. J. Am. Chem. Soc. 93:4940 (1971).
4. N.N. Shafranovskaya, E.N. Trifonov, Yu.S. Lazurkin and M.D. Franck-Kamenetskii, Clustering of thymine dimers in ultraviolet irradiated DNA and the long-range transfer of electronic excitation along the molecule. Nature New Biology 241:58 (1973).
5. H.J. Edenberg, Altered structure of ultraviolet-irradiated DNA: evidence for unwinding. Cold Spring Harbor Symp. Quant. Biol. 47:379 (1982).
6. H. Treibel, K.E. Reinert, H. Bar and H. Lang, Structural changes of ultraviolet-irradiated DNA derived from hydrodynamic measurements. Biochem. Biophys. Acta 561:59 (1979).
7. J.C. Wang, DNA supercoiling and DNA topoisomerases. In: "Nucleic Acid Research" K. Mizobuchi, I. Watanabe and J.D. Watons (eds.). Academic Press, Japan (1983).
8. J.C. Wang, Helical repeat of DNA in solution. Proc. Natl. Acad. Sci. USA 76:200 (1979).
9. G. Weisehahn and J.E. Hearst, DNA unwinding induced by photoaddition of psoralene derivatives and determination of dark-binding equilibrium by gel electrophoresis. Proc. Natl. Acad. Sci. USA 75:2703 (1978).
10. H.B. Gamber, K. Straub, M. Calvin and J.C. Bartholomew, DNA alkylation and unwinding induced by benza(a)pyrene diol epoxide: modulation by ionic strength and superhelicity. Proc. Natl. Acad. Sci. USA 77:2000 (1980).
11. G. Ciarrocchi and A.M. Pedrini, Determination of pyrimidine dimer unwinding angle by measurement of DNA electrophoretic mobility. J. Mol. Biol. 155:177 (1982).
12. G. Ciarrocchi, B.M. Sutherland and A.M. Pedrini, Photoreversal of DNA unwinding caused by pyrimidine dimers. Biochimie 64:665 (1982).
13. N. Camerman and A. Camerman, Photodimer of thymine in ultraviolet-irradiated DNA: proof of structure by X-ray diffraction. Science 160:1451 (1968).
14. S. Broyde, S. Stellman and B. Hingerty, DNA backbone conformation in cis-syn pyrimidine cyclobutane dimers. Biopolymers 9:1695 (1980).
15. D.A. Pearlman, S.R. Hoibrook, D.H. Pirkle and S.H. Kim, Molecular models for DNA damaged by photoreaction. Science 227:1304 (1985).
16. G.L. Cohen, R.W. Bauer, J.K. Barton and S.J. Lippard, Binding of cis- and trans- dichlorodiammineplatinum (II) to DNA: evidence for unwinding and shortening of the double helix. Science 203:1014 (1979).
17. M.E. Hogan, N. Dattagupta and J.P. Whitlock Jr., Carcinogen-induced alteration of DNA structure. J. Biol. Chem. 256:4504 (1981).

MISMATCH CORRECTION

Paul Modrich

Department of Biochemistry
Duke University Medical Center
Durham, NC 27710

Base pair mismatches result when coventional, but noncomplementary, Watson-Crick bases occur opposite each other on the two strands of the DNA helix. G-T and A-C mismatches are referred to as transition mispairs, while G-G, A-A, G-A, C-C, T-T, and C-T are called transversion mispairs. For the purposes of this review, the definition of mismatches will be extended to include structural anomalies due to insertion or deletion of one or more nucleotides within one strand of the helix. The study of homologous genetic recombination led to the initial suggestion that the occurrence and processing of mismatched base pairs may be of biological significance.[1,2] Genetic recombination occurs by breakage and rejoining of two homologous helics and is thought to proceed via an intermediate DNA spcies containing a heteroduplex joint, a region in which the two DNA strands are derived from different parental helices. If the heteroduplex joint spans genetic differences between the recombining molecules, then this region will contain one or more mispaired bases and will be genetically heterozygous. Repair of mismatches within the heteroduplex, and their consequent reduction to a genetically homozygous state, is a popular explanation for several marker effects associated with recombination such as gene conversion and map expansion in fungi,[1] and localized negative interference in bacteria.[3]

Meselson and colleagues[4,5] subsequently suggested that DNA biosynthetic errors may also contribute significantly to the generation of mispaired bases and that mismatch correction limited to the newly synthesized strand could serve to enhance replication fidelity. This hypothesis included the suggestion that at least in E. coli strand discrimination may be based on DNA methylation. Since polynucleotide methylation is a postsynthetic event, newly synthesized strands are expected to be transiently under-methylated and thus distinguishable from parental template strands. The idea that strand-specific mismatch correction contributes to replication fidelity has gained substantial support, at least in bacterial systems. Evidence to this effect will be summarized here.

MISMATCH CORRECTION IN VIVO

The best evidence that cells possess enzymatic machinery capable of processing mismatched base pairs has been provided by transfection experiments employing artificially constructed heteroduplex DNAs containing one or more mismatched base pairs. Transfection of E. coli with bacteriophage λ heteroduplexes under conditions of replication or recombination block[4,6] has clearly shown that this organism is capable of mismatch correction. This study also suggested that mismatch correction occurs by an excision repair mechanism involving rather large excision tracts, with an average size of several thousand nucleotides. Application of transfection assays for mismatch correction has led to identification of E. coli mutants defective in the process. Mutant strains defective in mutH, mutL, mutS, or urvD (also called urvE, mutU, or recL) exhibit reduced levels of mismatch correction as judged by this biological assay.[7,8,9] Since mutations in these loci also confer high spontaneous mutability,[10] the associated defects in mismatch correction are consistent with a role for this process in the enhancement of replication fidelity.

The evidence for mismatch correction in Streptococcus pneumoniae is somewhat more indirect but equally compelling. Transformation of this organism involves assimilation of single-strand fragments of donor DNA into the chromosome of recipient bacteria to generate a heteroduplex region. A particularly striking feature of this process is an associated marker specific variation in transformation efficiency (reviewed in 11,12). High efficiency markers yield transformants with an efficiency approaching one per genome equivalent of donor DNA entering the cell. In contrast, the transformation efficiency[11,12] of other markers is typically in the range of 0.05 to 0.5. It is now evident that marker discrimination in this system reflects mismatch repair on the donor strand within the heteroduplex recombination intermediate. Thus, when mismatches within the heteroduplex are subject to repair, the genotype of the donor strand is corrected to that of the recipient and consequently lost. Mutants defective in this mismatch correction system have been isolated,[13,14] and these so called hex⁻ strains behave as high efficiency hosts for transformation with all markers. Furthermore, as in the case of the E. coli mismatch repair deficient strains, hex⁻ mutants display high spontaneous mutability.

Transfection and DNA injection methods have also been employed to introduce mismatch heteroduplexes into mammalian cells. As in the case of E. coli, the results of such experiments suggest that higher cells are also capable of mismatch correction.[15,16,17] However, with the possible exception of the studies by Folger et al.[17] some events attributed to mismatch correction in such studies could have been the consequence of homologous rombination between heteroduplex substrates used.

MISMATCH SPECIFICITY

As may be surmised from the range of transformation efficiencies observed in S. pneumoniae, different types of mismatches are repaired with differing efficiencies. A number of S. pneumoniae markers of the various transformation classes have now been sequenced.[11,12] Similarly, Radman and colleagues[18,19] have examined correction of λ heteroduplexes containing defined mismatches in the E. coli system. The results with these two organisms are remarkably similar. In particular, the G-T and A-C transition mispairs are well repaired in both organisms. Similarly, A-A, G-G, and T-T transversion mismatches are also repaired in the two systems, as are single base pair deletions. (Ten nucleotide insertions and deletions have also been shown to be subject to correction in E. coli).[20] However, this class of mismatches appears to be less well recognized than the transition mispairs, at least in S. pneumoniae. Lastly, G-A, C-T, and C-C transversion mispairs and larger deletions or insertions appear to be weaker substrates for correction in both organisms. It is pertinent to note that data from the S. pneumoniae system indicates that the efficiency of recognition of some transversion mispairs may depend on the sequence environment in which the mismatch is embedded.[11]

STRAND SPECIFICITY

As discussed above, function of mismatch correction in enhancement of replication fidelity requires a system capable of discrimination of newly synthesized from parental DNA strands. Although several mechanisms for strand discrimination can be envisioned (see below), available evidence indicates that the strandedness of mismatch correction in the E. coli system can be dictated by the state of DNA methylation as initially hypothesized by Wagner and Meselson.[4] In this organism adenine methylation of d(G-A-T-C) sequences appears to govern the strand on which at least some repair events occur.

The evidence supporting involvement of d(G-A-T-C) methylation in mismatch correction in E. coli is several fold. Perhaps the most direct evidence has been provided by transfection with mismatch heteroduplexes in which the two DNA strands were in defined states of methylation at d(G-A-T-C) sites.[5,9,19,21] With hemimethylated heteroduplexes (methylated on only one DNA strand) mismatch repair is highly biased to the unmethylated strand, with the methylated strand serving as template for correction. Mismatch correction also occurs on heteroduplexes in which neither strand is modified, but in this case repair displays little strand bias. In contrast, the efficiency of mismatch correction has been found to be substantially reduced when both DNA strands are methylated.[9,19] However, the reduction in repair efficiency in the latter case is not as severe as that observed when a mutL strain is employed as host for transfection. These findings are in general accord with the Wagner-Meselson hypothesis.

Genetic evidence has also suggested a role in mismatch correction for the dam methylase, the enzyme responsible fr methylation of d(G-A-T-C) sequences. Thus, genetic analysis has implicated this enzyme in a pathway involving function of mutH, mutL and mutS, loci required for mismatch correction in E. coli.[22,23] Furthermore, strains which are deficient in the dam methylase and those which overproduce the enzyme have associated mutator phenotypes,[24,25,26] findings which can be explained in terms of the transfection results cited above. The mutator phenotype of dam^- strains can be understood in terms of a loss of strand bias, while that associated with overproduction of the methylase can be explained by a more rapid methylation of newly synthesized sequences coupled with a reduced efficienty of correction of symmetrically methylated regions. In sum, several independent lines of evidence have implicated d(G-A-T-C) methylation in mismatch correction in a manner that is consistent with the Wagner-Meselson hypothesis. It is important to note, however, that the mechanism by which methylation dictated the strandedness of mismatch correction is not understood. In particular, it has not been known whether the effects of methylation are direct or indirect, nor in fact, whether d(G-A-T-C) sequences oer se are required for mismatch correction by this pathway in vivo.

Although the evidence for methyl-directed mismatch correction appears compelling in the E. coli system, it is evident that this cannot be a truly general phenomenon since the DNA of some organisms does not appear to be methylated. For example, there is no reason to believe that DNA methylation has a role in hex mediated mismatch correction in S. pneumoniae.[11] Nevertheless, mismatch correction in this system is a stranded process, with repair apparently limited to the incoming donor strand during transformation. As pointed out by Lacks and coworkers,[11] this strand bias could be dictated by the presence of breaks at the end(s) of the donor strand withe heteroduplex recombination intermediate. Such an idea is readily extended to replication associated mismatch correction since newly synthesized strands are elongated by virtue of a free end.

In summary, analysis of these two prokaryotc organisms has provided compelling evidence for the existence of enzyme systems capable of strand specific mismatch correction. In both cases, the ss directing the strand specificity of repair events are envisioned to be some distance from the mismatch. It seems likely that reconstitution of the reaction in a defined in vitro system will be required in order to deduce the mechanism of this action at a distance as well as the nature of protein-DNA interactions involved in mismatch recognition.

METHYL-DEPENDENT MISMATCH CORRECTION IN VITRO

In order to approach the mechanism of mismatch correction in E. coli, we have developed a cell-free system which supports the reaction in vitro.[27,28] The assay employed in this work monitors repair of mismatched base pairs within the single EcoRI site of bacteriophage f1 R229 heteroduplexes. Since this DNA also contains four d(G-A-T-C) sites, the state of methylation of which can be controlled, this molecule can also be used to address the - methyl-dependence in in vitro correction.

As observed in vivo, in vitro mismatch correction is dependent on the products of mutH, mutL, mutS and uvrD loci, with repair in mutant extracts deficient in such proteins being less than 10% of that in wild type fractions.[27] In addition, cell-free repair also requires ssb, the E. coli single strand DNA binding protein,[28] a requirement that has not been tested in vivo. The reaction exhibits a specific requirement for ATP (GTP will not suffice), and is reduced somewhat in the absence of exogenous deoxyribonucleoside-5'-triphosphates.

As observed in λ transfection experiments,[9,19] in vitro mismatch repair is reduced when both DNA strands are methylated, with the extent of reduction being 60 to 85% depending on the mismatch.[28] This reduction is substantial, bu not as great as that observed when ATP is omitted or wen wild type extracts are replaced by fractions from mut strains. As mentioned above, similar observations have been made in vivo.[19] The in vitro reaction is also similar to the cellular syin the sense that it supports good repair of both G-T and A-C transition mispairs.[28] However, since correction of transversion or insertion/deletion mismatches has not been tested in vitro, it is not yet clear to what extent mismatch specificity will parallel that of the cellular system.

As anticipated from the transfection results of Wagner and Meselson,[4] in vitro mismatch correction is accompanied by repair DNA synthesis.[27,28] This synthesis, which occurs predominantly on the unmethylated strand of hemi-methylated heteroduplexes, is dependent on the presence of a mismatch, d(G-A-T-C) sequences unmethylated on at least one strand, and function of mutH, mutL and mutS loci. In highly concentrated extracts, which are required for mismatch correction, the strand-specific repair tracts appear to be quite long since synthesis occurs over much of the 6 kilobase substrate. However, some strand-specific, mut-dependent repair synthesis is also observed in more dilute lysates,[28] and in this case is localized to regions near d(G-A-T-C) sequences. This latter observation suggests that some mismatch repair events may involve scission of the unmethylated strand in the vicinity of d(G-A-T-C) sites.

Although individual mut^- extracts are defective in cell-free mismatch correction, activity is restored upon mixing extracts.[27,28] This has provided a biological assay for purification of the products of mutH, mutL and mutS genes, and the three proteins have now been purified to near homogeneity in biologically active form (K. Welsh, M. Su, A.-L. Lu and P. Modrich, unpublished). Purified mutH, mutL and mutS polypeptides are of M_r 25,000, 70,000, and 97,000 respectively. These values are in agreement with the results of

Grafstrom et al.[29] and Pang et al.[30] who have identified and partially purified radiolabeled forms of the E. coli mutH and the S. typhimurium mutH, mutL and mutS polypeptides. Preliminary attempts to reconstitute mismatch correction with purified biologically active mut gene products, single strand binding protein, DNA helicase II (the product of the uvrD gene,[31,32] and DNA polymerase I have not been successful (R. Lahue, M. Su, K. Welsh, and P. Modrich, unpublished). This suggests that DNA polymerase 1 may not be the synthetic enzyme involved in mismatch correction and/or that other as yet unidentified factors are also required. Nevertheless, the availability of proteins known to be involved in mismatch repair should facilitate attempts to reconstitute the reaction in a defined system.

OTHER MISMATCH CORRECTION SYSTEMS

As described above, the efficiency of mismatch correction of heteroduplexes in vivo of f1 heteroduplexes in vitro is substantially reduced if the DNAs are methylated at d(G-A-T-C) sites on both strands. In both cases, however, significant repair is observed with fully methylated molecules. Utilizing plasmid pBR322 heteroduplexes containing several 10 base pair insertion mismatches, Fishel and Kolodner[20,33] have examined both methyl-dependent and methyl-independent correction in E. coli hosts. With hemi-methylated plasmid heteroduplexes introduced by transformation, mismatch correction in vivo was methyl-dependent, involved long repair tracts, and required mutH, mutL, mutS and uvrD products. This correction system is presumably identical to that reponsible for correction of hemi-methylated or unmethylated heteroduplexes as discussed above.

However, two alternate pathways of mismatch correction were identified when the transforming heteroduplex was symmetrically modified at d(G-A-T-C) sites. One pathway involved long excision tracts and did not require mutH or mutL function, but was reduced by 50 to 60% in mutS or uvrD hosts.[33] A second and very weak system was characterized by short repair tracts (probably less than 300 base pairs). Short patch correction was reduced 50 to 100-fold by mutations in recF or recJ loci, and 4-fold by ssb mutations. In contrast to the methyl-directed system described above, correction by these two pathways can occur on either DNA strand.

Fishel and Kolodner[33] have also applied in vitro methods to initial analysis of these systems. Application of a transformation assay method allowed demonstration of the low level, short patch correction in crude fractions, and as observed in vivo, this reaction was dependent on recF function. They also demonstrated higher efficiency mismatch correction of fully modified DNA by restriction endonuclease assay. In this case, however, results with the cell-free system differed from those observed in vivo. In particular, high level mismatch correction was not observed in extracts prepared from mutS or recF mutants, suggesting possible overlap of short and long patch repair pathways in the cell-free system. In addition, the repair defects associated with mutS and recF mutant extracts was complex in the sense complementations not observed in vitro. Nevertheless, these findings clearly show that some mismatch correction can occur on templates that are initially fully methylated. It is not yet clear whether these activities can account for the residual mismatch correction observed with symmetrically modified heteroduplexes observed in vivo or that found with f1 heteroduplexes in vitro (above).

A third, and quite distinct alternate pathway for mismatch repair in E. coli has been suggested by the work of Lieb.[34,35] In her elegant analysis of the fine structure genetics of the bacteriophage λ cI gene, she has identified two exceptional mutations which yield excess recombinants in four factor crosses over short intervals. These two mutations both involve C-G --> T-A transition mutations at the second position within the sequence 5'-C-C-A/T-G-G.

Lieb has suggested that excess recombination of these markers reflects very short patch correction (spanning a distance of less than 20 bp on either side of the marker) of mismatches within this sequence. Furthermore, correction by this system appears to be unidirectional with the strand containing the 5'-C-C-A/T-G-G sequence serving as template for repair. This directionality appears to be dictated by sequence as opposed to methylation since the unidirectional nature of correction was found to be independent of the state of adenine or cytosine DNA methylation.[35]

These finding are of particular interest since in E. coli K strains, 5'-C-C-A/T-G-G is methylated at the internal C. Such sequences are hot spots for mutation due to spontaneous deamination of 5-methylcytosine to thymine.[36] Although the E. coli uracil DNA glycosylase can recognize and eliminate the spontaneous deamination product of cytosine, this activity cannot excise thymine arising from 5-methylcytosine deamination.[37] The very short patch mismatch correction system described by Lieb provides a simple mechanism for circumventing this problem by virtue of its sequence specificity. Given the frequency of methylated C_pG sequences in mammalian DNA, an analogous system may play a signficant role in maintenance of genetic stability in this class of organism.

REFERENCES

1. Holliday, R., Molecular aspects of genetic exchange and gene conversion, Genetics 78:273 (1974).
2. Radding, C.M., Genetic recombination: Strand transfer and mismatch repair, Ann. Rev. Biol. 47:847 (1978).
3. White, R.L. and Fox, M.S., On the molecular basis of high negative interference, Proc. Natl. Acad. Sci. USA 71:1544 (1874).
4. Wagner, R. and Meselson, M., Repair tracts in mismatched DNA heteroduplexes, Proc. Natl. Acad. Sci. USA 73:4135 (1976).
5. Radman, M., Wagner, R.E., Glickman, B.W., and Meselson, M., DNA methylation, mismatch correction and genetic stability, in: "Progress in Environmental Mutagenesis," M. Alacevic, ed., Elsevier/North Holland, Amsterdam. p 121 (1980).
6. White, R.L. and Fox, M.S., Genetic consequences of transfection with heteroduplex bacteriophage lambda DNA, Mol. Gen. Genet. 141:163 (1975).
7. Nevers, P. and Spatz, H., Escherichia coli mutants UvrD uvrE deficient in gene conversion of lambda heteroduplexes, Mol. Gen. Genet. 139:233 (1975).
8. Bauer, J., Krammer, G., and Knippers, R., Asymmetric repair of bacteriophage T7 heteroduplex DNA, Mol. Gen. Genet. 181:541 (1981).
9. Pukkila, P.J., Peterson, J., Herman, G., Modrich, P. and Meselson, M., Effects of high levels of DNA adenine methylation on methyl-directed mismatch repair in Escherichia coli, Genetics 104:571 (1983).
10. Cox, E.C., Bacterial mutator genes and the control of spontaneous mutation, Ann. Rev. Genet. 10:135 (1976).
11. Lacks, S.A., Dunn, J., and Greenberg, B., Identification of base mismatches recognized by the heteroduplex DNA repair system of Streptococcus pneumoniae 31:327 (1982).
12. Claverys, J-P., Mejean, V., Gasc, A-M, and Sicard, A.M., Mismatch repair in Streptococcus pneumoniae: Relationship between base mismatches and transformation efficiencies, Proc. Natl. Acad. Sci. USA 80:5956 (1983).
13. Lacks, S., Mutants of Diplococcus pneumoniae that lack deoxyribonucleases and other activitics possibly pertinent to genetic transformation, J. Bacteriol. 101:373 (1970).
14. Tiraby, J-G. and Fox, M.S., Marker discrimination in transformation and mutation of pneumococcus, Proc. Natl. Acad. Sci. USA 70:3541 (1973).
15. Miller, L.K., Cooke, B.E., and Fried, M., Fate of mismatched base-pair regions in polyoma heteroduplex DNA during infection of mouse cells,

Proc. Natl. Acad. Sci. USA 73:3073 (1976).
16. Abastado, J-P., Cami, B., Dinh, T.H., Igolen, J., and Kourilsky, P., Processing of complex heteroduplexes in Escherichia coli and Cos-1 monkey cells, Proc. Natl. Acad. Sci. USA 81:5792 (1984).
17. Folger, K.R., Thomas, K., and Capecchi, M.R., Efficient correction of mismatched bases in plasmid heteroduplexes injected into cultured mammalian cell nuclei, Mol. Cell Biol. 5:70 (1985).
18. Dohet, C., Wagner, R., and Radman, M., Repair of defined single base-pair mismatches in Escherichia coli, Proc. Natl. Acad. Sci. USA 82:503 (1985).
19. Wagner, R., Dohet, C., Jones, M., Doutriaux, M-P., Hutchinson, F., and Radman, M., Involvement of Escherichia coli mismatch repair in DNA replication and recombination, Cold Spring Harbor Symp. Quant. Biol. 49:611 (1984).
20. Fishel, R.A. and Kolodner, R., Gene conversion in Escherichia coli: The identification of two repair pathways for mismatched nucleotides, UCLA Symp. Mol. Cell Biol. 11:309 (1983).
21. Radman, M., Dohet, C., Bourguignon, M-F, Doubleday, O.P., and Lecomte, P., High fidelity devices in the reproduction of DNA, in: "Chromosome Damage and Repair," E. Seeberg and K. Kleppe, eds., Plenum Publishing Corp., p 431 (1981).
22. Glickman, B.W. and Radman, M., Escherichia coli mutator mutants deficient in methylation-instructed DNA mismatch correction, Proc. Natl. Acad. Sci. USA 77:1063 (1980).
23. McGraw, B.R. and Marinus, M.G., Isolation and characterization of dam^+ revertants and suppressor mutations that modify secondary phenotypes of dam^- strains of Escherichia coli K12, Mol. Gen. Genet. 178:309 (1980).
24. Marinus, M.G. and Morris, N.R., Biological function for 6-methyladenine residues in the DNA of Escherichia coli K12, J. Mol. Biol. 85:309 (1974).
25. Herman, G.E. and Modrich, P., Escherichia coli K-12 clones that overproduce dam methylase are hypermutable, J. Bacteriol. 145:644 (1981).
26. Marinus, M.G., Poteete, A. and Arraj, J.A., Correlation of DNA adenine methylase activity with spontaneous mutability in Escherichia coli K-12, Gene 28:123 (1984).
27. Lu, A-L., Clark, S., and Modrich, P., Methyl-directed repair of DNA base pair mismatches in vitro, Proc. Natl. Acad. Sci. USA 80:4639 (1983).
28. Lu, A-L., Welsh, K., Su, S-S., and Modrich, P., Repair of DNA base-pair mismatches in extracts of Escherichia coli, Cold Spring Harbor Symp. Quant. Biol. 49:589 (1984).
29. Grafstrom, R.H., Bear, J. and Hoess, R.H., Cloning, identification, and partial purification of mutH gene product, UCLA Symp. Mol. Cell Biol. 11:299 (1983).
30. Pang, P.P., Tsen, S-D., Lundberg, A.S., and Walker, G.C., The mutH, mutL, mutS and uvrD genes of Salmonella typhimurium LT2, Cold Spring Harbor Symp. Quant. Biol. 49:597 (1984).
31. Kumura, K. and Sekiguchi, M., Identification of the uvrD product of Escherichia coli as DNA helicase II and its induction by DNA-damaging agents, J. Biol. Chem. 259:1560 (1983).
32. Kushner, S.R., Maples, V.F., Easton, A., Farrance, I., and Peramachi, P., Physical, biochemical, and genetic characterization of the uvrD product, UCLA Symp. Mol. Cell Biol. 11:153 (1983).
33. Fishel, R.A. and Kolodner, R., An Escherichia coli cell free system that catalyzes the repair of symmetrically methylated heteroduplex DNA, Cold Spring Harbor Symp. Quant. Biol. 49:603 (1984).
34. Lieb, M., Specific mismatch correction in bacteriophage lambda crosses by very short patch repair, Mol. Gen. Genet. 191:118 (1983).
35. Lieb, M., Recombination in the lambda repressor gene: Evidence that very short patch (VSP) mismatch correction restores a specific sequence,

Mol. Gen. Genet. in press (1985).
36. Duncan, B.K. and Miller, J.H., Mutagenic deamination of cytosine residues in DNA, *Nature* 287:560 (1980).
37. Lindahl, T., DNA repair enzymes, *Ann. Rev. Biochem.* 51:61 (1982).

MOLECULAR APPROACHES TO THE STUDY OF NUCLEOTIDE EXCISION REPAIR IN EUKARYOTES

E.C. Friedberg, D.P. Barbis, J.M. Chenevert, R. Fleer, D. Kalainov, L. Naumovski, C.M. Nicolet, G.W. Robinson, R.A. Schultz, W.A. Weiss, and E. Yang

Department of Pathology
Stanford University Medical Center
Stanford, CA 94305

SUMMARY

Very little is known about the molecular mechanism of nucleotide excision repair in eukaryotes. Studies on human cells have been stimulated by the availability of excision repair-defective cell lines from patients suffering from the autosomal recessive disease xeroderma pigmentosum (XP). Such studies have contributed significantly to an understanding of the genetic complexity of excision repair in human cells. However, to date, no human excision repair genes or gene products known to complement the repair defect in XP cells have been isolated. The yeast *Saccharomyces cerevisiae* is an interesting model for exploring the molecular mechanism of nucleotide excision repair in eukaryotic cells. As is true in human cells, multiple yeast genes are involved and at least five genes are required for the specific incision of UV-irradiated DNA *in vivo*. These five genes have been isolated by molecular cloning and the nucleotide sequences of four of them have been determined. Each of these cloned genes is being used for overexpression of protein.

INTRODUCTION

Nucleotide excision repair defines a biochemical pathway by which base damage in DNA is excised as part of a nucleotide rather than as the free base; typically as part of an oligonucleotide fragment (see reference 1 for a recent review). It is now generally appreciated that this pathway is utilized by cells to excise base damage that causes significant conformational distortion of the secondary structure of DNA, e.g. the adduction to bases of large chemical substituents, or the covalent association of adjacent bases in the same polynucleotide chain (see reference 1 for a recent review).

The mechanism of such a biochemical pathway in eukaryotic cells is largely inferred from the results of studies using the prokaryote *E. coli* as a paradigm. Wherever possible, techniques successfully used for the study of DNA repair in *E. coli* have been applied to eukaryotes, with appropriate modification. By this strategy it has been demonstrated that eukaryotic cells can selectively remove certain forms of base damage from their DNA during incubation *in vivo*, and can also carry out repair synthesis and ligation of DNA.[1] Thus, the general model that nucleotide excision repair requires specific incision of DNA, excision of oligonucleotide fragments containing damaged bases, repair synthesis and DNA ligation seems universally applicable; however, at the present time *E. coli* is the only organism for which there is detailed information about the biochemistry of this process.[2-6] We are interested in understanding the mechanism of nucleotide excision repair in eukaryotic cells, particularly normal human cells, and the molecular basis for the defect(s) in DNA repair in the human disease xeroderma pigmentosum (XP). In this article we discuss the experimental approaches we are using in efforts to realize these goals.

NUCLEOTIDE EXCISION REPAIR IN HUMAN CELLS

E. coli is an attractive organism for the study of DNA repair because, in addition to its ease of handling, the availability of mutants with defined phenotypes has provided a genetic framework to evaluate the biochemical information obtained from cell-free studies. Hence, it is not surprising that the demonstration of defective DNA repair in XP[7] has made human cells a potentially attractive system for the study of nucleotide excision repair. Cell fusion studies have established the existence of eight complementation groups in XP, suggesting that the biochemistry of nucleotide excision repair in normal human cells may involve at least eight distinct genes.[7] Studies on XP cells from different complementation groups indicate that some of them (XP-A, -B, -C, -D and -G) are severely defective in the enzyme-catalyzed specific incision of damaged DNA.[8-10] On the other hand, cells from the XP complementation groups E, F and H are apparently deficient in DNA incision, but are clearly not totally defective in this process.[8-11] Further study is needed to consolidate these conclusions. In this regard, it would be helpful if some of the rarer XP complementation groups were represented by more than a single cell line,[7] so that trivial explanations such as leakiness of the mutation could be eliminated as a possible explanation for deficient DNA incision.

Such considerations notwithstanding, the dichotomous phenotype observed in XP bears an interesting resemblance to the situation in *E. coli*. Mutants in the *E. coli* *uvrA*, *uvrB* or *uvrC* genes are severely defective in a co-ordinated incision/excision reaction believed to occur during nucleotide excision repair in this becterium.[2-6] However, strains mutated in the *uvrD* or *polA* genes, while not totally defective in the incision/excision reaction, show some deficiency in this process.[1,12] These observations, coupled with recent demonstrations of the involvement of uvrD protein and of *E. coli* DNA polymerase I in incision/excision *in vitro*,[4-6] suggest to us the following model. We propose that in living cells the *uvrA*, *uvrB* and *uvrC* genes are absolutely required for nucleotide excision repair and *in vitro* the products of just these genes can catalyze the specific incision of UV-irradiated DNA. However, *in vivo* the optimal rate and efficiency of nucleotide excision repair require the products of the *uvrD* and *polA* genes (and perhaps others) (Table 1). Similarly, we suggest that in human cells the product of some genes may be absolutely required for nucleotide excision repair and perhaps are sufficient to catalyze specific biochemical events in cell-free systems. The products of other genes are important components of the complete human "repairosome" *in vivo*, but like the products of the *uvrD* and *polA* genes of *E. coli*, they may not be required for particular reactions *in vitro* (Table 1).

Given this analogy, to what may we attribute the apparent genetic complexity of nucleotide excision repair in human cells (in which at least eight genes may be involved, none of which is a known DNA polymerase), relative to *E. coli* (in which only five genes are involved, one of which is

Table 1. Genes Involved in Nucleotide Excision Repair

	E. coli	S. cerevisiae	Humans
DNA incision *in vitro*	uvrA uvrB uvrC	RAD1 RAD2 RAD3 RAD4 RAD10	XP-A XP-B XP-C XP-D XP-G
Nucleotide excision repair *in vivo*	uvrA uvrB uvrC uvrD polA	RAD1 RAD2 RAD3 RAD4 RAD10 RAD7 RAD14 RAD16 RAD23 MMS19	XP-A XP-B XP-C XP-D XP-G XP-E XP-F XP-H

DNA polymerase I)? Aside from the facile explanation of the greater complexity of proteins in human cells relative to their functional homologues in *E. coli*, it may be relevant that in eukaryotes DNA is structurally organized into nucleosomes, which in turn are subject to higher levels of organization yielding chromosomes. Hence, some of the genes involved in nucleotide excision repair in human cells may encode proteins that interact with chromosomes to provide access for catalytic components to sites of otherwise inaccessible base damage. Experimental support for this idea comes from experiments from this laboratory and that of Fujiwara [13,14].

It has proven extremely difficult to pursue studies with human cell-free systems. A major limitation is the extreme lability of one or more components in the extracts. Thus, for example, cell extracts of very high protein concentration are required to demonstrate the results just described [13,14]. In addition, freezing and thawing of the extracts results in a rapid loss of activity, as does the freezing of cells prior to their disruption.[15] In an effort to overcome the logistical restraints imposed by the use of conventional biochemical approaches, we are attempting to clone nucleotide excision repair genes from human cells with a view to overexpressing proteins for biochemical studies. However, molecular approaches to the study of nucleotide excision repair in humans are also problematic. Our recent studies on transfection of known genetic markers into normal and XP *diploid* fibroblasts have shown that the levels of stable expression of foreign genes attainable in such cells are far too low for the routine screening of genomic libraries for sequences that complement the UV sensitivity of XP cells (R.A. Schultz *et al*, manuscript submitted). On the other hand, only two of the eight existing XP complementation groups are currently represented by *immortalized* cell lines suitable for screening human genomic libraries. A second problem that we and others have encountered concerns reversion of the phenotype of UV sensitivity. Using stringent protocols that select for highly UV resistant cells, the reversion frequency of XP-A immortalized lines is low ($\sim 1 \times 10^{-7}$)[16] (R.S. Schultz *et al*, manuscript submitted). The failure to isolate true transfectants in the presence of such a low reversion background suggests that complementation of the phenomenon of UV sensitivity of XP-A cells by DNA transfection may be partial and hence would be missed by stringent selection. However, when we relaxed the stringency of the UV selection protocol, the frequency of UV resistant revertant cells was significantly higher, making it extremely difficult to detect true transfectants (R.A. Schultz et al, manuscript submitted). Clearly, alternative strategies are required for the molecular cloning of human nucleotide excision repair genes.

NUCLEOTIDE EXCISION REPAIR IN YEAST

The yeast Saccharomyces cerevisiae is much more amenable to the isolation of genes by molecular cloning. Furthermore, the genetics of nucleotide excision repair in yeast resemble the complexity observed in human cells, and yeast genes (called RAD) can also be segregated into those absolutely required for nucleotide excision repair and those which render mutants phenotypically deficient, but not defective in this process (Table 1). We have isolated five RAD genes required for the incision of damaged DNA in living cells. The RAD1, RAD2, RAD3 and RAD10 genes were cloned by screening a yeast genomic library for complementation of the UV sensitivity of the appropriate rad mutants.[17-20] The RAD4 gene was cloned by a gene rescue strategy to be described in detail elsewhere (R. Fleer and E.C. Friedberg, manuscript in preparation). The complete nucleotide sequence of four of these five genes has been determined.[17,20-22] As indicated in Table 2, the RAD1, RAD2 and RAD3 proteins are predicted to be of considerable size and all are significantly larger than RAD10 protein.

Regions of amino acid sequence homology have been detected between the RAD1, RAD3 and RAD10 polypeptides and known prokaryotic and eukaryotic proteins which bind and/or hydrolyze purine nucleotides such as ATP and GTP. Among these are the two E. coli nucleotide excision repair proteins uvrA and uvrD, both of which have ATPase activities[23-26] (Fig.1). The homology between the RAD3 and uvrD proteins is particularly interesting. As indicated earlier, the uvrD protein is involved in (but is apparently not essential for) nucleotide excision repair in E. coli. The uvrD gene is also involved in mismatch repair in E. coli and hence in addition to the phenotypes of UV sensitivity and deficient excision repair, uvrD mutants are spontaneous mutators (see reference 1 for a review). These phenotypes are consistent with the demonstration that the uvrD gene [which can encode a protein of calculated M_r~82kDa[24]] expresses an ATP-dependent DNA helicase called DNA helicase II of E. coli.[24-26] The RAD3 gene could encode a protein of calculated M_r~89kDa; a value close to that of uvrD protein (Table 2). Additionally, at least one rad3 mutant allele (rad3-12) has been reported to have an abnormally high frequency of spontaneous mutation at two independently tested loci, i.e., it has the phenotype of a spontaneous mutator.[27]

The significance of the sequence homologies shown in Fig 1 remains to be established by biochemical studies. One possibility is that, as is true in E. coli, in yeast some nucleotide excision repair proteins are ATPases. Another interesting possibility is that the RAD1, RAD3 and/or RAD10 proteins are G or G-like proteins which bind GDP and hydrolyze GTP.[28,29] A number of such proteins have been shown to be signal transducers for activation of other proteins[28,29] and it is worth considering that such a general role for one or more RAD proteins could result in alterations of the conformation of chromatin during nucleotide excision repair. Studies currently in progress are aimed at establishing whether the RAD1, RAD3 and RAD10 proteins bind and/or hydrolyze ATP or GTP. The RAD3 protein is also under investigation for DNA helicase activity. When placed under the control of the E. coli tac promoter, the RAD3 gene does not complement the UV

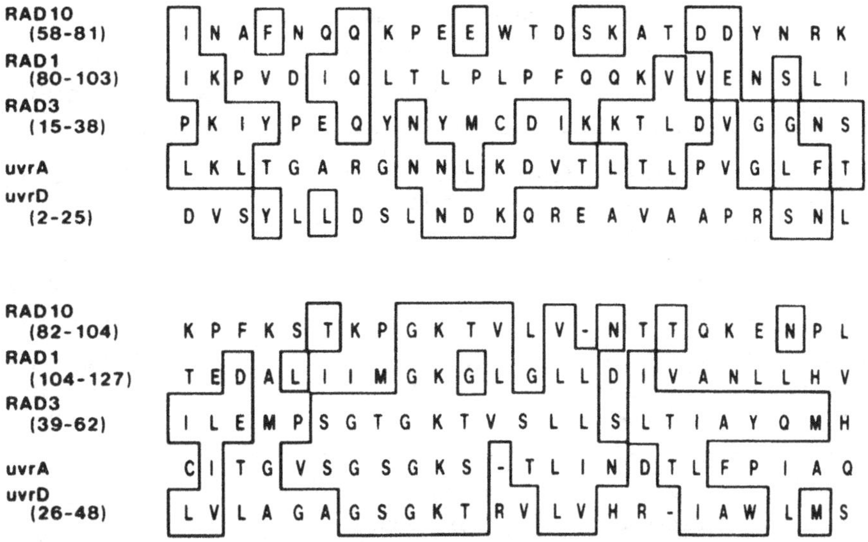

Figure 1. The uvrA data are derived from work in progress by Aziz Sancar and hence amino acid numbers are not yet assigned. The sequence GKS (T,G) is common to a number of purine nucleotide binding proteins.[24,33-37] The homologies shown include related as well as identical amino acids.

sensitivity of E. coli uvrA, uvrB, uvrC or uvrD strains.

In addition to these amino acid sequence homologies, the RAD1 and RAD2 polypeptides (which are virtually identical in size) (Table 2), share three limited domains of amino acid sequence homology.[21] Furthermore, the RAD3 polypeptide contains a stretch of twenty amino acids in which eleven are identical and two are related to those in a region of the yeast SIR3 polypeptide (Fig 2). The significance (if any) of these homologies also awaits further study.

Another interesting feature that has emerged from DNA sequence analysis is that all four sequenced RAD genes represent rare condons well, i.e., they do not show a particular bias for codon usage.[21] Codon usage has been correlated with levels of gene expression in yeast and in E. coli and it has been shown that many highly expressed genes have a significant bias for particular codons, while other codons are rarely used.[30] On the other hand, weakly expressed genes tend to utilize so-called rare codons. The indication from this analysis that the RAD1, RAD2, RAD3 and RAD10 genes are

Figure 2. Amino acid sequence homology between yeast RAD3 and SIR3 proteins.

Table 2. Properties of RAD Genes from S. cerevisiae

RAD1	RAD2	RAD3	RAD4	RAD10
Coding region = 2916bp	Coding region = 2925bp	Coding region =2334bp	---	Coding region =588bp
No introns	No introns	No introns	---	No introns
Calc. size of protein = 110kDa	Calc. size of protein =111kDa	Calc. size of protein = 89.8kDa	---	Calc. size of protein = 22.6kDa
Weakly expressed gene	Weakly expressed gene	Weakly expressed gene	---	Weakly expressed gene
---	---	Not inducible by DNA damage	---	---
Contains CS1, CS2, CS3	Contains CS1, CS2, CS3	Contains CS1, CS2, CS3	---	CS1, CS2, CS3 not present
No regulatory interaction with other RAD genes	---	No regulatory interaction other RAD genes	---	
Nonessential gene	Nonessential gene	Essential gene	Nonessential gene	Nonessential gene
---	---	Does not complement uvrA, uvrB, uvrC, uvrD or XP-A (?)	---	Does not complement uvrA, uvrB or uvrC

weakly expressed has been confirmed by direct experimentation. RAD3 mRNA is estimated to be present at levels of 1-5 copies/cell[19] and the RAD2 transcript at <1 copy/cell.[18] In addition, measurement of the levels of E. coli β-galactosidase expressed in yeast following transformation with single copy plasmids containing RAD-lacZ fusions indicates that the RAD1 gene is also weakly expressed (G.W. Robinson and E.C. Friedberg, manuscript in preparation). Collectively, these results suggest that like E. coli, yeast cells may normally contain very small amounts of the proteins required for nucleotide excision repair and have prompted us to examine the induction of RAD gene expression in cells exposed to DNA damage. These studies utilize cells transformed with single or multicopy plasmids carrying various RAD-lacZ fusions. Induction is detected by measuring increased levels of β-galactosidase activity in cells exposed to DNA damaging agents. To date, there is no evidence for induction of the RAD3 gene (Table 2)[22] (G.R. Robinson and E.C. Friedberg, manuscript in preparation). The RAD1, RAD2, RAD4 and RAD10 genes are still under study.

The RAD1, RAD2, RAD4 and RAD10 genes are not essential for the viability of haploid yeast cells. However, inactivation of the RAD3 gene is lethal[31,32] (Table 2). Hence, the RAD3 gene is required for an essential function in yeast. The observation that a number of viable rad3 allelic mutants are totally defective in nucleotide excision repair suggests that the essential and excision repair functions may be distinct. We have not yet defined the essential function of the RAD3 gene, however the results of both random and site-specific mutagenesis indicate that the linear sequence of the RAD3 gene is not organized into two discrete functional domains that specify these two functions (L. Naumovski and E.C. Friedberg, manuscript in preparation). We have recently isolated a temperature sensitive rad3 mutant (L. Naumovski and E.C. Friedberg, manuscript in preparation) and it is hoped that the detailed phenotypic characterization of this mutant will provide indications about the nature of the essential function of this gene.

The RAD1, RAD2, RAD3 and RAD10 genes are currently being tailored to bring them under the control of strong yeast or E. coli promoters. Overexpression of these genes should facilitate their isolation by protein fractionation and will set the stage for their detailed biochemical analysis.

ACKNOWLEDGEMENTS

We thank Dr. Aziz Sancar for communicating unpublished data and Dr. Russell Doolittle for pointing out the homology between the RAD3 and SIR3 polypeptide sequences. Studies from the senior author's laboratory are supported by research grants #CA-12428 from the USPHS and #1684 from the Council for Tobacco Research, and by contract #DE-AT03-76EV70032 with the U.S. Department of Energy.

REFERENCES

1. Friedberg, E.C. (1985) DNA Repair, W.H. Freeman and Co., New York.
2. Yeung, A.T., Mattes, W.B., Oh, E.Y. and Grossman, L. (1983) Proc. Natl. Acad. Sci. (USA) 80, 6157-6161.
3. Sancar, A. and Rupp, W.D. (1983) Cell 33, 249-260.
4. Kumara, K., Sekiguchi, M., Steinum, A-L. and Seeberg, E. (1985) Nucleic Acids Res. 13, 1483-1492.
5. Sancar, A., Franklin, K.A. and Sancar, G.B. (1984) Proc. Natl. Acad. Sci. (USA) 81, 7397-7401.
6. Caron, P.R., Kushner, S.R. and Grossman, L. (1985) Proc. Natl. Acad. Sci. (USA) (in press).
7. Cleaver, J.E. (1983) In: The Metabolic Basis of Inherited Disease, 5th ed. J.B. Stanbury, J.B. Wyngaarden, D.S. Fredrickson, J.L. Goldstein and M.S. Brown, eds. pp 1227-1248, McGraw-Hill, New York.
8. Zelle, B. and Lohman, P.H.M. (1979) Mutation Res. 62, 363-368.
9. Fornace, A.J., Jr., Kohn, K.W. and Kann, H.E., Jr. (1976) Proc. Natl. Acad. Sci. (USA) 73, 39-43.
10. Tanaka, K., Sekiguchi, M. and Okada, Y. (1975) Proc. Natl. Acad. Sci. (USA) 72, 4071-4075.
11. Moshell, A.N., Ganges, M.B., Lutzner, M.A., Coon, H.G., Barrett, S.F., Dupuy, J-M. and Robbins, J.H. (1983) In: Cellular Responses to DNA Damage. E.C. Friedberg and B.A. Bridges (eds) pp 209-213, Alan R. Liss, New York.
12. Friedberg, E.C. (1985) Cancer Surveys (in press).

13. Mortelmans, K., Friedberg, E.C., Slor, H., Thomas, G. and Cleaver, J.E. (1976) Proc. Natl. Acad. Sci. (USA) 73, 2757-2761.
14. Kano, Y. and Fujiwara, Y. (1983) Carcinogenesis 4, 1419-1424.
15. Slor, H., Lev-Sobe, T. and Friedberg, E.C. (1977) Mutation Res 45, 137-145.
16. Royer-Pokora, B. and Haseltine, W. (1984) Nature 311, 390-392.
17. Yang, E. and Friedberg, E.C. (1984) Mol. Cell. Biol. 4, 2161-2169.
18. Naumovski, L. and Friedberg, E.C. (1984) Mol. Cell. Biol. 4, 290-295.
19. Naumovski, L. and Friedberg, E.C. (1982) J. Bacteriol. 152, 323-331.
20. Weiss, W.A. and Friedberg, E.C. (1985) The EMBO J. (in press).
21. Nicolet, C.M., Chenevert, J.M. and Friedberg, E.C. (1985) Gene (in press).
22. Naumovski, L., Chu, G., Berg, P. and Friedberg, E.C. (1985) Mol. Cell. Biol. 5 17-26.
23. Seeberg, E. and Steinum, A-L. (1982) Proc. Natl. Acad. Sci. (USA) 79, 988-992.
24. Finch, P.W. and Emmerson, P.T. (1984) Nucleic Acids Res. 12, 5789-5799.
25. Maples, V.F. and Kushner, S.R. (1982) Proc. Natl. Acad. Sci. (USA) 79, 5616-5620.
26. Kumura, K. and Sekiguchi, M. (1984) J. Biol. Chem. 259, 1560-1565.
27. Brychcy, T. and Von Borstel, R.C. (1977) Mutation Res 45, 185-194.
28. Hughes, S.M. (1983) FEBS Lett. 164, 1-8.
29. Gilman, A.G. (1984) Cell 36, 577-579.
30. Bennetzen, J.L. and Hall, B. (1982) J. Biol. Chem. 257, 3026-3031.
31. Naumovski, L. and Friedberg, E.C. (1983) Proc. Natl. Acad. Sci. (USA) 80, 4818-4822.
32. Higgins, D.R., Prakash, S., Reynolds, P. Polakowska, R., Weber, S. and Prakash, L. (1983) Proc. Natl. Acad. Sci. (USA) 80, 5680-5684.
33. Gallwitz, D., Donath, C. and Sander, C. (1983) Nature 306, 704-707.
34. Powers, S., Kataoka, T., Fasano, O., Goldfarb, M., Strathern, J., Broach, J. and Wigler, M. (1984) Cell 36, 607-612.
35. Reymond, C.D., Gomer, R.H., Mehdy, M.C. and Firtel, R. (1984) Cell 39, 141-146.
36. Mozer, B., Marlor, R., Parkhurst, S. and Corces, V. (1985) Mol. Cell. Biol. 5 885-889.
37. Lochrie, M.A., Hurley, J.B. and Simon, M.I. (1985) Science 228, 96-99.

RESTORATION OF DNA REPAIR IN UV-SENSITIVE CHINESE HAMSTER OVARY CELL BY THE denV GENE FROM BACTERIOPHAGE T4

Kristoffer Valerie, Jon K. de Riel, and Earl E. Henderson

Fels Research Institute
Temple University School of Medicine
Philadelphia, PA 19140

The denV gene of bacteriophage T4, which is the principal determinant of resistance to ultraviolet light (UV) on the T4 genome,[1] codes for endonuclease V, a 16,000 dalton protein possessing two distinct enzymatic activities which can act sequentially to initiate repair of pyrimidine dimers. A pyrimidine dimer-DNA glycosylase activity first cleaves one of the two bonds attaching the pyrimidine dimer to the DNA backbone, leaving an apyrimidinic site, and an AP-endonuclease activity associated with the same polypeptide subsequently introduces a single-strand break at this site.[2,3] like the very similar enzyme from *Micrococcus luteus*, described by R. Grafstrom elsewhere in this volume, endonuclease V leaves residual damage products on both sides of the single-strand break, which presumably must be removed by other enzymes before repair synthesis can take place. The phage T4 and *M. luteus* enzymes catalyze a base excision process and are in this respect distinguishable from other known pyrimidine dimer repair systems, which proceed via a nucleotide excision repair mechanism. In spite of these differences, experiments with permeabilized cells have shown that endonuclease V is capable of restoring excision repair of UV damage not only in repair-deficient *E. coli*[4] but also in human fibroblasts from patients with the heritable UV repair disorder xeroderma pigmentosum.[5,6] Thus the denV gene is an interesting candidate for genetic correction of UV repair deficiency in both prokaryotic and eukaryotic cells.

We have cloned the denV gene[7] and shown that it does in fact partially restore UV resistance in a variety of repair-deficient *E. coli* strains.[8] Here we present data which show that the denV gene, coupled to an SV40 early gene promoter, restores repair of UV damage in the UV-sensitive Chinese hamster ovary cell line UV5.[9]

The plasmid construct used in these experiments, pneotk-denV, shown in Figure 1, contains the denV gene inserted into a truncated segment of SV40 DNA containing the promoter, splicing sites, and polyadenylation signal of the SV40 early gene.[10] The plasmid also contains the bla gene to permit ampicillin selection in *E. coli* and two markers for co-selection in mammalian cells: the *E. coli* neo gene, conferring resistance to the neomycin analogue G-418,[11] and the Herpes simplex virus thymidine kinase gene to permit selection in tk^- cells.[12] These markers were included to permit primary selection of transformants without exposure to UV. Construction of this vector will be described in more detail elsewhere (manuscript in preparation).

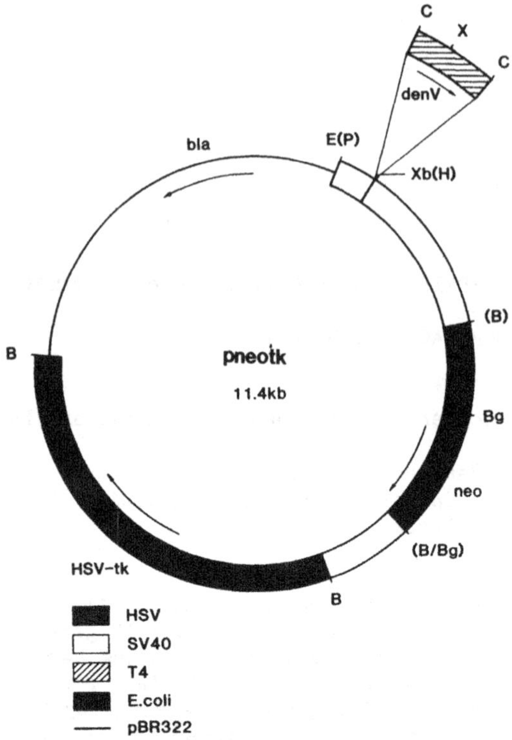

Figure 1. Plasmid map of pneotk and its denV⁺ derivative, pneotk-denV. Restriction sites are shown for EcoRI (E), HpaII (P), BamHI (B), BglII (Bg), XhoI (X), ClaI (C), XbaI (Xb), and HindIII (H). Orientations of transcription are shown with arrows.

Plasmid pneotk-denV was characterized by sequencing the denV gene insert, and then transfected into UV5 cells by calcium phosphate coprecipitation[13] followed by glycerol shock. UV5 cells and cells of the wild-type parent strain, AA8, were also transfected with the parent vector, pneotk. Transformants were selected with G-418 (1 mg/ml); clones formed with an efficiency of about 10^{-4}. Five neo^r clones from each transfection were subsequently screened for UV resistance. In the preliminary screen two of the pneotk-denV-transfected clones were clearly more resistant than UV5 (Figure 2), although one of these subsequently reverted to sensitivity. No UV-resistant clones were obtained from UV5 transfected with pneotk.

UV survival curves show that the stable transformant (I-A1) is substantially more resistant to UV than the UV5 mutant from which it was derived, but clearly not as resistant as wild-type. The approximate D_0 values calculated for the curves in Figure 2 are ~1.4 J/m^2 for the denV transformant, compared to ~2.0 J/m^2 for the wild-type AA8 cells and ~0.63 J/m^2 for mutant UV5 cells. Southern blots of DNA from the various lines (not shown) confirm that the stably resistant clone I-A1 contains a single copy of the denV gene integrated into host cell DNA and that phenotypic instability in clone I-C2 correlates with progressive loss of denV gene sequences from this line. There are a number of possible explanations for the failure of the denV gene in I-A1 to fully restore UV-resistance. (i) Residual sensitivity may be due to lesions other than pyrimidine dimers, which may be recognized by the endogenous repair system but not by endonuclease V. (ii) Repair initiated by endonuclease V may not be completed efficiently by the host cell. (iii) The effective level of endonuclease V in the cell nucleus may be inadequate, although the biochemical evidence presented below suggests this is not the case.

The level of repair in I-A1 cells has been characterized thus far by two biochemical measures: extent of single-strand breakage in response to UV irradiation and level of repair synthesis of DNA in response to UV. Controls for both these experiments were clones of AA8 and UV5 cells transformed with pneotk. Repair synthesis was quantitated on CsCl gradients as the specific activity of [^{3}H]thymidine incorporated into parental DNA following irradiation in the presence of 5-bromodeoxyuridine [14] (Figure 3). In unirradiated cultures, [^{3}H]thymidine was incorporated into

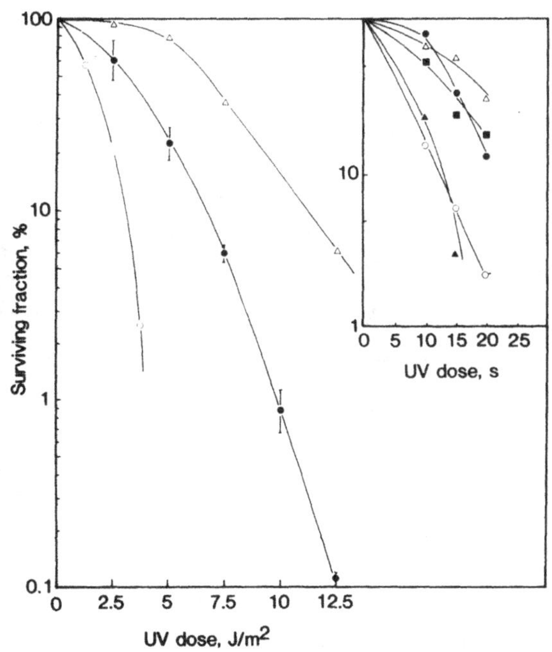

Figure 2. UV survival curves of CHO clones. Initial screens of G418-resistant clones (inset) were performed by serially irradiating cells suspended in a droplet. Subsequent curves were generated by irradiation of plated cells (UV fluence = 0.63 W/m^2). Δ, clone I-D5 (AA8/pneotk); ○, clone II-D3 (UV5/pneotk); ●, ▲, ■, clones I-A1, I-B5, and I-C2, respectively, all UV5 cells transformed with pneotk-denV.

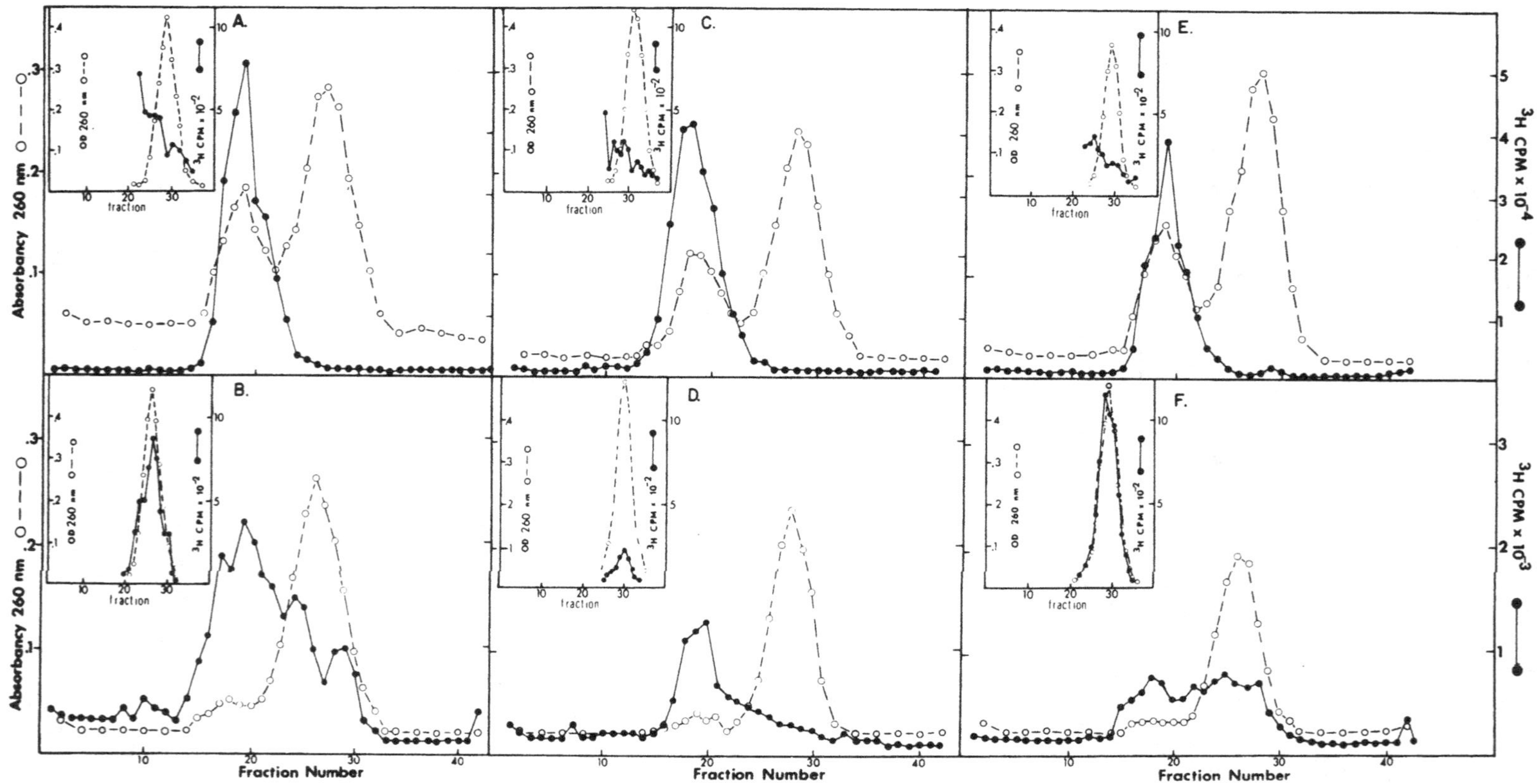

Figure 3. Measurement of DNA repair synthesis by equilibrium sedimentation. Cells were preincubated with BrdUrd (10 ug/ml) for 30 minutes, irradiated, incubated for 3 hours with BrdUrd and [methyl-^{3}H]thymidine, and harvested. DNA was extracted, sheared, and banded on CsCl (1.7 g/ml initial density). <u>Panels A</u> and <u>B</u>, clone I-D5 (AA8/pneotk); <u>panels E</u> and <u>F</u>, clone I-A1 (UV5/pneotk-denV). <u>Panels A</u>, <u>C</u>, and <u>E</u>: unirradiated cells; <u>panels B</u>, <u>D</u>, and <u>F</u>: irradiated with 40 J/M^2. (Insets) The five fractions of peak absorbancy were rebanded in CsCl. o, Absorbance at 260 nm; •, TCA-precepitable counts per minute (cpm). Increasing density from right to left.

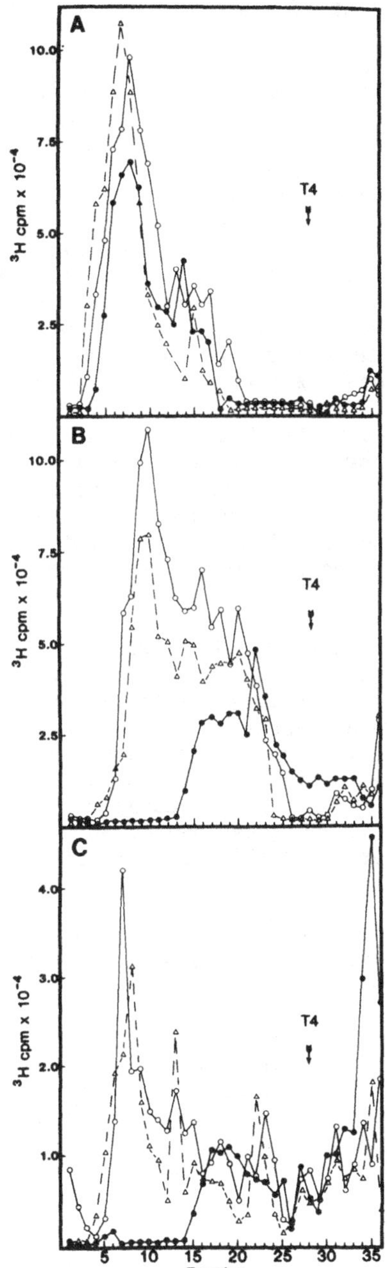

Figure 4. Alkaline sucrose gradients of [^{3}H]dt-labeled UV-irradiated CHO DNA. Cells were labeled for 24 hours, irradiated, incubated for 1 hour in fresh medium containing hydroxyurea (2mM and araC (10 um), and lysed directly on a 5-30% sucrose gradient (pH 12.1) prior to sedimentation (90 minute, 25,000 rpm). (A) Unirradiated cells. (B) ells irradiated with 5 J/m^2. (C) Cells irradiated with 30 J/m^2.

Δ - clone I-D5 (AA8/pneotk);
o - clone II-D3 (UV5/pneotk);
• - clone I-A1 (UV5/pneotk-denV).

Direction of sedimentation from right to left.

replicated DNA but not into the parental DNA (less dense peak). After 40 J/m^2, the amount of replication is greatly reduced in all three cell lines; AA8 and pneotk-denV transformed UV5 cells, but not the pneotk-transformed UV5 line, show incorporation of ^{3}H into parental DNA. Specific activities calculated from rebanded parental DNA peak fractions (insets) indicate that the level of repair synthesis in UV5 cells was not more than 29% of the level in wild-type AA8 cells, and jumped to 156% of the wild-type level in the denV$^+$ transformants.

Accumulation of single-strand breaks 1 hour after UV irradiation was measured by fractionation of uniformly labeled DNA on alkaline sucrose gradients.[15] Resealing of breaks after irradiation was prevented by inhibition of DNA polymerase and ligase with hydroxyurea and 1-β-D-arabinofuranosyl cytosine.[16] Breaks were quantitated by analysis of the distribution of ^{3}H across the gradient.[17] Such an analysis reveals that exposure to 5 J/m^2 induces essentially no breaks in UV5 cell DNA (relative to the unirradiated control), 2-4 breaks/10^9 daltons in AA8 cells, and 8-10 breaks/10^9 daltons in denV-transformed UV5 cells. More striking is the shape of the sedimentation profiles. Approximately 40% of the DNA in wild-type cells remains in the position of the main peak of unirradiated DNA after exposure to either 5 J/m^2 or 30 J/m^2, whereas the I-A1 DNA appears to move across the gradient in a more nearly unimodal fashion. The most straight-forward explanation for this observation is that hamster cell DNA is heterogeneous in susceptibility to the endogenous repair activity. This is consistent with other observations that active genes are repaired more rapidly than inactive genes in CHO cells[18] perhaps because the bulky endogenous repair complex has easier access to open DNA than to tightly packaged chromatin. Our results suggest that endonuclease V is not so constrained, perhaps because of its small size.

Comparison of relative levels of strand breakage and of repair synthesis suggests that the amount of ^{3}H incorporated per single-strand break is about half as much in denV-transformed UV5 cells as in wild-type cells. This could reflect smaller patch size of repair, or poor access by the repair polymerase to some single-strand breaks in tightly packaged DNA. However, it could also simply result from the inefficiency of the cell in converting an endonuclease V nick into a substrate for the repair polymerase. It should be possible to resolve these questions by more complete characterization of the repair process in denV$^+$ transformants of both UV5 and wild-type AA8 cells.

The reappearance of UV-induced excision repair in I-A1 cells indicates that endonuclease V can gain access to the nucleus, perhaps because of its small size, although the efficiency of nuclear uptake remains to be determined.

The data presented here suggest that the denV gene has created a base-excision repair pathway for pyrimidine dimers in pneotk-denV-transformed Chinese hamster ovary UV5 cells, and that this process is at least as active as the endogenous excision repair process although not as effective in maintaining viability. It should be of interest to determine its effect on UV mutagenesis in these cells.

ACKNOWLEDGEMENTS

This work was supported by a Basic Research Grant (1-899) to J.K. de Riel from the March of Dimes National Birth Defects Foundation.

REFERENCES

1. W. Harm, Mutants of phage T4 with increased sensitivity to ultraviolet, Virology 19:66-71 (1963).
2. E.H. Radany and E.C. Friedberg, A pyrimidine dimer-DNA glycosylase activity associated with the v gene product of bacteriophage T4, Nature 286:182-185 (1980).
3. Y. Nakabeppu and M. Sekiguchi, Physical association of pyrimidine dimer DNA glycosylase and apurinic/apyrimidinic DNA endonuclease essential for repair of ultraviolet-damaged DNA, Proc. Natl. Acad. Sci. USA 78:2742-2746 (1981).
4. K. Shimizu and M. Sekiguchi, Introduction of active enzyme into permeable cells of Escherichia coli: acquisition of ultraviolet light resistance by uvr mutants on introduction of T4 endonuclease V, Mol. Gen. Genet. 168:37-48 (1979).
5. K. Tanaka, H. Hayakawa, M. Sekiguchi and Y. Okada, Specific action of T4 endonuclease V on damaged DNA in xeroderma pigmentosum cells in vivo, Proc. Natl. Acad. Sci. USA 74:2958-2962 (1977).
6. C.A. Smith and P.C. Hanawalt, Phage T4 endonuclease V stimulates DNA repair replication in isolated nuclei from ultraviolet-irradiated human cells, including xeroderma pigmentosum fibroblasts, Proc. Natl. Acad. Sci. USA 75:2598-2602 (1978).
7. K. Valerie, E.E. Henderson and J.K. de Riel, Indentification, physical map location and sequence of the denV gene from bacteriophage T4, Nucl. Acids Res. 12:8085-8096 (1984).
8. K. Valerie, E.E. Henderson and J.K. de Riel, Expression of a cloned denV gene of bacteriophage T4 in Escherichia coli, Proc. Natl. Acad. Sci. USA, in press.
9. L.H. Thompson, D.B. Busch, K. Brookman, C.L. Mooney and D.A. Glaser, Genetic diversity of UV-sensitive DNA repair mutants of Chinese hamster ovary cells, Proc. Natl. Acad. Sci. USA 78:3734-3737 (1981).
10. V.B. Reddy, S.S. Tevethia, M.J. Tevethia and S.M. Weissman, Nonselective expression of simian virus 40 large tumor antigen fragments in mouse cells, Proc. Natl. Acad. Sci. USA 79:2064-2067 (1982).
11. P.J. Southern and P. Berg, Transformation of mammalian cells to antibiotic resistance with a bacterial gene under control of the SV40 early region promoter, J. Mol. Applied Genet. 1:327-342 (1982).
12. F. Colbere-Garapin, S. Chousterman, F. Horodniceanu, P. Kourilsky and A.-C. Garapin, Cloning of the active thymidine kinase gene of herpes simplex virus type 1 in Escherichia coli K12, Proc. Natl. Acad. Sci. USA 76:3755-3759 (1979).
13. F.L. Graham and A.J. van der Eb, A new technique for the assay of infectivity of human adenovirus 5 DNA, Virology 52:456-467 (1973).
14. C.A. Smith, P.K. Cooper and P.C. Hanawalt, Measurement of repair replication by equilibrium sedimentation, in: "DNA Repair: A Laboratory Manual of Reserach Procedures", vol. 1B, E.C. Friedberg, and P.C. Hanawalt, eds., Marcel Dekker, New York, pp. 289-305 (1981).
15. A.K. Ganesan, C.A. Smith and A.A. van Zeeland, Measurement of the pyrimidine dimer content of DNA in permeabilized bacterial or mammalian cells with endonuclease V of bacteriophage T4, in: "DNA Repair: a Laboratory Manual of Research Procedures", vol. 1A, E.C. Friedberg and P.C. Hanawalt, eds., Marcel Dekker, New York, pp. 89-97 (1981).
16. R.D. Snyder, B. van Houten, and J.D. Regan, The inhibition of ultraviolet radiation-induced DNA repair in human diploid fibroblasts by arabinofuranosyl nucleosides, Chem. Biol. Interactions 50:1-14 (1984).

17. A.R. Lehman, Measurement of postreplication repair in mammalian cells, in: "DNA Repair: A Laboratory Manual of Procedures", vol. 1B, E.C. Friedberg, and P.C. Hanawalt, eds., Marcel Dekker, New York, pp. 471-485 (1981).
18. V.A. Bohr, C.A. Smith, D.S. Okumoto and P.C. Hanawalt, DNA repair in an active gene: removal of pyrimidine dimers from the DHFR gene of CHO cells is much more efficient than in the genome overall, Cell 40:359-369 (1985).

THE ADAPTIVE RESPONSE OF MAMMALIAN CELLS TO ALKYLATING AGENTS

Leona Samson

Laboratory of Toxicology
Harvard School of Public Health
Boston, Massachusetts

Mutation, cell death and chromosome damage are the ultimate products of the response of mammalian cells to DNA-damaging agents. It is important to determine how cells respond to a continuous exposure to DNA damage (much like we experience in our environment) and to learn about the genetic and molecular events that lie between the incurrence of DNA damage and the final appearance of damaged chromosomes and dead or mutated cells. We now know that bacteria induce specific sets of genes in response to certain types of genetic damage;[1] evidence that eukaryotes employ similar strategies is rapidly accumulating. Among other things a detailed understanding of these mechanisms will provide a more sound scientific basis upon which to assess the risk to man from DNA-damaging agents in our environment.

We are continuously exposed to low levels of DNA-damaging agents from our environment, from our diet and from intermediary metabolites found inside every cell. Several years ago we devised a way to study continuous low-dose bacterial mutagenesis in an attempt to simulate this chronic environmental exposure to mutagens. Our objective was to determine whether mutation induction occurred with the frequency and the kinetics one would predict from the much more usual mutagenesis studies that involved exposing cells to extremely large doses of mutagen for a short period, i.e., we aimed to compare mutations induced during several days of mutagen exposure with mutations induced during several minutes of mutagen exposure. Since alkylating agents represent one of the major classes of DNA-damaging agents in our environment, the first mutagen we chose to study was the simple monofunctional methylating agent N-methyl-N'-nitro-nitrosoguanidine (MNNG). The results were totally unexpected. A growing E. coli population continuously exposed to an extremely low non-toxic MNNG dose (1 ug/ml) for three days did not continuously accumulate mutant cells because it had become refractory to mutation induction during the first 60 min of exposure (Fig. 1). It turned out that exposure of E. coli to sub-lethal levels of MNNG induced the de novo synthesis of at least two DNA repair enzymes that provide considerable protection against the killing and mutagenic effects of further alkylation damage even by very high doses of MNNG.[2] We have called this the E. coli adaptive response to alkylating agents and its characterization by several laboratories (using biochemistry, bacterial genetics and molecular biology) has allowed a better understanding of alkylation-induced cell death and mutation and of the mechanisms bacteria can use to avoid these fates.[3,4] The E. coli adaptive response to alkylating agents has been very well described elsewhere.[3,4] However,

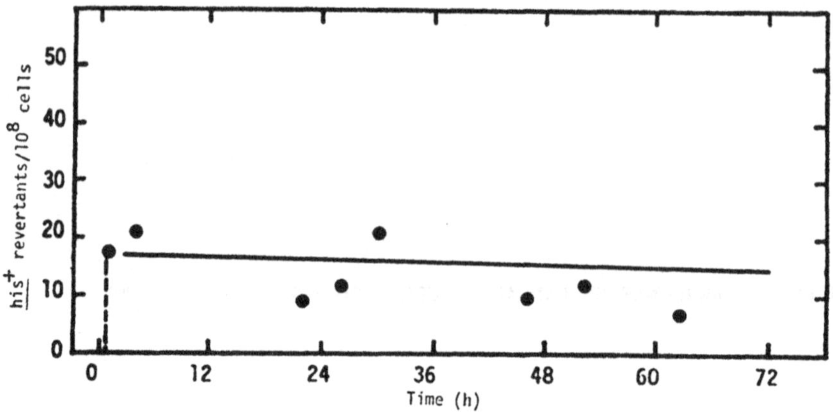

Figure 1. The accumulation of mutants in exponentially growing E. coli constantly exposed to 1 ug/ml MNNG.

here we describe two features of the response that are particularly relevant to the adaptive response of mammalian cells because they underscore the fact that, even in E. coli, MNNG adaptation and challenge doses must fall within narrow ranges in order to observe biologically an adaptive response.

Optimal doses for E. coli adaptation are shown in Figure 2, which shows killing and mutation induction by a 100 ug/ml MNNG challenge to cells

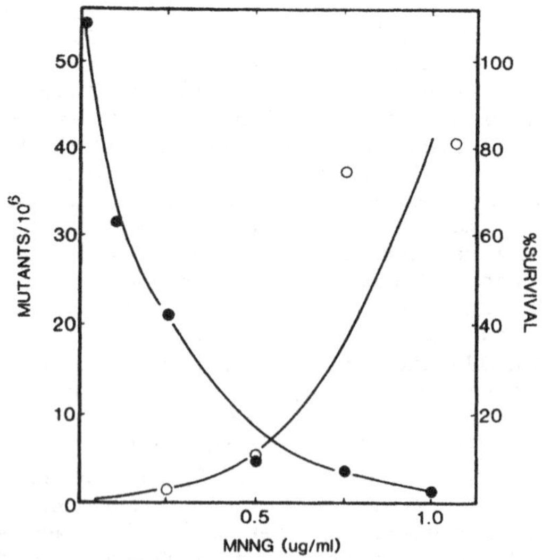

Figure 2. The effect of varying the adaptive dose of MNNG on the subsequent response of bacteria to a MNNG challenge. Six parallel cultures of E. coli were grown for 75 min in the presence of various low doses of MNNG and then challenged with 100 ug/ml MNNG for 5 min. The resulting mutation frequency (●) and cell survival (O) was measured.

pretreated with different low doses. We found that the induced resistance to killing and to mutation occurred with very different kinetics. For instance a 0.5 ug/ml MNNG adapting dose induced almost complete resistance to mutation but induced no resistance to cell killing by MNNG. This was the first indication that killing and mutagenic resistance are achieved via quite separate pathways and that cell death is not merely the result of mutations in essential genes. We now know, from the work of several laboratories, that the adaptive response involves the induction of a number of genes, at least two of which are DNA repair enzymes. For a detailed review of this work see references 3 and 4; a brief summary of the salient features is given below.

The induction of 3-methyladenine (3MeA) DNA glycosylase II, encoded by the alkA gene, provides killing resistance by repairing alkyl groups (on both purines and pyrimidines) that protrude into the minor groove of the DNA double helix. The induction of DNA methyltransferase, encoded by the ada gene, provides mutational resistance by removing methyl groups (that protrude into the major groove) from O6-methyl guanine (O6MeG) and O4-methyl thymine (O4MeT). Methyl groups at each of these positions interfere with base pairing and if left unrepaired cause misincorporations during DNA replication. The DNA methyltransferase directly transfers the offending methyl group from O6MeG or O4MeT to one of its own cysteine residues. This reaction is rather unusual because, once methylated, the DNA methyltransferase enzyme is inactivated, i.e., it is a suicide enzyme. The suicidal nature of the DNA methyltransferase repair mechanism accounts for the unusual kinetics of mutation-induction shown in Figure 3. As the MNNG challenge dose increases, the mutational resistance of adapted *E. coli* becomes saturated because all the methyltransferase molecules have been consumed. Thus, if we had originally chosen a very high MNNG challenge dose to test the resistance of adapted cells, the adaptive response of *E. coli* would have been completely missed. In summary, the adaptation and challenge doses of MNNG must fall within particular ranges in order to observe the adaptive response of *E. coli* to alkylating agents. It turns out that the same is true for the adaptive response of mammalian cells in tissue culture.

An awareness of the existence of a mammalian adaptive response could significantly influence our approach to assessing the mutagenic and carcinogenic risk of chemicals in our environment. We have found an

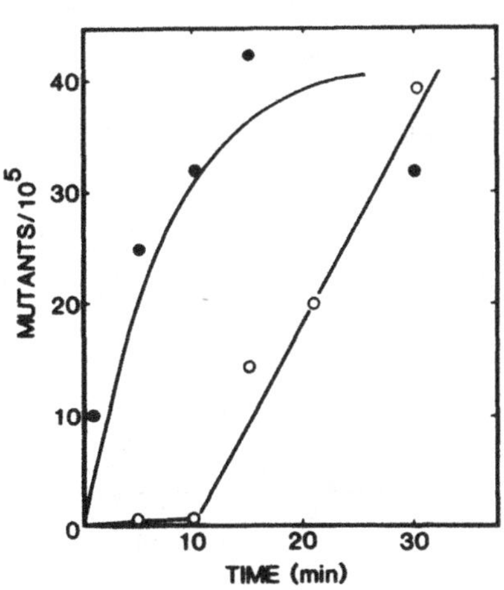

Figure 3. Mutation induction in adapted (O) and non-adapted (●) *E. coli* challenged with 10 ug/ml MNNG for various times.

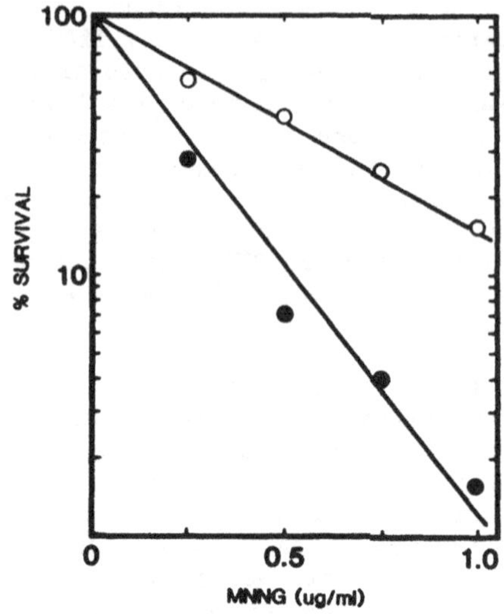

Figure 4. Sensitivity of adapted (O) and non-adapted (●) CHO cells to MNNG killing. Adapted cells received 0.01 ug/ml MNNG once every 6 hour for 48 hours.

adaptive response to alkylating agents in three mammalian cell types, namely Chinese hamster ovary (CHO), GM637 SV40 transformed normal human skin fibroblast[5] and HeLa CCL2 cells. We, and now others, find that chronic treatment with almost negligible levels of MNNG for at least one or two generation times (48 hr) renders mammalian cells highly resistant to the induction of sister chromatid exchange (SCE), mutation and cell death by further alkylation damage, even by very high doses of MNNG.[6,7] Resistance to cell killing and SCE is shown in Figures 4 and 5. The induced resistance decays once pretreatment is stopped and resistance to SCE and mutation induction[6] becomes saturated at high doses of alkyating agent, much like the characteristic saturation of mutational resistance in adapted bacteria. Resistance appears to be specific for alkylating agents because,

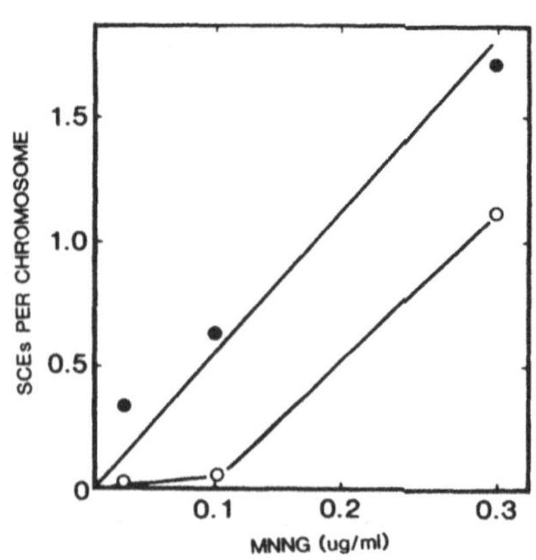

Figure 5. Sensitivity of adapted (O) and non-adapted (●) CHO cells to SCE induction by MNNG. Adapted cells received 0.01 ug/ml MNNG once every 6 hour for 48 hours.

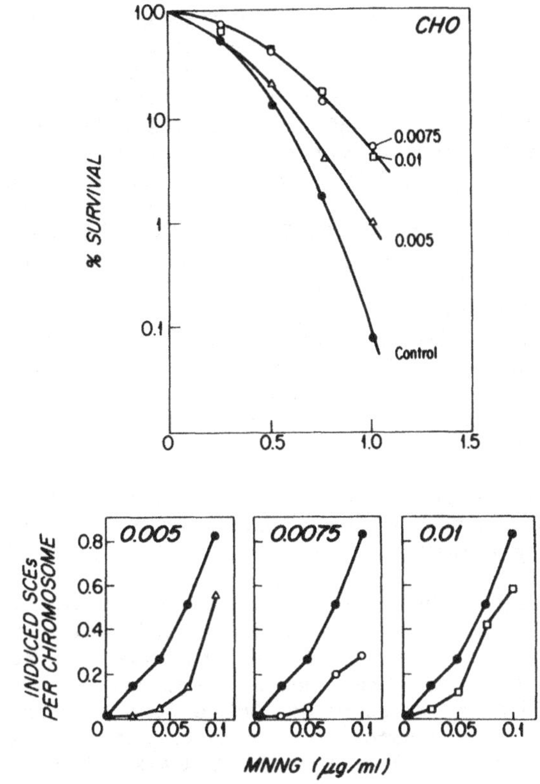

Figure 6. The effect of varying the adaptive dose of MNNG on the subsequent response of CHO cells to a MNNG challenge. The cells received 0, 0.005, 0.0075 or 0.01 ug/ml MNNG once every 6 hour for 48 hours.

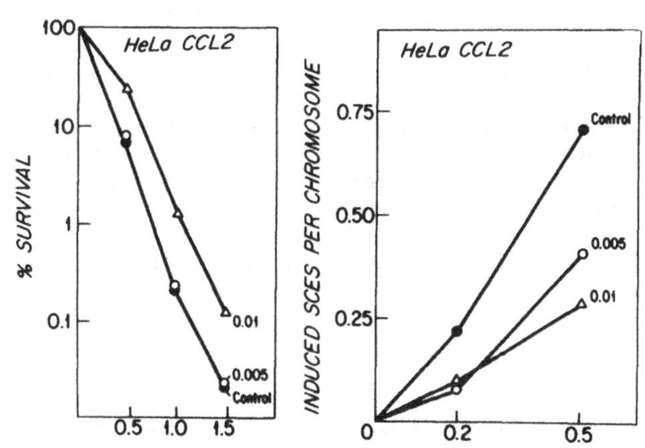

Figure 7. The effect of varying the adaptive dose of MNNG on the subsequent response of HeLa CCL2 cells to a MNNG challenge. The cells received 0, 0.005 or 0.01 ug/ml MNNG once every 6 hour for 48 hours.

while adapted cells are resistant to the SCE-inducing effects of three different alkylating agents, they are not resistant to SCE induction by UV light.[5] Having precluded explanations for the induced alkylation resistance such as the prevention of DNA alkylation and the redistribution of cells into resistant phases of the cell cycle, Jeffrey Schwartz and I deduced that adapted mammalian cells could repair DNA alkylation damage more efficiently. We have managed to observe a mammalian adaptive response by paying attention to its extreme dose dependency. This is exemplified in Figures 6 and 7. For CHO cells, 5, 7.5 or 10 ng/ml MNNG was added to the medium once every six hours for 48 hours. It is clear that the optimal adapting doses for killing resistance (7.5 and 10 ng/ml) are different from the optimal adapting doses for resistance to SCE induction (5 and 7.5 ng/ml). Separation of the adaptive resistance to killing and SCE induction is even more clearcut for HeLa cells pretreated with 5 and 10 ng/ml MNNG (Figure 7). Here, 5 ng/ml MNNG pretreatment induced considerable SCE resistance but absolutely no killing resistance, much like the results shown for E. coli in Figure 2.

In summary the adaptation and challenge doses of MNNG must fall within particular ranges in order to observe the adaptive response of mammalian cells to alkylating agents.

In the last few years it has become clear that mammalian cells possess some of the same armoury of DNA alkylation repair enzymes so far identified in E. coli, namely DNA methyltransferase and 3MeA and 3MeG-DNA glycosylases. Using our exact MNNG pretreatment regime, both adapted rodent and adapted human cells have been found to repair 06MeG more efficiently (3-5 fold) than non-adapted cells.[7,8] To date, no reports on the ability of the adapted mammalian cells to repair other types of DNA alkylation damage have been published.

In addition to learning about the events that lie between the incurrence of DNA damage and the final appearance of dead or mutated cells, we would also like to learn how mammalian cells respond at the genetic level to the kind of chronic low-dose exposure to DNA damage that we experience in our environment. A direct approach to learning how mammalian cells respond to alkylating agents at the genetic level is to isolate the genes involved in that response. The isolation of these genes will allow the study of their function, structure and in vivo expression. One of our approaches has been to attempt to directly isolate the human DNA methyltransferase gene.

A sensitive technique has been devised by Young and Davis for the purpose of cloning genes using antibodies as probe.[9] Thus, provided one can raise antibody to a particular protein, one can isolate the gene coding for that protein using the Young and Davis gt11 bacterial expression vector. gt11 permits the insertion of foreign DNA (in this case cDNA) into the - β galactosidase gene, lacZ, that has been incorporated in the genone. If - β galactosidase synthesis is induced by IPTG during infection, sufficient quantities of hybrid protein (coded by lacZ and the inserted foreign DNA) are produced for detection by antibodies. It is therefore possible to screen a gt11cDNA library using antibody as probe in much the same way as one would screen a cDNA library using nucleic acid as probe.

The mammalian DNA methyltransferase enzyme has been partially purified from several sources; their estimated molecular weights range from 18.5 - 24 kd. It turns out that the mammalian DNA methyltransferases have properties that are strikingly similar to the purified 18 kd E. coli enzyme. With this in mind we purified the bacterial enzyme to homogeneity and raised polyclonal antibodies to the purified protein in two rabbits. As expected, the antiserum from both rabbits specifically recognized the bacterial enzyme. Using the Western blot technique we showed that antiserum from one rabbit (but not the other) specifically recognized a single protein band among human cell extract proteins that were separated by SDS polyacrylamide gel electrophoresis. We were initially uncertain whether the crossreactive material (CRM) was indeed the human DNA methyltransferase because it had a molecular weight of 30 kd, substantially higher than expected. However, we found, in vitro, that methyl groups were transferred from a DNA substrate containing mainly 06MeG to a single protein species in human cell extracts that exactly co-migrated with the CRM on SDS polyacrylamide gels. This was true for five different human cell lines, namely HeLa CCL2, HT29, HepG2, HeLa S3 and Raji lymphoblastoid. Moreover, the CRM is significantly enhanced in adapted human cells. We therefore feel that the serum probably recognizes the human DNA methyltransferase enzyme. However, we observed occasionally that methyl groups were also transferred to a 20 kd protein, i.e., that our human cell extracts showed two peaks of methyltransferase activity. We subsequently found that serum taken from the rabbit at a later time recognized an additional protein band of 20 kd. The 20 kd CRM is also significantly enhanced in adapted human cells. This 20 kd DNA methyltransferase species could either be a product of processing of the 30 kd enzyme or a completely separate DNA methyltransferase. The processing of a DNA methyltransferase enzyme is precedented in E. coli.[4] The bacterial enzyme was initially purified as a homogeneous 18 kd protein.[10] However, a 37 kd precursor form of this enzyme was recently revealed by probing Western blots of E. coli cell extract with polyclonal antibodies raised to the pure 18 kd protein. It is not yet known if the cleavage of the 37 kd protein to yield the enzymatically active 18 kd species occurs in vivo or if this is merely a purification artifact.

We have used our human methyltransferase (HMT) antiserum to probe a gtll human cDNA library that was constructed from a human hepatoma cell line, namely HepG2. (This library was a kind gift from Dr. Harvey Lodish). We have screened 2 x 10^6 members of the library and have so far purified two gtll recombinant bacteriophage that produce a β-galactosidase fusion protein that is specifically recognized by the HMT antiserum. Recognition is entirely specific for the HMT antiserum because the preimmune serum and the antiserum which recognized only the E. coli methyltransferase, but not the human enzyme(s), did not recognize the fusion proteins made by these two gtll recombinants.

In summary, we have cloned a 2.5 kilobase (kb) cDNA and a 1.8 kb cDNA that each direct the synthesis of protein that is specifically recognized by our HMT antiserum. We have used Southern blot analysis to show that these two cDNAs are different from each other and Northern blot analysis to show that they were reverse transcribed from different mRNA species. The 2.5 kb cDNA hybridized to a 2.6 kb message and the 1.8 kb cDNA hybridized to a 3.2 kb message. That one or both of these cDNAs code for a DNA methyltransferase will only be confirmed when an active protein product is produced from the cDNA sequences.

REFERENCES

1. G.C. Walker, Mutagenesis and inducible response to deoxyribonucleic acid damage in Escherichia coli, Microbiol. Rev. 48:60 (1984).
2. L. Samson and J. Cairns, A new pathway for DNA repair in Escherichia coli, Nature 367:281 (1977).
3. T. Lindahl, DNA repair enzymes, Annu. Rev. Biochem. 51:61 (1982).
4. T. Lindahl, This volume, p. 335.
5. L. Samson and J.L. Schwartz, Evidence for an adaptive DNA repair pathway in CHO and human skin fibroblast cell lines, Nature 287:861 (1980).
6. B. Kaina, Enchanced survival and reduced mutation and aberration

frequencies induced in V79 Chinese hamster cells pre-exposed to low levels of methylating agents, Mutation Res. 93:195 (1982).
7. F. Laval and J. Laval, Adaptive response in mammalian cells: cross-reactivity of different pretreatments on cytotoxicity as contrasted to mutagenicity, PNAS 81:1062 (1984).
8. E.A. Waldstein, E.H. Cao, M.E. Muller, E.P. Cronkite, and R.B. Setlow, Adaptive resynthesis of O^6-methylguanine-accepting protein can explain the differences between mammalian cells proficient and deficient in methyl excision repair, PNAS 79:4786 (1982).
9. R. Young, and R.W. Davis, Efficient isolation of genes by using antibody as probes, PNAS 80:1194 (1983).
10. B. Demple, A. Jacobsson, M. Olsson, P. Robins, and T. Lindahl, Repair of alkylated DNA in Escherichia coli, physical properties of O^6-methylguanine-DNA methyltransferase, J. Biol. Chem. 257:13776 (1982).

DNA GLYCOSYLASES IN DNA REPAIR

Tomas Lindahl

Imperial Cancer Research Fund
Clare Hall Laboratories,
South Mimms, Herts EN6 3LD, U.K.

INTRODUCTION

The excision of potentially mutagenic and lethal lesions from DNA proceeds by one of two different routes. DNA damage which results in a major distortion of the DNA double-helix is generally recognized by a high-molecular weight nuclease that cuts two phosphodiester bonds in the altered strand, one on each side of the lesion. While the biochemical properties of the complex E. coli enzyme catalyzing this reaction are being elucidated rapidly,[1-3] the apparently analogous eukaryotic enzyme activity has not yet been demonstrated in a cell-free extract. The alternative pathway of excision-repair is initiated by one of several DNA glycosylases, which catalyze cleavage of the base-sugar bond of an altered nucleotide residue to release the damaged base component in free form and leave a repairable apurinic or apyrimidinic site in the DNA. The properties of DNA glycosylases have been reviewed.[4-7] Some recent observations on the multiplicity and substrate specificity of these enzymes are discussed here; references to previous work may be found in the review cited.

NUMBER OF DNA GLYCOSYLASES

Following the discovery of the uracil-DNA glycosylase in 1974, several additional DNA glycosylases were found in the late 1970's. Separate enzymes which catalyze the release of 3-methyladenine, hypoxanthine, formamidopyrimidine (an imidazole ring-opened purine), and thymine glycol from damaged DNA were characterized. These enzymes are widely or universally distributed, and they have been isolated from E. coli as well as from mammalian cells*. In addition, a cyclobutane pyrimidine dimer-DNA glycosylase has been found in Micrococcus luteus and in T4-infected E. coli, but not in other cells.[7] These observations, and the narrow substrate specificities of the enzymes

*Drosophila melanogaster cell extracts differ from those of many other sources in not containig detectable uracil-DNA glycosylase activity.[7] However, oxidatively damaged thymine residues are released by a Drosophila DNA glycosylase apparently analogous to the thymine glycol-DNA glycosylase of E. coli and mammalian cells (L. Breimer, unpublished data), so these insects are not deficient in all DNA glycosylases.

first discovered, indicated initially that many different DNA glycosylases might occur. However, no new glycosylases have been found during the last four or five years, so it now seems unlikely that many more such enzymes exist. (A recent preliminary report on an apparent hydroxymethyluracil-DNA glycosylase activity does not distinguish it clearly from uracil-DNA glycosylase). Instead, an unexpectedly broad substrate specificity of some of the more recently discovered DNA glycosylases has emerged, making it possible for some of these enzymes to remove a wide variety of lesions introduced by methylating or oxidizing agents. The two main examples are discussed below.

3-METHYLADENINE-DNA GLYCOSYLASES

The most abundant, and therefore potentially most important, cell-killing lesion introduced in DNA by simple methylating agents is 3-methyladenine. The presence of this modified base residue in a DNA template has been shown to block efficiently DNA polymerase I *in vitro*.[8,9] 3-Methyladenine is excised rapidly from DNA as a free base in *E. coli* and in mammalian cells. The *E. coli* enzyme mainly responsible for this reaction is the product of the *tag* gene, 3-methyladenine-DNA glycosylase I, a small constitutively expressed enzyme.[10] The *tag* gene has been cloned and sequenced; it encodes a cysteine-rich enzyme of molecule weight 20,000[11] (E. Seeberg, pers. commun.). The 3-methyladenine-DNA glycosylase I does not act on other methylated residues in DNA, including lesions as similar as 3-methylguanine and 7-methyladenine, but it excises at a slow rate 3-ethyladenine from ethylated DNA.[12,13] Recent genetic evidence from *tag*$^-$ combined with other mutations also strongly implicates the enzyme in the removal of a cell-killing but non-mutagenic lesion, i.e., 3-methyladenine, from DNA.

E. coli has a second DNA glycosylase that removes 3-methyladenine from DNA, the product of the *alkA* gene.[14] This enzyme of molecule weight 31,000 is inducible and it accounts for the increased resistance to the cell-killing activity of methylating compounds during the adaptive response to alkylating agents. The *alkA* gene has recently been cloned and sequenced.[14] The AlkA enzyme has a broader substrate specificity than the Tag enzyme in that it also removes 3-methylguanine from alkylated DNA.[10] In addition, the AlkA glycosylase has the surprising ability to excise efficiently two minor pyrimidine lesions, O^2-methylcytosine and O^2-methylthymine.[15] On the other hand, the enzyme does not remove the major alkylation product 7-methylguanine at a significant rate (<0.2% of that seen with 3-methyladenine), and 7-methylguanine does not appear to be actively repaired in *E. coli*. Thus, the *alkA* gene product, 3-methyladenine-DNA glycosylase II, can remove all alkylation lesions in which the methyl group protrudes into the minor groove of the double helix but not those with methyl groups located in the major groove.[15] This would provide one possible mechanism by which a single DNA glycosylase can remove several related but structurally dissimilar types of damage, such as N -alkylated purines and O -alkylated pyrimidines. *E. coli tag alkA* double mutants cannot excise 3-methyladenine from DNA after treatment with alkylating agents, and are very sensitive to killing by such agents.[16]

Mammalian cells apparently differ from *E. coli* in only having a single 3-methyladenine-DNA glycosylase, which is constitutively expressed.[17,18] enzyme of molecule weight 25,000 has a broader substrate specificity than the constitutively expressed *E. coli* enzyme. On the other hand, it does not recognize exactly the same lesions as the inducible *E. coli* DNA glycosylase. Thus, while all three enzymes remove 3-methyladenine efficiently, the mammalian enzyme also excises 3-methylguanine and 7-methylguanine from DNA; the latter lesion is not removed by the bacterial enzymes. Moreover, the mammalian 3-methyladenine-DNA glycosylase differs from the *E. coli* AlkA enzyme in that it does not excise 0 -methylpyrimidines.[19] It would appear that the mammalian enzyme employs a different strategy from the bacterial ones and removes purine residues in which a positive charge has been introduced by the alkylation event (3-methyladenine, 3-methylguanine, 7-methylguanine), while uncharged residues such as 1-methyladenine and O^6-methylguanine are not excised. The primary physiological role of the two *E. coli* enzymes as well as the mammalian enzyme is presumably the rapid removal of the important killing lesion 3-methyladenine; the additional removal of certain minor lesions could provide a slight advantage, and different mechanisms may have evolved secondarily to achieve the latter goal. Thus, the constitutive *E. coli* 3-methyladenine-DNA glycosylase does not have to be concerned with minor alkylation lesions, because even uninduced cells contain small amounts of the *alkA* gene product to remove such damage. Continuous or repeated exposure to methylating agents in an unfavorable physiological niche could lead to a relative accumulation of the minor alkylation lesions in the bacteria, but this would be counteracted by induction of the AlkA enzyme. Mammalian cells have a single, versatile enzyme that removes both 3-methyladenine and 3-methylguanine, and in this case it may not be necessary to provide for the repair of 0 -methylpyrimidines, which are very minor lesions. The slow but active repair of 7-methylguanine is presumably of little physiological relevance, since this base is neither a mutagenic nor a cell-killing lesion - the excision of 7-methylguanine may be regarded as a trivial consequence of the specific action mechanism employed by the mammalian enzyme to remove the more important 3-methylpurine residues.

The *E. coli* 3-methyladenine-DNA glycosylase I is strongly product-inhibited by free 3-methyladenine. Surprisingly, apparent non-competitive inhibition is observed; this would result if the free base prevents the enzyme from acting by a processive mechanism. The inducible *E. coli* enzyme and the mammalian glycosylase are not similarly inhibited. It is now of considerable interest to search for a specific inhibitor of the mammalian 3-methyladenine-DNA glycosylase, because an inhibitor would be expected to greatly sensitize mammalian cells to simple alkylating agents such as methyl methanesulfonate and dimethyl sulfate. These agents act by an S_N2 mechanism and consequently introduce as much 3-methyladenine but many fewer mutagenic (and potentially carcinogenic) O-alkylated lesions than nitrosoureas. They are now relatively inefficient cell-killing drugs because of the rapid excision-repair of 3-methyladenine in DNA.

THYMINE GLYCOL-DNA GLYCOSYLASE

The multitude of similar but non-identical base alterations caused by ionizing radiation makes it seem unlikely that enzymes of the narrow substrate specificity exhibited by uracil-DNA glycosylase or *E. coli* 3-methyladenine-DNA glycosylase I would be involved in their repair. The major type of purine lesion caused by such irradiation is the opening of the imidazole ring of adenine and guanine. A single enzyme, the formamidopyrimidine-DNA glycosylase, has been shown to excise both imidazole ring-opened adenine residues from γ-irradiated DNA and a secondary alkylation product, imidazole ring-opened 7-methylguanine residues, from alkylated DNA.[20] The latter may be regarded as a model compound and it is likely that the enzyme also removes ring-opened, unmethylated guanine residues from irradiated DNA, although for technical reasons a direct experimental demonstration has not been attempted.

Oxidative damage and ionizing radiation convert DNA pyrimidine residues to relatively labile hydroperoxides, which spontaneously decompose into more stable derivatives. In the case of thymine, the ring-saturated *cis* and *trans* forms of 6-hydro-5,6-dihydroxythymine (thymine glycol) and 5,6-dihydro-6-hydroxythymine, ring-opened derivatives such as methyltartronylurea, and hydroxymethylhydantoin obtained by ring contraction are reaction products in double-stranded DNA.[21,22] An E. coli thymine glycol-DNA glycosylase[23] has been shown to excise all the pyrimidine derivatives mentioned, as well as fragments of thymine residues such as unsubstituted urea.[22,24] The enzyme may generally recognize a local distortion caused by the loss of the planar pyrimidine ring structure, since non-conventional pyrimidine derivatives that have retained the 5,6 double bond (uracil O^2-methylthymine) are not removed, although they can be excised by other DNA glycosylases. Mammalian cells contain a thymine glycol-DNA glycosylase with very similar properties, although it is relatively more active on ring-fragmented base residues than on ring-saturated ones.[25,26]

The thymine glycol-DNA glycosylases of E. coli and mammalian cells differ from previously characterized enzymes such as the uracil- or 3-methyladenine-DNA glycosylases in that they have an associated AP endonuclease activity. However, they share this property with the pyrimidine dimer-DNA glycosylase of phage T4 and M. luteus. The thymine glycol-DNA glycosylase was first described as a nuclease and named E. coli endonuclease III, although the cleavage of the phosphodiester bond at the damaged site is a secondary reaction following the removal of the damaged base. The E. coli gene (nth) encoding the enzyme has recently been mapped and cloned, and enzyme-defective mutants isolated.[27] Surprisingly, they are not very sensitive to oxidative damage or ionizing radiaton.

CONCLUSIONS

Several different types of base damage in DNA may be excised by currently unknown DNA glycosylases. For example, it is unclear how oxidatively damaged cytosine residues are removed, because unusual base derivatives occur due to ring contraction. Further, the repair of whole groups of base lesions, such as N-oxides or naphtylamine adducts, has not been investigated. The limiting factor for experimental work in these cases is that the chemical characterization of the base derivatives is incomplete, and they are consequently not available as reference compounds for enzymological studies. The physiological roles of several of the known DNA glycosylases are being defined by isolation of enzyme-deficient mutants, and by detailed characterization of the substrate specificities of the enzymes. In this regard, the least well understood of the DNA glycosylases is the hypoxanthine-DNA glycosylase.[28,29] No genetic information is available on this enzyme, and while the apparently narrow substrate specificity indicates that the enzyme could serve to remove deaminated adenine residues from DNA, the very low activity in E. coli extracts implies that the enzyme might act more efficiently on a different, so far unrecognized, lesion. Interestingly, mammalian cell extracts contain markedly higher levels of hypoxanthine-DNA glycosylase activity than bacteria. The genes for most of the known DNA glycosylases of E. coli have already been cloned, and in some cases sequenced, and with time these studies will no doubt be extended to the analogous enzymes of mammalian cells.

REFERENCES

1. G.B. Sancar, A. Sancar and W.D. Rupp. Sequences of the E. coli uvrC gene and protein. Nucl. Acids Res. 12:4593 (1984).
2. K. Kumura, M. Sekiguchi, A-L. Steinum and E. Seeberg. Stimulation of the UvrABC enzyme-catalyzed repair reactions by the UvrD protein (DNA helicase II) Nucl. Acids Res. 13:1483 (1985).
3. P.R. Caron, S.R. Kushner and L. Grossman. Involvement of helicase II (urvD) and DNA polymerase I in the excision process mediated by the uvrABC protein complex. Proc. Natl. Acad. Sci. USA, in press.

4. T. Lindahl. DNA glycosylases, endonucleases for apurinic/apyrimidinic sites, and base excision-repair, in Progr. Nucl. Acid Res. Mol. Biol. 20:135 (1979).
5. B.K. Duncan. DNA glycosylases, in The Enzymes Vol. XIV. Academic Press Inc., N.Y. (1981).
6. T. Lindahl. DNA repair enzymes. Ann. Rev. Biochem. 51:61 (1982).
7. E.C. Friedberg. DNA repair. W.H. Freeman & Co., N.Y. (1985).
8. S. Boiteux, O. Huisman and J. Laval. The EMBO J. 3:2569 (1984).
9. K. Larson, J. Sahm, R. Shenkar and B. Strauss. Methylation-induced blocks to in vitro DNA replication. Mut. Res., in press (1985).
10. P. Karran, T. Hjelmgren and T. Lindahl. Induction of a DNA glycosylase for N-methylated purines is part of the adaptive response to alkylating agents. Nature 296:770 (1982).
11. N.D. Clarke, M. Kvaal and E. Seeberg. Cloning of Escherichia coli genes encoding 3-methyladenine DNA glycosylases I and II. Mol. Gen. Genet. 197:368 (1984).
12. S. Riazuddin and T. Lindahl. Properties of 3-methyladenine-DNA glycosylase from Escherichia coli. Biochemistry 17:2110 (1978).
13. L. Thomas, C.-H. Yang, and D.A. Goldthwait. Two DNA glycosylases in Escherichia coli which release primarily 3-methyladenine. Biochemistry 21:1162 (1982).
14. Y. Nakabeppu, H. Kondo and M. Sekiguchi. Cloning and characterization of the alkA gene of Escherichia coli that encodes 3-methyladenine DNA glycosylase II. J. Biol. Chem. 259:13723 (1984).
15. T.V. McCarthy. P. Karran and T. Lindahl. Inducible repair of O-alkylated DNA pyrimidine Escherichia coli. EMBO J. 3:545 (1984).
16. G. Evensen and E. Seeberg. Adaptation to alkylation resistance involves the induction of a DNA glycosylase. Nature 296:773 (1982).
17. P.E. Gallagher and T.P. Brent. Partial purification and characterization of 3-methyladenine-DNA glycosylase from human placenta. Biochemistry 21:6404 (1982).
18. R. Male, D.E. Helland and K. Kleppe. Purification and characterization of 3-methyladenine-DNA glycosylase from calf thymus. J. Biol. Chem. 260:1623 (1985).
19. J.A. Hall and P. Karran. Methylated pyrimidine lesions in the DNA of mammalian cells, in Repair of DNA lesions introduced by N-nitroso compounds, eds. H. Krokan and R. Myrnes. Oxford University Press, in press.
20. L.H. Breimer. Enzymatic excision from γ-irradiated polydeoxyribonucleotides of adenine residues whose imidazole rings have been ruptured. Nucl. Acids Res. 12:6359 (1984).
21. R. Teoule, C. Bert and A. Bonicel. Thymine fragment damage retained in the DNA polynucleotide chain after gamma irradiation in aerated solutions. II. Rad. Res 72:190 (1977).
22. L.H. Breimer and T. Lindahl. DNA glycosylase activities for thymine residues damaged by ring saturation, fragmentation, or ring contraction are functions of endonuclease III in Escherichia coli. J. Biol. Chem. 259:5543 (1984); Thymine lesions produced by ionizing radiation in double-stranded DNA. Biochemistry, in press.
23. B. Demple and S. Linn. DNA N-glycosylases and UV repair. Nature 287:203 (1980).
24. H.L. Katcher and S.S. Wallace. Characterization of the Escherichia coli X-ray endonuclease, endonuclease III. Biochemistry 22:4071 (1983).
25. L.H. Breimer. Urea-DNA glycosylase in mammalian cells. Biochemistry 22:4192 (1983).
26. M.C. Hollstein, P. Brooks, S. Linn and B.N. Ames. Hydroxymethyluracil DNA glycosylase in mammalian cells. Proc. Natl. Acad. Sci. USA 81:4003 (1984).
27. R.P. Cunningham and B. Weiss. Endonuclease III (nth) mutants of Escherichia coli. Proc. Natl. Acad. Sci. USA 82:474 (1985).

28. P. Karran and T. Lindahl. Enzymatic excision of free hypoxanthine from polydeoxynucleotides and DNA containing deoxyinosine monophosphate residues. J. Biol. Chem. 253:5877 (1978); P. Karran and T. Lindahl. Hypoxanthine in deoxyribonucleic acid: Generation by heat-induced hydrolysis of adenine residues and release in free form by a deoxyribonucleic acid glycosylase from calf thymus. Biochemistry 19:6005 (1980).

29. B. Myrnes, P-H. Guddal and H. Krokan. Metabolism of dITP in HeLa cell extracts, incorporation into DNA by isolated nuclei and release of hypoxanthine from DNA by a hypoxanthine-DNA glycosylase activity. Nucl. Acids Res. 10:3693 (1982).

APURINIC SITES AS COMMON INTERMEDIATES IN MUTAGENESIS

[1]Lawrence A. Loeb, Bradley D. Preston, Elizabeth T. Snow, and [2]Roeland M. Schaaper

[1]The Joseph Gottstein Memorial Cancer Research Laboratory University of Washington, Seattle, WA 98195 and [2]NIEHS Research Triangle Park, NC 27709

It is well known that the high fidelity of DNA replication in eucaryotes results in part from the accuracy by which DNA polymerases copy the base sequences in DNA. It is also well known that the reactions of many mutagens and carcinogens with DNA result in the formation of bulky adducts. Experimental evidence indicates that these bulky adducts block DNA synthesis _in vitro_ and _in vivo_.[1] A paradox in mutagenesis is how DNA polymerases, with their stringent requirement for correct base pairing, can accommodate bulky adducts to yield mutations in nascent DNA. One possibility is that the adducts responsible for mutagenesis are labile and form non-bulky intermediary products that are mutagenic. We have proposed that apurinic sites are mutagenic intermediates[2] and we have subsequently analyzed their mutagenic potential _in vivo_ and _in vitro_.

Apurinic sites result from the hydrolysis of the glycosylic bond between a purine base and the deoxyribose sugar leaving the DNA phosphodiester backbone intact. A landmark in quantitating the spontaneous rate of depurination of DNA was the study of Lindahl and Nyberg,[3] which yielded an _in vitro_ rate of approximately 4×10^{-11} depurinations/nucleotide residue/sec. This rate translates to about 10,000 depurinations per mammalian cell per day. In addition to spontaneous hydrolysis, depurination also results from labilization of the glycosylic bond as a result of chemical modification of purine bases[4] as well as enzymatic removal of altered bases by specific glycosylases.[5] If apurinic sites are present in DNA, bypass of these sites during DNA replication is likely to involve non-template directed incorporation of nucleotides by DNA polymerases. However, the concept of apurinic sites as promutagenic lesions was not supported by initial experiments. Only recently have studies documented the mutagenic potential of apurinic sites and pointed to the possibility that depurination is a major mechanism for spontaneous and induced mutagenesis.[6]

Effects of Apurinic Sites on DNA Synthesis In Vitro

A major effect of apurinic sites both in synthetic polynucleotide templates[2] and natural DNAs[7] is to inhibit chain elongation by DNA polymerases _in vitro_. This blockage of synthesis was most convincingly demonstrated by measuring the size distribution of nascent DNA synthesized on templates containing apurinic sites.[8] Nevertheless, bypass of apurinic sites must also occur, since when DNA polymerases copy synthetic polynucleotides containing apurinic sites, enhanced incorporation of nucleotides that are noncomplementary to the templates bases is found.[2] With each template studied, the number noncomplementary nucleotides incorporated relative to the number of complementary nucleotides incorporated (error rate) increased in direct proportion to the extent of depurination. Abolishment of enhanced misincorporation by pretreatment of the depurinated template with alkali, at concentrations known to hydrolyze apurinic sites, suggests that the apurinic sites themselves are mutagenic. These initial studies with synthetic polynucleotides demonstrated the potential mutagenicity of apurinic sites; however, the types of misincorporations varied with different polynucleotide templates.[2,9]

A genetic system has been developed to analyze the frequency of misincorporation by DNA polymerase on a natural biologically active DNA template by measuring the frequency of revertants produced as a result of copying a mutant DNA containing a single base substitution.[10] With _am3_ ØX174 DNA as a template, all revertants result from misincorporations opposite deoxyadenosine at position 587.[11] Treatment of ØX174 DNA with acid and heat causes depurination throughout the template, including the adenine residue at this position. Therefore, mutagenesis as a result of depurination at this position can be measured as an increase in reversion at the _am3_ locus.

The copying of depurinated _am3_ ØX174 single stranded DNA with purified DNA polymerases _in vitro_ is mutagenic.[12] The frequency of revertants correlates positively with the ability of different DNA polymerases to copy past apurinic sites. In general, procaryotic DNA polymerases exhibit high fidelity and copy past apurinic sites only with difficulty. In contrast, purified eucaryotic DNA polymerases are error-prone and readily copy past apurinic sites. DNA sequence analysis of DNA synthesized on ØX templates containing apurinic sites yielded an unexpected result. Instead of random misincorporations, as one might expect in the absence of a template base, substitutions opposite presumed apurinic sites were specific; dAMP was most frequent, followed by dGMP. A similar preferential incorporation of dAMP as a terminal nucleotide was observed during the _in vitro_ replication of ØX174 or M13 templates containing apurinic sites.[8,13] The mechanism for dAMP incorporation opposite these sites remains to be determined. The suggestion that dAMP incorporation is determined by the inherent binding specificity of DNA polymerase in the absence of a template[8] is not adequately supported by the only published measurements for binding; the order of deoxynucleoside triphosphate binding to _E. coli_ DNA polymerase I is G>A>T>C.[14]

Mutagenic Potential of Apurinic Sites In Vivo

Transfection assays using single stranded phage DNA containing graded amounts of apurinic sites have been used to measure the _in vivo_ mutagenic potential of apurinic sites. Transfection of upcopied, closed-circular, single-stranded _am3_ ØX174 DNA containing varying numbers of apurinic sites into wild type _E. coli_ spheroplasts failed to produce increased titers of revertant phage.[15] However, a large increase in mutagenicity was observed when the same depurinated DNA was introduced into spheroplasts obtained from _E. coli_ that were previously exposed to UV irradiation. Pretreatment of the depurinated template with purified HeLa apurinic endonuclease abolishes mutagenicity, indicating that apurinic sites are the responsible promutagenic lesions. The SOS response is required for the expression of mutations induced by apurinic sites. The mutagenic response is not observed in UV irradiated spheroplasts derived from _E. coli_ containing defective _rec_A, _rec_F or _umu_C genes.[16] Mutagenesis is also observed in wild type _E. coli_ pretreated with nalidixic acid or UV irradiation.[16] The types of mutations produced in UV-treated _E. coli_ were determined by sequencing the DNA from the amber revertants.[15] Similar to the results produced in vitro, the predominant substitution occurring opposite presumed apurinic sites is dAMP. Kunkel[17] utilized a forward mutation assay to measure the entire spectrum of mutations produced by depurination of M13mp2 DNA containing the lacZ gene. Since this gene is nonessential, all classes of mutations can be detected. In this system, mutations resulting from depurination are also SOS dependent, and are predominantly single base substitutions with a strong preference for dAMP.

Apurinic sites might also be mutagenic in eucaryotes; transfection of depurinated SV-40 DNA into monkey kidney cells results in increased SV-40 mutagenesis.[18] Since this mutagenesis is not increased by prior treatment of the host cell with UV irradiation, the results suggest that mutagenesis by depurination of DNA from animal viruses does not require the induction of an error prone DNA repair pathway.

Spontaneous Mutagenesis

Little is known about the molecular basis for spontaneous mutagenesis. Two likely sources are misincorporations during DNA replication and damage to DNA by endogenous chemical reactions; although other possibilities, such as fluctuations in nucleotide pools and alteration in DNA polymerases, can be invoked. The possibility that a significant fraction of spontaneous mutations result from replication past cryptic apurinic sites has been raised by the studies of Miller and Low.[19] examination of over 600 independently occurring amber and ochre mutations in the lacI gene in SOS induced bacteria revealed predominantly G:C->T:A and, to a lesser extent, A:T->T:A transversions. Mutagenesis occurs only at certain sites, and the types of substitutions are different from normal replication errors,[19] which appear to be predominatly transitions.[20] The transversions observed in SOS induced cells are likely to have resulted from the insertion of dAMP opposite purines or apurinic sites, since the other pathways leading to these transversions involve pyrimidine-pyrimidine base pairs which are seldom observed.[21] An alternate mechanism, not involving apurinic sites, is that the SOS induction process itself alters the spectrum of misincorporations by the DNA replication apparatus to favor the misincorporation of dAMP opposite purines. This seems less likely when one considers that most mutations caused by UV or chemicals in SOS induced cells are targeted, i.e., occur opposite sites of damage.[20]

Chemically-induced Mutagenesis

A large number of chemical agents damage DNA to produce bulky adducts[23] that block chain elongation during DNA replication.[1] In order for mutagenesis to occur, these presumably lethal blocks to DNA replication must be removed and mutations must accumulate in the nascent DNA. Many of the adducts produced by mutagens involve covalent binding to the N-3 and N-7 positions of purines, causing labilization of the glycosylic bond.[4] For N-acetoxy-N-acetyl-2-aminofluorene and benzo[a]pyrene diolepoxide, labile adducts comprise a major fraction (up to 40%) of the total DNA adducts.[23,24] On this basis, it has been postulated that apurinic sites might be key intermediates in mutagenesis by some chemical agents.[2] This concept is supported by studies in the lacI gene of E. coli which have revealed a category of bulky mutagens (e.g., benzo[a]pyrene diolexpoxide, cyclopenta[c,d]pyrene epoxide and aflatoxin B1) that are dependent on the SOS system for mutagenesis and which induce primarily targeted G:C->T:A transversions.[25,26] Studies on single-stranded phage DNA provide more direct evidence that apurinic sites may play an important role in mutagenesis by some chemicals.[27] Modification of ØX174 DNA in vitro by B-propriolactone, N-acetoxy-N-acetyl-2-aminofluorene or benzo[a]pyrene diolepoxide induces mutations after transfection of the modified DNA into E.

coli spheroplasts. Mutagenicity of these chemicals is also SOS dependent. In the case of B-propiolactone, mutagenesis is increased by heating the modified DNA and can be abolished by incubation with an apurinic endonuclease. These data provide evidence that apurinic sites formed from labile adducts can contribute to mutagenesis.

In addition to bulky adducts, other base modifications may enhance depurination. For example, Cu^{2+} interacts preferentially with the N-7 position of purines in DNA, and increases the rate of depurination.[29] After heat treatment of ØX am3 DNA in the presence of Cu^{2+}, transfection of the modified DNA into E. coli yielded revertants which contained predominantly dAMP at position 587. Thus, depurination may also play a role in mutagenesis by certain metals.

Possible Mechanism of Chemically Induced Mutagenesis via Apurinic Sites

Based on the data from the replication of depurinated DNA in vitro, and the above studies on the specificity of mutagenesis by chemicals that form bulky DNA adducts on DNA, we have proposed a model to account for both spontaneous and chemically-induced mutagenesis via apurinic sites.[6] In formulating this model, there are two important considerations: 1) Since the number of DNA adducts (potential sites for depurination) resulting from chemical treatment of intact cells is likely to be less than the number of apurinic sites arising from spontaneous depurination, many of the chemical adducts must have special properties that increase mutagenesis above that which can be attributed to spontaneous depurination. 2) Apurinic sites, once formed, are likely to be rapidly hydrolyzed by cellular apurinic endonuclease; thus, a mechanism must exist for the protection of newly formed apurinic sites. The key concepts in this model are that DNA replicating complexes are completely or transiently blocked by bulky adducts. When depurination occurs, they are poised to immediately copy past the apurinic site by incorporating dAMP. Thus, DNA polymerases themselves may serve to protect newly formed apurinic sites from apurinic endonucleases.

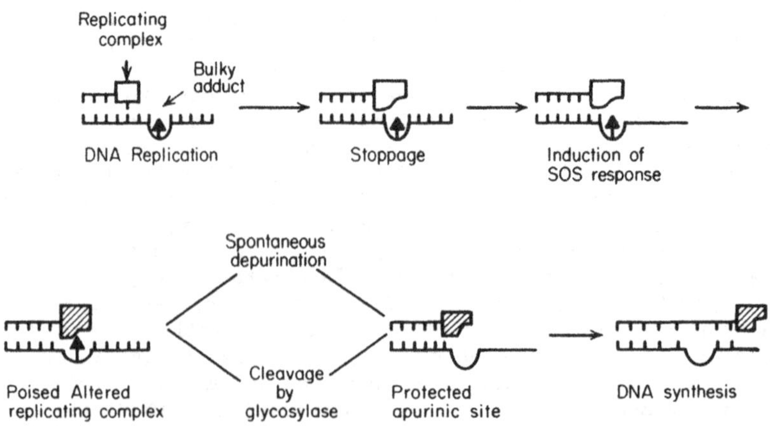

Figure 1. (reprinted from Ref. 6, with permission)

In procaryotes, bulky agents block DNA replication and induce the SOS response, which is believed to involve alterations in the DNA replication complex.[20] Depurination could result from labilization of the glycosylic bond by certain adducts or by enzymatic excision of the altered base. The apurinic site on the template provides a smaller obstruction than does the original bulky adduct. The altered DNA replicating complex is poised at the apurinic site, and may be able to protect and copy past the site prior to cleavage by cellular apurinic endonucleases. Synthesis past apurinic sites results in the insertion of dAMP as a single base substitution. In eucaryotes, bulky base modifications are also likely to block DNA replication. Since eucaryotic DNA polymerases are able to copy past apurinic sites _in vitro_,[12] it is not necessary to postulate the induction of an SOS response involving an altered replicating complex. The normal DNA polymerase would be transiently poised at these potential sites for depurination. In eucaryotes, the apurinic sites in DNA may also be protected from apurinic endonucleases by histones and other chromosomal proteins.

In order to test this model for chemically-induced mutagenesis, it will be necessary to prepare biologically active DNA containing a single apurinic site at a defined nucleotide position. We have recently developed a rapid procedure for the site-specific insertion of nucleotide analogues into ØX174 DNA.[30] This procedure is based on the observation that, under specified conditions _in vitro_, purified DNA polymerases will insert single nucleotides into DNA.[31] Using this procedure, apurinic sites could be site-specifically generated by: 1) insertion of dUMP followed by removal of the uracil with purified uracil glycosylase; 2) insertion of 7-methyl-dGMP followed by treatment with mild heat to hydrolyze the labile glycosylic bond[32] and/or; 3) direct insertion of deoxyribose-5'-triphosphate. Site-specific apurinic DNA would permit a comparison of procaryotic and eucaryotic DNA polymerases for their abilities to protect and replicate past these lesions in the presence of apurinic endonucleases _in vitro_. Moreover, the mutagenic spectrum of apurinic sites could be unambiguously determined in both eucaryotes and procaryotes _in vivo_ following replication of site-specifically depurinated viral and bacteriophage DNAs in the appropriate hosts.

One of the major problems to quantitating accumulated exposure of humans to mutagens has been the diversity of DNA damaging agents. This is dramatically emphasized by the presence of over 300 mutagenic compounds in cigarette smoke. As a result, purified human DNA is likely to contain a spectrum of base adducts. Exposure of purified DNA to elevated temperatures would convert many of these altered bases to apurinic sites. Thus, measurements of potential apurinic sites may be indicative of accumulated exposure to certain chemical mutagens and could be prognostic for susceptibility to malignancy.

It is difficult to measure small numbers of apurinic sites in DNA using chemical degradation and analysis of the size distribution of nonspecific DNA sequences. However, we have recently developed a technique to measure the presence of apurinic sites in specific DNA sequences such as mitochondrial DNA (mtDNA) using purified apurinic endonuclease to cut the DNA at the sites of damage and then analyze the size distribution of the denatured DNA on agarose gels using the Southern procedure. By using mtDNA-specific probes and a single site restriction endonuclease to linearize the mtDNA prior to scission with apurinic endonuclease (thereby avoiding the complications of the D-loop region) it should be possible to quantitate less than one apurinic site in 3×10^4 nucleotides. With the addition of the cloning and sequencing techniques we have developed for mtDNA,[33] we will be able to quantitate the number of apurinic sites necessary to produce mutagenesis of specific segments of the mitochondrial genome. It will also be possible to measure the rate of formation and repair of apurinic sites in both mtDNA and any genomic sequences for which appropriate hybridization probes are available. Other types of damage could also be analysed by using the appropriate damage specific endonuclease, a combination of glycosylase and apurinic endonuclease.

We have also considered several alternative techniques for measuring apurinic sites in eucaryotic DNA. Our biochemical assay, utilizing apurinic endonuclease and DNA polymerase I to specifically incorporate radiolabeled nucleotides into closed circular plasmid DNA, is sensitive, with a limit of detection of approximately one apurinic site per 5 X10^5 nucleotides, and highly specific because it utilizes the inherent specificity of a purified apurinic endonuclease. We have also investigated the production and use of monoclonal antibodies specific for the apurinic sites in DNA. This approach has been successfully used by others to assay specific DNA damage such as O^6methylguanine-, 2-acetyl-aminofluorine-, benzo[a]pyrenediolepoxide-, and cis-diamminedichloroplatinum-DNA adducts.[3,4] These multiple techniques should make it possible to quantitate the contribution of apurinic sites to both spontaneous and induced mutagenesis.

ACKNOWLEDGEMENTS

This work was supported by the following Grants: CA 24845 and AG 00057 from National Institutes of Health; R809623010 from Environmental Protection Agency.

REFERENCES

1. P. Moore and B.S. Strauss. Sites of inhibition of in vitro DNA synthesis in carcinogen- and UV-treated ØX174 DNA. Nature, 278:664, (1979).
2. C.W. Shearman and L.A. Loeb. Effects of depurination on the fidelity of DNA synthesis. J. Mol. Biol., (1979).
3. T. Lindahl and B. Nyberg. Rate of depurination of native deoxyribonucleic acid. Biochemistry, 11:3610-3618, (1972).
4. B. Singer and D. Grunberger, Molecular Biology of Mutagens and Carcinogens, Plenum Press, New York, (1983).
5. T. Lindahl, Annu. Rev. Biochem., 51:61-87, (1982).
6. L.A. Loeb. Apurinic sites as mutagenic intermediates. Cell, 40:483-484, (1985).
7. M.L. Lockhart, J.F. Deutsch, I. Yamanra, L.F. Cavalieri and B.H. Rosenberg. Termination of DNA synthesis in vitro at apurinic sites but not at ethyl adducts on the template. Chem. Biol. Interact., 42:85-95, (1982).
8. B. Strauss, S. Rabkin, D. Sagher and P. Moore. The role of DNA polymerase in base substitution mutagenesis and non-instructional templates. Biochimie 64:829-838, (1982).
9. S. Boiteux and J. Laval. Coding properties of poly (deoxycytidylic acid) templates containing uracil or apyrimidinic sites: In vitro modulation of mutagenesis by DNA repair enzymes. Biochemistry, 21:6746-6751, (1982).
10. L.A. Weymouth and L.A. Loeb. Mutagenesis during in vitro DNA synthesis. Proc. Natl. Acad. Sci. USA, 75:1924-1928, (1978).
11. T.A. Kunkel and L.A. Loeb. On the fidelity of DNA replication: The accuracy of Escherichia coli DNA polymerase I in copying natural DNA in vitro. J. Biol. Chem., 255:9961-9966, (1980).
12. T.A. Kunkel, R.M. Schaaper and L.A. Loeb. Depurination-induced infidelity of deoxyribonucleic acid replication proteins in vitro. Biochemistry, 22:2378-2384, (1983).
13. D. Sagher and B. Strauss. Insertion of nucleotides opposite apurinic/apyrimidinic sites in deoxyribonucleic acid during in

vitro synthesis: Uniqueness of adenine nucleotides. Biochemistry, 22:4518,4526, (1983.)
14. P.T. Englund, J.A. Huberman, T.M. Jovin and A. Kornberg. Enzymatic synthesis of deoxyribonucleic acid. XXX. Binding of triphosphates to deoxyribonucleic acid polymerase. J. Biol. Chem., 244:3038-3044, 1969.
15. R.M. Schaaper, T.A. Kunkel and L.A. Loeb. Infidelity of DNA synthesis associated with bypass of apurinic sites. Proc. Natl. Acad. Sci. USA, 80:487-491, 1983.
16. R.M. Schaaper, B.W. Glickman and L.A. Loeb. Mutagenesis resulting from depurination is an SOS process. Mutat. Res., 106:1-9, 1982.
17. T.A. Kunkel. The mutational specificity of depurination. Proc. Natl. Acad. Sci. USA, 81:1494-1498, 1984.
18. Gentil, Margot and Sarasin, Mutat. Res., (in press).
19. J.H. Miller and K.B. Low. Specificity of mutagenesis resulting from the induction of the SOS system in the absence of mutagenic treatment. Cell, 37,675-682, 1984.
20. G.C. Walker. Mutagenesis and inducible responses to deoxyribonucleic acid damage in Escherichia coli. Microbiol. Rev., 48:60-93, 1984.
21. M.D. Topol and J.R. Fresco. Complementary base pairing and the origin of substitution mutations. Nature, 263:285-289, 1976.
22. L.A. Loeb and T.A. Kunkel. Fidelity of DNA synthesis. Annu. Rev. Biochem., 52:429-459, 1982.
23. J.A. Miller and E.C. Miller. Searches for ultimate chemical carcinogens and their reactions with cellular macromolecules. Cancer, 47:2327-2345, 1981.
24. E. Sage and W.A. Haseltine. High-rato of alkali-sensitive lesions to total DNA modification iduced by benzo[a]pyrene diolepoxide. J. Biol. Chem., 259,11098-11102, 1984.
25. E. Eisenstadt, A.J. Warren, J. Porter, D. Atkins and J.H. Miller. Carcinogenic epoxides of benzo[a]pyrene and cyclopenta[cd]pyrene induce base substitutions via specific transversions. Proc. Natl. Acad. Sci. USA, 79:1945-1949, 1982.
26. P.L. Foster, E. Eisenstadt and J.H. Miller. Base substitution mutations induced by metabolically activated aflatoxin B_1. Proc. Natl. Acad. Sci. USA, 80:2695-2698, (1983).
27. R.M. Schaaper, B.W. Glickman and L.A. Loeb. Role of depurination in mutagenesis by chemical carcinogens. Cancer Res., 42:3480-3485.
28. W.G. Tarpley, J.A. Miller and E.C. Miller. Rapid release of carcinogen- guanine adducts from DNA after reaction with N-acetoxy-2-acetylamino- fluorene or N-benzyoyloxy-N-methyl-4-aminoazobenzene. Carcinogenesis 3:81-88, (1982).
29. L.A. Loeb and R.M. Schaaper, unpublished results.
30. B.D. Preston, B. Singer and L.A. Loeb. Site specific mutagenesis by polymerase-catalyzed incorporation of alkyl-deoxyribonucleoside triphosphate (dNTP) adducts into ØX174 DNA. Proc. 76th Ann. Meeting, Amer. Assoc. Cancer Res., 26:96, (1985).
31. R.A. Zakour, E.A. James and L.A. Loeb. Site specific mutagenesis: Insertion of single noncomplementary nucleotides at specified sites by error-directed DNA polymerization. Nucl. Acid Res., 12:6615-6628, (1984).
32. J.H. Fleissbach, J.M. Keith and J.B. Guttenplan. Incorporation of 7-methylguanine into a known position of pBR322: A potential method for the site-specific introduction of apurinic sites. Proc. 76th Ann. Meeting Amer. Assoc. Cancer Res., 26:88, (1985).
33. R.J. Monnat, Jr., C.L. Maxwell and L.A. Loeb. Nucleotide sequence preservation of human leukemic mitochondrial DNA. Cancer Res., 45:1809-1814, (1985).
34. M.C. Poirier. The use of carcinogen-DNA adduct antisera for quantitation and localization of genomic damage in animal models and the human population. Environ. Mutagen., 6:879-887, (1984).

THE REPAIR OF URACIL-CONTAINING DNA

Bernard Weiss and Hiyam H. El-Hajj

Department of Molecular Biology and Genetics
The Johns Hopkins University School of Medicine
Baltimore, MD 21205

For what purpose does DNA contain thymine rather than uracil? This question has been approached by studying the consequences of the misincorporation of large amounts of uracil into DNA in place of thymine. Such studies have been facilitated by the isolation of special bacterial mutants. Let us first consider how uracil residues can appear in DNA.

ORIGINS OF DNA URACIL

Uracil may arise in DNA either from the utilization of dUTP by DNA polymerases or from the chemical alteration of DNA cytosine residues. The latter may occur by the spontaneous hydrolytic deamination of cytosine to uracil,[1] or it may be effected by the action of bisulfite, a common environmental mutagen.[2] In the bacterium Escherichia coli, uracil is frequently misincorporated in place of thymine, perhaps as often as once per every 300 thymines in DNA.[3] This occurs because dUTP is a major metabolic intermediate and because it can be utilized by the DNA polymerases.

Figure 1 outlines the metabolism of dUTP in E. coli. As reviewed by O'Donovan and Neuhard,[5] dUTP is an obligatory intermediate in the biosynthesis of dTMP. An estimated 75% of it arises from the deamination of dCTP, and the remainder from the phosphorylation of dUDP. dUDP arises, as do the other deoxyribonucleotides, from the reduction of the corresponding ribonucleoside diphosphates. The dUTP may be incorporated into DNA by the DNA polymerases of E. coli, which do not distinguish between dUTP and dTTP, but this is usually prevented by the enzyme dUTPase (deoxyuridine triphosphatase). dUTPase is a very active enzyme in crude extracts of E. coli. It specifically hydrolyzes dUTP to pyrophosphate and dUMP, which is the substrate for thymidylate synthetase. dUTPase is therefore not only required for the biosynthesis de novo of thymidylate, but it is the cell's first line of defense against the incorporation of uracil into DNA.

Mutants defective in dUTPase activity were expected to accumulate dUTP and to have a concomitant reduction in the synthesis of dTMP, both of which should contribute to the extensive substitution of uracil for thymine during DNA synthesis. Surprisingly, their chromosomes contained no detectable uracil.[6] An explanation was offered by the discovery of uracil-DNA glycosylase,[7] the existence of which suggested that there is a pathway for the base-excision repair of uracil-containing DNA.

Uracil can appear in DNA in chiefly one of two ways: (a) it can be incorporated in place of thymine, giving rise to an A:U base pair as shown (Fig. 2), or the uracil can arise from the spontaneous deamination of cytosine residues in DNA, producing a mutagenic G:U base pair. In either case, it is removed by uracil-DNA glycosylase, which releases the free base, leaving an apyrimidinic site in the DNA. At this point the phosphodiester backbone of the DNA is still intact. The next steps result in the excision of the base-free sugar. In E. coli there are several AP endonucleases, i.e., DNases that cleave at apurinic and apyrimidinic sites.[7] The major AP endonuclease activity belongs to exonuclease III, a multifunctional enzyme that is also a 3'-->5' exonuclease. It is responsible for about 85% of the AP endonuclease activity measurable in crude extracts. It cleaves 5' to the apyrimidinic site. Unfortunately, it cannot then remove the lesion exonucleolytically because it is a 3'-->5' exonuclease. DNA polymerase I, however, can remove the base free sugar by its 5'-->3' exonucleolytic activity while it repairs the resulting gap via nick translation. Finally, the nick is sealed by DNA ligase. It has been suggested[9] that the exonucleolytic activity of exonuclease III might also play an important role by functioning as an "anti-ligase". By producing a small gap adjacent to the base-free sugar, it might prevent premature sealing of the nick by DNA ligase. The gap might also provide a better priming site for DNA polymerase I.

The evidence for this pathway comes from the analysis of mutants. We can increase uracil misincorporation by mutating the dut gene, which codes for dUTPase. When we do so, we find an excessive fragmentation of nascent DNA that is consistent with very active excision-repair.[4] It was found that nonlethal mutations affecting the polymerase or the ligase became conditionally lethal when combined with a dut mutation,[10] indicating that these enzymes are needed for the repair of uracil-containing DNA.

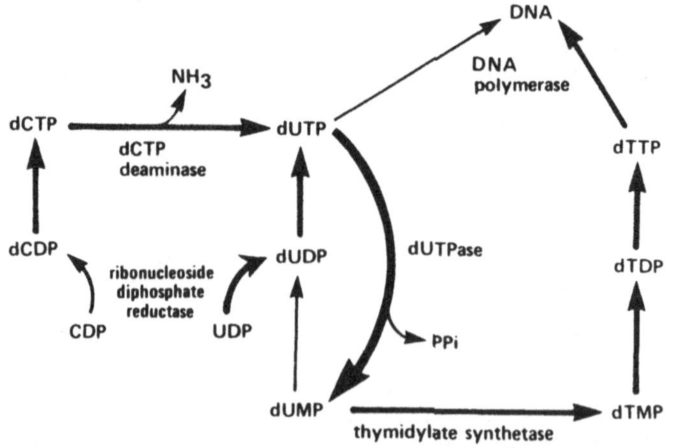

Figure 1. The metabolism of dUTP and the role of dUTPase in the biosynthesis de novo of dTMP in E. coli. (From ref. 4).

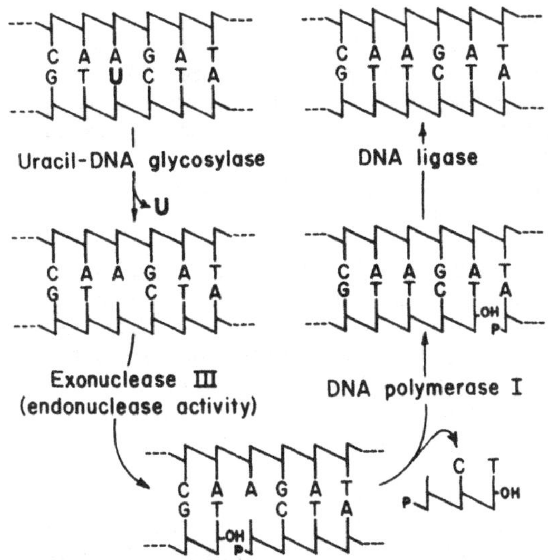

Figure 2. The major pathway in E. coli for the base-excision repair of uracil-containing DNA. (From ref. 8).

Similarly, nonlethal exonuclease III (xth) mutations became conditionally lethal when combined with a dut mutation;[8] moreover, this lethality could be reversed by a mutation in ung, the glycosylase gene. In other words, when the cell does not generate apyrimidinic sites by removing the uracil, it does not need exonuclease III.

There other AP endonucleases in E. coli, which although less abundant than exonuclease III, might also be active in this repair pathway.[7] The residual AP endonuclease activity in exonuclease III mutants is due almost exclusively to endonucleases III and IV. The former is an activity associated with thymine glycol-DNA glycosylase, and it cleaves 3' to the AP site. It has been proposed[11] that if this enzyme acts in conjunction with an AP endonuclease that cleaves 5' to the site (e.g., exonuclease III or endonuclease IV), the lesion might be excised without the need for an exonuclease. Tight insertion mutations have recently been isolated for endonuclease III[12] and for endonuclease IV (S. Goijman, B. Weiss, and R.P. Cunningham, unpublished). Unlike exonuclease III mutations, they are not lethal in combination with a dut mutation. Therefore, they do not appear to be essential for the repair of uracil-containing DNA. Perhaps this is due to their lesser activity. Because these genes have also been cloned, it will be interesting to see now if overproduction of either of these enzymes can compensate for an exonuclease III mutation.

THE CONSEQUENCES OF URACIL IN DNA

Many of the effects of uracil incorporation are secondary to the repair process itself, which breaks the DNA. Thus, cells that carry the dut mutation alone display an excessive fragmentation of newly synthesized DNA.[4] When DNA phages are grown on a dut mutant, their progeny are reduced in number and their DNA's contain many single strand breaks[8]. Agents that lead to cleavage of DNA are known to activate the "SOS" regulon,[13] producing a series of effects similar to those we observe in the dut mutants; the latter, therefore, might also be SOS-mediated. These properties include elevated rates of spontaneous mutation (or weak mutator phenotype), elevated rates of intrachromosomal recombination (or hyper-Rec phenotype), a reduced ability to form stable lysogens of bacteriophage λ, and a reduced ability to undergo cell division (mild filamentation).[6,8] In addition, dut mutants are also thymidine auxotrophs,[8] as expected.

In ung mutants we can observe the effects of the retention of uracil in DNA. These mutants have an elevated rate of spontaneous mutations, specifically of G:C --> A:T transitions.[14] This is the result we expect from cells that are unable to remove from DNA the uracil residues that arise from deaminated DNA cytosine residues. Unlike the uracil that replaces thymine and is therefore properly paired with adenine, that which arises from DNA cytosine is in a mutagenic G:U pair.

The dut ung double mutation results in a relatively stable DNA containing high levels of uracil.[15] The dUTPase deficiency enhances uracil misincorporation while the glycosylase mutation blocks its removal. Such mutants replace over 10% of DNA thymine with uracil, and they appear to be quite healthy. Higher levels (up to 18%) were associated with a prolonged generation time that may have been secondary to the growth conditions, which entailed thymine deprivation.[16] A greater replacement was not obtained, probably because of the "leakiness" of the mutations; the mutant used in these studies, for example, had about 1% to 5% residual dUTPase activity. There was, however, one observable abnormality that was a premonition that a greater replacement of thymine might be deleterious: colonies of dut ung mutants appeared to be constitutive for the expression of the lac operon. This can be explained by the observation that within the lac operator region, there is a critical thymine methyl group required for the binding of the repressor.[17] If only a fraction of these sites are modified within a colony, it will appear to be lac-constitutive. If essential genes could be similarly affected, we shall never be able to obtain a very high level of uracil misincorporation.

As stated above, many of the phenotypic abnormalities of the dut mutants may be secondary to the repair process itself. Extensive repair, because it involves breakage of DNA, is lethal. This is best demonstrated with dut ung double mutants. A bacteriophage grown on such a strain may have had 10 to 20% of its DNA thymine replaced with uracil. These uracil-containing phages can be propagated only on ung mutants; they will not grow efficiently in an ung^+ host.[15] Similarly, gene transfer (by conjugation, transduction, or transformation) from a dut ung mutant is virtually impossible unless the recipient strain carries an ung mutation; otherwise the transferred uracil-containing DNA would be destroyed through the combined action of uracil-DNA glycosylase and AP endonucleases. This is an unusual example of enzymatic restriction.

THE dut OPERON

The dut region of the E. coli has been sequenced.[18] It appears to contain an operon consisting of two genes; the first is dut and the second is a gene for a 23-kDa polypeptide of unknown function. Deletions and large insertions into the dut gene prevent the expression of the 23-kDa polypeptide, confirming that it is co-transcribed with dut (ref. 19 and E. Spitzer and B. Weiss, unpublished). The existence of an operon raises the suspicion that it must be under some type of control, but if so, the signals are not well understood. No more than about a 2-fold variation in dUTPase activity has been demonstrated in our lab under a variety of conditions, including thymine starvation. The 23-kDa protein also remains a mystery. Recently, we constructed an insertion mutation in a plasmid copy of the gene and then transferred it to the chromosome. The mutant has no apparent defect; it is viable and displays no auxotrophy.

THE dut GENE IS ESSENTIAL

The dut mutations were produced by mutagenesis with nitrosoguanidine, an agent that primarily produces base substitution mutations that result in altered protein products. A strain carrying the dut-1 allele, the tightest dut mutation, had about 5% of normal dUTPase activity at 25°C and 0.8% at 42°C. The mutation behaves as if it is temperature-sensitive. Thus, the dut-1 mutation specifies a 60%-reduced viability at 42°C, a temperature-sensitive thymidine auxotrophy, and in the presence of an xth deletion, a temperature-sensitive conditional lethality.[8] The dut-1 mutation is therefore certainly leaky at 25°C. Perhaps its residual activity at 42°C may also be significant. Would a tighter mutation be lethal? If not, it might be used in conjunction with an ung mutation to study the consequences of obtaining even higher levels of uracil in chromosomal DNA. Tight (insertion) mutations for the ung gene, isolated by Bruce Duncan, already exist. In our earlier attempts, using conventional genetic manipulations *in vivo*, we were unable to obtain dut mutations by transposon insertion or by deletions extending from nearby genes. Our failure to obtain deletion mutants, however, might have been due to the nature of the neighboring genes. On one side of the dut operon is an essential gene that affects DNA synthesis;[20] on the other side is the pyrE (orotidylate pyrophosphorylase) gene, mutations in which are incompatible with dut mutations for unknown reasons.[6]

Recently, we constructed a dut insertion mutation by cloning a DNA segment into the coding sequence for dUTPase on a plasmid carrying the *E. coli* dut gene. The cloned segment specifies chloramphenicol resistance, thus providing a convenient marker for following the mutation during gene transfer and for detecting it even in the presence of a functional dut gene. The insertion mutation was then transferred to the chromosome via genetic recombinaion. Subsequently, we transferred the gene to other strains via generalized transduction with bacteriophage P1, selecting for chloramphenicol resistance, and always propagating our cells in the presence of thymidine. This procedure should result in a recombinational substitution of a functional copy of the dut gene with a mutant copy. We found that the insertion mutation could be transferred only to those cells that carry at least two copies of dut. For example, it could not be transferred to wild type *E. coli* cells, which have only one (chromosomal) copy of the gene, but it could be transduced into cells carrying a second copy of the gene either on a plasmid or on a λ prophage, or in what may be a spontaneous tandem duplication of the chromosomal dut region. A strain carrying both the insertion mutation on the chromosome and the intact gene on a λ prophage could not be cured of the λ by a procedure that cured wild type cells with greater than 50% efficiency. Therefore, the insertion is lethal; it can exist only in cells that also have at least one functional copy of the dut gene.

Why is a tight dut mutation lethal? One possibility was that it causes a high degree of uracil incorporation that leads to breakdown of DNA during repair. We should be able to block the repair by an ung mutation. However, a strain carrying insertions in both dut and ung was still inviable. A second possibility is that there is a limit to the amount of uracil a cell will tolerate in its DNA. On this assumption, we tried to reduce the level of uracil misincorporation with a mutation in the gene that governs the production of most of the dUTP, namely, the dcd (dCTP deaminase) gene. It is known, for example, that dcd mutations can reverse the lethality of dut xth double mutants.[8] Nevertheless, we were unable to introduce the dut insertion into a strain carrying a dcd deletion. We are continuing experiments in this vein.

DNA-URACIL IN OTHER ORGANISMS

As reviewed by others[7,15] most of the enzymes described above, although first discovered in E. coli, were later found in mammalian cells. These include dUTPase, uracil-DNA glycosylase, thymine glycol-DNA glycosylase (endonuclease III), and simple AP endonucleases, resembling endonuclease IV, that cleave 5' to the AP site. An exception is exonuclease III; the only similar bifunctional AP endonuclease-exonuclease that we know of is in Hemophilus influenzae. It has been suggested, however, that its exonuclease activity might merely be a manifestation of a degenerate specificity of the AP endonuclease rather than the function of a separate catalytic domain.[21]

Unlike E. coli, mammalian cells make most of their dUMP by deaminating dCMP. Therefore, dUTP is not a necessary intermediate in the biosynthesis of thymidylate. Nevertheless, they have the capacity to make dUTP and therefore possess dUTPase. Inhibition of dihydrofolate reductase by methotrexate blocks the synthesis of thymidylate, leading to an increase in the dUTP pool; the resulting misincorporation of uracil into DNA might contribute to the cell death produced by such compounds.[22]

SOME GENERAL OBSERVATIONS

A major unresolved question is why E. coli goes out of its way to make dUTP. The most direct pathway for the biosynthesis of dTMP should be dUDP --> dUMP --> dTMP, instead of dCTP --> dUTP --> dTMP. In the latter pathway not only is a high energy phosphate bond and an amino group wasted, but a potentially harmful intermediate, dUTP, is generated. Additional energy will be required for the repair of the DNA into which some of this dUTP will be accidentally incorporated. As discovered in the studies of dut mutants, the turnover of DNA uracil enhances genetic recombination and mutation, which may be desirable for the cell population, and this may be why E. coli makes large amounts of dUTP. Another possibility is that dUTP serves another important role that is yet to be discovered. Further studies with strains that underproduce or that overproduce dUPTase may help to resolve these questions.

Bacteriophage PBS2 of Bacillus subtilis normally contains uracil instead of thymine in its DNA. To synthesize such DNA, it produces inhibitors of its host's dUTPase and uracil-DNA glycosylase.[15] The existence of such an organism argues that there is no natural law prohibiting the replication and transcription of uracil-containing DNA. The findings of Lindahl's laboratory[1,7] offer a probable explanation for the almost universal choice of thymine over uracil as a component of DNA. It was found that DNA-cytosine residues undergo spontaneous hydrolytic deamination to uracil, and that although the reaction is very slow, it can be a significant source of mutation in a large genome. Uracil-DNA glycosylases were probably evolved to deal with these lesions. On the other hand, there was no need to deal with the A:U base pairs generated by uracil misincorporation. Even though the latter are much more common, they are not mutational. A glycosylase that could specifically recognize a G:U base pair would probably not be very efficient because of the transient nature of the unstable base pair. Consequently, there evolved uracil-DNA glycosylases that do not distinguish between uracil in A:U or G:U configurations; they even operate on single-stranded DNA. Such enzymes preclude the existence of uracil as a stable component of DNA.

ACKNOWLEDGEMENTS

Work of the authors' laboratory was supported by research grants from the American Cancer Society (#MV205) and National Cancer Institute (#5P01 CA16519).

REFERENCES

1. T. Lindahl, and B. Nyberg, Heat-induced deamination of cytosine residues in deoxyribonucleic acid, Biochemistry 13:3405-3410 (1974).
2. H. Hayatsu, Bisulfite modification of nucleic acids and their constituents, Prog. Nucl. Acid. Res. 16:75-124 (1976).
3. J. Shlomai and A. Kornberg, Deoxyuridine triphosphatase of Escherichia coli. Purification, properties, and use as a reagent to reduce uracil incorporation into DNA, J. Bacteriol. 253:3305-3312 (1978).
4. B.-K. Tye, P.O. Nyman, I.R. Lehman, S. Hochhauser, and B. Weiss, Transient accumulation of Okazaki fragments as a result of uracil incorporation into nascent DNA, Proc. Natl. Acad. Sci. USA 74:154-157 (1977).
5. G.A. O'Donovan and J. Neuhard, Pyrimidine metabolism in microorganisms, Bacteriol. Rev. 34:278-243 (1970).
6. S.J. Hochhauser and B. Weiss, Escherichia coli mutants deficient in deoxyuridine triphosphatase, J. Bacteriol. 134:157-166 (1978).
7. T. Lindahl, DNA repair enzymes, Ann. Rev. Biochem. 51:61-87 (1982).
8. A.F. Taylor and B. Weiss, Role of exonuclease III in the base-excision repair of uracil-containing DNA, J. Bacteriol. 151:351-357 (1982).
9. C. Goffin and W.G. Verly, T4 DNA ligase can seal a nick in double-stranded DNA limited by a 5'-phosphorylated base-free deoxyribose residue, Nucleic Acids Res. 11:8103-8109 (1983).
10. B.-K. Tye and I.R. Lehman, Excision repair of uracil incorporated in DNA as a result of a defect in dUTPase, J. Mol. Biol. 117:293-306 (1977).
11. H.R. Warner, B.F. Demple, W.A. Deutsch, C.M. Kane, and S. Linn, Apurinic/apyrimidinic endonucleases in repair of pyrimidine dimers and other lesions in DNA, Proc. Natl. Acad. Sci. USA 77:4602-4606 (1980).
12. R.P. Cunningham and B. Weiss, Endonuclease III (nth) mutants of Escherichia coli, Proc. Natl. Acad. Sci. USA 82:474-478 (1985).
13. G.C. Walker, Mutagenesis and inducible responses to deoxyribonucleic acid damage in Escherichia coli, Microbiol. Rev. 48:60-93 (1984).
14. B.K. Duncan and B. Weiss, Specific mutator effects of ung (uracil-DNA glycosylase) mutations in Escherichia coli, J. Bacteriol. 151:750-755 (1982).
15. B.K. Duncan, DNA glycosylases, p. 565-586, in: "The Enzymes," P.D. Boyer, ed., Academic Press, New York, (1981).
16. H.R. Warner, B.K. Duncan, C. Garrett, and J. Neuhard, Synthesis and metabolism of uracil-containing deoxyribonucleic acid in Escherichia coli, J. Bacteriol. 145:687-695. (1981).
17. E.F. Fisher and M.H. Caruthers, Studies on gene control regions XII. The functional significance of a Lac operator constitutive mutation, Nucleic Acids Res. 7:401-416 (1979).
18. L.G. Lundberg, H.-O. Thoresson, O.H. Karlstrom, and P.O. Nyman, Nucleotide sequence of the structural gene for dUTPase of Escherichia coli K-12, The EMBO Journal, 2:967-971 (1983).
19. L.G. Lundberg, O.H. Karlstrom, P.O. Nyman, Isolation and

characterization of the dut gene of Escherichia coli. II. Restriction enzyme mapping and analysis of polypeptide products, Gene 22:127-131 (1983).

20. E.D. Spitzer and B. Weiss, The dna-707 mutation of Escherichia coli, Fed. Proc. 41:1194 (1982).
21. B. Weiss, Exodeoxyribonucleases of Escherichia coli, p. 203-231, in: "The Enzymes," vol. 14, P.D. Boyer, ed., Academic Press, New York (1981).
22. M. Goulian, B. Bleile, and B.Y. Tseng, Methotrexate-induced misincorporation of uracil into DNA, Proc. Natl. Acad. Sci. USA 77:1956-1960 (1980).

METABOLIC CONSEQUENCES OF DNA DAMAGE: THE ROLE OF POLY (ADP-RIBOSE)POLYMERASE AS MEDIATOR OF THE SUICIDE RESPONSE

Nathan A. Berger and Sosamma J. Berger

Hematology/Oncology Division
Departments of Medicine and Biochemistry
Case Western Reserve University, School of Medicine
Cleveland, Ohio 44106

DNA damage such as adducts, strand breaks and cross-links may have direct toxic or mutagenic effects. DNA damage may also initiate metabolic alterations that have indirect toxic or mutagenic effects. These metabolic changes, consequent to DNA damage, can alter the fidelity of DNA replication,[1] alter the regulation of macromolecular synthesis[2,3] or alter the cell's ability to repair damage and maintain its viability.[4,5,6]

Recent studies show that DNA damage can produce rapid alterations in steady state levels of deoxynucleoside triphosphate pools.[1-3] For example, Das et al.[1] showed that treatment of Chinese Hamster V79 cells with MNNG or UV-irradiation caused rapid increases in dATP and dTTP pools without significant changes in dGTP or dCTP pools. These changes occurred at MNNG doses with relatively high survival rates and were accompanied by a marked increase in mutation frequency.[1] In vitro, studies with purified eukaryotic DNA polymerases showed that the frequency of nucleotide misincorporation was affected by alterations in relative concentrations of the deoxynucleoside triphosphates.[7,8] Thus the alterations in dNTP pool sizes that occur consequent to DNA damage may contribute, in part, to an increased mutagenic frequency.

Alterations in deoxynucleoside triphosphate pools also affect the regulation of DNA synthesis. Using Chinese hamster ovary cells,[2,3] Newman and Miller showed that within ten minutes of UV irradiation there was a 3 to 5 fold increase in cellular dTTP which then returned to control values within 3 to 5 hours. In contrast, dCTP levels decreased within the first ten minutes to 10% of the level present in the unirradiated cells.[2,3] Using permealilized cells to simulate these alterations in deoxynucleotide pools sizes, the authors showed that a 20 fold excess of DTTP resulted in 95% inhibition of DNA synthesis whereas a 23 fold excess of dCTP resulted in a 2 fold increase in DNA synthesis.[2] These alterations in deoxynucleoside triphosphate pool sizes, consequent to DNA damage, appear to vary in different cell types and to contribute to alterations in both mutation and replication.[1-3]

DNA damage also causes alterations in pyridine nucleotide pools. A decrease in cellular NAD^+ content occurs following treatment of cells with gamma radiation, alkylating agents or a number of other DNA damaging agents[4-6,9] The fall in NAD^+ levels is due to activation of poly(ADP-ribose)

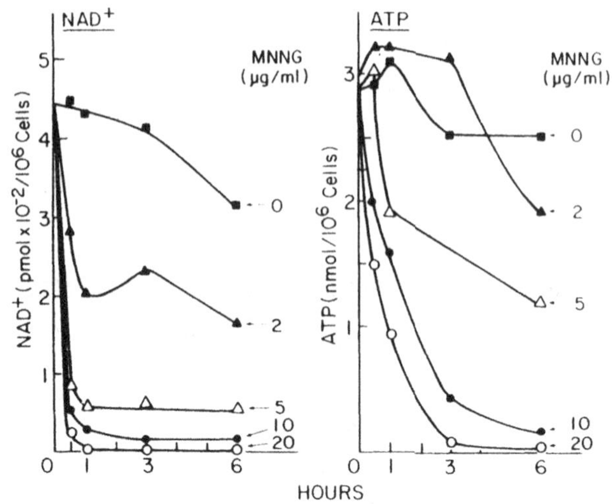

Figure 1. Effects of MNNG concentration on rates of NAD^+ (left) and ATP (right) lowering in L1210 cells. Final concentrations of MNNG are indicated at right of each curve.

polymerase in response to the formation of DNA strand breaks.[10,11] Thus, UV-irradiation activates poly(ADP-ribose) synthesis in cells competent for UV excision repair but not in cells from patients with xeroderma pigmentosum which are incompetent for UV excision repair processes.[11,12] Poly(ADP-ribose) polymerase uses NAD^+ as substrate to synthesize poly(ADP-ribose) covalently attached to chromosomal proteins.[13] Since the enzyme is activated by DNA strand breaks, this protein modification usually occurs during the cellular response to DNA damage. In addition, activation of the enzyme can consume large enough quantities of NAD^+ to totally deplete the cellular pools of this nucleotide.[4-6]

Figure 1 (left) shows that in L1210 lymphocytes, increasing doses of MNNG cause more rapid and drastic depletion of cellular NAD^+ pools. Figure 1 (right), shows that similar doses of MNNG also cause depletion of cellular ATP pools. Smaller doses of MNNG cause partial depletion of NAD^+ and ATP whereas the higher doses cause total depletion of these nucleotide pools. The smaller doses of MNNG (13 uM) that cause partial decreases of NAD^+ and ATP are in the range of those concentrations (5 uM) that were found to increase the dTTP or dATP pools in V79 cells.

As shown in Figure 2, both the decrease in NAD and the decrease in ATP can be blocked by the presence of the poly(ADP-ribose) polymerase inhibitors nicotinamide or 3-amino benzamide. These results indicate that depletion of both pools is dependent on the activation of poly (ADP-ribose) polymerase by DNA strand breaks. Since NAD is required for oxidation reduction metabolism leading to the synthesis of ATP, the depletion of NAD by activation of poly(ADP-ribose) polymerase results in the cessation of ATP synthesis and consequent depletion of cellular ATP levels.[5]

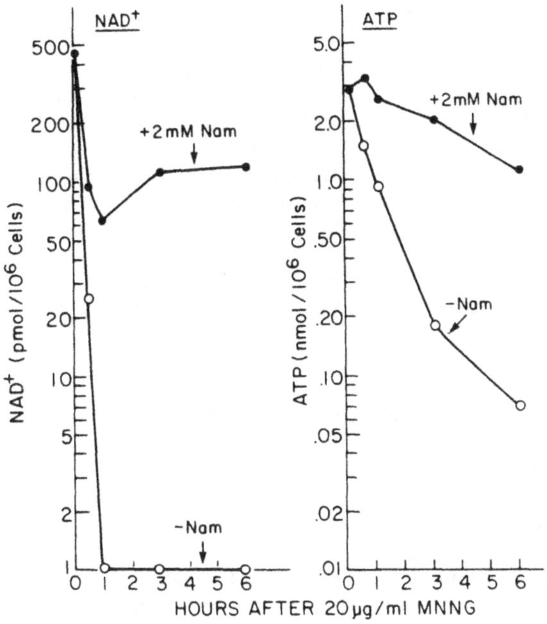

Figure 2. Effect of nicotinamide in MNNG-induced NAD^+ (left) and ATP (right) depletion. Cells were treated with 20 ug/ml MNNG (0) and then a portion of the culture was adjusted to contain 2 mM nicotinamide.

Table 1. METABOLITE POOLS (pmol/10^6 CELLS) IN L1210 CELLS FOLLOWING TREATMENT WITH 20 ug/ml MNNG

HOURS	NAD^+	ATP	G-6-P	$NADP^+$
0	595	3,730	163	23.4
0.5	55	750	151	19.7
1	21	964	104	17.1
3	0	29	300	12.1
6	0	0	0	3.6

Table 1 shows the effect of treating L1210 cells with 20 ug/ml MNNG on several different metabolic processes. At this dose of MNNG, cell survival is reduced by several logs as determined by clonogenic assay; however, viability at 6 hours is 90% as determined by trypan blue exclusion. These cells show the characteristic depletion of NAD due to maximal activation of poly(ADP-ribose) polymerase and the consequent depletion of ATP pools. They also show a characteristic depletion pattern for glucose-6-phosphate indicating that the cells lose the ability to take up and use glucose. Thus, there is an initial decrease in glucose-6-phosphate levels, followed by an increase at 3 hours after damage, followed by a complete disappearance of this metabolite. Both the transient increase in glucose-6-phosphate and its subsequent decrease can be partially blocked by inhibitors of poly(ADP-ribose) polymerase, again indicating the role of poly(ADP-ribose) polymerase in the alterations of the metabolite pools.

It is interesting to note that while NAD^+ pools are totally depleted in response to 20 ug/ml MNNG, the much smaller $NADP^+$ pools are reduced but not depleted. This probably reflects the inability of poly(ADP-ribose) polymerase to use $NADP^+$ as substrate as well as the lack of significant conversion of $NADP^+$ to NAD^+.

As a consequence of the drastic depletion of NAD^+ and ATP pools and the decrease in glucose utilization and energy production, the cell with high levels of DNA damage loses its ability to conduct energy dependent processes such as DNA, RNA or protein synthesis.[4-6] Table 2 shows the drastic decrease in thymidine, uridine and leucine incorporation that occurs when cells are treated with 20 ug/ml MNNG. The relation of this reduction in macromolecular synthesis to the depletion of NAD^+ as a result of the activation of poly(ADP-ribose) polymerase is shown by the partial restoration of macromolecular synthesis when the damaged cells are treated

Table 2. EFFECTS OF POLY(ADP-RIBOSE) POLYMERASE INHIBITORS ON MACROMOLECULAR SYNTHESIS

Treatment	$[^3H]$dTMP	$[^3H]$UMP	$[^3H]$Leu
		(dpm/10^6 cells/6hr)	
Control	140,000	27,400	400,000
20 ug/ml MNNG	18,300	3,500	5,100
20 ug/ml MNNG + 2mM nicotinamide	77,300	13,300	20,900
20 ug/ml MNNG + 2mM benzamide	109,000	18,700	30,000

Note: Logarithmically growing L1210 cells were treated with MNNG and where indicated were adjusted to contain 2 mM nicotinamide or benzamide. Different samples were labeled to assess DNA, RNA or protein synthesis by the addition of $[^3H]$dThd, $[^3H]$Urd, or $[^3H]$Leu.

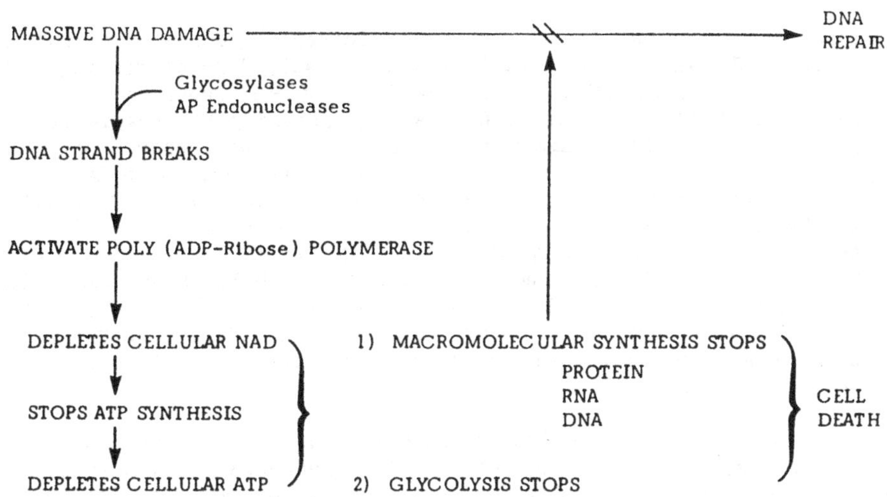

Figure 3. Schematic proposal for the involvement of poly(ADP-ribose) polymerase in the suicide response to massive DNA damage.

with the poly(ADP-ribose) polymerase inhibitors nicotinamide or benzamide.[4-6]

These indirect metabolic consequences of high levels of DNA damage are summarized in Figure 3. Activation of poly(ADP-ribose) polymerase by massive DNA damage is actually a suicide response that causes rapid NAD and ATP depletion. This results in cessation of all energy dependent processes. Macromolecular synthesis stops. Glucose utilization stops. And cells die before they have the opportunity to repair the DNA damage. Thus, indirect metabolic consequences cause cell death rather than death resulting from specific genotoxic events.

Does the suicide pathway outlined above have any physiologic or pathophysiologic relevance? One potentially important function of the suicide mechanism is to limit the possibility that massively damaged cells will be able to repair and survive with a high mutation frequency. Thus, activation of poly(ADP-ribose) polymerase by DNA damage may provide a safety mechanism to remove massively damaged cells from a population to decrease the likelihood of preserving cells with highly mutated phenotypes.

Another area where the suicide pathway may contribute to pathophysiologic pathways occurs in cells with intrinsically low NAD^+ levels. In this situation, activation of the suicide effect may account for increased susceptibility to the cytotoxic effects of DNA damaging agents. For example, cells from patients with Fanconi's anemia show marked sensitivity to many DNA damaging agents. We found that cells from these patients had NAD^+ levels that were below those in normal cells.[14] Furthermore, when the cells from Fanconi's anemia patients were treated with DNA damaging agents, the consequent activation of poly(ADP-ribose) polymerase resulted in lower NAD^+ levels in the FA cells than in the normal cells. This reduction of NAD^+ to lower levels in the FA cells may contribute to the increase in cytotoxicity of the DNA damaging agents in FA cells compared to normals.

The suicide effect mediated by activation of poly(ADP-ribose) polymerase may also explain some previous observations of the interaction of different types of DNA damage.[15] For example, Strauss has shown that increased levels of DNA damage induced by increasing doses of methyl methane sulfonate (MMS), in the range of 1 to 4 mM, resulted in decreasing levels of DNA repair.[15] Similarly, increasing doses of MMS resulted in inhibition of the cell's ability to conduct DNA repair following UV-irradiation. When the same levels of DNA damage were carried out in the presence of 3-amino benzamide, the level of DNA repair increased in proportion to the dose of MMS and the MMS treatment no longer interfered with the cell's ability to repair DNA in response to UV-irradiation.[15] In these cases, the treatment with MMS apparently activated poly(ADP-ribose) synthesis and the suicide response to deplete the cell's energy metabolites and limit the repair of MMS and UV induced DNA damage. When the activity of poly(ADP-ribose) polymerase was inhibited by 3-amino benzamide, the energy metabolites were preserved and the cells were able to repair both the MMS and the UV-induced DNA damage.[15]

An interesting observation has recently been made which implicates the poly(ADP-ribose) polymerase mediated suicide mechanism in the toxicity of adenosine deaminase deficiency and severe combined immune deficiency disease in humans.[16] Individuals with this disease suffer severe lymphopenia due to the toxic effects of deoxyadenosine. The lymphocytotoxic effect of adenosine deaminase deficiency can be simulated in lymphocyte cell lines from normal individuals by incubating them with the adenosine deaminase inhibitor, deoxycoformycin. Incubation of such leukocytes with deoxycoformycin and deoxyadenosine results in the gradual accumulation of DNA strand breaks and the depletion of NAD^+ leading to cell death over a period of several days.[16] This depletion of NAD and loss of cell viability were effectively blocked by nicotinamide or 3-amino benzamide.[16] Thus, persistant activation of poly(ADP-ribose) polymerase by unrepaired or recurrent DNA strand breaks may activate the suicide mechanism of cell death. This study provides a basis for the interesting suggestion that treatment with nicotinamide could block the persistent activity of poly(ADP-ribose) polymerase and may help preserve lymphocyte function in patients with adenosine deaminase deficiency.[16]

In summary, metabolic changes in response to DNA damage may cause rapid and drastic alterations in metabolic pool sizes. These alterations result in significant, indirect changes in viability and in the fidelity and regulation of replication. Indirect metabolic consequences of DNA damage may account for increasing susceptibility to DNA damage of certain individuals with pre-existing abnormalities of metabolic pool sizes.

ACKNOWLEDGEMENTS

This research was supported by the National Institutes of Health grants CA-35983 and GM-32647. Some of the studies were performed in facilities supported by the Sohio Foundation.

REFERENCES

1. S.K. Das, E.P. Benditt and L.A. Loeb, Rapid changes in deoxynucleoside triphosphate pools in mammalian cells treated with mutagens. Biochem. Biophys. Res. Communs. 114:458-464 (1983).
2. C.N. Newman and J.H. Miller, Mutagen-induced changes in cellular deoxycytidine triphosphate and thymidine triphosphate in Chinese hamster ovary cells. Biochem. Biophys. Res. Commun. 114:34-40 (1983).
3. C.N. Newman and J.H. Miller, Kinetics of UV-induced changes in deoxynucleoside triphosphate pools in Chinese hamster ovary cells and their effect on measurements of DNA synthesis. Biochem.

Biophys. Res. Commun. 116:1064-1069 (1983).
4. J.L. Sims, G.W. Sikorski, D.M. Catino, S.J. Berger and N.A. Berger, Poly(adenosinediphosphoribose) polymerase inhibitors stimulate unscheduled deoxyribonucleic acid synthesis in normal human lymphocytes. Biochemistry 21:1813-1821, 1982.
5. J.L. Sims, S.J. Berger and N.A. Berger, Poly(ADP-ribose) polymerase inhibitors preserve nicotinamide adenine dinucleotide and adenosine 5'-triphosphate pools in DNA-damaged cells: Mechanism of stimulation of unscheduled DNA synthesis. Biochemistry 22:5188-5194 (1983).
6. N.A. Berger, Poly(ADP-ribose) in the cellular response to DNA damage. Rad. Res. 101:4-15 (1985).
7. N. Battula, D.K. Duke and L.A. Loeb, Avian nyeloblastosis virus DNA polymerase: Kinetic studies of the incorporation of non-complimentary nucleotides. J. Biol. Chem. 250:8404-8408 (1975).
8. T.A. Kunkel, J.R. Silber and L.A. Loeb, The mutagenic effect of deoxynucleotide substrate imbalances during DNA synthesis with mammalian DNA polymerases. Mutat. Res. 94:413-419 (1982).
9. C.J. Skidmore, M.I. Davies, P.M. Goodwin, H. Halldorsson, P.J. Lewis, S. Shall and A-A Zia'ee, The involvement of poly(ADP-ribose) polymerase in the degradation of NAD caused by gamma-radiation and N-methyl-N-nitrosurea. Eur. J. Biochemistry 101:135-142 (1979).
10. R.C. Benjamin and D.M. Gill, Poly(ADP-ribose) synthesis in vitro programmed by damaged DNA. A comparison of DNA molecules containing different types of strand breaks. J. Biol. Chem. 255:10502-10508 (1980).
11. N.A. Berger and G.W. Sikorski, Poly(adenosine diphosphoribose) synthesis in ultraviolet-irradiated xeroderma pigmentosum cells reconstituted with micrococcus luteus UV endonuclease. Biochemistry 20:3610-3614 (1981).
12. N.A. Berger, G.W. Sikorski, S.J. Petzold and K.K. Kurohara, Defective poly(adenosine diphosphoribose) synthesis in xeroderma pigmentosum. Biochemistry 19:289-293 (1980).
13. C.S. Surowy and N.A. Berger, Unique acceptors for poly(ADP-ribose) in resting, proliferating and DNA-damaged human lymphocytes. Biochim. Biophys. Acta 740:8-18 (1983).
14. N.A. Berger, S.J. Berger and D.M. Catino, Abnormal NAD^+ levels from patients with Fanconi's anemia. Nature 299:271-273 (1982).
15. B.S. Strauss, The interaction of UV- and methyl methane sulfonate-induced DNA repair synthesis: A role for poly(ADP-ribose). Carcinogenesis 5:527-582, (1984).
16. S. Sato, C.J. Carrera, H. Kubota, D.B. Wasson and D.A. Carson, Mechanism of deoxyadenosine and 2-chlorodeoxyadenosine toxicity to nondividing human lymphocytes. J. Clin. Invest. 75:377-383 (1985).

DNA DAMAGE, DNA REPAIR AND INDUCED MUTAGENESIS:

SOME ENZYMOLOGICAL AND STRUCTURAL CONSIDERATIONS*

Miroslav Radman

Laboratoire de Mutagenes, Institut Jacques Monod
C.N.R.S. - Universite Paris VII
2, Place Jussiev, Tour 43
75251 Paris, Cedex 05, France

INTRODUCTION

The particular interest of this symposium is to review all relevant knowledge which may help in the assessment of carcinogenic and mutagenic risk, in particular the risk arising from diverse environmental and metabolic conditions. Of special interest are the origins of somatic and germ line mutations. A priori, the majority of deleterious mutations are recessive, which as germ line mutations may cause more victims than the dominant mutations. The reason is that the majority of dominant mutations are readily recognized in the first carrier; either their propagation is directly stopped in the case of lethal mutations or they can be controlled in the case of non-lethal mutations. Recessive mutations are usually not recognized either in the first carrier or in later heterozygous carriers and they thus spread in the population repeatedly causing victims among the homozygous progeny. Homozygosity arises in the offspring of heterozygous parents by the Mendelian rule of random chromosome assortment. Homozygous somatic cell lineages can arise in heterozygous individuals by chromosomal rearrangements (mitotic recombination, deletion, non-disjunction). Since the normal human population contains roughly 1% heterozygous carriers for any given recessive gene mutation, a normal individual may carry on the order of 10^3 diverse recessive mutations, some of which when homozygous may cause severe cellular metabolic defects including cancerous growth.[1] Therefore, when trying to estimate risk from carcinogenic and other deleterious mutations, one cannot underestimate the role of chromosomal rearrangement in expression of pre-existing as well as of newly arising recessive mutations.[1] Early routine detection of recessive (and other) mutations in humans may become possible through development of new sensitive molecular probes.

*This paper is dedicated to Professor Maurice Errera on the occasion of his 70th birthday and of his retirement as the director of the Laboratoire de Biophysique et Radiobiologie of the Universite Libre de Bruxelles. The majority of the work reviewed here was carried out in the "Service Errera".

PATHWAYS OF MUTAGENESIS

Mutation can be defined in molecular terms as any alteration in some original genomic nucleotide sequence. (Traditionally, the terms mutagenesis and mutation referred to relatively rare stochastic events which are stable, i.e., revert back to the original sequence as rarely as the occurence of the initial wild type -> mutant alteration. This traditional notion of mutation is distinct from the pre-programmed DNA rearrangments which are genetically controlled and which can occur frequently and reversibly and fulfill a defined biological function).[1] Thus some mutations can profoundly affect a biological function whereas others can be phenotypically neutral.

Mutations arise either "*de novo*", as errors in DNA replication, or "*de antiquo*" by the scrambling of pre-existing sequences in the course of diverse DNA recombination events.[1,2]

I shall consider here only the "*de novo*" mutagenesis which can arise either as errors in the course of replication of unmodified DNA templates or as modified-template-directed errors. *Errors in the course of replication of unmodified templates* can be inducible (e.g., by the presence of chemically modified deoxyribonucleoside triphosphates or by unbalanced pools of the four normal deoxyribonucleoside triphosphates or by the presence of some divalent cations such as Mn^{++} and Be^{++})[3] or they may arise simply as the consequence of the limited fidelity of the DNA replication machinery and/or as the consequence of the limited efficiency of postreplicative mismatch correction.[4] *Template-directed* errors may occur either by direct miscopying of the modified (*miscoding*) base in the DNA or indirectly as a complex metabolic consequence of the presence of *non-coding* DNA lesions in replicating DNA. A number of DNA lesions are known to cause such a stereochemical change of the original base as to make its hydrogen bonding and stacking characteristics resemble those of another normal base more than the intact original base (e.g., O^6-methyl-guanine pairs better with thymine than with cytosine, see Reference 5 for this and more examples). Base analogs and small alkylating agents are typical for this group of mutagens.

A majority of the known potent chemical mutagens/carcinogens cause non-coding lesions in DNA which under normal high fidelity replication conditions block the elongation of the newly synthesized strands. These agents are potent mutagens because they induce a cellular mutator activity which is required for fixation of mutations opposite non-coding lesions (so called *targeted* mutations) but which appears to cause mutations also in undamaged DNA (so called *untargeted* mutations). This discussion will deal with these inducible mutagenic processes, which are just one part of a complex cellular metabolic response to DNA damage, the so-called SOS induction.[6-8] Since the *E. coli* mismatch repair system has been used to characterize SOS-induced mutagenesis, a brief review of mismatch repair is presented below.

DIRECTED AND UNDIRECTED MISMATCH REPAIR IN *E. COLI*

Correction of replication errors

In order for a process of mismatch repair to be able to correct replication errors, it must be able to distinguish newly synthesized strands from parental strands and must selectively correct the copy according to the original and not *vice versa*. In *E. coli*, a means of strand discrimination for the mismatch repair system is provided by adenine methylation in GATC sequences.[4] Mismatch repair is greatly reduced in regions of DNA in which GATC sequences are fully adenine methylated and repair occurs preferentially, if not exclusively, on unmethylated strands of hemimethylated heteroduplexes.[9,10] Because newly synthesized strands are transiently undermethylated, i.e., methylation lags somewhat behind replication, mismatch repair in E. coli is believed to occur primarily on newly synthesised strands immediately behind the replication fork. The mismatch repair system eliminates over 99 percent of replication errors.[11]

Specificity of mismatch repair

The repair of all classes of non-complementarities between the two strands was tested in heteroduplex DNA's of bacteriophage λ transfecting E. coli. The transition mismatches (G:T and A:C) appear to be efficiently repaired, whereas three of the six transversion mismatches (G:A, C:T and C:C) are not.[10,12,13] All four single unpaired bases in frameshift:wild type heteroduplexes are well recognized and repaired on either of the two strands.[10] Thus, it appears that replication errors involving either deletion or insertion of a single base are corrected.[10,14] However, the system does not appear to act on large unpaired regions (about 800 base pairs) suggesting that there is a maximum size of unpaired region that the system can recognize as a mismatch.[10] Further experiments are required to determine what that size is.

Mismatches in GC rich regions are more efficiently repaired than mismatches in AT rich regions (M. Jones, R. Wagner and M. Radman, unpublished). Petruska and Goodman have shown that the in vitro error rate of T4 phage DNA polymerase is proportional to the GC content of the environment.[15] Therefore, it appears that the mismatch repair system acts most efficiently on the most frequent replication errors (e.g., G:T, A:C, frameshift and transversions, Pu:Pu and Py:Py, in GC rich regions). The final effect of mismatch repair on DNA replication is the elimination of over 99 percent of mutations resulting in equilibration of the occurence of different types of mutations along the genome.

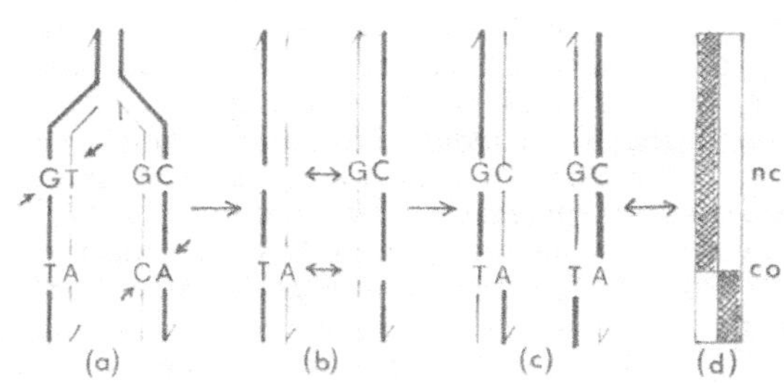

Figure 1. Postreplicative mismatch repair by recombination resulting in sister chromatid exchange.

Heavy lines represent parental DNA strands, thin lines newly synthesized strands. (a) T and C residues in the newly synthesized strands are the newly arising mutations, i.e. replication errrors creating a G:T and an A:C mismatch which are attacked by the mismatch repair enzymes on both DNA strands leading to (b) double-strand breaks which are repaired (c) by recombination with the intact region of the sister duplex (see reference 16 for details of the molecular mechanism). The G:T mismatch is repaired without accompanying crossing over (NCO, for non-crossover); the A:C mismatch is repaired with accompanying crossing over (CO, for crossover) and therefore will be visualised as sister chromatid exchange in (d). See text.

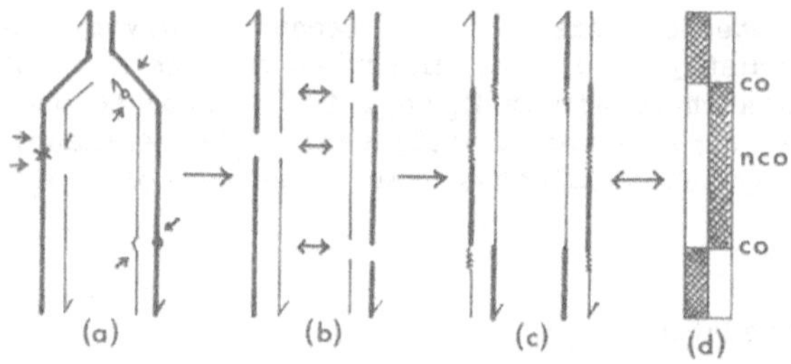

Figure 2. Postreplicative repair of DNA damage by recombination resulting in sister chromatid exchange.

Symbols are as in Figure 1. The cross in the parental strand represents a non-coding DNA lesion such as a pyrimidine dimer, which blocks DNA chain elongation resulting in gaps in the newly synthesized DNA. The open dot is a lesion in the newly synthesized strand, the dark dot is a lesion in the parental strand which has been copied ("bypassed"). All these structures can be attacked by single-strand specific (S1-like) endonucleases resulting in double-strand breaks (b), which can be repaired as in Figure 1 with the same consequences (c,d) (see Figure 1 and text for further explanations).

Undirected mismatch repair

When both strands of a DNA duplex bearing one or more repairable mismatches are undermethylated, such a heteroduplex is subject to attack by mismatch repair enzymes on both strands, frequently resulting in the killing of DNA, presumably by double-strand breaks.[10] Double-strand breaks can be repaired in *E. coli*, and even more efficiently in eukaryotic cells, by a recombination process which apparently involves copying of the intact homologous DNA. The repair of a double-strand break or gap is thus accompanied by information transfer (gene conversion) from the intact duplex to the broken duplex.[16] We have therefore proposed that in eukaryotes mismatch repair of replication errors may occur by recombination between sister chromatids following double-strand breakage at the site of the replication error.[10] Figure 1 shows schematically that recombination repair by gene conversion could remove replication errors via creation of double strand breaks. Figure 2 shows that DNA damage could cause double-strand breaks in newly replicated daughter duplexes and thus be repaired by the same mechanism as the mismatch repair in Figure 1 (see Legend).

GENETIC DIAGNOSIS OF UNTARGETED AND TARGETED MUTATIONS

(1) The *E. coli* mismatch repair system normally operates by excision-resynthesis and thus can correct replication errors only if the parental template strand is intact. When the parental DNA strands are damaged, and when the SOS system is induced, two kinds of mutations can occur: targeted mutations, which arise opposite DNA lesions by the error-prone copying of the damaged template, and untargeted mutations, which arise in the course of error-prone replication of intact DNA.[8] It was shown that, as expected,

only untargeted mutations (which arise as genuine mismatches) are subject to mismatch correction.[17] Excision of the targeted mutation could not lead to its correction as long as the initial lesion remains in the template. Indeed, it was found that the E. coli mismatch repair system decreases untargeted mutagenesis but has no effect on targeted mutagenesis.[17]

(2) Bacteriophage λ provides a morphological distinction between targeted and untargeted mutations arising in the λ immunity region.[17-19] Clear plaque mutants arising in irradiated phage genomes, following infection of SOS-induced bacteria, occur in infective centers containing only (or predominantly) mutant phages ("pure" clear plaques). This is an unexplained curiosity because we do not know the mechanism of the loss of the two damaged wild-type strands (or of their phage progeny). On the other hand, untargeted mutations, like spontaneous mutations, occur in infective centers in which the clear plaque mutant phages are a minority ("mixed" clear plaques).[17-19]

The fact that mismatch repair deficient bacteria yield more "mixed" mutant plaques than mismatch repair proficient bacteria (with no difference in the yield of "pure" mutant plaques) suggests strongly that the UV-induced untargeted mutations are genuine replication errors.[17]

(3) Targeted and untargeted mutations have different genetic requirements, e.g., the umuC,D gene mutation abolishes only the targeted mutagenesis of bacteriophage λ.[18,19] The fact that E. coli umuC,D mutants are practically non-mutable by radiations and many chemical mutagens (when testing for mutagenesis of the bacterial chromosome) suggests that untargeted mutations either do not occur in bacterial DNA or (more likely) that they are efficiently mismatch corrected.[17]

COPYING OF THE NON-CODING DNA LESIONS: "BYPASS REPLICATION" IN VIVO AND IN VITRO

The notion of mutagenic lesion-bypass replication was proposed along with the notion that such a mutagenic process is part of the inducible SOS induction program triggered by DNA damage.[6] Sedgwick's analysis of the replication of UV irradiated E. coli DNA supported this model but the mechanism was not clear (e.g., recombination repair and bypass replication were equally plausible).[20] Caillet-Fauquet et al.[21] analysed the fate of UV-irradiated single-strand ØX174 phage DNA in single infection and demonstrated that UV-induced SOS-mutagenesis in E. coli is correlated with acquisition of the capacity to elongate growing DNA chains across the sites of initial blocking lesions. However, their work did not address the still unknown mechanism of the "bypass", e.g., did the "bypassed" lesions (pyrimidine dimers) remain in the unaltered form?

In vitro studies. The difficulty of approaching the question of the enzymology of "bypass" replication in vivo led to the search for "bypass" replication in vitro and to attempts to stimulate such replication in simple defined systems.[22,23]

The first report on inducible error-prone replication of homopolymer templates in cell extracts showed that both UV irradiation of polypyrimidine templates and SOS-induction prior to extract preparation increased synergistically the in vitro misincorporation frequency.[22] However, we failed to purify the error-prone polymerase activity which, in retrospect, is likely to be related to Pol I* purified from induced E. coli.[24] (The biological role of Pol I* enzyme is not clear at present because UV-induced mutagenesis of E. coli is not Pol I-dependent and the umuC mutant shows normal Pol I* levels).

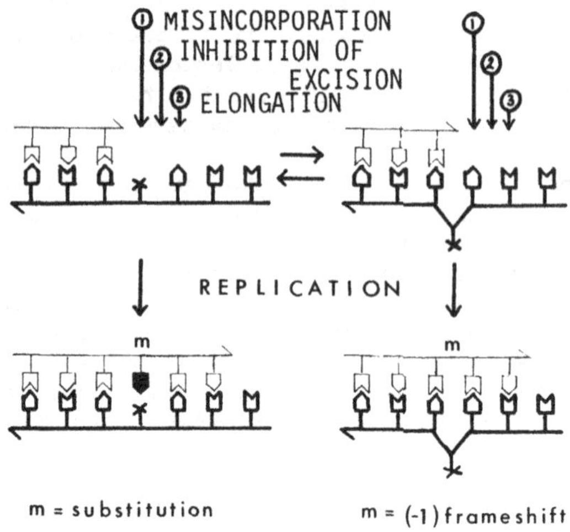

Figure 3. Rate-limiting steps in "bypass" replication of non-coding DNA damage: alternative structures for substitution and deletion frameshift mutagenesis.

Postulated rate-limiting steps in "bypass" replication of non-coding DNA damage such as an apurinic/apyrimidinic site (cross). Left side: pathway for substitution mutagenesis (misincorporated base is dark). Right side: extrahelical conformation of the damaged site may facilitate "bypass" replication by single base deletion mutagenesis. See text for further discussion and references.

Yet another approach was to try to define the rate limiting steps in the copying of non-coding lesions in vitro using purified DNA polymerases (hoping that the SOS-induced factors could act in vivo on thus defined steps). It appears that both fidelity functions of DNA polymerases limit the efficiency of "bypass replication" in vitro. It was first demonstrated that the 3' to 5' "proofreading" exonuclease appears to diminish the efficiency of "bypass" of pyrimidine dimers by excision of (mis)incorporated residues.[23] The turnover of dNTP substrates to free dNMPs in the absence of significant chain elongation was called "polymerase idling".[23]

However, the lack of efficient stimulation of the bypass replication of pyrimidine dimers by the inhibition (or absence) of the 3' to 5' exonuclease [25,26] led to the discovery that nucleotide selection by DNA polymerases is probably the predominant rate-limiting step in "bypass replication" in vitro [25-28] The use of sequencing techniques by P. Moore, and B. Strauss elegantly demonstrated that DNA chain elongation in vitro usually stops one base before a pyrimidine dimer in the template.[26] Manganese, the mutagenic divalent cation which appears to inhibit nucleotide selection rather than the proofreading exonuclease,[3] permits the incorporation of one more base; yet for the majority of DNA polymerases (including those devoid of exonucleases), elongation at mismatches opposite a pyrimidine dimer is rather inefficient.[25-28] The manganese-stimulated "bypass replication" of apurinic sites is much more efficient than the "bypass" of pyrimidine dimers.[26-28]

In conclusion, we can postulate at least three limiting steps in "bypass replication" of non-coding lesions (see Figure 3):

(1) misincorporation opposite the non-coding lesion,
(2) inhibition of the exonucleolytic removal of the newly incorporated mismatched base, and
(3) elongation past DNA damage.

WHAT HAVE WE LEARNED FROM SEQUENCING OF INDUCED MUTATIONS?

The first sequences of UV-induced mutations in E. coli DNA clearly indicated that adjacent pyrimidines are frequent sites of UV-induced mutations.[29] There is still some controversy as to the relative contribution of pyrimidine dimers versus 6, 4 photoproducts (see contributions by Drs. Hutchinson and Glickman and their coworkers, this volume). The analysis of mutations induced by carcinogens which attack predominantly guanines also supported the notion of targeted mutations.[30]

Targeted (UV and gamma ray) and untargeted substitution mutations were first sequenced in M13 phage[32] and then extensively analyzed in a forward mutation system.[33-35] Now, a large body of data in phage λ and E. coli DNA allows the conclusions that both targeted and untargeted mutations do occur, that the nature of the damage determines the specificity of targeted mutations and that, in general, frameshift (one base deletion) mutations are produced in both targeted (e.g., UV and apurinic site induced) and untargeted (e.g., UV-induced) mutagenesis[32-35] in addition to base substitutions.

MODELS FOR "BYPASS REPLICATION"

Models, or working hypotheses, for "bypass replication" in E. coli must account for the following reviewed characteristics: (1) targeted mutations are generated by the replication of damaged parental DNA, such that the viable progeny molecules contain a high proportion of (sometimes only) the mutant sequence (viz., strand loss), (2) the nature of the targeted mutation is influenced by the nature of the lesion, thus diverse base substitution mutations should be possible as well as one base deletion (frameshift) mutations, (3) low-fidelity replication of undamaged DNA may accompany the "bypass".[32]

Recent genetic studies clearly show that the enzymology of "bypass replication" is complex. The initial idea that untargeted mutagenesis is the necessary consequence of the low-fidelity requirement for the "bypass" may not hold (G. Maenhaut-Michel, A. Brotcorne-Lannoy, and P. Caillet-Fauquet, personal communication), because mutants exist which affect only targeted mutagenesis (e.g., umuC,D) or only untargeted mutagenesis (polA, some din mutations, etc...). Two studies implicate DNA polymerase III as being involved, directly or indirectly, in at least the untargeted mutagenesis.[37,38] Furthermore, the genetically-induced SOS mutator effect (e.g., recA441 mutation) and the UV-induced mutator effect appear to be different (G. Maenhaut-Michel, A. Brotcorne-Lannoy, and P. Caillet-Fauquet, as above).

Hence, there are not sufficient elements available to draw a detailed model for SOS mutagenesis. Figure 3 illustrates a simplified model which, although deficient in enzymology, makes a few specific suggestions about the possible structures of the damaged single-stranded template and of the damaged double-stranded DNA, which are compatible with all three requirements stated above.

The principal suggestion is that the misincorporation event opposite DNA damage (step 1, Figure 3) may be facilitated both by the decrease in the intrinsic fidelity of DNA polymerases and by the modification of the DNA structure at the site of damage. There is considerable experimental evidence that the inhibition of both base selection[26-29] and the 3' to 5' exonuclease facilitates the "bypass". It was recently suggested that the SOS mutator effect[18] and the misincorporation event (1 in Figure 3)[38] do do not require functional UmuC,D proteins, but appear to require some other inducible activity. Thus the umuC,D gene products may be involved in steps 2 and/or 3 in Figure 3. However, attempts to complement the umuC,D gene requirement in targeted mutagenesis by the mutD mutation (which severely inhibits the 3' to 5' exonuclease proofreading associated with DNA polymerase III)[39] has given negative results so far (A. Brotcorn-Lannoy, personal communication). Since the *in vitro* "bypass" replication of apurinic sites is facilitated by pyrophosphate depletion, it was suggested that polymerase-associated pyrophosphate metabolism may be affected by SOS induction.[38,40]

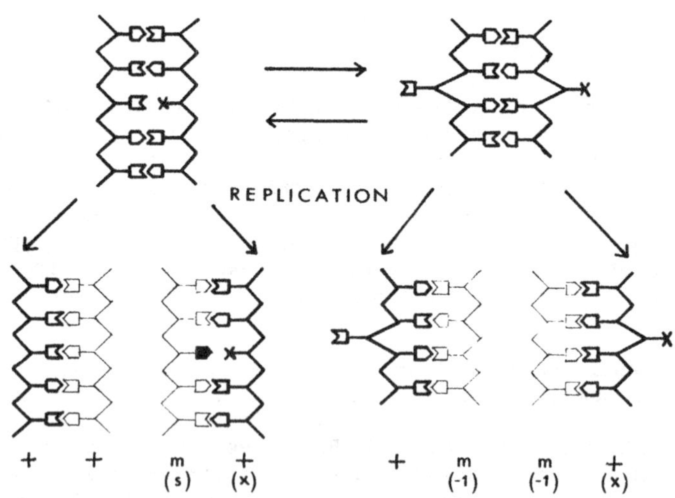

Figure 4. Hypothetical extrahelical structures of a damaged DNA duplex.

The lack of stacking and hydrogen bonding may facilitate an extrahelical conformation of the damaged base (or base loss site) and of the undamaged base in the complementary strand (see the text). The intrahelical conformation may be involved predominantly in substitution mutagenesis, extrahelical conformation in single base deletion (-1) mutagenesis. The extrahelical conformation (left side) offers an intriguing possibility for template-directed (-1) replication errors on both damaged and undamaged DNA strands. The cross represents a non-coding lesion (e.g. a base loss site); m stands for mutation, s for substitution, (-1) for deletion frameshift.

I propose that all three steps in Figure 3 could be facilitated, and some circumvented, if the damaged base (or a base-loss site) could swing out of the helix, allowing the stacking of the two undamaged flanking bases in the template. Thus facilitated "bypass" replication would produce a one-base deletion mutation. There is considerable evidence (from model building (reference 41 and M. Radman, unpublished), photochemistry (reference 40 and K. Minton, personal communication) and N.M.R. studies)[42-44] that such extrahelical structures can occur both in single-stranded and in double-stranded DNA, even in the absence of damage.

If the extrahelical conformation of the damaged base could induce the extrahelical conformation of the base opposite the damage in double-stranded DNA (as illustrated in Figure 4), then it is conceivable that parental DNA strands could strand-direct mistakes at the same site, presumably involving single-base detection. In other words, the damaged strand could affect the complementary strand in its coding properties. I presume that special inducible proteins could be responsible for the stabilization of the extrahelical base (or damaged base) conformations in Figures 3 and 4. UmuC,D and RecA proteins are obvious candidates. These and probably some other proteins could affect steps 1 and 2 in the "bypass" (Figure 3). The RecA protein was shown to inhibit the polymerase III-associated 3' to 5' exonuclease activity and to cause decreased fidelity of DNA polymerase III in vitro.[45] On the other hand, the single-strand binding protein (SSB) is expected to prevent the extrahelical base conformation by its cooperative binding to the phosphate moiety.

The intention of this speculation is just to create hypotheses to explain how damaged double-stranded DNA could yield (largely) pure mutant progeny in the pathway of targeted mutagenesis. A less speculative phenomenon which can contribute to pure mutant progeny of damaged DNA is the strand loss phenomenon, which occurs with phage λ DNA even in the absence of damage.[46] Strand loss is of regular occurence when only one of the two phage λ DNA strands is UV irradiated (M. Radman, R. Wagner, and Z. Trgovevic, unpublished experiments). The genetics of targeted mutagenesis dictate that the UmuC,D proteins should be involved in the strand loss phenomenon and/or in the stabilization of the extrahelical conformation in Figure 4, if these speculations are relevant to the molecular mechanism of targeted mutagenesis.

ACKNOWLEDGEMENTS

I thank my colleagues G. Maenhaut-Michel, P. Caillet-Fauquet, and A. Brotcorne-Lannoye for their collaboration and many discussions on SOS mutagenesis. The mismatch repair studies reviewed were carried out by my colleagues C. Dohet, M. Jones, P. Doutriaux, and R. Wagner. The "Laboratoire de Mutagenese" is supported by grants from the C.N.R.S., Association pour la Recherche sur le Cancer and Ligue Natinale Francaise contre le Cancer.

REFERENCES

1. M. Radman, R. Wagner and P. Jeggo, Mutation Res. 92:101 (1982).
2. M. Radman, in "Cellular Responses to DNA Damage", E.C. Friedberg and B.A. Bridges ed, Plenum Press, New York, pp. 287-298 (1983).
3. L.A. Loeb and T.A. Kunkel, Annu. Rev. Biochem. 52:429 (1982).
4. M. Radman and R. Wagner, Curr. Topics Microbiol. Immun. 108:23 (1984).
5. J.W. Drake and R.H. Baltz, Annu. Rev. Biochem. 45:11 (1976).
6. M. Radman, in "Molecular and Environmental Aspects of Mutagenesis", eds L. Prakash et al., C.C. Thomas Publ. Spring Field, III. pp. 127

(1974).
7. E.M. Witkin, Bacteriol. Rev. 40:869 (1976).
8. G.C. Walker, Microbiol. Rev. 48:60 (1984).
9. P.J. Pukkila, J. Peterson, G. Herman, P. Modrich and M. Meselson, Genetics 104:571 (1983).
10. R. Wagner, C. Dohet, M. Jones, M.P. Doutriaux, F. Hutchinson and M. Radman, Cold Spring Harbor Symp. Quant. Biol. 49:611 (1984).
11. B.W. Glickman and M. Radman, Proc. Natl. Acad. Sci. USA 77:1063 (1980).
12. C. Dohet, R. Wagner and M. Radman, Proc. Natl. Acad. Sci. USA 82:503 (1985).
13. B. Kramer, W. Kramer and H.J. Fritz, Cell 38:879 (1984).
14. C. Dohet, R. Wagner and M. Radman, submitted.
15. J. Petruska and M.F. Goodman, J. Biol. Chem. (1985) in press.
16. J.W. Szostak, T.L. Orr-Weaver, R.J. Rothstein and F.W. Stahl, Cell 33:25 (1983).
17. P. Caillet-Fauquet, G. Maenhaut-Michel and M. Radman, EMBO J. 3:707 (1984).
18. G. Maenhaut-Michel and P. Caillet-Fauquet, J. Mol. Biol. 177:181 (1984).
19. R.D. Wood and F. Hutchinson, J. Mol. Biol. 173:293 (1984).
20. S. Sedgwick, Proc. Natl. Acad. Sci. USA 72:2753 (1975).
21. P. Caillet-Fauquet, M. Defais and M. Radman, J. Mol. Biol. 117:95 (1977).
22. M. Radman, P. Caillet-Fauquet, M. Defais and G. Villani, in "Screening Tests in Chemical Carcinogenesis" eds R. Montesano, H. Bortsch and L. Tomatis, IARC Sci. Publ. No. 12, Lyon, pp. 537-545 (1976).
23. G. Villani, S. Boiteux and M. Radman, Proc. Natl. Acad. Sci. USA 75:3037 (1978).
24. D. Lackey, S.W. Krauss and S. Linn, Proc. Natl. Acad. Sci. USA 79:330 (1982).
25. O.P. Doubleday, G. Maenhaut-Michel, A. Brandenburger, P. Lecomte and M. Radman, in "Chromosome Damage and Repair", eds. E. Seeberg and K. Kleppe, Plenum Press, New York, pp. 447-459 (1981).
26. P.D. Moore, K.K. Bose, S.D. Rabkin and B.S. Strauss, Proc. Natl. Acad. Sci. USA 78:110 (1981).
27. S. Boiteux and J. Laval, Biochemistry 22:6746 (1982).
28. S.D. Rabkin, P.D. Moore and B.S. Strauss, Proc. Natl. Acad. Sci. USA 80:1541 (1983).
29. O.P. Doubleday, P. Lecomte, A. Brandenburger, W.P. Diver and M. Radman in "Induced Mutagenesis", ed. C.W. Lawrence, Plenum Press, New York, pp. 115-134 (1983).
30. C. Coulondre, J.H. Miller, P.J. Farabaugh and W. Gilbert, Nature 274:775 (1978).
31. L.A. Loeb, Cell 40:483 (1985).
32. A. Brandenburger, G.N. Godson, M. Radman, B.W. Glickman, C.A. Van Sluis and O.P. Doubleday, Nature 294:180 (1981).
33. T.A. Kunkel, Proc. Natl. Acad. Sci. USA 81:1494 (1984).
34. J.H. Miller, J. Mol. Biol. 182:45 (1985).
35. E.J. Le Clerc and N.L. Istock, J. Mol. Biol. 180:217 (1984).
36. B.A. Bridges, R.P. Mottershead and S.G. Sedgwick, Mol. Gen. Genet. 144:53 (1976).
37. A. Brotcorne-Lannoye, G. Maenhaut-Michel and M. Radman, Mol. Gen. Genet. 199:64 (1985).
38. B.A. Bridges and R. Woodgate, Mol. Gen. Genet. 196:364 (1984).
39. R. Scheuermann, T. Schuman, P.M.J. Burgess, C. Lu and H. Echols, Proc. Natl. Acad. Sci. USA 80:7085 (1983).
40. O.P. Doubleday, P. Lecomte and M. Radman, in "Cellular Responses to DNA Damage", eds. E.C. Friedberg and B.A. Bridges, Alan R. Liss, Inc., New York, pp. 489-499 (1983).
41. A.J. Lomant and J.R. Fresco, Progr. Nucl. Acids Res. Mol. Biol. 15:185, (1975).

42. A. Pardi, K.M. Morden, D.J. Patel and I. Tinoco, Biochemistry 21:6567 (1982).
43. K.M. Morden, G.Y. Chu, F.H. Martin and I. Tinoco, Biochemistry 22:5557 (1983).
44. Farzakerley, R. Teoule, A. Guy and W. Guschlbauer, FEBS Letters 176:449 (1984).
45. A.R. Fersht and J.W. Knill-Jones, J. Mol. Biol. 165:669 (1983).
46. R. Wagner and M. Meselson, Proc. Natl. Acad. Sci. USA 73:4135 (1976).

MECHANISMS OF MUTAGENESIS OF E. COLI BY ULTRAVIOLET LIGHT

Franklin Hutchinson and Richard D. Wood

Department of Molecular Biophysics and Biochemistry
Yale University
New Haven, CT 06511 U.S.A.

This paper summarizes information concerning mechanisms of mutagenesis by ultraviolet light (UV), one of the most extensively studies mutagenic agents. The mechanisms are those in E. coli, one of the most thoroughly studied living systems. E. coli is a host for a number of bacteriophages (lambda, M13, etc.) which rely mainly on host enzymes for DNA replication and repair, and thus should be mutagenized by processes very similar to those which affect the host cell.

If irradiated lambda phage are absorbed to E. coli cells, extensive mutagenesis is usually seen only if the host cells are also irradiated.[1] The greater part of this mutagenesis is caused by activation of the "SOS system".[2,3] UV-induced DNA damage activates the RecA protein to catalyze the cleavage of the LexA repressor and derepress 20 or more genes whose activities, to the extent they have been identified, are related to DNA repair. Of these, the umuD and umuC gene products are needed for at least some UV mutagenic processes,[2,3] and RecA protein has a second role in mutagenesis in addition to its part in control of the SOS system.[4,5]

Unirradiated lambda phage shows a marked increase in mutation when the host cells are irradiated.[6,7] In this case the number of mutant phage in a single burst from one infected bacterium is typically one or two in a total burst of about 100. In contrast, irradiated phage typically show 30 to 50 mutants per burst;[7,8] the mutation occurs when repair or replication enzymes first encounter a photoproduct, so many copies of the mutated DNA are made. Because such a mutation requires a lesion in or near the mutated gene, it is sometimes referred to as "targeted". For unirradiated phage, the small number of mutants per burst is consistent with a loss of fidelity in DNA replication,[7] so the mutation usually occurs late in the lambda DNA replication cycle when most of the DNA is made;[9] such a mutation is sometimes referred to as "nontargeted".

MUTATIONAL SPECTRA INDUCED BY ULTRAVIOLET LIGHT

In this paper we concentrate on studies of the changes in base sequence induced by UV, and on the locations of the mutations in various base sequences in mutated genes. Extensive studies of this kind have been made in the E. coli lacI gene, using the ingenious approaches devised by J.H. Miller,[10-12] and in the E. coli lacZ(alpha) fragment cloned into the single-strand DNA phage M13.[13] These studies have been invaluable in providing new insights and in confirming conclusions derived from our results in the lambda phage cI gene.

Table I summarizes some results of UV mutagenesis of the cI repressor gene of lambda phage under various conditions.

The mutational spectra are rather similar for phage mutagenized as prophage integrated into the host cell genome and as free phage, except there are rather more transversions and fewer transitions in the former case; at present we have no good explanation for this difference. A really interesting feature is the similar mutation rate and spectrum for lambda phage DNA which had been irradiated as DNA in solution, photoreactivated with an E. coli enzyme,[14] repackaged into phage heads and assayed. The ability of phage from photoreactivated DNA to form plaques was much greater than for samples of the same irradiated DNA packaged without photoreactivation. However, the number of clear-plaque mutants per plaque-forming unit were the same for photoreactivated and for non-photoreactivated DNA;[15] thus the cyclobutyl pyrimidine dimers, the only photoproducts known to be acted on by photoreactivation, are not the major premutagenic lesions. Photoreactivation of irradiated E. coli cells does reduce the induced mutation rate,[16] but this is because photoreactivation reduces induction of the SOS functions[17,15] required for UV mutagenesis.

MECHANISMS OF MUTAGENESIS BY ULTRAVIOLET LIGHT

In this section we summarize conclusions we have drawn concerning the types of mutations listed in Table I.

Transitions. This is the most common mutation induced by UV in cI (Table I), in lacI (10) and in lacZ.[13] We conclude that it is formed predominantly at the 3' end of a 4-(pyrimidin-6-yl)-pyrimidinone, the "6-4 photoproduct",[18] for the following reasons.

1. Transitions occur overwhelmingly at -Py-Py- sequences in the cI, lacI and lacZ genes; the results for lacZ in the single-strand DNA phage M13 rule out a possible association with -Pu-Pu- sequences (13). Since transitions are also rare at the middle Py in a 5'-Pu-Py-Py- sequence,[9,10,13] they must be mostly at the 3' end of a -Py-Py sequence.

When an -A-C- sequence in the lacI gene was converted to a -T-C- sequence, UV-induced transitions at the C position were increased 15 fold,[12] vividly demonstrating the importance of a -Py-Py sequence.

2. The most common UV photoproduct is the cyclobutyl pyrimidine dimer and the next most common is the 6-4 photoproduct, with a yield about 10% that of the dimer.[18] Both products are formed in DNA only at -Py-Py- sequences. Since photoreactivation does not decrease the yield of mutation in the cI gene,[15] the dimer cannot be the premutagenic lesion. Since there is roughly one mutation per 100 dimers in the cI gene,[19] the premutagenic lesion must occur at a -Py-Py- sequence with a yield at least 1% that of the dimer, and the 6-4 photoproduct is the only known product meeting the requirements.

3. Another reason for excluding the dimer as the premutagenic lesion is that most mutations in phage irradiated with 313 nm light in the presence of acetophenone (which produces mostly T-T, with some T-C, dimers)[20] are transversions, with very few transitions.[19]

4. The group attached to the 4 position of the pyrimidine on the 3'

Table 1. UV MUTATIONAL SPECTRA IN THE LAMBDA cI GENE, DETERMINED BY DNA SEQUENCING

Mutation	Prophage in uvr^+ cells[a]	SOS-induced uvr^- host cells: phage $+30 J/m^2$, no PR[b]	phage $+30 J/m^2$, with PR[c]	phage no UV[d]
Transitions				
G:C to A:T	16	23	9	1
A:T to G:C	8	18	7	0
Transitions	21	8	10	2
Frameshifts	3	7	1	10
Double events (see text)	7	6	1	0
Deletions/insertions greater than 3 bp	1	0	0	0
	---	---	---	---
Total mutants sequenced	56	62	28	13

[a]Lambda cI857 indl Oam Ap^R as a prophage in AB1157 uvr^+ cells given 40 J/m^2 (R.D. Wood, unpublished).
[b]Lambda cI857 indl Oam irradiated with 30 J/m^2 and absorbed to AB1886 uvrA cells given 3 J/m^2 (ref. 19).
[c]DNA from lambda cI857 indl Oam Ap^R irradiated with 30 J/m^2 in solution, photoreactivated, packaged, and absorbed to AB1886 uvrA cells given 3 J/m^2 (ref. 15).
[d]Unirradiated lambda cI857 indl Oam absorbed to AB2497 uvr^+ cells previously given 100 J/m^2 (ref. 8).

side is lost in forming the 6-4 photoproduct. Since this group on the 4 position participates in Watson-Crick base pairing, the 3' pyrimidine cannot supply coding information to the DNA polymerase; this accounts neatly for the location of transitions at the 3' end of -Py-Py- sequences. The groups on the 5' pyrimidine which participate in Watson-Crick pairing are intact.

5. There is a rough correlation between the level of mutations at T-C, C-C, T-T, and T-C sites in lacZ [10] and in lambda cI [19] and the yields of 6-4 photoproducts at such sites.[21,22]

Since transitions are half or more of all UV-induced mutations, the dependence of UV mutagenesis on umuDC and recA shows that these gene products are required to form most transitions.

Transversions. Unirradiated E. coli cells in which the SOS functions were induced by genetic manipulation showed an order of magnitude increase in mutation, mostly transversions.[23] Nontargeted base substitution mutations in the O_R2 locus in unirradiated lambda phage were caused by UV induction of the SOS system in the host cells;[24] presumably, most of these were transversions. In both cases, mutagenesis was abolished in umuDC cells.

The level of transversions shown in the first three columns of Table I, 15% to 35% of the total, is higher than can be ascribed to the nontargeted component, so transversions must also be caused by targeted mutagenesis. The correlation with -Py-Py- sequences is much poorer than for transitions,[19,12] so it is not known whether transversions occur directly at photoproducts, in the vicinity of photoproducts, or both. Mutagenesis in umuDC cells is reduced to less than 15% of that in $umuDC^+$ cells, so these gene products must be essential for at least some targeted transversions.

Frameshifts. Nontargeted mutations in heavily irradiated host cells are mostly frameshifts (Table I), and the process is independent of umuDC.[8] The data available are inadequate to determine whether or not targeted mutagenesis also induces frameshifts. Most frameshifts induced by UV are loss or insertion of a single base pair and occur in runs of three or more identical base pairs.[9,13,12] Miller[12] fused lacI gene with a known +1 or -1 frameshift to the lacZ gene, and investigated the ability of UV to insert a compensating frameshift in lacI such that transcription would read through into lacZ and produce beta-galactosidase; screening thousands of UV-induced frameshifts in this way, he could show that -1 frameshifts are 20-fold more common than +1 frameshifts.

Double events. By this is meant two adjacent or nearby (1 to 4 nucleotides) base substitutions, or a base substitution plus a nearby frameshift. These double events occur at a level roughly 10% that of total UV mutagenesis in lambda cI (Table I), in lacI[10] and in lacZ in M13.[13] They occur at a frequency much too high to be ascribed to coincidences between two independent events (see also (25)): furthermore, direct sequencing in the lambda cI gene[8,19] and in lacZ[13] have not shown mutations 5 to 100 base pairs apart, as would be expected if the double events above were caused by tow independent events.

The double events require photoproducts in the mutated gene, and are not caused just by inducing the SOS system. (1) In hundreds of lacI mutations in cells with a genetically induced SOS system, no double events were found.[23] (2) In 73 lambda cI gamma ray mutations in SOS-induced cells, no double events of the type described were found (K.R. Tindall, unpublished); several pairs of gamma-ray-induced mutations 6 to 20 base pairs apart were found, but the mechanism is clearly different. Thus double events from UV must be targeted and must require a characteristic lesion(s) to induce them, a lesion not formed by ionizing radiation. They also must require umuDC activity, since lack of umuDC gene products decreases mutation by more than 90%.

Double events would be expected to have a low reversion rate by any agent, and so may account for some UV-induced "deletions" which have been characterized as such by the absence of reversion.

Deletions and Insertions more than 3 Base Pairs. Only one such mutation in the lambda cI gene, an 8 base pair deletion in a prophage, is listed in Table I. The lambda cI system apparently discriminates against the detection of large deletions, since about 5% of the UV mutations in lacI on an F' plasmid are of this type,[12] and large deletions involving two or more genes in the E. coli chromosome are well documented-- e.g. in the lac[26] and his[27] operons.

One mechanism for deletion and insertion is presumed to be recombination between segments of DNA having the same base sequences. UV is well known to stimulate recombination.[28,29] Very little is currently known about how these recombinational events lead to those DNA rearrangements which are scored as mutations.

CONCLUSIONS

This summary shows that UV mutagenesis involves several processes and several types of mutations. It is important to know, if some step or event affects, say, UV-induced reversion of a his mutant, what kinds of mutation cause the reversion. More, if reversion of the mutant is not affected, it is essential to know what kinds of mutation are involved, because statements can only be made about these mutations, and not about UV mutagenesis in general.

It is also clear that a plot of mutagenesis versus UV fluence will be complex, as has been shown,[30,31] and that the spectrum of mutations will depend on dose. Thus, extrapolation from experimental data at high dose to low dose situations involve considerations both of numbers and of kinds of mutations.

Finally, extrapolation of these results to other organisms may be particularly difficult because the SOS functions play such a large role in UV mutagenesis of E. coli. At least some bacteria apparently do not have an SOS system.[32] It is unclear whether any SOS-like system has been found in yeast,[33] although gene transcription induced by DNA damage has recently been demonstrated.[34] Published accounts showing, or not showing, SOS-like activities in mammalian cells in culture are too numerous to mention here; it is fair to say that about the same number of papers take each side, and that even the positive results are considerably smaller than those routinely found in E. coli.

ACKNOWLEDGEMENTS

This research was supported in part by grant GM28297 from the U.S. National Institutes of Health and by contract EVO 3571 from the U.S. Department of Energy.

REFERENCES

1. J. Weigle, Induction of mutations in a bacterial virus, Proc. Natl. Acad. Sci. USA 39:628 (1953).
2. J.W. Little and D.W. Mount, The SOS regulatory system of Escherichia coli, Cell 29:11 (1982).
3. G.C. Walker, Mutagenesis and inducible responses to deoxyribonucleic acid damage in Escherichia coli, Microbiol. Rev. 48:60 (1984).
4. D.G. Ennis, B. Fisher, S. Edmiston and D. Mount, Dual role for E. coli RecA protein in SOS mutagenesis, Proc. Natl. Acad. Sci. 82:3325 (1985).
5. E. Witkin and T. Kogoma, Involvement of the activated form of RecA protein in SOS mutagenesis and stable DNA replication in Escherichia coli, Proc. Natl. Acad. Sci. 81:7539 (1984).
6. R. Devoret, Influence du genotype de la bacterie hote sur la mutation du phage lambda produite par la rayonnement ultraviolet, C.R. Acad. Sci. Paris 260:1510 (1965).

7. H. Ichikawa-Ryo and S. Kondo, Indirect mutagenesis in phage lambda by ultraviolet preirradiation of host bacteria, J. Mol. Biol. 97:77 (1975).
8. R.D. Wood and F. Hutchinson, Nontargeted mutagenesis induced by ultraviolet light in Escherichia coli, J. Mol. Biol. 173:293 (1984).
9. A.M. Skalka, DNA replication--bacteriophage lambda, Curr. Topics Microbiol. Immunol. 78:201 (1977).
10. C. Coulondre and J.H. Miller, Genetic studies of the lac repressor IV. Mutagenic specificity in the lacI gene of Escherichia coli, J. Mol. Biol. 117:577 (1977).
11. P.A. Todd and B.W. Glickman, The mutational specificity of UV-light in Escherichia coli: indications for a role of DNA secondary structure, Proc. Natl. Acad. Sci. 79:4123 (1982).
12. J.H. Miller, On the mutagenetic specificity of ultraviolet light, J. Mol. Bol. 182:45 (1985).
13. J.E. LeClerc, N.L. Istock, B.R. Saran and R. Allen Jr., Sequence analysis of ultraviolet-induced mutations in M13lacZ hybrid phage DNA, J. Mol. Biol. 180:217 (1984).
14. R.M. Snapka and B.M. Sutherland, Escherichia coli photoreactivating enzyme: Purification and properties, Biochemistry 19:4201 (1980).
15. R.D. Wood, Pyrimidine dimers are not the principal pre-mutagenic lesions induced in lambda phage DNA by ultraviolet light, J. Mol. Biol. in press (1985).
16. E.M. Witkin, Ultraviolet mutagenesis and inducible repair in E. coli, Bacteriol. Rev. 40:869 (1976).
17. W.A. Haseltine, Site specificity of ultraviolet light induced mutagenesis. In "Cellular Responses to DNA Damage," E.C. Friedberg and B.A. Bridges, ed. A.R. Liss Inc. NY (1983).
18. M. Patrick and R. Rahn, Photochemistry of DNA and polynucleotides: Photoproducts, in "Photochemistry and Photobiology of Nucleic Acids," S.Y. Wang ed. Academic Press, NY II:35 (1976).
19. R.D. Wood, T.R. Skopek, and F. Hutchinson, The changes in DNA base sequence induced by targeted mutagenesis of lambda phage by ultraviolet light, J. Mol. Biol. 173:273 (1984).
20. R. Rahn and M.H. Patrick, Photochemistry of DNA, in "Photochemistry and Photobiology of Nucleic Acids," S.Y. Wang ed. Academic Press, NY II:97 (1976).
21. D.E. Brash and W.A. Haseltine, UV-induced mutation hotspots occur at DNA dosage hotspots, Nature 298:189 (1982).
22. W.A. Franklin, K.M. Lo, and W.A. Haseltine, Alkaline lability of fluorescent photoproducts produced in ultraviolet light-irradiated DNA, J. Biol. Chem. 257:13535 (1982).
23. J.H. Miller and K.B. Low, Specificity of mutagenesis resulting from the induction of the SOS system in the absence of mutagenic treatment, Cell 37:675 (1984).
24. P. Quillardet and R. Devoret, Damaged-site independent mutagenesis of phage lambda produced by inducible error-prone repair, Biochimie 64:789 (1982).
25. R.W. Dunst, R.D. Coleman, B.W. Harnish, and L.W. Hill, A system for measuring the frequencies of tandem and non-tandem double base substitutions induced by ultraviolet irradiation, Mol. Gen. Genet. 184:445 (1982).
26. D.O. Schwartz and J.R. Beckwith, Mutagenesis which causes deletions in Escherichia coli, Genetics 61:371 (1969).
27. E.P. Goldschmidt, M.S. Cater, T.S. Matney, M.A. Butler and A. Greene, Genetic analysis of histidine operon in Escherichia coli K12, Genetics 66:219 (1978).
28. P.F. Lin and P. Howard-Flanders, Genetic exchanges caused by ultraviolet photoproducts in phage lambda DNA molecules: The role of DNA replication, Mol. Gen. Genet. 146:107 (1976).

29. J.B. Hays, T.A.G. Smith, S.A. Friedman, E. Lee and G.L. Coffman, RecF and RecBC function during recombination of nonreplicating, UV-irradiated phage lambda DNA and during other recombination processes. Cold Spring Harbor Symp. Quant. Biol. 49:475 (1984).
30. C.O. Doudney, Complexity of the ultraviolet mutation frequency response curve in Escherichia coli B/r: SOS induction, one lesion and two lesion mutagenesis, J. Bacteriol. 128:815 (1976).
31. N.J. Sargentini and K.C. Smith, Multiple independent components of ultraviolet radiation mutagenesis in Escherichia coli K-12 uvrB5, J. Bacteriol. 140:436 (1979).
32. S.G. Sedgwick and P.A. Goodwin, personal communication.
33. W. Siede and F. Eckardt, Inducibility of error-prone DNA repair in yeast?, Mutat. Res. 129:3 (1984).
34. S.W. Ruby and J.W. Szostak, Specific Saccharomyces cerevisiae genes are expressed in response to DNA-damaging agents, Mol. Cell. Biol. 5:75 (1985).

ACTIVATION OF H-ras-1 ONCOGENES BY CHEMICAL CARCINOGENS

Helmut Zarbl, Saraswati Sukumar, Anne L. Arthur, Dionisio Martin-Zanca, and Mariano Barbacid

Basic Research Program
Frederick Cancer Research Facility
Frederick, Maryland 21701

INTRODUCTION

Accumulating evidence indicates that oncogene activation is intimately associated with the development of at least certain human cancers. Genes capable of inducing malignant transformation of normal, non-tumorigenic cells in culture have been identified in a significant fraction of human tumors (for a review, see Ref. 1). Most of these transforming genes are members of the ras gene family,[2-5] which are known to acquire their neoplastic properties by single point mutations within their coding sequences.[6-8] Recent studies have demonstrated that these critical mutations are somatic events associated with tumor development.[9] Furthermore, ras oncogenes are known to efficiently induce tumors in animals upon incorporation into retrovial vectors.[10-15] Yet these observations cannot be considered as sufficient evidence to establish a causal relationship between oncogene activation and the onset of human malignancies.

Efforts aimed at establishing the role of ras oncogenes in human neoplasia are complicated by the unknown etiology of most human cancers and the impossibility of experimental manipulations. We have, therefore, directed our research towards animal model systems that closely parallel human neoplasia. We have shown that transforming ras genes can be reproducibly identified in mammary carcinomas of rats induced by the chemical carcinogen nitroso-methyl-urea (NMU).[16,17] Similar results have also been obtained in other laboratories utilizing other carcinogen-induced animal tumor systems.[18-20] Moreover, carcinogen-activated ras oncogenes acquired their malignant properties by the same types of mutations as those present in human tumors.[16,17,21,22] These findings attest to the validity of animal tumor systems as models to study the role of ras oncogene activation in human neoplasia.

In this article we will discuss the specific activation of the H-ras-1 locus in mammary carcinomas of rats induced by different chemical carcinogens. More importantly, we provide experimental evidence indicating that the mutagenic activity of the carcinogens utilized is directly responsible for the malignant activation of the H-ras-1 oncogenes.

MATERIALS AND METHODS

Mn1 I Restriction Fragment Length Polymorphisms

In order to detect Mn1 I restriction fragment length polymorphisms (RFLP) in H-ras-1 oncogenes, forty micrograms of DNA isolated from normal breasts and carcinogen-induced mammary carcinomas were digested for sixteen hours with 40 units of Mn1 I using conditions recommended by the supplier (New England Biolabs, Inc.). Digests were electrophoresed (18 hours at 40 V) in horizontal 2.8% agarose gels (w/v) and blotted to either zeta-probe™ blotting membranes (Bio-Rad Laboratories) or nitrocellulose as described by Southern[28]. Membranes were hybridized under stringent conditions (50% formamide, 5 x SSC and 42°C) for 72 hours with 10^7 cpm/ml of ^{32}P-labelled probe obtained by nick translation of the 120 bp Hpa II-Sac I DNA fragment described in Figure 1A. Blots were exposed for 24 hours to Kodak XAR-5 film at -70°C in the presence of intensifier screens.

Synthetic Oligonucleotide Probes

Synthetic oligonucleotide probes were used to determine the point mutations responsible for malignant activation of the H-ras-1 locus in carcinogen-induced rat mammary carcinomas. DNAs were isolated from NIH/3T3 cells, representative NIH/3T3 transformants derived from mammary carcinomas, and NIH/3T3 cells co-transfected with the normal H-ras-1 gene and pSV2-neo[24] and selected for growth in the presence of G418. Fifty micrograms of each DNA were digested with 400 units of Hind III using conditions recommended by the supplier (New England Biolabs, Inc.). Digests were electrophoresed (12 h at 50 V) in horizontal agarose (0.7% w/v) gels which were subsequently prepared for hybridization as previously described by Kidd et al.[25] Dried gels were prehybridized for 1 h in hybridization buffer[25] containing 500 μg/ml of sonicated herring sperm DNA and subsequently hybridized[17] for 18 at the discriminating temperature previously determined for each individual oligonucleotide probe. [^{32}P]-labelled oligonucleotide probes were used at a concentration of 5 ng/ml (specific activity of ~2 x 10^9 cpm/μg) or 1.5 ng/ml (specific activity of ~7 x 10^9 cpm/μg) depending on the labelling protocol utilized (see below). Hybridized gels were then washed with constant agitation as follows: 300 ml (four changes) of 6 x SSC for 15 min each at room temperature and 400 ml of 6 x SSC for 30 min at room temperature. Then 500 ml of 6 x SSC at 2°C above the hybridization temperature were poured onto the gel and allowed to cool to room temperature for 30 min. Finally the gel was rinsed with 200 ml of 6 x SSC at room temperature, dried and exposed to Kodak XAR-5 film for 18 h at -70°C in the presence of intensifier screens.

Oligonucleotide Labelling I: Polynucleotide Kinase Method

Labelling of oligonucleotide probes was performed by two different procedures. Specific activities of about 2 x 10^9 cpm/μg were obtained by quantitative phosphorylation of 5' termini with polynucleotide kinase. Reactions (300 μl) containing 0.01 nmole of individual nonadecamers, 1 nmole of [α-^{32}P]ATP (7000 ci/mole, New England Nuclear), 60 mM Tris-HCl (pH 7.8), 15 mM 2-mercaptoethanol, 10 mM $MgCl_2$ and 60 units of T4 polynucleotide kinase (New England Biolabs) were incubated for 30 minutes at 37°C and were stopped by making mixtures 20 mM with respect to EDTA (pH 7.5). Labelled oligonucleotides were separated from unincorporated [α-^{32}P]ATP by ion exchange chromatography on Elutip™-d columns (Schleicher & Schuell, Inc.) according to manufacturers protocol. [^{32}P]-oligonucleotides were eluted with 0.5 ml of 1 M NaCl, 20 mM Tris-HCl (pH 7.5) 1 mM EDTA and added directly to hybridization mixtures.

Oligonucleotide Labelling II: Primer Extension Method

High specific activity probes (~7 x 10^9 cpm/μg) were synthesized as previously described.[26,27] An octamer complementary to the 3' end of the template nonadecamers was quantitatively phosphorylated at its 5' end under conditions described above. After phenol extraction, the phosphorylated primer was purified by gel filtration on Sephadex G-25 columns (Superfine) eluted in water. The octamer primer was then annealed to the nonadecamer templates in 50 mM NaCl-10 mM Tris-HCl (pH 8.0) - 1 mM EDTA at final concentrations of 50 ng/ml of template (7.5 μM) and 100 ng/ml of primer (30 μM) by heating to 80°C and cooling slowly to 4°C. Primer extension reactions were done in a final volume of 10 μl containing 0.1 mM dATP, 0.1 mM TTP, 5 mM β-mercaptoethanol, 50 mM NaCl, 25 mM Tris-HCl (pH 7.5), 10 mM $MgCl_2$, 1 μl of annealed primer and template mixture (30 pmoles and 7.5 pmoles, respectively), 30 pmoles each of [α^{32}P]dCTP and [α^{32}P]dGTP (Amersham, at 3000 Ci/mole, evaporated to dryness), and DNA Polymerase I Klenow Fragment at a final concentration of 0.5 U/μl (New England Biolabs). Reaction mixtures were incubated at 0°C for 90 min. 10 μl of formamide containing xylene cyanol and bromophenol blue marker dyes were then added and the samples were boiled for 5 min. The labelled nonadecamer probes were then separated from the unlabelled template by electrophoresis in a denaturing 20% polyacrylamide gel.[26] The positions of nonadecamer probes were determined by autoradiography, the bands excised and eluted by washing gel slices three times for 1-3 hours each with 300 μl of 0.9 M NaCl, 0.18 M Tris-HCl (pH 8.0) and 6 mM EDTA. Eluted probes were added directly to hybridization mixtures.

RESULTS

H-ras-1 Oncogenes in Carcinogen-Induced Mammary Carcinomas

Treatment of female rats with a single dose of carcinogen during their sexual development leads to the efficient and reproducible induction of mammary carcinomas.[28,29] We have previously shown that each of nine mammary carcinomas induced by NMU in Buf/N rats contained a transforming H-*ras*-1 oncogene.[16] These observations have now been generalized. 79% (30 of 38) of those DNAs isolated from NMU-induced mammary carcinomas of Buf/N rats were capable of inducing morphological transformation of NIH/3T3 cells in gene transfer experiments. Similar results (18 of 26 or 70%) were obtained with tumors induced in Sprague-Dawley rats, a random-bred strain. Transforming genes were also identified in mammary tumors induced by DMBA (7,12-dimethylbenz(a)anthracene), a different carcinogen. In this case, however, only five of twenty-six DNAs were capable of inducing transformation of NIH/3T3 in parallel gene gransfer experiments. The results with DMBA-induced tumors contrast the high frequency of oncogene activation observed in NMU-induced tumors. The reason for this difference remains to be determined.

The transforming genes present in these carcinogen-induced mammary carcinomas were characterized by submitting DNAs isolated from representative NIH/3T3 transformants to Southern blot analysis utilizing DNA probes specific for known oncogenes. Each of the transforming genes detected in the present studies was also identified as a transforming allele of the H-*ras*-1 locus, independently of the carcinogen utilized to induce the mammary carcinomas. The specificity and reproducibility of these results strongly support the concept that H-*ras*-1 oncogenes play an important role in mammary carcinogenesis.

Specific Mutagenesis Within the H-ras-1 Locus

ras oncogenes in animal tumors acquire their transforming properties by the same missense mutations previously identified in human tumors.[16,17,21,22] Some of these mutations alter sequences specifically recognized by restriction endonucleases, leading to the generation of diagnostic RFLPs which have been utilized to identify oncogenes in human tumor DNA.[6,7,9,30] We have utilized this experimental approach to identify the point mutations responsible for the specific activation of the H-*ras*-1 locus in NMU and DMBA-

Table 1. Restriction Enzyme Polymorphism Created by Point Mutations in Codons 12 and 61 of the Rat H-ras-1 Locus

Codon	Mutation	Polymorphism
12	G^{34} -> A	None
	G^{34} -> C	New Taq I, Ava I, Xho I sites
	G^{35} -> N	Lost Mnl I site
	G^{35} -> C	New Bbv, I, Fnu 4HI, Pst I sites
61	C^{181} -> A	None
	C^{181} -> G	New Hph I site
	A^{182} -> C	New Ava II, Sau 96I sites*
	A^{182} -> G	New Taq I site
	A^{182} -> T	New Xba I site
	A^{183} -> C	None
	A^{183} -> T	New Nla III site*

Positions 34-36 and 181-183 correspond to codons 12 and 61, respectively. A^{36} -> N and A^{183} -> G do not alter the coding properties of their respective condons. G^{34} -> T and C^{181} -> T lead to the generation of termination codons. *Due to multiple cleavage sites in the vicinity of codon 61, these polymorphisms cannot be detected by Southern blot analysis of genomic DNA.

induced mammary carcinomas. Table 1 summarizes the polymorphic restriction endonuclease cleavage sites that will be created or eliminated as a consequence of mutations affecting the coding properties of the 12th or 61st codon of the H-ras-1 gene. Digestion of tumor DNAs with Taq I, Pst I, Hph I, and Xba I failed to reveal any polymorphisms indicating that the point mutations G^{34} -> C, G^{35} -> C, C^{181} -> G, A^{182} -> T, and A^{182} -> G were not responsible for activation of the H-ras-1 locus in these mammary carcinomas (Table 1).

In contrast, most of the DNAs isolated from NMU-induced mammary carcinomas exhibited a polymorphic Mnl I DNA fragment (Figure 1). This Mnl I RFLP is not diagnostic of one specific mutation but rather identifies alterations that affect the GAGG sequence located in positions 35 to 38 of the first exon of the H-ras-1 locus. This sequence includes the second and third nucleotides of codon 12 (residues 35 and 36) and the two coding deoxyguanosines of codon 13 (residues 37 and 38). As illustrated in Figure 1A, the first exon of the normal H-ras-1 allele spans two Mnl I DNA fragments of 206 and 74 base pairs (bp). Elimination of this Mnl I cleavage site generates a single 280 bp Mnl I DNA fragment. Wild type (206 bp) and polymorphic (280 bp) DNA fragments can be identified by Southern blot analysis utilizing the 120 bp Hpa II-Sac I DNA fragment depicted in Figure 1A as a radioactive probe. Figure 1B shows a representative experiment. Most of the tumors tested exhibited the polymorphic 280 bp Mnl I DNA fragment diagnostic of a mutated H-ras-1 gene. In addition to this polymorphic fragment, each tumor DNA exhibited the normal 206 bp Mnl I DNA fragment, indicative of heterozygocity at this locus. The stronger intensity of the latter (Figure 1B) is probably contributed by the normal H-ras-1 alleles of the accompanying stroma.

In total 60 NMU-induced tumors of 71 tested, scored as positive in the Mnl I RFLP assay, indpendent of the strain of rats utilized (Table 2). In contrast, none of twenty-six normal breasts obtained from eighteen Buf/N and six Sprague-Dawley rats exhibited the Mnl I RFLP (Table 2). More importantly, four of these normal breasts were obtained from animals carrying one or more

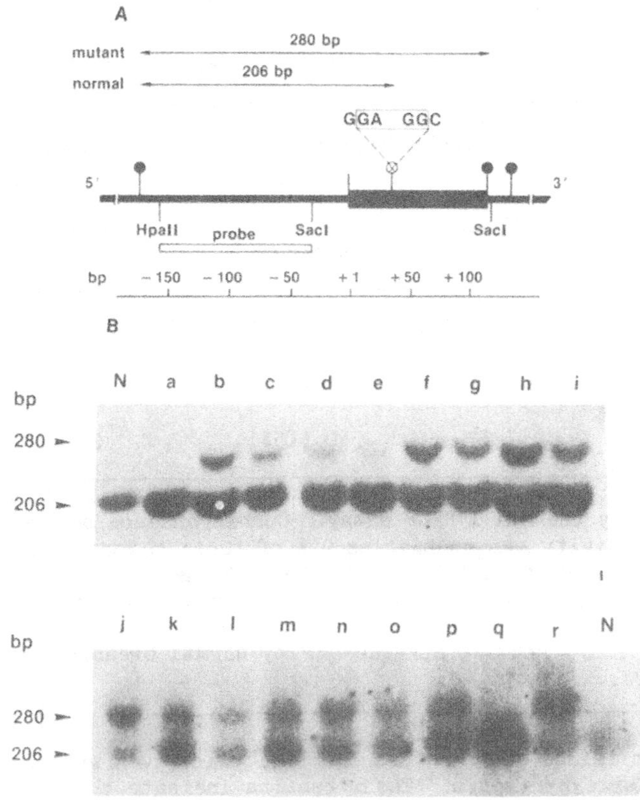

Figure 1. A) Strategy for detection of Mnl I restriction fragment length polymorphisms (RFLP) created by mutations within the twelfth or thirteenth codons of the rat H-ras-1 locus. Wild-type Mnl I restriction sites (●) in the vicinity of the first exon (solid box, nucleotides +1 to +111) are indicated. The position of the polymorphic Mnl I site (GAGG, ⊗ , nucleotides +35 to +38) relative to the sequences of the normal twelfth (GGA) and thirteenth (GGC) codons is shown. Arrows span the normal (206 bp) and polymorphic (280 bp) Mnl I restriction fragments detected by the 120 bp Hpa II-Sac I probe.
B) DNA isolated from normal breasts (N) and NMU-induced mammary carcinomas (a-r). Arrows indicate the positions of the normal (206 bp) and the polymorphic (280 bp) Mnl I restriction fragments.

mammary carcinomas. These observations rule out the possibility that the high frequency of Mnl I RFLPs might be due to random mutagenesis, and thus, suggest a close association between mutagenesis in this specific domain of the H-ras-1 locus and tumor development.

Definitive evidence correlating the Mnl I RFLP with malignant activation of the H-ras-1 locus was provided by analysis of representative NIH/3T3 transformants derived from NMU-induced mammary carcinomas. Each of the tranformants tested exclusively exhibited the polymorphic 280 bp Mnl I DNA fragment (data not shown), confirming the transforming properties of H-ras-1 alleles carrying the Mnl I RFLP. Those tumor DNAs that did not possess the Mnl

Table 2. Malignant Activation of the H-ras-1 Locus in NMU-Induced Mammary Carcinomas of Rats[a]

Strain	Mammary carcinomas		Normal breasts	
	Tested	H-ras-1 oncogene	Tested[b]	H-ras-1 oncogene
Buf/N	38	31 (81%)	18	0
Sprague-Dawley	26	23 (88%)	8	0
Fischer 344	7	6 (86%)	-	-
TOTAL	71	60 (85%)	26	0

[a]As determined by the Mnl I RFLP assay. Correspondence between Mnl I RFLP and transforming activity was as high as 92% of those tumor DNAs that were tested in more than one gene transfer assay. None of the normal breast DNAs was capable of inducing malignant transformation upon transfection into NIH/3T3 cells.
[b]Includes four Buf/N and two Sprague-Dawley normal breasts obtained from tumor-bearing animals.

I polymorphism failed to transform NIH/3T3 cells, even when repeatedly tested in gene transfer experiments. These results indicate that (i) malignant activation of the H-ras-1 oncogene in NMU-induced mammary carcinomas results from mutations specifically localized in a region no larger than four nucleotides, and (ii) oncogenes not capable of transforming NIH/3T3 cells may exist in a small percentage of NMU-induced mammary carcinomas.

NMU-Induced H-ras-1 Oncogenes are Activated by the Same G^{35} -> A Mutation

To establish the role of ras oncogenes in the development of neoplasia, it is necessary to define the stage at which they become activated. Recently, Balmain et al. have demonstrated the presence of transforming ras genes at the papilloma stage of DMBA-induced skin carcinomas.[31] Induction of rat mammary carcinomas by a single dose of NMU, however, does not proceed through identifiable pre-neoplastic stages in which the presence of ras oncogenes could be demonstrated. However, this animal tumor system possesses two important properties. First, initiation of carcinogenesis is a well defined process that is completed within hours following administration of NMU.[16,29] Second, the mutagenic properties of NMU have been well characterized. NMU is a direct acting alkylating agent that preferentially induces G -> A mutations as a consequence of its ability to methylate the O^6 position of deoxyguanosine residues (for reviews, see ref. 32, 33). In an effort to establish whether H-ras-1 oncogenes were directly activated by NMU during initiation of the carcinogenesis we examined whether the diagnostic Mnl I RFLP's shown above were generated by G -> A mutations.

For this purpose, we utilized oligonucleotide probes capable of detecting specific point mutations in genomic DNA.[34] We synthesized nonadecamers to identify substitutions in position 35. This residue is the only nucleotide of the Mnl I cleavage site that can alter the coding properties of the critical twelfth codon of the H-ras-1 gene. Oligomers included Ha19-G^{35} [5'TGGGCGCTGG*AGGCGTGGG-3', where the underlined nucleotides define the diagnostic Mnl I cleavage site and G* is the deoxyguanosine residue located in position 35]; while Ha19-A^{35} and Ha19-T^{35} had position 35 (G*) substituted by either a deoxyadenosine (Ha19-A^{35}) or a thymidine (Ha19-T^{35}). The corresponding Ha19-C^{35} oligonucleotide was not synthesized because a G^{35} -> C^{35} transversion would have created a polymorphic Pst I cleavage site (Table I) which, as indicated above, could not be detected in any of the tumors tested. The specificity of each of these three nonadecamers were verified by hybridization to plasmids containing the normal rat H-ras-1 gene (pH-ras-1) and a transforming allele (pNMU-1) previously shown to carry an activating G^{35} -> A mutation.[16]

The oligonucleotide probes were hybridized to Hind III cleaved DNAs isolated from the representative NIH/3T3 transformants obtained from each of the NMU-induced mammary carcinomas that scored positive in gene transfer experiments. NIH/3T3 transformants were selected because they contain slightly amplified oncogene sequences and lack the normal H-ras-1 rat allele. Results of representative experiments are depicted in Figure 2. At the discriminating temperature, each of the NIH/3T3 transformants tested hybridized to Ha19-A^{35} but not to Ha19-G^{35} or Ha19-T^{35}, indicating that their transforming H-ras-1 oncogenes carried identical G -> A transitions in position 35. As expected, DNA isolated from NIH/3T3 cells containing multiple copies of the normal H-ras-1 proto-oncogene (which was co-transfected with a selectable neo gene marker) hybridized to Ha19-G^{35}, but not to the Ha19-A^{35} or Ha19-T^{35} oligonucleotide probes (Figure 2).

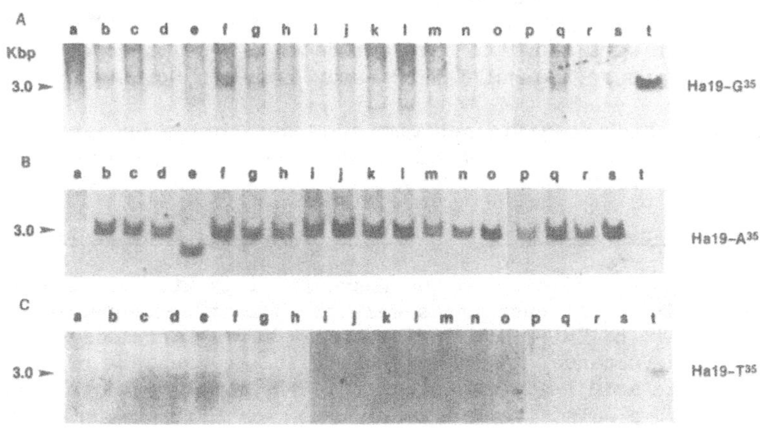

Figure 2. Use of synthetic oligonucleotide probes to determine the specific point mutations responsible for malignant activation of the H-ras-1 locus in NMU-induced rat mammary carcinomas. DNAs were isolated from NIH/3T3 cells (a), representative NIH/3T3 transformants derived from NMU-induced mammary carcinomas (b to s), and NIH/3T3 cells co-transfected with the normal H-ras-1 gene and pSV2-neo and selected for growth in the presence of G418 (t). A) Hybridization with probe Ha19-G^{35} (5'TGGGCGCTGGAGGCGTGGG 3'); B) Hybridization with probe Ha19-A^{35} (5'TGGGCGCTGAAGGCGTGGG 3'); C) Hybridization with probe Ha19-T^{35} (5'TGGGCGCTGTAGGCGTGGG 3'). Arrowheads indicate the expected 3.0 kbp Hind III DNA fragment of H-ras-1, which contains all coding sequences except the first 12 nucleotides.[16]

Detection of H-*ras*-1 oncogenes directly in tumor tissue was made possible by using high specific activity oligonucleotide probes The results of such an analysis is presented in Figure 3. Panel A depicts the results obtained when Bam HI digested tumor DNA were hybridized with the nonadecamer probe Ha19-G^{35}* (whose sequence is complementary to that of the nonadecamer probe Ha19-G^{35} described in the legend to Figure 2), which specifically detects the normal H-*ras*-1 locus. This oligonucleotide probe hybridizes with the expected 10 kbp Bam HI fragment in DNA isolated from both normal breast tissue (N) and mammary carcinomas (a-k). The detection of a normal H-*ras*-1 allele in tumor DNA corroborates the results obtained in the Mnl I RFLP assay (see Figure 1) and are indicative of heterozygocity of the H-*ras*-1 locus in these tumors.

Panel B shows the results obtained when Bam HI digested tumor DNAs were hybridized with the nonadecamer probe Ha19-A^{35}* (whose sequence is complementary to that of the nonadecamer probe Ha19-A^{35} described in the legend to Figure 2), which specifically detects H-*ras* oncogenes harboring G -> A transitions to the second deoxyguanosine residue of codon 12. The Ha19-A^{35}* probe hybridized with two Bam HI fragments. The first of these is a 12 kbp fragment that was detected in DNA from both normal breast and breast carcinomas. Hybridization of the probe with the 12 kbp fragment was more frequently detected in DNA from Fischer 344 rats (N and h-k) than in Sprague-Dawley rats (N and d-g) or Buf/N rats (N and a-c), and probably represents a polymorphism in the H-*ras*-2 pseudogene locus. The second fragment detected by the Ha19-A^{35}* probe is the expected 10 kbp Bam HI fragment containing the H-*ras*-1 oncogene.[16] This fragment was not detected in DNA from normal breast (N) or in DNA from mammary tumors that had been previously shown not to have activated H-*ras*-1 oncogene (lanes, a, d, and h) using DNA mediated NIH/3T3 cell transfection assays and the Mnl I RFLP assay. The Ha19-A35* probe did, however, detect the 10 kbp fragment in those tumors that were shown to have activated H-*ras*-1 oncogenes by transfection assays or the Mnl I RFLP assay (lanes b, c, e-g, and i-j).

Using these high specific activity oligonucleotide probes, we can now identify the activating G^{35} -> A mutations of H-*ras*-1 oncogenes in DNA isolated directly from tumor tissues.

DMBA-Induced H-ras-1 Oncogenes are Activated by Mutations in the 61st Codon

The striking specificity of the above findings strongly suggests that the activating G^{35} -> A transitions were the direct consequence of the mutagenic activity of NMU. However, alternative explanations must also be considered. For instance, it is possible that the presence of p21 proteins carrying glutamic acid residues in position 12 (coded by the mutated GAA codon) may confer a selective growth advantage to the neoplastic mammary cells. Alternatively, mammary cells may have repair systems that preferentially introduce deoxyadenosine residues. To address this fundamental question, we examined the mutations responsible for the activation of H-*ras*-1 oncogenes in mammary carcinomas induced by DMBA. This carcinogen forms large adducts with deoxyguanosine and dexoyadenosine residues leading to the induction of excision repair mechanisms, which occassionally generate point mutations of undefined specificity.[32] DNA from representative NIH/3T3 transformants derived from each of five DMBA-induced mammary carcinomas known to contain a transforming H-*ras*-1 oncogene, efficiently hybridized to the Ha19-G^{35} probe but not to the Ha19-A^{35} or Ha19-T^{35} probes.[17] These results indicate that H-*ras*-1 oncogenes in DMBA-induced tumors were not activated by point mutations in either the twelfth or adjacent codons In agreement with these observations, none of these oncogenes exhibit the polymorphic Mnl I RFLP.[17]

Previous studies have indicated that *ras* genes may also acquire

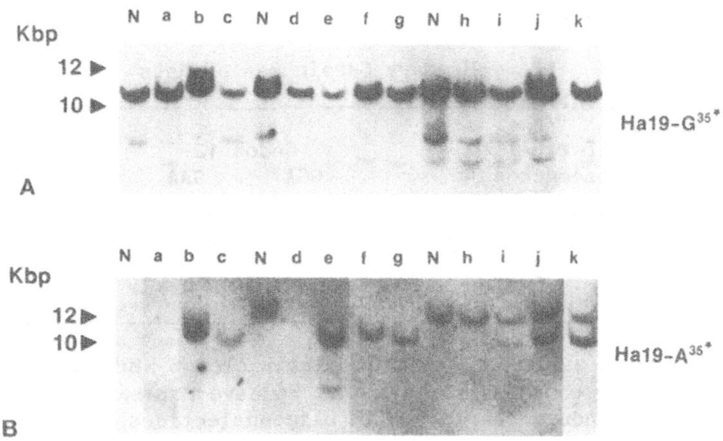

Figure 3. Use of high specific activity nonadecamer probes to determine specific point mutations responsible for the malignant activation of the H-ras-1 locus in NMU-induced rat mammary carcinomas. DNAs were isolated from normal breast tissue (N), from NMU-induced tumors that did not have activated H-ras-1 oncogenes (a, d, and h) and tumors which did have activated H-ras-1 oncogenes (b, c, e-g, and i-j). A) Hybridization with probe Ha19-G^{35*} (~7 x 10^9 cpm/µg), and B) hybridization with probe Ha19-A^{35*} (~6 x 10^9 cpm/µg). Arrowheads indicate the expected 10 kbp and 12 kbp Bam HI fragments which contain all of the coding sequences of the H-ras-1 gene and H-ras-2 pseudogene, respectively.[16]

transforming properties as a result of missense mutations in the sixty-first codon.[35,36] It was, therefore, plausible that H-ras-1 oncogenes present in DMBA-induced mammary tumors were activated as a result of point mutations within the sixty-first codon. In order to examine this possibility, we utilized a nonadecamer probe complementary to the sixty-first codon (CAA) and flanking sequences of the normal rat H-ras-1 locus. DNA from transformants derived from DMBA-induced tumors failed to hybridize to the nonadecamer probe using hybridization conditions that will not allow for detection of sequences with single base mismatches. These results indicated that each of the five H-ras-1 oncogenes present in the DMBA-induced tumors harboured mutations in the region of the sixty-first codon.

In order to localize these mutations we utilized mixed sequence nonadecamer probes. These oligomers consist of a mixture of sequences with all possible activating base substitutions at the position of interest. Base substitutions that lead to the generation of termination codons and those that do not lead to amino acid substitutions were omitted. Hybridization with these probes allows one to determine whether a mutation has occurred at a particular nucleotide, but does not allow identification of the particular mutation involved. Mixed sequence oligonucleotides capable of detecting all mutations in the deoxycytosine residue of codon sixty-one (CAA), failed to hybridize to DNA from any of the five transformants. Four of the transformants hybridized to the oligomer probe which detected point mutations in the second residue (deoxyadenosine) of codon sixty-one. One transformant hybridized to the oligomer probe which detected transversions in the third residue (deoxyadenosine) of codon sixty-one.

Table 3. Mutagenesis of H-ras-1 Oncogenes in Carcinogen-Induced Mammary Carcinomas of Rats

Carcinogen	H-ras-1 oncogenes	Codon 12 GGA	Codon 12 GAA	Codon 61 CAA	Codon 61 CNPy⁺
NMU	60	0	60	NT*	NT
DMBA	5	5	0	0	5

*Nucleotide sequence analysis of a molecularly cloned NMU-induced H-ras-1 oncogene revealed no changes in codon 61. We have not systematically hybridized the NMU-induced H-ras-1 with oligonucleotides specific for the 61 codon.
⁺N is deoxyguanosine, deoxycytosine, or thymidine. Py is deoxycytosine or thymidine.

These results indicate that H-ras-1 oncogenes of DMBA-induced tumors were activated by point mutations in either of the deoxyadenosine residues of codon 61 (CAA). These findings, summarized in Table 3, rule out the possibility that the G^{35} -> A mutations present in each of the NMU-induced H-ras-1 oncogene are the result of either positive growth selection or specific repair systems. Instead they indicate that malignant activation of the H-ras-1 locus in NMU-induced mammary carcinomas is the result of the direct mutagenic effect of NMU on this locus.

DISCUSSION

In spite of the wealth of information regarding ras oncogenes, their role in human neoplasia remains to be defined. Based on the apparent simplicity with which these genes acquire transforming properties[6-8,35,36] and on the fact that ras gene activation does not correlate with the histopathology of any type of human neoplasia,[37-39] it has been speculated that ras oncogenes may only play a circumstantial role in carcinogenesis. The reproducible and specific activation of H-ras-1 oncogenes in 60 of 71 NMU-induced mammary carcinomas of rats strongly argues against this possibility. Instead, it favors the concept that malignant activation of this ras locus is a necessary step in the development of these mammary carcinomas.

Additional support for the hypothesis that ras oncogenes play a major role in carcinogenesis has been provided by similar studies in other animal tumor systems.[16-22,40-42] For instance, ras oncogenes have been reproducibly detected in DMBA-induced mouse skin carcinomas[18] as well as in X-ray and NMU-induced mouse thymomas.[20] More recently, an EGF receptor-related oncogene, designated neu, has been identified in each of four ethyl-nitroso-urea induced neuroblastoma cell lines,[43] indicating that reproducible activation of specific oncogenes is not limited to members of the ras gene family, but instead might be a general property of tumors sharing a common genetic background, and perhaps more importantly, a defined etiology. Although the molecular bases for these observations are unknown, it is likely that malignant activation of a given locus might be determined by its role in those cells that become the target for carcinogenesis.

Induction of mammary carcinomas in rats by injection of a single dose of NMU is one of the animal tumor systems most amenable to experimental analysis by molecular approaches.[16,17] This is due in part to the fact that the mechanisms by which NMU induces carcinogenesis have been extensively studied (for reviews, see ref. 32, 33). NMU preferentially interacts with the N^7 and O^6 positions of deoxyguanosine (G) and with the N^3 position of deoxyadenosine (A). Whereas N^7MeG and N^3MeA elicit repair responses, O^6MeG acts as an efficient template for DNA polymerases. However O^6MeG frequently directs the incorporation of thymidine instead of deoxycytosine residues. Thus, if DNA replication occurs prior to the removal of O^6Me group by O^6MeG DNA methyl transferases, G -> A transitions will be generated. Our observations that each of 60 NMU-induced H-ras-1 oncogenes examined carried the same G^{35} -> A activating mutation suggest that NMU might be directly responsible for the malignant activation of these oncogenes.

Alternative explanations such as the existence of G -> A error prone repair systems in neoplastic mammary cells or a selective growth advantage of cells carrying ras p21 proteins with glutamic acid residue in position 12, were ruled out by experiments utilizing DMBA-induced mammary tumors. Each of five H-ras-1 oncogenes identified in these tumors became activated by single point mutations in either of the two deoxyadenosine residues of codon 61 (CAA). DMBA is known to form adducts with deoxyadenosine as well as deoxyguanosine residues. However, it has been proposed that deoxyadenosine adducts are most likely to be responsible for the carcinogenic effect of DMBA.[44] This hypothesis is based on the fact that whereas all benzanthracine derivatives react equally well with deoxyguanosine residues their carcinogenic potential correlates with their ability to form deoxyadenosine adducts. Taken together, our findings strongly implicate chemical carcinogens in the generation of point mutations that activate ras oncogenes during mammary carcinogenesis.

Our results have further biological implications regarding the role of ras oncogenes in the multistep process of carcinogenesis. Several laboratories have shown that ras oncogenes cannot transform normal (primary) rat embryo cells unless these cells also acquire the capability of self-proliferation either spontaneously, by carcinogen treatment, or by transfection of "immortalizing" oncogenes such as c-myc.[45-47] Thus, it might be postulated that initiation of carcinogenesis by NMU requires the concomitant activation of two cooperating oncogenes only one of which, H-ras-1, can be detected in gene transfer assays. However, we favor a somewhat different hypothesis. The inability of ras oncogenes to transform primary embryo cells might be a mere consequence of the limited proliferation properties of these cultures. Gullino et al.,[28] have demonstrated that induction of mammary carcinomas in rats requires that the carcinogenic insult must take place during the development of the mammary gland. Thus, it is very likely that hormone-dependent proliferation and/or differentiation, a normal physiological process, might provide conditions adequate for the mutation and/or phenotypic expression of the carcinogen-activated H-ras-1 oncogenes. Although it is evident that many questions remain to be answered, our studies illustrate how molecular characterization of transforming genes in experimentally induced tumors can provide important clues in unveiling the molecular mechanisms of carcinogenesis.

ACKNOWLEDGEMENTS

Dr. Helmut Zarbl is the recipient of a Postdoctoral Fellowship from the Fonds de la Recherche en Sante du Quebec, Quebec, Canada.

Research sponsored by the National Cancer Institute, DHHS, under contract No. N01-CO-23909 with Litton Bionetics, Inc. The contents of this publication do not necessarily reflect the views or policies of the

Department of Health & Human Services, nor does mention of trade names, commercial products, or organizations imply endorsement by the U.S. Government.

REFERENCES

1. Cooper, G.M. Science 217:801-806 (1982).
2. Der, C.J., Krontiris, T.G., and Cooper, G.M. Proc. Natl. Acad. Sci. USA 79:3637-3640 (1982).
3. Parada, L.F., Tabin, C.J., Shih, C., and Weinberg, R.A. Nature 297:474-478 (1982).
4. Santos, E., Tronic, S.R., Aaronson, S.A., Pulciani, S., and Barbacid, M. Nature 298:343-347 (1982).
5. Shimizu, K., Goldfarb, M., Suard, Y., Perucho, M., Li, Y., Kamata, T., Feramisco, J., Stavnezer, E., Fogh, J., and Wigler, M.H. Proc. Natl. Acad. Sci. USA 80:2112-2116 (1983).
6. Tabin, C.J., Bradley, S.M., Bargmann, C.I., Weinberg, R.A., Papageorge, A.G., Scolnick, E.M., Dhar, R., Lowy, D.R., and Chang, E.H. Nature 300:143-149 (1982).
7. Reddy, E.P., Reynolds, R.K., Santos, E., and Barbacid, M. Nature 300:149-152 (1982).
8. Taparowsky, E., Suard, Y., Fasano, O., Shimizu, K., Goldfarb, M., and Wigler, M. Nature 300:762-765 (1982).
9. Santos, E., Martin-Zanca, D., Reddy, E.P., Pierotti, M.A., Della Porta, G., and Barbacid, M. Science 223:661-664 (1984).
10. Harvey, J.J., Nature 204:1104-1105 (1964).
11. Kirsten, W.H., and Mayer, L.A. J. Natl. Cancer Inst. 39:311-335 (1967).
12. Peters, R.L., Rabstein, L.S., Louise, S. Van Vleck, R., Kellof, G.F. and Heubner, R.J. J. Natl. Cancer Inst. 53:1725-1729 (1979).
13. Rasheed, S., Gardner, M.B., and Huebner, R.J. Proc. Natl. Acad. Sci. USA 75:2972-2976 (1978).
14. Ellis, R.W., DeFeo, D., Shih, T.Y., Gonda, M.A., Young, H.A., Tsuchida, N., Lowy, D.R., and Scolnick, E.M. Nature 292:506-511 (1981).
15. Andersen, P.R., Devare, S.G., Tronick, S.R., Ellis, R.W., Aaronson, S.A., and Scolnick, E.M. Cell 26:129-134 (1981).
16. Sukumar, S., Notario, V., Martin-Zanca, D., and Barbacid, M. Nature 300:658-661 (1983).
17. Zarbl, H., Sukumar, S., Arthur, A.V., Martin-Zanca, D., and Barbacid, M. Nature 315:382-385 (1985).
18. Balmain, A., and Pragnell, I.B. Nature 303:72-74 (1983).
19. Eva, A., and Aaronson, S.A. Science 220:506 (1983).
20. Guerrero, I., Calzada, P., Mayer, A., and Pellicer, A. Proc. Natl. Acad. Sci. USA 81:202-205 (1984).
21. Notario, V., Sukumar, S., Santos, E., and Barbacid, M., In: "Cancer Cells" Vol. 2, pp. 425-432 (A. Levine, G. Vande Woude, and J. Watson, eds). Cold Spring Harbor Laboratory Press (Cold Spring Harbor, NY).
22. Guerrero, I., Villasante, A., Corces, V., and Pellicer, A. Science 225:1159-1162 (1984).
23. Southern, E.M. J. Mol. Biol. 98:503-517 (1985).
24. Southern, P.J., and Berg, P. J. Molec. Appl. Genetics 1:327-341 (1982).
25. Kidd, V.J., Wallace, R.B., Itakura, K., and Woo, S.L.C. Nature 304:230-234 (1983).
26. Studnecki, A.B., and Wallace R.B. DNA 3:7-15 (1984).
27. Bos, J.L., Verlaan-de Vries, M., Jansen, A.M., Veeneman, G.H., van Boom, J.H., and van der Eb, A.J. Nucleic Acids Res. 12:9155-9163 (1984).
28. Gullino, P.M., Pettigrew, H.M., and Grantham, F.H. J. Natl. Cancer Inst. 54:401 (1975).

29. McCormick, D.L., Adamowski, C.B., Fiks, A., and Moon, R.C. Cancer Res. 41:1690 (1981).
30. Fujita, J., Yoshida, O., Yuasa, Y., Rhim, J.S., Hatanaka, M., and Aaronson, S.A. Nature 309:464-466 (1984).
31. Balmain, A., Ramsden, M., Bowden, G.T., and Smith, J. Nature 307:658-660 (1984).
32. Singer, B., and Kusmierek, J.T. Ann. Rev. Biochem. 51:655-693 (1982).
33. Pegg, A.E. Cancer Investigation 2:223-231 (1984).
34. Conner, B.J., Reyes, A.A., Morin, C., Itakura, K., Teplitz, R.L. and Wallace, R.B. Proc. Natl. Acad. Sci. USA 80:278-282 (1983).
35. Yuasa, Y., Srivastava, S.K., Dunn, C.Y., Rhim, J.S., Reddy, E.P., and Aaronson, S.A. Nature 303:775-779 (1983).
36. Taparowsky, E., Shimizu, K., Goldfarb, M., and Wigler, M. Cell 34:581-586 (1983).
37. Murray, M.J., Shilo, B.-Z., Shih, C., Cowing, D., Hsu, H.W., and Weinberg, R.A. Cell 25:335-361 (1981).
38. Perucho, M., Goldfarb, M., Shimizu, K., Lama, C., Fogh, J., and Wigler, M. Cell 27:467-476 (1981).
39. Pulciani, S., Santos, E., Lauver, A.V., Long, L.K., Aaronson, S.A., and Barbacid, M. Nature 300:539-542 (1982).
40. Padhy, L.C., Shih, C., Cowing, D., Finkelstein, R., and Weinberg, R.A. Cell 23:865-871 (1982).
41. Parada, L.F., and Weinberg, R.A., Mol. Cell Biol. 3:2298-2301 (1983).
42. Sukumar, S., Pulciani, S., Doniger, J., DiPaolo, J.A., Evans, C.H., Zbar, B., and Barbacid, M. Science 223:1197-1199 (1984).
43. Schechter, A.L., Stern, D.F., Vaidyanathan, L., Decker, S.J., Drebin, J.A., Greene, M.I., and Weinberg, R.A. Nature 312:513-516 (1984).
44. Dipple, A., Pigott, M., Moschel, R.C., and Costantino, N. Cancer Res. 43:4132-4135 (1983).
45. Land, H., Parada, L.F., and Weinberg, R.A. Nature 304:596-602 (1983).
46. Ruley, H.E. Nature 304:602-606 (1983).
47. Newbold, R.F., and Overell, R.W. Nature 304:648-651 (1983).

ACTIVATION OF THE c-myc ONCOGENE

Michael D. Cole

Department of Molecular Biology
Princeton University
Princeton, NJ 08544

INTRODUCTION

The cellular proto-oncogenes were first identified as the sequences within the DNA of normal cells that were homologous to the transforming genes of the acute oncogenic RNA tumor viruses.[1] Numerous studies have linked neoplastic transformation to genetic changes in these evolutionarily conserved cellular genes. However, the function of the cellular proto-oncogenes in normal cell growth and differentiation is still largely unknown, but it is almost certain that these genes will represent important control elements in the cell. The mechanisms by which these essential genes become activated to induce abnormal growth are quite varied, ranging from point mutations [2-4] and chromosomal translocation [5-7] to gene amplification.[8-10] In some instances, the resultant protein is altered in structure, while in others the major consequence of the activating lesion appears to be the overexpression or deregulation of the cellular gene. Studies of the cellular oncogenes promise to provide insight not only into the process of neoplastic transformation, but also into the fundamental control of normal cell growth.

The c-myc oncogene has been implicated in many tumors

One of the cellular proto-oncogenes, c-*myc*, has been found to be activated in a wide range of tumors. Activation, in this case, refers to an identifiable genetic lesion that alters the cellular DNA at or near the c-*myc* gene. High levels of c-*myc* RNA have been found in many tumor cells[11] but in most cases this is probably a secondary consequence of the rapid rate of cell growth and not attributable to an altered c-*myc* gene. The first, and most striking, examples of activated c-*myc* genes were found in B cell neoplasia of three different species; avian bursal lymphomas,[12] mouse plasmacytomas [5,13,14] and human Burkitt lymphomas.[6,7] Virtually every tumor of these three types has an activated c-*myc* gene, either by proviral insertion in the case of avian lymphomas or chromosomal translocation in plasmacytomas and Burkitt lymphomas. The c-*myc* oncogene has also been linked to the conversion of small cell lung carcinomas to a more malignant phenotype by gene amplification.[10] Other examples of c-*myc* activation have been found in murine T cell lymphomas,[15] a promyelocytic leukemia[8,9] and a neuroendocrine tumor.[16] Interestingly, none of these neoplastic diseases are identical to the leukemias, carcinomas and fibrosarcomas that were originally described for MC19,[1] the prototype avian RNA tumor virus from which the c-myc gene was first identified.

Tumors linked to c-myc Oncogene Activation

Avian B Cell Lymphomas
Murine Plasmacytomas and T Cell Lymphomas
Human Burkitt Lymphomas
Small Cell Lung Carcinomas

However, subsequent experiments have demonstrated that MC29 virus can induce B cell lymphomas.[18] It is clear from all of the studies above that activation of the c-myc oncogene can be linked to the transformation of a wide range of cell types.

The function of the c-myc oncogene is presently unknown. The protein is localized in the nucleus and has been reported to have DNA binding activity.[19,20] However, the majority of the protein remains tightly bound to the nuclear matrix in fractionated cells.[21] The level of expression of the gene has been tightly linked to the proliferative state of the cell; the gene is not transcribed in quiescent cells and can be induced by mitogens in many cell types.[22,23] Constitutive expression of c-myc has been shown to induce growth in the absence of certain growth factors such as PDGF.[24]

The role of c-myc activation in cell transformation remains controversial. One model suggests that c-myc activation provides an immortalizing function to primary cells, while activated ras (or ras-like) genes induce transformation.[25] However, activated ras genes have not been found in plasmacytoma DNAs that have been tested in the NIH3T3 focus assay (unpublished results from our laboratory) and we have been unable to detect transforming activity with plasmacytoma DNA similar to that described in some studies.[26] Furthermore, our laboratory has recently found that immortalized cell lines which express viral promoter-linked c-myc genes are not morphologically transformed but become tumorigenic in animals[27] (see below).

Models of c-myc oncogene activation

All of the mechanisms that have been found to activate the c-myc oncogene seem to alter the regulation or level of expression of the gene. This is true for chromosomal translocation, gene amplification and proviral insertion. Conversely, only rare instances have been found in which any of the ras oncogenes have been activated by mechanisms other than mutation.[28] One potential explanation for this may be the lack of a functional assay for point mutations within the c-myc gene, either within the protein coding sequences or the control region. However, the frequency with which c-myc activation has been associated with overexpression or deregulation of the gene, as compared to ras, suggests that changes in the amount of myc protein in the cell may have a more profound influence on cell growth and subsequent transformation than alterations in protein structure.

The finding of chromosomal translocations and proviral insertions that activate the c-myc gene without altering the protein coding regions[29] suggests that regulation of the gene in normal cells is under stringent control. Further support for the importance of c-myc regulation comes from the observation that c-myc is induced in response to mitogens and shut off in quiescent cells.[22,23] Hence, it has been of great interest to study the mechanism by which the immunoglobulin (Ig) domain alters expression of c-myc after translocation of the gene in plasmacytomas and Burkitt lymphomas. Initial studies demonstrated that the c-myc and Ig genes were in opposite transcriptional orientation (head to head),[5] making it clear that no known promoter from the Ig gene was directly responsible for transcription of the translocated gene. It was also found that the normal c-myc gene, which co-exists with the translocated gene in the tumor cells, is not expressed.[29] These observations have given rise to several models to account for the transcriptional activation of the gene. For example, the Ig domain may enhance the expression of c-myc by either the presence of specific sequences that promote transcription or an altered chromatin configuration that increases the accessibility of the gene to transcription factors. An alternative model suggests that a negative regulator controls c-myc transcription and that displacement of, or mutations within, the 5' end of the gene disrupt the binding of this repressor leading to deregulated gene expression.[30] Another model suggests that there is an increased translational efficiency of the c-myc RNA associated with displacement of the first untranslated c-myc exon, which is frequently displaced by chromosomal translocation.[31] None of these models are mutually exclusive.

TRANSLOCATION AND TRANSCRIPTIONAL ACTIVATION OF THE C-MYC ONCOGENE

Translocation involves an abortive immunoglobulin gene rearrangement

Chromosomal translocation links the c-myc oncogene to the immunoglobulin alpha heavy chain gene (Ca) in the majority of mouse plasmacytomas.[5] The breakpoint within the c-myc gene usually occurs in the first intron, displacing the first, untranslated exon.[29] The c-myc and Ca genes are in opposite transcriptional orientation, with the breakpoint within the alpha switch sequences in the majority of tumor lines.[5] The Ig sequences immediately adjacent to the breakpoints often contain the short repeat elements that are characteristic of Ig switch regions[32-34] indicating that the translocation probably results from a random mistake in the switching process that fortuitously activates the c-myc gene and leads to clonal outgrowth of tumor cells. However, there are no sequences within the c-myc gene that have any significant homology to switch repeats and that would promote an abortive rearrangement.[33] The translocation breakpoints in Burkitt lymphomas are similar to those in plasmacytomas, with c-myc most frequently linked in a head-to-head configuration with the Cu gene.[35] The positions of the breakpoints within the c-myc locus are more variable, often occuring on the 5' or 3' sides of the gene.

Activation of a bidirectional promoter within the first c-myc intron

Studies of c-myc expression in plasmacytomas demonstrated that there were heterogenous RNAs that were often shorter than the normal transcript.[5] Since the translocation displaces the 5' end and normal promoters of the c-myc gene,[29] it was of interest to determine where the RNAs transcribed from the translocated gene initiated. Northern blot analysis demonstrated that the tumor-specific RNAs initiated within the first intron, utilizing sequences adjacent to the second c-myc exon.[23] S1 nuclease studies indicated that the RNAs utilized many different initiation sites, with no requirement for a TATA sequence like that found upstream from most eukaryotic promoters. The tumor-specific RNAs have heterogenous 5' untranslated sequences but still preserve the same AUG start codon as found in the normal c-myc RNA. Thus, there is no difference in the c-myc protein produced from the translocated allele.

Somewhat surprisingly, single stranded probes detected a second heterogenous set of RNA transcripts that initiated within the c-myc intron, but which are transcribed in the opposite direction, toward Ca.[23] These RNAs were also found to have hetergeneous 5' ends which lacked TATA sequences. Sandwich hybridization demonstrated that RNAs initiating within the c-myc intron were transcribed through the translocation breakpoint and into the Ca gene itself.

The bidirectional nature of the transcription from the translocated c-myc oncogene suggests the presence of sequences within the Ig domain that promote transcription from the c-myc intron. These transcriptional enhancing sequences within the Ig gene do not involve the 'heavy chain enhancer' that has been shown to promote transcription of productively rearranged Ig genes.[36,37] The heavy chain enhancer moves to the reciprocal of the translocation and does not influence transcription of the c-myc gene in plasmacytomas. Furthermore, analysis of the tumor line MOPC 104E, where c-myc is joined to the sequence 3' of the Ca gene, suggests that the Ca coding regions are not required for transcriptional activation of c-myc.[34] Somatic cell hybrid experiments also suggest that the Ig locus influences the expression of the c-myc gene after chromosomal translocation.[38]

Similar c-myc RNA Levels in Tumor and Proliferating Normal Cells

Since no changes in the structure of the c-myc protein are expected from the translocation, the major change must be in the regulation of the gene. In particular, it was essential to compare the level of c-myc expression between normal and tumor cells. We found that the level of c-myc RNA in plasmacytoma cells was at least 30X higher (on an RNA transcripts/cell basis) than that found in normal, quiescent lymphocytes.[23] Yet, when the lymphocytes were stimulated with bacterial lipopolysaccharide (a polyclonal mitogen), there was a dramatic induction of c-myc expression, to a level similar to that in the tumor cells. The difference observed between the tumor cells and quiescent normal cells probably represents the primary influence of the translocation, namely constitutive transcription of the c-myc oncogene at a level similar to that found in growing cells.

BIOLOGICAL ACTIVITY OF THE C-MYC ONCOGENE; ITS ROLE IN TUMORIGENESIS

The finding that the c-myc oncogene is translocated in virtually every plasmacytoma clearly links the transcriptional activation of the gene to the growth of the tumor cells. A major question that remains is what role c-myc activation plays in cell transformation, more specifically whether c-myc transforms cells or only extends their growth potential. To test for biological activity of c-myc, we have linked viral promoters to the gene, which should give constitutive expression when introduced into any cell type[27]. Initial attemps were made to induce foci or morphologically transformed cells by transfection of these activated c-myc genes onto fibroblast monolayers. No foci could be detected, while control plates transfected with the EJ ras gene gave numerous dense foci. This result demonstrates that the focus-formation assay using fibroblast monolayers has a limited ability to detect activated cellular oncogenes, since, as discussed below, activated c-myc genes induce cell transformation when other assays are used.

Tumorigenicity of fibroblast lines expressing activated c-myc genes

To assay more directly for c-myc induced alterations in cellular growth properties, established fibroblast lines were selected which express activated c-myc genes. Quite strikingly, all of the c-myc transfected lines formed tumors with a relatively short talent period (1-3 weeks), while the controls gave no tumors.[27] c-myc transfected Rat2 cells were also tumorigenic in syngeneic rats. These results demonstrate that the c-myc oncogene can transform previously immortalized cells, as defined by the acquisition of a tumorigenic phenotype. However, none of the tumorigenic cells were altered significantly in their morphology and they did not form foci in monolayers.

It was possible that the tumors that arose did not derive from the c-myc transfected cells as a whole, but from some rare phenotypic or genetic variant. Cells from one of the tumors were re-established in culture for comparison to the parental cells. No differences were observed either in morphology, growth rate or in anchorage-independent growth.[27] Furthermore, DNA from the tumors did not induce foci of transformed cells when transfected onto monolayers. These results indicate that the tumorigenicity of the c-myc transformed cells almost certainly derives from the expression of the transfected c-myc oncogene and not from some secondary genetic change.

Subtle alterations in cellular growth induced by c-myc

Studies of fibroblast lines expressing activated c-myc oncogenes showed several subtle alterations in growth properties.[27] First, the morphology of the c-myc transfected cells was not altered significantly when compared to controls, but there was a slight increase in refractility and a decrease in 'swirling' of the monolayers at confluence. The growth rate of the cells in 10% serum was similar to that of the controls, but a major difference was noted in the growth rate in low serum concentrations (0.5%). Cells expressing an activated c-myc gene grew at a significant rate, while the controls grew very slowly (Rat2) or not at all (NIH/3T3). The c-myc-transfected cells had a low, but significant, increase in the efficiency of anchorage-independent growth (1-2%). This is substantially lower than the anchorage-independent growth of ras-transformed cells, in which >90% of the cells form relatively large colonies.

Expression of the transfected and endogenous c-myc genes

The c-myc RNA levels in the fibroblast lines were analysed and compared to the levels found in normal cells and in plasmacytomas, where the c-myc gene is activated by chromosomal translocation. All of the cell lines which had been transfected with activated c-myc genes expressed novel transcripts which were consistent with the introduced constructs.[27] Studies of the expression of the exogenous versus endogenous c-myc genes yielded two very interesting results. The first is that expression of the exogenous c-myc gene(s) does not influence the expression of the endogenous gene, even though the former is expressed at 4-6 fold higher levels. Thus, if it is assumed that high levels of c-myc RNA yield high levels of the c-myc protein, this protein does not feed back and block expression of the endogenous gene.

The second observation is that the endogenous c-myc gene is differentially expressed in the c-myc-induced tumors as compared to the cell lines. As noted above, the endogenous gene is transcribed in the c-myc transfected cell lines at the same level as in the parental cells. However, in the tumors that arose from these cell lines, the endogenous c-myc gene was shut off. This result suggests that the endogenous c-myc gene is still responding to the control signals that normally govern the gene, even though the exogenous genes are constitutively expressed and the cells are growing as part of a tumor mass. Furthermore, when the tumor cells were reestablished in culture, the endogenous c-myc gene is reactivated (unpublished observations).

There are two likely explanations for this down regulation of the endogenous gene. The major influence may be in the accessibility of growth factors to the cells after injection into the animal. It has been shown that the c-myc gene is induced by platelet-derived growth factor (PDGF) in quiescent fibroblasts in culture.[22] Furthermore, the accessible concentration of PDGF in the animal is much less than that in the in vitro culture medium.[39] Thus, the endogenous c-myc gene may turn off because of the lack of exogenous growth factor stimulation, with a resulting inability of the parental fibroblast lines to grow in vivo. Cells expressing a viral promoter-linked c-myc gene have constitutive c-myc expression and grow continuously. A second signal that may regulate c-myc expression is contact inhibition. We have found that expression of the c-myc gene is decreased in confluent monolayers, while exogenously introduced genes are expressed at the same level (unpublished results). It is possible that the dense cell mass found in a tumor leads to decreased c-myc expression in a similar manner. Both of these factors may contribute to the low level of endogenous c-myc gene expression in the fibroblast tumors.

Biological Activity of the c-myc Oncogene

Transformation of Established Fibroblast Lines
Immortalization and Transformation of Primary Cells
Reduction in Growth Factor Requirements
Cooperation with the ras and Polyoma MT Oncogenes
in Primary Cell Transformation
Sensitivity to Tumor Promoters

Immortalizing activity of activated cellular oncogenes

In addition to the ability of activated cellular oncogenes to transform previously established cell lines, oncogenes can also immortalize primary cells. We, and others, have found that activated c-myc genes can induce the establishment of permanent cell lines from several types of early passage rodent cell cultures.[40,41] Immortalization requires constitutive expression of the c-myc gene, since viral promoter-linked genes are active while the c-myc gene with its normal promoter and 5' flanking sequences is inactive (A. Kelekar, unpublished results). It has also been shown that the H-ras gene can immortalize primary cells when it contains a point mutation at amino acid 12,[42] a change that has been linked to the genesis of cancer cells.[2-4] Thus, both c-myc and ras genes can rescue early passage cells from senescence as well as transform established lines. Moreover, we have recently found that transfecting additional copies of an activated c-myc oncogene into primary cells previously immortalized by c-myc induces anchorage-independent growth and tumorigenicity in nude mice (work in progress). It has also been shown that treatment of activated c-myc oncogene-transfected primary cells with the tumor promoter TPA induces foci of transformed cells.[43]

Cooperation of c-myc with ras in cell transformation

An additional activity associated with activated c-myc genes is the cooperation with mutated forms of the ras oncogene in the transformation of early passage rodent cells in a focus assay.[25] It was found that neither activated c-myc nor ras genes alone could transform primary cells, but that cotransfection of the two genes gave fully transformed and tumorigenic cells. The transformed cells have multiple copies of both the c-myc and ras genes and express high levels of each gene. However, it was subsequently found that when the mutant ras gene was linked to a strong promoter, fully transformed cells could be induced in one step with ras alone.[42] Thus, activated forms of either c-myc or ras can transform primary cells under the appropriate conditions.

SUMMARY

Activated forms of the c-myc oncogene have been linked to cell transformation in a variety of mammalian and avian tumors. Several mechanisms of gene activation have been described, but each mechanism gives rise to constitutive levels of expression, usually no higher than that found in proliferating normal cells. Thus, activation of the c-myc oncogene appears to require high levels of the c-myc protein, in particular at stages of cell growth and differentiation during which the gene is supposed to be inactive.

ACKNOWLEDGEMENTS

This work was supported by grants from the National Institutes of Health and the American Cancer Society.

REFERENCES

1. Bishop, J.M., Ann. Rev. Biochem. 52:301-354 (1983).
2. Tabin, C.J., S.M. Bradley, E.I. Bargmann, R.A. Weinberg, A.G. Papageorge, E.M. Scolnick, R. Dhar, D.R. Lowy and E.H. Chang, Nature 300:143-148 (1982).
3. Reddy, E.P., R.K. Reynolds, E. Santos and M. Barbacid, Nature 300:149-152 (1982).
4. Taparowsky, E., Y. Suard, O. Fusano, K. Shimizu, M. Goldfarb, and M. Wigler, Nature 300:762-765 (1982).
5. Shen-Ong, G.L.C., E.J. Keath, S.P. Piccoli and M.D. Cole, Cell 31:443-452 (1982).
6. Taub, R., I.Kirsh, C. Morton, G. Lenoir, D. Swan, S. Tronick, S. Aaronson and P. Leder, Proc. Natl. Acad. Sci. USA 79:7837-7841 (1982).
7. Erikson, J., A. ar-Rushdi, H.L. Drwinga, P.C. Nowell and C.M. Croce, Proc. Natl. Acad. Sci. USA 80:820-824.
8. Collins, S., and M. Groudine, Nature 298:679-681 (1982).
9. Dalla-Favera, R., F. Wong-Staal and R.C. Gallo, Nature 299:61-63 (1982).
10. Little, C.D., M.M. Lau, D.N. Carney, A.F. Gazdar and J.D. Minna, Nature 306:194-196 (1983).
11. Eva et al. Nature 295:116-119 (1982).
12. Hayward, W.S., B.G. Neel and S.M. Astrin, Nature 290:475-480 (1981).
13. Marcu, K.B., L.J. Harris, L.W. Stanton, J. Erikson, R. Watt and C.M. Croce, Proc. Natl. Acad. Sci. USA 80:519-524 (1983).
14. Adams, J., S. Gerondakis, E. Webb, L.M. Corcoran and S. Cory, Proc. Natl. Acad. Sci. USA 80:1982-1986 (1983).
15. Steffen, D., Proc. Natl. Acad. Sci. USA 81:2097-2101 (1984).
16. Alitalo, K., M. Schwab, C.C. Lin, H.E. Varmus and J.M. Bishop, Proc. Natl. Acad. Sci. USA 80:1707-1711 (1983).
17. Graf, T. and H. Geug, Biochim. Biophys. Acta 516:269-299 (1978).
18. Hayward, W.S., C.-K. Shih and C. Moscovici in Tumor Viruses and Differentiation, Alan R. Liss, Inc., New York, pp. 279-287 (1982).
19. Donner, P., I. Greiser-Wilke and K. Moelling, Nature 296:262-266, (1982).
20. Abrams, H.D., L.R. Rohrschneider and R.N. Eisenman, Cell 29:427-439 (1982).
21. Eisenman, R.N., C.Y. Tachibana, H.D. Abrams and S.R. Hann, Mol. Cell.

Biol. 5:114-126 (1985).
22. Kelly, K., B.H. Corcoran, C.D. Stiles and P. Leder, Cell 35:603-610 (1983).
23. Keath, E.J., A. Kelekar and M.D. Cole, Cell 37:521-528 (1984).
24. Armelin, H.A., M.C.S. Armelin, K. Kelly, T. Stewart, P. Leder, B.H. Cochran and C.D. Stiles, Nature 310:655-660 (1984).
25. Land, H., L.F. Parada and R.A. Weinberg, Nature 304:596-602 (1983).
26. Lane, M.A., A. Sainten and G.M. Cooper, Cell 28:873-880 (1982).
27. Keath, E.J., P.G. Caimi and M.D. Cole, Cell 39:339-348, (1984).
28. Schwab, M., K. Alitalo, H.E. Varmus, J.M. Bishop and D. George, Nature 303:497-501 (1983).
29. Stanton, L.W., R. Watt and K.B. Marcu, Nature 303:401-406 (1983).
30. Leder, P., J. Battey, G. Lenoir, C. Moulding, W. Murphy, H. Potter, T. Stewart and R. Taub, Science 222:765-771 (1983).
31. Saito, H., A.C. Hayday, K. Wiman, W.S. Hayward and S. Tonegawa, Proc. Natl. Acad. Sci. USA 80:7476-7480 (1983).
32. Calame, K., S. Kim, P. Lalley, R. Hill, M. Davis and L. Hood, Proc. Natl. Acad. Sci. USA 79:6994-6998 (1982).
33. Gerondakis, S., S. Cory and J.M. Adams, Cell 36:973-982 (1984).
34. Piccoli, S.P., P.G. Caimi and M.D. Cole, Nature 310:327-330 (1984).
35. Klein, G., Cell 32:311-315 (1983).
36. Gillies, S.D., S.L. Morrison, V.T. Oi and S. Tonegawa, Cell 33:717-728 (1983).
37. Banerji, J., L. Olson and W. Schaffner, Cell 33:729-740 (1983).
38. Nishikura, K., R. Abbas, J. Erikson, R. Watt, G. Rovera and C.M. Croce, Proc. Natl. Acad. Sci. USA 80:4822-4826 (1983).
39. Huang, J.S., S.H. Huang and T.F. Deuel, J. Cell Biol. 97:383-388 (1983).
40. Ruley, H.E., J.F. Moomaw and K. Maruyama in Cancer Cells (eds. Vande Woude, G.F., Levine, A.J., Topp, W.C. and Watson, J.D.) Cold Spring Harbor Press, pp. 481-486.
41. Mougneau, E., L. Lemieux, M. Rassoulzadegan and F. Cuzin, Proc. Natl. Acad. Sci. USA 81:5758-5762 (1984).
42. Spandidos, D.A. and N.M. Wilkie, Nature 310:469-475 (1984).
43. Connan, G., M. Rassoulzadegan and F. Cuzin, Nature 314:277-279 (1985).

SALMONELLA TESTER STRAINS: MUTATIONAL TARGETS AND CORRELATION WITH ANIMAL CARCINOGENICITY AND TERATOGENICITY

Philip E. Hartman[1] and Sharon Lea Aukerman[1,2]

The Johns Hopkins University
Department of Biology
Baltimore, MD 21218

INTRODUCTION

Other presentations in this session summarize molecular changes that appear to be primary participants in the activation of cellular proto-oncogenes. These mutations involve chromosome rearrangements and simple base substitutions. Mutations in human ras genes have been detected that affect codon positions 12 or 61 and involve five of the six possible kinds of base substitutions; only an A/T -> C/G transversion had not been detected as of 1984.[1] This absence is not surprising because of the limited cases available, the fact that 4 of the 6 base pairs in the first two positions of the normal target codons happen to be G/C base pairs, and the possibility that only a limited set of substitutions may activate the protein product.

Recent experience in this exciting field also instructs us that multiple genetic events leading to activations of multiple cellular oncogenes probably are necessary for fully expressed tumors.[2-4] Single mutations in individual genes certainly are not sufficient for full cancer induction in one step,[5] but single mutations could well form critical stages in tumor "progression" and in the formation of lethal metastases (cf. 6; other situations are reviewed in 5).

An important point to consider is that multistep mutational processes in progression to cancer allow great impacts to be made by defects in DNA repair processes. For example, if four independent mutational events are required to generate an ultimate tumor and the appropriate mutation frequency is increased 40X in a repair-defective clone, the probability of reaching the ultimate combination of genetic changes is increased by a factor of over two million-fold. Therefore, it is not surprising that roughly 20% of tumors examined exhibit defective O^6-alkyltransferase activity[7] or that defective excision repair underlies oncogenesis in xeroderma pigmentosum.[8] It is to be expected that a very wide variety of molecular changes could lead to inactivation of DNA repair activities; these mutations could include frameshifts, base substitutions, deletions, and probably more rarely, translocations.

---

[2] Present address: The University of Texas System Cancer Center, M.D. Anderson Hospital and Tumor Institute, Department of Cell Biology, Houston, Texas 77030 USA.

A critical role of somatic mutation in cancer causation is supported by a broad variety of evidence [9] beyond the now established fact of sequenced mutations affecting the structure and function of cellular oncogenes. Even the so-called "promotion" phase of tumor induction probably has mutation as a critical component in multi-step (not just two-step) progression.[10] It is no wonder, then, that short-term tests for mutagens constitute an essential element in a broadening array of valuable tools available today to fight cancer before it begins, namely by lessening human exposure to mutagens.[11-14] Results from a wide variety of short-term tests have been summarized to 1979,[11] and the distribution of the genetic toxicology data base among assay systems portrayed to 1983.[14] We pick out for description here the most widely used of these tests, the "Ames test." Aspects of the utility of the Salmonella tests have been summarized previously.[15-18]

EVOLVING SALMONELLA TESTER STRAINS

The Salmonella mutagen tester strains are sometimes spoken of as if they were part of the fossil record. In fact, there has been continual development of the strains, and development continues today. Salmonella histidine-requiring mutants were first used to assay for mutagenicity of cycasin aglycone (methylazoxymethanol), a naturally occurring carcinogen and methylating agent,[19] and to determine that covalent linkage to DNA produced much stronger frameshift mutagenesis than did mere intercalation or stacking.[20] Particular histidine-requiring mutants were carefully selected from among over 1,000 characterized mutants [21] in searches for those with conveniently low background levels of spontaneous mutability and optimal responses to particular known mutagens. Examples are the response of hisG46 to a methylating agent.[19] and the response of hisD3052 to a variety of frameshift mutagens.[22] We were particularly fortunate to have a series of histidine-negative frameshift mutants induced by ICR 191 and ICR 364-OH.[23] This is because this set included about 50% "plus" frameshifts, a property typical of the ICR compounds and not merely peculiar to the histidine operon.[24] In fact, our collection of 140 histidineless frameshift mutants that arose spontaneously or after x-ray, fast neutron, nitrous acid, or ultraviolet light treatments included only one classified as a "plus" frameshift.[21] "Plus" frameshifts in repetitive sequences are much more readily mutable by exogenous agents than are "minus" frameshifts. Reasons for this behavior are reviewed by Roth [25] and are apparent in the mutagen-susceptible frameshift "hot-spots" described below.

The Ames group sequentially developed derivatives that were: (a) lacking in complete cell walls (gal, rfa) and thus more penetrable by some mutagens, (b) defective in the uvr system of DNA repair, a repair system active on pyrimidine dimers and other helix-distorting lesions (eg. DNA adducts the size of a propyl group or larger), and (c) containing plasmid pKM101 that sensitizes the bacteria to some mutagens, such as DNA interstrand cross-linking agents, to which the original strains were poorly responsive[26] (reviewed in 27). The molecular activity of the genes (mucA, mucB) on pKM101 active in enhancing mutagenesis is only partially understood [28] Plasmid pKM101 may allow apurinic and apyrimidinic sites to be much more mutagenic than would be the case in its absence. Strains containing pKM101 are more mutagen sensitive, have high spontaneous backgrounds, and differentiate base substitution and frameshift mutagens less well than do their plasmidless counterparts. The addition of microsomal extract to activate promutagens was a critical innovation [29,30] (reviewed in 27). Table 1 shows how some of the above steps influenced the Salmonella response to a pharmaceutical, hycanthone, which is an agent used to treat schistosomal infections in the human.[31]

Table 1. Effects of DNA repair (uvrB) and cell wall (gal, rfa) defects on the mutagenic response to hycanthone furoate (100 µg per plate) of Salmonella frameshift mutations hisC3076 and hisD3052 Spontaneous reversions were subtracted in all cases.

Strain	Induced Reversions per plate	Relative No. of Induced Reversions
hisC3076	19	=1
TA1952 (hisC3076 uvrB)	63	3
TA1532 (hisC3076 uvrB gal)	218	11
TA1537 (hisC3076 uvrB rfa)	2145	112
hisD3052	10	=1
TA1534 (hisD3052 uvrB)	165	17
TA1964 (hisD3052 uvrB gal)	349	35
TA1538 (hisD3052 uvrB rfa)	2420	242
TA1978 (hisD3052 rfa)	379	38

One important aspect of the work of the Ames laboratory, not often appreciated, was a critical attempt to assess the utility and validity of the test system by the examinaion of a wide spectrum of chemical structures encompassing almost 300 compounds.[32,33] A short-term test is not useful if it takes extreme effort to obtain positive results with only a few potent mutagens.

Since the classic summary of the tester strains in 1975,[27] emphasis has been on the procuring of histidine-requiring mutants with broad specificities for mutagens not readily diagnosed by the original strains (eg. introduction of hisG428) and in adapting the test procedure. The latter includes the important introduction of a preincubation step (reviewed in 17, 34).

Other Salmonella mutagenicity tests have been described utilizing forward-mutation assays to 8-azaguanine resistance,[35-37] a tryptophan-requiring mutant with broad specificity,[38] mutations to streptomycin independence,[39] and double mutants carrying both frameshift (histidine-requirement) and base substitution (leucine-requirement) mutations.[40] A simple taped-plate assay has been developed for detection of volatile mutagens.[41]

Although each system may possess particular advantages in research experimentation, the histidine-requiring tester strains stand apart for routine testing.[14,17] A strategy for sequential use of individual Salmonella strains to make them even more cost-effective is being developed[42]. In such a case, it often would be necessary to store chemicals in solution between assays, a procedure that appears to be effective for a wide array of chemicals.[43]

MOLECULAR TARGETS FOR MUTAGENESIS

The Salmonella mutants, selected from among a large number of mutants for broad mutagen sensitivities, have interesting molecular targets for mutagens. Figures 1 through 5 and the respective figure legends describe the primary mutagen-sensitive nucleotide sequences of some widely used and/or particularly interesting mutants. The sequences depicted include frameshifts hisD3052, hisD6610, and hisD6580 as well as base substitutions hisG428 and hisG46. It is apparent that multiple modes of reversion to a wildtype phenotype are possible in each instance; this fact is sometimes forgotten in the mutation literature where it is assumed that reversion automatically signifies a very specific kind of molecular change. In fact, increasingly frequent misinterpretation of specificities in the literature is one primary reason the authors decided to assemble these sequences here. It is true, though, that there are particular highly mutagen-susceptible nucleotide stretches in each target sequence. These stretches are the hisD3052 GCGCGCGC sequence, the hisD6610 CCCCCC sequence, the hisD6580 AAAAA sequence, as well as the A/T and G/C-containing codons of hisG428 and hisG46, respectively.

Between them, hisG428 and hisG46 are reverted by all of the six possible base substitution mutations, including both transitions and all four transversions. The frameshift strains contain critical G/C-rich and A/T-rich regions. Regarding hisD3052, it has been speculated that mutagens that form adducts at the C8 position of guanine may preferentially act at repeating GC sequences [64] and that DNA repair may be decreased.[65] N-methyl-N-nitrosourea has been shown to form a ratio of N^7 to O^6-methylguanine of 0.1 with poly(dG).poly(dC) versus 0.2 with poly(dG-dC).poly(dG-dC) in solution, again indicating that an alternating GCGCGCGC sequence may be a true "hot spot" for the detection of mutagens.[66] While there are suitable explanations for induction of frameshift mutations in repetitive sequences by intercalating agents (see ref. 25), currently we know of no good explanations for the ability of methylating agents to induce minus frameshift mutations (examples for N-methyl-N'-nitro-N-nitrosoguanidine, NG in Figure 1).

Frameshift hisC3076 was induced with ICR 364-OH in S. typhimurium LT-2 strain ara-9 (non-fermentation of arabinose) and found to revert well with the same compound.[23] Strains carrying frameshift hisC3076 were originally recommended for mutagenicity testing[27] and have been widely used; however, they have been replaced by strains carrying hisD6610 in the Ames laboratory's most recent recommendations.[34] The sequences surrounding the hisC3076 mutation are not known, but the mutation is assumed to be a + G/C base pair in a sequence of G/C's, akin to other strains that might have been developed into equally useful tester strains, such as hisA3043, hisD3018, and hisF3031 (see ref. 22, 67). The mutagen-susceptible sequence is presumed through the pattern of reversion with known mutagens [22,31,67] and on studies of frameshift suppressor genes that read four-base GGGG or CCCC as glycine or as proline codons, respectively [47,48,52,53] (reviewed in 25). Strains carrying hisD6610 reveal several mutagens not active on strains carrying hisC3076;[50] on the other hand, hisD6610 is more difficult to handle and the existing data base on hisC3076 is extremely broad.

MONOSPECIFIC TESTERS

The recommended Salmonella tester strains have been designed to allow detection of broad ranges of mutagens. They each have multiple modes of reversion that detect a variety of changes in the nucleotide sequence. On the other hand, strains are being developed that respond specifically to just one kind of base pair substitution.[61] These are strains presumably containing mutations in critical codons, for example in a codon affecting a particular amino acid essential in the active site of an enzyme and for which multiple suppressor loci are not present. One such strain, TR5071,

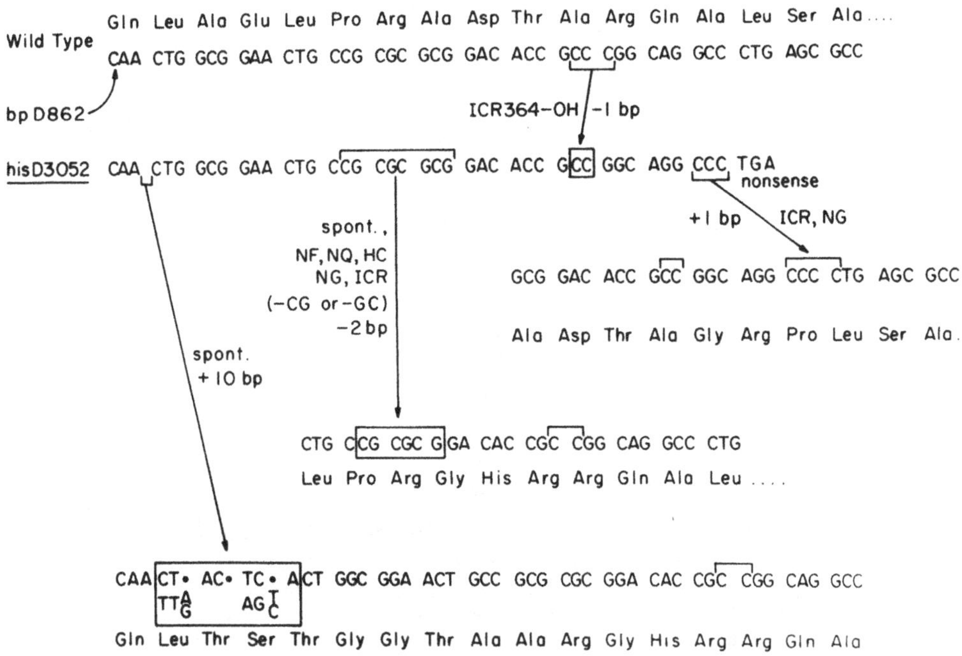

Figure 1. Mutant hisD3052 was derived from an ara-9 (no arabinose fermentation) mutant of *Salmonella typhimurium* LT-2. A minus C frameshift was induced in the hisD gene by the mutagen ICR 364-OH.[23] Nucleotide sequences are provided by the data of Isono and Yourno,[44] derived from amino acid sequencing of peptides of wild type and revertant proteins, and from the DNA sequence data of the wild type gene (W.M. Barnes and E. Tuley, personal communication). The strand sequence shown is that which directly reflects the messenger RNA sequence. Standard abbreviations are used for amino acids (Gln = glutamine, etc.), and ... indicates that translation of the polypeptide continues in-phase to its normal terminus. "Nonsense" indicates premature termination of the polypeptide chain. The UGA codon is predominantly expressed as nonsense, but a low level of translation occurs in many *Salmonella* strains.[45] In those rare instances where translation does continue, another nonsense codon would terminate the translation "downstream." There are at least 6 classes of revertants of which only three classes are diagrammed. Nucleotide changes in the other 3 classes have not been determined. The predominant class of revertants (center of Figure) lack a CG or a GC from a redundant sequence (boxed). This class includes all of the 2-nitrosofluorene (NF), nitroquinoline-*N*-oxide (NQ), and hycanthone (HC) induced revertants, about half of the spontaneous (spont.) and *N*-methyl-*N*'-nitro-*N*-nitrosoguanidine (NG) induced revertants, and a smaller proportion of the ICR-induced revertants analyzed.[44] There is ambiguity in the sequence of the +10 base pair (bp) insertion (left side of Figure) since the nucleotide sequence has been deduced only by peptide mapping.[44] The wild type sequence depicted starts at base pair #862 of the hisD gene. Mutation hisD3052 is present in the tester strains TA1534, TA1538 and TA98.[27,34]

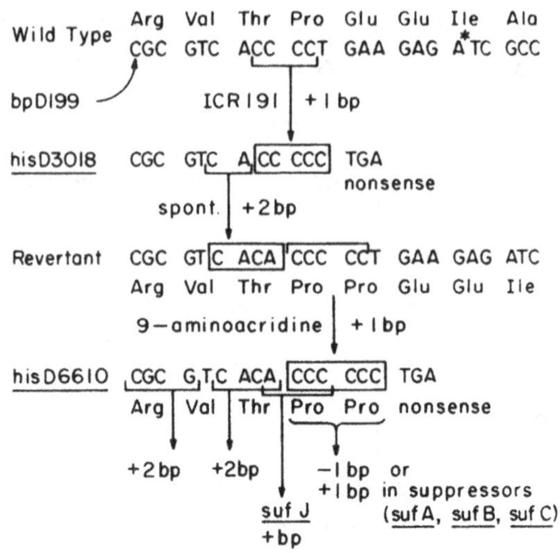

Figure 2. Mutant hisD6610 was derived in multiple steps [50] starting with an ICR-induced mutant, hisD3018 in *S. typhimurium*.[23] Mutation hisD3018 (boxed) was determined to be a +C frameshift based on the amino acid sequence of revertant proteins [46] and on ability of an unlinked suppressor mutation (sufB) to insert proline and concomitantly restore the reading frame [47-49] (summarized in 25). The nucleotide sequence of wild type has been detailed by DNA sequencing (W.M. Barnes and E. Tuley, personal communication). A spontaneous (spont.) revertant of hisD3018 contained a two-base pair (bp) repeat of the CA doublet, yielding an extra proline (Pro) in the amino acid sequence.[46,50] A new histidine mutation, hisD6610, was introduced into a derivative strain by treatment with 9-aminoacridine. The mutation was located to the hisD3018 region of the hisD gene by genetic mapping into one of the 79 deletion segments of the histidine operon (cf. 21). The pertinent change (boxed) was confirmed by DNA sequencing.[50] The original derivative strain, TA88, also contained a hisO1242 mutation that eliminated attenuation of his messenger RNA formation, leading to high constitutive levels of histidine enzymes. Comparison of this strain with a hisO$^{\pm}$ derivative showed that reversion of hisD6610 induced by several diagnostic mutagens was slightly enhanced in the presence of the hisO1242 mutation.[50] This enhanced mutagenicity may be due to the effect of derepression of the histidine operon in inducing an error-prone ("SOS") DNA repair system [51] or merely to enhanced transcription activity in this genetic region. TA97, the recommended tester strain containing hisD6610,[34] also contains the hisO1242 mutation and plasmid pKM101 which carries genes for a very active SOS system (reviewed in 28). Mutant hisD6610 has a mutation-prone sequence of 6 C's in a row wherein a -C would restore the reading frame and the wild type phenotype. Mutant hisD6610 also may be pictured as reverting intragenically by + 2bp additions to either of the two redundant sequences, CGCG or CACA. In addition, a presumed + 1bp addition to any one of three tRNA suppressor genes (suf) is pictured as capable of inserting proline,

using a 4-base sequence and, thus, restoring the reading frame[52,53] (also see 25). Finally, sufJ suppressors recognize ACCC as a "triplet" codon.[53] Strains TA89, TA90 and TA110 carry hisD6610.[34] The asterisk indicates a 6-methyldeoxyadenosine residue in the sequence GA*TC, involved in strand recognition during mismatch repair.[54,55]

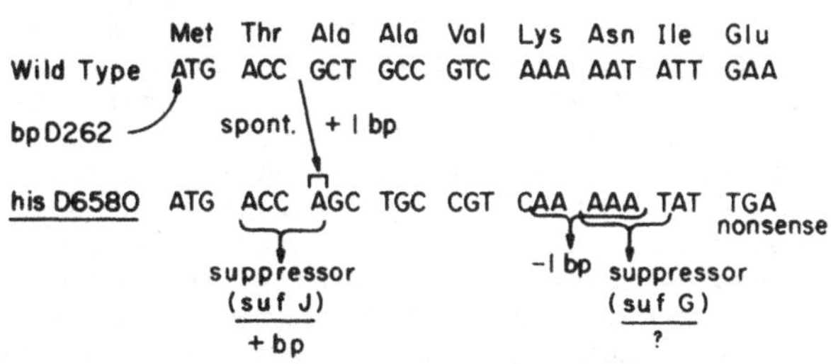

Figure 3. Mutant hisD6580 was obtained in S. typhumurium LT-2 as a spontaneous (spont.) frameshift mutant with an insertion of a single A/T base pair (bp) in the hisD gene as determined by DNA sequencing.[53] Although not in a set of Salmonella tester strains, hisD6580 is of interest for it contains a sequence of 5 contiguous A residues that may be susceptible to -1 frameshifts and thus restore the reading frame. In fact, a strain containing hisD6580 responds very clearly to the mutagen malondialdehyde.[56] Mutant hisD6580 can also revert by an +C frameshift mutation added to a stretch of 3 C residues in the frameshift suppressor (sufJ) tRNA gene. This permits reading of the 4-base sequence ACCX by allowing insertion of a threonine residue.[53,57,58] Suppression is also achieved by an unclassified mutation in another suppressor gene, sufG, known to act at runs of A residues in the message[53,57]. Derivatives of hisD6580 (strains TA95 and TA96) containing rfa and uvrB mutations as well as plasmid pKM101 have been constructed.[56]

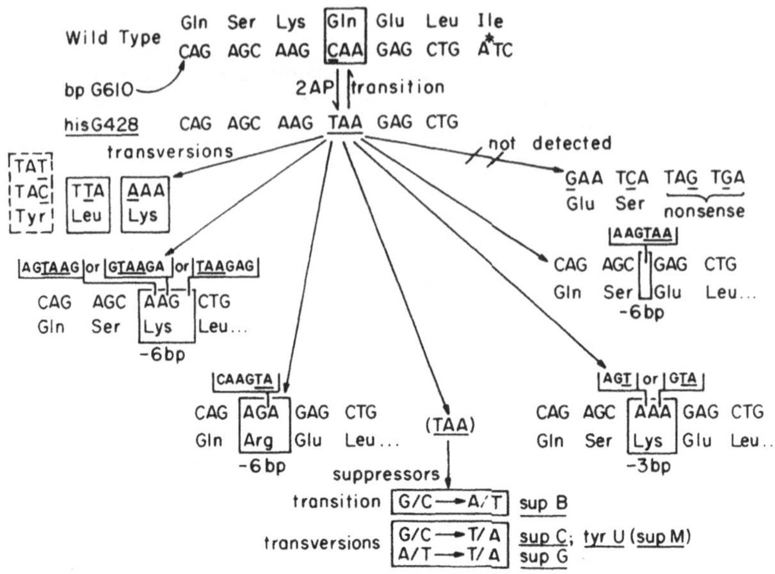

Figure 4. Mutant hisG428 was isolated after treatment of an ara-9 (arabinose fermentation negative) strain of *S. typhimurium* LT-2 with 2-aminopurine (2AP). The mutation was classified as a nonsense ochre mutation in the hisG gene.[21] The nucleotide sequence around the mutant site was determined by DNA sequencing, a method also applied in determining nucleotide changes in intragenic revertants.[59] The mutation affects codon 207 of the hisG gene (W.M. Barnes and E. Tuley, personal communication). Mutant hisG428 was examined because it was found to be the most responsive to oxidative mutagens of a number of ochre histidineless mutants surveyed.[56] It can revert by AT -> GC transition mutations and AT -> TA transversion mutations at the mutant codon, giving rise to a hisG enzyme that is sensitive to feedback inhibition by the histidine analogue 2-thiazolealanine.[59] Tyrosine (Tyr) revertants are shown in a dashed box since they have not been detected among revertants sequenced to date; however, tyrosine-inserting suppressors are active (below). Mutant hisG428 also can revert intragenically by 3- or 6-base pair in-phase deletions of which 7 possible types have been observed. All gave rise to a thiazolealanine-resistant phenotype.[59] Levin *et al.*[59] have pointed out that there is an AGAGC....AGAGC direct repeat in the target region (see Wild Type sequence); whether this repeat is involved in mutagen susceptibility is not clear, but short deletions are a common class of spontaneous revertant.[59,60] Reversion also is accomplished by transition and transversion mutations in any one of 4 characterized suppressor tRNA genes (bottom of figure).[59,61] At least one additional suppressor locus exists.[61] Suppressor revertants may be distinguished readily becouse of their hypersensitivity to the histidine· analogues thiazolealanine and D,L-1,2,4-triazole-3-alanine. Suppressors may be further subdivided by more laborious tests involving sensitivity to mutants of

Salmonella phage P22 or by genetic mapping.[61] Suppressor revertants cannot readily be distinguished merely on the basis of colony size. The recommended tester strain, TA102,[34] carries hisG428 on a multicopy plasmid; this offers multiple copies of targets of the hisG sequence as opposed to a unitary target for each of the chromosomal suppressor loci. Strains such as TA104 exist that carry hisG428 only in the chromosome.[56,62] The asterisk (*) indicates a 6-methyldeoxyadenosine residue in the sequence GA*TC, involved in strand recognition during mismatch repair.[54,55]

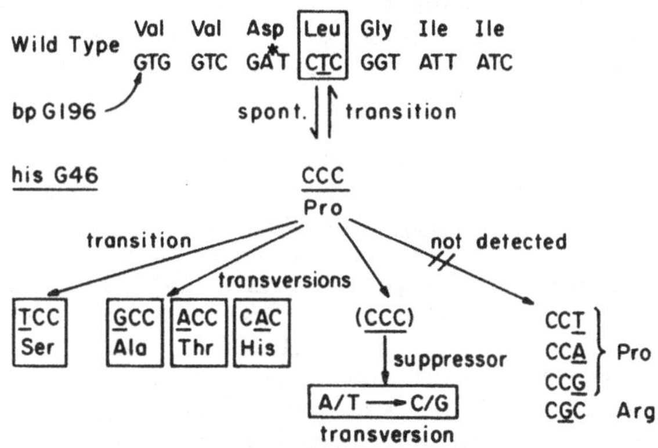

Figure 5. Mutant hisG46 is a spontaneous (spont.) base substitution mutant of S. typhimurium LT-2 that was mapped in the proximal portion of the first structural gene of the histidine operon.[21,63] DNA sequencing by W.M. Barnes and E. Tuley[60] demonstrated that an A/T -> G/C transition mutation had created hisG46 at hisG codon 69, revertible by both transition and transversion mutations at the mutant codon. Revertants carrying arginine (Arg) at position 69 have not yet been detected. The pertinent nucleotide sequence is being synthetized and inserted at the proper location in order to determine precisely the phenotype of the resultant strain (W.M. Barnes, personal communication). The different classes of intragenic revertants of hisG46 exhibit varying sensitivities to the histidine analogue D,L-1,2,4-triazole-3-alanine (see Fig. 1 in ref. 60), presumably due to differing specific enzyme activities. However, differences are not great enough to allow valid assignment of codons among revertants. Reversions also occur by virtue of suppressor mutations; suppressor-containing revertants can clearly be distinguished by increased sensitivity to histidine analogues;[60] suppression is brought about exclusively by an A/T -> C/G transversion in a tRNA gene.[61] Tester strains carrying hisG46 are described in publications from the Ames laboratory.[27,34] The asterisk (*) indicates a 6-methyldeoyadenosine residue in the sequence GA*TC, involved in strand recognition during mismatch repair.[54,55]

containing hisG8646 has been used,[60] and a set of four of the possible six base substitution-specific strains has been developed to date (61 and personal communication). Such strains may find use as second-step checks on compounds that might give only weak, questionable responses in hisG46 or in hisG428 due to extreme mutagen specificity. One example is 5-azacytidine which is a weak mutagen for hisG46 (W.M. Barnes, personal communication) and is thought to cause predominantly G/C -> C/G transversions.[61] Monospecific testers also may be of use in differentiating contrasting mutagens present in different fractions of complex mixtures and in detecting minor modes of mutagenesis by a single mutagen.

FINDING CARCINOGENS

Salmonella testing has proved itself in practical usage by detecting mutagens later found to be carcinogens and by providing insights into mutagenic mechanisms. The latter provide rationale for selective elimination of genotoxic chemicals as opposed to structurally related analogues whose use may be economically advantageous. A few of an increasing list of examples of Salmonella mutagens later found to be carcinogens are hycanthone[22] (reviewed in 68, 69) furylfuramide (AF-2), ethylene dichloride, ethylene dibromide, and tris (2,3-dibromopropyl) phosphate (reviewed in 15, 17). Salmonella tests on analogues led to the formulation of a mechanism for the mutagenicity of hycanthone and the development of a non-mutagenic and non-carcinogenic, yet pharmacologically active analogue.[31,67-70] Salmonella was used to detect and allow subsequent elimination of mutagenic nitroarenes in xerographic toners.[71] More importantly from a long-range viewpoint, Salmonella may help provide predictability as to biological activity based on chemical structure alone of this important class of environmental contaminants,[72-75] an approach that we anticipate will be used more widely in the future as our knowledge advances. Salmonella also has played an important role in the detection and the characterization of mutagens in charred foods.[76,77]

Brusick[16] has concluded that "the Salmonella/microsomal assay meets the criteria of an established predictive test" for potential carcinogens. Detailed data on the reproducibility of the test on 59 chemicals in four different laboratories[78] and on 250 chemicals encompassing 370 tests[79] have been published in the open literature. An analysis of extensive data in the literature has also appeared.[80] In 12 studies of 1975-1982 with a mean sample size of 150 chemicals the sensitivity of the test (mutagenic carcinogens/total carcinogens) was 79%; the sensitivity varied from 54% (in a study biased by a high percent of halogenated organics known to give false negative tests) to 93%.[16] Salmonella tests detected 24/30 (80%) of chemicals either identified or suspected to be carcinogens in humans.[81,82] The six compounds missed (arsenicals, asbestos, benzene, chloroform, DDT, and diethlstilbesterol) also were negative in mutagenicity tests in mammalian cells.[82] Of these six compounds, four (benzene, arsenicals, asbestos and diethylstilbesterol) were detected as positive in tests for cell transformation; in the case of the latter three compounds, concomitant mutagenicity tests were negative.[83] Salmonella also has been used to examine 61 suspect chemicals that eventually proved negative in long-term, two-species rodent carcinogenicity studies.[84] Of these 61 compounds, 38 (62%) were also negative with Salmonella, 21 (34%) were positive, and two compounds gave conflicting results.[84]

Examination is under way to discern the ability of various short-term tests to detect carcinogens[85] that are inactive or cannot readily be detected by the Salmonella test. The test chemicals are acrylonitrile, benzene, diethylhexylphthalate, diethylstilbestrol, hexamethylphosphoramide, phenobarbital, safrole and o-toluidine.[86] Adjunct tests are clearly needed;[87] the major problem is to find the most appropriate test battery.

MUTAGENS AND OTHER FOCAL LESIONS

Elsewhere, the view has been expressed that somatic mutations initiate numerous focal lesions other than cancers, lesions that accumulate and develop during the lifespan.[88,89] While individual lesions generally can be considered as benign, an array of such lesions may ultimately have health impacts. Also, an occasional focus can progress in a gradual, step-wise fashion under appropriate environmental circumstances to develop into a truly deleterious individual lesion. Thus, the impacts of somatic mutations on the human condition are much more important than implied by the current emphasis on the mutagen/cancer relationship. Consequences extend into diverse conditions such as atherosclerosis, duodenal ulcers and gallbladder disease, i.e. many common diseases of the elderly (reviewed in 88, 89).

IDENTIFYING TERATOGENS

Besides prospective carcinogens, Salmonella would seem to be an efficient detector of teratogens and fetotoxic agents, exceeding cytogenetic tests in this regard in the small sample assembled.[90] The total correlation is about 75% (Table 2). "Exceptions" have been listed in the table because it is these compounds that are of most interest both from the standpoint of validity of test procedures and from the standpoint of understanding mechanisms involved with regard to chemical classes of compound.

Table 2. Correlation of Salmonella test results with results from tests for teratogenicity (adapted from 90).

	Teratology	Salmonela	# Compounds
A.	+	+	35
B.	-	-	9
C.	+	-	11 [actinomycin D, aldrin, aminopterin, carbon tetrachloride, dieldrin, diethylstilbesterol, griseofulvin, lead, mirex, procarbazine, urethan]
D.	-	+	3 [benzo(a)pyrene*, captan*, 2-methyl-4-nitro-1-(4-nitrophenyl)imidazole]
E.	Total compounds		58

Total correlation [(A + B) / E x 100] = 75%

*Fetotoxic activity observed

IDENTIFYING GERM LINE MUTAGENS

Comparisons of results from a number of short-term tests and from studies of germ-cell mutagenesis in mammals are summarized by a committee of the International Commission for Protection Against Environmental Mutagens and Carcinogens.[91] Their conclusions on germ-line mutagenicity may be exaggerated,[92] and hycanthone is listed as a non-mutagen for Salmonella in spite of extensive evidence to the contrary[22,31,67,70] (see legend to Figure 1). If we correct for the latter obvious discrepancy, *in vitro* *Salmonella* tests detected 19 of 23 compounds scored as positive or questionably positive in the dominant lethal assay.[91] Only methotrexate (amethopterin), procarbazine-HCl (Natulan), DDT, and saccharin were listed as positive or questionably positive in the dominant lethal assay and negative with *Salmonella*. In one report, saccharin did produce demonstrable mutagenic activity for *Salmonella* strain TA100 in the urines of mice to which it was orally administered.[93] Induced mutations in germ line cells do seem very low except for the nitrosoureas; this may be a prime location to look for antimutagenic compounds and novel DNA repair mechanisms.

Data compiled by Heddle and Bruce[94] on 58 compounds tested for induction of sperm abnormalities in mice showed 20 to be positive in both tests (this includes Mitomycin C and 5-bromo-2'-deoxyuridine which are, in fact, mutagens for *Salmonella*) while 16 were negative in both tests. Twelve compounds produced sperm abnormalities but were negative in the *Salmonella* tests that the authors compiled (Actinomycin D, aminopterin, butylated hydroxytoluene, colchicine, cycloheximide, griseofulvin, hydroxyurea, 5-iodo-2-deoxyuridine, lead acetate, procarbazine, γ-rays, and vinblastine). Among ten compounds positive in the *Salmonella* test, but negative in the sperm abnormality assay were aflatoxin B1, chloral hydrate, *N*-methyl-*N'*-nitro-*N*-nitrosoguanidine and quinacrine).

INFORMATION SOURCES

Citations of articles relating to specific agents and mutagenesis or teratogenesis are available from the Environmental Mutagen Information Center (EMIC) and the Environmental Teratology Information Center, Oak Ridge National Laboratory, P.O. Box Y, Oak Ridge, Tennessee 37831 USA; phone (615) 574-7871. The three most frequently cited publications in the EMIC file of nearly 60,000 citations from over 3,000 sources are *Mutation Research*, *Environmental Mutagenesis* and *Cancer Research*.

INSTRUCTIVE DISCREPANCIES

The addition of mammalian activation systems[29,30] and modifications in test procedures and *Salmonella* tester strains[17,27,34] (also consult Figure 3) have greatly strengthened the ability of *Salmonella* to detect carcinogens and procarcinogens as mutagens. Nevertheless, discrepancies remain; some of these have apparent reason. Other qualitative and quantitative discrepancies exist for which initially there is no apparent reason based on current knowledge. We believe that investigation of such discrepancies can contribute to fundamental knowledge concerning the process of carcinogenesis and the carcinogenic potency of environmental agents. *Salmonella* can contribute substantially to such analyses, as illustrated in the following example.

Mutagenic potency in *Salmonella* strain TA100 fails to correspond to carcinogenic potency of a wide variety of methylating and ethylating agents. In contrast, there is a relationship between carcinogenic potency versus the initial ratio of N^7 versus O^6-alkylguanine DNA which is related to the

Swain-Scott constant.[95] Other factors may be important in mutagenicity and carcinogenicity, however. We have recently studied with Dr. R.B. Brundrett a series of alkylating agents that include nitrosocarbamates, true nitrosamides and nitrosoureas; quantitative aspects of mutagenesis in Salmonella contrast sharply with antitumor activity and carcinogenesis for compounds in which these attributes have been studied.[96-100] The decomposition of the compounds studied leads to common reactive chemical species for homologous sets of compounds (Figure 6). Figure 7 shows results of mutagenesis tests on hisG46 performed with 7 different methylating agents, including methylnitrosocarbamate and dimethylnitrosourea. The carbamate is two orders of magnitude more mutagenic on a molar basis than is the urea, and two other carbamates are even more genetically active. Some attributes of nitrosamides and nitrosocarbamates, perhaps a structural conformation,[101] serve to make them susceptible to cysteinyl-sulfhydryl catalysis of breakdown. Intracellular glutathione is an important activator of the mutagenicity of N-methyl-N'-nitro-N-nitrosoguanidine.[102,103] We hypothesize that the nitrosoureas spontaneously decay at about the same rate within and outside the bacteria. In contrast, activation to mutagens of the nitrosocarbamates and nitrosamides is preferentially brought about intracellularly by the near-millimolar levels of reduced glutathione present [98,100]. We conclude that the mutagenic potency of the nitrosocarbamates and nitrosamides selectively is enhanced by active creation of reactive chemical derivatives by intracellular glutathione in contrast to nitrosoureas that decay in a more random fashion, and thus, predominantly decay extracellularly.

The question remains as to why the nitrosocarbamates and nitrosamides are relatively impotent in the rodent species tested (reviewed in 96-98). After all, N-methyl-N'-nitro-N-nitrosoguanidine appears to require sulfhydryl activation for gastric carcinogenesis.[104] At least a major component in the answer appears to be the high content in rodent plasma and tissues of one or more carboxylesterases that lead to rapid hydrolysis, and thus inactivation, of the nitrosocarbamates and nitrosamides but do not

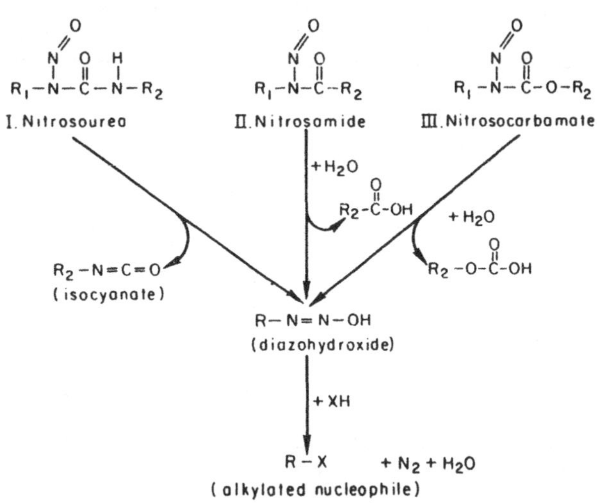

Figure 6. General structures for three closely related classes of N-nitroso compounds, members of which are direct-acting mutagens for Salmonella. A decomposition scheme leading to common electrophilic species for homolgous compounds is also diagrammed (from ref. 97).

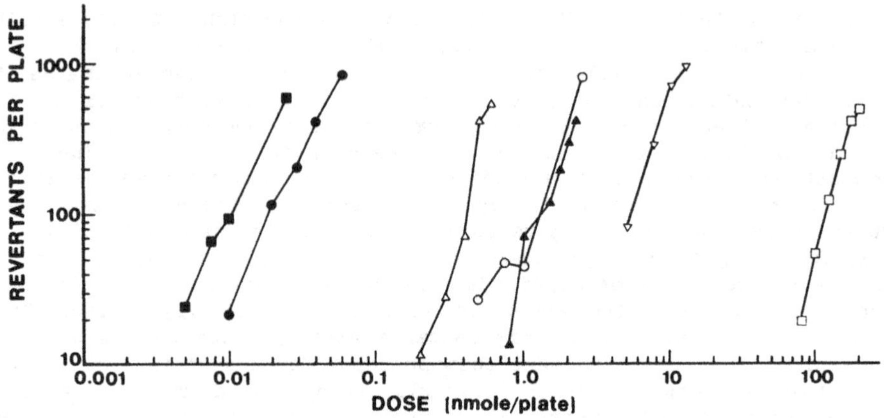

Figure 7. Dose-response curves for seven methylating agents in reversion of S. typhimurium strain hisG46. N-nitroso compounds were dissolved in 100% ethanol (or sulfolane in the case of nitrosobaygon), serially diluted in 100% ethanol, and then diluted 50-fold into bacteria suspended in 0.1M phosphate buffer, pH 7.4. After 10 minutes at 37°C, duplicate 0.1 ml samples were plated for detection of revertants Nitrosocarbaryl was also tested without preincubation; in this instance nitrosocarbaryl was diluted 50-fold into bacteria and plated immediately.
● = nitrosocarbaryl (1-napthyl-N-nitrosomethylcarbamate);
○ = nitrosocarbaryl without preincubation;
■ = nitrosobaygon (σ-isopropoxyphenyl-N-nitrosomethylcarbamate);
▽ = methyl-p-tolyltriazene (data from ref. 96);
△ = N-methyl-N'-nitro-N-nitrosoguanidine;
▲ = N-methyl-N-nitrosoethylcarbamate; and
□ = dimethylnitrosourea. Preincubation increased the sensitivity of the assay for nitrosocarbaryl mutagenicity by a factor of about 100X. Approximate slopes of the dose-response curves, determined by least squares, range from 2.4 to 3.8.[100]

inactivate the nitrosoureas tested.[97-100] Thus, carboxylesterases, whose function in plasma of particular animal species is unknown, could comprise a critical host defense mechanism against particular reactive environmental agents.[98,100] The content of plasma carboxylesterases varies widely among animals (eg. high in rodents and low in man) and parallels inactivating capacity for nitroso compounds (Table 3). These observations indicate that extrapolation to man of rodent test results (as is currently done) with nitrosocarbamates and nitrosamides may not be valid. Additionally, since rodent microsomes contain high esterase activity, Salmonella mutagenicity tests on these classes of compounds should be performed in the absence of S-9 mix.[98]

Some Salmonella tester strains and E. coli release reduced glutathione into the medium, amounting to an extracellular accumulation in the micromolar range in fully saturated cultures (105 and unpublished results). These low levels do not affect ordinary mutagen testing where they can be

Table 3. N-(2-Chlorethyl)-N-nitrosocarbamate half lives in plasma from different species*.[98]

Species	Mutagenic[a] half-life (# samples tested)		Chemical[b] half-life (# samples tested)	
Mouse	15.3 ± 6	12	9.5 ± 0.5	9
Rat	18	2	27.5	3
Rabbit[c]	14.5	2	18.7	4
Human	159	2	239 ± 25	32
Hog[c]	230	2	312	
Fetal bovine[c]	272	2	Not tested	

*Plasma concentrations were 5% (approximately 2.5 to 3 mg/ml total plasma protein).
[a]Half-lives are in minutes ± standard deviation from the mean; tested versus hisG46.
[b]Half-lives were determined by a colorimetric procedure and are reported in minutes ± standard deviation from the mean.
[c]Serum was obtained from these species, rather than plasma.

swamped out by excess mutagen, but they are effective cellular protectors against micromolar levels of toxic agents such as N-methyl-N'-nitroso-N-nitrosoguanidine and some heavy metals (R.A. Owens, personal communication).

It is important that we become familiar with the nitrosamides, for they may be endogenously synthesized agents involved in gastric, esophageal and bladder cancer among non-smokers worldwide.[106-109]

Our major conclusion is a general one: Nature has evolved numerous mechanisms, some quite novel and unexpected, as defenses against potential mutagens and carcinogens. Salmonella cannot alone pinpoint and unravel these mechanisms, but the bacteria complement well other laboratory methodologies in tracking down "antimutagens" and "anticarcinogens."

ACKNOWLEDGEMENTS

This paper is fondly dedicated to Drs. Bruce N. and Giovanna Ames for numerous inspirations and insights provided to the senior author over many years with complete freedom and grace. Bruce's associates, most recently David E. Levin, Dorothy M. Maron, and Robin W. Morgan, also have contributed substantially. We thank Wayne M. Barnes and Elodee Tuley for supplying unpublished nucleotide sequence information for genes hisG and hisD. Supported in part by grant ES3217 from the National Institute of Environmental Health Sciences to P.E.H. Contribution #1285 from the Department of Biology, The Johns Hopkins University.

REFERENCES

1. H.E. Varmus, Annu. Rev. Genet. 18:553 (1984).
2. H. Land, L.F. Parada and R.A. Weinberg, Science 222:771 (1983).

3. D.J. Slamon, J.B. deKernion, I.M. Verma, and M.J. Cline, Science 224:256 (1984).
4. H.E. Ruley, Nature 304:602 (1983).
5. P.H. Duesberg, Science 228:669 (1985).
6. A. Kasid, M.E. Lippman, A.G. Papageorge, D.R. Lowy and E.P. Gelmann, Science 228:728 (1985).
7. R.S. Day III, Bioscience 31:807 (1981).
8. R.B. Setlow, Nature 271:713 (1978).
9. D.S. Straus, JNCI 67:233 (1981).
10. H. Hennings, R. Shores, M.L. Wenk, E.F. Spangler, R. Tarone and S.H. Yuspa, Nature 304:67 (1983).
11. M. Hollstein, J. McCann, F.A. Angelosanto and W.W. Nichols, Mutat. Res. 65:133 (1979).
12. D. Brusick, "Principles of Genetic Toxicology," Plenum Press, New York (1980).
13. J.F. Douglas (editor), "Carcinogenesis and Mutagenesis Testing," Humana Press, Clifton, NJ (1984).
14. J.F. Crow, S. Abrahamson et al., "Identifying and Estimating the Genetic Impact of Chemical Mutagens," pp. 136-143, National Academy Press, Washington, DC (1983).
15. B.N. Ames, Science 204:587 (1979).
16. D. Brusick, Ann. N.Y. Acad. Sci. 407:164 (1983).
17. M.J. Prival, Ann. N.Y. Acad. Sci. 407:154 (1983).
18. B.N. Ames, Cancer 53:2034 (1984).
19. D.W.E. Smith, Science 152:1273 (1966).
20. B.N. Ames and H.J. Whitfield, Cold Spring Harbor Symp. Quant. Biol. 31:211 (1966).
21. P.E. Hartman, Z. Hartman, R.C. Stahl and B.N. Ames, Adv. Genet. 16:1 (1971).
22. P.E. Hartman, K. Levine, Z. Hartman and H. Berger, Science 172:1058 (1971).
23. N.S. Oeschger and P.E. Hartman, J. Bacteriol. 101:490 (1970).
24. M.P. Calos and J.H. Miller, J. Mol. Biol. 153:39 (1981).
25. J.R. Roth, Annu. Rev. Genet. 8:319 (1974).
26. J. McCann, N.E. Spingarn, J. Kobori and B.N. Ames, Proc. Natl. Acad. Sci. USA 72:979 (1975).
27. B.N. Ames, J. McCann and E. Yamasaki, Mutat. Res. 31:347 (1975).
28. G.C. Walker, Microbiol. Rev. 48:60 (1984).
29. B.N. Ames, W. Durston, E. Yamasaki and F. Lee, Proc. Natl. Acad. Sci. USA 70:2281 (1973).
30. H.V. Malling, Mutat. Res. 13:425 (1971).
31. P.E. Hartman, H. Berger and Z. Hartman, J. Pharmacol. Exp. Ther. 186:390 (1973).
32. J. McCann, E. Choi, E. Yamasaki and B.N. Ames, Proc. Natl. Acad. Sci. USA 72:5135 (1975).
33. J. McCann and B.N. Ames, Proc. Natl. Acad. Sci. USA 73:950 (1976).
34. D.M. Maron and B.N. Ames, Mutat. Res. 113:173 (1983).
35. T.R. Skopek, H.L. Liber, J.J. Krolewski and W.G. Thilly, Proc. Natl. Acad. Sci. USA 75:410 (1978).
36. T.R. Skopek, H.L. Liber, J.J. Krolewski and W.G. Thilly, Proc. Natl. Acad. Sci. USA 75:4465 (1978).
37. M.G. Bignami and R. Crebelli, Toxicol. Letters 3:169 (1979).
38. E. Balbinder and D. Kerry, Mutat. Res. 130:315 (1984).
39. T. Kada, K. Aoki and T. Sugimura, Environ. Mut. 5:9 (1983).
40. D.G. MacPhee and M.F. Leyden, Mutat. Res. 143:1 (1985) and in press.
41. L.M. Distlerath, J.C. Loper and C.R. Dey, Mutat. Res. 136:55 (1984).
42. E. Zeiger, K.J. Risko and B.N. Margolin, Environ. Mut. 7:Suppl. 3:45 (1985).
43. D.A. Pagano and E. Zeiger, Environ. Mut. 7:293 (1985).
44. K. Isono and J. Yourno, Proc. Natl. Acad. Sci. USA 71:1612 (1974).
45. J.J. Ferretti, J. Bacteriol. 106:691 (1971).

46. J. Yourno and S. Heath, J. Bacteriol. 100:460 (1969).
47. J. Yourno and S. Tanemura, Nature 225:422 (1970).
48. J. Yourno, J. Mol. Biol. 66:483 (1972).
49. J. Yourno and T. Kohno, Science 175:650 (1972).
50. D.E. Levin, E. Yamasaki and B.N. Ames, Mutat. Res. 94:315 (1982).
51. D. Downs and J. Roth, Abstr. H59, 85th Annu. Mtg. Amer. Soc. Microbiol., Las Vegas, p. 118 (1985).
52. D. Riddle and J. Roth, J. Mol. Biol. 66:483 (1972).
53. L. Bossi and J.R. Roth, Cell 25:489 (1981).
54. M.G. Marinus, in "DNA Methylation" ed. by A. Razin, H. Cedar and A.D. Riggs, Springer-Verlag, New York, pp. 81-109 (1984).
55. A.-L. Lu, S. Clark and P. Modrich, Proc. Natl. Acad. Sci. USA 80:4639 (1983).
56. D.E. Levin, M. Hollstein, M.F. Christman, E.A. Schwiers and B.N. Ames, Proc. Natl. Acad. Sci. USA 79:7445 (1982).
57. T. Kohno and J.R. Roth, J. Mol. Biol. 126:37 (1978).
58. L. Bossi and D.M. Smith, Proc. Natl. Acad. Sci. USA 81:6105 (1984).
59. D.E. Levin, L.J. Marnett and B.N. Ames, Proc. Natl. Acad. Sci. USA 81:4457 (1984).
60. Z. Hartman, P.E. Hartman, W.M. Barnes and E. Tuley, Environ. Mut. 6:633 (1984).
61. D.E. Levin and B.N. Ames, Environ. Mut. 7 (#6): in press (1985).
62. L.J. Marnett, H.K. Hurd, M.C. Hollstein, D.E. Levin, H. Esterbauer and B.N. Ames, Mutat. Res. 148:25 (1985).
63. I. Hoppe, H.M. Johnston, D. Biek and J.R. Roth, Genetics. 92:17 (1979).
64. A.H.-J. Wang, G.J. Quigley, F.J. Kolpak, J.L. Crawford, J.H. van Boom, G. van der Marel and A. Rich, Nature 282:680 (1979).
65. C. Lagravere, B. Malfoy, M. Leng and J. Laval, Nature 310:798 (1984).
66. W.T. Briscoe and L.E. Cotter, Chem-Biol. Interact. 52:103 (1984).
67. D.S. Straus, P.E. Hartman and Z. Hartman, in "Molecular and Environmental Aspects of Mutagenesis" ed. by L. Prakash, F. Serman, M.W. Miller, C.W. Lawrence and H.W. Baber, C.C. Thomas Publ., Springfield, IL pp. 73-88 (1974).
68. P.E. Hartman and P.B. Hulbert, J. Toxicol. Environ. Hlth. 1:243 (1975).
69. H.Tsuda, D.S.R. Sarma, S. Rajalakshmi, J. Zubroff, E. Farber, R.P. Batzinger, Y.-N. Cha and E. Bueding, Cancer Res. 39:4491 (1979).
70. P.B. Hulbert, E. Bueding and P.E. Hartman, Science 186:647 (1974).
71. H.S. Rosenkranz, E.C. McCoy, D.R. Sanders, M. Butler, D.K. Kirizides and R. Mermelstein, Science 209:1039 (1980).
72. T. Sugimura and S. Takayama, Environ. Hlth. Perspect. 47:171 (1983).
73. H.S. Rosenkranz, G. Klopman, V. Chankong, J. Pet-Edwards and Y.Y. Haimes, Environ. Mut. 6:231 (1984).
74. E.C. McCoy, P. Zimmerman and H.S. Rosenkranz, Environ. Mut. 7:Suppl. 3:44 (1985).
75. W.A. Vance and D.E. Levin, Environ. Mut. 6:797-811 (1984).
76. T. Sugimura and S. Sato, Cancer Res. 43:Suppl:2415S (1983).
77. L.L. Vuolo and G.J. Schuessler, Environ. Mut. 7 (#4): in press (1985).
78. V.C. Dunkel, E. Zeiger, D. Brusick, E. McCoy, D. McGregor, K. Mortelmans, H.S. Rosenkranz and V.F. Simmon, Environ. Mut. 6:Suppl. 2:1-254 (1984).
79. S. Haworth, T. Lawlor, K. Mortelmans, W. Speck and E. Zeiger, Environ. Mut. 5:Suppl. 1:3-142 (1983).
80. J. McCann, L. Horn and J. Kaldor, Mutat. Res. 134:1 (1984).
81. H. Bartsch, L. Tomatis and C. Malaveille, Regulatory Toxicol. Pharmacol. 2:94 (1982).
82. H. Bartsch, Ann. N.Y. Acad. Sci. 407:351 (1983).
83. J.C. Barrett, T.W. Hesterberg and D.G. Thomassen, Pharmacol. Rev. 36:53S (1984).
84. M.D. Shelby and S. Stasiewicz, Environ. Mut. 6:871 (1984).
85. E.S. Von Halle, Progr. Mutat. Res. 5:699 (1985).
86. J. Ashby, F.J. deSerres, M. Draper, M. Ishidate, Jr., B.H. Margolin,

B.E. Matter and M.D. Shelby (Editors), Progr. Mutat. Res. 5:1-725 (1985).
87. R.J. Pienta, L.M. Kushner, and L.S. Russell, Regulatory Toxicol. Pharmacol. 4:249 (1984).
88. P.E. Hartman, Environ. Mut. 5:139 (1983).
89. P.E. Hartman and R.W. Morgan, in "Molecular Biology of Aging: Gene Stability and Gene Expression" ed. by R.S. Sohal, L. Birnbaum and R.G. Cutler, Raven Press, in press (1985).
90. C.A. Schreiner and H.E. Holden, Jr. in "Teratogenesis and Reproductive Toxicology" ed. by E.M. Johnson and D.M. Kochhar, Springer-Verlag, New York, pp. 135-168 (1983).
91. D. Brusick, B.J. Kilbey, et al., Mutat. Res. 114:117 (1983).
92. W.L. Russell, Environ. Mut. 7:Suppl. 3:78 (1985).
93. R.P. Batzinger, S.-Y.L. Ou and E. Bueding, Science 198:944 (1977).
94. J.A. Heddle and W.R. Bruce, in "Origins of Human Cancer" ed. by H.H. Hiatt, J.D. Watson and J.A. Winsten, Cold Spring Harbor Laboratory, New York, pp. 1549-1557 (1977).
95. H. Bartsch, B. Terracini, C. Malaveille, L. Tomatis, J. Wahrendorf, G. Brun and B. Dodet, Mutat. Res. 110:181 (1983).
96. R.B. Brundrett, M. Colvin, E.H. White, J. McKee, P.E. Hartman and D.L. Brown, Cancer Res. 39:1328 (1979).
97. S.L. Aukerman, R.B. Brundrett, J. Hilton and P.E. Hartman, Cancer Res. 43:175 (1983).
98. S.L. Aukerman, R.B. Brundrett and P.E. Hartman, Environ. Mut. 6:835 (1984).
99. R.B. Brundrett and S.L. Aukerman, Cancer Res. 45:1000 (1985).
100. S.L. Aukerman, "The Selective Degradation of Nitrosamides, Nitrosocarbamates and Nitrosoureas by Blood Plasma, Tissue Homogenates and Purified Esterase." Ph.D. Thesis, The Johns Hopkins University, 256 pp (1983).
101. A.M. Sapse, G. Snyder and L. Osorio, Cancer Res. 44:1904 (1984).
102. B. Sedgwick and P. Robbins, Mol. Gen. Genet. 180:85 (1980).
103. G. Mohn, P. deKnijff and A. Baars, Mutat. Res. 111:25 (1983).
104. O. Wiestler, A. von Deimling, O. Kobori and P. Kleihues, Carcinogenesis 4:879 (1983).
105. R.A. Owens and P.E. Hartman, Environ. Mut. 7:Suppl. 3:47 (1985).
106. P.E. Hartman, in "Chemical Mutagens, Vol. 7" ed. by F.J. de Serres and A. Hollaender, Plenum Publ. Corp., pp. 211-294 (1982).
107. P.E. Hartman, in Banbury Report 12: "The Possible Role of Nitrosamines in Human Cancer" ed. by P.N. Magee, Cold Spring Harbor Laboratory, N.Y. pp. 415-431 (1982).
108. P.E. Hartman, Environ. Mut. 5:111 (1983).
109. S.S. Mirvish, JNCI 71:630 (1983).

MECHANISMS OF SPONTANEOUS MUTAGENESIS:

CLUES FROM MUTATIONAL SPECIFICITY

Barry W. Glickman, Douglas F. Fix, Fumio Yatagai,
Philip A. Burns, and Roel M. Schaaper*

Biology Department, York University, Toronto, Ontario
M3J 1P3 and *National Institute of Environmental Health
Sciences, Research Triangle Park, N.C. 27709

INTRODUCTION

Spontaneous mutation can be viewed as the sum total of everything that can go wrong with DNA during the life cycle of a cell. Yet, the student of mutation only sees a small fraction of the events; only those changes that produce a selectable and hence observable alteration in phenotype. Moreover, our observations are made through a filter; the majority of errors made by a cell are corrected by the plethora of mechanisms that have evolved to maintain the accurate transmission of genetic material. Our increasing knowledge about mechanisms of DNA repair and their influence on mutation, in combination with the newly developed ability to clone and sequence mutants, provide an opportunity to explore the sources of spontaneous mutation.

Early speculation on spontaneous mutation suggested that its source might be natural background radiation. However, Muller and Mott-Smith [1] showed that only a negligible fraction of spontaneous mutation could have arisen in this manner. Current views suggest that spontaneous mutation is essentially the product of intracellular events. These range from errors of DNA metabolism (either during its replication or repair); the incorporation of unusual nucleotides (base analogs); or represent the consequences of DNA damage resulting from naturally occurring cellular components. In addition, some inherent properties of DNA sequences may contribute to spontaneous mutation. These ideas are reviewed by Von Borstel; [2] Loeb and Kunkel; [3] Drake *et al.*; [4] and Sargentini and Smith. [5]

In this paper we present the nucleotide sequences of 270 spontaneous *lacI* mutants. These mutants include examples of frameshift, deletion, duplication and base substitution events as well as mutagenesis due to the transposition of the mobile genetic element, IS1. The relative importance of each class of mutation will be discussed as well as some of the sources and mechanisms responsible.

THE APPROACH

The *lacI* gene of *E. coli*, which codes for the repressor of the *lac* operon, has been used extensively for the analysis of mutational specificity. Initially, nonsense mutations were analyzed by classical genetic techniques to provide information on base substitution mutation at more than 70 sites within the gene.[6-11] This same gene has been the subject of direct analysis by DNA sequencing.[12,13] Recently, we developed a rapid approach to the cloning and sequencing of *lacI* mutants.[14] The approach involves the recovery of *lacI* mutations by recombination *in vivo* onto a bacteriophage M13 derivative which is $lacI^+$-$lacZ^-$. The double selection for $lacI^-$-$lacZ^+$ recombinants insures that the desired *lacI* mutation is recovered. Sequencing is done by the dideoxy method using synthetic oligomers as primers. The methodology is described in detail by Schaaper *et al.*[14,15] Here we present a spontaneous mutational spectrum based upon the sequences of 270 *lacI* mutants.

RESULTS

The results of sequencing the 270 cloned mutants are given in Tables 1, 2 and 3. About 66% of all the mutations recovered occur at the hotspot site previously reported by Farabaugh *et al.*[16] This site involves position 620-631 where the sequence 5'-CTGG-3' is repeated three times (figure 1). The hotspot event involves either the addition of a CTGG sequence (53%), or the loss of a CTGG sequence (13%). The fraction of hotspot mutations, as well as the ratio of addition versus deletion events determined here by DNA sequencing, are identical to the values obtained from mapping and reversion studies.[16]

The remaining 93 mutations are distributed among (larger) deletions (10.5%), base substitutions (16%), insertion events involving IS1 (3%), single base frameshifts (4.5%) and duplications (1.1%). These mutants are described in detail in Table 2. The base substitution mutations are summarized in Table 3. The deletion, insertion and duplication events are presented schematically in Figure 2. Each class of mutation is considered further below.

```
         600          610          620          630          640
          |            |            |            |            |
5'-CTGTCTCGGCGCGTCTGCGTCTGGCTGGCTGGCATAAATAT-3'
```

Figure 1. The sequence surrounding the frameshift hotspot. The mutations are the result of the gain or loss of this 4 base pair repeat which is repeated three times (positions 620-632). The repeated sequences are underscored.

Table 1. The Distribution of Spontaneous $lacI^-$ Mutants by Class

Class of Mutation	Number of Occurrences	Frequency X 10^{-7}
Base Substitutions	43 (16%)	3.2
Single Base Frameshifts	12 (4.5%)	0.9
Deletions	28 (10.5%)	2.1
Duplications	3 (1.1%)	0.2
Insertions (IS1)	7 (2.6%)	0.5
(+) 5'-CTGG-3'[1]	142 (53%)	10.6
(-) 5'-CTGG-3'[1]	35 (13%)	2.6
TOTAL	270	20 X 10^{-7}

[1]Frameshift hotspot at position 620-631

Table 2. Description of All Non-hotspot Mutations

A. Base Substitutions

Position	Change	Occurrences	Amino Acid Change
23	G => C	1	val => met
56	G => A	1	ala => thr
80	C => T	1	gln => amb (am5)
83	A => G	1	thr => ala
93	G => A	3	arg => his
104	C => T	1	gln => amb (am6)
108	C => T	1	ala => val
140	G => A	1	val => met
141	T => A	1	val => glu
150	C => A	1	ala => glu
158	G => T	1	glu => amb (am7)
183	T => A	1	val => glu
213	T => A	1	leu => amb (am11)
222	G => A	1	gly => asp
293	C => A	1	gln => och (oc13)
419	C => T	1	gln => amb (am15)
582	T => G	1	leu => opal
768	A => T	1	asp => ser
777	C => A	1	ala => glu
845	T => G	1	try => asp
885	T => G	1	ser => opal
926	A => C	2	ser => arg
1085	G => C	1	val => leu
+6(lacO)	T => C	17	O^c

B. Non-hotspot Frameshifts

Position	Alteration	Occurrences	Sequence Alteration
52	(-)T	1	CGA T GTC => CGAGTC
125-128	(-)A	1	CG AAA GC => CGAAGC
286-287	(+)C	1	TCG C GCC => TCGCCGCC
290	(-)G	1	GCC G ATC => GCCATC
388	(-)T	1	CAG T GGG => CAGGGG
394	(-)G	1	GCT G ATC => GCTATC
397	(-)C	1	GAT C ATT => GATATT
562	(+)TC	1	GG TC ACC => GGTCTCACC
581-582	(-)TT	1	CTG TT AG => CTGAG
620-621	(+)C	1	CGT C TGG => CGTCCTGG
646	(-)T	1	CAC T CGC => CACCGC
855-856	(-)C	1	ATA CC GA => ATACGA

C. Insertion Elements

Insertion Element	Position [1]	Orientation[2]
IS1	42-50	II
IS1	53-61	I
IS1	148-156	II
IS1	222-230	II
IS1	488-496	II
IS1	496-504	I
IS1	670-678	I

[1]Nine base-pair repeat sequence generated upon IS1 insertion; only the 3' end of the IS1 insertion was sequenced. [2]Orientation I is that of S114.

(continued)

Table 2. (Continued)

D. Deletions

Class I

Position	Size (bp)	Repeated Bases	Sequence
-33- 18	52	none	
-10-137	148	none	
91-177	87	8	CCGCGTGG
114-127	14	3	ACG
146-268	123	8	GCGGCGAT
218-262	45	4	ATTG
272-286	15	2	GC
267-282	16	5	TCGCG
281-369	89	8	TCTCGCGC
313-334	22	7/8	GG^{T}GTCGA
507-524	18	none	
610-615	6	6	GCGTCT
691-768	78	none	
536-617	81	7/8	CGCG^{A}CTG
790-806	117	none	
814-1046	233	9	GCGCGTTGG
919-1029	111	7	GCAAACC
928-1014	87	10/12	AAACCA^{G}C^{G}TGG
928-1013	86	10/12	AAACCA^{G}C^{G}TGG

Class II

Position	Size (bp)	Ocurrences	Repeated Bases	Sequence
824-(+)8	380	1	7	GCGGATA
900-(+)4	300	1	none	-
916-(+16)	296	1	none	-
929-(+20)	287	1	none	-
954-(+9)	251	1	none	-
1108-(+4)	92	2	7	GTGAGCG
(-64)-(+7)	71	1	none	-
(-64)-(+12)	76	1	none	-

E. Tandem Duplications

Mutant	Duplicated Bases	Size (bp)
1-8	362-619	258
1-28	368-824	457
3-25	380-534	155

Table 3. The Nature of Spontaneous Base Substitutions

Transitions	Occurrences	Transversions	Occurrences
A:T => G:C[1]	19	A:T => T:A	4
G:C => A:T	9	A:T => C:G	5
		G:C => T:A	4
		G:C => C:G	2
TOTAL	28		15

[1]Includes 17 mutations at (+6) of lacZ messenger (O^{C} mutation).

DISCUSSION

The sequencing of 270 spontaneous mutations in the lacI gene of E. coli yielded representatives of six different classes of mutational events, presumably representing a wide variety of mutational mechanisms. The spectrum is dominated by addition and deletion events at the site of a repeated sequence (Table 1), the spontaneous lacI hotspot first reported by Farabaugh et al.[16] Together, these account for about 66% of all spontaneous mutations in the lacI gene. The question arises of how representative the prevalence of such a hotspot is for mutation in an average gene. Roughly calculated, the probability of a direct triple repeat of a four base sequence is only 1 in about 65,000. Thus, for genes of the size of lacI, only 1 out of roughly 50 might be expected to have a similar hotspot sequence. However, considerable evidence indicates that DNA sequences generally are not random and therefore similar kinds of repeated sequences may exist more often than might be expected.

The Frameshift Hotspot

As described above, a large fraction of spontaneous lacI mutations arise as the consequence of the gain or loss of one copy of the four base-pair sequence, 5'-CTGG-3' tandemly repeated three times at nucleotides 620-631 (figure 1). The ratio of the addition of four base-pairs compared to their loss was 4:1, similar to that previously determined by genetic analysis.[16] Frameshift mutations involving repeated sequences are thought to occur via slipped or misaligned intermediates mediated by the repeats as proposed by Streisinger et al.[17] The model therefore predicts the occurrence of these mutations. However, additional factors may also contribute to their predominance. The region neighboring the frameshift hotspot is extremely rich in both palindromic and repeated sequences. In fact, the four base-pair repeats are part of two, more extensive, nine base-pair repeats which predict the same mutational outcome. In addition, two seven base-pair repeats are immediately 3' to, and partially overlapping with, the nine base pair repeats; these neighboring repeats jointly form nearly perfect repeats of 19 and 17 base pairs also capable of templating the gain or loss of the four bases at the hotspot site. Moreover, palindromic DNA sequences are present that may help to stabilize certain misaligned intermediates. Hence, the predominance of mutations at this hotspot site may reflect the accumulative effect of several structural components which may cooperate in enhancing the formation and stability of the structural misalignments templating the mutation. A discussion of how misalignments can generate mutation is found in Ripley and Glickman.[18]

Base Substitutions

Base substitutions accounted for 16% of the LacI$^-$ mutants and were recovered at a frequency of 2 X 10^{-7} (Table 2). Taking into account the number of generations the bacterial population had undergone and estimating the number of sites in the gene that can yield a lacI mutation as roughly 200, the mutation rate can be estimated as 10^{-10} mutations per cell per round of replication. Since 17 of the mutations occurred at an apparent hotspot site in the operator region of the lacZ, the actual mutation rate in lacI is slightly lower. This estimate agrees well with earlier estimates by Drake.[19]

The nature of the base substitutions was also examined. The ratio of transversion to transition mutations is shown in Table 3. Subtracting the 17 lacO^c mutations (A:T => G:C), no single event seems grossly favored. There are four possible transversion and two transition pathways. The ratio of transition and transversion events may not necessarily reflect their occurrence during DNA replication. Instead they are more likely to be a reflection of the specificity of DNA repair. For example, mismatch repair may preferentially correct transitional mismatch intermediates.[20] In addition, the apparent recovery of transverion events may also be enhanced by the action of uracil DNA-glycosylase. The removal from DNA of uracil, the deamination product of cytosine, specifically reduces the frequency of G:C => A:T events.[6] Not only is the frequency of transitions reduced by DNA repair mechanisms, attempts at repair can also alter a mutational spectrum. Indeed, mutation at depurinated sites favors transversions.[21,22] This makes any simple correlation between the relative recovery of transitions and transversions and their formation during DNA replication impossible.

The A:T => G:C transition at position +6 of the lacZ messenger has been shown previously to result in the 0^c (operator constituitive) phenotype.[23] The frequency of this base substitution is 23 X 10^{-8}. This is at least 50-fold greater than the frequency of base substitution at an average site within the lacI gene and greater even than the frequency of mutation at the deamination hotspots.[6,24] We note that Cheung et al.,[25] in an NMR study, observed the enhanced imino proton exchange for this particular base pair compared to other base pairs in this region. This would be expected to result in an enhancement of the deamination of this adenine residue. In the absence of repair, the resulting hypoxanthine subsequently would mispair with cytosine and A:T => G:C transition would result. We suggest that this mechanism is the basis for this base substitution hotspot.

Single Base Frameshifts

Twelve independent frameshift mutations have been so far recovered in this study. Since frameshifts severely disrupt protein structure, they are more likely to be recovered than, for example, base substitution mutations. Thus, correcting for effective target size, single base pair frameshift events appear to occur at least one order of magnitude less frequently than single base pair substitutions. As can be seen in Table 2, 8 of the frameshifts involved a single base deletion and 2, a single base addition. There was one occurrence each of a frameshift involving the gain or loss of two base pairs.

The DNA sequence information provides details of two critical aspects of mutation, the identification of the sequence change at the site of the mutation and the larger DNA context in which the mutational site resides. This information provides an opportunity to examine the mechanisms by which these frameshift mutations may have occurred. Such an analysis provided the basis for the proposal of Streisinger et al.[17] who suggested that frameshift mutations were the consequence of slippage during strand extension at sites of repetitive sequences. This model was later supported by several instances where the DNA sequence of frameshift sites could be resolved[26,27] and by the sequencing of the lacI frameshift hotspot. The Streisinger model has been extended to slippage schemes involving non-adjacent repeats. For example, the reversion of the T4 rIIB frameshift mutation, FC47, is thought to be templated by a 16 base pair repeat 256 base pairs downstream from the mutated site.[27,28] More recent models of frameshift mechanisms provide a role for not only directly repeated sequences, but also palindromic sequences in producing misaligned intermediates for frameshift mutagenesis.[29,18] In the case of direct repeats, inter-strand misalignments result from the hydrogen bonding of one copy of the repeat onto the complement of the second copy of the repeat whereas in the case of inverted repeats, intra-strand misalignments are potentiated by the palindromic nature of the sequence. An interesting aspect of mutation directed by such misalignments is that not only frameshift, but also base substitution mutations can result.[29,18,30]

Table 4. Properties of Spontaneous Frameshift Mutations

Mediated by Direct Repeats	3[1]
Mediated by Palindromic Repeats	4
In Runs	3
5'-GATC-3' (*dam* methylation sites)	3
Without Obvious Template	2

[1]The total is greater than the number of frameshifts recovered since some fall into more than one category. Repeats and palindromes that template these events are all within 25 bp of the site of mutation.

An analysis of the origin of the frameshift mutations is presented in Table 4. Structural intermediates could be proposed for most of the mutations (10/12). Of these, in only three cases are repeated sequences available that could have directed a slippage event in the manner proposed by Streisinger.[17] Most (10/12) of the sequenced frameshifts could have been templated by a nearby sequence, either a direct or an inverted repeat. In two cases, templates capable of producing the mutations and located within 25 base pairs were lacking. It is possible that these may reflect events templated by more remote sequences. In any case, our results indicate that a diverse source of sequences are available to direct the

```
A.
        120       130                        120       130
         |         |            =>            |         |
    5'-TTTCTGCGAAAACGCGG-3'              5'-TTTCTGCGAAACGCGG-3'

B.
  840       850         860               840       850       860
   |         |           |      =>         |         |         |
5'-TGGGATACGACGATACCGAAAG-3'             5'-TGGGATACGACGATACGAAAG-3'

C.
                      A
                   A     A
                T           T
                   T     C
                    A•T
                    G•C
                    C•G
                    G +C  <=
                    G•C
            265     C•G          295
             |      G•C     *     |
         5'-T-T-G-T-C C-G-A-T-C-A-A-3'

D.
 280        290      300                  280                 300
  |          |*       |         =>         |                   |
5'-ATCTCGCGCCGATCAACTGGG-3'              5'-ATCTCGCGCCATCAACTGGG-3'
```

Figure 2. DNA sequences surrounding frameshift mutations. A. The loss of an A from a repeat of four A's. B. A frameshift (loss of a C from position 855/856) templated by a nearby direct repeat. C. The insertion of a C directed by a palindromic DNA sequence. D. The loss of a G at a *dam* methylation site. Underscoring defines direct repeats. The asterisk indicates 6-MethylAdenine residues.

templating of frameshift events. Figure 2 includes an example of each of the above.

Frameshift mutations however, do not appear to occur at random throughout the gene. Even though there are only three dam-sites (5'-GATC-3') in the lacI gene, three of the 12 mutations involve the loss of a G:C base pair within the dam recognition sequence. The prevalence of frameshift mutations at dam methylation sites is unlikely to be due to chance. More likely, they reveal a "DNA context" effect reflecting metabolic activities involving the 5'-GATC-3' sequence. Indeed, these sequences may represent metabolic hotspots. Several years ago, Lark and Gomez-Eichelmann[31] concluded that dam methylation sites were non-random with respect to Okazaki fragments. They suggested that the 5'-GATC-3' sequences were at the ends of the Okazaki pieces. More recently, evidence from the study of the mechanism of methylation-instructed mismatch repair[32] and from mutational spectra in mismatch repair defective strains (in preparation) indicate that during mismatch repair, the incision occurs not at the site of the mismatch but rather at the 5'-GATC-3' sequences. An example of a frameshift at a dam methylation site is given in figure 2D.

The identification of 5'-GATC-3' sites as sites predisposed to single-strand breakage likely accounts for the DNA contextual effect. Single-strand breaks can be expected to provide potential substrates for the formation of misaligned intermediates, or sites for the action of DNA exonucleases leading to frameshift errors. The prediction that DNA context could modulate frameshift frequencies has been suggested by Ripley and Shoemaker[33] and Ripley and Glickman[28], in the case of bacteriophage T4, where the reversion rate of different runs of A's appears to be related to their proximity to the dinucleotide AC, the sequence where Okazaki fragments are known to be initiated in the case of bacteriophage T4.

Deletions

The mechanism of deletion formation has been the subject of much discussion. In many cases short stretches of repeated bases are found at the deletion endpoints.[16,12] This observation has lent support to the hypothesis that deletion mutation occurs via structural intermediates in which one copy of the repeat is misaligned on the complement of the other copy of the repeated sequence. The deletion then encompasses one copy of the repeat and the sequence between the repeats. This model is an extension of the Streisinger model for the formation of frameshift mutations by misalignment on repeated sequences.[17] Secondary structure may also play a role in deletion formation, either by bringing together otherwise greatly separated sequences[12] or juxtaposing deletion endpoints in deletion intermediates in which misaligned hydrogen-bonding is mediated by inverted or quasi-palindromic DNA sequences.[18,34]

A total of 28 deletion mutations were sequenced in this study. One deletion was isolated twice and another (146-268 in Section D of Table 2) is identical to S23 isolated earlier.[16] The recovery of mutants more than once among a relatively small sample confirms that deletion formation is not a random event. For convenience, deletion events have been classified into two distinct groups. Class I deletions events (19/28) lie within the lacI gene while Class II events (9/28) extend into the lacO region. Class II events are $lacO^{c}$.

We will first consider Class I events. Class I events tend to be relatively short, and are most often flanked by repeated sequences (14/19). An analysis of the sequences surrounding these endpoints indicate that in 3 of the 5 cases where repeated sequences do not flank the deletion, quasi-palindromic sequences are available which could have directed the mutation.

In the case of the two deletions for which neither direct or inverted repeats were identified, it should be pointed out that another mechanism need not be invoked; perhaps the templating sequence is further away than the rather arbitrary 50 base pairs scanned. The most interesting aspect of deletion specificity is just how often both palindromes and direct repeats cooperate to produce a deletion. In most of the Class I deletions analyzed, the misalignments potentiated by direct repeats were further stabilized by inverted repeats. In 12 of these, the quasi-palindromic sequences alone

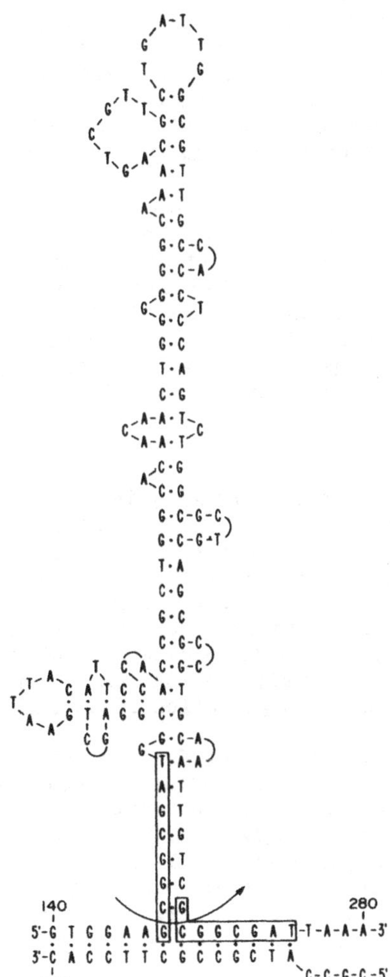

Figure 3. Sequence-directed deletion mutation isolated more than once in the <u>lacI</u> gene. This 123 bp deletion is flanked by direct repeats (boxed) and inverted repeats (i.e., quasi-palindromic sequences). The direct or inverted repeats are capable of accounting for the specificity of the deletion endpoints. However, both repeats can cooperate in the formation and stabilization of the proposed structural intermediate.

could account for the specificity of the deletion endpoints. These included both deletions which have now been recovered more than once [16] and for which structural intermediates have been described. [34] This deletion is presented in figure 3 as an example of a deletion event in which both repeated and palindromic sequences play a role. We have suggested that the presence of both repeated and palindromic sequences enhances the stability of the intermediate, hence making it more likely that such mutations will be recovered. [34]

How significant is the recovery of deletions whose endpoints can be juxtaposed by inverted as well as directly repeated sequences? In this study, 12 out of 14 mutants lying within the _lacI_ gene (i.e. Class I mutants) fell into this category, while five of the seven from the study of Farabaugh _et al_. [16] demonstrated the concurrent occurrence of both palindromic and repeated sequences at their endpoints. [34] In contrast, an analysis of 25 simulated deletions ranging from 10 to 100 base pairs and contained entirely within the _lacI_ gene, revealed only one case where their termini could be juxtaposed by palindromic sequences of four or more base pairs. Clearly, the endpoints of the deletions formed _in vivo_ are not random. They must reflect the mechanism responsible for their creation.

An examination of the Class II deletions revealed a number of commonalities. Class II events are generally longer and often lack repeated sequences at their endpoints. Few of the Class II mutants can be explained on the basis of misalignments. This may be related to the fact that selection for the _lacI_ mutants requires that the cell remain _lacZ_$^+$. In addition, many of the Class II deletion have their endpoints in the CAP-binding site. We do not yet know what influence the binding of the CAP protein to this region might have on mutation.

Duplications

Three tandem duplications, ranging in size from 155 to 457 base pairs were recovered (Table 2, Section E). The origin of duplication mutations is not well understood. They can potentially be created by the processing of structural intermediates similar to those envisioned for the production of deletion events (see 18). None of the duplications recovered in this study however, were flanked by repeated sequences capable of directing their formation. Nor were palindromic sequences readily able to account for the duplication endpoints. A similar situation was noted by Calos _et al_.[35] who recovered a spontaneous duplication of 88 base pairs in the _lacI_ gene.

While we are unable to account for the specific endpoints of these mutants, each of the duplicated sequences contain within them a region of very extensive secondary structure potential. As a consequence, structural intermediates formed in a single-strand region including the duplication would be expected to be stabilized. It is also possible that such structures may interfere with normal repair and replication processes and thereby enhance the opportunity for errors during subsequent repair or replication. DNA context effects may also play an important role. Duplications would be expected to occur where DNA breakage was frequent. Potential candidates are sites of Okazaki primer synthesis or sites of frequent mismatch repair. We note with interest that all three of the duplications have one endpoint near the second 5'-GATC-3' sequence in the _lacI_ gene, the furthest being only 29 base pairs removed. Possibly the concentration of endpoints in the vicinity of this _dam_ methylation site is suggestive of a role for such sites in providing metabolic 'hotspots' where repair is ultimately responsible for the formation of the duplication events. In addition, mismatch repair tracts are thought to extend over extensive distances, easily more than 1000 base-pairs.

Transposable Elements

Transposable elements are normal constituents of most bacterial genomes and many extrachromosomal elements. The E. coli chromosome contains between 8 and 14 copies of IS1 (c.f. 36). They can alter the organization and expression of procaryotic genes at frequencies similar to, and often greater than, spontaneous mutation. Thus, among the collection of spontaneous $lacI^-$ mutations, representatives of this class of events can be expected. Indeed, seven independent occurrences of insertion elements were observed. In each case the insertion was of IS1, a 768 base pair insertion element characterized by short inverted terminal repeats and producing a 9 base pair direct repeat of the 9 base pair segment at the site of insertion.[37,38]

The seven IS1 insertions occurred at different sites in the lacI gene (Table 2). Two insertions of IS1 into lacI reported earlier by Calos et al.[39] occured at sites identical to our mutants 3-19 and 3-24. However, in each case, the insertion had the reverse orientation to that found here. The insertion of IS1 is apparently not random[40] but demonstrates what has been termed regional rather than site specificity.[36] The recovery of IS1 at seven different sites in the lacI gene would confirm the relaxed aspect of integration specificity. G:C base pairs were found at one, and often at both ends of the 9-base pair duplication (10 out of 14), a feature of IS1-mediated insertion noted previously by Galas et al.[41] It has been reported that insertions of IS1 and also transposon Tn9, which carries IS1 as its flanking elements, occur preferentially in A/T rich regions.[41] The sites at which IS1 insertions were recovered in the lacI gene were not particuarly rich at A/T. In fact, 64% of the bases in the duplicated 9 base pair repeated sequence at the sites of insertion were G:C base pairs.

The seven IS1 insertions recovered in this study represent 12% of the non-hotspot mutations. These insertions are therefore not rare events. In addition, due to the lack of A/T rich regions, the contribution of IS1-mediated mutagenesis in lacI might be an underestimate of the general role IS1-insertional mutagenesis plays in E. coli.

CONCLUSIONS

The sequencing of 270 spontaneous mutants in the lacI gene of E. coli has provided insights into the relative role of the various pathways of mutagenesis. In addition, the analysis of the specificity of mutation has provided insights into their origin. Deletions, insertions, duplications, frameshift and base substitution events were recovered. Within the various classes of mutational events, often evidence for different mechanisms emerged. For example, frameshifts can result from several diverse mechanisms and most intriguingly, insights into DNA "context" effects is emerging. In the future specific hypotheses concerning the origin of specific classes of mutation will be further examined in strains with altered DNA repair capabilities, and with the aid of specifically altered DNA sequences. The ability to clone and sequence large numbers of mutants opens the door to a radical new approach to our understanding of DNA repair and mutagenesis.

REFERENCES

1. H.J. Muller and L.M. Mott-Smith, Evidence that natural radioactivity is inadequate to explain the frequency of "natural" mutations, Proc. Natl. Acad. Sci. USA 16:277-285 (1930).
2. R.C. von Borstel, On the origin of spontaneous mutations, Japan. J. Genetics 44:102-105 (1969).

3. L.A. Loeb and T.A. Kunkel, Fidelity of DNA synthesis, Annu. Rev. Biochem. 51:429-457 (1982).
4. J.W. Drake, B.W. Glickman, and L.S. Ripley, Updating the theory of mutation, American Scientist 71:621-630 (1983).
5. N.J. Sargentini and K.C. Smith, Spontaneous mutagensis: the roles of DNA repair, replication, and recombination, Mut. Res. 154:1-27 (1985).
6. C. Coulondre, J.H. Miller, P.J. Farabaugh, and W. Gilbert, Molecular basis of base substitution hotspots in Escherichia coli, Nature 274:775-780 (1978).
7. B.K. Duncan and J.H. Miller, Mutagenic deamination of cytosine residues in DNA, Nature 287:560-561 (1980).
8. P.A. Todd and B.W. Glickman, Mutational specificity of UV-light in Escherichia coli: Indications for a role of DNA secondary structure, Proc. Natl. Acad. Sci. USA 79:4123-4127 (1982).
9. P.L. Foster, E. Eisenstadt, and J.H. Miller, Base substitution mutations induced by metabolically activated aflatoxin B1, Proc. Natl. Acad. Sci. USA 80:2695-2698 (1983).
10. J.H. Miller, Mutational specificity in bacteria, Ann. Rev. Genet. 17:215-238 (1983).
11. B.A. Kunz and B.W. Glickman, The role of pyrimidine dimers as premutagenic lesions: A study of targeted vs. untargeted mutagenesis in the lacI gene of Escherichia coli, Genetics 106:347-364 (1984).
12. A.M. Albertini, M. Hofer, M.P. Calos, and J.H. Miller, On the formation of spontaneous deletions: The importance of short sequence homologies in the generation of large deletions, Cell 29:319-328 (1982).
13. M.P. Calos and J.H. Miller, Genetic and sequence analysis of frameshift mutations induced by ICR-191, J. Mol. Biol. 153:39-66 (1981).
14. R.M. Schaaper, B.N. Danforth, and B.W. Glickman, Rapid repeated cloning of mutant lac repressor genes, Gene (in press) (1985).
15. R.M. Schaaper, B.N. Danforth, and B.W. Glickman, Mechanisms of spontaneous mutagenesis: an analysis of spectrum of spontaneous mutation in the E. coli lacI gene, J. Mol. Biol. (in press) (1985).
16. P.J. Farabaugh, U. Schmeissner, M. Hofer, and J.H. Miller, Genetics studies of the lac repressor, VII. On the molecular nature of spontaneous hotspots in the lacI gene of Exherichia coli, J. Mol. Biol. 126:847-863 (1978).
17. G. Streisinger, Y. Okada, J. Emrich, J. Newton, A. Tsugita, E. Terzaghi, and M. Inouye, Frameshift mutations and the genetic code, Cold Spring Harbor Symp. Quant. Biol. 33:77-84 (1966).
18. L.S. Ripley and B.W. Glickman, Unique self-complementarity of palindromic sequences provides DNA structural intermediates for mutation, Cold Spring Harbor Symp. Quant. Biol. 47:851-861 (1983).
19. J.W. Drake, Comparitive rates of spontaneous mutation. Nature 221:1182 (1969).
20. B.W. Glickman and M. Radman, Escherichia coli mutator deficient in methylation-instructed DNA mismatch correction. Proc. Natl. Acad. Sci. USA 77:1063-1067 (1980).
21. T.A. Kunkel, The mutational specificity of depurination. Proc. Natl. Acad. Sci. USA 81:1494-1498 (1984).
22. R.M. Schaaper, T.A. Kunkel, and L.A. Loeb, Infidelity of DNA synthesis associated with bypass of apurinic sites, Proc. Natl. Acad. Sci. USA 80:487-491 (1983).
23. L. Macquat, K. Thornton, and W. Reznikoff, J. Mol. Biol. 139:537-549 (1980).
24. B.W. Glickman, Spontaneous mutagenesis in Escherichia coli strains lacking N6-methyladenine residues in their DNA. Mutat. Res. 61:153-159 (1979).

25. S. Cheung, K. Arndt, and P. Lu, Correlation of lac operator DNA imino protein exchange kinetics with its function, Proc. Natl. Acad. Sci USA 81:3665-3669 (1984).
26. Y. Okada, G. Streisinger, J. Owen, J. Newton, A. Tsugita, and M. Inouye, Molecular basis of a mutational hotspot in the lysozyme gene of bacteriophage T4, Nature 263:338-341 (1972).
27. D. Pribnow, D.C. Sigurdson, L. Gold, B.S. Singer, J. Brosius, T.J. Dull, and M.F. Noller, rII cistrons of bacteriophage T4: DNA sequence around the intercistronic divide and positions of genetic landmarks, J. Mol. Biol. 149:337-376 (1981).
28. L.S. Ripley and B.W. Glickman, In cellular to DNA damage (E. Friedberg, ed.) Alan Liss, New York (1983).
29. L.S. Ripley, Model for the participation of quasipalindromic DNA sequences in frameshift mutation, Proc. Natl. Acad. Sci. USA 79:4128-4132 (1982).
30. J.G. DeBoer and L.S. Ripley, Demonstration of the production of frameshift and base substitution mutation by quasipalindromic DNA sequences, Proc. Natl. Acad. Sci. USA 81:5528-5531 (1984).
31. M.C. Gomez-Eichelmann and K.G. Lark, Endo R Dpn I restriction of Escherichia coli DNA synthesized in vitro. Evidence that the ends of Okazaki pieces are determined by template deoxyribonucleotide sequence, J. Mol. Biol. 117:621-635 (1977).
32. A.L. Lu, K. Welsh, S. Clark, S.S. Su, and P. Modrich, Repair of DNA basepair mismatches in extracts of Escherichia coli, Cold Spring Harbor Symp. of Quant. Biol. 49:589-596 (1984).
33. L.S. Ripley and N.B. Shoemaker, A major role for bacteriophage T4 DNA polymerase in frameshift mutagenesis, Genetics 103:353-366 (1983).
34. B.W. Glickman and L.S. Ripley, Structural intermediates of deletion mutagenesis: A role for palindromic DNA, Proc. Natl. Acad. Sci. USA 81:512-516 (1984).
35. M.P. Calos, D. Galas and J.H. Miller, Genetic studies on the lac repressor. VIII. DNA sequence change resulting from an intragenic duplication. J. Mol. Biol. 126:865-869 (1978).
36. K. Kleckner, Ann. Rev. Genet. 15:391-404 (1981).
37. L. Johnsrud, DNA sequence of the transposable element IS1, Mol. Gen. Genet. 169:213-218 (1979).
38. M.P. Calos and J.H. Miller, Transposable elements, Cell 20:579-595 (1980).
39. M.P. Calos, L. Johnsrud and J.H. Miller, DNA sequence at the integration sites of the insertion element IS1, Cell 13:411-418 (1978).
40. S. Kuhn, P. Starlinger and H.J. Fritz, Mol. Gen. Genet. 167:235-242 (1979).
41. D.J. Galas, M.P. Calos and J.H. Miller, Sequence analysis of Tn9 insertions in the lacZ gene, J. Mol. Biol. 144:19-41 (1980).

USE OF GRADIENT DENATURING GELS TO DETERMINE MUTATIONAL SPECTRUM IN HUMAN CELLS

Neal F. Cariello and William G. Thilly

Massachusetts Institute of Technology
Department of Applied Biology
Cambridge, Massachusetts 02139

ABSTRACT

Based on the fact that mutagens induce specific patterns of gene mutations, this paper outlines a method to allow discrimination among mutagen-treated populations. The technique should allow direct screening of human tissue for genetic change, using human peripheral blood lymphocytes deficient in the enzyme hypoxanthine guanine phosphoribosyl transferase. The method is based on gradient denaturing gel electrophoresis, which separates short DNA molecules according to their melting properties. The melting behavior of DNA fragments is extremely sequence-dependent, and DNAs with single basepair substitutions often migrate differently. Even DNA fragments with the same basepair substitutions at different locations in the molecule have been resolved. Gradient-denaturing gel electrophoresis has the capacity to separate mutant DNA on the basis of the nature and position of the mutation.

INTRODUCTION

Genetic changes occur in human offspring, but the specific causes of the changes remain unknown. Nearly 1500 specific locus mutations have been identified in the human population[1], and it has been estimated that 1-2% of all liveborns suffer a specific locus mutation associated with significant pathology. Now, at the other end of the spectrum there exists a plethora of tests showing the mutagenicity of various compounds in bacteriophage, bacteria, yeast, and Drosophila. Many of these compounds have been detected in the environment; however, no causal link has been made between environmental mutagens and human genetic changes. If such a link exists, the development of a sensitive yet rapid assay for DNA alteration may permit the tracing of a particular genetic change to a specific environmental exposure.

The problem of discerning the relationship between environmental mutagens and human genetic change is clearly difficult given the minute concentrations of mutagens found in the environment and the low frequency of any specific genetic change. However, one biological phenomenon, the appearance of a specific set of mutations correlated with a given mutagen, offers a solution to this complex problem. The key to any such analysis of genetic change is the concept of mutational spectrum[2]. Mutagens do not act randomly on DNA, rather they produce unique alterations at certain locations. In short, mutagens show specificity with regard to the kind and position of mutation; this specificity can be thought of as a chemical fingerprint on the DNA, this fingerprint can identify the specific mutagen to which a population of cells has been exposed. Mutational analysis of cells sampled from the human population could reveal which segments of the population have suffered genetic damage from a particular mutagen.

MUTATIONAL SPECTRUM

The formation of a chemically-induced mutation is a complex process. In some instances, the compound reacts directly with DNA, forming a premutagenic lesion. This premutagenic lesion can then be acted upon by repair processes and finally acted upon by the replicative machinery. The opportunity for mutagenic specificity exists at every step. The chemical may: 1) react only with certain nucleotides 2) react preferentially with nucleotides in a specific sequence, or 3) react preferentially with DNA in a particular chromatin formation. The repair enzymes may correct certain lesions at a greater frequency, and finally, the replicative machinery may differentially misincorporate nucleotides opposite the lesion, depending on the surrounding sequence. The result is a pattern of genetic change induced by a particular chemical. The pattern of genetic change, with regard to the kind and position of mutation, is termed the mutational spectrum. Importantly, the pattern of induced genetic changes is non-random, and varies from chemical-to-chemical. Three important experimental studies will now be examined.

Pioneering work was done by Benzer and Freese[2]. They mapped spontaneous and bromouracil-induced mutations in the rII region of the bacteriophage T4. 126 spontaneous and 67 bromouracil-induced mutations were mapped, using recombination with a known set of deletion mutants. In this way, the mutations could be localized to a certain portion of the genome. Hotspots for both spontaneous and bromouracil-induced mutations were found, and, importantly, the hotspots were found at different locations in the rII region. Therefore, if an experimenter was blindly given two sets of rII mutants, one spontaneous and the other bromouracil-induced, it would be possible for him to determine which population had been exposed to bromouracil. It should be stressed that the mutational spectrum has another layer of complexity superimposed, namely the selective conditions. Not only must the premutagenic lesion be processed as described above, but in this case the mutation must produce an altered plaque morphology. Any mutation that does not score positively under the selective conditions will be missed. Benzer and Freese were therefore examining a subset of the total genetic changes induced by bromouracil, yet this subset possessed sufficient specificity to allow discrimination among spontaneous and bromouracil-induced mutations.

The site-specificity of mutagens has been extensively studied in the lacI gene of *E. coli* by Miller and co-workers[3,4,5]. Supressible amber and ochre mutants in the lacI gene were selected and their location determined by crosses with deletion mutants. Since the sequence of the lacI gene was known, the precise location of each mutation and the particular basepair change could be determined. The pattern of mutation was determined for spontaneous change and for 4-nitroquinoline-1-oxide (NQO), ethylmethanesulfonate (EMS), 2-aminopurine (AP), N'-methyl-N'-nitro-N-nitrosoguanidine (MNNG) and ultraviolet light (UV). These results are summarized in Figure 1. Benzer's earlier observations were confirmed, namely, that mutagens act specifically with regard to kind and position of mutation.

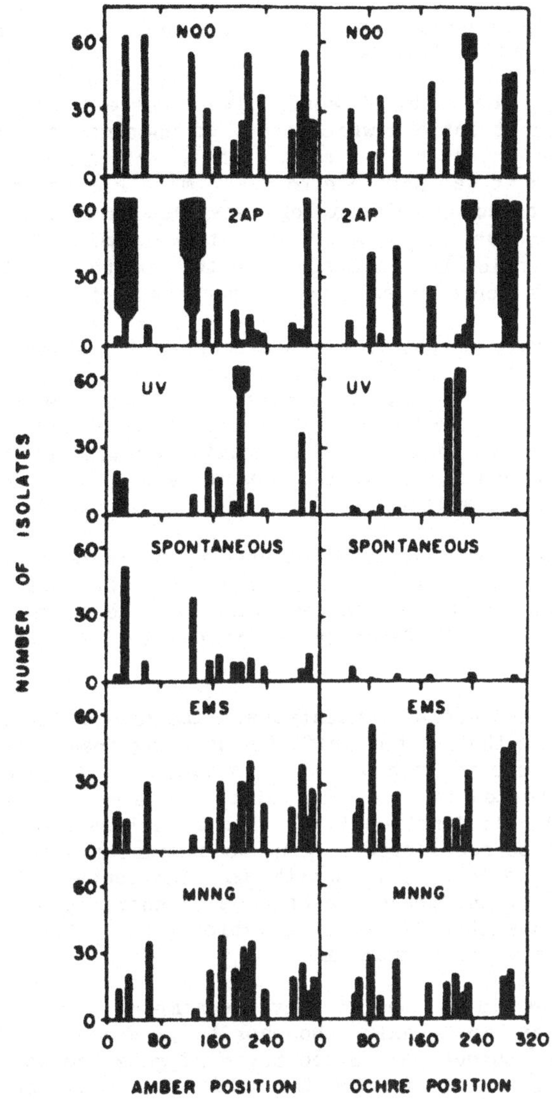

Figure 1. Distribution of 3738 independent E. coli mutations arising from the GC->AT transition. The number of independent occurrences at each site is indicated by bar height.

From reference[5].

In these studies, 72 amber and ochre sites which can arise from a single basepair change were utilized; only the AT->GC transition was not detected. This system essentially examined the mutational spectrum at 72 noncontigous nucleotides, representing about 7% of the target, the lacI gene. The selection system involved two steps: first screening for lacI mutants, then examining these mutants for amber and ochre suppressibility.

The nonsense mutants were then mapped to a particular codon and the exact nature of the mutation deduced.

Importantly, the mutagens were shown to have different hotspots. Only those basepair substitutions which produced nonsense codons in the 72 bp "window" could be detected, yet even with these limitations of target size and mutation type it was possible to discriminate among mutagens. The foundation for molecular epidemiology has thus been laid. Although a great number of mutants must still be analyzed to determine the mutational spectrum for a particular compound, once the spectrum is known, a smaller number of mutants could reveal the "fingerprint" of a particular compound.

Another study made use of certain forward selection systems which monitor mutations at relatively few nucleotide pairs. These systems are all similar, in that the selective agent used seems to be toxic to wild-type cells because it binds to, and inactivates, essential proteins. Resistance presumably occurs from a specific modification of the target protein which renders it functional but refractory to selective agent binding and/or inactivation.

This approach was developed and used by Phaik-Mooi Leong who examined mutational spectra in a human lymphoblast line[6]. She examined the ability of different mutagens to induce resistance to the drugs 6-thioguanine (TG), ouabain (OUA), dichloroibofuranosylbenzimidazole (DRB), and podophyllotoxin (PPT).

Only restricted kinds of mutations, missense or small addition/deletions that do not shift the reading frame, would be expected to confer resistance to such toxins. Nonsense and frameshift mutations should not be detected in this assay since these mutations would presumably destroy the protein's functionality. The mutations producing toxin resistance would be restricted in <u>kind</u> to certain missense mutations and restricted in <u>place</u> to certain positions. The acutal kinds and positions would be defined by the structure of each essential protein and defined by the binding sites within it for an existing toxin. OUA, PPT and DRB appear to act via this type of mechanism.

TG^r, however, is not mediated by the same mechanism. Loss of functional hypoxanthine guanine phosphoribosyl transferase (HPRT) will result in the TG^r phenotype. Since the HPRT gene product is nonessential for cell survival, it is expected that this locus will be sensitive to all forms of gene mutation, including deletions and frameshifts.

In this assay, the data were normalized by expressing the ratio of $DRUG^r$ to TG^r. The TG^r fraction is a reflection of the frequencies of all types of mutation, while the $DRUG^r$ fraction only reveals that portion of cells that have suffered a missense mutation at the drug binding site. The TG^r fraction is taken as a general indication of the frequency of mutation, and if a mutational hotspot exists for a particular drug-binding site, then an elevated $DRUG^r/TG^r$ ratio is expected. In this system, the "window" for mutational spectrum is the different drug-binding sites. The data are shown in Figure 2; the results are normalized with respect to TG^r.

The results show that the spontaneous mutant fraction is different from the EMS-, MNNG- or NQO-induced mutant fraction. Furthermore, the system can discriminate between EMS- and MNNG-induced mutants and between EMS- and NQO-induced mutants. The system cannot discriminate among spontaneous and ICR-191-induced mutants. This latter finding is particularly noteworthy, since it indicates that the system is behaving as expected. ICR-191 generates essentially only frameshift mutations in bacteria[7]. But the "small marker" assay used here is not expected to be

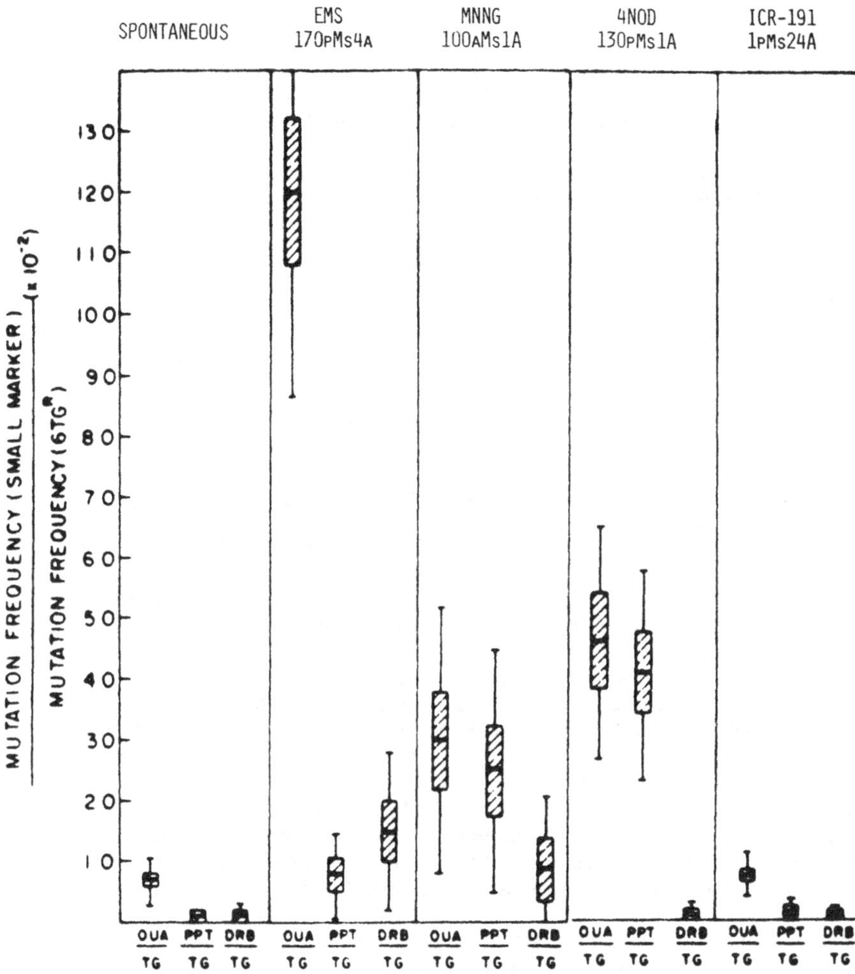

Figure 2. Normalized small marker spectrum from human lymphoblasts. $DRUG^r/TG^r$ is determined at three $DRUG^r$ loci for spontaneous, EMS-, MNNG- and ICR191-induced mutants.

From reference [6].

sensitive to frameshift mutations since these mutations would destroy the essential gene product. The results suggest that ICR-191 is not causing basepair substitution mutations in the human lymphoblast line; extrapolating from the prokaryotic system, we can predict that ICR-191 is causing frameshift mutations. The results also suggest that EMS, MNNG and NQO are causing basepair substitution mutations. Furthermore, it appears that there is specificity of mutation induced by EMS versus MNNG, and EMS versus NQO. This work is the first report of mutational spectrum in human cells, and this work shows that human cell populations can be used to derive mutational spectra of sufficient precision to resolve differences among populations mutated spontaneously and those populations mutated by certain mutagens.

What are the possible approaches for determining mutational spectrum in human samples? Several criteria must be met before a method would be deemed appropriate for human sampling:

1) the sampling method should be reasonable; that is, it should not cause the donor undue pain or discomfort.
2) a sufficient number of genetically detectable events must be present in a sample.
3) the events detected must be of sufficient diversity so that a mutational spectrum can be determined.
4) the protocol, when extended to human sampling, should not require excessive laboratory time or cost.

A relatively new approach to studying DNA double-stranded structure, called gradient-denaturing gel electrophoresis, may allow the resolution of mutational spectrum from human cells. This system, developed by Fischer and Lerman[8] should allow the detection and enumeration of point mutations in specific sequences within total human genomic DNA.

GRADIENT-DENATURING GEL ELECTROPHORESIS

The use of gradient denaturing gels is based on the fact that the mobility of DNA in polyacrylamide gels is sensitive to the secondary structure of the molecule. As the concentration of a denaturant is increased, DNA in solution passes through three discrete states: 1) totally helical 2) partially helical-partially random coil (melted), and 3) strand separation. As the concentration of denaturant is increased, certain sections of the DNA will make the transition from helix to random coil. Therefore, under certain denaturing conditions, a section of the DNA molecule can be expected to be helical while another section will be melted. This type of DNA molecule will show a markedly reduced migration rate on a polyacrylamide gel. In practice, one often finds several isomelting regions of a few hundred basepairs, separated by very steep temperature gradients. The mobility of the molecule will be severely retarded when the lowest-temperature isomelting domain becomes random coil.

Fischer and Lerman have developed a gel system to examine the melting behavior of DNA. The system employs an increasing gradient of urea and formamide. When the DNA enters the gel, it is completely helical, and it will migrate to that concentration of urea/formamide where its lowest-melting domain becomes random coil; when this occurs the DNA's migration rate is reduced.

The melting of a domain is <u>extremely</u> sequence-dependent. The melting of a domain has been described by Fischer and Lerman as a cooperative phenomenon, in which a single basepair substitution can affect the temperature at which the entire domain melts. Fischer and Lerman have examined the effect of known point mutations on the mobility of a 536 bp lamda fragment. The results of this experiment are shown in Figure 3. To understand the power of these gels, it is necessary to examine this figure in some detail. In some cases, a <u>single</u> basepair substitution can shift the fragment about one centimeter; for example, mutation cy3019, which is a CG to TA substitution at position 129, results in retardation of the fragment about 0.9 cm above wild-type. Even two mutants identical in overall DNA composition can be resolved; for example, a TA-to-GC or a TA-to-CG at the <u>same</u> lambda position can be separated. cy2001 is a TA-to-GC at position 136, while cy3071 is a TA to CG at the same position, yet these fragments are separated by about 2mm on a gradient denaturing gel. It should be noted that the four mutants with substitutions above basepair 153 migrate as wild-type. The basepair substitutions for the lambda fragments are also given in Figure 3.

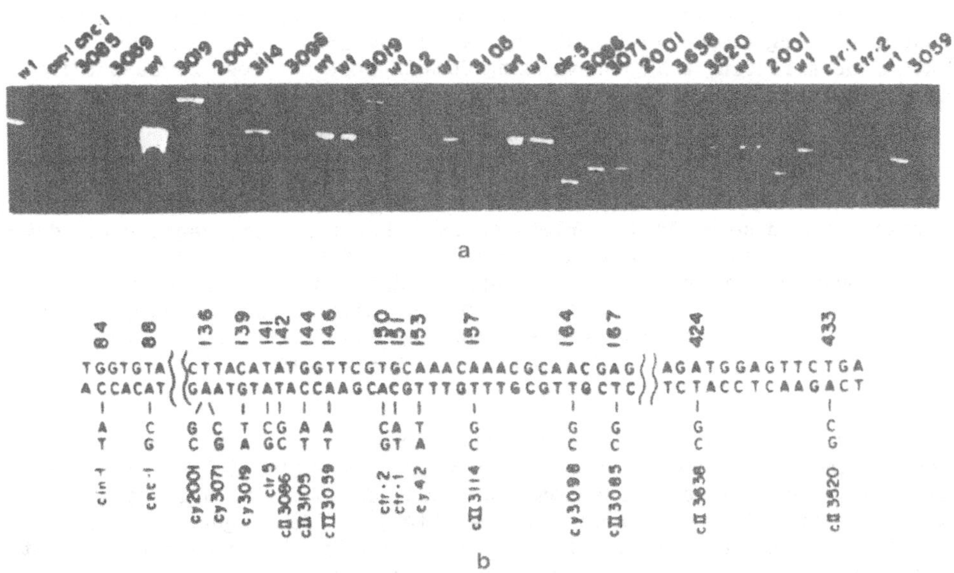

Figure 3. a) Effect of mutation on fragment depth in gradient denaturing gel. Size of lambda fragment is 535 bp. b) Sites and nature of lambda mutants.

From reference[8].

Fischer and Lerman have also developed an algorithm which predicts the melting behavior of DNA solely from its sequence. They have used the Fixman and Freire[9] modification of Poland's[10] algorithm for calculation of the equilibrium melting transition as a function of sequence. Fischer and Lerman have replaced the two-valued stability parameters for basepairs by the set of 10 values for nearest-neighbor doublets as suggested by Gotoh and Tagashira[11]. The resulting algorithm can predict DNA melting solely from the DNA sequence. The melting map of the wild-type 536 bp lambda fragment is shown in Figure 4. Note that the DNA is organized into

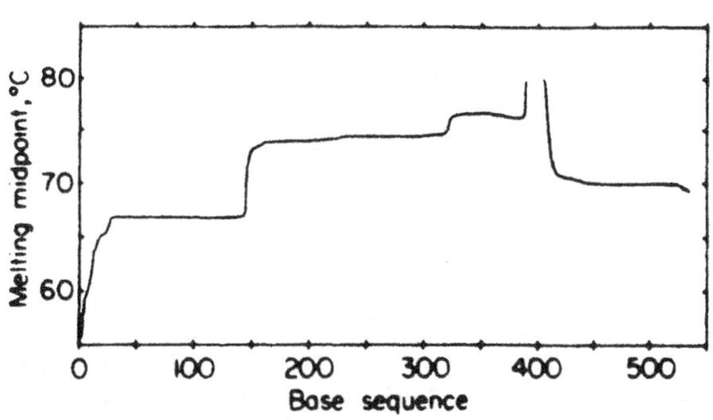

Figure 4. Calculated melting map of wild-type lambda fragment.

From reference[8].

isomelting domains of a few hundred basepairs. Only mutations in the lowest-melting domain (1-153 bp) affected the mobility in the gel. Mutations affecting a higher-melting domain were not seen because this higher melting-domain did not get the opportunity to denature.

Up to now, gradient-denaturing gels have been used only to study the behavior of homoduplex DNA (perfectly paired). How would heteroduplex (mismatched) DNA run on a gradient-denaturing gel? It is expected that the mismatch would severely destablize an isomelting domain, causing the domain to melt at reduced temperature. As is shown in Figure 5, wild-type and mutant DNAs containing a single basepair substitution were allowed to anneal, forming four DNA species: wild-type, mutant, AC mismatch, and GT mismatch. Four discrete bands representing the different DNA duplexes are resolved; it is assumed that the least stable species is the AC mismatch, followed by the GT mismatch DNA.

This type of heteroduplex analysis has recently been extended to detect single basepair changes in human genomic DNA[12]. A ^{32}P wild-type single-stranded cloned probe was hybridized to total DNA with known point mutations in IVS1 of the human beta globin gene (See Figure 6). Four mutations were examined; all four species were displaced from the wild-type and, more importantly, all four mutants were resolved from one another, as is shown in Figure 7. The specific mutations are:

IVS1, 1 is a G->A at basepair position 1
IVS1, 6 is a C->T at position 6
IVS1, 110 is G->A at position 110
IVS1, 5 is G->T at position 5.

Basepair position 1 refers to the 5' nucleotide of IVS-1. Note that the same substitution, G->A in mutants 1 and 110, produces mismatched DNA which runs to discrete locations on a gradient-denaturing gel. An individual heterozygous for the beta-thalessemia mutation is shown in lane 2; both alleles are resolved. A homozygous patient is given in lane 3; as expected a single band is seen.

Gradient-denaturing gel electrophoresis seems to be a reasonable method for detection of mutational spectra. The spectra would indicate the relative position and frequency of the mutants on a gradient-denaturing

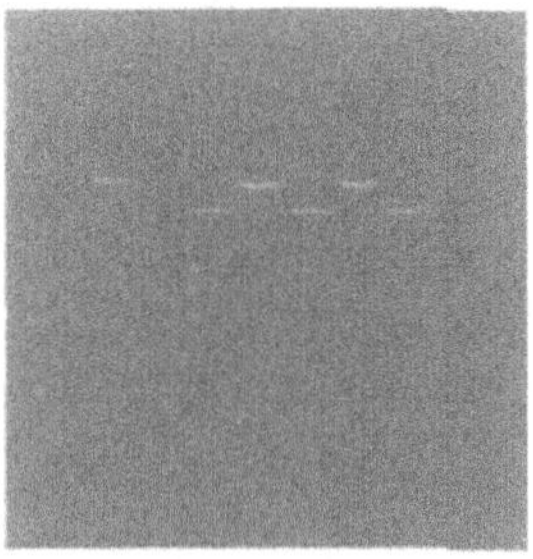

Figure 5. Effect of mispairing, using cloned DNA in gradient-denaturing gel. Left to right: 1) wild-type, 2) G to A substitution, 3 and 4) wild-type and mutant heated and reannealed separately, 5) wild-type and mutant heated and reannealed together, 6 and 7) other samples.

From reference[12].

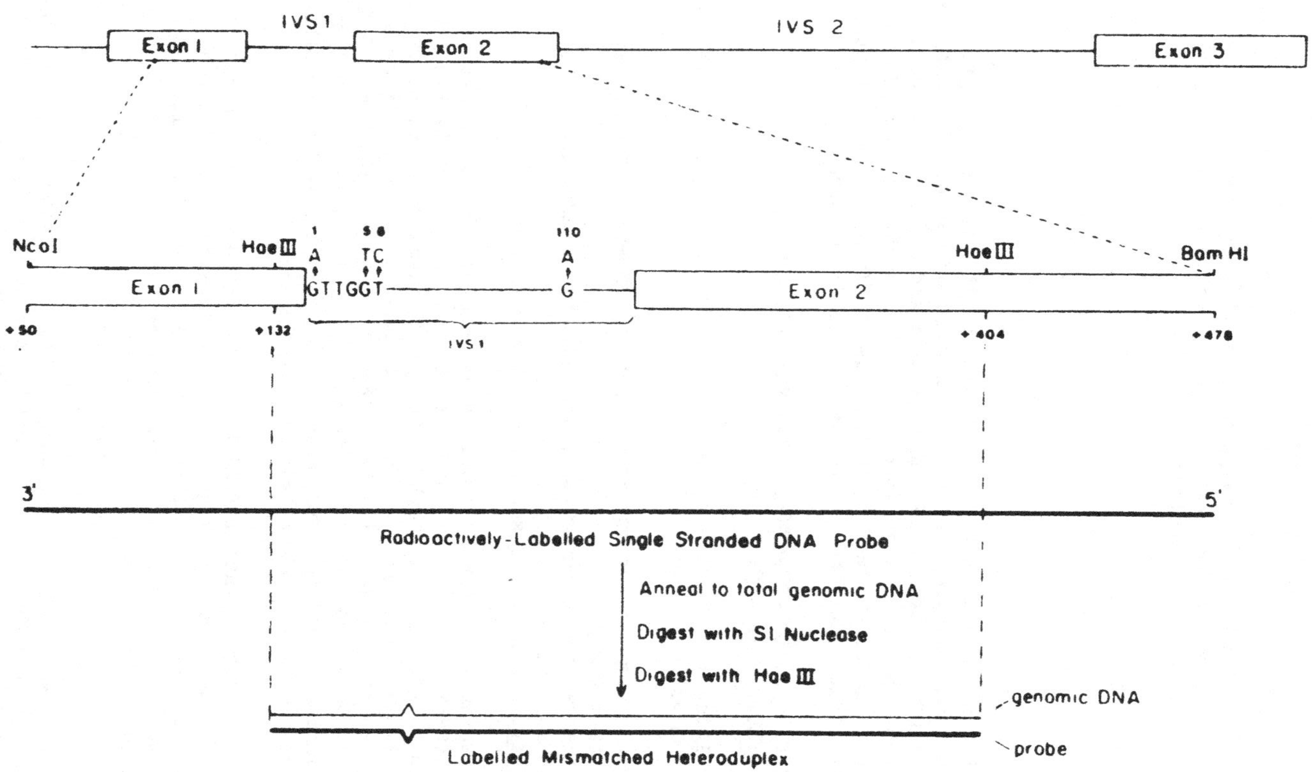

Figure 6. Diagram showing the location of the four beta thalassemia mutations in the human beta globin gene. Basepair substitutions are shown.

From reference 13.

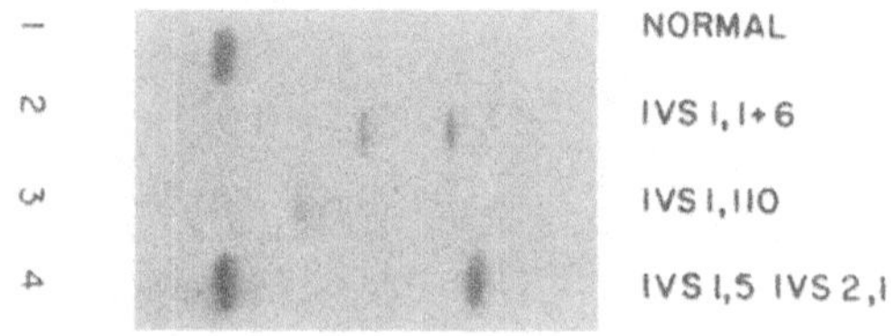

Figure 7. Autoradiogram analyzing hetreoduplex formed between single-strand wild-type probe and human genomic DNA. Nature of mutations given in Figure 6.

From reference 13.

gel. A number of mutants could be examined by the method developed by Myers, et al. For example, if a GC->AT transition hotspot for MNNG existed at position 1 of IVS1 of the beta-globin gene and a GC- AT>EMS hotspot existed at position 110, these species could be resolved (See Figure 6). By analyzing a sufficient number of the EMS-and MNNG-induced mutants, the mutagenic specificity of these compounds could be discerned.

APPLICATION OF GRADIENT-DENATURING GEL ELECTROPHORESIS TO EXAMINE MUTATIONAL SPECTRUM IN HPRT

A gene locus must meet several criteria if it is to be useful for mutational analysis, including:

1) mutants must be present in sufficient numbers in the sample for spectral analysis.
2) a nonessential locus is desired, allowing accumulation of mutations without killing the cell.
3) _in vivo_ and _in vitro_ selection systems must exist.

The nonessential X-linked enzyme hypoxanthine guanine phosphoribosyl transferase (HPRT) is a likely candidate for further investigation by gradient-denaturing gel electrophoresis. HPRT mutants can be selected and _cloned_ from human peripheral blood, and the mutant fraction appears to be on the order of 10^{-5} to 10^{-6} [14,15,16]. Therefore, a 100 ml blood sample would yield between 100 and 1000 mutants for further spectral analysis. To devise a concrete method for obtaining spectra from human HPRT we will have to examine the structure of the gene in some detail.

The cDNA coding for human HPRT has been cloned and sequenced[17]. The organization of the human HPRT genome is now known; the entire gene spans about 35 kb, and the locus is split into nine exons. The length of the intervening sequences (IVS) and exons are shown in Figure 8 (P.I. Patel, personal communication). Many of the exons are small; the largest coding portion of a single exon is contained in exon 3.

Gradient-denaturing gel analysis requires contigious genomic DNA. The most straightforward approach is to make restriction enzyme cuts _within_ exons of individual mutants and to probe this section of the exon with the wild-type radiolabeled probe. In this way, a mismatch at the site of the mutation would be formed. The heteroduplex may run differently than wild-type on a gradient-denaturing gel. However, many of the exons are small and do not contain the appropriate restriction enzyme sites. Exon three lends itself to this direct approach; it is the largest exon, at 184 bp,

Exon	Size(bp)	Intron	Size(kb)
1	about 115	1	13.8
2	107	2	1.6
3	184	3	8.3
4	66	4	3.6
5	18	5	3.5
6	83	6	4.0
7	47	7	0.17
8	77	8	1.6
9	about 900		

Figure 8. Organization of the human HPRT Organization. Total size is about 35 kb. Exon and intron sizes are given.

comprising 28% of the reading frame. Restriction sites exist near the 5' and 3' ends of the exon, such that a 161-bp fragment can be freed; this fragment represents 25% of the coding region.

Another possible approach would be to make the cuts in the mutant's DNA 5' and 3' of each exon and then hybridize this fragment to a wild-type fragment of the same size. In this way, all of the coding portion for a particular exon and some flanking sequence would be available for analysis. The exons are closely spaced at the 3' end of the gene. Here the proper 400 bp fragment would contain exons 7 and 8 with IVS.

Using the melting algorithm of Lerman, we generated the melting profile of the fragment of exon 3 that can be cleaved by restriction enzymes; this is given in Figure 9. The easily-cleavable 161 bp fragment shows a lower-melting domain from 300 to 385 bp; this represents 13% of the coding region. It is assumed that mutations in this section will all be displaced from wild-type using heteroduplex formation. Perhaps even mutations in the highest-melting domain can be resolved. A frameshift mutation in the high-melting domain should destabilize the heteroduplex even more than a simple point mutation, and it may be possible to resolve a frameshift in the high-melting domain, whereas a point mutation may not be seen. To recapitulate, the 161-bp section of exon 3 appears to have, at minimum, about 85 bp that may be useful for spectrum; these 85 bp comprise 13% of the entire coding region of HPRT.

At this point, two related approaches to mutational spectrum could be tried, a clone-by-clone approach and an approach that polls the mutants. For the clone-by-clone approach, the DNA from each mutant would be individually isolated, and one mutant per lane would be run on a gel. If 100 total mutants were isolated, perhaps 10 would run differently than wild-type. If a mutational hotspot exists, several of the mutants should run to the same spot on the gel. In the mass approach, the DNA from some number of mutants would be pooled and run in a single lane. A densitometer trace should reveal any hotspots.

Clearly the pooled approach is the more desirable. The DNA from a large number of mutants would be pooled, cut with the appropriate restriction enzymes, hybridized to single-stranded radioactive probe and run on a gradient denaturing gel. The majority of the DNA should run as wild-type but mutational hotspots should produce a non-wild-type signal. The intensity of a non-wild-type band on an autoradiogram would be proportional to the frequency of its occurrence. In this way, the mutational spectrum could be generated by a scanning densitometer trace of the film. The distance from wild-type would represent the specificity of

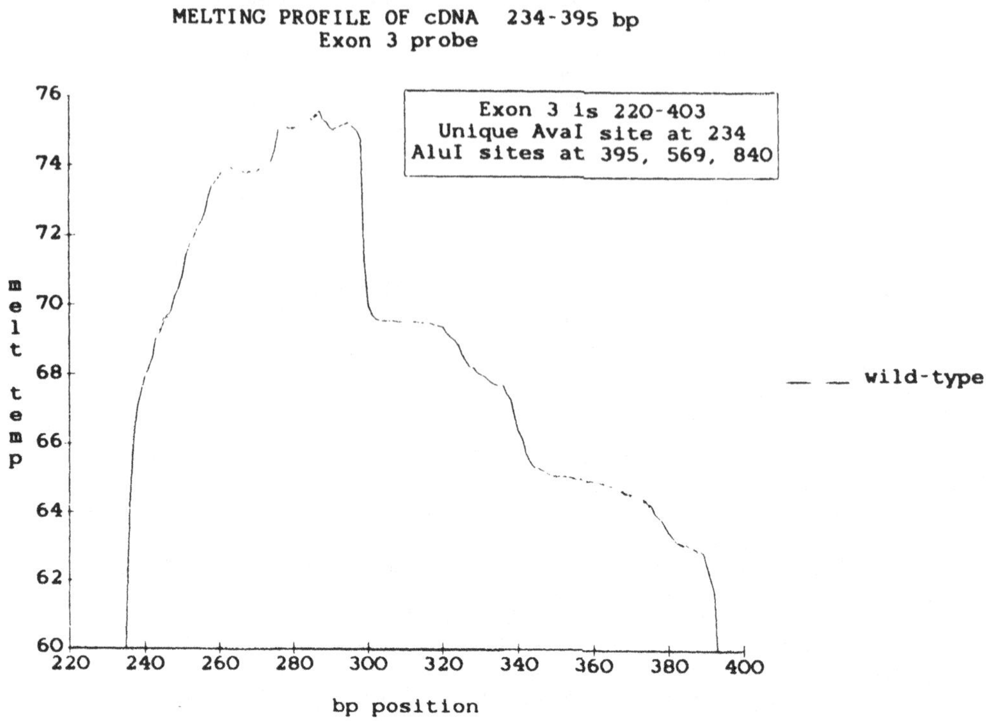

Figure 9. Melting profile of human HPRT exon 3 fragment. 161-bp fragment can be generated by restriction enzyme digestion. This 161-bp fragment represents about 25% of the coding region.

the mutation, while the intensity of the band would indicate the frequency of occurrence. This is summarized in Figure 10. If a group of high-risk individuals -- for example, chemical workers --showed identical mutational spectrum, it could be argued that something common to the experience of these workers is causing the pattern of genetic change.

A central problem is the generation of a spontaneous spectrum for a given individual. For example, a person with a defect in a DNA repair enzyme may produce a pattern of mutation suggesting that he or she has been heavily exposed to environmental mutagens; such a conclusion would be in error. It is possible that each person can be his own control. A relatively small number of wild-type T-lymphocytes, say 10^4, could be expanded *in vitro* to 10^8; during this time spontaneous mutagenesis will occur. Very few previously existing mutants would be present in the starting population of 10^4 lymphocytes. HPRT clones from the expanded population would then be analyzed; this pattern would comprise the spontaneous mutational spectrum. This assumes that spontaneous *in vitro* mutagenesis closely resembles spontaneous *in vivo* mutagenesis.

The spontaneous spectrum would then be compared to the spectrum found in a large sample of the person's blood, say 10^8 lymphocytes; this is summarized in Figure 11. For the latter spectrum, selective pressure would be applied immediately, selecting HPRT mutants generated *in vivo*. If the spontaneous spectrum is not identical to the spectrum found in the large-

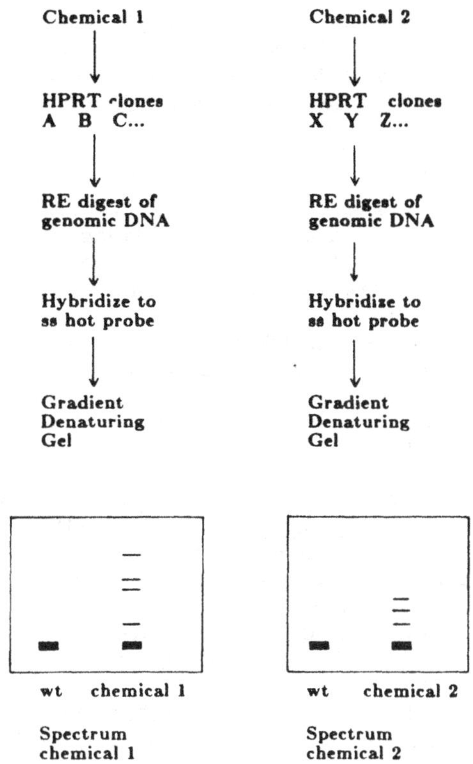

Figure 10. Schematic representation of mutational spectra derived from two mutagens, using gradient-denaturing gels.

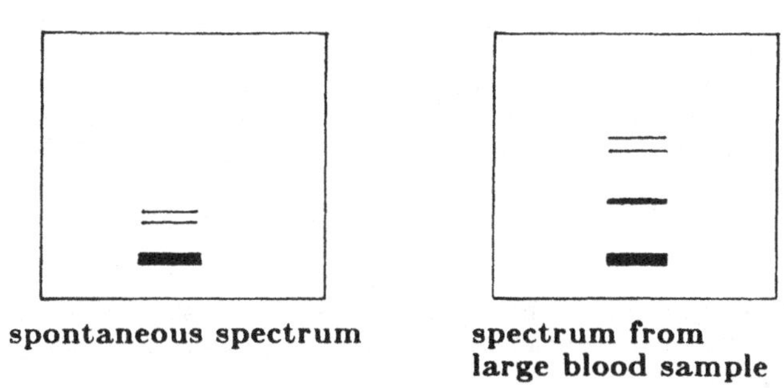

Figure 11. Schematic representation of spontaneous and non-spontaneous spectra from a human donor. Spontaneous spectrum is generated from the <u>in vitro</u> expansion of a small number of cells. The non-spontaneous spectrum is generated by selection of existing <u>in vivo</u> mutants.

scale sample, it can be argued that this individual has suffered induced mutations. The final step would be to match the spectrum of an environmental mutagen with that of the non-spontaneous spectrum found in an individual; the individual might have been exposed to this mutagen.

REFERENCES

1. E. B. Hook, The value and limitations of clinical observations in assessing chemically-induced genetic damage in humans, in: "Indicators of Genotoxic Exposure", Banbury Report 13. B. A. Bridges, B. E. Butterworth, and I. B. Weinstein, eds., Cold Spring Harbor (1982).
2. S. Benzer, and E. Freese, Induction of specific mutations with 5-bromouracil, Proc. Natl. Acad. Sci. 44:112-119 (1958).
3. J. H. Miller, D. Garrem, P. Lu, and A. Schmitz, Genetic studies of the lac repressor: I. Correlation of mutational sites with specific amino acid residues. J. Mol. Biol. 109:275-301 (1977).
4. C. Coulondre, and J. H. Miller, Genetic studies of the lac repressor: III. Additional correlation of mutational sites with specific amino acid residues, J. Mol. Biol. 117:525-575 (1977).
5. C. Coulondre, and J. H. Miller, Genetic studies of the lac repressor. IV: mutagenic specificity in the lac I gene of E. coli, J. Mol. Biol. 117:577-606 (1977).
6. P. M. Leong, Means of recognizing the causes of mutation in human cell populations: the normalized small marker mutational spectra. Ph.D. Thesis, MIT (1984).
7. M. P. Calos, and J. H. Miller, J.H., Genetic and sequence analysis of frameshift mutations induced by ICR-191, J. Mol. Biol. 135:39-66 (1981).
8. S. G. Fischer, and L. S. Lerman, DNA fragments differing by single base-pair substitutions separated in denaturing gradient gels: correspondence with melting theory, Proc. Natl. Acad. Sci. 80:1579-1583 (1983).
9. M. Fixman, and J. J. Freire, Theory of DNA melting curves, Biopolymers 16:2693-2704 (1977).
10. D. Poland, Recursive relation generation of probability profiles for specific-sequence macromolecules. Biopolymers 13:1859-1871 (1974).
11. O. Gotoh, and Y. Tagashira, Stabilities of nearest-neighbor doublets in double-helical DNA determined by fitting calculated melting profiles to observed profile, Biopolymers 20:1033-1042 (1981).
12. L. S. Lerman, S. G. Fischer, I. Hurley, K. Silverstin, and N. Lumelsky, Sequence-determined DNA separations, Ann. Rev. Biophys. Bioeng. 13:339-423 (1984).
13. R. M. Myers, N. Lumelsky, L. S. Lerman, and T. Maniatis, Detection of single base substitutions in total genomic DNA. Nature 313:495-498 (1985).
14. R. J. Albertini, K. L. Castle, and W. R. Borcherding, T-cell cloning to detect the mutant 6-thioguanine-resistant lymphocytes present in human peripheral blood, Proc. Natl. Acad. Sci. 79:6617-6621 (1982).
15. A. A. Morley, K. J. Trainor, R. Seshadri, and R. G. Ryall, Measurements of in vivo mutations in human lymphoyctes. Nature 302:155-156 (1983).
16. G. H. Strauss, and R. J. Albertini, Enumeration of 6-thioguanine-resistant peripheral blood lymphocytes in man as a potential test for somatic cell mutations arising in vivo. Mut. Res. 61:353-379 (1979).
17. D. J. Jolly, H. Okayama, P. Berg, A. C. Esty, D. Filpula, O, Bohlen, G. G. Johnson, J. E. Shively, T. Hunkapillar, T. Friedmann, Isolation and characterization of a full-length expressible cDNA for human hypoxanthine phosphoribosyl transferase. Proc. Natl. Acad. Sci. 80:477-481 (1983).

URINARY N-NITROSAMINO ACIDS AS INDICES OF ENDOGENOUS FORMATION OF N-NITROSO COMPOUNDS

H. Ohshima[1], B. Pignatelli[1], J. Nair[1], N. Munoz[1], S. Calmels[1], M. Crespi[2], S.H. Lu[3], S.V. Bhide[4], P. Vincent[5], H. Leclerc[6], S. Kamiyama[7], and H. Bartsch[1]

1. International Agency for Research on Cancer, Lyon, France; 2. The Regina Elena Institute for the Study and Therapy of Tumors, Rome, Italy; 3. Cancer Institute, Chinese Academy of Medical Sciences, Beijing, People's Republic of China; 4. Cancer Institute Tata Memorial Center, Bombay, India; 5. Universite Claude Bernard, Lyon, France; 6. INSERM, Villeneuve d'Ascq, France; 7. Akita University, Akita, Japan

SUMMARY

Exposure to their precursors (e.g., amines, nitrate/nitrite, NO_x) can lead to formation in the human body of N-nitroso compounds (NOC), a class of potent animal carcinogens, which are also suspected of being carcinogenic in man. A non-invasive method, the 'N-nitrosoproline (NPRO) test', for estimating endogenous nitrosation in man was developed in our laboratory. This test, which monitors 24-hr-excretion of urinary N-nitrosamino acids. is now applied in clinical and field studies, with the aim of measuring nitrosamine exposure and of identifying dietary, life-style, and host factors, or disease states, that affect nitrosation in man. Results from such studies are used to identify populations/individuals at high risk for cancers of the stomach, oesophagus, and oral cavity possibly caused by endogenous nitrosamines, and to indicate preventive measures by which the body burden of endogenous nitroso carcinogens can be lowered efficiently.

INTRODUCTION

Endogenous formation of N-nitroso compounds (NOC), a versatile class of carcinogens[1], from ingested precursors appears to be the largest source of exposure to these compounds for the general population[2]. Both nitrosatable amino compounds and nitrate are commonly present in various foodstuffs, the latter being reduced to nitrite in the body. Endogenous formation of NOC has been suspected to be associated with an increased risk of cancer of the stomach, oesophagus, and bladder, but convincing epidemiological evidence is lacking.

Because of the lack of non-invasive methods, our laboratory has recently developed a simple and sensitive method (NPRO-test) for quantitative estimation of endogenous nitrosation in man[3-5]. It is based on the fact that certain non-carcinogenic N-nitrosamino acids (NAA), such as NPRO, are excreted quantitatively in the urine[6-8], while concomitantly formed carcinogenic nitrosamines are not readily detectable, due to fast metabolism and reaction with cellular material. Thus, the amount of urinary NPRO and of other NAA is excreted per 24 hrs per person following ingestion of nitrate and/or proline, is an index of endogenous nitrosation.

The general objective of our studies is to collect more data on endogenous nitrosation in: a) human subjects with precancerous lesions of the oesophagus and the stomach, b) asymptomatic subjects from high and low risk areas for these cancers, and c) subjects exposed to different levels of precursors (amines, nitrosating agents) and nitrosation modifiers. Ongoing studies and results obtained to-date are summarized below.

METHODS

The NPRO-test[3,9]

(i) Human subjects are given 200 ml beetroot juice (260 mg nitrate); (ii) 30 minutes later they are given L-proline (500 mg); (iii) the subjects fast for a further 2 hrs; (iv) 24-hr urine samples are collected in plastic bottles containing 10 g NaOH (during urine collection, foodstuffs rich in nitrate and cured meat, smoked fish and beer are avoided); (v) 100-ml urine aliquots are stored at -20°C prior to analysis; no artefact formation or degradation of NPRO and nitrate/nitrite was shown to occur. Alternatively, three 24-hr urine samples may be collected from each subject according to the following protocols: (i) undosed, (ii) after intake of 100 mg L-proline 3-times a day after each meal, and (iii) after intake of 100 mg L-proline and 100 mg vitamin C 3 times a day. Whenever possible, information is being collected from each study subject on demographic data, smoking, drinking and dietary habits and clinical findings.

The rationale for applying this procedure in field or clinical studies is based on the following observations: (i) NPRO has been reported not to be carcinogenic or mutagenic[9-10]; (ii) after gavage of rats with [^{14}C]NPRO, the $^{14}CO_2$ production and DNA alkylation were negligible[7], but urinary excretion of NPRO (as the unchanged compounds) was rapid and almost complete[6,7,8]; (iii) urinary levels of NPRO in rats gavaged with proline and nitrite reflected well endogenous nitrosation of proline[8]; (iv) in humans, preformed NPRO ingested in the aqueous extract of broiled dried squids was also rapidly and almost quantitatively eliminated in the urine within 24 hrs after ingestion[4]. Thus, the amount of NPRO excreted in the 24-hr urine was an indicator of daily endogenous nitrosation[3].

Analysis of N-nitroso compounds in the urine

Urine samples were spiked with N-nitrosopipecolic acid as internal standard and analysed for NPRO, N-nitrososarcosine (NSAR), and other nitrosamino acids, according to a published procedure[3,11] after conversion to their methyl esters by diazomethane. Samples were analysed on a tracor 550 gas chromatograph (GC), which was interfaced to a Thermal Energy Analyzer (TEA), a nitrosamine specific detector.

Monitoring of excreted N-nitrosamino acids as a new method to quantify endogenous nitrosation in rodents and in humans

Human urine contains several N-nitrosamino acids (NAA), among which the major ones have been identified as NPRO, NSAR, N-nitrosothiazolidine-4-carboxylic acid (NTCA), and *trans*- and *cis*-isomers of N-nitroso-2-methylthiazolidine 4-carboxylic acid (NMTCA)[3,11,12,13], all of which are currently analysed as an indicator of human exposure to exogenous and endogenously formed N-nitroso compounds.

These N-nitrosamino acids, when administered orally to rats, are excreted unchanged in the urine and feces almost quantitatively[6,11]. Most N-nitrosamino acids have been reported to be weakly carcinogenic (NSAR) or non-carcinogenic (e.g. NPRO)[9,10].

Although the potential carcinogenicity of these N-nitrosamino acids is weak or negligible, the kinetics for formation of these compounds from the reaction of amino acids with nitrite have been shown to be similar to those for carcinogenic N-nitrosamines[14]. Furthermore, endogenous formation of N-nitrosamino acids *in vivo* in rats, estimated from the amounts excreted in the urine, was shown to occur via a similar mechanism to that observed *in vitro*[6,11]. Thus, the amount of N-nitrosamino acids excreted in the 24-hr urine was an indicator of daily exposure to N-nitroso compounds formed endogenously. The kinetic model derived from rat studies[14] can be used to predict nitrosamine formation in humans[15].

Kinetic studies on the endogenous nitrosation in human subjects

In the absence of adverse biological effects of NPRO[9,10], a number of kinetic studies were previously carried out on the formation of NPRO *in vivo* in human volunteers who had ingested vegetable juice (as a source of nitrate) and an aqueous solution of proline with or without dietary nitrosation modifiers. Urinary excretion of NPRO was monitored as an index for endogenous nitrosation. Results allowed three major conclusions to be drawn: (i) an unequivocal demonstration that N-nitroso compounds are formed in the human body, even after ingestion of amounts of precursors (amine, nitrate) that are considered as normal daily intake; the amounts of nitrosated amino acids formed *in vivo* were around 25 μg per day per person, when subjects were living on an uncontrolled western diet: (ii) Nitrosation inhibitors, like vitamin C, vitamin E, and polyphenolic compounds[3,6,16,17] markedly reduced the yield of nitrosamino acids formed in healthy human subjects (see Fig. 1).

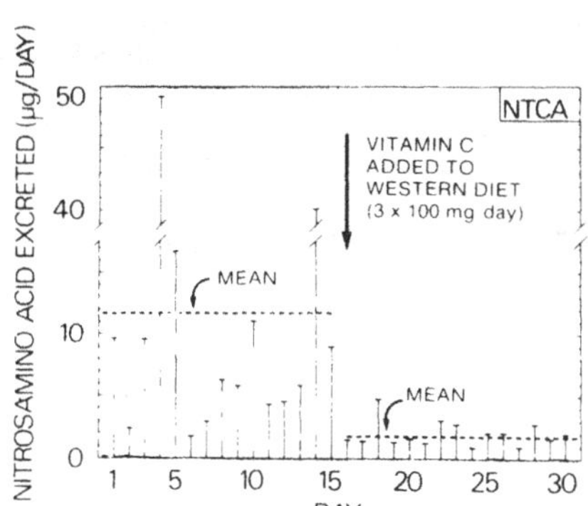

Figure 1. Endogenous formation of N-nitrosothiazolidine-4-carboxylic acid (NTCA) in one human subject living on an uncontrolled western diet and its inhibition by vitamin C. 24-hr urine samples were collected each day during a 30 day period from one healthy subject. From days 16 to 30 the diet was supplemented with vitamin C (as indicated by the arrow)[18].

Catalysis of nitrosation *in vivo* was observed in cigarette-smoking subjects attributable to high level thiocyanate in the saliva of smokers and to higher exposure to nitrosating agents like NO_x present in cigarette smoke[19,20]; (iii) It was found that the intake of nitrate above a dose of 260 mg per day per person led to sharp increase in the amount of *N*-nitrosated products (NPRO) formed *in vivo*[3]. Taken together our data indicate that, because of the effectiveness of the nitrosation catalysts/inhibitors and other modifying factors, individual monitoring will be required, rather than measuring the intake of precursors (amine and nitrite), to associate endogenous formation of *N*-nitroso compounds with cancer at specific sites.

Monitoring of endogenous nitrosation in human subjects at high risk for stomach cancer

In humans an association between relative gastric anacidity and increased rates for stomach cancer is well documented. It has been postulated that such an achlorhydric environment may provide a suitable milieu for intragastric formation of *N*-nitroso compounds by the presence of a large number of bacteria, which convert nitrate to nitrite and also may catalyse nitrosation. Four groups of subjects, in which gastric anacidity is found as a common factor, have been included in our pilot studies: individuals with chronic atrophic gastritis (CAG), those with pernicious anaemia, those who underwent partial gastrectomy, and subjects with different gastric secretory functions or those receiving cimetidine treatment. All the study subjects have been selected from those attending the gastroenterology department of collaborating hospitals.

Interim results[21] indicated that after ingestion of nitrate and proline, urinary NPRO levels in patients with chronic atrophic gastritis were dependent on gastric pH, showing maximal yields at about pH 2, with large interindividual variations of excreted nitrosated amino acids, ranging from 0 to 120 μg per day per person. Patients, as compared to healthy controls, however, excreted no apparent excess of NPRO.

These data clearly indicate that endogenous nitrosation does occur in the human stomach, but its relation to the induction of upper gastrointestinal cancer remains to be proven. Final comparison of subjects with and without precancerous lesions of the stomach is pending until more data are collected, in particular from subjects with normal stomach mucosa. In addition, the levels of total nitroso compounds, measured by the method of Walters *et al.*[22], and the bacterial count in fasting gastric juice obtained from each subject are being determined.

Studies on N-nitrosation catalysis by bacteria

Since endogenous nitrosation catalyzed by bacteria has been suggested to play a role in the stomach and the urinary bladder[23], studies on the microbial formation of NOC have been initiated[24]. A total of 38 strains of various bacteria and one strain of candida from human sources were examined for their ability to form *N*-nitrosomorpholine (NMOR) from morpholine and nitrite at neutral pH. Twenty five bacterial strains exhibited nitrosation activity. Kinetic studies were conducted with resting cells of *E. coli* A 10. The formation of NMOR was optimal at pH 7.25, and the reaction followed Michaelis-Menten kinetics. Substrate specificity for several amines was found (Table 1), with proline being a poor substrate. These results indicate that nitrosation is catalyzed by a bacterial enzyme(s), and attempts to characterize this process in *E. coli* are in progress. Bacterial strains isolated and identified in gastric juice samples from patients with chronic atrophic gastritis, and urine samples from subjects

Table 1. Substrate specficity of E. coli strain A10 to catalyse nitrosation of various amines[24].

Amine	pKa	Specific nitrosation rate	
Morpholine	8.7	154[a]	(100)[b]
Thiazol	-	42	(27.3)
N-Methylbenzylamine	9.54	14	(9.0)
Aminopyrine	5.04	17	(11.0)
Proline	-	5.7	(3.7)
Dimethylamine	10.72	5	(3.2)
Piperidine	11.2	1.4	(0.9)
Pyrrolidine	11.27	1.6	(1)

[a] nmol of nitrosamine formed per mg bacterial protein per hour
[b] relative activity in parentheses, taking NMOR as 100.
Nitrosation reaction was measured in 0.1 mol/l Tris-HCl buffer pH 7.2, in the presence of 25 mmol/l sodium nitrite and 25 mmol/l of each amine, for 1 hr at 37°C.

having urinary bladder infections are now being tested for their ability to catalyse nitrosation.

Study in a high- and a low-incidence area for gastric cancer in Northern Japan

In order to study the role of endogenous nitrosation and nutritional deficiency in the aetiology of stomach cancer, a pilot study was initiated in 1983. Samples of 24-hr urine and blood were collected from 100 subjects living in high- (Akita) and low- (Iwate) incidence areas for stomach cancer in the northern part of Japan. Three different urine samples were collected from each subject: (a) undosed, (b) after ingestion of 100 mg proline three times a day after each meal, and (c) after ingestion of 100 mg proline three times a day together with 100 mg vitamin C. These samples were analysed for nitrate and chloride ions excreted and for some NAA such as NPRO, NTCA and NMTCA. Blood samples are being analysed for the levels of vitamins (A, B_2 and E), trace elements (Zn, Fe, Se), uric acid, and some enzyme activities as an index of nutritional status. In addition, dietary questionnaires were obtained from each subject. Initial results revealed that Akita subjects had a two-fold higher exposure to NAA (median value: 47 µg NAA per day per person) then did those in Iwate (25 µg per day per person). Vitamin C led to a reduction of endogenous NAA (16 µg per day). Potential of endogenous nitrosation in each individual is now being correlated with the results obtained from the analysis for nutritional status and food habits.

Evaluation of human exposure to N-nitroso compounds and nitrate in high-risk areas for oesophageal cancer

Oesophageal cancer is a prevalent disease in northern China and the Linxian county in Henan province has the highest age-adjusted mortality rate, with 151/100,000 in men and 115/100,000 in women. An extensive search for the causative factors of this cancer was begun in 1972[25]. Those studies suggest that tobacco and alcohol, which account for over 70% of the risk in populations of Europe and America, appear to play a minor role in China, but other environmental factors may be more important. N-Nitroso compounds and their precursors (secondary amine, nitrate, nitrite) have received the most attention.

In order to evaluate human exposure to N-nitroso compounds through either ingestion of preformed compounds in foods or endogenous formation,

24-hr urine samples were collected from about 250 subjects living in Linxian (high-risk)area) and Ganxian (low-risk area with a mortality rate for oesophageal cancer of 35/100,000 in men and 16/100,000 in women) according to three different protocols: (a) undosed subjects, (b) subjects who had ingested 100 mg L-proline three times a day, after each meal, and (c) subjects who had ingested 100 mg L-proline together with 100 mg vitamin C three times a day. The urine samples were analysed for the presence of NPRO, NTCA and NSAR (µg/24-hr per person), and nitrate (mg/24-hr per person). Interim results have been published[26].

The conclusions that can be drawn from this study are: (i) in the high-incidence area (Linxian), exposure to NPRO, NTCA, NSAR and nitrate is higher than in Fanxian county; (ii) intake of L-proline three times a day further increased the urinary excretion of NPRO in these subjects; the potential for endogenous nitrosation appears to be at least twice as high in the inhabitants of the high-risk area as in those of the low-risk area; (ii) ingestion of vitamin C reduced the amount of NPRO and NTCA formed *in vivo* in the high-risk subjects to levels seen in the low-risk subjects. Although it remains to be demonstrated whether endogenous nitrosation is one of the risk factors related to the development of oesophageal cancer in certain provinces in northern China, intervention studies can be considered, as our results[26] have demonstrated the efficiency of vitamin C in blocking N-nitrosation *in vivo*.

The significance of these findings is being verified in studies which involve different areas in the People's Republic of China: The excretion of urinary nitrosamino acids is being correlated with mortality figures for cancer of the oesophagus, stomach, and liver in different provinces; a number of risk factors for malignant diseases and nutritional indices which had also been analysed will be included in the comparisons (in collaboration with Dr. J. Chen, National Center for Preventive Medicine, Beijing; and Dr. T.C. Campbell, Cornell University, Ithaca, N.Y., USA).

Nitrosamine exposure of betel quid chewers

The correlation between oral cancer and chewing of betel quid often containing tobacco in India and other South East Asian countries is well established[27]. Nitrosation *in vivo* of areca nut-specific alkaloids, in particular arecoline, leads to the formation of NOC of which N-nitrosomethylpropionitrile was found to be carcinogenic in experimental animals[28]. The extent to which this nitrosation of betel nut constituents is taking place *in vivo*, -- i.e., the oral cavity of chewing subjects -- has not been investigated.

Saliva and urine samples were collected from chewers of betel quid with or without tobacco, from tobacco chewers, from cigarette smokers and from people with no such habits. In order to evaluate exposure of betel quid chewers to NOC, these samples were analysed for tobacco-specific nitrosamines (TSNA) and betel nut-specific nitrosamines (BSNA)[29].

TSNA and BSNA were detected (ng/ml) in saliva of chewers of betel quid with tobacco (BQT); BSNA were detected in chewers of betel quid without tobacco (BQ). Precursor tobacco alkaloids (i.e. nicotine) and betel nut alkaloids (i.e. arecoline), nitrite and thiocyanate were detected (µg/ml) in the saliva. A good positive correlation was found between TSNA and nicotine in saliva; no correlation was seen between arecoline and BSNA. Urinary NPRO levels tended to be higher in chewers of BQT, BQ, or tobacco and in smokers as compared to the habit group, suggesting an increased endogenous nitrosation. Urinary levels of tobacco-related compounds were similar in chewers of BQT and tobacco when compared with smokers, implying similar uptake. Certain TSNA and BSNA can be formed while chewing betel quid with or without tobacco, as demonstrated by an increase in their levels after *in vitro* nitrosation of betel quid extract at pH 7.4. Up to 250 times higher concentrations of TSNA and BSNA are formed at pH 2.1 (simulated gastric condition) when compared with pH 7.4 after nitrosation of BQT and BQ. These levels of NOC could be formed intragastrically after ingestion of betel quid. Two new *N*-nitrosamino acids were detected in the acidic fraction of the saliva in BQT chewers and were identified as 3-(*N*-nitroso-*N*-methylamino)propionic acid and 4-(*N*-nitroso-*N*-methylamino)butyric acid. These NOC were shown to occur ubiquitously in all tobacco products.

It is not yet known which of these *N*-nitroso compounds and precursor alkaloids induced cancer of the oral cavity and oesophagus in betel quid chewers. However, there is a firmly established association between buccal cancer and snuff dipping in humans[30]; these tobacco products contain TSNA at high concentration[31].

TSNA, known to be potent animal carcinogens may therefore also be causative agents in cancers associated with betel quid chewing, either alone or in combination with other betel quid ingredients, some of which exert genetic and toxic effects.

CONCLUSION

Our results demonstrate unequivocally that NOC are formed in the human body even after ingestion of levels of precursors (nitrate, amines), which are considered to be normal daily intake, and that *in vivo* nitrosation is modified by many factors. In view of the evidence accumulated to date, there is little reason to believe that humans are resistant to the carcinogenic action of NOC, from either preformed or endogenous sources. Now that highly exposed populations/individuals can be identified, for which the 'NPRO-test' is currently used, there are various ways by which exposure to NOC can be lowered: modification of habits (e.g., stop smoking or chewing of tobacco); reduction of *in vivo* nitrosation (e.g., by dietary inhibitors like vitamin C and lowered precursor intake); and removal of potential risk factors, such as micro-nutrient deficiences (e.g., vitamins). Such measures should reduce a portion of diet-related and environmental cancers.

ACKNOWLEDGEMENTS

We wish to thank J.C Bereziat, M.C. Bourgade and J. Michelon for technical assistance and Y. Granjard for secretarial work. A part of these studies was supported by Contract No. 9304498 from the Delegation Generale a la Recherche Scientifique et Technique (DGRST), France. One of the TEA detectors was provided on loan by the National Cancer Institute of the United States under Contract NO1 CP-55715. The work reported on betel quid was undertaken during the tenure of a Research Training Fellowship awarded to J. Nair by the International Agency for Research on Cancer.

REFERENCES

1. H. Druckrey, R. Preussmann, S. Ivankovic, and D. Schmahl, Oraganotrope carcinogene Wirkungen bei 65 verschiedenen *N*-Nitroso-Verbindungen and BD-Ratten. *Z. Krebsforsch.* 69:103 (1967).
2. Coordinating Committee for Scientific and Technical Assessment of Environmental Pollutants - Nitrates: An Environmental Assessment, National Academy of Sciences (1978).
3. H. Ohsima, and H. Bartsch, Quantitative estimation of endogenous nitrosation in humans by monitoring *N*-nitrosoproline excreted in the urine. *Cancer Res.* 41:3658 (1981).

4. H. Ohsima, and H. Bartsch, Quantitative estimation of endogenous nitrosation in humans by measuring excretion of N-nitrosoproline in the urine, in "Environmental Mutagens and Carcinogens", T. Sugimura, S. Kondo and H. Takabe, eds., University of Tokyo Press, pp. 577 (1982).
5. H. Bartsch, H. Ohshima, N. Munoz, M. Crespi, and S, H, Lu, Measurement of endogenous nitrosation in humans: Potential applications of a new method and initial results, in Human Carcinogenesis, C. C. Harris, H. N. Autrup, eds., Academic Press, pp. 833 (1983).
6. H. Ohsima, J. C. Bereziat, and H, Bartsch, Monitoring N-nitrosamino acids excreted in the urine and feces of rats as an index for endogenous nitrosation, Carcinogenesis 3:115 (1982).
7. C. Chu, and P. N. Magee, The metabolic fate of nitrosoproline in the rat, Cancer Res. 41:3653 (1981).
8. R. E. Dailey, R.C. Braunberg and A. M. Blaschka, The absorption distribution and excretion of (^{14}C)-nitrosoproline by rats, Toxicology 3:23 (1975).
9. International Agency for Research on Cancer, "Monographs on the Evaluation of the Carcinogenic Risk of Chemicals to Humans, Vol.17. Some N-nitroso Compounds" (1978).
10. S. S. Mirvish, O. Bulay., R. G. Runge and K. Patil, Study of the carcinogenicity of large doses of diemthylnitramine, N-nitroso-L-proline and sodium nitrite administered in drinking water to rats, J. Natl. Cancer Inst. 64:1435 (1980).
11. H. Ohshima, I. K. O'Neill, M. Friesen, J.C. Bereziat and H. Bartsch, Occurrence in human urine of new sulphur-containing N-nitrosamino acids N-nitrosothiazolidine 4-carboxylic acid and its 2-methyl derivative, and their formation, J. Cancer Res. Clin. Oncol. 108:121 (1984).
12. H. Ohshima, M. Friesen, I.K. O'Neill and H. Bartsch Presence in human urine of a new N-nitroso compound, N-nitrosothiazolidine 4-carboxylic acid, Cancer Letters 20:183 (1983).
13. H. Tsuda, T. Hirayama and T. Sugimura, Presence of N-nitrosoL-thioproline and N-nitroso-L-methylthioproline in human urine as major N-nitroso compounds, Gann 74:331 (1983).
14. H. Ohshima, G. A. T. Mahon, J. Wahrendorf and H. Bartsch, Dose-response study of N-nitrosoproline formation in rats and a deduced kinetic model for predicting carcinogenic effects caused by endogenous nitrosation, Cancer Res. 43:5072 (1983).
15. H. Ohshima and H. Bartsch, Monitoring endogenous formation in man, in: "Monitoring human exposure to carcinogenic and mutagenic agents", A. Berlin, M. Draper, K. Hemminki and H. Vainio, eds., IARC Scientific Publications No. 57, International Agency for Research on Cancer, p. 233 (1984).
16. B. Pignatelli, R. Scriban, G. Descotes and H. Bartsch, Modifying effects of polyphenols and other constituents of beer on the formation of N-nitroso compounds, Amer. Soc. Brew. Chem. J. 42:18 (1984).
17. H. F. Stich, H. Ohshima, B. Pignatelli, J. Michelon and H. Bartsch, Inhibitory effect of betel nut extracts on endogenous nitrosation in humans, J. Natl. Cancer Inst. 70:1047 (1983).
18. H. Ohshima, I. K. O'Neill, M. Friesen, J. C. Bereziat and H. Bartsch, Occurrence in human urine of new sulphur-containing N-nitrosamino acids N-nitrosothiazolidine 4-carboxylic acid and its 2-methyl derivative, and their formation, J. Cancer Res. Clin. Oncol. 108:121 (1984).
19. D. Hoffmann and K. D. Brunnemann, Endogenous formation of N-nitroso-proline in cigarette smokers. Cancer Res. 43:5570 (1983).
20. K. F. Ladd, H. L. Newmark and M. C. Archer, N-Nitrosation of proline in smokers and non-smokers, J. Natl. Cancer Inst. 73:83 (1984).

21. H. Bartsch, H. Ohshima, N. Munoz, M. Crespi, V. Cassale, V. Ramazotti, R. Lambert, Y. Minaire, J. Forichon and C. L. Waters, In-vivo nitrosation, precancerous lesions and cancers of the gastrointestinal tract. On-going studies and preliminary results, in "N-Nitroso Compounds: Occurrence, Biological Effects and Relevance to Human Cancer", I. K. O'Neill, R. C. von Borstel, C. T. Miller, J. Long and H. Bartsch, eds., Scientific Publications No. 57, International Agency for Research on Cancer, 955 (1984).
22. C. L. Walters, L. J. Downes, L. W. Edwards and P. L. R. Smith, Determination of a non-volatile N-nitrosamine on a food matrix, Analyst 103:1127 (1978).
23. G. Hawksworth and M. Hill, Bacteria and the N-nitrosation of secondary amines, Br. J. Cancer 25:520 (1971).
24. S. Calmels, H. Ohshima, P. Vincent, A.M. Gounot and H. Bartsch, Screening of microorganisms for nitrosation catalysis at pH 7 and kinetic studies on nitrosamine formation from secondary amine by E. Coli strains, Carcinogenesis 6: in press (1985).
25. C. S. Yang, Research on oesophageal cancer in China: a review, Cancer Res. 40:2633 (1980).
26. S. H. Lu, H. Ohshima and H. Bartsch, Recent studies on N-nitroso compounds as possible etiological factors in oesophageal cancer, in: "N-Nitroso Compounds: Occurrence, Biological Effects and Relevance to Human Cancer", I. K. O'Neill, R. C. von Borstel, C. T. Miller, J. Long and H. Bartsch, eds., Scientific Publications No. 57, International Agency for Research on Cancer, 947 (1984).
27. S. V. Bhide, S. S. Shah, J. Nair and D. Nagarajarao, Epidemiological and experimental studies on tobacco related oral cancer in India, in: N-Nitroso Compounds: Occurrence, Biological Effects and Relevance to Human Cancer", I. K. O'Neill, R. C. von Borstel, C. T. Miller, J. Long and H. Bartsch, eds., IARC Scientific Publications No. 57, International Agency for Research on Cancer, 851 (1984).
28. G. Wenke and D. Hoffmann, A study of betel quid carcinogenesis 1. On the in vitro N-nitrosation of arecoline, Carcinogenesis 4:169 (1983).
29. J. Nair, H. Ohshima, M. Friesen, M. Croisy, S. V. Bhide and H. Bartsch, Tobacco-specific and betel nut-specific N-nitroso compounds: Occurrence in saliva and urine of betel quid chewers and formation in vitro by nitrosation of betel quid, Carcinogenesis 6:295 (1985).
30. D. M. Winn, Tobacco chewing and snuff dipping: an association with human cancer, in: "N-Nitroso Compounds: Occurrence, Biological Effects and Relevance to Human Cancer, I. K. O'Neill, R. C. von Borstel, C. T. Miller, J. Long and H. Bartsch, eds., IARC Scientific Publications No. 57, International Agency for Research on Cancer, Lyon, 837 (1984).
31. D. Hoffmann, K. D. Brunnemann, J. D. Adams and S. S. Hecht, Formation and analysis of N-nitrosamines in tobacco products and their endogenous formation in consumers, in: "N-Nitroso Compounds: Occurrence, Biological Effects and Relevance to Human Cancer", I. K. O'Neill, R. C. von Borstel, C. T. Miller, J. Long and H. Bartsch, eds., Scientific Publications No. 57, International Agency for Research on Cancer, Lyon, 743 (1984).

APPLICATION OF MONOCLONAL ANTIBODIES TO

MONITOR HUMAN EXPOSURE TO AFLATOXIN B1

John D. Groopman

Boston University School of Public Health
Environmental Health Section
80 E. Concord Street
Boston, MA 02118

Human exposure to aflatoxins occurs because these compounds are the products of mold-related spoilage of foods. The aflatoxins are highly toxic and carcinogenic compounds, which are consistently found to contaminate human food supplies in many areas of the world, and are epidemiologically linked to increased incidences of human liver cancer in Asia and Africa[1]. The biological effects, metabolism, and DNA adducts of aflatoxin B1 have recently been reviewed by Busby and Wogan[1]. Adverse biological effects of aflatoxin B1 (AFB1) are thought to be related to its ability to be metabolized oxidatively by constitutive cellular enzymes. Of particular importance may be the ability of AFB1 to form covalent adducts with cellular macromolecules, including DNA. It is well established that AFB1 forms covalent DNA adducts following enzymatic oxidation to a highly reactive 2,3-epoxide, which nucleophilically attacks the N7 atom of guanine[2-4]. This reaction produces the major DNA lesion 2,3-dihydro-2-(N7-guanyl)-3-hydroxyaflatoxin B1 (AFB1-N7-Gua) and in secondary reactions the other aflatoxin B1 DNA adducts. With this chemical structural knowledge, we initiated a series of projects to begin to produce monoclonal antibodies which recognize these aflatoxin derivatives. The goal was to monitor the exposure of people to this environmental chemical carcinogen.

Since 1979, we have produced various monoclonal antibodies that recognize aflatoxins. The antigens used to produce these monoclonal antibodies have ranged from aflatoxin-modified DNA to aflatoxin-adducted proteins. The antibodies are being used, in conjunction with other chemical analytic techniques, as non-invasive screening methodologies for monitoring human exposure to these environmentally occurring mycotoxins. These screening methods require the ability to quantify aflatoxin and its metabolites, including DNA adducts, in readily accessible compartments, such as serum and urine. Earlier research by Donahue et al[5] produced a chemical protocol for quantifying AFB1-N7-Gua in urine, based on the isolation of aflatoxin-DNA adducts from human urine by preparative and analytical liquid chromatography. A radiometric labeling technique using tritiated dimethyl-sulfate was used in these procedures to tag the aflatoxin-DNA adduct at the 9-position of the guanine moiety. This post-labeling technique is specific for this DNA adduct, and it provided *in vitro* a limit of detectability in spiked samples of 1 pg AFB1-N7-Gua/ml urine; however, when this method was applied to *in vivo* urine samples, non-specific interfering materials present in urine sample often prevented the attainment of this level of sensitivity. Therefore, we have undertaken efforts in collaboration with this group to combine the chemical analytic and radiometric procedures with a monoclonal antibody-affinity chromatography column, in order both to purify the aflatoxin adducts from urine and to have a confirmation analysis for the DNA adduct.

The development of a reusable monoclonal antibody-affinity chromatographic column has required the production of high-affinity monoclonal antibodies recognizing aflatoxins. Our initial endeavors to produce aflatoxin-specific monoclonal antibodies culminated in the production of five antibodies which were found to recognize aflatoxin B1-adducted DNA; eventually, in a competitive ELISA (using USERIA methodology) a limit of detectability of one AFB1 residue per 1,355,000 nucleotides was thus obtained[6,7]. These five monoclonal antibodies were obtained after fusion of mouse P3x63 myeloma cells with spleen cells isolated from BALB/c mice that had been immunized with aflatoxin B1-adducted DNA complexed with methylated bovine serum albumin. Selected hybridomas were found to produce monoclonal antibodies specific for modified DNA containing both the AFB1-N7-Gua and the 2,3-dihydro-2-(N5-formyl-2',5',6'triamino-4'-oxo-N5-pyrimidyl)-3-hydroxy-aflatoxin B1 (AF-FAPyr) adducts, suggesting that these DNA adducts share a common antigenic determinant.

Further experiments were performed to ascertain if these monoclonal antibodies were specific for the major AFB1 guanine adducts, AFB1-N7-Gua and AF-FAPyr, and other oxidative metabolites of AFB1. These monoclonal antibodies apparently recognize only AFB1-bound to DNA, since the following concentrations of compounds resulted in less than 2% inhibition in a competitive ELISA: 11,000 pmoles AFB1; 530 pmoles AFB2; 32,000 pmoles AFP1; 5,100 pmoles AFM1; 10,000 pmoles AFB2a; 260 pmoles AFG1; 2,000 pmoles AF-Diol; 2,100 pmoles AFB1-N7-Gua and 2,100 pmoles AF-FAPyr. Finally, we also determined that 5.5 pmoles of B[a]P metabolite bound to DNA at a level of 1.24 adducts per 100 nucleotides was not recognized by either antibody. These data support the hypothesis that these monoclonal antibodies are specific for an altered conformation in DNA resulting from the covalent binding of AFB1.

These antibodies were used for the quantification of _in vivo_ rat liver adducted AFB1-modified DNA. [3H]-AFB1was administered to rats at dosages ranging from 0.01-1.0 mg AFB1/kg. These dosages represent sub-acute toxic regimens for these animals[1]. The adducted DNA was isolated from the livers of these animals 2 hours following exposure. There was one AFB1 residue per 35,000, 251,000, and 1,355,000 nucleotides, respectively, in these DNAs, and HPLC analysis of these nucleic acids revealed that 75-80% of the adducts bound were chromatographically identical to the AFB1-N7-Gua adduct. These nucleic acids were exposed to 0.1 N KOH, 37°C for 10 minutes to quantitatively convert the AFB1-N7-Gua adducts to the AF-FAPyr derivative. Representative results of inhibition experiments using the USERIA and monoclonal antibody I-A-2 are shown in Figure 1. The 50% inhibition level obtained with these _in vivo_ modified DNAs is 350 fmol AFB1 bound to DNA, as compared with 250 fmol AFB1 bound to DNA found for the _in vitro_-modified DNA. This discrepancy is attributed to the necessity of applying 3-90 µg of _in vivo_-modified DNA to the micro-wells, as compared with 5 ng of _in vitro_-adducted DNA for the competition analyses. Therefore, monoclonal antibodies directed towards AFB1-modified DNA can be used to quantify AFB1 bound _in vivo_ in rat liver to DNA at levels of one AFB1 residue per 1,355,000 nucleotides. However, the disadvantage of these initial antibodies was their lack of cross reactivity with other aflatoxin derivatives which are expected to be present in environmental samples.

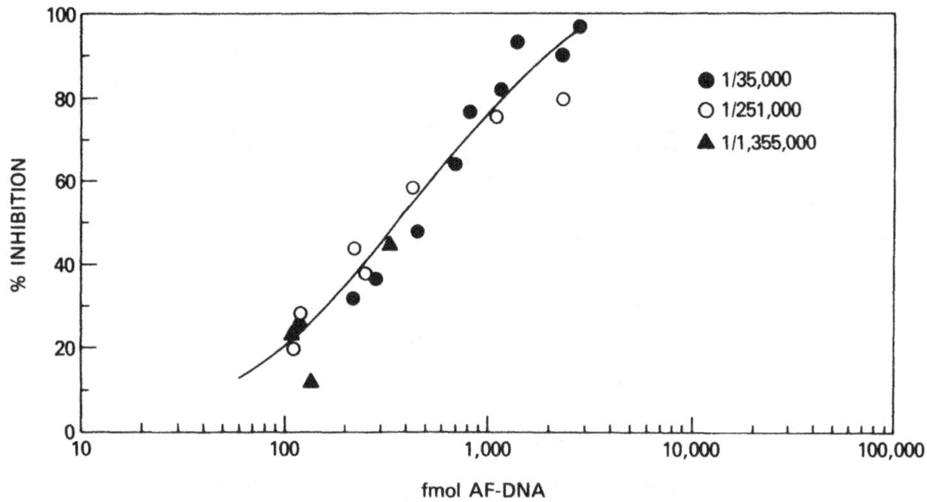

Figure 1. Competitive inhibition USERIA for in vivo rat liver AFB_1-DNA.

Since these initial antibodies produced from animals immunized with aflatoxin modified DNA did not cross react with the aflatoxin derivatives expected to be present in human urine and serum samples, we continued attempts to produce high-affinity monoclonal antibodies with the desired properties. To accomplish this, female BALB/By CJ mice were immunized with AFB1-adducted BGG in complete Freund's adjuvant[8]. Seven of the ten mice injected were found to produce significant anti-aflatoxin B1 serum titers, as measured by noncompetitive ELISA. Spleen cells from five of these mice were fused with SP-2 myeloma cells, and a number of monoclonal antibodies, grown as ascites fluid, were obtained. One of these monoclonal antibodies proved to be a high-affinity IgM (2B11), and the specificity and application of this antibody to affinity chromatography are described in the following.

Values for the affinity constant for 2B11 were determined to be 8x10E8 and 1x10E9 liters pmole, by Scatchard plot analysis and the method described by Muller[9], respectively. These values were derived from data obtained by the competitive RIA described above. The specificity of 2B11 in the competitive RIA for aflatoxin B1 and its metabolites, including the major aflatoxin B1-DNA adducts, are depicted in Figures 2 and 3. The 50% inhibition levels for aflatoxin B1, aflatoxin B2, and aflatoxin M1 were found to be 3.0 pmoles (1 ng), whereas those for AFG1, AFG2 and AFQ1 were 60.0, 84.0 and 275.0 pmoles, respectively. These data indicate clearly that the major epitope for 2B11 recognition of aflatoxin lies in the coumarin ring and the cyclopentenone ring of the aflatoxin molecule.

Figure 3 depicts the specificity and 50% inhibition in the competitive RIA of 2B11 for aflatoxin B1 and the two major aflatoxin-DNA adducts, AF-FAPyr and AFB1-N7-Gua. The 50% inhibition values were 3.0 pmoles, 24.0 pmoles and 89.0 pmoles, respectively. Based on the data in Figure 2, we expected high recognition by the antibody for the two aflatoxin-DNA adducts. In addition, data in Figure 3 indicate that the antibody is about 4-times more sensitive in detecting the AF-FAPyr adduct than the AFB1-N7-Gua adduct (24.0 pmoles versus 89.0 pmoles). This finding was unanticipated, since the aflatoxin moiety is identical in these adducts

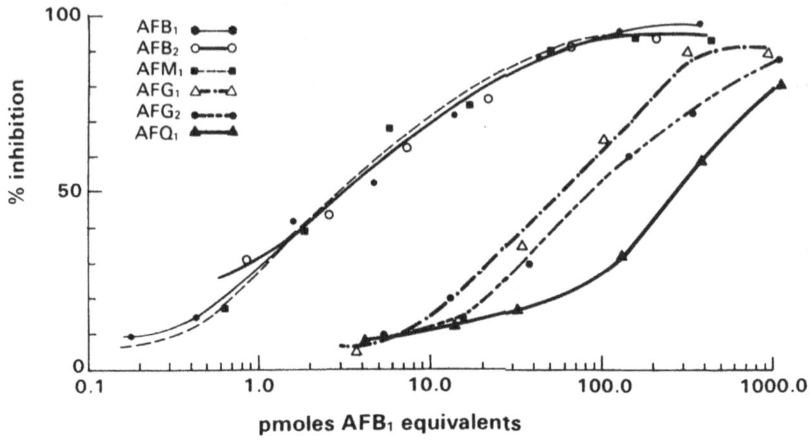

Figure 2. Competitive RIA for Mab 2B11.

when compared to the parent molecule. These observations suggest that the epitope for the antibody may be partially obstructed in these DNA adducts, due to the stereochemistry of covalent binding of aflatoxin to guanine.

At this time, we began our efforts to produce a preparative monoclonal antibody affinity column to isolate aflatoxins from complex biological fluids. Ascites fluid containing antibody 2B11 was initially fractionated by precipitation with saturated ammonium sulfate, then purified by HPLC in a steric exclusion chromatographic mode. The purified antibody was bound to a solid phase matrix of Sepharose 4B. The columns made for these experiments contained 1 mg of antibody per ml of column volume.

Initial experiments were performed to determine the ability of the antibody column to bind [3H]-aflatoxin B1. In these studies, the concentration of [3H]-AFB1 was 1 ng (3 pmoles) in 10 ml PBS, a level which would approximate concentrations expected in body fluids of people exposed to contaminated foods. When the sample containing the [3H]-AFB1 was applied to the column, about 10% of the radioactivity washed through in the loading and PBS wash phase. The column was then washed with 2M potassium

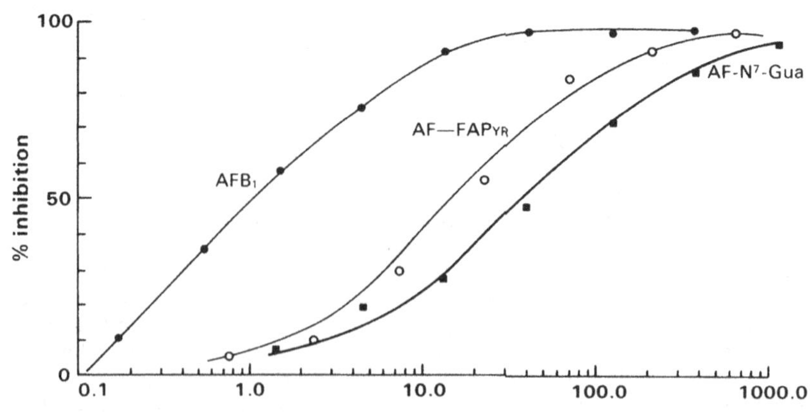

Figure 3. Competitive RIA for Mab 2B11 and DNA adducts.

thiocyanate, a commonly used elution buffer for antibody affinity chromatography, which eluted less than 2% of the applied [3H]-AFB1. Sequential washes with phosphate buffer, pH 3.0, diethanolamine buffer, pH 9.9, and phosphate buffer, pH 2.0, removed less than 1% of the applied [3H]-AFB1. However, quantitative elution of the bound [3H]-AFB1 was achieved using 50% dimethylsulfoxide (DMSO) in phosphate buffer, pH 7.4 (50% DMSO buffer). The antibody column was regenerated by washing the column with PBS, pH 7.4, and we have used and regenerated this column more than 100 times to date, with no apparent loss of activity.

To determine the optimal DMSO concentration for elution of AFB1, the affinity column prepared as above was eluted sequentially with 1%, 5%, 10%, 20%, 30% and 40% DMSO-PBS. Less than 4% of the applied aflatoxin was eluted by all of these DMSO solutions and quantitative recovery was obtained only with 50% DMSO. These properties emphasize the potential of this column for use as a preparative tool in recovery of aflatoxins from complex biological mixtures.

The capacity of the antibody affinity column to bind AFB1 was determined by radiometric and absorbance techniques, and the column was found to be capable of binding 1-1.3 μg of AFB1 from 10 ml of PBS per 1 ml of column bed volume.

Two adult male Fisher 344 rats were each injected with 1 mg [14C]-AFB1 per kg body weight and their urine collected for 20 hours, by which time 10-12% of the [14C] from the initial dose had been excreted into the urine. These urine samples were fractionated initially on the affinity column and then analyzed by competitive RIA, and the results from this assay were compared to the amount of aflatoxin determined from the radioactivity data. A precise correspondence (data not shown) was obtained when these data were compared, and these results indicated that the majority of the aflatoxin derivatives in the rat urine would be recognized by the monoclonal antibody. These data also indicated that the competitive RIA has the sensitivity required to determine aflatoxin content in biological samples.

Aliquots (100 μl) of each urine containing 290 and 310 ng of [14C]-aflatoxin equivalents, respectively, were diluted with 1.9 ml of PBS and applied to the antibody affinity column, and 65% of the applied [14C] became bound to the affinity matrix. We also recycled the initially unretained aflatoxin moieties back through the column. However, this second passage through the column failed to result in the binding of more aflatoxin to the column. These data indicate that the unretained aflatoxins are not immunologically recognized by the antibody and thus are consistant with the specificity reported for this monoclonal antibody. The retained aflatoxins were eluted from the column with 50% DMSO-PBS and analyzed by analytical reversed-phase HPLC[10]. Representative HPLC chromatograms of the UV and radioactivity aflatoxin profiles in the rat urine are found in Figures 4 and 5. Of the retained radioactivity, the predominant metabolite is aflatoxin M1, accounting for between 41% and 50% of the recovered [14C]. Aflatoxin P1 and B1 are detected and in total account for less than 10% of the radioactivity and/or UV absorbance. A major metabolite detected was the AFB1-N7-Gua adduct, which accounted for 16% of the applied radioactivity. The level of AFB1-N7-Gua in the urine corresponded to the amount calculated from the pharmacokinetic data of Bennett, _et al._, 1981[11]. The overall recovery of the radioactive aflatoxins applied to the HPLC column was greater than 95%. The unretained material from the affinity column (PBS washes) was analyzed by preparative HPLC procedures. All of the radiolabeled aflatoxin chromatographed from this fraction chromatographed as unretained polar derivatives of aflatoxin. These data show that the major metabolites isolated from the urine of rats dosed _in vivo_ are aflatoxin M1, P1, and AFB1-N7-Gua. This information

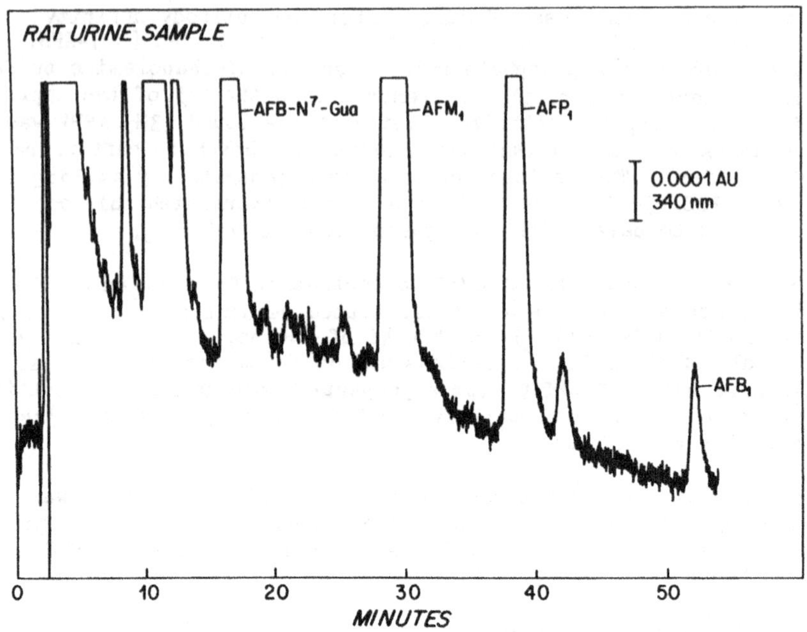

Figure 4. HPLC profile of UV absorbance from rat urine sample.

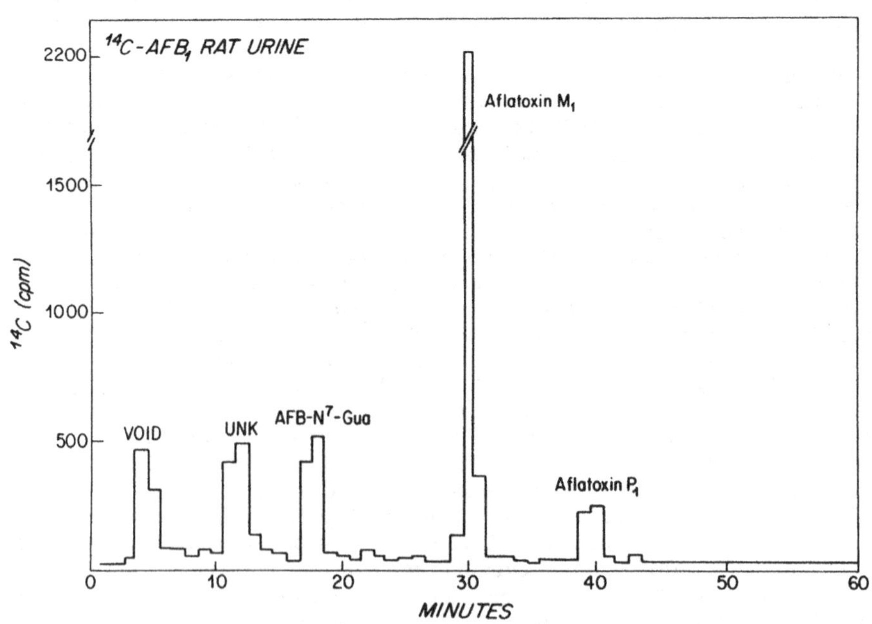

Figure 5. HPLC profile of radioactivity from rat urine sample.

indicates that by using monoclonal antibodies having different specificities, we should be able to quantify other urinary aflatoxin derivatives, such as the oxidative conjugates.

People exposed to aflatoxin B1 in their diet were identified in the Guangxi Province of the People's Republic of China. Their urine samples were used to gain preliminary evidence of the applicability of the monoclonal antibody affinity column and HPLC analysis procedures for monitoring people for exposure to aflatoxins. For this study, 20 individuals were selected, and the intake of aflatoxin B1 from the diet on the previous day was calculated to be from 13.4 to 87.5 μg. Competitive RIA were run on the human urine samples eluted from the monoclonal antibody affinity column. The resulting data indicate that the concentrations of aflatoxin in the urine collected were in the 0.1 to 10 ng/ml range. These data were calculated using a linear extrapolation of the RIA data. Further experiments need to be performed to verify that this linear extrapolation is consistent with the pharmacokinetics of aflatoxin excretion into human urine. Therefore, these data should be interpreted as representing the range of aflatoxin content in the human samples. Urine samples from individuals who had been exposed to the highest dose (87.5 μg) the previous day were run on the antibody affinity column and then analyzed by analytical HPLC. An HPLC profile of one of these individuals is depicted in Figure 6. This sample shows the occurrence of the major aflatoxin-DNA adduct, AFB1-N7-Gua, in the urine. This peak represents between 7 and 10 ng of the DNA adduct. These data indicate that the monoclonal antibody columns coupled with HPLC can quantify the levels of aflatoxin-DNA adducts in human samples obtained from environmentally exposed people in less than one hour.

Summary of Aflatoxin Monoclonal Antibody Research

We have been applying monoclonal antibody technology to produce antibodies which recognize aflatoxins. These antibodies are being used in conjunction with other chemical analytic techniques to develop non-invasive screening methodologies for monitoring human exposure to environmentally occurring carcinogens. These screening methods require the ability to quantify aflatoxin and its metabolites, including DNA adducts, in readily accessible compartments such as serum and urine. Another important requirement of these procedures is that they have to be rapid enough to measure many samples in a short period of time. Earlier measurements of aflatoxin metabolites, including the DNA adducts, have used techniques such as thin layer chromatography, high performance liquid chromatography, and immunological assays to prepare and analyze aflatoxin-containing samples[12-14]. However, a fundamental disadvantage of many of these procedures was the lengthy preparation time required before the sample could be submitted to analytical analysis. We consider our new methods to be rapid enough (less than one hour) to be applied to large-scale epidemiological studies to measure aflatoxin exposure in populations. Furthermore, these methods permit the resolution by HPLC of aflatoxin mixtures into their individual components. These components, such as the DNA adducts, can be submitted to chemical procedures to enhance the ultimate sensitivity of the analysis. An example of this was reported earlier, using a chemical procedure to quantify AFB1-N7-Gua in urine[5]. Taken together, it now appears that a combination of monoclonal antibody technology and analytical chemistry is providing a rapid and sensitive means for the initiation of epidemiologic pilot studies to assess the exposure status of populations to the environmentally occuring carcinogen aflatoxin B1.

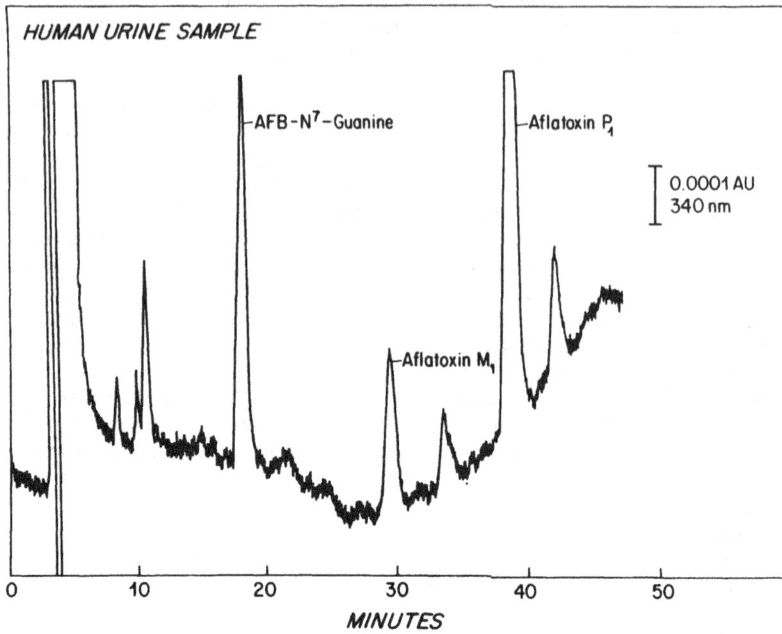

Figure 6. HPLC profile of UV absorbance from human urine sample.

REFERENCES

1. W. F. Busby, and G. N. Wogan, "Aflatoxins in Chemical Carcinogens," 2nd Ed. C.E. Searle, ed., pp. 945-1136 (1985).
2. J. M. Essigmann, R. G. Croy, A. M. Nadzan, W. F. Busby, Jr., V. N. Reinhold, G. Buchi, G. N. Wogan, Structural identification of the major DNA adduct formed by aflatoxin B1 in vitro, Proc. Natl. Acad. Sci. USA 74, 1870-1874 (1977).
3. R. G. Croy, J. M. Essigmann, V. N. Reinhold, G. N. Wogan, Identification of the principle aflatoxin B1-DNA adduct formed in vivo in rat liver, Proc. Natl. Acad. Sci. USA 75, 1745-1749 (1978).
4. J. K. Lin, J. A. Miller, E. C. Miller, (1977) 2,3-Dihydro-2-(guan-7-yl)-3 hydroxy-aflatoxin B1, a major acid hydrolysis product of aflatoxin B1-DNA or-ribosomal RNA adducts formed in hepatic microsome mediated reactions in rat liver in vivo, Cancer Res. 37, 4430-4438 (1977).
5. P. R. Donahue, J. M. Essigmann, and G. N. Wogan, Aflatoxin-DNA adducts: Detection in Urine as a dosimeter of exposure Banbury Report 13: Indicators of Genotoxic Exposure (Cold Spring Harbor Laboratory) pp. 221-229 (1982).
6. A. Haugen, J. D. Groopman, I. C. Hsu, G. R. Goodrich, G. N. Wogan, and C. C. Harris, Monoclonal antibody to aflatoxin B1-modified DNA detected by enzyme immunoassay, Proc. Natl. Acad.Sci. USA, 78, 4124-4127 (1981).
7. J. D. Groopman, A. Haugen, G. R. Goodrich, and C. C. Harris, Quantitation of aflatoxin B1-modified DNA using monoclonal antibodies, Cancer Res., 42, 3120-3124 (1982).
8. J. D. Groopman, L. J. Trudel, P. R. Donahue, A. Rothstein, and G. N. Wogan, High affinity monoclonal antibodies for aflatoxins and their application to solid phase immunoassay, Proc. Nat'l Acad. Sci., USA, 81:7728-7731 (1984).

9. R. Muller, Calculation of average antibody affinity in anti-hapten sera from data obtained by competitive radioimmunoassay, J. Immunological Methods, 34, 345-352 (1980).
10. J. D. Groopman, P. R. Donahue, J. Zhu, J. Chen, and G. N. Wogan, Aflatoxin metabolism in humans: Detection of metabolites and nucleic acid adducts in urine by affinity chromatography, Proc. Nat'l. Acad. Sci., USA.82: 6492-6497 (1985).
11. R. A. Bennett, J. M. Essigmann, and G. N. Wogan, Excretion of an aflatoxin-guanine adduct in urine of aflatoxin B1-treated rats, Cancer Res., 41, 650-654 (1981).
12. S. Tsuboi, T. Nakagawa, M. Tomita, T. Seo, H. Ono, K. Kawamura, and N. Iwamura, Detection of aflatoxin B1 in serum samples of male Japanese subjects by radioimmunoassay and high-perforance liquid chromatography, Cancer Res. 44, 1231-1234 (1984).
13. T. C. Campbell, J. P. Caedo, J. Bulatoa-Jayma, L. Salamat, and R. W. Engel, Aflatoxin M1 in human urine Nature, 227, 403-404 (1970).
14. Z. Sun, Y. Wu, and S. Wu, Monoclonal antibody against aflatoxin B1 and its potential applications, Chin. J. Oncol. 5, 401-405 (1983).

BLOOD PROTEINS AS CARCINOGEN DOSIMETERS

Steven R. Tannenbaum and Paul L. Skipper

Massachusetts Institute of Technology
Department of Applied Biological Sciences
Cambridge, MA 02139

INTRODUCTION

The widespread application of bacterial and other bioassays for the mutagenicity of environmental chemicals has shown that humans regularly encounter large numbers of chemical substances with genotoxic activity. These observations have led to concern over the potential impact of environmental mutagens on the incidence of cancer and genetically-related pathologic states in humans, and have emphasized the need for risk assessment.

Risk assessment can be considered at two levels -- qualitative and quantitative. In a qualitative sense, knowledge of the existence of risk involves two elements: evidence that the substance under consideration is mutagenic and evidence that significant human exposure takes place. Thus, for a large number of environmental chemicals, qualitative evidence of genetic risk already exists. Extending the assessment to a quantitative level is a much more complex undertaking. For this purpose, in addition to the qualitative evidence mentioned above, at least the following are needed:

1. Information on dose to susceptible cells and on dose-response relationships;

2. knowledge of metabolism and mechanisms of action;

3. information on patterns and levels of exposure; and

4. methods of risk extrapolation to relate mutagenic properties to specific disease states in humans.

Risk assessment has typically been based on exposure as determined by estimation or direct measurement of average ambient levels of carcinogens. This approach is necessarily defective in several ways. It ignores interindividual differences in exposure resulting from differences in, for example, consumption of contaminated foods, or consumption of products with differing levels of contamination. It ignores interindividual variations that operate at the post-consumption point such as different activities of constitutive or inducible metabolic enzymes. To a considerable extent, these difficulties can be ameliorated by measuring the end-product of carcinogen exposure -- the macromolecular adduct.

Because of the complexity of the chemical environments to which human populations are exposed, the problem of quantifying exposure to genotoxins in a given individual represents a formidable challenge. Many possible approaches have been suggested and are currently under active investigation[1]. In this paper we will discuss methods which rely on the covalent binding of carcinogens and their metabolites to blood proteins, an approach which was pioneered by Ehrenberg and co-workers in Sweden[2].

Background on Protein Adducts

Carcinogens interact with proteins as well as with DNA, although whether protein-carcinogen adducts can result in genetic damage has not been established. It has been shown, however, that the amount of a protein carcinogen adduct formed may be used as a quantitative measure of exposure to a carcinogen[2,3]. Such a measure presumably is reflective of the absorption, metabolism, and excretion of the compound in an exposed individual. DNA adduction and mutation have generally been considered to be the most relevant indicators of oncogenesis, but may be difficult to measure. DNA adduct levels, in addition, may reflect only exposure experienced during the several-day period immediately preceding measurement. Protein adduction may, in some cases, be more readily measurable and may reflect exposure in a time-frame of weeks to months rather than days. Thus, protein adduct measurement is a form of human chemical dosimetry which may be used with DNA adducts and mutation to represent both the recent and longer term history of an individual's exposure to genotoxic substances.

Proteins that might prove generally useful for revealing the historical record of exposure should: (1) react with a given compound with predictable dose-dependency to form covalent, relatively stable, and reliably detectable products, (2) be present in easily accessible compartments such as saliva, blood, or urine, and (3) have long biological lifetimes.

Hemoglobin and albumin are two promising candidates for such dosimeters. Both are accessible with minimally invasive sampling techniques. Each has advantages, and may provide information complementary to that obtained from the other. Hemoglobin has a lifetime of about 120 days in humans; thus, circulating levels of carcinogen-modified hemoglobin will reflect the level of carcinogen exposure during a period of nearly four months. It also possesses some metabolic competence, particularly, the ability to oxidize aromatic hydroxylamines to nitroso compounds which react quite efficiently with sulfhydryl groups. Albumin also has a relatively slow turnover, which leads to a half-life of 20-25 days in man. This protein does not possess metabolic capacity other than, perhaps, some esterase activity. In contrast to hemoglobin, though, it is not protected by the erythrocyte membrane and might be the target for a greater number of carcinogens. It is present and is synthesized in the same cells in which the reactive metabolic intermediates of carcinogens are mostly formed -- the hepatocytes. Also, albumin has a number of high-affinity binding sites for a broad spectrum of xenobiotics and endobiotics[4]. This suggests the possibility of a generic assay, one in which defining elements of the binding site are used to isolate a variety of bound carcinogens. Subsequent identification of an unknown member of the group would be a major step in identifying previously unrealized risks.

At least two relevant types of reactions with hemoglobin or albumin have been investigated.

1. The electrophilic interaction of carcinogenic epoxides or other alkylating agents with nucleophilic sites on hemoglobin -- primarily

residues of cysteine, methionine, and histidine -- results in covalent modification of the protein. This work was pioneered in the early 1970's by Ehrenberg and co-workers, using the carcinogen ethylene oxide[2]. The generality of hemoglobin alkylation by a variety of environmentally important carcinogens has since been shown[5-8]. Covalent modification of albumin by potent mutagens found in pyrolyzed foods, such as Trp-P-2[9] and IQ (our work, unpublished), have been demonstrated.

2. Some xenobiotics, or their metabolites, interact first with the oxyheme site of a hemoprotein, resulting in an oxidation, followed by the binding of the oxidized, highly reactive metabolite either to the heme moiety itself or to the attendant protein. This mechanism is the basis of the suicide inactivation by certain drugs of cytochrome P450, as studied by Ortiz de Montellano _et al._[10,11]. The aromatic amines, some of which are known human carcinogens for the bladder, interact with the oxyheme site of hemoglobin in a similar fashion, as shown by Kiese, Eyer and others[12-14].

Thus, ample evidence exists to confirm the concept of using albumin and hemoglobin as traps for covalent binding of electrophilic metabolites of carcinogens. The properties of these proteins are sufficiently different so that both could be exploited for studies on molecular epidemiology of environmental carcinogens.

Current Research at M.I.T

Current studies in our laboratory on a number of carcinogens demonstrate the extent of differences that might be expected for compounds of different structure binding to different proteins. The results of these studies are summarized in Table 1 and discussed below.

Table 1. Comparison of Albumin and Hemoglobin Binding in the Rat

	Fraction of Dose Associated with Total Protein			
	Albumin		Hemoglobin	
Carcinogen	Total Bound (% of Dose)	Specific Adduct (% of Total Bound)	Total Bound (% of Dose)	Specific Adduct (% of Total Bound)
4-Aminobiphenyl	0.1	TRP-214 (20%)	8	CYS-93β (90%)
IQ	0.2	CYS-34 (15%)	0.05	currently unknown
Aflatoxin B_1	1.5	currently unknown	0.03	currently unknown
Fluoranthene	0.04	currently unknown	0.02	currently unknown

1. 4-Aminobiphenyl is a typical aromatic amine, which will be discussed more extensively in the following section. Briefly, it interacts through its hydroxylamine with hemoglobin and through its hydroxamic acid with albumin. For purposes of measuring exposure, hemoglobin gives an overwhelmingly higher yield of adduct than does albumin.

2. IQ (2-amino-3-methyl[4,5-f]-imidazoquinoline) is a potent food mutagen and carcinogen present in cooked flesh foods and in commercial beef extract[15-18]. The nature of its binding is currently under study (Robert Turesky, unpublished observations). The only adduct that has yet been characterized is to the sole free sulfhydryl in rat albumin. The majority

of covalently bound adducts to both albumin and hemoglobin are of unknown structure, but the quantity of binding to both proteins is similar.

3. Aflatoxin B_1 is a potent hepatocarcinogen the metabolism of which has been extensively characterized[19]. Binding to DNA occurs via the very reactive 2,3-epoxide, and binding to proteins has generally been suggested to proceed via either the epoxide or the dialdehyde generated by hydrolysis of the epoxide, to form the dihydrodiol, followed by tautomerization of the hemiacetal. The relatively high binding to albumin, compared to hemoglobin, may be indicative of a highly reactive electrophile (such as epoxide) which has a relatively short half-life after its escape from the hepatocyte. Characterization of the albumin adducts is currently underway by Dr. Gabriele Sabbioni of our laboratory.

4. Fluoranthene is a typical polycyclic aromatic hydrocarbon, which is both a mutagen and a carcinogen. The structure of its principal DNA adduct has been determined by John Babson and Gerald Wogan of M.I.T. as being derived from fluoranthene diol epoxide (1,2,3,10b-tetrahydro-2,3-dihyroxy-fluoranthene-1, 10b oxide). Binding to proteins could occur either via the epoxides, the dihydrodiol epoxides, or the quinones derived from the dihydrodiols. Current research by Deborah Hutchins suggests the <u>syn</u> and <u>anti</u> diol epoxides as the principal reactants for albumin binding. Although albumin binding predominates, the hemoglobin adducts could also prove useful for further investigation.

In sum, the nature of the binding of carcinogens to proteins depends upon the selectively of the electrophile for a nucleophilic site on the protein. Reactivity factors, steric factors, and the affinity of the carcinogen for a binding site in the vicinity of the nucleophile all play a role in determining the extent of binding.

<u>4-Aminobiphenyl (ABP)</u>

Our most extensive results are with ABP for which we have characterized the major hemoglobin and albumin adducts. This aromatic amine was chosen as a model compound because it is a known human bladder carcinogen for which there might be quantifiable human exposure through cigarette smoking[20].

The feasibility of monitoring exposure to ABP via quantification of the hemoglobin binding has now been established. Studies with the rat have shown that greater than 90% of ABP bound to hemoglobin is in the form of a cysteine sulfinic acid amide, which can be released by mild hydrolysis and be detected as the parent amine. Between 5 and 10% of a single dose becomes bound to the total hemoglobin. During chronic administration of ABP, the level bound to hemoglobin increases to a plateau value about 30-times greater than that resulting from a single dose. Bound ABP persists in circulation until a full erythrocyte lifetime has elapsed. The dose-response relationship for a single dose is linear between 500 ng/kg to 5 mg/kg[21]. The structure of the principal hemoglobin adduct as a sulfinamide of cysteine β-93 has now been confirmed by x-ray crystallography[22].

A method was developed by Matthew Bryant to quantify nonradioactive ABP adducted to hemoglobin. It utilizes basic hydrolysis of the isolated hemoglobin, followed by extraction of the regenerated amine into hexane. After clean-up, the amine is derivatized with pentafluoropropionic anhydride and analyzed on a capillary gas chromatograph coupled to a mass spectrometer. An ABP analog, 4'-F-4-ABP, is added at the beginning to serve as an internal standard. The method is sensitive to levels below 10 pg (or 50 fmol) per 10 ml blood. Preliminary studies on blood samples taken from humans showed variable levels of up to 400 pg/10 ml blood.

Serum albumin was isolated from rats at 27 hours after administration of the carcinogen [2,2'-^{3}H]-4-aminobiphenyl[23]. Pronase digestion of the albumin fraction yielded a mixture of radiolabelled materials which was resolved into 5 major components by reverse-phase HPLC. From detailed UV, proton NMR, and mass spectral analyses, four of these were determined to be ABP, 4'-OH-4-Acetylaminobiphenyl, and two other metabolites, all of which are presumed to be non-covalently associated with the protein. The fifth component, however, resulted from covalent bond formation and was identified as a tetrapeptide containing 3-(tryptophan-N1-yl)-4-acetylaminobiphenyl, whose amino acid sequence was H_2H-ala-trp-ala-val. Rat serum albumin contains only a single tryptophan residue in a hydrophobic drug-binding site. Its high selectivity for binding of this metabolite suggests a unique role for this protein in the detoxification of ultimate carcinogenic metabolites.

The example given to us by ABP shows how the two blood proteins can provide information that is mutually complimentary from the perspective of molecular epidemiology. The risk for bladder cancer has been related to formation of the β-glucuronide of the hydroxylamine in the liver followed by decomposition of this conjugate in the bladder[24]. The risk for liver, intestinal, and mammary cancer may be related to formation of a reactive hydroxamic acid ester with subsequent decomposition in the target organ[25]. If the hemoglobin adduct is proportional to the hydroxylamine, while the albumin adduct is proportional to the reactive hydroxamate, a useful differential measure of organ risk may be available for epidemiological studies.

REFERENCES

1. Banbury Report 13. "Indicators of Genotoxic Exposure", B. Bridges, B. Butterworth, and I. B. Weinstein, eds., Cold Spring Harbor Laboratory, New York (1982).
2. L. Ehrenberg, K. D. Hiesche, S. Osterman-Golkar and I. Wennberg, Evaluation of genetic risks of alkylating agents: Tissue dose in the mouse from air contaminated with ethylene oxide, Mutat. Res. 24:83-103 (1974).
3. C. J. Calleman, L. Ehrenberg, B. Jansson, S. Osterman-Golkar, D. Segerback, K. Svensson, and C. A. Wachtmeister, Monitoring and risk assessment by means of alkyl groups in hemoglobin in persons occupationally exposed to ethylene oxide, J. Environ. Pathol. Toxocol. 2:427-442 (1978).
4. J. R. Brown and P. Shockly, Serum albumin: Structure and characterization of its ligand binding sites, in: "Lipid-Protein Interactions", Vol. 1, P. C. Jost and O. H. Griffith, eds., John Wiley and Sons, New York (1982).
5. L. Ehrenberg and S. Osterman-Golkar, Alkylation of macromolecules for detecting mutagenic agents, Tetrato. Carcino. Mutag. 1:105-127 (1980).
6. P. B. Farmer, E. Bailey, and J. B. Campbell, Use of alkylated proteins in the monitoring of exposure to alkylating agents, in: "Monitoring Human Exposure to Carcinogenic and Mutagenic Agents", A. Berlin, M. Draper, K. Hemminki, and H. Vaino, eds., IARC Scientific Publication No. 59, Lyon, France (1984).
7. P. B. Farmer, E. Bailey, J. H. Lamb, and T. A. Connors, Approach to the quantitation of alkylated amino acids in haemoglobin by gas chromatography-mass spectrometry, Biomed. Mass Spectrometry 7:41-46 (1980).
8. M. A. Pereira and L. W. Chang, Binding of chemical carcinogens and mutagens to rat hemoglobin, Chem.-Biol. Interact. 33:301-305 (1981).

9. M. Yamada, M. Mori, and T. Sugimura, Myeloperoxidase-catalyzed binding of 3-amino-1-methyl-5H-pyrido[4,3-b]indole, a tryptophan pyrolysis product, to protein, Chem.-Biol. Interact. 33:19-33 (1980).
10. P. R. Ortiz de Montellano, K. L. Knuze, G. S. Yost, and B. A. Mico, Self-catalyzed destruction of cytochrome P-450: Covalent binding of ethynyl sterols to prosthetic heme, Proc. Natl. Acad. Sci., USA, 76:746-749 (1979).
11. P. R. Ortiz de Montellano, G. S. Yost, V. A. Mico, S. E. Dinizo, M. A. Correia, and H. Kumbara, Destruction of cytochrome P-450 by 2-isopropyl-4-pentinamide and methyl 2-isopropyl-4-pentenoate: Mass spectrometric characterization of prosthetic heme adducts and non-participation of epoxide metabolites, Arch. Biochem. Biophys. 197:524-533 (1979).
12. M. Kiese, "Methemoglobinemia: A Comprehensive Treatise", CRC Press, Cleveland (1974).
13. P. Eyer and E. Lierheimer, Biotransformation of nitrosobenzene in the red cell and the role of glutathione, Xenobiotica 10:517-526 (1980).
14. H. G. Neumann, Analysis of hemoglobin as a dose monitor for alkylating and arylating agents, Arch. Toxicol. 56:1-6 (1984).
15. M. Takahashi, K. Wakabayashi, M. Nagao, M. Yamamoto, T. Masui, T. Goto, N. Kinae, I. Tomita, and T. Sugimura, Quantification of 2-amino-3-methylimidazo[4,5-f]quinoline (IQ) and 2-amino-3,8-dimethylimidazo-[4,5-f]quinoxaline (MeIQx) in beef extracts by liquid chromatography with electrochemical detection (LCEC), Carcino. 6:1195-1199 (1985).
16. W. A. Hargraves and M. W. Pariza, Purification and mass spectral characterization of bacterial mutagens from commercial beef extract, Cancer Res. 43:1467-1472 (1985).
17. R. J. Turesky, J. S. Winshnok, S. R. Tannenbaum, R. A. Pfund, and G. H. Buchi, Qualitative and quantitative characterization of mutagens in commercial beef extract, Carcino. 4:863-866 (1983).
18. J. S. Felton, M. G. Knize, C. Wood, B. J. Wuebbles, S. K. Healy, D. H. Stuermer, L. F. Bjeldanes, B. J. Kimble, and F. T. Hatch, Isolation and characterization of new mutagens from fried ground beef, Carcino. 5:95-102 (1984).
19. A. A. Stark, J. N. Essigman, A. L. Demain, T. R. Skopek, and G. N. Wogan, Aflatoxin B_1 mutagenesis, DNA binding and adduct formation in Salmonella typhimurium, Proc. Natl. Acad. Sci. USA 76:1343-1347 (1979).
20. C. Patrianakos and D. Hoffmann, Chemical studies on tobacco smoke. LXIV. On the analysis of aromatic amines in cigarette smoke, J. Anal. Toxicol. 3:150-154 (1979).
21. L. C. Green, P. L. Skipper, R. J. Turesky, M. S. Bryant, and S. R. Tannenbaum, In vivo dosimetry of 4-aminobiphenyl via a cysteine adduct in hemoglobin, Cancer Res. 44:4254-4259 (1984).
22. D. Ringe, R. J, Turesky, P. L. Skipper, and S. R. Tannenbaum, Structure of the single stable hemoblogin adduct formed by 4-aminobiphenyl in vivo. Manuscript submitted to Science.
23. P. L. Skipper, M. W. Obiedzinski, S. R. Tannenbaum, D. W. Miller, R. K. Mitchum, and F. F. Kadlubar, Identification of the major serum albumin adduct formed by 4-aminobiphenyl in vivo. Cancer Res. (in press).
24. F. F. Kadlubar, J. A. Miller, and E. C. Miller, Hepatic microsomal N-glucuronidation and nucleic acid binding of N-hydroxy arylamines in relation to urinary bladder carcinogenesis, Cancer Res. 37:805-814 (1977).
25. C. A. McQueen, C. J. Maslansky, and G. M. Williams, Role of the acetylated polymorphism in determining susceptibility of cultured rabbit hepatocytes to DNA damage by aromatic amines, Cancer Res. 43:3120-3123 (1983).

DEVELOPMENT OF AN AVIDIN-BIOTIN AMPLIFIED ENZYME-LINKED IMMUNOASSAY FOR DETECTION OF DNA ADDUCTS OF THE HUMAN BLADDER CARCINOGEN 4-AMINOBIPHENYL

Dean W. Roberts, R. Wayne Benson, Thomas J. Flammang, and Fred F. Kudlubar

Division of Biochemical Toxicology
National Center for Toxicological Research
Jefferson, AR

ABSTRACT

4-Aminobiphenyl (ABP) is a known human urinary bladder carcinogen which is present in tobacco smoke and may be ubiquitous in the environment. As a biological monitor of carcinogen exposure, we have developed an immunological method for measuring the predominant carcinogen-DNA adduct of ABP, N-(deoxyguanosin-8-yl)-ABP (dG-C8-ABP). Rabbits were immunized with keyhole limpet hemocyanin (KLH) conjugate prepared by a periodate oxidation and coupling of N-(guanosin-8-yl)-ABP (rG-C8-ABP) to the protein. The resulting polyclonal antisera was systematically characterized using dual inhibitor methodology augmented by specialized computer and software support; and a competitive avidin-biotin enzyme-linked immunoassay (A-B ELISA) assay employing polyclonal rabbit anti-KLH-(rG-C8-ABP) was developed. Under the assay conditions described, the detection limit for dG-C8-ABP was 18 fmol/well. The relative lack of reactivity toward ABP, N-acetyl-4-aminobiphenyl, N-(deoxyadenosin-8-yl)-ABP, N-(deoxyguanosin-8-yl)-2-aminofluorene and deoxyguanosine as inhibitors indicated that primary specificity involves epitopes found on the purine and biphenyl rings. Results emphasize the need to define polyclonal anti-adduct sera operationally in the context of the antigen/assay system used to evaluate it. Assay sensitivity was achieved by decreasing the amount of antibody and solid-phase antigen in the competitive portion of the assay and the use of avidin-biotin as well as enzymatic amplification. This methodology is a useful alternative to other ultrasensitive techniques and should be directly applicable to the detection of ABP-DNA adducts in exposed human populations.

INTRODUCTION

The human bladder carcinogen 4-aminobiphenyl (ABP) requires metabolic activation in order to form arylamine-DNA adducts which can give rise to mutations in the genome and to malignant transformation[1]. Metabolic factors such as the capacity to N-acetylate arylamines, which shows a genetic polymorphism in humans, and the levels of liver monooxygenases, which are subject to induction by environmental agents, affect the balance between N-acetylation and N-hydroxylation and are probably important modifiers of susceptibility to aromatic amine-induced bladder cancer[2].

Physiological factors such as urine pH and frequency of urination[3] also determine the extent of conversion of the probable proximate carcinogenic metabolite, the N-hydroxy arylamine N-glucuronide[4], in the bladder lumen to a DNA-binding electrophile[3]. Recent evidence indicates that ABP may be a ubiquitous environmental carcinogen and a significant component in tobacco smoke[5]. The probable occurrence of ABP-DNA adducts in humans would be expected to depend on circumstances of environmental exposure combined with heritable individual differences in the metabolism and pharmacodynamics of aromatic amines. Together, these factors dictate the need for sensitive biomonitoring methods to better estimate possible health risks associated with environmental exposure to ABP. In particular, there is a specific requirement for ultrasensitive and non-invasive screening methods to detect and quantify non-radiolabeled ABP-DNA adducts in small amounts obtained from potentially exposed individuals.

Methods to detect DNA modification by chemical carcinogens have been reviewed recently[6] and include photon-counting fluorescence spectroscopy, [^{32}P] post-labeling analysis, and immunological detection techniques including radioimmunoassay (RIA), enzyme-linked immunosorbent assay (ELISA) and ultrasensitive enzymatic radioimmunoassay (USERIA). There have been a number of important advances in the development of sensitive immunological techniques to detect DNA components structurally modified by mutagenic and DNA-reactive carcinogens[7-12]. Although RIA techniques remain especially useful for the determination of affinity constants of anti-adduct antibodies[13], they have been largely superceded by more sensitive (10-100 fold) solid-phase enzyme immunoassays[6,8]. These assays typically employ either polyclonal rabbit antisera or murine monoclonal antibodies. Although monoclonal antibodies have obvious potential to be more specific than polyclonal antisera, they are not generally observed to be more sensitive or more specific than corresponding conventional antisera[14]. The single epitope specificity of monoclonal antibodies may be an advantage in structural analysis studies, but can be a limitation in assay situations where there is a requirement for immuno-detection of antigens after electrophoresis and electroblotting or procedures such as histological fixation which may denature the unique determinant recognized by a monoclonal antibody. We report here the production of polyclonal rabbit antisera with primary specificity for N-(deoxy-guanosin-8-yl)-ABP, the major DNA adduct formed by ABP *in vivo*[1] along with sensitive methodology useful in the characterization of antibodies and reduction of cross-reactivity. The assay described employs an avidin-biotin amplification system which is an alternative to other ultrasensitive detection techniques and does not require chromatographic isolation of substrate or the use of radioisotopes.

MATERIALS AND METHODS

Preparation of Modified Nucleic Acid Components

N-(Deoxyguanosin-8-yl)-4-aminobiphenyl (dG-C8-ABP), N-(doexyadenosin-8-yl)-4-aminobiphenyl (dA-C8-ABP), N(deoxyguanosin-8-yl)-2-aminofluorene (dG-C8-AF), N-(doexyguanosin-8-yl)-benzidine (dG-C8-BZ) and N-(deoxyguanosin-8-yl)-N'-acetylbenzidine (dG-C8-AcBZ) were all synthesized and then purified by high pressure liquid chromatography (HPLC) exactly as described[15-17]. Hydrolyzed DNA was prepared by enzymatic digestion of mixtures of calf thymus DNA (1 mg/ml, Sigma Chemical Co., St. Louis, MO) as described[18].

Preparation of Nucleoside-Protein Conjugates and Immunization

N-Hydroxy-4-aminobiphenyl was prepared from 4-nitrobiphenyl (Aldrich Chemical Co., Milwaukee, WI) by reduction with ammonium sulfide light

(Fisher Scientific, St. Louis, MO) as described[19]. The ribonucleoside adduct, N-(guanosin-8-yl)-4-aminobiphenyl (rG-C8-ABP), was prepared essentially by the method of Lee and King[20] except that after the removal of ethanol the reaction mixture was extracted 3 times with n-butanol, the combined extracts dried under reduced pressure, and the resulting residue was dissolved in methanol. The adduct was then purified by Sephadex LH-20 column chromatography by stepwise elution with 25%, 50%, and 75% aqueous methanol.

The rG-C8-ABP was coupled to either bovine serum albumin (BSA, Sigma, or KLH (Sigma) by a modified procedure of Erlanger and Beiser[21]. Five mg of rG-C8-ABP were combined with 2.0 ml of 0.05 M $NaIO_4$ and stirred in the dark for 45 minutes at room temperature. The reaction was stopped by the addition of 10 µl ethylene glycol. This mixture was added dropwise to 250 µl of 20 mg/ml BSA in 0.05 M $NaCO_3$, pH 9.5 and stirred for 45 minutes in the dark at room temperature during which time the pH was maintained at 9.5 by addition of 2 M $NaCO_3$. The protein-(rG-C8-ABP) conjugate was then stabilized by reduction with 8 mg $NaBH_4$ for 3 hours at 4°C and the reaction mixture was then dialyzed 3x against 4 liters of pH 7 phosphate-buffered saline (PBS). Protein concentrations were determined by the method of Lowry. KLH-(rG-C8-ABP) conjugates were prepared similarly except 7.5 mg of rG-C8-ABP were oxidized in 3 ml of 0.05 M $NaIO_4$, combined with 15 mg of KLH, and then reduced with 12 mg $NaBH_4$.

Rabbits were immunized with KLH-(rG-C8-ABP) in two series of immunizations two years apart. For each series, the initial immunization employed Freund's complete adjuvant and subsequent booster immunizations employed Freund's incomplete adjuvant. PBS containing 500 µg KLH-(rG-C8-ABP)/ml was emulsified with an equal volume of adjuvant and 2 ml of this immunogen was divided between two 0.5-ml subcutaneous injection sites and two 0.5-ml intramuscular injection sites. Data presented here relate to sera obtained during the second series of immunizations from rabbit BH2 eight days after the third immunization.

<u>Avidin-Biotin Amplified Enzyme-Linked Immunosorbent Assay (A-B ELISA)</u>

A-B ELISA assays were conducted in 96-well polystyrene plates (Nunc Immuno Plate I, Vangard International Inc., Neptune, NJ) which were coated by allowing 25 ng (protein) BSA-(rG-C8-ABP) in 60 µl saline/well to dry overnight at 37°C. Wells were washed 5x with PBS containing 0.05% Tween-20 (Sigma), 0.1% gelatin (Difco, Detroit, MI) and 1% fetal calf serum (Gibco, Grand Island, NY) (assay buffer). To block remaining free binding sites, wells were filled with assay buffer and incubated 70 minutes at 37°C. For non-competitive A-B ELISA's, dilutions of rabbit anti-KLH-(rG-C8-ABP) were added to appropriate wells and incubated at 37°C for 1.5 hours. For competitive A-B ELISA's rabbit antiserum was diluted 1:1000 in assay buffer containing 40 µM deoxyguanosine (dG), combined with an equal volume of an appropriate concentration of inhibitor diluted in assay buffer, and incubated overnight at 4°C. 100 µl of this mixture was added to duplicate wells and incubated for 1.5 hours at 37°C. Plates were washed 5x as before and 100 µl of a 1:3000 dilution of biotinylated goat anti-rabbit IgG (Enzo Biochemicals, Inc., New York, NY) was added to each well. After incubation for 1.5 hours at 37°C in a humidifed chamber, plates were washed 5x and 100 µl of a 1:250 dilution of avidin-biotin-horseradish peroxidase (HRP) complex (Enzo) was added to each well.

After a 15-minute incubation at 37°C, plates were washed 5x and 100 µl of a 0.16% solution of 5-amino-2-hydroxybenzoic acid (5-ASA, Sigma, recrystallized as described in[22]) containing 3% H_2O_2 was added to each well. Reactions were stopped after 30 minutes by addition of 50 µl of 1N NaOH; and thirty minutes later, substrate to chromophore conversion was

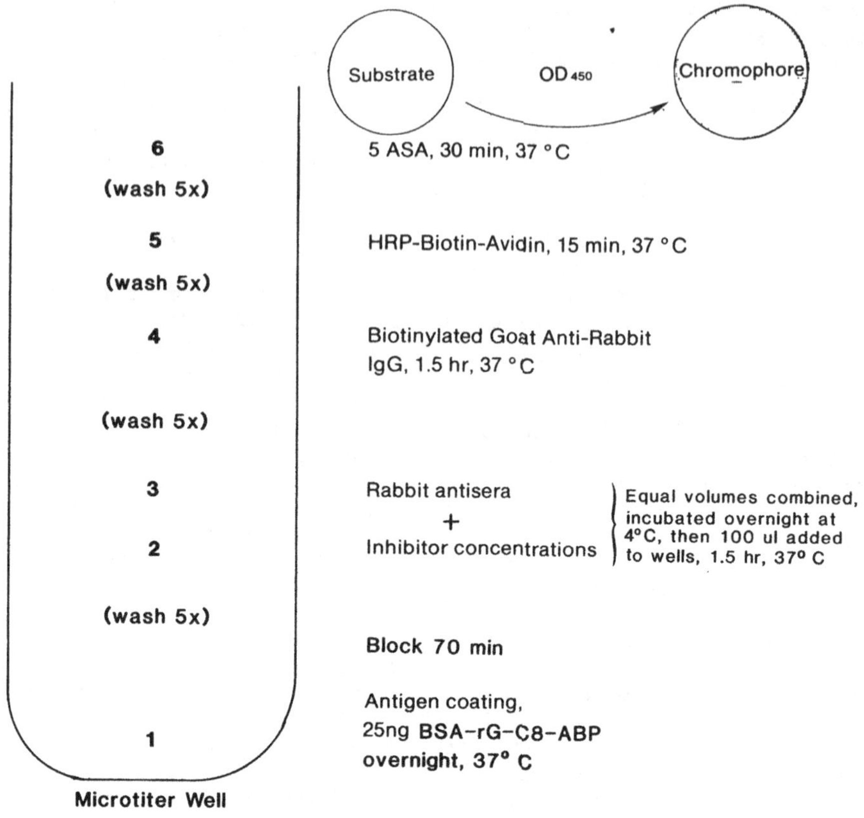

Figure 1. Schematic representation of A-B ELISA. Numbers in the microtiter well represent, in order of addition to the well, the stack of reagents which comprise the A-B ELISA: 1 = solid-phase antigen coating; 2 = inhibitor; 3 = rabbit anti-KLH-(rG-C8-ABP); 4 = biotinylated goal anti-rabbit IgG; 5 = avidin-biotin-HRP complex; calculated according to the formula:

$$\%I = \frac{(\text{ODmax BKG}) - (\text{OD individual-BKG})}{(\text{ODmax-BKG})} \times 100$$

By number, ODmax = 1,3,4,5,6; OD individual = 1,2,3,4,5,6; Background = 3,4,5,6. Controls include 1,4,5,6 to rule out second (detection) antibody recognition of coating antigen; and 6 only, which is used to zero the ELISA reader.

quantified at OD_{450} with an ELISA plate reader (Chromoscan, Bio-tek Instruments, Inc. Burlington, VT).

Conventional ELISA

Conventional competitive ELISA assays were performed in a manner similar to the competititve A-B ELISA except that the amount of coating antigen was 50 ng BSA-(rG-C8-ABP)/well, rabbit antiserum was diluted to a final dilution of 1:225 in buffer that did not contain dG, and a 1:500 dilution of HRP-conjugated goat anti-rabbit IgG (Cappel, Cochranville, PA) was used to detect rabbit antibody bound to slide-phase antigen.

Computer Support

Data from the ELISA plate reader were collected on an Apple 2E computer via RS232c serial interface and stored on diskettes for analysis using software and graphics rountines modified or developed in our laboratory. Standard curves for competititve A-B ELISAs were evaluated and plotted using a weighted, iterative least squares regression method for four-parameter curve-fitting[23]. Inhibition assays were evaluated using a program which reads data stored on floppy disks and allows flexible user input of the location of up to six inhibition curves by row and column, legend information, inhibitor concentration, log scale, dilution increment and location of background wells. The program calculates percent inhibition for each inhibitor concentration, computes scale, plots inhibition curves fitted by polynomial regression, and prints hard copy reports and graphics.

RESULTS AND DISCUSSION

Assay Development

A preliminary series of competitive and non-competitive ELISA assays were conducted to titer antisera and to determine optimal assay conditions. Strategy for increasing the sensitivity of the competitive ELISA assay involved decreasing the quantity of reactants in the competitive portion of the assay [BSA-(rG-C8-ABP), anti-KLH-(rG-C8-ABP), and inhibitor (No.'s 1-3 in Fig. 1)] and the use of an antibody dilution such that solid-phase antigen was in substantial excess. The capacity of an inhibitor to compete with solid phase antigen for a limited amount of antibody is subject to the law of mass action and, in theory, a 10-fold reduction in reaction size should be accompanied by a 10-fold decrease in the amount of inhibitor required to effect 50% inhibition. Realization of theoretical sensitivity by decreasing reaction size while increasing amplification is limited by a concomitant decrease in the signal to noise ratio due to increased prominence of non-specific background. Amounts of biotinylated goat anti-rabbit conjugate and HRP-biotin-avidin complex were systematically adjusted to achieve a high signal to noise ratio and obtain bound enzyme activity such that ODmax (no inhibitor) was approximately 1:10 after incubation for 30 minutes. Similarly, conditions for blocking after adsorption of antigen, and changes in incubation conditions, incubation time, washing technique, etc., also influence signal to noise ratio and were systematically optimized.

Dual Inhibitor Assays

In this assay system, only antibodies capable of binding the solid-phase coating antigen, BSA-(rG-C8-ABP), potentially contribute to total binding and thus are subject to inhibition. Antibodies with other specificities such as to the immunogenic KLH carrier protein do not bind to the solid-phase coating antigen, are removed during washing, and are therefore not relevant except as possible contributors to non-specific background. Only those antibodies which react or cross-react with the coating antigen are retained in the assay.

In order to evaluate the specificities of polyclonal rabbit antisera raised by immunization with KLH-(rG-C8-ABP), a series of competitive ELISA assays were conducted. These assays employed a dual inhibitor methodology in which three different inhibitors, dG-C8-ABP, dG, and N-acetyl-4-amino-biphenyl (AABP), were each added in decreasing increments to produce standard curves in the presence of a constant amount of a second inhibitor which was included in the assay buffer used to dilute polyclonal anti-KLH-(rG-C8-ABP). The normal assay buffer was spiked with dG or ABP to

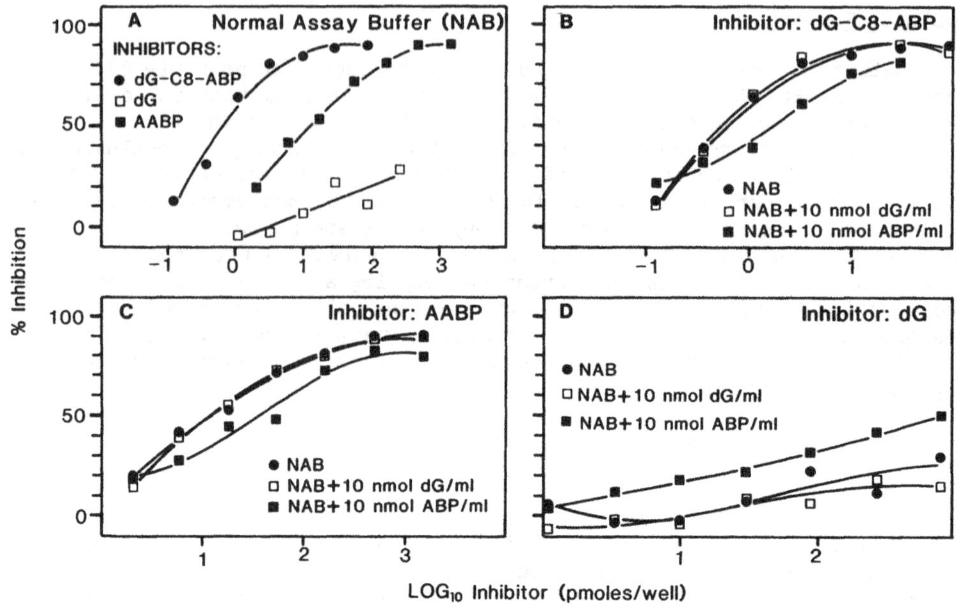

Figure 2. Dual inhibitor analyses of polyclonal rabbit anti KLH-(rG-C8-ABP). Antiseri diluted in normal assay buffer (NAB) plus 10 μg dG or 10 μM ABP were used to prepare standard curves constructed using serial 3-fold concentration increments of the following three inhibitors: dG-C8-ABP (90, 30, 10, 3, 1, 0.3, 0.1 pmole/well; dG (810, 270, 90, 30, 10, 3, 1 pmole/well); and AABP (1500, 500, 166, 55, 18, 6, 2 pmole/well). Assay wells were coated with 50 mg BSA-(rG-C8-ABP) and bound rabbit antibodies were detected with HRP conjugated goat anti-rabbit IgG.

determine if absorption of anti-dG or anti-ABP reactivity within the population of polyclonal KLH-(rG-C8-ABP) antibodies would alter the capacity of selected inhibitors to competitively displace antibody from solid-phase BSA-(rG-C8-ABP). The constant amounts of dG and ABP in the spiked assay buffer were equivalent to 500 pmol ABP/well for the buffer containing 10 μM ABP and 500 pmol dG/well for the buffer containing 10 μM dG. In comparison to the normal (unspiked) assay buffer, 10 μM ABP resulted in 63% inhibition and 10 μM dG resulted in 10.4% inhibition.

In Fig. 2, Panel A shows standard inhibition curves in normal buffer. Note that polyclonal antisera reacted with all three inhibitors; and 50% inhibition levels for dG-C8-ABP, AABP, and dG were 0.48 pmol/well, 11.7 pmol/well, and >> 800 pmol/well, respectively. Panel B shows three inhibition curves for identical concentrations of dG-C8-ABP assayed using serum diluted in each of the three different assay buffer solutions. The concentrations of dG-C8-ABP which caused 50% inhibition of binding were 488 fmol/well for antibody diluted in normal buffer, 490 fmol/well for antibody diluted in buffer containing dG, and 1700 fmol/well for antibody diluted in buffer containing ABP. This indicates that absorption of anti-dG reactivity has little effect on detection of HPLC- purified dG-C8-ABP and that ABP in the dilution buffer neutralizes some of the antibody which would otherwise be available to react with dG-C8-ABP.

Panel C shows the inhibition curves for identical concentrations of AABP assayed using sera diluted in the three buffers. Note that, as expected, absorption of anti-ABP reactivity but not anti-dG reactivity affects the detection of AABP, indicating that the population of antibodies which can be neutralized with dG does not effect inhibition by AABP. The observation that 500 pmol ABP (alone)/well resulted in 63% inhibition while about 30 pmol AABP (alone)/well was required to give comparable assay inhibition indicated that, on a molar basis, AABP was at lease 10-fold more efficient than ABP as an inhibitor.

Dual inhibition assays in which dG was included in the assay buffer suggested that the level of anti-dG reactivity was probably negligible when purified samples of dG-C8-ABP were assayed as inhibitors. These data suggested that the relative proportion of antibodies with anti-dG reactivity was very small compared to the population recognizing AABP or dG-C8-ABP. The importance of evaluating low levels of anti-dG reactivity lies in the potential contribution to total binding of even low levels of anti-nucleoside antisera when assaying samples such as hydrolyzed DNA that contain a 10^5 - to 10^8 - fold preponderance of unadducted bases as available antigen. Panel D shows the inhibition curves for dG assayed using serum diluted in the three buffers. When 10 μM ABP was included in the assay buffer, the result was a very low amplitude response (ODmax = 0.25) in which dG served as an inhibitor in the relative absence of anti-ABP reactivity. Under these special circumstances, 50% inhibition could be achieved with 842 pmol dG/well. As a result of these studies, dG was routinely included in the buffer for the more sensitive A-B ELISA in order to make the assay relatively refractive to dG as an inhibitor.

The desirability of including dG in the assay buffer does not imply extensive cross-reactivity or lack of specificity of the rabbit antiserum since analogous techniques to absorb cross-reactivity are routinely employed in USERIA techniques using polyclonal rabbit antisera when 1 μg DNA is included in the antigen-antibody reaction mixture[10]. Muller and Rajewsky[14] showed that antibodies specific for O^6-ethyl-dG have high affinity for the alkylated deoxyribonucleoside as compared to 10^6-fold lower affinity for unmodified dG. The problem of minor cross-reactivity is not unique to polyclonal antisera. The highly sensitive and specific immuno-slot-blot assay for carcinogen-modified nucleosides in DNA is also limited in sensitivity by low-level, non-specific binding of monoclonal antibodies to DNA[12]. The population of antibodies elicited by immunization with the antigenic mosaic comprised of rG-C8-ABP coupled to KLH includes relevant antibodies with specificity for dG-C8-ABP, ABP, and dG. This is consistent with the hypothesis that these molecules contain antigenic determinants which are shared by the immunogen. However, the poor efficiency of AABP, ABP, and dG as inhibitors as compared to dG-C8-ABP indicates that the major epitope(s) recognized by the polyclonal serum is unique to dG-C8-ABP.

Structurally Related Inhibitors

The sensitivity and specificity of the competititve A-B ELISA using polyclonal rabbit anti KLH-(rG-C8-ABP) are presented in Figure 3. As an internal standard, dG-C8-ABP was routinely assayed as inhibitor over 6-serial, 3-fold dilutions of inhibitor from 500 fmol/well to 2 fmol/well. The standard curve in Figure 3A was analyzed using a weighted, iterative, least squares regression method for four-parameter curve-fitting[16]. For this assay, the minimum detectable inhibitor concentration for duplicate samples at the 95% confidence level was 18 fmol/well. Detection limits are influenced not only by the affinity and specificity of the antisera, but also by the reproducibility and number of replicate samples. The dG-C8-ABP standard curve in Panel A represents the same data presented in Panel B for dG-C8-ABP plotted as % inhibition.

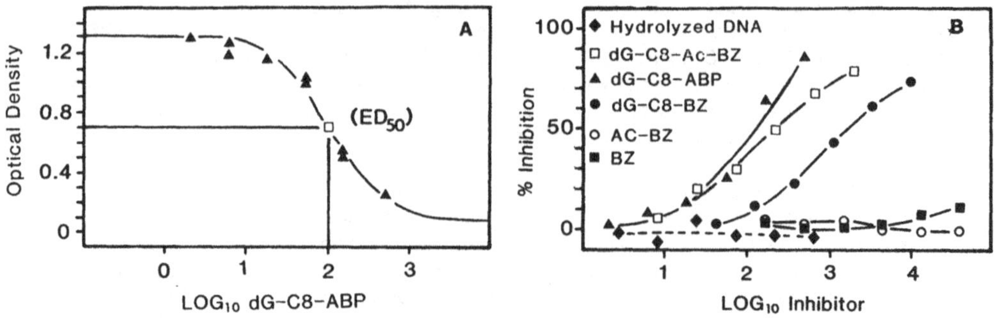

Figure 3. Competitive A-B ELISA's. The standard curve (Panel A) for the A-B competitive ELISA for dG-C8-ABP inhibitor was evaluated using a weighted, iterative least squares regression method for four-parameter curve-fitting[16] wherein the relationship between OD response (y) to inhibition concentration in fmol/well (x) is described by the formula:

$$y = \frac{a - d}{1 + (x/c)^{b}} + d$$

Where a = OD max, b = (-1) slope, c = ED_{50}, and d = OD min. For this standard curve using BH2.3 antisera and dG-C8-ABP as inhibitor, slope = 1.21, ED_{50} = 106 fmol/well, OD max = 1.32 and OD min = 0.07.

The relative efficiences of several structurally related molecules for inhibition of antibody binding to solid phase BSA-(rG-C8-ABP) are presented in Figure 3B and Table 1. Under the described assay conditions, polyclonal rabbit antisera detects fmol quantities of dG-C8-ABP, dG-C8-AcBZ and dG-C8-BZ, weakly reacts with AABP and dA-C8-ABP, and does not recognize N-acetyl-benzidine (AcBZ), benzidine (BZ), ABP, or hydrolyzed DNA. Note that the scale for hydrolyzed DNA in Figure 3B is in pmol/well while the scale for other inhibitors is in fmol/well. These data demonstrate that the polyclonal antiserum recognizes dG-C8 adducts that are structurally related to dG-C8-ABP but with less efficiency. The relative inefficiency or total inability of hydrolyzed DNA and haptenic groups AABP, ABP, BZ and AcBZ to function as inhibitors in the assay further demonstrate that primary antibody specificity involves epitopes not found on the haptenic groups or on the dG-C8-ABP adduct. The 10-fold greater inhibitory capability of dG-C8-ABP as compared to dG-C8-AF suggests that the biphenyl ring structure is also an important component of the recognized antigenic determinant. The critical importance of the purine ring structure to antibody recognition is also demonstrated by the 58-fold greater inhibitory efficiency of dG-C8-ABP as compared to dA-C8-ABP.

Present data indicate that polyclonal rabbit antisera raised against KLH-(rG-C8-ABP) recognize the major DNA adduct (dG-C8-ABP) of the human bladder carcinogen ABP. The avidin-biotin-amplified competitive ELISA provides a sensitive and reproducible means to detect fmol-quantities of carcinogen-DNA adducts and may be considered a useful alternative to other ultrasensitive techniques. Our results demonstrate the importance of operationally defining polyclonal anti-adduct antisera in the assay context as well as a means of diminishing the influence of low levels of anti-dG cross-reactivity. This methodology is capable of detecting ABP-DNA adducts

Table 1. Specificity and Cross-Reactivity of Polyclonal Rabbit Antisera

Inhibitor	Relative Inhibition	50% Inhibition (fmol/well)[a]
dG-C8-ABP	1 (relative standard)[b]	148
dG-C8-AcBZ	1.42	208
dG-C8-BZ	3.22	432
dG-C8-AF	11.67	1,622
AABP	58	10,242
dA-C8-ABP	626	100,208
ABP, BZ, AcBZ	---[c]	---
Hydrolyzed DNA	---	---

[a] Percent inhibition of binding of solid-phase BSA-(rG-C8-ABP) was calculated according to the formula:

$$\%I = \frac{(\text{ODmax-BKG})-(\text{OD Individual-BKG})}{(\text{ODmax-BKG})} \times 100$$

using software developed in our laboratory which fits a 3° polynomial to the inhibition curve and calculates concentration at 50% inhibition.

[b] The capacity of each adduct to inhibit binding was also expressed relative to inhibition of binding by dG-C8-ABP, which was conducted on each assay plate as an internal standard (50% inhibition = 148 $\pm$ SEM 11, n = 8) where relative inhibition was calculated by dividing the concentration of the sample at 50% inhibition by the concentration of dG-C8-ABP resulting in 50% inhibition on the same plate.

[c] Did not inhibit binding of antibody with solid-phase BSA-(rG-C8-ABP) under the assay conditions.

at levels of biological significance and should be applicable as a biomarker in the estimation of health risk associated with exposure to the carcinogen ABP.

REFERENCES

1. F. A. Beland, D. T. Beranek, K. L. Dooley, R. H. Heflich, and F. F. Kadlubar, Arylamine-DNA adducts in vitro and in vivo: their role in bacterial mutagenesis and urinary bladder carcinogenesis, Environ. Health Perspect. 49:125 (1983).
2. G. J. Hammons, F. P. Guengerich, C. C. Wesi, F. A. Beland, and F. F. Kudlubar, Metabolic oxidation of carcinogenic arylamines by rat, dog, and human hepatic microsomes and by purified flavin-containing and cytochrome P-450 monooxygenases, Cancer Res. 45:3578 (1985).
3. J. F. Young and F. F. Kadlubar, A pharmacokinetic model to predict exposure of the bladder epithelium to urine pH, voiding interval, and resporption, Drug Metab. Disp. 6:641 (1982).
4. F. F. Kadlubar, J. A. Miller, and E. C. Miller, Hepatic microsomal N-glucuronidation and nucleic acid binding of N-hydroxy arylamines in relation to urinary bladder carcinogenesis, Cancer Res. 37:805 (1977).

5. Tannenbaum, et al., this volume.
6. E. Kriek, L. D. Engelse, E. Scherer, and J. G. Westra, Formation of DNA modifications by chemical carcinogens: Identification, localization, and quantitation, Biochem. Biophys. Acta in press (1985).
7. M. C. Poirier, Antibodies to carcinogen-DNA adducts, J. Natl. Cancer Inst. 67:515 (1981).
8. A. Haugen, J. D. Groopman, I. Hsu, G. R. Goodrich, G. Wogan, and C. Harris, Monoclonal antibody to aflatoxin B_1-modified DNA detected by enzyme immunoassay, Proc. Natl. Acad. Sci. 78:4124 (1981).
9. R. Muller and M. F. Rajewsky, Immunological quantiation by high-affinity antibodies of O^6-ethyldeoxyguanosine in DNA exposed to N-ethyl-N-nitrosourea, Cancer Res. 40:887 (1980).
10. I. Hsu, M. C. Poirier, S. H. Yuspa, D. Grunberger, I. B. Weinstein, R. H. Yolken, and C. C. Harris, Measurement of benzo(a)pyrene-DNA adducts by enzyme immunoassays and radioimmunoassay, Cancer Res. 41:1091 (1981).
11. J. D. Groopman, L. J. Trudel, P. R. Donahue, A. Marshak-Rothstein and G. A. Wogan, High-affinity monoclonal antibodies for aflatoxins and their application to solid-phase immunoassays, Proc. Natl. Acad. Sci. 81:7728 (1984).
12. P. Nehls, J. Adamkiewicz, and M. F. Rajewsky, Immuno-slot-blot: a highly sensitive immunoassay for the quantitation of carcinogen-modified nucleosides in DNA, J. Cancer Res. Clin. Oncol. 108:23 (1984).
13. R. Muller, Calculation of average antibody affinity in anti-hapten sera from data obtained by competitive radioimmunoassay, J. Immuno. Methods 34:345 (1980).
14. R. Muller and M. R. Rajewsky, Antibodies specific for DNA components structurally modified by chemical carcinogens, J. Cancer Res. Clin. Oncol. 102:99 (1981).
15. F. F. Kadlubar, F. A. Beland, D. T. Beranek, K. L. Dooley, R. H. Heflich, and F. E. Evans, Arylamine-DNA adduct formation relation to urinary bladder carcinogenesis and Salmonella typhimurium, in: "Environmental Mutagens and Carcinogens, T. Sugimura, S. Kondo, H. Takabe, eds., Alan R. Liss, Inc., New York, NY pp. 385 (1982).
16. F. A. Beland, W. T. Allaben, and F. E. Evans, Acyltransferase-mediated binding of N-hydroxy-arylamides to nucleic acids, Cancer Res. 40:834 (1980).
17. C. N. Martin, F. A. Beland, R. W. Roth, and F. F. Kadlubar, Covalent binding of benzidine and N-acetylbenzidine to DNA at the C-8 atom of deoxyguanosine in vivo and in vitro, Cancer Res. 42:2678 (1982).
18. Y. Yamazoe, R. W. Roth, and F. F. Kadlubar, Reactivity of benzidine diimine with DNA to form N-(deoxyguanosin-8-yl)-benzidine, Carcinogenesis in press (1985).
19. M. R. Thissen, R. W. Roth, and W. P. Duncan, Convenient synthesis of selected ^{14}-C- and ^{3}-H-labeled aromatic hydroxylamines, Organic Prep. Procedures Int. 12:337-334 (1980).
20. M. -S. Lee and S. M. King, New synthesis of N-(guanosin-8-yl)-4-aminobiphenyl and its 5'-monophosphate, Chem. Biol. Interactions 34:239 (1981).
21. B. F. Erlanger and S. M. Beiser, Antibodies specific for ribonucleosides and ribonucleoties and their reaction with DNA, Proc. Natl. Acad. Sci. USA 52:768 (1964).
22. D. J. Ellens and L. J. Geilkens, A simple method for the purification of 5-aminosalicylic acid, application of the product as substrate in enzyme-linked immunosorben assay (ELISA), J. Immunol. Methods 37:325 (1980).
23. D. Rodbard, Statistical quality control and routine data processing for radioimmunoassay and immunoradiometric assays, Clin. Chem. 20:1255 (1974).

INTRAGENOMIC HETEROGENEITY IN DNA DAMAGE PROCESSING: POTENTIAL IMPLICATIONS FOR RISK ASSESSMENT

Philip C. Hanawalt

Department of Biological Sciences
Stanford University
Stanford, California 94305

INTRODUCTION

There have been many attempts to analyze the complex relationships between measured DNA repair levels in mammalian cells and various biological end-points such as survival, mutagenesis, and transformation. In such studies the repair proficiency has been assumed to be uniform throughout the genome and, in fact, cell population homogeneity is also generally assumed. However, it is quite likely that DNA damage in some domains of the genome is processed more efficiently than in others and that such heterogeneity in repair would result in corresponding differences in the responses seen for particular biological effects. The genome includes many different functional classes of DNA of which some are "silent" (e.g., repetitive sequences in heterochromatin, unexpressed genes) while others are active (e.g., expressed genes, elements required for the translation machinery, and various regulatory regions). The consequences of unrepaired or misrepaired damage in DNA will most certainly depend upon the precise location of the damage with respect to these functional classes. This fact is well-documented at the nucleotide sequence level in bacteria by correlations of the spectrum of particular lesions with "hot spots" for mutagenesis. However, we have obtained little information on the specificity of the mutagenic response in relation to damage and repair in mammalian systems, analyzed at the various levels of genomic organization. Differences in the repair response to damage in selected genomic regions may account for some of the profound differences seen in the carcinogenic response in different tissues or when different organisms are compared. Therefore, it may be important to understand the "fine structure" of DNA repair in mammalian genomes in order to assess carcinogenic risks.

Most of the methods utilized to study DNA repair are not intrinsically sensitive to intragenomic heterogeneity in distribution of repair sites[1]. Measurement of repair replication by the 5-bromodeoxyuridine density labeling protocol[2,3] is insensitive to heterogeneity in DNA repair as is the direct determination of adduct release from genomic DNA. However, the single cell autoradiographic method for determining "unscheduled DNA synthesis" should at least resolve gross cellular population heterogeneity in the repair response[4]. The methods that utilize lesion-specific endonucleases, such as the T4 endonuclease V to produce strand breaks at sites of UV induced cyclobutane pyrimidine dimers in DNA, should, in principle, be able to detect heterogeneities in DNA repair within a population of DNA molecules[5]. Thus, any DNA molecules selectively repaired should retain their integrity upon exposure to such endonucleases while those molecules inaccessible to repair should be cleaved at the sites of persisting lesions. The bromouracil photolysis method might also be expected to reveal gross heterogeneity in repair within a population of DNA molecules[6]. In that case DNA molecules should become selectively sensitive to fragmentation by 313nm irradiation as repair patches containing 5-bromodeoxyuridine are introduced. Finally, the approach of using monoclonal antibodies to 5-bromouracil in DNA developed in our laboratory by S. A. Leadon, to selectively isolate those molecules containing repair patches provides a promising new method for detection of heterogeneity in repair[7]. Cohn and Lieberman[8,9] have recently reported use of a polyclonal antibody to 5-bromouracil in DNA to demonstrate that the frequency of repair patches in ribosomal gene sequences in human cell DNA corresponds with that of pyrimidine dimers introduced by UV and is about 3 to 4 fold lower than that in the bulk DNA. Most if not all of the difference may be explained by the high guanine-cytosine content of the ribosomal gene sequences. In contrast, Irvin and Wogan[10] have detected enhanced binding of activated tritium-labeled aflatoxin B_1 within ribosomal gene sequences in rat liver DNA and the guanine enrichment in those sequences appeared insufficient to explain the preferential interaction.

An early indication of DNA repair heterogeneity in mammalian cells was the observation of a biphasic time course for the repair of pyrimidine dimers[1]. Rapid removal of dimers in the first few hours after UV irradiation was followed by a gradually decreasing rate, approaching a plateau by 24 hours. Also indirectly suggestive of heterogeneity was the report of the rapid recovery of RNA synthesis in UV irradiated mammalian cells, consistent with the preferential repair of actively transcribed domains of the genome[11]. However, attempts to demonstrate preferential repair (or not) of pyrimidine dimers in general classes of DNA, such as repetitive DNA sequences isolated from satellite DNA in rodent cells, did not reveal significant variation in repair efficiencies[12]. More recent investigation of repair in repetitive DNA in monkey cells has shown that whereas pyrimidine dimers are proficiently repaired certain bulky adducts are refractory to repair in alpha DNA sequences relative to that in the bulk genomic DNA[13].

We will consider in this review the three systems in which we have discovered intragenomic heterogeneity in the processing of damaged DNA. Alpha DNA in primates represents the first example of deficiency in the repair of a unique DNA species within otherwise repair-proficient cells[13]. In contrast, the demonstrated proficient repair of an active, "housekeeping" gene in Chinese hamster ovary cells may explain how these cells are able to tolerate high UV doses in spite of an overall inefficient excision-repair system[14]. In the third example, we have shown that in at least one human genetic deficiency in DNA repair, the defect results in partial loss of repair efficiency in some domains of the genome. Thus, in xeroderma pigmentosum, complementation group C (XP-C) some DNA sequences are repaired normally while others are evidently less accessible to repair enzymes[15].

DEFICIENT REPAIR OF BULKY ADDUCTS IN ALPHA DNA

Primate genomes contain appreciable amounts of a highly repetitive species termed alpha DNA. In the African green monkey line, BS-C-1, alpha DNA comprises nearly 20 percent of the genome; most of it in long tandem arrays of a repeating unit of 172 base pairs and localized in heterochromatin near centromeres on the chromosomes. Recent analyses in our laboratory have demonstrated tract lengths of over 450 tandem repeats (Madhani et al. J. Biol. Chem., in press). Since most of the monomer units contain a single recognition site for the restriction endonuclease HindIII, a large fraction of the alpha DNA can be isolated by separating the restriction products of cellular DNA by gel electrophoresis. The formation and repair of a number of DNA lesions in alpha DNA (isolated as the 172 bp monomer and its dimer) have been compared with that in the high-molecular weight bulk DNA. Markedly reduced repair of adducts of the furocoumarins (angelicin and psoralens), aflatoxin B_1, and N-acetoxy-2-acetylaminofluorene (NA-AAF) is measured in alpha DNA as compared to that in bulk DNA. In contrast, roughly the same amount of repair of damage produced by UV (254 nm), thymine glycols produced by ionizing radiation, and certain methylated bases formed by treatment with methyl methane sulfonate (MMS) or dimethyl sulfate (DMS) occurs in alpha and in non-alpha DNA. For a number of the agents studied the adduct frequencies were determined directly at different times after treatment. Thus, a sensitive monoclonal antibody to thymine glycol in DNA[16] was used to determine that there are no significant differences in the repair of this lesion in alpha and in non-alpha DNA. Loss of tritium-labeled adduct from the respective classes of DNA was used to follow repair of aflatoxin B_1, NA-AAF, and hydroxymethyl trimethyl psoralen (HMT)[17,18,19].

For adducts of the furocoumarin, HMT, the possibility was ruled out that the repair deficiency in alpha DNA might be due to the formation of a class of adducts that were inherently refractory to repair[19]. However, the quantum efficiency for photochemical conversion of psoralen monoadducts to interstrand crosslinks in alpha DNA was shown to be much lower than that for bulk DNA[19]. That result suggests that the alpha DNA is in a more constrained state or a different secondary structure than is non-alpha DNA and this may be related to its location in constitutive heterochromatin and/or the possible short repeat-length "phasing" of nucleosome cores in alpha sequences. Our working hypothesis is that the chromatin structure of alpha reduces its accessibility to the enzymes that initiate the repair of bulky adducts in DNA.

Our hypothesis of unique chromatin structure in alpha is also supported by our results with activated aflatoxin B_1[18]. In addition to being substrates for the bulky adduct incision system, the primary adducts of this agent can also be lost in a spontaneous chemical reaction in which modified guanine is removed from the DNA, leaving an apurinic (AP) site. It is thought that this spontaneous process is responsible for the residual loss of aflatoxin adducts in XP-A cells and that repair of the AP sites accounts for the slight amount of repair replication observed in XP-A cells treated with aflatoxin[20,21]. The residual loss of aflatoxin adducts in alpha DNA was shown to be the same as that previously observed in total DNA in XP-A cells[20]. Thus for aflatoxin in alpha DNA, the adducts are not repaired by the system that recognizes bulky adducts, but are lost more slowly by a spontaneous chemical reaction, producing AP sites that are repaired. However, we have found that UV irradiation of the cells increases the repair of aflatoxin adducts in alpha DNA to the levels seen in bulk DNA[18]. Thus, it is likely that DNA distortion by pyrimidine dimers or repair at dimer sites results in opening up of alpha chromatin allowing access of repair enzymes to other types of bulky lesions. Some other types of bulky lesions such as NA-AAF may result in partial disruption of alpha chromatin and that may account for the intermediate levels of repair deficiency in alpha DNA for repair of these lesions[13,18]. We have recently found that repair of bulky adducts in alpha DNA can occur at certain phases of the cell cycle (Leadon and Hanawalt, submitted). It will be important to correlate such variations in repair efficiency in unique domains with changes in chromatin structure during the cell cycle.

The overall conclusion from our analysis of DNA repair in alpha sequences is that the unique chromatin structure for certain nontranscribed regions of the genome, such as alpha DNA, may result in the deficient repair of some types of lesions in those regions. However, it is not possible to assess the significance of persisting damage in those regions since we don't yet know the function of alpha DNA. For purposes of risk assessment it may be more relevant to understand the processing of lesions in active or potentially active genes.

PREFERENTIAL REPAIR IN ACTIVE GENES

There have been a number of recent attempts to determine DNA damage and repair levels in defined, active genes in mammalian cells. Thus, evidence has been presented for increased binding and removal of benzo(a)pyrene adducts in the transcribed (i.e., DNase I sensitive chromatin) regions of the genome[22]. The production of alkaline labile sites by the alkylating agent MNNG in a 4kb fragment of the collagen gene and in a 5 kb fragment of the beta globulin gene in contact inhibited human WI38 cells was studied by Nose and Nikaido[23]. Roughly similar rates of disappearance of intact fragments analyzed on alkaline gels were noted with increasing dose of carcinogen, and the change in the number of alkaline labile sites with time after treatment also appeared similar for the two fragments investigated. Although the lesions were uncharacterized it is likely that repair was initiated by the glycosylase - AP endonuclease pathway rather than that responsible for direct incision at bulky adducts in DNA.

We have recently quantified the levels of pyrimidine dimers and their repair in the active dihydrofolate reductase (DHFR) gene of Chinese hamster ovary (CHO) cells in which the gene is amplified 50 fold, using T4 endonuclease V as a sensitive probe for dimers[14]. Restricted genomic DNA was treated or not with T4 endonuclease V, electrophoresed under denaturing conditions, transferred to nitrocellulose paper and hybridized to a ^{32}P-labelled probe specific for a restriction fragment in the gene. The proportion of fragments free of endonuclease sensitive sites (ESS) in each sample was determined from the difference in the amount of probe hybridized at the position of full length fragments for T4 endonuclease-treated and untreated samples, measured directly by scintillation counting. The use of a cell line amplified for a roughly 1000 kb region containing the DHFR gene helped greatly with the sensitivity of the analysis.

Roughly 70% of the pyrimidine dimers were removed from a restriction fragment within the gene in 24 hr while little repair was detectable in a fragment 30 kb upstream from the gene and only 15% of the dimers were lost from the genome overall. We have now confirmed these results in CHO cells containing only single copies of the DHFR gene , so the observed microheterogeneity in repair is clearly unrelated to the gene amplification phenomenon (Bohr et al., submitted). Rodent cell lines typically exhibit UV resistance similar to that of repair proficient human cell lines while carrying out much lower levels of repair[24]. On the basis of our results we have suggested that the preferential repair of vital DNA sequences facilitates the UV resistance of the CHO cells in spite of their low overall repair levels. This is a plausable explanation but it leaves unanswered the question of how the replication machinery in these cells is able to overcome many persisting lesions in the DNA. Nevertheless, viable progeny do arise so it is clear that bypass replication or some recombinational process must have occurred to generate functional daughter genomes.

In our analyses it was essential to determine the contribution of semiconservative replication to the appearance of dimer-free DNA in the restriction fragment bands in the gels. The addition of hydroxyurea to the cell cultures to suppress replicative DNA synthesis did not appreciably affect the apparent level of repair and therefore the generation of dimer-free DNA must have been mostly due to repair. The small amount of replicated DNA was categorically eliminated in some experiments in which 5-

bromodeoxyuridine was added to density-label the replicated DNA in the cultures. Following enzymatic restriction the replicated DNA was separated from the unreplicated DNA on CsCl equilibrium density gradients so that the T4 endonuclease V sensitive site analysis could be carried out on parental DNA that had not undergone replication. The results were consistent with those obtained on total DNA. The density labeling methodology when combined with the restriction fragment analysis also provided an approach to study the extent of replication of the different genomic regions.

In contrast to the results with CHO cells the analysis of repair in human cells in which the DHFR gene is amplified has not revealed significant differences in the extent of repair within the gene and in nearby sequences. Of course, normal human cells are repair proficient overall for pyrimidine dimers so one would not expect to find striking differences in repair in the various genomic domains. A careful comparison of the time course of repair indicates that pyrimidine dimers in the DHFR region are repaired earlier than those in the bulk of the genome (Mellon et al, to be published). Differences might be more prominent for certain bulky adducts like the furocoumarins since human cells do not efficiently remove those adducts from their DNA[25]. In principle, the methodology we have developed could be used to look at repair at the gene fragment level, of any lesion that can be quantified by site-specific nicking of the DNA. For example, the uvrABC repair endonuclease complex from _E. coli_ might be used as lesion probe for a variety of bulky adducts. The use of lesion specific glycosylases followed by treatment of the DNA with an AP endonuclease would also serve to quantify repair of those lesions for which glycosylases have been isolated[26].

DOMAIN-LIMITED REPAIR OF PYRIMIDINE DIMERS IN XERODERMA PIGMENTOSUM, GROUP C

Different complementation groups of XP are characterized by the fact that they exhibit varying degrees of deficiency at an early step in excision-repair. Thus, XP-A cell lines are nearly totally deficient in the removal of UV induced pyrimidine dimers, XP-C cell lines typically exhibit about 10-40% of normal levels of dimer removal and repair synthesis, while the XP variants are evidently normal with respect to damage removal and repair synthesis[27]. We know that the deficiency in excision-repair in the "classical" XP lines affects an early step since it can be corrected by the introduction of T4 endonuclease V which nicks the DNA at each dimer site, thereby permitting subsequent steps of excision-repair to occur[28,29,30,31]. The suggestion has been made that the deficiency in repair in XP is due to an inaccessibility of the damage to endogenous repair endonucleases, since dimer excising activities appear to be present in the extracts from XP cell lines[32,33]. Mansbridge and Hanawalt[15] set out to test the hypothesis that the partial deficiency in repair in certain XP complementation groups was due to exclusion of repair endonucleases from some domains of the genome. If this were the case one might find heterogeneity in the distribution of the remaining pyrimidine dimers in the DNA during the course of repair. Thus, repair patches would appear in some regions, replacing the dimers, while other regions would be devoid of repair patches and would still contain the initial concentration of dimers. A means was devised to test this hypothesis taking advantage of the high specific activity labeling of repair patches with ^{3}H-thymidine and the specific cleaving of DNA strands containing pyrimidine dimers by the T4 endonuclease V. The size distribution of ^{3}H-containing DNA after treatment with the enzyme provided a measure of the frequency of dimers remaining in regions containing repair patches while the size distribution of bulk DNA (prelabeled by growing cells in ^{14}C thymidine prior to irradiation) gave an indication of overall repair. As expected the ^{3}H and ^{14}C sedimentation profiles in alkaline sucrose gradients for DNA from normal human fibroblasts were congruent, with or without T4 endonuclease V treatment, at all times during the course of repair. In contrast, for XP-C, a marked disparity was seen, with the ^3H labeled DNA exhibiting less sensitivity to T4 endonuclease V after longer incubation times while most of the ^{14}C DNA retained the original sensitivity to the enzyme. The low overall level of repair could be explained by the exclusion from repair of a large number of genomic domains[15]. Control experiments using density-labeling with 5-bromodeoxyuridine to resolve repair synthesis from normal semiconservative replication verified that the ^3H label was mostly in repair patches, and determinations of unscheduled DNA synthesis by single cell autoradiography showed that population heterogeneity could not be responsible for the result. The same heterogenous pattern was observed with three different XP-C cell lines. This discovery represented an important new development in our understanding of the basis for the repair deficiency in XP. In addition, the procedure physically separated the repaired "domains" from the bulk DNA, thereby facilitating the eventual characterization of the DNA in these domains. The phenomenon has been confirmed by Mullenders et al.[34] who have indicated further that the repairable regions in XP-C are preferentially associated with the nuclear matrix, suggestive that they may include some actively transcribed genes.

The XP-C cell lines exhibit overall dimer repair levels comparable to those for the CHO cells and yet they are much more sensitive to UV. We have now compared repair of pyrimidine dimers in the single-copy DHFR gene in several XP-C cell lines with that in normal human fibroblasts and normal human epidermal keratinocytes. While the normal cells removed over 80% of the dimers from the gene in 24 hours following a UV dose of 20 J/m^2 the XP-C cells exhibited no more than 30% repair within the DHFR gene fragment studied (Bohr, Okumoto and Hanawalt, submitted). Thus, the efficiently repairable "domains" in XP-C evidently do not include this essential "housekeeping" gene. It remains to be determined which domains in XP-C are still repaired normally. In fact it is not clear whether the <u>same</u> domains are excluded from repair in all cells in the XP-C population, or whether the results represent an average over a population in which the excluded domains are different in different cells. Comparative analysis of repair levels in a number of different genes should resolve this question.

DISCUSSION AND CONCLUSIONS

In summary, we have documented three examples of heterogeneity in DNA repair in mammalian genomes. In African green monkey kidney cells certain bulky adducts are poorly repaired in the heterochromatic alpha DNA sequences relative to that in the bulk DNA. In Chinese hamster ovary cells an active "housekeeping" gene is proficiently repaired in a background of overall inefficient repair in the genome. In both of these cases the heterogeneity was illustrated in normal cells. In the third example, a hereditary defect in humans has been shown to cause a deficiency in DNA repair in some domains of the genome selectively. In this case the affected "domains" have yet to be identified and characterized, but our results so far indicate that an essential gene is poorly repaired in that system.

We can consider potential heterogeneity of repair in the genome at several different organizational levels in chromatin[1,35]. At the primary structure level in DNA the nucleotide sequences flanking the lesion might affect enzymatic recognition of damage, although there is as yet no experimental documentation of such bias. It is clear that the secondary structure of DNA can affect repair, however, as illustrated by the preferential action of the T4 endonuclease V on dimers in native vs. denatured DNA[36]. It is possible that lesions in Z-DNA and other rare secondary structures may be repaired differentially in relation to those introduced into the B-DNA conformation. It is also possible, indeed likely, that the superhelical tension in genomic DNA in mammalian cells affects the rate of repair[35]. Evidence that chromatin structure interferes with access of repair enzymes to lesions was first reported by Wilkins and Hart[37] who showed that many of the pyrimidine dimers in permeabilized human fibroblasts were inaccessible to exogenously supplied *M. luteus* UV endonuclease. Kantor and Setlow[38] attributed the observed differential rates of removal of pyrimidine dimers from DNA in cultured human fibroblasts to different classes of accessibility to dimers by repair enzymes. At the level of the nucleosome there is evidence that some types of bulky adducts are excluded from "core" regions and that the repair process itself causes some disruption of the normal nucleosome structure[1,35]. Furthermore, the nucleosome "cores" do not necessarily reform at their original locations along the DNA following repair[39]. At higher levels of chromatin structure and DNA condensation we have even less understanding of the rules governing accessibility of genomic domains to damage and repair. In fact, the term, "domain," is still only operationally defined in terms of its structural and functional significance[35].

This area of research deserves further study to learn how chromatin structure in the mammalian genome affects accessibility of the DNA to repair in the various functional domains. In the case of the CHO cells the resistance to UV may be achieved through the selective repair of active, "housekeeping" genes while adjacent nontranscribed regions of the genome are ignored by the repair system. However, the genetic defect in XP-C clearly renders those cells less able to cope with damage in some vital sequences in their DNA. It is possible that the domains inaccessible to repair in XP-C are much larger than are those represented by active genes but that they include many essential genes. Thus, the heterogeneity in repair due to the genetic defect in XP-C may be non-selective for active genes and therefore of little value to cellular survival. This hypothesis is presently under test in our laboratory.

It is now clear that repairability of damage in mammalian chromatin depends upon the type of lesion, its location in the genome, and the functional state of the DNA at the particular site. Information obtained on the processing of damage overall or in one domain of the genome may not be relevant to an understanding of a biological response that is dependent upon damage and/or repair activity in another domain. Thus, for example the question of whether proto-oncogenes are located in efficiently repairable domains may have significance for risk assessment. We have preliminary evidence that the expressed *c-abl* proto-oncogene is proficiently repaired while the unexpressed *c-mos* proto-oncogene is poorly repaired in UV-irradiated mouse 3T3 cells (Madhani, Bohr and Hanawalt, submitted). At this point we don't have enough information to even guess whether the likelihood of malignant transformation is reduced or enhanced when a relevant proto-oncogene containing potentially activating damage is located in a genomic domain that is excluded from repair. The repair process itself introduces strand breaks that might enhance the probability of intragenomic transpositions that could result in activation of oncogenes like c-*myc*, while unrepaired damage could activate others like the c-*ras* proto-oncogene. Reaction of the c-*ras* proto-oncogene with benzo(a) pyrene diol-epoxide *in vitro* has been shown to generate a transforming oncogene when the modified DNA is introduced into NIH 3T3 cells[40]. The fate of the cell and ultimately the organism of which it is a part may depend upon the processing of an altered DNA region near or within an oncogene.

Since rodents are used widely in carcinogen testing for human risk assessment it is imperative that we learn the unique features of DNA damage processing in the respective systems. Various reports have indicated that levels of DNA repair in mouse embryo cells decline with successive passage in culture[41,42] and accompanying their transformation to established heteroploid cultures[43]. Do some domains become selectively excluded from repair and if so, which? Do similar changes in DNA repair occur with development and/or aging in the cells *in vivo*? Does the spectrum of proficiently repaired genes and/or domains vary from tissue to tissue in the same organism? Evidence has been presented for higher rates of nucleotide substitution and consequent evolution in rodents than in humans[44]. Can this be explained in part by accumulation of mutagenic lesions in underrepaired domains in rodent cell genomes? Finally, what are the features of mammalian chromatin structure and/or DNA sequences that determine whether a particular region of the genome should be accessible or not to repair enzymes? I would conclude with the predictive speculation that we may find a correlation of repairability with the replication timing of genes. Those genes that replicate early in S-phase, including the "housekeeping" genes, will be found in the proficiently repaired class, while those that replicate late will include the permanently inactive genes[45] and will turn out to be less accessible to repair.

ACKNOWLEDGEMENTS

Our research on DNA repair in defined DNA sequences has been supported by grants from the National Institutes of Health and the American Cancer Society and a contract with the Department of Energy. The work benefits from ongoing discussion with many individuals in my group, particularly Ann Ganesan and Charles Allen Smith.

REFERENCES

1. P. C. Hanawalt, P.K. Cooper, A.K. Ganesan and C.A. Smith, DNA repair in bacteria and mammalian cells, Annu.Rev.Biochem. 48:783-836 (1979).
2. D. E. Pettijohn, and P.C. Hanawalt, Evidence for repair replication of ultraviolet damaged DNA in bacteria, J.Mol.Biol. 9:395-410 (1964).
3. C. A. Smith, P. K. Cooper, and P. C. Hanawalt, Measurement of repair replication by equilibrium sedimentation, in E. C. Friedberg and P.C. Hanawalt (Eds.), DNA Repair: A Laboratory Manual of Research Procedures, Marcel Dekker Inc., Vol I, part B 289-305 (1981).
4. J. E. Cleaver, and G. H. Thomas, Measurement of unscheduled synthesis by autoradiography, in E. C. Friedberg and P. C. Hanawalt (Eds.), DNA Repair: A Laboratory Manual of Research Procedures Vol I, part B pp.277-287 (1981).
5. A. K. Ganesan, C. A. Smith and A. A. van Zeeland, Measurement of the pyrimidine dimer content of DNA in permeabilized bacterial or mammalian cells with endonuclease V of bacteriophage T4, in E. C. Friedberg and P. C. Hanawalt (Eds.), DNA Repair: A Laboratory Manual of Research Procedures, Marcel Dekker Inc., Vol 1, part A, pp.89-97 (1980).
6. R. B. Setlow and J. D. Regan, Measurement of repair synthesis by photolysis of bromouracil, in E. C. Friedberg and P. C. Hanawalt (Eds.), DNA Repair: A Laboratory Manual of Research Procedures, Marcel Dekker Inc., Vol 1, part B, pp.307-318 (1981).
7. A. S. Leadon and P. C. Hanawalt, Differential repair of DNA damage in specific nucleotide sequences in mammalian cells, Radiation Research Society Abstracts Gg-6 (1985).
8. S. M. Cohn, and M.W. Lieberman, The use of antibodies to 5-bromo-2'-deoxyuridine for the isolation of DNA sequences containing excision-repair sites, J. Biol. Chem. 259:12456-12462 (1984).
9. S. M. Cohn, and M.W. Lieberman, The distribution of DNA excision-repair sites in human diploid fibroblasts following ultraviolet irradiation, J. Biol. Chem. 259:12463-12469 (1984).

10. T. R. Irvin, and G. N. Wogan, Quantitation of aflatoxin B_1 adduction within the ribosomal RNA gene sequences of rat liver DNA, Proc.Natl. Acad. Sci. U.S.A. 81:664-668 (1984).
11. L. V. Mayne, Inhibitors of DNA synthesis (aphidicolin and araC/HU) prevent the recovery of RNA synthesis after UV-irradiation, Mutat. Res. 131:187-191 (1984).
12. M. W. Lieberman and M.C. Poirier, Intragenomal distribution of DNA repair synthesis: repair in satellite and mainband DNA in cultured mouse cells, Proc. Natl. Acad. Sci. U.S.A. 71:2461-2465 (1984).
13. M. E. Zolan, G. A. Cortopassi, C. A. Smith, and P. C. Hanawalt, Deficient repair of chemical adducts in alpha DNA of monkey cells, Cell 28:613-619 (1982).
14. V. A. Bohr, C. A. Smith, D. S. Okumoto, and P. C. Hanawalt, DNA repair in an active gene: removal of pyrimidine dimers from the DHFR gene of CHO cells is much more efficient than in the genome overall, Cell 40:359-369 (1985).
15. J. N. Mansbridge, and P.C. Hanawalt, Domain-limited repair of DNA in ultraviolet irradiated fibroblasts from xeroderma pigmentosum complementation group C, in E.C. Friedberg and B.R. Bridges (Eds.) Cellular Responses to DNA Damage, UCLA Symp on Mol. and Cellu. Biol. Vol II, Alan R. Liss, Inc. New York, pp.195-207 (1983).
16. S. A. Leadon and P. C. Hanawalt, Monoclonal antibody to DNA containing thymine glycol, DNA Repair Reports: Mutation Res 112:191-200 (1983).
17. S. A. Leadon, M. E. Zolan, and P. C. Hanawalt, Restricted repair of aflatoxin B_1 induced damage in alpha DNA of monkey cells, Nucl. Acids Res. 11:5675-5689 (1983).
18. S. A. Leadon, and P. C. Hanawalt, Ultraviolet irradiation of monkey cells enhances the repair of DNA adducts in alpha DNA, Carcinogenesis 5: 1505-1510 (1984).
19. M. E. Zolan, C. A. Smith, and P. C. Hanawalt, Formation and repair of furocoumarin adducts in alpha DNA and bulk DNA of monkey cells, Biochemistry 23:63-69 (1984).
20. S. A. Leadon, P. A. Amstad, and P. A. Cerutti, Repair and expression of aflatoxin B_1-induced DNA damage, in A. Castellani (Ed), Proc. NATO Adv. Study Inst. on use of human cells for assessment of risk from physical and chemical agents., Plenum Press, 60:105-117 (1983).
21. A. Sarasin, C. A. Smith, and P. C. Hanawalt, Repair of DNA in human cells after treatment with activated aflatoxin B_1, Cancer Res.37:1786-1793 (1977).
22. J. E. Arrand, and A. M. Murray, Benzpyrene groups bind preferentially to the DNA of active chromatin in lung cells, Nucl. Acids Res.10: 1547-1555 (1982).
23. K. Nose, and O. Nikaido, Transcriptionally active and inactive genes are similarly modified by chemical carcinogens and X-ray in normal human fibroblasts, Biochim. Biophys. Acta 781:273-278 (1984).
24. A. Ganesan, G. Spivak, and P.C. Hanawalt, Expression of DNA repair genes in mammalian cells, in P. Nagley, A. W. Linnane, W. J. Peacock, and J. A. Pateman (Eds.), Manipulation and Expression of Genes in Eukaryotes, Academic Press, Sidney, Australia, pp.45-54 (1983).
25. P. C. Hanawalt, J. Kaye, C. A. Smith, and M. Zolan, Cellular responses to psoralen adducts in DNA, in Psoralens - in Cosmetics and Dermatology Pergamon Press, pp.133-142 (1982).
26. T. Lindahl, DNA repair enzymes, Annu. Rev. Biochem 51:61-87 (1982).
27. J. H. Robbins, K. H. Kraemer, M. A. Lutzner, B. W. Festoff, and H. G. Coon, Xeroderma Pigmentosum: An inherited disease with sun sensitivity, multiple cutaneous neoplasms, and abnormal DNA repair, in Annals of Internal Medicine 80:221-248 (1974).

28. K. Tanaka, S. Sekiguchi, and Y. Okada, Restoration of ultraviolet-induced unscheduled DNA synthesis of xeroderma pigmentosum cells by the concomitant treatment with bacteriophage T4 endonuclease V and HVJ (Sendai virus), Proc. Natl. Acad. Sci. USA 72:4071-4075 (1975).
29. K. Tanaka, H. Hayakawa, M. Sekiguchi, and Y. Okada, Specific action of T4 endonuclease V on damaged DNA in xeroderma pigmentosum cells in vivo, Proc. Natl. Acad. Sci. USA 74:2958-2962 (1977).
30. G. Ciarrocchi, and S. Linn, A cell-free assay measuring repair DNA synthesis in human fibroblasts, Proc. Natl. Acad. Sci. USA 75:1887-1891 (1978).
31. C. A. Smith, and P.C. Hanawalt, Phage T4 endonuclease V stimulates DNA repair replication in isolated nuclei from ultraviolet-irradiated human cells, including xeroderma pigmentosum fibroblasts, Proc. Natl. Acad. Sci. USA 75:2598-2606 (1978).
32. K. Mortlemans, E.C. Friedberg, H. Slor, G. Thomas, and J.E. Cleaver, Defective thymine dimer excision by cell-free extracts of xeroderma pigmentosum cells, Proc. Natl. Acad. Sci. USA 73:2757-2761 (1976).
33. Y. Fujiwara and Y. Kano, Characteristics of thymine dimer excision from xeroderma pigmentosum chromatin, in E. C. Friedberg and B. A. Bridges (Eds.) Cellular Responses to DNA Damage, UCLA Symp. on Mol. and Cellu.Biol. Vol II, Alan R. Liss, Inc. New York, pp. 215-224, (1983).
34. L. H. F. Mullenders, A. C. van Kesteren, C. J. M. Bussmann, A. A. van Zeeland, and A. T. Natarajan, Preferential repair of nuclear matrix associated DNA in xeroderma pigmentosum complementation group C, Mutation Res. 141:75-82 (1984).
35. V. Bohr and P. Hanawalt, Factors that affect the initiation of excision-repair in chromatin, in A. Collins, C. S. Downes, and R. T. Johnson (Eds), DNA Repair and its Inhibition, IRL Press, Oxford pp.109-125 (1984).
36. L. K. Gordon and W. A. Haseltine, Comparison of the cleavage of pyrimidine dimers by the bacteriophage T4 and Micrococcus luteus UV-specific endonucleases, J.Biol.Chem. 255:12047-12050 (1980).
37. R. J. Wilkins and R. W. Hart, Preferential DNA repair in human cells, Nature 247:35-36 (1974).
38. G. J. Kantor and R. B. Setlow, Rate and extent of DNA repair in nondividing human diploid fibroblasts, Cancer Res. 41:819-825 (1981).
39. M. E. Zolan, C. A. Smith, N. M. Calvin, and P. C. Hanawalt, Rearrangement of mammalian chromatin structure following excision repair, Nature 299:462-463 (1982).
40. C. J. Marshall, K. H. Vousden, and D. H. Phillips, Activation of c-Ha-ras-1 proto-oncogene by in vitro modification with a chemical carcinogen, benzo(a)pyrene diol-epoxide, Nature 310:586-589 (1984).
41. R. Ben-Ishai and L. Peleg, Excision-repair in primary cultures of mouse embryo cells and its decline in progressive passages and established cell lines, in P. C. Hanawalt, and R. B. Setlow (Eds.) Molecular Mechanisms for Repair of DNA Part B Plenum Press, New York, pp.607-610 (1975).
42. T. Yagi, DNA repair ability of cultured cells derived from mouse embryos in comparison with human cells, Mutation Res. 96:89-98 (1982).
43. G. C. Elliott and R. T. Johnson, DNA repair in mouse embryo fibroblasts - I. Decline in ultraviolet-induced incision rate accompanies cell transformation, J. Cell Sci. 60:267-288 (1983).
44. C. Wu and W. Li, Evidence for higher rates of nucleotide substitution in rodents than in man, Proc. Natl. Acad. Sci. USA 82:1741-1745 (1985).
45. M. A. Goldman, G. P. Holmquist, M. C. Gray, L. A. Caston, and A. Nag, Replication timing of genes and middle repetitive sequences, Science 224:686-692 (1984).

THE ROLE OF PHARMACOKINETICS IN RISK ASSESSMENT

Richard H. Reitz, Tony R. Fox, and Philip G. Watanabe

1803 Building
Dow Chemical Company
Mammalian and Environmental Toxicology Laboratory
Midland, MI 48640

Pharmacokinetics can aid in the formulation of risk estimations in several ways: (a) by selection of doses for toxicity studies, (b) by distinguishing between "internal dose or toxifor concentration" and "applied dose", (c) by providing a physiological basis for extrapolating between species, and (d) by helping us to visualize the toxicological consequences of processes which we cannot (yet) quantify.

(A) Selection of Doses for Toxicity Studies. One of the most significant uses of pharmacokinetics in risk assessment is in the setting of dose levels for studies from which risk estimates are to be derived. This is particularly true when chronic studies are involved. The objective of chronic studies (defined as studies in which chemicals are administered to animals for the majority of their lifetimes) is the discovery of toxic effects which are not manifested during short-term toxicity tests. Obviously, one of the most significant effects which can be discovered in such tests is an increase in the incidence of a specific type of tumor. However, the current practice of employing very high doses, often frankly toxic to the organism, in such studies makes risk extrapolation difficult or impossible. Many of the processes which function to eliminate toxic chemicals from the body have limited capacity. This may cause the occurrence of unexpected toxicity once these capacities are exceeded. Pharmacokinetics can play a valuable role in defining the areas where such discontinuities can be expected. Experiments can then be designed so that the toxic potential of materials at nonsaturating levels can be realistically evaluated. A general approach for generating and utilizing pharmacokinetic data is outlined below:

The first step is the development of an analytical method for quantification of the parent compound in biological media. Alternatively,

Table 1. Approach to Gathering Pharmacokinetic data for Use in Dose Selection during Chronic Studies.

A. Develop an Analytical Method or Synthesize Radiolabeled Compound.
B. Admininster Intravenously - Follow Parent Compound.
C. Administer by Other Appropriate Route(s).
D. Study in the Other Sex/Species, (Limited Evaluation).
E. Evaluate Repeatedly Dosed Animals (2 Weeks).

a radiolabeled form of the chemical may be synthesized to provide greater sensitivity, provided that sufficient attention is paid to separation of the parent compound from its radioactive metabolites.

Next, an intravenous dose of the material is administered to one sex of the test species, and the time course of elimination of the radioactivity is measured. At least two doses should be evaluated. The highest dose should be just below that needed to cause frank toxicity, and the lowest dose should be determined by the limits of analytical sensitivity.

Then the material is administered to animals by the route of administration to be employed in the chronic study, and the time course of elimination is redetermined. Two doses are chosen, by the same criteria employed for the intravenous studies, although the absolute value of these doses will undoubtedly be different. This allows an estimation of the rates and major routes of elimination, as well as the extent of absorption of the test material. If examination of the results reveals non-linear behavior (for examples of non-linear behavior see Levy[1]), then intermediate doses should be evaluated in order to define the regions where these deviations occur. If radiolabeled material is available, it is usually helpful to determine the major routes of elimination, although this is not essential.

Once these studies have been completed, it should be possible to determine whether significant non-linearities occur in the disposition of the material, and whether the material has potential for accumulation during repeated dosing.

Selected parameters should then be monitored in the other sex and species to be tested. It is not anticipated that every parameter mentioned above will be monitored; the major objective is to determine whether there are major differences between sex/species.

Finally, limited studies should be conducted in animals dosed for approximately two weeks with the material of interest. This period is a compromise between the need to investigate changes in Pharmacokinetics and the expense of obtaining such animals. It may be possible to coordinate subchronic toxicity evaluations with pharmacokinetic studies for this purpose. These data allow verification of any predictions about bioaccumulation, as well as provide time for induction/repression of enzyme systems and readjustment of metabolic pools (e.g. glutathione) to occur.

The next step is to incorporate this information into the dose-selection process. This is accomplished in the following manner.

(1) If the studies show no indication of non-linear behavior, and if the material is completely cleared from the animals during the period between administration of doses, then pharmacokinetics will play no role in selection of doses. Doses will be chosen upon pathological criteria alone.

(2) If accumulation occurs, then doses based upon pathological criteria should be reduced by the accumulation factor (e.g. steady state levels 10X higher than single-dose levels, multiple-dose levels reduced by 10X).

(3) If non-linearities are observed, then anticipated human exposures should be considered in determining the setting of dose levels. If anticipated human exposures occur well below the area where non-linear pharmacokinetics predominate, then the chronic toxicity study <u>must</u> contain at least two dose levels in the linear region if extrapolation is to be

possible. This will usually mean utilizing at least three, and often four, doses in such studies, and these doses will often be widely spaced. Conversely, if human exposures are in the region where non-linear pharmacokinetic behavior is anticipated, then the chronic study should place major emphasis upon defining the shape of the dose-response curve in this area, and the doses may be more closely spaced. For risk assessment purposes, however, a study will always be of limited usefulness if it does not employ at least three treatment levels and span at least one order of magnitude between high and low dose levels.

(B) Distinguishing between "Internal Dose or Toxifor Concentration" and "Applied Dose": Another important use of pharmacokinetics is understanding the complex relationships between the dose administered to the test animals and the concentration of the toxic species within the animal at the target organs. Pharmacokinetics may also be useful in helping to formulate a hypothesis about the mechanisms of toxicity of individual agents. As an example, it is instructive to consider two recent chronic studies of methylene chloride (Dichloromethane, DCM).

One chronic study of DCM toxicity in B6C3F1 mice was conducted by dissolving DCM in the drinking water at levels sufficient to give 250 mg/kg/day[2]. In a second study, B6C3F1 mice were exposed to DCM vapor, 6 hr/day, 5 days/week. The concentrations of DCM in this study were 2000 and 4000 ppm[3].

Assuming that 2/3 of the inhaled DCM was absorbed by the animals, we can calculate that the dose of DCM to animals exposed to 2000 ppm was between 850 and 1700 mg/kg/day, depending upon the value used for the ventilation rate of mice. Thus, the administered dose in the drinking water study was 1/3 to 1/7th of the dose in the inhalation study. However, the biological results were dramatically different. DCM failed to produce a significant increase in tumors at any site in the drinking water study, but it caused a significant increase in lung and liver tumors in the inhalation study.

Knowledge of the metabolism of DCM helps provide an explanation for the results of these two studies. Data have been collected in the rat, hamster, human, and mouse. In each of these species, DCM is extensively (>95%) metabolized at low doses. The products of this metabolism are CO, CO_2, formate, and formaldehyde, and there appear to be at least two pathways of metabolism: one dependent upon mixed function oxidases, and one dependent upon a source of reduced glutathione[4]. However, in each of the species studied, saturation of at least one pathway of metabolism occurs at higher doses. This means that the toxicity produced by DCM is not a simple linear function of administered dose.

With the use of a pharmacokinetic model[5] with kinetic constants measured by Andersen et al., (personal communication), the internal dose of DCM (estimated as the area under the concentration/time curve for blood, liver, or lung) may be calculated in each assay. When this is done, one possible explanation for the different results becomes apparent: the internal dose of DCM is at least 100X higher in the inhalation study than it is in the drinking water study. The higher dose in the inhalation study appears to result from two factors: (1) saturation of metabolism by the high blood levels achieved during inhalation; (2) a "first pass" effect, caused by the fact that most of the methylene chloride absorbed from the G.I. tract goes directly to the liver, which is the principal site of metabolism for DCM.

Another valuable insight may be gained from the pharmacokinetic analysis of this compound. In describing the metabolic pathways for DCM, various investigators have speculated that an intermediate formed by MFO enzymes (formyl chloride) might be reactive enough to cause toxicity. However, analysis of the data indicates that both exposure protocols result in the same amount of MFO metabolite formation. This is because the MFO pathway saturates at low concentrations (~200-300 ppm in the atmosphere) so that increasing the concentration to 2000 ppm does not cause the accumulation of additional MFO metabolites. Consequently, it appears that the MFO pathway may play a protective role in animals exposed to DCM.

(C) Providing a Physiological Basis for Extrapolating between Species: It is almost always necessary to perform an extrapolation between species as well as an extrapolation between doses in conducting risk estimations. Occasionally it is necessary to extrapolate between different routes of exposure as well. One technique that has proven very useful in accomplishing this in recent years is the technique of physiological modeling. In this technique, a computer model is constructed with compartments which correspond to actual tissues or tissue groups of the body. The size of these compartments, as well as the blood and air flowing through these compartments, may be obtained from the medical literature. Once partition coefficients have been determined, differential equations describing the dynamic mass balance in each compartment may be written. An example of such a model is given in Figure 1, which shows a physiologically-based pharmacokinetic model (PBPM) for 1,1,1-trichloroethane (methyl chloroform, MC).

Several pharmacokinetic studies of MC have been reported, (two from our own laboratory), and we were interested in seeing whether a single model could accurately describe the data gathered by inhalation exposure in rats, mice, and humans. In addition, we were interested in seeing whether it was possible to predict the fate of ^{14}C-MC administered in the drinking water of rats with a model developed to describe inhalation exposures. The success that we had with this approach is described in Figures 2-5.

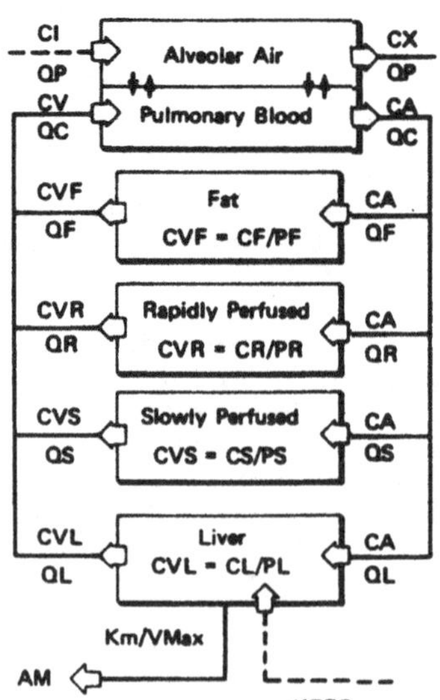

Figure 1. A physiologically based pharmacokinetic model (PB-PM) for methylchloroform capable of simulating inhalation and drinking water exposures.

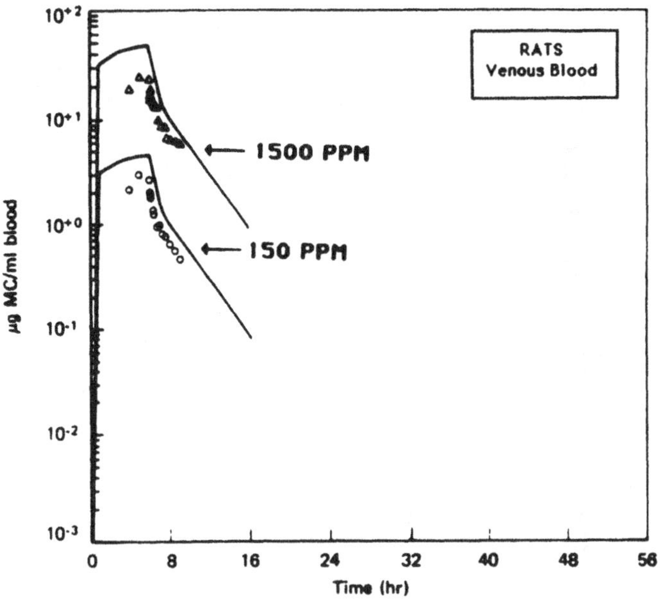

Figure 2. Predicted and observed venous blood concentrations of methylchloroform during inhalation exposures.

—— = computer predictions, Δ = 1500 ppm exposure, o = 150 ppm exposure

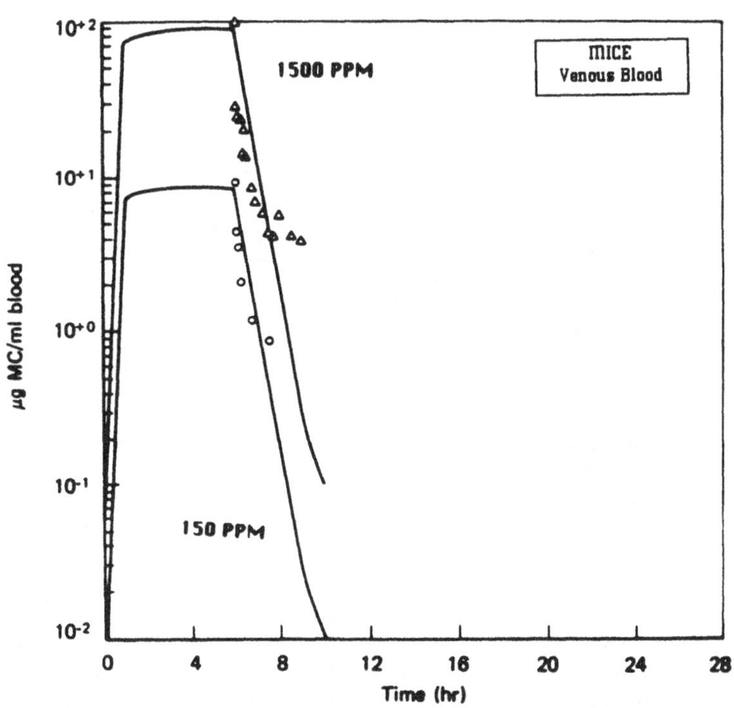

Figure 3. Summarizes the same type of data for the B6C3F1 mouse. No adjustments to the model were made, other than those appropriate to the altered physiology. Excellent agreement between data and the model are apparent.

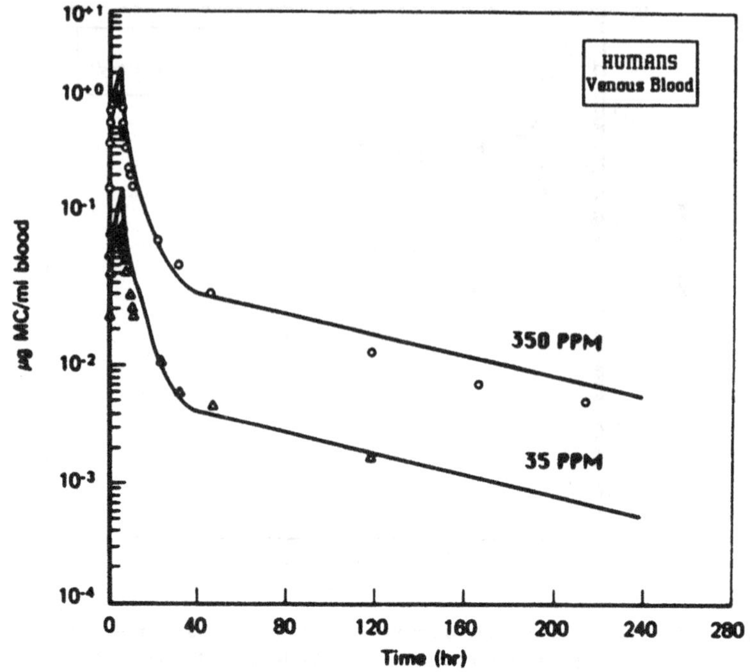

Figure 4. Summarizes the same type of data from humans. This is the bottom line - extrapolating from rodents to humans - and it was very successful.

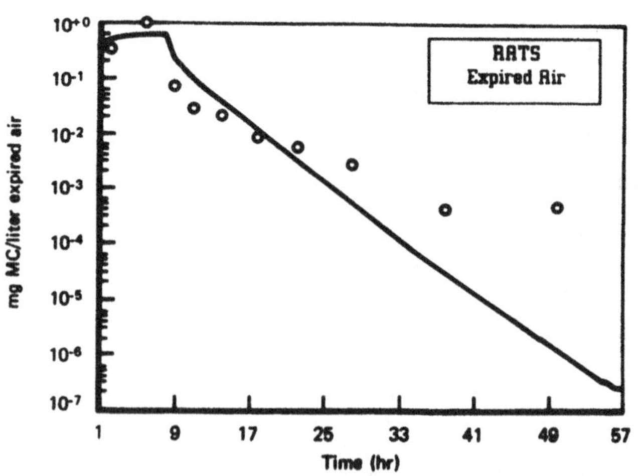

Figure 5. Depicts the elimination of MC in expired air after rats were given access to ^{14}C-MC in drinking water for 8 hr. The only change made to the model used for inhalation data with rats was to change the input function for MC from inhalation to drinking. Good fit was obtained for the first 30 hr, during which 99.9% of the administered radioactivity was recovered.

The data gathered by Schumann _et al._[6] in F344 rats were used as the basis for extrapolation to the other species. Actual venous blood levels of MC are shown as either circles (150 ppm exposure) or triangles (1500 ppm exposure). The computer stimulation is depicted as a solid line. Good fit was obtained quickly with this type of model.

In addition to predicting the time course of MC in expired air and blood, the model did an excellent job of predicting total body burden, concentration of MC in liver and fat tissue, and production of non-volatile metabolites for all the species studied. The remarkable success of this exercise suggests that PBPM will be an important tool for risk assessment in the future.

(D) Visualizing the Toxicological Consequences of Processes which we Cannot Quantify: The mathematical exercises conducted for describing the PBPM can be employed for predicting the qualitative nature of inaccessible processes as well. Gehring and Blau[7] described a procedure for visualizing the effects of saturable DNA repair systems upon the production of genetic damage and subsequent consolidation of this damage by cellular replication. The results of this simulation are shown in Figure 6. Chemical (C) bioactivated by metabolism may be detoxified by a saturable process (CaD2), bind to non-critical macromolecules (CaM), or bind to genetic targets (CaG). If cellular replication occurs and "consolidates" the damage so that it is no longer susceptible to DNA repair systems, it produces a species designated as CaG*. To demonstrate the nonlinearity of these processes, realistic values for rate constants were chosen, and then a normalized plot of CaM, CaD2, and CaG* was prepared. If all processes were operating linearly, each of these curves would be flat. Instead it can be seen that as the initial concentration rises from 10^{-4} to 0.1 molar, the relative amount of CaD2 first rises, then falls. In contrast, the amount of CaG* formed rises 1000X more than would be expected from the change in initial concentration.

Reitz[8] extended this model to show that a similar disproportionate increase in replicated genetic lesions can result when a chemical stimulates cellular regeneration, even if this chemical has no direct effect on DNA itself. This follows from the assumption that endogenous mutagens normally produce a low amount of genetic damage which is subject to the action of DNA repair systems. Consequently, altering the ratio between DNA repair and DNA replication rates has the effect of speeding accumulation of CaG*.

Detection of a Cellular Oncogene in Tumors from B6C3F1 Mouse Liver:

Although we have discussed several ways in which pharmacokinetics can aid in the formulation of risk estimations, many other factors need to be considered as well. Fundamental to the choice of a model for risk estimation is an assumption as to whether the chemical event associated with increased frequencies of tumorigenesis involves isolated "hits" on unique targets such as DNA, or indirect effects on non-genetic targets within the cell.

While considerable effort has been expended in evaluating the role of oncogenes in human cancer, suprisingly little is known about the presence or absence of oncogenes in the animals used for long term studies of carcinogenicity. The male B6C3F1 mouse utilized by the NCI and NTP for long term bioassays is prone to develop liver tumors in the absence of chemical treatment. The average incidence of liver tumors in control animals was 25% in a recent compilation of studies conducted under the auspices of the NTP[9]. The detection of cellular oncogenes in a variety of presumably spontaneous human tumors suggested that oncogenes might also be

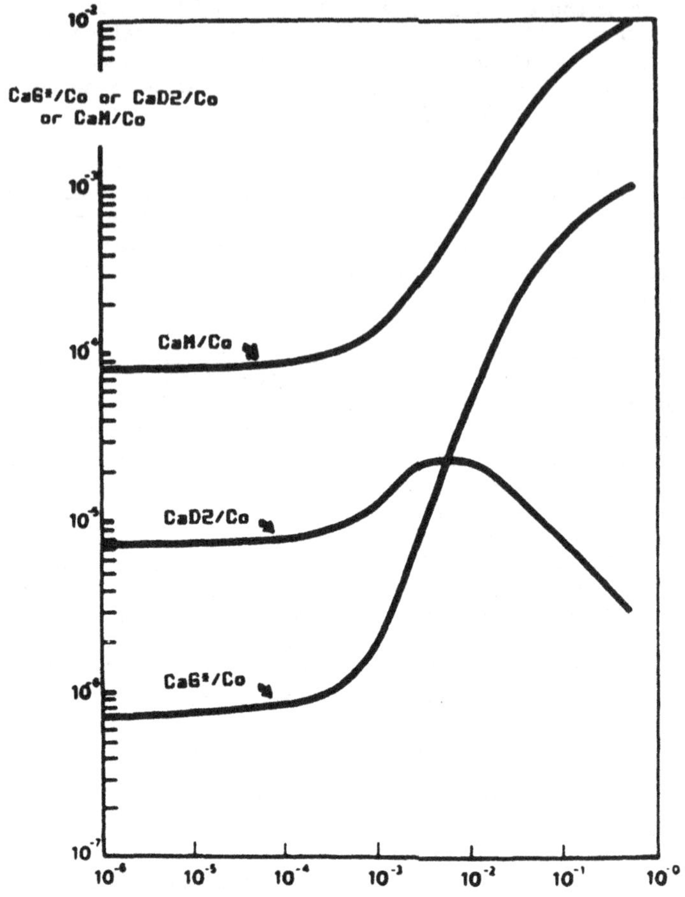

Figure 6. Simulated effects upon replicated kinetic material CaG*, metabolite bound to macromolecules (CaM) and metabolite detoxified by a saturable process (CaD_2). All parameters are normalized to initial concentration (C_o).

involved in the usually high incidence of liver tumors in this mouse. We have confirmed this hypothesis by demonstrating the presence of an active cellular oncogene in DNA isolated from the tumor tissue of 24-month old control mice in a bioassay conducted in our laboratory[10]. The results of these experiments are shown in Table 2.

The specific transforming frequencies ranged from 0.005 to 0.02 focus-forming units/μg DNA. No foci were observed with NIH-3T3 cells transfected with DNA from calf thymus, or DNA isolated from non-tumor liver tissue of either old or young mice (Table 2). Nine of the eleven tumor-bearing animals tested contained an active oncogene. This suggests that this strain of mouse may be unusual in containing a genetic predisposition toward activation of this particular gene segment. In any case, a better understanding of the process(es) involved in expression of this oncogene may have important implications for the interpretation of animal bioassay data.

Table 2: Transforming Activity in DNA from Liver Tumors of B6C3F1 Assayed in NIH-3T3 Cells.

DNA Source	Age (Months)	# of FOCI/ over # Plates Observed	# of Animals whose DNA induced Transformation
		Experiment #1	
Mice with Tumors	24	10 of 44	6 of 11
Mice without Tumors	3	0 of 4	0 of 1
Calf Thymus		0 of 4	
		Experiment #2	
Mice with Tumors	24	14 of 44	7 of 11
Mice without Tumors	3-5	0 of 43	0 of 2
Mice without Tumors	24	0 of 44	0 of 10

REFERENCES

1. G. Levy, Dose-dependent effects in pharmacokinetics. in: "Importance of Fundamental Principles in Drug Evaluation", D.H. Tedeschi, and R.E. Tedeschi, eds., Raven Press, N.Y. pp. 141-172, (1968).
2. National Coffee Association (NCA), Twenty-four month oncogenicity study of methylene chloride in mice. Prepared by Hazelton Laboratories America, Inc., Vienna, VA. Unpublished, (1983).
3. National Toxicology Program (NTP), NTP Technical report on the toxicology and carcinogenesis studies of dichloromethone in F344/N rats and B6C3F1 mice (inhalation studies). NTP TR 306. Board Draft (1985).
4. M. W. Anders, V. L. Kubic, R. A. Jones, and L. J. Jenkins, Jr., Metabolism of halogenated methanes and macromolecular binding. J. Environ. Pathol. Toxicol. 1:117-124 (1977).
5. J. C. Ramsey, and M. E. Andersen, A physiologically-based description of the inhalation pharmacokinetics of styrene monomer in rats and humans. Toxicol. Appl. Pharmacol. 73:175 (1984).
6. A. M. Schumann, T. R. Fox, and P. G. Watanabe, [^{14}C]-Methylchloroform (1,1,1-trichloroethane): Pharmacokinetics in rats and mice following inhalation exposure. Toxicol. Appl. Pharmacol. 62:390-401 (1982).
7. P. J. Gehring, and G. E. Blau, Mechanisms of carcinogenesis: Dose response. J. Environ. Pathol. 1:163 (1977).
8. R. H. Reitz, Mechanistic considerations in the formulation of carcinogenic risk estimations. Banbury Report on Risk Quantitation and Regulatory Policy, May 13-16 (1984) (in press) (1985).
9. K. C. Chu, J. Cueto, and J. Ward, J. Toxicol. Environ. Health 8:251 (1981).
10. T. R. Fox, and P. G. Watanabe, Detection of a Cellular Oncogene in Spontaneous Liver Tumors of B6C3F1 Mice. Science, 228:596-597 (1985).

IMMUNOLOGIC QUANTIFICATION OF CARCINOGEN-DNA ADDUCTS

Regina M. Santella, Ling Ling Hsieh, and Frederica Perera

Cancer Center/Insititute of Cancer Research
School of Public Health
Division of Environmental Sciences
Columbia University
New York, New York 10032

ABSTRACT

Sensitive immunological methods for the detection of carcinogen-DNA adducts have recently been developed. These techniques are particularly useful for screening human populations for exposure to environmental carcinogens. Measurement of the biologically effective dose in humans may be useful in detecting carcinogenic hazards and carrying out risk estimates. We have developed monoclonal antibodies to several carcinogen-DNA adducts. These have included DNA modified by a benzo[a]pyrene diol epoxide (BPDE-I), 1-aminopyrene (1-AP) and 8-methoxypsoralen (8-MOP). BALB/cCr mice were immunized with the modified DNAs complexed electrostatically to methylated bovine serum albumin. Several stable clones have been isolated for each of the modified DNAs and characterized by enzyme-linked immunosorbent assay (ELISA). All antibodies are highly specific for the appropriate modified DNA and do not cross-react with nonmodified DNA. The antibodies to BPDE-I-DNA have significant cross-reactivity with DNAs modified by similar antitrans diol epoxides of benz[a]anthracene and chrysene. These DNAs all contain N-2 of guanine adducts. The antibody probably recognizes a shared determinant encompassing the guanine base and the hydrocarbon ring containing the hydroxide groups. The antibodies cross-react with BPDE-I-dG, the monoadduct isolated from DNA, but with lower sensitivity than for the intact modified DNA. They do not react with acetylaminofluorene (AAF) or 1-AP modified DNA, both of which contain C-8 of guanosine adducts. The antibodies to 1-AP-modified DNA demonstrate cross-reactivity with 8-nitro-1-aminopyrene- and 6-nitro-1-aminopyrene-modified DNA, as well as some slight cross-reactivity with BPDE-I-DNA and AAF-DNA. In general, all the antibodies have their highest reactivity with the original antigen and do not cross-react with the free carcinogens. Competitive ELISA with these antibodies can quantify modification levels with a sensitivity of about one adduct per 10^7-10^8 nucleotides. The antibodies are currently being used to measure adduct levels in animals exposed to carcinogens. In addition, DNA isolated from the lung, placenta, and pheripheral white blood cells of humans with various environmental exposures are being tested.

INTRODUCTION

Methods are needed to improve hazard identification and risk assessment for humans exposed to environmental carcinogens. Markers of the dose of carcinogen to the target cells or markers of preclinical response/risk could be valuable tools in this regard. Integration of such laboratory markers into studies of human populations, termed "molecular cancer epidemiology", has many advantages including the timely identification of carcinogenic hazards, more precise estimation of human cancer risk, and the ability to provide evidence of causal relationships. For example, the use of the same methods in different species can facilitate extrapolation of risks from animals to humans and between human populations with varying levels of exposure to carcinogens.

It is becoming feasible to measure the amount of substance that has interacted with the critical cellular targets or surrogates in humans and animals; i.e., the biologically effective dose of carcinogens. (For review, see[1]). Indicators of biological dose include somatic cell mutation, cytogenetic damage, DNA damage (including adduct formation), and DNA repair. At the present time, these markers demonstrate that a biological response has occurred as a result of exposure to a carcinogen; hence there is a presumption of potentially increased risk. However, use of these or other markers as predictors of the actual magnitude of risk requires future validation in prospective or prospective-retrospective experimental or epidemiological studies[2]. Hence available markers are currently useful as dosimeters, not as precise indicators of future risk. Most methods are in the experimental or validation stage and are not yet applicable to routine studies. In addition, available markers reflect genetic toxicity and hence will be most relevant to initiating-type carcinogens. Finally, the human samples most readily accessible for study are usually remote from actual target tissues. Therefore, information on the relationship between the endpoint measured in the target tissue and that in the sampled tissue is needed. For example, in the studies discussed below carcinogen adduct levels are measured in placental or peripheral white blood cell DNA, while the target may be the lung, liver, etc.

It is now possible to detect and quantify carcinogen adducts by immunological methods. This technique should be especially useful in monitoring humans with exposure to potential environmental carcinogens to provide information about the biologically effective dose[1,3]. A large number of polyclonal and monoclonal antibodies to various carcinogen DNA adducts have been developed. Two types of antigens have been used, carcinogen-modified mononucleosides (monoadducts) covalently coupled to protein (bovine serum albumin or keyhole limpet hemocyanin), or carcinogen-modified DNA electrostatically complexed to methylated bovine serum albumin. Table 1 gives a list of some of the antibodies that have been developed (for a review see[4,5]). In general, antibodies generated to the monoadducts show their highest reactivity with the monoadducts and lower reactivity for the adduct when it is present in DNA. In contrast, antibodies generated to intact modified DNA react best with DNA and have lower cross-reactivity with the isolated adducts. Depending on the application of the antibody, one approach or another may be preferable. One of the advantages of having an antibody specific for the monoadduct is that large amounts of DNA can be digested and the adduct isolated by various chromatographic procedures before quantification by immunoassay. Sensitivities and specificities may sometimes be increased by this procedure and cleaner samples obtained, since all the nonmodified bases are eliminated. For other types of studies (e.g., immunohistochemical studies) antibodies to the intact DNA may be preferable.

Table 1. Polyclonal and Monoclonal Antibodies Against Carcinogen-DNA Adducts.

Antigen	Ref	Antigen	Ref
O^6Me dGuo	(6,7)	AAF-DNA	(13)
O^2Me dThd	(5)	BPDE-I-DNA	(14-16)
O^6Et dGuo	(8-10)	cis Pt-DNA	(17,18)
O^4Et dThd	(6,9)	1AP-DNA	(19)
O^6-Bu Guo	(6,9,12)	8-MOP-DNA	(20,21)
		AFL-B_1-DNA	(22,23)
Guo-C8-AAF	(10,11)	OsO_4-oxidized poly (dT) (thymiine glycol)	(24,25)
		UV-DNA (thymidine dimer)	(26)

Monoclonal Antibodies to Benzo[a]pyrene diol epoxide I modified DNA

A total of five stable clones producing antibody specific for BPDE I-DNA have been isolated and characterized[16,27]. One clone designated 8E11 was obtained from the spleen cells of a BPDE-I-G-BSA immunized animal, while four clones designated 5D11, 5D2, 1D7, and 4C2 were from the spleen cells of several BPDE-I-DNA immunized animals. Only antibody 5D11 is of the IgG_2, kappa isotype; all others are IgG_1, kappa.

Competitive ELISAs were used to determined the sensitivity of the antibodies and to obtain additional information about their specificity. These data are summarized in Table 2. Antibody 8E11, which was generated against monomeric adduct coupled to BSA, has a 50% inhibition value with BPDE-I-DNA of 350 fmole. It has better reactivity against the monoadduct BPDE-I-dG (50% inhibition at 146 fmole) than with DNA, as is typical of antibodies generated to monoadducts. In addition, it shows good cross-reactivity against BPDE-I tetraols (50% at 250 fmole), the hydrolysis products of BPDE-I. This suggests that the antibody recognizes mainly determinants present on the BP ring. It is specific for BPDE-I since it does not react with BPDE-II-DNA. Antibody 5D11 has better sensitivity for BPDE I-DNA (50% inhibition at 19 fmole) than for the monoadduct BPDE-I-dG (50% inhibition at 21000 fmole). There is no cross-reactivity with BPDE-I-tetraols or BPDE-II-dG, but there is some low cross-reactivity with BPDE-II-DNA (50% inhibition at 4400 fmole). Clones 5D2, 4C2 and 1D7, produced from the spleen cells of animals immunized with BPDE-I-DNA, all produce antibodies with characteristics similar to those of 5D11. The higher reactivity with DNA, compared to monoadduct suggests that they also recognize antigenic determinants in the surrounding DNA structure as well as the adduct itself. Neither N-2-acetylaminofluorene- nor 1-aminopyrene-modified DNA (AAF-DNA or 1-AP-DNA) showed any cross-reactivity with any of the antibodies when tested at high concentrations. Both of these carcinogens produce adducts at the C-8 position of guanine instead of the N-2 position, to which BPDE binds. This can be contrasted with studies of the reactivity of antibody 5D11 with DNAs modified by other trans diol epoxides. When tested with DNAs modified by benz[a]anthracene trans-8,9-diol-10, 11-epoxide, benz[a]anthracene, trans3,4-diol- 1,2-epoxide, or chrysene trans-1,2-diol-3,4-epoxide, 50% inhibition values of 140, 145 and 345 fmole, respectively, were obtained. Similar cross-reactivites were obtained with a polyclonal antibody to BPDE-I-DNA which we developed in collaboration with Dr. Miriam Poirier of the NCI. Both the monoclonal and polyclonal antibodies can therefore recognize N-2 guanine adducts that are formed from trans diol epoxides with similar stereochemistry. This fact must be kept in mind when quantifying adducts from complex exposures such as cigarette smoke.

In a competitive ELISA, it is possible to test a maximum of 50 μ of DNA in a well. If 20% inhibition (about 5 fmole with antibody 5D11) is considered the lower limit of sensitivity in the assay, it is possible to quantify levels of adducts in the range of one per 10^7 to 10^8 bases. This high level of sensitivity is especially important to quantify DNA damage in human tissues, where the levels of DNA modification are likely to be extremely low. Several recent studies have begun to utilize this approach in human monitoring. One study[28] reported that lung tissue and/or white blood cell DNA from five of 27 patients contained significant levels of adducts, but they were not related to current smoking. The five subjects, lung cancer patients, were former or passive smokers. Another study using the same polyclonal antibody found similar frequencies of positive samples in the white blood cell DNA of roofers, foundry workers and volunteer laboratory workers[29]. The two positive volunteers were smokers. We are currently continuing this type of study, using monoclonal antibodies as well as the polyclonal antibody, to quantify adducts in a variety of human samples, including lung, placenta, and white blood cells.

Table 2. Competitive inhibition of monoclonal antibody binding (fmole causing 50% inhibition).

	8E11	5D11	5D2	4C2	1D7
BPDE-I-DNA	350	19	17	160	370
BPDE-I-dG	145	21000	$>10^{5a}$	$>10^{5a}$	4000
BPDE-I-tetraols	250	$>10^{5a}$	90000	64000	1200
BPDE-I-RNA	580	18000	20000	210	900
BPDE-II-DNA	$>3x10^{4c}$	4400	$>10^{5a}$	$>10^{5a}$	8800
BPDE-II-dG	$>10^{5a}$	$>10^{5a}$	$>10^{5a}$	$>10^{5a}$	$>10^{5a}$
AP-DNA	$>1.5x10^{4b}$	$>1.5x10^{4b}$	$>10^{5a}$	$>10^{5a}$	$>10^{5a}$

[a] no inhibition detected with 10^5 fmole per assay
[b] no inhibition detected with $1.5x10^4$ fmole per assay
[c] no inhibition detected with $3x10^4$ fmole per assay

<u>Monoclonal Antibodies to DNA Modified by 1-Aminopyrene</u>

1-Nitropyrene is responsible for a significant portion of the mutagenic activity of diesel exhaust particulates. The structure of the DNA adduct formed when both <u>Salmonella typhimurium</u>[30] and rats[31] are treated with 1-NP has been identified as N-(deoxyguanosin-8-yl)-1-aminopyrene. To quantify adduct levels in human and animal samples we have developed a panel of monoclonal antibodies specific for 1-aminopyrene modified DNA (1-AP-DNA). A total of ten stable clones were isolated from animals immunized with AP modified denatured DNA[19]. Since the AP adduct is unstable to heat denaturation, the DNA was denatured before modification. The most sensitive clone, 11H2, has been completely characterized by competitive ELISA. It reacts best with AP modified denatured DNA (50% inhibition at 18 fmole), while with native DNA the 50% inhibition value is 250 fmole (Table 3). This is similar to the results seen with the BPDE-I-DNA antibodies and is probably due to the adduct being more accessible to antibody-binding when present in single-stranded DNA. Antibody 11H2 also cross-reacts with other C-8 modified DNAs. With denatured 8-nitro-1-AP DNA, 50% inhibition is at 34 fmole but with denatured 6 nitro-1-AP DNA it is at 3200 femole. Denatured AAF-modified DNA also cross-reacts with a 50% inhibition at 2200 fmole. Surprisingly, denatured BPDE-I-DNA also cross-reacts with the antibody (50% inhibition at 3000 fmole), even though it binds to the N-2 position of guanine. Since 1-AP-DNA cannot be denatured before assaying, we have recently developed a more sensitive ELISA utilizing native DNA as the competitor, with a 50% inhibition at 48 fmole. This assay has been validated by quantifying adducts in Salmonella typhimurium treated with [^{3}H] 1-NP[19] and is currently being used to quantify adducts in animals treated with 1-NP.

Table 3. Competitive Inhibition of Monoclonal Antibody 11H2 Binding (fmole Causing 50% Inhibition).

1-AP-d-DNA	18
1-AP-n-DNA	250
8-nitro-1-AP-d-DNA	34
6-nitro-1-AP-d-DNA	3200
AAF-d-DNA	2200
BPDE-I-d-DNA	3000

Monoclonal Antibodies to 8 Methoxypsoralen-Modified DNA

8-Methoxypsoralen (8-MOP) plus ultraviolet-A light (320-400 nm) is used clinically in the treatment of psoriasis[32] and, more recently, extracorporeally as an experimental cytoreductive treatment for the leukemic phase of cutaneous T-cell lymphoma. 8-MOP forms two monoaddition adducts on thymine, as well as a cross-linked derivative[33]. Several case reports in the literature have associated treatment with 8-MOP plus UVA with cutaneous carcinoma[35], and with leukemia[34]. Patients undergoing this treatment provide an ideal population for studying adduct levels, since a defined quantifiable exposure has occurred. We have therefore developed a panel of monoclonal antibodies which recognize 8-MOP modified DNA, to quantify adduct levels in patients. A total of 14 stable clones have been isolated and four characterized completely[21]. They are all highly specific for 8-MOP modified DNA and do not cross-react with nonmodified DNA. Competitive inhibition curves of antibody 8G1 binding to plates coated with 8-MOP-DNA are shown in Figure 1. While 1 μg of 0.50% modified DNA (17 fmole 8-MOP adduct/well) inhibits antibody binding by 50%, the addition of up to 50 μg of nonmodified DNA per well does not inhibit antibody binding. To determine the antibody specificity, reactivity with 8-MOP modified poly(dA-dT).poly(dA-dT) -- which forms both monoadducts and cross-links (50% inhibition at 17 fmole) -- and poly(dA).poly(dT) --which forms only monoadducts (50% inhibition at 13 fmole) -- was also tested (Figure 1). DNA was also modified by two other psoralen derivatives, 4 aminomethyl-4,5', 8-trimethylpsoralen (50% inhibition at 580 fmole) and dimethylangelicin (50% inhibition at 104 fmole). Neither BPDE-I- or AAF-modified DNA or free 8-MOP showed any cross-reactivity with up to 10^5 fmole/well. Further characterization of antibody 8G1 with fractions from an HPLC digest of modified poly(dA-dT).poly(dA-dT), as well as the data in Figure 1, suggest that the antibody has its highest specificity for the monoadduct. DNA adduct levels in myeloma cells treated with several doses of 8-MOP and a UVA dose of 14 J/cm^2 were quantified. The data in Table 4 demonstrate that this antibody can be used to quantify adduct lines.

We have recently begun to monitor adduct levels in patients undergoing extracorporeal treatment with 8-MOP and UVA for cutaneous T-cell lymphoma, and plan to extend these studies to patients undergoing treatment for psoriasis.

DISCUSSION

A large number of polyclonal and monoclonal antibodies are now available for detecting and quantifying carcinogen-DNA adducts. These antibodies can be utilized in immunohistochemical studies to localize

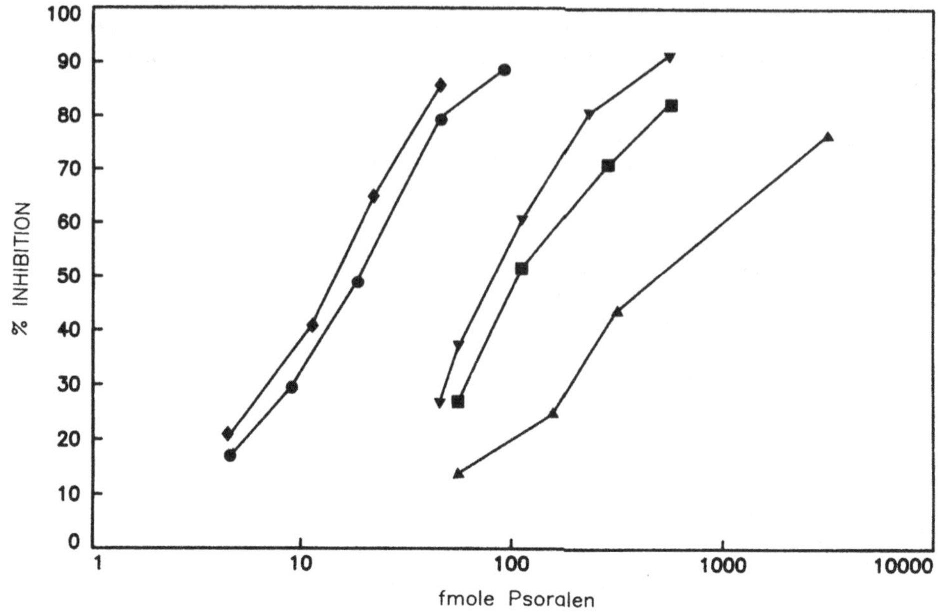

Figure 1. Competititve inhibition of monoclonal antibody 8G1 binding to 8-MOP-DNA. The competitors were 0.50% modified 8-MOP-DNA (●), 7.0% modified 8-MOP-poly(dA-dT).poly(dA-dT) (▼), 1.3% modified 8-MOP-poly(dA).poly(dT) (◆), 3.4% modified DMA-DNA (■) and 1.9% modified AMT-DNA (▲). The antibody was used at a 1:60,000 dilution.

carcinogen adducts in various tissues or cell types. In addition, they are an essential tool for quantifying adduct levels in exposed populations by RIA or ELISA. This approach, termed "molecular cancer epidemiology", can help to identify at a molecular level specific exogenous agents or host factors that play a role in human cancer causation. A limited number of studies have already begun to apply these techniques to human monitoring. Two published studies using the antibodies to BPDE-I-DNA were mentioned earlier. In addition to our current studies using the BPDE-I-DNA and 8-MOP-DNA antibodies, several other studies are using antibodies to quantify adduct levels. Dr. M. Poirier is conducting a study of the DNA adduct levels in patients undergoing treatment with cis platinum[36]. This study group, similar to that receiving 8-MOP treatment, is a population in which the exposure to the DNA-damaging agent is well defined. Finally, a study has been carried out on esophageal DNA from a high-risk group in China. In this study, the DNA was digested and the O^6-alkylguanine adducts isolated by HPLC before quantification by ELISA[37]. Such molecular epidemiology studies should eventually provide valuable information on human cancer causation.

Table 4. ELISA Quantification of 8-MOP-DNA Adducts.

ng 8-MOP/ml	fmole 8-MOP adducts	adducts/10^7 bases
50	.54	1.8
250	.87	2.8
500	.98	3.2

The potential advantages for risk assessment have been briefly described. However, with respect to determining the absence or presence of safe levels or "thresholds" for carcinogens, not accomplished by bioassays or epidemiology[38,39,40,41], molecular dosimetry methods are unlikely to be a panacea. Even if experimental thresholds are shown for specific compounds and markers, population thresholds are highly unlikely, given the demonstrated variability in human response.

ACKNOWLEDGEMENTS

The authors wish to thank Drs. I.B. Weinstein, A. Jeffrey, F. Beland, and F. Gasparro for samples and helpful discussion, and N. Dharmaraja and B. Silber for excellent technical assistance. This work was supported by grants from The Council for Tobacco Research, U.S.A., Inc., #1483A; Health Effects Institute; NIH CA 21111, CA 31696, and CA 39174.

REFERENCES

1. F. P. Perera, and I. B. Weinstein, Molecular epidemiology and carcinogen-DNA adduct detection: New approaches to studies of human cancer causation, J. Chron. Dis 35:581-600 (1982).
2. Committee: "Report of the Ad Hoc Panel on Chemical Carcinogenesis Testing and Evaluation of the National Toxicology Program Board of Scientific Counselors", National Toxicology Program, Aug. (1984).
3. G. Theall, G. Hatch, I. B. Weinstein, D. Grunberger, and S. Nesnow, Quantitative relationships between DNA adduct formation and biological effects, in: "Banbury Report: Indications of Genotoxic Exposure", B. A. Bridges, B. E. Butterworth, and I. B. Weinstein, eds., Cold Spring Harbor Lab., Cold Spring Harbor, pp. 231-244 (1982).
4. E. Kriek, L. P. Engelse, E. Scherer, and J. G. Westra, Formation of DNA modifications by chemical carcinogens identification, localization and quantification, Biochemin. Biophys. Acta 738:181-201 (1984).
5. P. T. Strickland, and J. M. Boyle, Immunoassays of carcinogen modified DNA, Prog. in Nucl. Acids Res. 31:2-201 (1984).
6. R. Muller, and M. F. Rajewsky, Antibodies specific for DNA components structurally modified by chemical carcinogens, J. Cancer Res. Clin. Oncol. 102:99-113 (1981).
7. C. P. Wild, G. Smart, R. Saffhill, and J.M. Boyle, Radioimmunoassay of O^6 methyldeoxyguanosine in DNA of cells alkylated in vitro and in vivo, Carcinogenesis 4:1605-1609 (1983).
8. R. Muller, and M. F. Rajewsky, Immunological quantification by high-affinity antibodies of O^6-ethyldeoxyguanosine in DNA exposed to N-ethyl-N-nitrosourea, Cancer Res. 40:887-896 (1980).
9. M. F. Rajewsky, R. Miller, J. Adamkiewica, and W. Drosdziok, Immunological detection and quantification of DNA components structurally modified by alkylating carcinogens ethylnitrosourea, in: "Carciongenesis: Fundamental Mechanisms and Environmental Effects", B. Pullman, P. O. Ts'o, and H. V. Gelboin, eds., Reidel Press, Dordrecht, Holland, pp. 207-218 (1980).
10. C. J. Van der Laken, A. M. Hagenaars, G. Hermsen, E. Kriek, A. J. Kuipers, J. Nagel, E. Scherer, and M. Welling, Measurement of O^6-ethyl-deoxyguanosine and N-(deoxyguanosine-8-yl)-N-acetyl-2-aminofluorene in DNA by high-sensitive enzyme immunoassays, Carcinogenesis 3569-572 (1982).
11. M. C. Poirier, S. H. Yuspa, I. B. Weinstein, and S. Blobstein, Detection of carcinogen-DNA adducts by radioimmunoassay, Nature, 270:186-188 (1977).

12. R. Saffhill, P. T. Strickland, and J. M. Boyle, Sensitive radioimmunoassays for O^6 -n-butyldeoxyguanosine O^2-n-butylthymidine and O^4-n-butylthymidine, Carcinogenesis, 3:547-552 (1982).
13. E. Sage, R. P. Fuchs, and M. Leng, Reactivity of the antibodies to DNA modified by the carcinogen N-acetoxy-N-acetyl-2-aminofluorene, Biochemistry 18:1328 (1979).
14. M. C. Poirier, R. Santella, I. B. Weinstein, D. Grunberger, and S. H. Yuspa, Quantitation of benzo[a]pyrene-deoxyguanosine adducts by radioimmunoassay, Cancer Res. 40:412-416 (1980).
15. H. Slor, N. Mizusawa, T. Nechart, R. Kakefuda, R. S. Day, and M. Bustin, Immunochemical visualization of binding of the chemical carcinogen benzo[a]pyrene diol epoxide to the genome, Cancer Res. 41:3111-3117 (1981).
16. R. M. Santella, C. D. Lin, W. L. Cleveland, and I. B. Weinstein, Monoclonal antibodies to DNA modified by a benzo[a]pyrene diol epoxide, Carcinogenesis 5:373-377 (1984).
17. B. Malfoy, B. Hartmann, J. P. Macquet, and M. Leng, Immunochemical studies of DNA modified by cis dichlorodiammine platinum (II), Cancer Res. 41:4127-4131 (1981).
18. M. C. Poirier, S. Lippard, L. A. Zwelling, M. Ushay, D. Kerrigan, R. M. Santella, D. Grunberger, and S. H. Yuspa, Antibodies elicited against cis -diamminedichloroplatinum(II)-modified DNA are specific for cis- diamminedichloroplatinum(II)-DNA adducts formed in vivo and in vitro, Proc. Natl. Acad. Sci. 79:6443-6447 (1982).
19. L. L. Hsieh, A. M. Jeffrey, and R. M. Santella, Monoclonal antibodies to 1-aminopyrene-DNA, Carcinogenesis In press (1985).
20. Z. Zarebska, M. Jarbabek-Chorzelska, T. Chorzelski, and S. Zablonska, Immune serum against anti DNA-8-methoxypsoralen photoadduct, Z. Naturforsch. 39:136-140 (1984).
21. R. M. Santella, N. Dharmaraja, F. P. Gasparro, and R. L. Edelson, Monoclonal antibodies to DNA modified by 8-methoxypsoralen and ultraviolet A light, Nucl. Acids Res. 13:2533-2544 (1985).
22. A. Haugen, J. D. Groopman, I. C. Hau, G. R. Goodrich, G. W. Wogan, and C. C. Harris, Monoclonal antibody to aflatoxin B1-modified DNA detected by enzyme immunoassay, Proc. Natl. Acad. Sci. 78:4124-4127 (1981).
23. P. J. Hertzog, J. R. L. Smith, and R. C. Garner, Production of monoclonal antibodies to guanine imidazole ring opened aflatoxin B1 DNA, the persistent DNA adduct in vivo, Carcinogenesis 3:825-828 (1982).
24. S. A. Leadon, and P. C. Hanawalt, Monoclonal antibodies to DNA containing thymine glycol, Mut. Res. 112:191-200 (1983).
25. R. Raghavan, R. J. Melamede, M. F. Laspia, B. F. Erlanger, and S. S. Wallace, Properties of antibodies to thymine glycol a product of the radiolysis of DNA, Rad. Res. 97:499-510 (1984).
26. P. T. Strickland, and J. M. Boyle, Characterisation of two monoclonal antibodies specific for dimerised and non-dimerised adjacent thymidines in single stranded DNA, Photochem. Photobiol. 34:595-601 (1981).
27. R. M. Santella, L. L. Hsieh, C. D. Lin, S. Viet, and I. B. Weinstein, Quantitation of exposure to benzo[a]pyrene with monoclonal antibodies, in: "Environmental Health Perspectives", in press (1985).
28. F. P. Perera, M. C. Poirier, S. H. Yuspa, J. Nakayama, A. Jaretzki, M. M. Curnen, D. M. Knowles, and I. B. Weinstein, A pilot project in molecular cancer epidemiology: determination of benzo[a]pyrene-DNA adducts in animal and human tissues by immunoassays, Carcinogenesis 3:1405-1410 (1982).

29. A. K. M. Shamsuddin, N. T. Sinopoli, K. Hemminki, R. R. Boesch, and C. C. Harris, Detection of benzo[a]pyrene-DNA adducts in human white blood cell DNA, Cancer Res. 45:66-68 (1985).
30. D. C. Howard, R. H. Heflich, F. E. Evans, and F. A. Beland, Formation of DNA adducts in vitro in Salmonella typhimurium upon metabolic reduction of the environmental mutagen 1-nitropyrene, Cancer Res. 43:2052-2058 (1983).
31. C. A. Stanton, F. L. Chow, D. H. Phillips, P. L. Grover, R. C. Garner, and C. N. Martin, Evidence for N-(deoxyguanosin-8-yl)-1-aminopyrene as a major DNA adduct in female rats treated with 1-nitropyrene, Carcinogenesis 6:535-538 (1985).
32. J. A. Parrish, T. B. Fitzpatrick, M. A. Pathak, and L. Tannenbaum, Photochemotherapy of psoriasis with oral methoxsalen and long wave ultraviolet light, New Engl. J. Med. 291:1207-1211 (1974).
33. P. S. Song, and J. K. Tapley, Photochemistry and photobiology of psoralens, Photobiol. 29:1177-1197 (1979).
34. A. Eskelinen, K. Halme, and A. Lassus, Risk of cutaneous carcinoma in psoriatic patients treated with PUVA, Photodermatology 2:10-14 (1985).
35. K. Freeman, and A. P. Warin, Acute myelomonocytic leukemia developing in a patientwith psoriasis treated with oral 8-methoxypsoralen and longwave ultraviolet light, Clinical & Exper. Derm. 10:144-146 (1985).
36. M. C. Poirier, E. Reed, L. A. Zwelling, R. F. Ozols, and S. H. Yuspa, The use of polyclonal antibodies to quantitate cis-platinum drug DNA adducts in cancer patients, Environ. Health Prosp., in press (1985).
37. D. Umbenhauer, C. P. Wild, R. Montesano, R. Saffhill, J. M. Boyle, N. Huh, V. Kirstein, and M. F. Rajewsky, Detection of alkylated bases and alkyltransferase activity in human tissues, in: "The Role of Cyclic Nucleic Acid Adducts in Carcinogenesis and Mutagenesis," B. Singer, and H. Bartsch, eds., IARC Scientific Publ. New York, p. in press (1985).
38. N. A. Littlefield, J. H. Farmer, D. W. Gaylor, and W. G. Sheldon, Effects of dose and time in long-term, low-dose carcinogenic study, J. Environ. Path. & Toxic. 3:17-34 (1979).
39. D. W. Gaylor, C. H. Frith, and D. L. Greenman, Urinary bladder neoplasms induced in BALB/C female mice with low doses of 2-acetylaminofluorene, J. Environ. Path & Toxic., in press (1985).
40. U. H. Ehling, D. Averbeck, P. A. Cerutti, J. Friedman, H. Greim, A. C. Kilbye, and M. L. Mendelsohn, Review of the evidence for the presence or absence of thresholds in the induction of genetic effects by genotoxic chemicals, International Comm. for Protection Against Environmental Mutagenesis and Carcinogens, ICPEMC Publication No. 10. Mutat. Res. 123:281-341 (1983).
41. E. Anderson, Quantitative approaches in use to assess cancer risk, Risk Analysis 3277-295 (1983).

INTERPRETATION OF GENOTOXICITY DATA IN ASSESSING THE RISKS OF GENETIC DISEASES

Carter Denniston

University of Wisconsin
Madison, Wisconsin 53706

INTRODUCTION

Risk assessment may be either qualitative or quantitative. The screening of chemicals for mutagenicity is a qualitative matter, akin to a test of hypothesis in statistical analysis. Monitoring a population for an increase in the mutation rate also falls into this category. On the other hand, determining dose-response curves and estimating the impact on the population of an increase in exposure to some mutagenic agent is a quantitative process, involving statistical estimation.

I will discuss one example of each of these two kinds of assessment: the logic of screening programs as an example of qualitative analysis, and the estimation of the impact of an increase of exposure to a mutagen on the frequencies of genetic disease.

QUALITATIVE RISK ASSESSMENT: Screening Program Theory

The screening of chemicals for possible mutagenic or carcinogenic effects is an awesome task facing both government and industry. I would like to describe some of the formal and logical aspects of such screening. Obviously, no amount of formalism can, by itself, be sufficient; knowledge of the biology and tests involved is mandatory. Nevertheless, a formal look at the problem may be useful to those charged with the testing and regulation of chemicals.

The primary problem is simply stated: Given a set of test results, how are these best combined to help reach a final determination regarding the mutagenicity or nonmutagenicity of a tested agent?

An Example

A simple example may help to fix the ideas. Recently, the Committee on Chemical Environmental Mutagens of the National Research Council proposed a screening program for chemical mutagenesis (Table 1). The program was described as having two "tiers". Tier 1 consists of three tests, A, B and C; tier 2 of one test, D. Tier 1 tests are done first. If any two of them is positive, the chemical is called a mutagen, if all three are negative, it is called a nonmutagen; otherwise the tier 2 test is done to decide the issue.

Table 1. A Proposed Screening Program

Tier 1	Tier 2	A	B	C	D	Decision
		Outcome of test				
A. Salmonella	D. Drosophila	+	+	+		m
B. Mammalian cell	X-linked lethal	+	+	-		m
gene mutation		+	-	+		m
C. Mammalian cell		-	+	+		m
chromosomal breakage		+	-	-		?
		-	+	-		?
m = mutagen n = nonmutagen		-	-	+		?
? = go to Tier 2		-	-	-		n
					+	m
					-	n

From reference [1].

The way I have described the proposal -- essentially the same way the Committee described it -- is common but tends to blur the important distinction between a decision function and a protocol. In this example, the decision function may be written as AB+AC+AD+BC+BD+CD. That is, we are to label the agent a mutagen if, and only if, any two of the four tests turn out positive. Although not explicitly stated, this is what is implied by the description. A protocol, the order, and the conditions under which the tests are to be done, was not completely specified by the Committee. A possible protocol is shown in Figure 1. In it we do A and B first, if both are positive, we call the agent a mutagen, if either or both are negative, we do C. If all are negative, we call the agent a nonmutagen; otherwise we do D.

The decision function for this program is uniquely determined by the Committee's description and is one of 116 possible decision functions using four binary tests. The protocol suggested is one of 288 possible protocols for this particular decision function. The 116 decision functions on four binary tests fall into 20 decision structures. These are shown in Table 2 (from Denniston[2]).

Let us pursue this example a bit further by examining it as a lattice. Any decision function may be represented as a lattice of its possible outcomes, the set of outcomes partitioned into two or more decisions (Figure 2). In this case each of the four tests has two possible results, + and -, giving 2^4=16 possible outcomes. Each of these 16 outcomes is assigned either m or n (mutagen or nonmutagen). The decision function is depicted in two ways, on the left as a single latice, on the right, as two

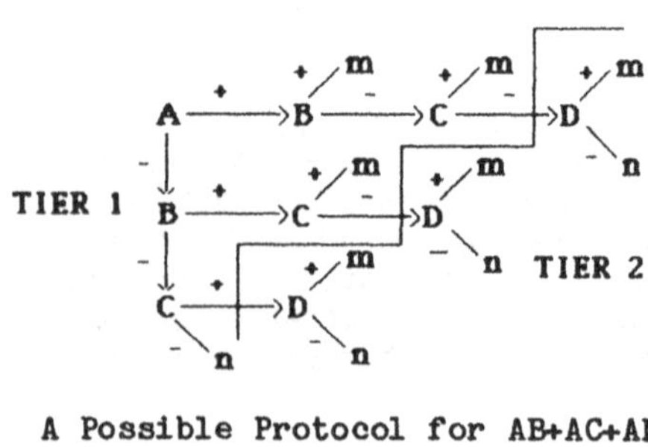

Figure 1. A Possible Protocol for AB+AC+AD+BC+BD+CD.

Table 2. Decision Structures Utilizing Four Tests[1]

Structure/Dual	ID	Protocols	ID	Structure/Dual
WXYZ	4.1.1.0[2]	24	4.15.1.0	W+X+Y+Z
WXY+WXZ	4.3.6.0	36	4.13.6.0	W+X+YZ
WXY+WXZ+WYZ	4.4.4.0	120	4.12.4.0	W+XY+XZ+YZ
WXY+WXZ+WYZ+XYZ	4.5.1.0	288	4.11.1.0	WX+WY+WZ+XY+XZ+YZ
WX+WYZ	4.5.12.0	36	4.11.12.0	W+XY+XZ
WX+WYZ+XYZ	4.6.6.0	120	4.10.6.0	WX+WY+WZ+XY+XZ
WX+YZ	4.7.3.0	48	4.9.3.0	WY+WZ+XY+XZ
WX+WY+WZ	4.7.4.0	24	4.9.4.0	W+XYZ
WX+WY+XYZ	4.7.12.0	108	4.9.12.0	WX+WY+WZ+XY
WX+WY+WZ+XYZ	4.8.4.0	144		
WX+WY+XZ	4.8.12.0	80		

[1]From reference 2.
[2]Identification number = n.r.p.i., where n= no. of tests, r = no. of outcomes that yield m, p = size of permutation class for that structure, and i = arbitrary index to distinguish structures with identical n, r and p.

lattices to represent the two "tiers". In the latter case, the Tier 1 lattice is partitioned into three parts, m, n, and "go to tier 2". The Tier 2 lattice is very simple, of course, involving only one test, two possible outcomes and two decisions. This second way of depicting the situation is a sort of mixture of decision function and protocol. There is, of course, nothing inherently wrong about describing a screening program in terms of "tiers", but neither is the idea at all fundamental. The fundamental ideas are those of the decision function and the protocol. One should be able to think about screening programs without tiers!

THE IDEA OF CONSISTENCY

Table 2 requires some explanation. How were the 20 structures arrived at? The question: "How many decision functions are there using four binary tests?" may be restated as: "How many ways can one "color" a 2x2x2x2 lattice using two colors (e.g., m and n). Without any constraints the answer is obviously 2^{16} ways. However, for our application we must impose a restriction that I have chosen to call consistency. All nodes above a

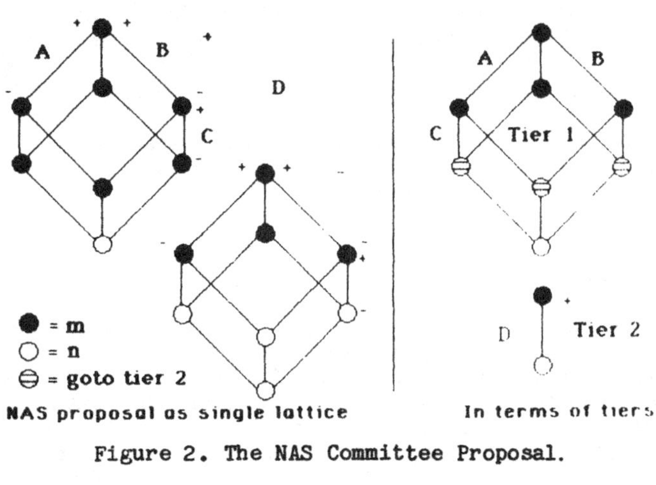

Figure 2. The NAS Committee Proposal.

node colored m must also be colored m. The idea of consistency is a natural one. If we are willing to conclude our test chemical is a mutagen based on a certain set of positive results, we should certainly conclude the same if even more results turn out to be positive. With this constraint, there are 116 distinct decision functions with four tests. If, for convenience, we chose to consider the four tests indistinguishable, these 116 functions fall into 20 structures as shown in Table 2. For example, structure 4.9.4.0 W+XYZ represents four functions: A+BCD, B+ACD, C+ABD and D+ABC. It may also be noted that the structures come in pairs (duals) which correspond to turning the lattice upside-down.

A PROBLEM WITH TIERS

Although I have made some critical remarks about tiers, they are, nonetheless a useful device for describing a screening program. In our example there were three results in Tier 1 that lead to Tier 2. Let us call them exit points. Here all three exit points lead to the same second tier, but this is not necessary. It is quite possible for different exit points to lead to different second tiers, even involving different sets of tests. In such a case, inconsistencies can arise. In our example, the three exit points are mutually noncomparable, that is, no exit point is above or below any other exit point in the Tier 1 lattice. In such a situation, as long as the second tiers are themselves internally consistent, the entire decision function will be consistent. But problems can arise when exit points are comparable. Then the following rule must be obeyed.

The nodes of a decision lattice may be ranked in two ways: with regard to their positions in the lattice and with regard to their "color". Given two nodes, a and b, if a is above (or equal to) b, we write $a \geq b$; if a is of higher or equal color rank (e.g. m is higher than ? and ? is higher than n) than b, we write $a \gg b$. Our rule of consistency may be simply stated: For any two nodes, a and b, if $a \geq b$ then $a \gg b$. That is, higher nodes cannot be of lower color rank. Now consider two exit points, a and b, where a is above b in a Tier 1 lattice. Let the second tiers associated with these points be called T(a) and T(b), respectively, (in general, these are identical lattices with different colorings). Then the entire decision function will be consistent if the following holds: For any two nodes, $a' \in T(a)$ and $b' \in T(b)$, if $a' \geq b'$ then it must be that $a' \gg b'$.

ERROR RATES AND COSTS

Any screening program has two additional properties which I shall only mention. If each test has a false positive and false negative error rate, then so does the entire program. These error rates depend entirely on the decision function but not at all on what protocol is used. A general method for obtaining program error rates may be found in Denniston[2].

In addition, a program has two costs: the average cost of testing a mutagen and the average cost of testing a nonmutagen. These costs depend, of course, on the costs of the individual tests, the error rates and the protocol used.

EXTENSIONS OF SCREENING THEORY

Our example screening program was, of course, very specific: It used four tests; each test had two possible outcomes, + and -; and the entire program had but two decisions, m and n. This immediately suggests three kinds of extensions to our theory: vary the number of tests; vary the number of outcomes or levels per test; e.g., +, - and ?; and vary the number of decisions; e.g., m, n, and ?. The first varies the number of dimensions in the lattice, the second the number of levels per dimension, and the third the number of "colors"; i.e., the number of parts into which we must partition the lattice, subject to the consistency principle.

For example, consider a program with two tests, A and B; three levels per test, +, -, and ? (indeterminate); and three decisions, m, n and ? (Figure 3). There are 62 decision functions and 38 structures (Table 3). Each function may have many possible protocols.

The body of Table 3 should be interpreted as follows. Each structure is of the form M:N. M is the condition for concluding that the agent tested is a mutagen, N is the condition for concluding that it is a nonmutagen. For example, consider the structure Xy: xy. The outcome Xy, X positive and Y negative, and all outcomes above this one in the lattice, lead to the conclusion m. The outcome xy, X negative and Y indeterminate, and all outcomes below this one in the lattice, lead to the conclusion n. All other outcomes lead to ? no conclusion. The three dicisions, m, n and ?, could, of course, have other meanings without affecting the validity of Table 3.

I will conclude this brief introduction to screening theory by pointing out some of its potential. From the examples I have given, it is clear that with more than a very few tests the number of different decision structures and protocols becomes much too large to enumerate. One way to cut down on the number of possibilities and, at the same time, aid the process of decision-making is to assign numerical weights to each test and a score (some nondecreasing function of the test weights) to each node of the outcome lattice. The partitioning of the lattice can then be done numerically; e.g., if the score is $\geq x$ conclude that the chemical is a mutagen. This approach looks promising.

For the general problem it seems, from what we already know, that it would not be difficult to write computer programs that can: 1) generate decision functions and protocols with specified attributes, 2) calculate error rates and costs, and 3) search for decision functions and protocols with certain optimal properties. I wish to emphasize again, however, that such mechanical procedures can never substitute for scientific judgement based on an intimate knowledge of the tests utilized and the biological problems involved.

QUANTITATIVE RISK ASSESSMENT: Estimating Impact

The basic equation used to assess the impact on a population of increased exposure to a mutagen has been (BEIR[3], UNSCEAR[4], Denniston[5]):

$$I = S \cdot R \cdot M \tag{1}$$

where I = impact per unit dose
S = the spontaneous burden of genetic disease
R = the relative mutation risk
M = the mutation component.

The impact is usually number of additional cases of genetic disease in the population per additional unit of dose, but it could just as well be additional number of years of life lost (Jones[6]) or even extra hospital beds needed or funds expended.

The relative mutation risk is defined as $\Delta m/m$, the relative change in the mutation rate, m, per unit of dose. The mutation component is $(\Delta S/S)/(\Delta m/m)$, the elasticity of response to an increase in the

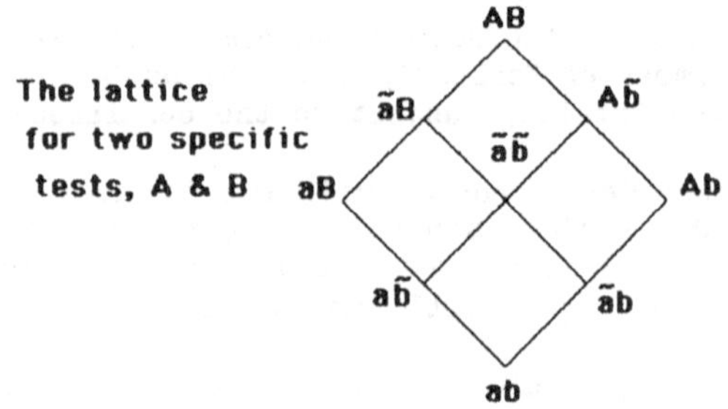

Figure 3. The Outcome Lattice for Two Trinary Tests

mutation rate, to borrow a term from the field of economics. Our basic equation then becomes

$$I = S\cdot[\Delta m/m] \cdot [(\Delta S/S)/(\Delta m/m)] = \Delta S. \qquad (2)$$

However, we must add some specificity to the formulation!

In radiation work, two approximations are generally made. First, it is assumed that S is a linear function of the mutation rate, $S = A+Bm$, so that the mutation component is then $M=Bm(A+Bm)$. Second, especially when effects at low dose are of interest, it is assumed that the mutation rate is also approximately linear in dose, $m = a+bD$, in which case $R = b/a$. In this special case, R is the reciprocal of the doubling dose.

The main problem of quantitative risk assessment, then, is obtaining reliable estimates of S, R and M. How are such estimates obtained? How well we measure I depends, of course, on the precision of our estimates of S, R and M. In fact, the variance of I is given by

$$V^I = (S^2 + V_S)(R^2 + V_R)(M^2 + V_M) - (I^2). \qquad (3)$$

Table 3. Decision Structures: Two Tests, Trinary Outcomes*

Structure/Dual		Structure/Dual	
Xy : xy	XY : xy	Xy+xY : xY	Xy : xy+xy
Xy : xy	XY : xY	Xy+xY : xY	Xy : xY+xy
Xy+xY : xy	XY : xy+xy	Xy+xY : Xy	xY : xy+Xy
Xy+xY : xy	XY : xY+xy	Xy+xY : xy+xy	xY+Xy : xY+xy
Xy+xY : xy	XY : xY+Xy	Xy+xY : xy+xy	xY+Xy : xY+Xy
Xy : xy	Xy : xY	XY : xy	
xY : xy	Xy : Xy	Xy : xy	
Xy+xY : xy	Xy : xy+xy	xY : xy	
Xy+xY : xy	Xy : xY+xy	Xy+xY : xy+xy	
Xy+xY : xy	Xy : xy+Xy	Xy+xY : xy+xY	
Xy+xY : xy	Xy : xY+Xy	Xy+xY : Xy+xy	

*X, Y = positive results; x, y = negative results; x, Y = indeterminate; structures are of form M:N, where M is condition for decision m, and N is condition for decision n. See text for further discussion.

THE SPONTANEOUS BURDEN

Table 4 shows various estimates of the frequencies of genetic diseases among newborn. Not all of the columns are, of course, independent. The Stevenson[7] data overestimate dominants because of the inclusion of many mild disorders; the Trimble & Doughty[8] data underestimate them because of the failure to count adult diseases. The Table demonstrates how imprecise are our estimates of the background level of human genetic disease.

The biggest, and possibly most important, uncertainty lies in the category of "irregularly inherited" traits (malformations and multifactorial in the table). Estimates of this category have been going up over time, but without a corresponding increase in data. In addition, our lack of knowledge of the mutation component for this category aggrevates the situation.

THE RELATIVE MUTATION RISK

Estimation of this parameter depends upon knowledge of the background mutation rate and the relevant dose-response curve for the particular trait being measured. Committees concerned with the effects of radiation on human populations have been forced to use data from the mouse, in particular the specific locus test, to estimate R. The Hiroshima-Nagasaki data give us only upper bounds for the induction rates due to radiation.

Chemicals raise a new host of problems. Each chemical is likely to have its own dose-response characteristics, and individual variability in sensitivity may well be greater for chemicals than for radiation. Quantitative progress in this area is now beginning, as the work reported in this Symposium attests.

THE MUTATION COMPONENT

For simple Mendelian and chromosomal traits, the mutation component is one, because their equilibrium frequencies are directly proportional to the mutation rate. For more complex traits and for simple ones with phenocopies, the problem is a difficult one. Some information may be obtained by ascertaining the heritabilities of a trait. The relation between the mutation component and heritability is shown in Table 5.

The broad heritability is that proportion of the total phenotypic variance that is attributable to genetic factors; in simple models it is the regression of genotype on phenotype. The narrow heritability is that proportion of the total phenotypic variance attributable to the

Table 4. Estimates of the Spontaneous Genetic Burden*

Genetic Category	Stevenson[7]	UNSCEAR[9]	Trimble & Doughty[8] min	Trimble & Doughty[8] adj**	UNSCEAR[4]	BEIR[3]
Dominant & X-linked	336	99	9	12	100	100
Recessive	21	21	9	11	10	25
Chromosomal	__	42	16	20	40	40
Malformations	141	250	358	428	430	
Multifactorial	148	150	158	473	470	900

*Frequencies per 10,000 live births; modified from Crow & Denniston[10].
**Min=minimum; adj=adjusted for under-ascertainment.

Table 5. Heritabilities and the Mutation Component*

Broad sense Heritability	Narrow sense Heritability	Mutation Component	Example
High	High	High	Rare Dominant
High	Low	High	Rare recessive
High	Low	Low	Overdominant
Low	Low	Low	Environmental

From reference 11.

"additive" effects of genes; in simple models it is the regression of additive genotype on phenotype. One sees from Table 5 that if both heritabilities are high the mutation component is high, and if both are low, the mutation component is low. On the other hand, if they are far apart, indicating genetic interactions, nothing can be inferred about the size of the mutation component. Luckily, we can draw one important conclusion in this case. If the broad heritability is high, indicating strong genetic influence on the trait, but the narrow heritability is low, indicating little correlation between parents and offspring, we are dealing with a trait that is not likely to be affected by a change in the mutation rate, or if affected, not very quickly. In either case, such traits will contribute little in the short term to our estimates of total impact following exposure to a mutagen.

SUMMARY

Two aspects of genetic risk assessment have been discussed briefly. Some new results concerning the logical development of mutagenic screening programs have been reviewed. In addition, a brief introduction to quantitative risk estimation has been given.

REFERENCES

1. NAS, Identifying and Estimating the Genetic Impact of Chemical Mutagens. Report of the Committee on Chemical Environmental Mutagens, National Academy of Sciences, Washington, D.C. p. 223 (1982).
2. C. Denniston, Screening program theory: Decision functions and protocols, Environ. Mutagen. 7:53-72 (1985).
3. BEIR, The Effects on Populations of Exposure to Low Levels of Ionizing Radiation. National Academy of Sciences, Washington, D.C. (1980).
4. UNSCEAR, Genetic effects of radiation, in: Ionizing Radiation: Sources and Biological Effects, United Nations Scientific Committee on the Effects of Ionizing Radiation, United Nations, Twenty-Seventh Session, Supplement No. 45 (A/37/45), pp. 425-569 (1982).
5. C. Denniston, Low level radiation and genetic risk in man, Ann. Rev. Genet. 16:329-355 (1982).
6. M. B. Jones, Years of life lost due to Huntington's disease, Am. J. Hum. Genet. 31:711-717 (1979).
7. A. C. Stevenson, The load of hereditary defects in human populations, Rad. Res. Suppl. 1:306-325 (1959).
8. B. K. Trimble and J.H. Doughty, The amount of hereditary disease in human populations, Ann. Hum. Genet. 38:199-223 (1974).
9. UNSCEAR, Genetic effects of radiation, in: Ionizing Radiation: Levels and Effects, United Nations Scientific Committee on the Effects of Ionizing Radiation, United Nations, Twenty-Seventh Session, Supplement No. 25 (A/8725), pp. 199-302 (1972).

10. J. F. Crow, and C. Denniston, Mutation in human populations. in: Advances in Human Genetics, H. Harris and K. Hirschhorn, eds., Plenum Publishing Co., Chapter 2 (1985).
11. J. F. Crow and C. Denniston, The mutation component of genetic damage, Science 212:888-893 (1981).

BACKGROUND LEVELS OF DNA DAMAGE IN THE POPULATION

Robert L. Saul & Bruce N. Ames

Biochemistry Department
University of California
Berkeley, California 94720

ABSTRACT

In addition to the dangers of man-made mutagenic substances, there exist a variety of naturally-occurring agents and processes than can cause DNA damage in man. These include exogenous agents, such as sunlight and dietary mutagens, as well as endogenous agents, such as reactive oxygen species which are formed within the cell via normal metabolism. These processes can lead to at least six major types of DNA damage: base loss, base deamination, base alkylation, base dimerization, base oxidation, and single-strand breakage. Based upon *in vitro* measurements in cell-free model systems or cultured cells, estimates of the rates of production of these lesions have been made for the average human cell: about 26,000 base losses per day, primarily purines lost because of spontaneous hydrolysis of the glycosyl bond; about 350 cytosine deaminations per day; tens of thousands of base alkylations per day, primarily methylations of guanine caused by S-adenosylmethionine; up to 50,000 pyrimidine dimerizations per day, dependent upon exposure to sunlight; and about 100,000 single-strand breaks per day. Based upon measurements of DNA oxidation products in human urine, we estimate that the total rate of formation of all oxidative DNA damage products in man is on the order of a few thousand events per cell per day. In addition, base mispairing can take place in the absence of DNA damage because of tautomeric shifts, base ionization, or base rotation. Such estimates of DNA lesion formation are only very approximate because they are based on limited data, and in most cases have not been corrected for the many possible influences of tertiary and quaternary DNA structure and the presence of histones, divalent cations, and other factors *in vivo*. The biological importance of each particular lesion is difficult to access since this depends not only on the rate of formation, but also on the persistence and chemical nature of that lesion within the DNA. The steady-state concentrations of most lesions are probably very low *in vivo* because of a variety of efficient DNA repair enzymes that repair damage, and "proof-reading" enzymes that correct base mismatching. Although the importance of this background of damage is difficult to evaluate *per se*, it may be possible to assess the relative importance of environmental mutagens by comparison to this background.

INTRODUCTION

The efficient DNA repair systems present in modern day organisms have evolved from ancestral repair systems dating back several billion years.

The evolution of such repair enzymes has come in response to the existence of DNA damaging agents throughout the earth's history. It is important for those who evaluate the dangers of industrially produced mutagens to recognize that some DNA damage has always been with us, and that the effects of man-made DNA-damaging agents are superimposed upon a pre-existing background rate of DNA damage product formation.

Accurate estimation of the magnitudes of the rates of formation of DNA damage products in man is made difficult by (i) the analytical obstacles involved in measuring very sparce concentrations of analyte in complex biological matrices; (ii) the large variety of possible damage products, many of which are not yet well characterized; and (iii) the uncertainty in extrapolating from chemical model systems, whole cell systems, or animal model systems to the human situation. Faced with these difficulties, only a very rough estimation (at best) is possible for the background levels of the major types of DNA damage in man. In the following review, we discuss the sources of background DNA damage, estimate the rates of formation of such damage, and assess the possible significance of these rates.

SOURCES OF DNA DAMAGE

Hanawalt _et al._[1] defined DNA damage as "any modification of DNA that alters its coding properties or its normal function in replication or transcription". Under this rather broad definition, DNA damage can be categorized into three main types: (i) induced chemical damage resulting from interaction of DNA with chemical agents, UV light or ionizing radiation; (ii) spontaneous chemical damage caused by the intrinsic thermodynamic instability of DNA; and (iii) base mispairing caused by the innate physicochemical properties of the nucleoside components of DNA. The sources of background DNA damage have been summarized in Table 1 under these three categories. Agents that induce chemical change in DNA have been subclassified into exogenous agents and endogenous agents. Listed here as exogenous sources of DNA damage are UV radiation which comes from sunlight, and background ionizing radiation, which comes from cosmic rays and decay of naturally occurring radioisotopes in the earth's crust. Another source of exogenous mutagens is the wide variety of naturally-occurring toxins present in plants[2], many of which are present in the human diet.

Table 1. Sources of Background Levels of DNA Damage

INDUCED CHEMICAL CHANGES
- Exogenous Agents
  - Solar UV Radiation
  - Background Ionizing Radiation
  - Natural Mutagens in Food
- Endogenous Agents
  - Reactive Oxygen Species
  - Endogenous Alkylating Agents

SPONTANEOUS CHEMICAL CHANGES
- Hydrolysis
  - Base Loss
  - Strand Breakage
- Deamination

PHYSICOCHEMICAL MISPAIRING
- Tautomerization
- Ionization
- Base Rotation
- Natural Base Analogs

Endogenous processes can also lead to DNA damage. Considerable interest has grown over the possibility that reactive oxygen species can be formed in the cell as a consequence of normal metabolism[2-4]. Superoxide anion and hydrogen peroxide are known to be formed in vivo, and these may further react to yield more reactive species such as hydroxyl radical, which can produce numerous types of oxidative damage in DNA. Several endogenous processes may also lead to DNA alkylation. Magee and Barrows [5] proposed that the normal metabolite S-adenosylmethionine may methylate DNA bases in vivo. Another class of endogenous alkylating agents are the N-nitroso compounds, which can be formed in the human gastrointestinal tract from dietary precursors[6].

The spontaneous chemical changes and physicochemical properties listed in Table 1 are based on a similar table by Singer and Kusmierek[7]. Proposed roles for hydrolysis and deamination as damage processes have been discussed by Lindahl and co-workers[8-10] and Shapiro[11].

The concept of physicochemical mispairing was first advanced by Watson and Crick[12]. Errors in base incorporation may occur during replication because either the template base or the incoming nucleotide is in a rare tautomeric, ionized or syn-rotated form. Topal and Fresco[13] have advanced this theory by using model building, the use of base analogs, and chemical considerations of enol or imino tautomers. One other type of natural DNA lesion is that caused by the presence of natural base analogs in the cell; in particular, dUTP, which may be mistaken for TTP by the DNA polymerase[14].

ESTIMATES OF BACKGROUND LEVELS OF DNA DAMAGE IN MAN

Tables summarizing estimates of the rates of formation of various types of DNA damage have recently been compiled by Shapiro[11], Tice and Setlow[15], and Hartman[16]. In Table 2, we have combined and expanded these tables. Our values are expressed in events/cell/day, and assume average human cells to contain 5.6×10^{12} Daltons of DNA (which is 1.8×10^{10} residues, or about 4.5×10^9 of each of the four main types of residues). Specific DNA repair enzymes are known for each type of damage product listed[17]. For example, AP endonucleases initiate the repair of baseless sites, and uracil-DNA glycosylase removes the uracil produced by cytosine deamination. The evolution of these specific enzymes, in itself, is evidence for the natural existence of the corresponding damage products in vivo.

The base loss numbers in Table 2 are from Shapiro[11], who based his numbers on chemical rate constants for DNA in aqueous solution. The base deamination value in Table 2 is also from Shapiro[11], again based on chemical rate constants. Deamination of adenine residues is also possible[10], but is expected to be a very rare event, and not biologically important[11]. The rate given for cytidine deamination is based on a hypothetical unbuffered solution, and may be very different in vivo in the presence of buffers[11].

The values for the background rate of base alkylation were calculated assuming that the principle source is methylation by S-adenosylmethionine. We used the data of Barrows and Magee[5] who methylated calf thymus DNA with methyl-tritiated-S-adenosylmethionine in an in vitro system. Our estimates were calculated using their rate constants, and assuming a concentration of S-adenosylmethionine in the cell nucleus of 0.065 mM[18]. The products N-7-methylguanine and N-3-methyladenine are repaired by DNA-glycosylases.

Table 2. Estimation of Rates of Formation of DNA Damage Products in Man

Damage	Events/cell/day	Primary damage source	Experimental system tested
BASE LOSS			
Depurination	25,000	Spontaneous	Chemical model[11]
Depyrimidination	1,300	Spontaneous	Chemical model[11]
BASE DEAMINATION			
Cytosine	350	Spontaneous	Chemical model[11]
BASE ALKYLATION			
N-7-Me-G	84,000	S-Adenosyl-Met.	Chemical model[5]
N-3-Me-A	840	S-Adenosyl-Met.	Chemical model[5]
BASE DIMERIZATION			
Cyclobutane type	up to 37,500	U.V. Light	Bacteria[15]
Pyr(6-4)pro type	up to 12,500	U.V. Light	Bacteria[15]
BASE OXIDATION			
Thymine Glycol	400	Ox. Metabolism	Human Urine[20]
5-Hydroxymethyluracil	600	Ox. Metabolism	Human Urine[22]
Other oxidized prod.	2,000	Ox. Metabolism	Estimates
SINGLE STRAND BREAKS			
Direct	100,000	Spontaneous	Chemical model[11]

Ultraviolet light cannot penetrate more than a few millimeters of tissue or clothing, and so the only cells with significant formation of pyrimidine dimers are exposed skin cells. The amount of dimer formation per exposed cell per day is dependent upon numerous factors including the number of hours of sun exposure per day, the time of day, the time of year, latitude, weather conditions, and the racial group of the exposed individual. Tice and Setlow[15] used data from bacteria to estimate that human cells exposed to 1 hour of Texas noon sun would have about 50,000 pyrimidine dimers formed per cell. For Table 2, we have assumed that this value is about the maximum number of photoproducts that exposed skin would receive in an entire day. We have separated this value into two parts, for the two major types of dimers; the pyr-pyr-cyclobutanes and the pyr(6-4)pyo-photoproducts, assuming that the former is formed in about three times greater yield than the latter[19].

Our lab has used a novel approach to estimate the rate of formation of oxidative DNA damage <u>in vivo</u>[20]. Our approach has been to assay human urine for the presence of certain DNA damage products that are removed from DNA by repair and are excreted. This method has the advantages that it is non-invasive and is highly sensitive because of the bioconcentration of damage products from the entire body into a relatively small volume of urine. This approach is not, however, generally applicable to all types of DNA damage products because of numerous associated technical problems. We have found one particular oxidative DNA damage product, cis-5,6-dihydroxydihydrithymidine (cis-thymidine glycol) which can be used reliably as a tool for monitoring oxidative DNA damage, without technical problems. This compound has the following useful properties: (i) it is rapidly repaired and excreted; (ii) it is chemically and biologically stable; (iii) it is relatively easy to purify and detect; (iv) it is not formed artifactually during collection, storage, or analysis; (v) it is not absorbed from the gut, which means that urinary levels are unaffected by performed dietary levels of the compound; and (vi) it has two different possible stereochemical forms (cis-(5R,6S)-thymidine glycol and cis-(5S,6R)-thymidine glycol) and the relative abundance of these stereoisomers can provide information on the chemical mechanism(s) of formation. We estimate that about 100 cis-thymidine glycol molecules are formed per cell per day in the average adult human, almost exclusively as only one of the two possible cis-isomers. Two other oxidative DNA damage products investigated by our lab are the bases cis-thymine glycol (TG) and 5-hydroxymethyluracil (HMU), which would presumably be removed from DNA *in vivo* by TG-DNA glycosylase and HMU-DNA glycosylase, respectively[21]. Both TG and HMU are present in human urine, at levels which correspond to formation rates of about 300 events per cell per day for TG and 600 events per cell per day for HMU[22], although these numbers are less certain, because of possible contributions of exogenous sources of these two compounds. Rats, which have a higher specific metabolic rate and a shorter life span, have about a 15-fold higher rate of formation of TG and HMU per cell. Based on the levels in human urine, and on estimated relative abundances of other types of oxidative DNA damage, we suggest that the total rate of formation of all oxidative DNA damage products in man is on the order of a few thousand events per cell per day.

Single strand breaks are thought to arise primarily via two mechanisms: directly, by spontaneous hydrolysis of the phosphodiester bond, and indirectly, as a secondary result of depurination. The value given in Table 2 is for the direct mechanism and was calculated from estimates of Shapiro[11]. It has been suggested that the rate of single strand breakage via the indirect mechanism should be equal to the total rate of base loss[11], since strand breakage is thought to proceed quite easily at an AP site[23]. However, *in vivo*, AP repair processes would interrupt this two step process and so make this indirect mechanism less important.

RELEVANCE OF BACKGROUND LEVELS OF DNA DAMAGE TO MUTAGENESIS

DNA damage can produce mutations directly or indirectly. In the case of the direct mechanisms, the damage causes a change in the hydrogen bonding properties of the template, leading to base mispairing to produce point mutations. An example of this is cytosine deamination to uracil, leading to a CG to TA transition mutation. Finally, as noted in Table 1, base mispairing can occur in the absence of covalent changes in the bases, as a result of errors by the polymerase caused by tautomerism, base ionization, or base rotation. Mathematical models, based on known physicochemical properties of the bases, and assuming a "proofreading" step during errors/replicated base pair in bacteria[13], although this frequency in human cells must be much lower.

Factors influencing the biological effects of a particular type of DNA damage product are (i) the rate at which it is formed; (ii) the length of time it persists in the DNA (dependent upon the efficiency of repair); and (iii) the chemical nature of the lesion. The "steady state" level of a particular lesion in DNA is equal to the first fraction (the formation rate) multiplied by the second factor (the persistence time). It should be noted that the values in Table 2 indicate only the formation rates, and cannot be used by themselves to predict biological effects without considering the persistence of each particular lesion.

Most of the lesions listed in Table 2 (and most of those not listed) are probably not important to mutagenesis, either because their steady state level is numerically insignificant (because of slow formation or rapid repair), or because the lesion itself is benign (e.g., substitution of uracil for thymine is probably relatively harmless, since uracil will continue to base-pair with adenine). In the case of methylated adducts, numerous products are known, but only two relatively rare products, O-6-methylguanine and O-4-methylthymine, are thought to be mutagenic, presumably because these particular lesions have altered hydrogen bonding properties. The possibility remains, however, that many more types of DNA damage products may cause mutation indirectly as a result of errors that arise during their repair.

At this time, it is difficult to assess the absolute importance of the various rates discussed above. However, these values, representing background levels of damage and mutation, may be useful for providing some feeling for the relative importance of various mutagenic hazards.

ACKNOWLEDGEMENTS

This work was supported by Department of Energy Contract DE-AT03-76EV70156, by a contract from the National Foundation for Cancer Research, and by National Institute of Environmental Health Sciences Center Grant ES01896. Robert Saul was supported by a postdoctoral fellowship from the American Health Association, California Affiliate.

REFERENCES

1. P. C. Hanawalt, P. K. Cooper, A. K. Ganesan, and C. A. Smith, DNA repair in bacteria and mammalian cells, Ann. Rev. Biochem. 48783-836 (1979).
2. B. N. Ames, Dietary carcinogens and anticarcinogens: Oxygen radicals and degenerative diseases, Science 221:1256-1264 (1983).
3. D. Harman, Role of free radicals in mutation, cancer, aging, and the maintenance of life, Radiat. Res. 16:753-763 (1962).
4. J. R. Trotter, Spontaneous cancer and its possible relationship to oxygen metabolism, Proc. Natl. Acad. Sci. USA 77:1763-1767 (1980).
5. L. R. Barrows, and P. N. Magee, Nonenzymatic methylation of DNA by S-adenoxylmethionine in vitro. Carcinogenesis 3:349-351 (1982).
6. M. C. Archer, Hazards of nitrate, nitrite, and nitrosamines in human nutrition, in: "Nutritional Toxicology", J. N. Hathcock, ed., Vol. 1, pp. 327-381, Academic Press, New York (1982).
7. B. Singer, and J. T. Kusmierek, Chemical mutagenesis, Ann. Rev. Biochem. 51:655-693 (1982).
8. T. Lindahl, and B. Nyberg, Rate of depurination of native deoxyribonucleic acid, Biochem. 19:3610-3618 (1972).
9. T. Lindahl, and B. Karlstrom, Heat induced depyrimidation of deoxyribonucleic acid, Biochem. 25:5151-5154 (1973).
10. P. Karran, and T. Lindahl, Hypoxanthine in deoxyribonucleic acid: Generation by heat-induced hydrolysis of adenine residues in free form by a deoxyribonucleic acid glycosylase from calf thymus, Biochem. 19:6005-6011 (1980).
11. R. Shapiro, Damage to DNA caused by hydrolysis, in: "Chromosome Damage and Repair", E. Seeberg and K. Kleppe, eds., pp. 3-18, Plenum Press, New York (1981).
12. J. D. Watson, and F. H. C. Crick, Genetical implications of the structure of deoxyribonucleic acid, Nature 171:964-967 (1953).
13. M. D. Topal, and J. R. Fresco, Complementary base pairing and the origin of substitution mutations, Nature 263:285-289 (1976).
14. B. M. Olivera, P. Manlapaz-Ramos, H. R. Warner, and B. K. Duncan, DNA intermediates at the Escherichia coli replication fork. II. Studies using dut and ung mutants in vitro, J. Mol. Biol. 128:265-275 (1979).
15. R. R. Tice, and R. B. Setlow, DNA repair and replication in aging organisms and cells, in: "Handbook of the Biology of Aging" (in press).

16. P. E. Hartman, Bacterial mutagenesis: Review of new insights, Environ. Mutagen. 2:3-16 (1980).
17. T. Lindahl, DNA repair enzymes, Ann. Rev. Biochem. 51:61-87 (1982).
18. T. O. Eloranta, Tissue distribution of S-adenosylmethionine and S-adenosylhomocysteine in the rat, Biochem. J. 166521-529 (1977).
19. D. E. Brash, and W. A. Haseltine, UV-induced mutation hotspots occur at DNA damage hotspots, Nature 298:189-192 (1982).
20. R. Cathcart, E. Schwiers, R. L. Saul, and B. N. Ames, Thymine glycol and thymidine glycol in human and rat urine: A possible assay for oxidative DNA damage, Proc. Natl. Acad. Sci. USA 81:5633-5637 (1984).
21. M. C. Hollstein, P. Brooks, S. Linn, and B. N. Ames, Hydroxymethyluracil DNA glycosylase in mammalian cells. Proc. Natl. Acad. Sci. USA 81:4003-4007 (1984).
22. B. N. Ames, R. L. Saul, E. Schwiers, R. Adelman, and R. Cathcart, Oxidative DNA damage as related to cancer and aging: the assay of thymine glycol, thymidine glycol, and hydroxymethyluracil in human and rat urine, in: Proceedings of the Symposium on "Molecular Biology of Aging: Gene Stability and Gene Expression," held on Oct. 19, 1984, New York, NY, Raven Press, New York (in press).
23. T. Lindahl, and A. Andersson, Rate of chain breakage at apurinic sites in double-stranded deoxyribonucleic acid, Biochem. 19:3618-3623 (1972).

USE OF MODELS IN LOW-DOSE EXTRAPOLATION*

John Van Ryzin

Columbia University
New York, New York 10032

INTRODUCTION

A major problem faced by regulatory agencies today is establishing acceptably low exposure levels to known or potential human carcinogens in the environment. Such regulations are required of the Environmental Protection Agency for toxic substances, of the Food and Drug Administration for food contaminants, of the Occupational Safety and Health Administration for workplace environments, and of the Consumer Products Safety Commission for hazardous consumer products.

Sometimes a given chemical is determined to be a human carcinogen based on high-dose epidemiological findings (e.g., asbestos in workers) or an animal carcinogen based on high-dose animal chronic bioassay data (e.g., saccharin in male rats). Such high-dose carcinogenic results immediately raise the question of how much risk from these carcinogens may be present at the low-dose levels commensurate with human environmental exposure.

Since the majority of such carcinogenic studies are based on chronic animal bioassays, the problem of high-dose to low-dose extrapolation is accompanied by the problem of conversion from animal data to human data. This is particularly true since the National Toxicology Program currently tests 20 to 30 chemicals per year for carcinogenicity in both rats and mice. Faced with positive carcinogenic results in animals, what should a regulatory agency do?

Putting aside the problem of converting to humans a low-dose level in the animal assessed to be a relevant risk, we discuss here the problem of extrapolating downward in the animal from high-dose to low-dose. In doing such extrapolations, mathematical models are used. We will discuss the use of such models. For a discussion of the conversion problem see the Office of Technology Assessment report[1].

One option to avoid the extrapolation problem would be simply to ban the use of such chemicals. This has been done for DDT and cyclamates. However, other known carcinogens cannot be so simply banned. Aflatoxin in peanuts, corn and wheat cannot be totally banned without seriously compromising our food supply. Past use of asbestos cannot be banned retrospectively. We must now assess the environmental risk of existing

*Research supported by National Institute of Environmental Health grant ES-02257

asbestos exposures and balance this against their costly removal. These are but two examples among many of important societal problems requiring low-dose risk assessment.

A considerable amount of work has gone into the problem of low-dose risk assessment in recent years. Overview reports exist in the Office of Science and Technology Policy statement[2], the Food Safety Council Report[3], and the Office of Technology Assessment report[1]. In this article, the problem of high-dose to low-dose extrapolation from animal carcinogenic bioassays is discussed.

LOW-DOSE EXTRAPOLATION IN ANIMAL BIOASSAYS.

The usual animal bioassay uses groups of animals (mice and/or rats typically) exposed to a chemical for a lifetime (24 to 30 months) at a variety of dose levels. These include: one group at a control dose of zero, one at the maximum tolerated dose (MTD), and other groups at various dose levels between the control and MTD, but usually within an order of magnitude or two of the MTD. At the end of the study, the number of animals exhibiting tumors of interest are recorded at each experimental dose level. Examples of such dose-response data are given in Table 1 for vinyl chloride (liver angiosarcoma in rats), for DDT (liver hepatoma in mice) and for saccharin (bladder tumors in male rats). See Van Ryzin and Rai[4] for the sources of these data.

With the use of such data one can fit a postulated mathematical dose-response curve to the data, using appropriate statistical procedures. A number of such mathematical models have been hypothesized for carcinogenic endpoints. These include the one-hit, multistage, and multihit models, based on modeling carcinogenesis through stochastic biologically occurring events over time; and the Weibull, probit, and logit models, based on tolerance distribution theory. A review of such modeling and statistical procedures can be found in Krewski and Van Ryzin.[5] However, the biological arguments favoring any one model over any other model for carcinogenic endpoints are by no means established. Hence, at this stage of knowledge of modeling and carcinogenesis none of these models can be taken as absolutely scientifically correct; nor can any of these models be ruled out.

This uncertainty in models causes a severe problem in low-dose extrapolation. Typically, several of the models will adequately fit the observed data in the experimental range, but they differ widely in their

Table 1. Three Sets of Dose-Response Data

Chemical	Dose	Dose Levels: No. of animals with Tumor/No. of Animals on Test						
DDT	parts per million in diet	0	2	10	50	250		
		4/111	4/105	11/124	13/104	60/90		
Vinyl Chloride	parts per million in air	0	50	250	500	2500	6000	
		0/58	1/59	4/59	7/59	13/59	13/60	
Saccharin	percentage in diet	0	1	3	4	5	6.25	7.5
		0/324	5/658	8/742	12/189	15/120	20/120	37/118

estimates of risk when extrapolated downward to the low-dose range. Thus, the question is: "How should one adequately use models for low-dose extrapolation?"

To illustrate this model variability in risk estimates, consider the low-dose extrapolation problem as illustrated in Figure 1 for a typical dose-response curve. The curve in Figure 1 represents a typical estimated dose-response curve using a specified mathematical model and the data from an animal experiment. The data at high-doses and control are represented by dots in Figure 1. On the horizontal axis are plotted the dose units d and on the vertical axis is the probability of tumor at dose level d. The problem of low-dose extrapolation is to come down the estimated dose-response curve to a prescribed low level of risk or probability of tumor, denoted by P in Figure 1. The chosen risk level P is usually 10^{-4} to 10^{-7} for most regulatory purposes. For example, if $P = 10^{-5}$ is chosen, then the goal is to find that dose level d which gives a probability of having one animal in 100,000 with the tumor occurrence over a lifetime. Such a dose level d leading to a prespecified risk level of, say, $P = 10^{-5}$ has been referred to in the literature as the virtually safe dose (VSD) at the risk level 10^{-5}. See Figure 1 for illustration. A lifetime risk of 10^{-5} corresponds roughly to the occurrence of 30 such tumors per year in the US population, assuming that the entire population is exposed at a dose level equivalent to that to which the animal is exposed.

Such VSD estimates are highly dependent on the mathematical model chosen for extrapolation, as well as the observed data. To illustrate this, the first four rows of Table 2 give the VSD estimates at the $P = 10^{-5}$ risk level for four models (one-hit, multistage, Weibull and a kinetic model discussed below) for the three data sets of Table 1.

To put at least one of these estimates into perspective, consider the kinetic model extrapolation given for saccharin in Table 2: 1.4×10^{-3} percent in the diet. Let us assume that a person has an exposure of 130 mg of saccharin per day (roughly the equivalent of two cans of diet soda per day, for many of the currently used diet sodas). Furthermore, assume that the person eats on the average, say, two pounds of food per day; that is, 907×10^{3} mg per day. Hence the percent of saccharin in such a person's diet would be .014 percent, that is, 1.4×10^{-2}. Since the kinetic model is linear at low doses, this exposure would lead to a risk of 1.0×10^{-4}, or one case of bladder cancer per 10,000 persons so exposed. The conversion from rat to man used here is based on an equivalent percentage in the diet, a conversion factor often used by the FDA.

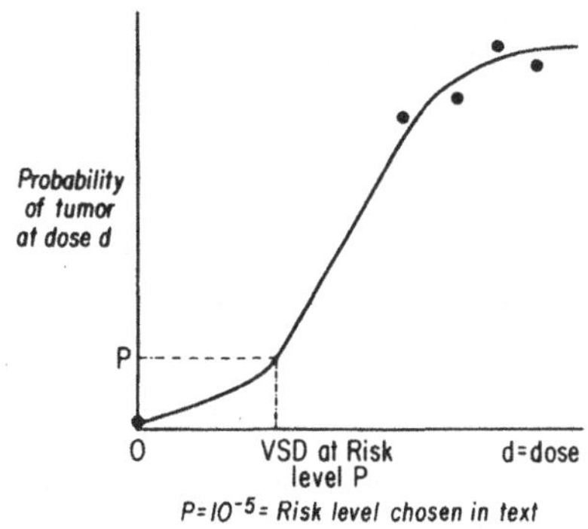

Figure 1. Low-Dose Extrapolation

Table 2. Estimated Virtually Safe Doses at the Prescribed Risk Level of $P = 10^{-5}$ for Various Methods of Low-Dose Extrapolation for the Three Data Sets of Table 1

VSD at $P = 10^{-5}$ risk

Chemical*	One-hit Model	Multistage Model	Weibull Model	Kinetic Model	Linear Extrapolation from $P = 10^{-2}$
DDT	2.8×10^{-3}	6.4×10^{-3}	8.3×10^{-2}	5.9×10^{-3}	6.2×10^{-3}
Vinyl Chloride	2.0×10^{-1}	2.0×10^{-1}	3.8×10^{-7}	2.5×10^{-2}	2.0×10^{-1}
Saccharin	4.8×10^{-4}	2.3×10^{-1}	2.8×10^{-1}	1.4×10^{-3}	2.1×10^{-3}

*For appropriate dose units for each chemical see the second column of Table 1.

As can be seen from the data of Table 2, the VSD estimates at $P = 10^{-5}$ differ considerably (up to 6 orders of magnitude) depending on which of the mathematical models is used for extrapolation. This vast uncertainty in estimates is <u>not</u> due primarily to data variability, which could be accounted for in any model by confidence interval techniques (Krewski and Van Ryzin).[5] It is inherent in the use of different dose-response models each having its unique low-dose mathematical behavior.[6] Such mathematical behavior, when extrapolated to low-dose levels, orders of magnitude below the lowest dose level above control, leads to extremely varying VSD estimates. This is because the models project the observed nonlinearity in the data to the low-dose range. Steeply upward curving data, such as for saccharin in Table 1, sharply orders the VSD estimates from lowest (one-hit) to highest (Weibull); while downward curving data, such as for vinyl chloride in Table 1, sharply orders the VSD estimates from lowest (Weibull) to highest (one-hit and multistage) (see Table 2). Roughly linear data, such as for DDT in Table 1, do not so sharply order the VSD estimates. For further discussion of this phenomenon, see Krewski and Van Ryzin[5] and Van Ryzin[6,7].

One possible way of choosing which VSD estimate to use would be to choose arbitrarily that mathematical model which fits the observed data the best and to extrapolate with that model. However, this is an inadequate solution. For the three data sets of Table 1, the multistage, Weibull, and kinetic models <u>all</u> give adequate fits to the data, based on chi-square goodness-of-fit tests. Furthermore, these three models are statistically indistinguishable from each other for these three data sets, based on likelihood ratio statistics. Yet, as can be seen from Table 2, the VSD estimates are clearly very different when these three models are extrapolated to the risk level of 10^{-5}.

Given these extremes in plausible VSD estimates based on mathematical models which are statistically indistinguishable, what should the regulator do? One solution is to always do linear extrapolation. Such an extrapolation seems defensible from a public health viewpoint as is argued here.

LINEAR EXTRAPOLATION AND ITS JUSTIFICATION

Linear extrapolation to low-dose levels can be done by using the data and models to estimate a dose level leading to an intermediate risk level, say 10^{-2}, near the experimental range followed by linear extrapolation downward to, say, 10^{-5}. For example, to extrapolate linearly from the dose level $d_{.01}$ yielding a risk of $10^{-2} = .01$ to that dose level $d_{.00001}$ yielding a risk level of $10^{-5} = .00001$, one merely divides $d_{.01}$ by 1000; that is, $d_{.00001} = (d_{.01})/1000$. The last column of Table 2 shows such calculations based on the Weibull model, which adequately fits all three data sets of Table 1, to obtain the estimate $d_{.01}$ at the 10^{-2} risk level followed by extrapolation to $d_{.00001}$ at 10^{-5} risk.

Linear extrapolation often leads to low VSD estimates compared to nonlinear models. However, it is justifiable from a public health perspective based on a variety of arguments. Peto[8] shows that any smooth dose-response curve which is increasing in dose and does not have a threshold (believable for most carcinogenic responses) will lead to low-dose linearity provided the environmental doses encountered add to similar naturally occurring background doses. Crump et al.[9] show that low-dose linear extrapolation always results from a multistage model of carcinogenesis when data uncertainty is accounted for by upper confidence limits on risk. Finally, Van Ryzin and Rai[4] show that low-dose linearity is a natural consequence of pharmacokinetic modeling of carcinogenic responses, which has recently been much supported by the literature. See, for example, Gehring and Blau[10], Hoel, et al.[11] and Van Ryzin and Rai[12]. The second last column of Table 2 gives the VSD estimates at the $P = 10^{-5}$ risk level for the three data sets of Table 1 resulting from the pharmacokinetic model introduced in the latter paper. This model fits all three of these data sets and indicates that the nonlinearities in the observed data are completely consistent with nonlinear pharmacokinetics operative at high-doses, but linear at low-doses. Note that the VSD estimates with this kinetic model in Table 2 agree to within an order of magnitude with the VSD estimates based on linear extrapolation from the 10^{-2} to the 10^{-5} risk level (last column of Table 2). For a fuller discussion of the consequences of this kinetic model on low-dose extrapolation see Van Ryzin[13].

The linear type VSD estimates in Table 2 are orders of magnitude different from those given for the highly nonlinear data sets (vinyl chloride and saccharin) by the nonlinear models (e.g., Weibull). The estimates one gets from other nonlinear models (probit, logit, and multihit) suffer similar discrepancies as the Weibull model when compared to linear extrapolation. See Krewski and Van Ryzin[5] for a further discussion of this point.

CONCLUSION

Based on the above three arguments favoring low-dose linearity, it appears that regulatory agencies in using mathematical models should use some form of linear low-dose extrapolation in assessing low-level environmental risk based on high-dose animal carcinogenicity data. That such estimates may overstate risk is quite plausible. However, to substantiate scientifically and rigorously that these low-dose risks do not exist appears beyond the scope of current animal bioassays. Thus, for the immediate future low-dose linear extrapolation in the regulatory setting will, I predict, become the accepted methodology to use.

REFERENCES

1. Office of Technology Assessment (OTA), Assessment of Technologies for Determining Cancer Risks from the Environment, Washington: U.S. Government Printing Office, (1981).
2. Office of Science and Technology Policy, Chemical carcinogens; notice of review of the science and its associated principles, Federal Register, May 22, pp. 21594-21661 (1984).

3. Food Safety Council, Proposed System for Food Safety Assessment, Washington: Food Safety Council, (1980).
4. J. Van Ryzin and K. Rai, A dose-response model incorporating Michaelis-Menten kinetics, American Statistical Assoc. Proc. Biopharmaceutical Section American Statistical Association, Washington, D.C. pp. 59-64 (1984).
5. D. Krewski and J. Van Ryzin, Dose-response models for quantal response toxicity data, in: "Statistics and Related Topics", M. Csorgo, D. Dawson, J. N. K. Rao and E. Saleh, eds., New York:North Holland, pp. 201-231 (1981).
6. J. Van Ryzin, Discussion: The assessment of low-dose carcinogenicity, Biometrics 38:130-139 (Suppl.) (1982).
7. J. Van Ryzin, Quantitative risk assessment, J. Occupat. Med. 22:321-326 (1980).
8. R. Peto, Carcinogenic effects of chronic exposure to very low levels of toxic substances, Environ. Health Perspect. 22:155-159 (1978).
9. K. S. Crump, H. Guess, and K. Deal, Confidence intervals and tests of hypotheses concerning dose response relations inferred from animal carcinogeicity data, Biometrics 33:437-451 (1977).
10. P. J. Gehring and G. E. Blau, Mechanisms of carcinogenesis: dose response, J. Environ. Pathol. Toxicol. 1:581-591 (1978).
11. D. G. Hoel, N. L. Kaplan, and M. W. Anderson, Implication of nonlinear kinetics on risk estimation in carcinogenesis, Science 219:1032-1037 (1983).
12. J. Van Ryzin and K. Rai, The use of quantal response data to make predictions, in: "The Scientific Basis of Toxicity Assessment", H. R. Witschi, ed., Amsterdam: Elsvier/North Holland Biomedical Press, pp. 273-290 (1980).
13. J. Van Ryzin, Consequences of nonlinear kinetic dose-response models on carcinogenic risk assessment, in: Banbury Report 19: "Risk Quantitation and Regulatory Policy", in press (1985).

PROBLEMS IN INTERSPECIES COMPARISONS

Richard Wilson and Edmund A. C. Crouch

Energy and Environmental Policy Center
and Physics Department
Harvard University
Cambridge, MA 02138

In discussing the risk of cancer, we are concerned with the risk to man. But we rarely have data on man - and hope to get little more, because to have data implies that exposure has been excessive. Accordingly we must use data on animals, and it is then necessary to find out how to use these data to predict human cancers.

We usually use data on rodents for two reasons. Their lifetime is short -- 2 years -- and we can get data in a reasonable time. Secondly, they are mammals, and the cell structure is in many ways similar to the cell structure in man, so that some similarity in cancer incidence can be expected.

Rodents have been used for a century of toxicological research. The threshold for acute toxicological response has been compared between animals and man, and it has been found to be similar when expressed as a fraction of body weight. Then, for purposes of prediction for public policy, we can estimate the LD_{50} in man from measurement of the LD_{50} in rodents. It seems natural to attempt a similar type of comparison for other types of response. However, when considering a chronic response such as cancer, another variable enters - time. Should we compare lifetime cancer incidence in two species, when the animals are dosed daily for life, or only compare incidence for equal elapsed time? If the lifetime of the two species is the same, it makes no difference. But cancer incidence varies as a high power of age, the fourth or higher power, so that the difference in these two recipes for comparing cancer incidence in a man and in a mouse is $(70 \text{ years}/2\text{years})^5$ or about a factor of a hundred million.

The problem of how to use the animal data is often mentioned but rarely discussed. We first realized the importance of the problem when we became dissatisfied with the logic, or rather the lack of logic, of the cancer risk assessments being performed by FDA and EPA. This led us to consider this logic from scratch. The only approach that we have found to be intellectually satisfying so far, is to use the few chemicals where we have data in both animal and man, and to ask how well someone using the animal data would have predicted the human data (Crouch and Wilson[1]; Crouch[2]). Although the data are far from precise, the lifetime cancer incidence in man is similar to the lifetime cancer incidence in rodents, if both are fed the same fraction of their body weight daily for life. Then we can make the same argument that we have used for toxic responses for a century; if the carcinogenic potency is measured in a rodent, it may be estimated in man by using this correlation.

The data can be much improved, and in this part of the paper we make some suggestions on the improvement. Since we depend upon these very few studies for calibration of our bioassay results, we must be sure that the calibration is well done.

The animal bioassays with which the human data are compared are old, and not according to the newer protocols. Until the bioassays are repeated with the newer strains and protocols, an additional source of error is introduced into the intercomparison. Another unnecessary complication is that the human exposures were often occupational, and were not uniform throughout life. The necessary extrapolation to a continuous lifetime exposure leads to additional error. This extra error could be avoided by a special bioassay where the time of the animal exposure was deliberately adjusted to match the time of the occupational exposure expressed as a fraction of a lifetime. We made these suggestions 6 years ago, and no doubt we are not unique. But the toxicological community is obsessed with the dichotomous distinction between a carcinogen and a non-carcinogen, and once a chemical is found to be carcinogenic in an animal test, no further bioassays are deemed to be worthwhile. We have repeatedly expressed our view that this is very short-sighted.

The number of chemicals in which the carcinogenic effects can be compared can be increased by considering negative epidemiology. This is because the primary purpose of a risk assessment, and the interspecies comparison included therein, is a cautious protection of the public. We want, in most cases, to be sure that we do not underestimate human risk. For example, we now know that inhalation of unleaded gasoline causes kidney tumors in rats. By examining refinery workers, we can obtain an upper limit on the number of kidney tumors in man, an upper limit which is just at the value expected by using the usual extrapolation from animals. This same argument can, no doubt, be applied to many more chemicals.

Crouch and Wilson[1] have noted, further, that better comparisons are possible between two laboratory animals, and have proceeded to make them. They have found that if the carcinogenic potency in the mouse is related to that in man by an interspecies comparison factor K, K is lognormally distributed, with a standard deviation of 1.3. Thus 2% of all values of K are greater than 20 and 2% less than 1/20. However, laboratory animals tend to have the same lifetime of 2 years, so the time comparison was not made thereby . But even if the data were improved in this respect, this "kill'em and count'em" approach would remain insensitive because of the statistical error associated with the numbers of animals used. This has led many people to search for another approach. The next part of the paper addresses this question. At the risk of losing our readers, we state at the outset that we know of no magic formula, but we outline some of the issues involved, and some attempts at their solution.

The principle behind many such attempts is to search for an intermediate biological marker to show that a cancer is being developed, and to measure this marker by chemical means which avoid the statistical uncertainties. However, many people have searched for such markers for many other purposes, of which the most important perhaps is the early prediction of a cancer with the possibility of an increased cure rate.

A study, by Armitage and Doll[3], of the age-distribution of cancer incidence and mortality led to the idea of multistage models. Overt cancers appear to occur only after a sequence of events in a single cell. The simplest assumptions about the probability per unit time of such events

- that they are constant -- lead to expected power law variations of cancer incidence with age. Roughly what is observed can be expressed as:

$$I(t) = A * t^n \tag{1}$$

where I is the incidence rate at age t, and A is a constant dependent on the tissue and tumor type. The exponent n lies in the range 2 to 11 for human cancers, and is typically 4 to 6. There are some cancers, notably hormone-dependent ones, for which this summary formula does not work, but the underlying multistage model still seems appropriate when the age dependence of hormone levels is taken into account (Pike, et al[4]).

This leads to an immediate question, how do the parameters A and n vary between species? The experimental data are sufficient to determine one of these (usually A) and it is assumed to be the same.

A simple application of these multistage ideas can lead to remarkable quantitative differences between species, but the differences depend crucially on the mechanisms assumed to be acting. If we compare a mouse and man, for which species the lifetime risks of cancer are not substantially different (less than a factor 10), we see that the constant A has to differ radically because of the differences in lifetime. If we work at the cellular level and assume that the numbers of cells in human and mouse differ by the same factor as their bodyweights, the disparity is even greater. For the same probability of tumor, we have:

$$\frac{A_{man}}{A_{mouse}} \times \frac{\text{human life}^{\,n}}{\text{mouse life}} \times \frac{\text{human mass}}{\text{mouse mass}} = 1 \tag{2}$$

With n typically about 5, this leads to a ratio of (A_{man}/A_{mouse}) of 2×10^{-11}. It seems unlikely that humans are really 10^{11} times more resistant to cancer than are mice. Referring back to individual "stages", we have to take the 5th root of this quantity, indicating that for each stage human cells are about 100 times as resistant as mouse cells. Such a factor is plausible, but even this would be reduced if the unit of time that should be considered is metabolic rate, or cell division rate, in which case the ratio is reduced to simply a factor 3 or 4.

Whatever the theoretical justification for such ideas, experimental evidence is vital. Can the rate or absolute number of DNA adducts formed be correlated with the metabolic rate of the animal tested, or with the cell division rate of the organ?

The Armitage and Doll model[3] can be extended to incorporate several further ideas about carcinogenesis. Assume that a cancer arises after a sequence of n events have occurred in a cell; e.g., a sequence of DNA damage events incorporated into the genome. If we allow for four possibilities for a given cell -- mitosis, death, another event in the sequence occurring, and repair of the previous event -- we can write the probability of one or more cells being in the state in which k events have occurred by age t as $Q_k(s,t)$ evaluated at s=0, where $Q_k(s,t)$ satisfies the partial differential equation:

$$\frac{\partial Q_k}{\partial t} - [\lambda_k (1 - e^{-s}) + \xi_k + \mu_k] \frac{Q_k}{s} = - \xi_{k-1} \frac{\partial Q_{k-1}}{\partial s}$$

$$k = 0, 1, 2 \ldots n \tag{3}$$

Here λk is the probability per unit time for binary division
μk is the probability per unit time for death
ξk is the difference between the transition rate from state k to state k+1 and the repair rate from state k+1 to state k

$$(\xi_{-1} = \xi_n = 0) \tag{4}$$

This model assumes that repair only occurs between adjacent states, but it could clearly be extended to incorporate repair of two types of damage at the same time. The transition rates between states may depend on the dose of carcinogen, the division rate, and other factors.

We present this model simply to show how many parameters are needed even to approach realism. Here we need information on 3n-2 parameters, all as a function of time (and in the case of the transition rates, as a function of dose also). With n typically about 5, there is a paucity of experimental data to fit all these 13 parameters.

Formulae including these complexities have been presented by Crump and Howe[5] and by Zeise and Crouch[6]. There is no a priori reason why any of these parameters should vary between species in any simple way. However, some of them might be related to other processes and measured indirectly[7,8].

The basis for the assumption that cancer incidence in man is related to cancer incidence in animals is that mechanisms for carcinogenesis are believed to be similar at the cellular or sub-cellular levels -- so that the same mechanisms found to operate in animals will also operate in humans. Insofar as certain parts of some mechanisms are known, this assumption appears to be borne out in practice.

The following steps must be taken into account in order to extrapolate between species, (although not all these steps may be present for any particular carcinogen):

(1) Entry of the original material into extracellular fluids and transport throughout the body.
(2) Transport through cell membranes.
(3) Receptors within the cytoplasm, and transport within the cell.
(4) Reactions producing the actual carcinogen - e.g. cytochrome P450 enzyme activation reactions.
(5) Transport of the carcinogen throughout the cell.
(6) Transport of the carcinogen across cell membranes.
(7) Transport of the carcinogen through extracellular fluids.
(8) Transport of the carcinogen through "target tissue" cell membranes.
(9) Receptors within the "target" cells, and transport throughout such cells.
(10) Action on the DNA and other cellular components.
(11) DNA repair mechanisms and their failures.
(12) Cellular division rates and damage incorporation mechanisms.
(13) Other necessary effects before a cancerous cell is generated.
(14) Immune system responses of the whole organism to the altered cells.
(15) Growth rates of damaged cells.

(1) to (3) of these represent initial entry and transport of the material of interest. (4) is the activation step (if necessary). (5) to (9) are the stages of the transport of the actual carcinogen which may differ significantly between species. (10)-(14) represent the mechanisms which go into the "microscopic" dose-response relationship.

Each one of the steps listed above might differ between species, although similar mechanisms of DNA damage and repair appear to be active in cells from all eukaryotic organisms -- albeit with differences of detail -- and it appears that DNA damage lies at the heart of carcinogenesis.

We would expect to see quantitative differences in rates of transport and in activation rates, for example, but there may also be qualitative differences. The stages (1) through (9) may combine, so that the effective "microscopic" dose to the affected cell has a diffferent relation to the externally applied dose in the two species of interest; e.g., if the enzymes necessary for the activation step are available at much lower levels in one species, they may be saturated in that species at much lower doses.

Much of the work now in progress on pharmokinetics is focussed on the study of steps 1-9. We would urge here the importance of studying them differentially and noting differences between species. In Crouch and Wilson[1], we had already noted that carcinogenic potency compares more favorably between species if the site of the tumor was not taken into account. This is particularly evident for amines, where rodents get liver tumors and dogs and men get bladder tumors. This is clearly because of differences in one or other of these steps.

The studies of DNA adduct formation indicate that the dose-response relation for adduct formation is linear, or very nearly so, over the wide range of doses from those at which tumors may be observed in experimental animals down to those to which humans may be exposed in the environment (Dunn[9]). Such a result indicates linearity in the combination of steps 1-10 above. But the adducts are not themselves on the chain of the cancer process. They are indicators of damaged DNA in cells, whereas it is far less numerous cells that go on to form cancer. Therefore, to extend the argument to the complete description of carcinogenesis requires the widely contested assumption that the fraction of cells in which DNA is repaired is independent of dose. Very small variations between the (mostly reparable) DNA adduct formation and the unrepairable DNA damage can lead to very different dose-response relations for tumor formation versus dose. Without a direct measurement of such unrepairable (or unrepaired) damage, observation of total adduct formation cannot provide sufficient sensitivity to detect possible non-linearity.

Nontheless, it would be interesting to compare the production of DNA adducts in different species. It is possible, for example, that adduct formation captures much of the variation between species.

If an animal is fed a dose of carcinogen (say 10 microgram of benzo(a)pyrene) that may be expected to produce observable numbers of tumors in some target organ (Peirce[10]; Zeise and Crouch[6]), it is observed that the number of DNA adducts formed within that target organ is of the order of 10^{10}. The fraction of animals getting tumors in the target organ at such doses is of the order of 0.1, so there is a ratio of about 10^{11} between adduct formation and cancer as an end point (Dunn[9]).

A simplified example should make this clear. Suppose that there is one particular non-repairable site on the DNA, at which adduct formation necessarily leads to a tumor (or which is linearly related to tumor formation). The dose-response relation for adduct formation at this site (and hence for tumor formation) could be highly non-linear, for example:

$$n = (d/d_0)^3 \qquad (5)$$

where n is the total number (in the target organ) of such sites affected at dose d, and d_0 is a typical dose which causes observable tumors. If there are also other, repairable sites with a linear dose-response relation:

$$N = 10^{11} * (d/d_0) \qquad (6)$$

where N is the total number of sites in the organ affected, then the overall dose-response relation between total adducts and dose would be simply:

$$\text{Total adducts} = 10^{11} * (d/d_0) + (d/d_0)^3 \qquad (7)$$

which is experimentally indistinguishable from linearity (for doses below $10^4 * d_0$, which would probably kill the animal outright).

This example is given solely to point out the possibility that there may be non-linear dose-response relations hidden in the overall linear observations of adduct formation. It should be evident that one could substitute practically any non-linear term for the cubic term above, with the same result. The enormous factor of 10^{11} between the total adduct formation and the final response of interest can hide practically anything. In particular the mechanism of this example is not the only one which might be operating to cause a non-linear carcinogenic response even with linear DNA adduct formation. There is some evidence for a two step mechanism*, with cell division fixing the DNA damage in some way. If the unrepairable DNA damage is produced linearly but at a low rate, but the rate of cell division is affected by dose, then one could also obtain a non-linear dose-response relation for carcinogenesis.

There are many mechanisms involved in getting the ultimate carcinogen to the target sites on the DNA, many of which could be non-linear. Active transport within the cell, and especially any activation steps, are two mechanisms which immediately spring to mind. The observed linearity of adduct formation can be interpreted as showing that in practice the "effective dose" is indeed proportional to the dose applied to the animal. These ideas should be contrasted with those suggested by Hoel *et al.*[11], who suggest that the relation between DNA adducts (possibly specific adducts rather than total adducts) is likely to be linear, but that the relation between applied dose and the "effective dose" (measured by DNA adduct concentration) might be non-linear because of non-linear kinetic mechanisms.

As we consider step 10 by itself, we note that it is widely believed that this first step in forming a cancer, usually called initiation, is damage of DNA, and therefore is akin to mutagenesis. This has led to a hope that there could be a correlation between carcinogenic potency, and mutagenic potency in an *in vitro* test, such as that developed by Ames. There was an initial success along these lines by Meselson and Russell[12], but it now appears that the correlation exists only for a limited class of chemicals. Although disappointing, this does not invalidate a comparison between the step 10 of carcinogenesis noted above and mutagenesis, because the other stages may be very different for the two species.

Other biological markers may be located further along the cancer process. A study of oncogenes suggest that they may, at some future time, become a good marker.

The interspecies comparison problem would then take on a different color. If, for example we can convince ourselves that cancer incidence is proportional to the average oncogene concentration, we could measure this concentration in both animal and man at doses which are low enough to be ethical to inflict upon people, and use this comparison to predict human cancers from animal ones.

In conclusion, we note that attempts to find better ways of interspecies conversion all imply a model, from which some of the parameters can be determined by experiments which do not have the usual statistical error.

REFERENCES

1. E. A. C. Crouch, and R. Wilson, Interspecies comparison of carcinogenic potency, J. Tox. Env. Hlth. 5:1095-1118 (1979).
2. E. A. C. Crouch, Uncertainties in interspecies extrapolation of carcinogenicity, Env. Hlth. Persp. 50:327 (1983).
3. P. Armitage, and R. Doll, The age distribution of cancer and a multi-stage theory of carcinogenesis, Brit. J. Cancer 8:1-12 (1954).
4. M. C. Pike, M. D. Krailo, B. C. Henderson, J. T. Casagrande, and D. G. Hoel, Hormonal risk factors, 'Breast tissue age' and the age of incidence of breast cancer, Nature 303:767 (1983).
5. K. S. Crump, and R. B. Howe, The multistage model with a time-dependent dose pattern: Applications to carcinogenic risk assessment, Risk Analysis 4163 (1984).
6. L. Zeise, and E. A. C. Crouch, Benzo(a)pyrene paper, in preparation (1985).
7. E. Farber, The multistep nature of cancer development, Cancer Res. 44:4217-4223 (1984).
8. E. Farber, Cellular biochemistry of the stepwise development of cancer with chemicals, C. H. A. Clowes Memorial Lecture, Cancer Res. 44:5463-5474 (1984).
9. B. P. Dunn, Wide-range linear dose-response curve for DNA binding of orally administered benzo(a)pyrene in mice, Cancer Res. 43:2654-2658 (1983).
10. W. E. H. Peirce, Tumour-promotion by lime oil in the mouse forestomach, Nature 189:497-498 (1961).
11. D. G. Hoel, N. L. Kaplan, and M. W. Anderson, Implication of nonlinear kinetics on risk estimation in carcinogenesis, Science 219:1032-1037 (1983).
12. M. Meselson, and K. Russell, Comparison of carcinogenic and mutagenic potency, in: "Origins of Human Cancer", H. H. Hiatt, J. D. Watson, and J. A. Winstein, eds., pp. 1473-1482, Cold Spring Harbor, NY (1977).

THE TIME TO TUMOR APPROACH IN RISK ASSESSMENT

Roy E. Albert

Institute of Environmental Medicine
New York University Medical Center
550 First Avenue
New York, N.Y. 10016 USA

ABSTRACT

The multistage model is currently a widely used mathematical tool in carcinogen risk assessment to obtain a low-dose linear non-threshold slope for estimating cancer risks and comparing carcinogens with respect to potency. However, the multistage model is a single pathway model, whereas biological evidence indicates that carcinogenesis proceeds through multiple pathways. Furthermore, recent studies suggest that carcinogens induce a generalized increase in the susceptibility to neoplastic transformation triggered as rare events by cell proliferation. Such evidence supports the $dt^n = c$ time to tumor model in a modified form.

When assessment of carcinogenic chemicals was begun by the Carcinogen Assessment Group (CAG) at the U.S. Environmental Protection Agency in 1976, the use of low-level non-threshold linear extrapolation was taken over from the field of ionizing radiation on the basis of the mechanistic linkage between carcinogenesis and mutagenesis, the linearity of dose-response for mutagenesis, and the consistency with linearity of at least some epidemiological dose-response data for cancer[1]. Additional arguments were added in support of low-dose linearity for chemicals: non-threshold dose-response linearity for tumor initiation[2] and virtual low-dose linearity when the mode of action of the agent in question and the background causes are the same; as indicated below, this is inherent in the multistage model of carcinogenesis.

In recent years, there has been growing skepticism of the rationale for non-threshold low-dose linearity on the grounds that much of it is speculative. Indeed, Federal Agencies have taken the position that low dose linearity is a fall back position when no more plausible mechanistic rationale exists[3,4]. However, the lack of any other accepted approach has left the dominance of non-threshold low-dose linear extrapolation unchallenged.

About 5 years ago, the practice of drawing a straight line through the origin of the dose-response plot from the lowest statistically significant response was dropped. The multistage model was adopted by the CAG primarily on the grounds that it made better use of the data in establishing the low-dose linear slope. However, the multistage model does have biological implications. It was proposed more than 30 years ago as an explanation for the observation that the incidence rate of many cancers in human populations is proportional to integer powers of age[5]. According to the multistage theory, a certain number of cellular changes are necessary to transform a normal cell into a malignant one. The transformed cells clone out to become a detectable cancer. Armitage and Doll introduced the assumption that the cellular changes had to occur in a predetermined order[6]. Fisher pointed out that clonal expansion of partially transformed cells reduces the number of cellular transitions that are needed for neoplastic transformation[7]. As the multistage model is currently used, each stage has a transition rate which is the sum of the transition rate from the background causes of cancer and a transition rate which is linearly dependent on carcinogen dose; thus, the carcinogen and background processes are working together. The shape of the dose-response curve with this form of the multistage model can have linear, quadratic, and higher order components.

In practice, the multistage model continues to be a widely used mathematical tool: it is flexible and can accommodate most any data set. The low-dose linear slope is a simple way to compare potencies of different carcinogens and to estimate individual excess cancer risks and numbers of excess cancer deaths for given exposures. Technical details are still being ironed out. For example, the number of stages to be used in any given data set: Should there be one stage less than the number of doses or the fewest stages that provide a reasonable fit to the data? Should constraints that are inherent in the mathematics of the multistage model be applied to linear term; for example, if two stages are affected by the carcinogen, the linear term must be greater than two times the square root of the background term times the quadratic term[8]. There is also the issue of whether to use maximum likelihood, average likelihood, or upper 95% confidence limit for the linear slope.

Although the multistage model is in widespread use for risk assessment, there is a concern about whether the model is biologically sound. The model postulates a single pathway from normal to neoplastic transformation, yet biologically there can be multiple pathways. This is evident from our own studies with benzo(a)pyrene (B(a)P) in the mouse skin, where chronic application produces carcinomas most all of which appear de novo, that is, without first appearing as benign papillomas; however, when B(a)P is combined with the promoting agent, phorbol myristate acetate, the first lesions are all papillomas which take months to become cancerous[2]. This is a case where there is not one but two pathways to cancer. Indeed, Foulds whose classic studies on the multistaged nature of cancer progression emphasized that cancer usually progresses along multiple pathways[9]. According to Foulds, structure and behavior of tumors are determined by numerous characters, that within wide limits are independently variable, capable of highly varied combinations and assortments, and liable to independent progression. This accounts for the extraordinary variety of tumors of any given type that are commonly seen and the fact that generally the first clinical manifestation of a tumor may be at any stage of progression. Foulds illustrated the idea of multiple pathways in Figure 1, where lesions of increasing gravity x, y and z arise from a region of "incipient neoplasia"; most lesions arise concurrently but there can be consecutive progression[9]. Foulds invoked his concept of regional incipient neoplasia to deal with the clinically silent latent period. His is basically a multiple pathway model which is different from the single pathway multistage model used for risk assessment.

Kennedy and Little[10,11] have obtained experimental support for Foulds' concept of regional incipient neoplasia and for the role of cell proliferation in neoplasia. They have demonstrated in a series of studies with cultured mouse embryo cells that radiation induces a frequent altera-

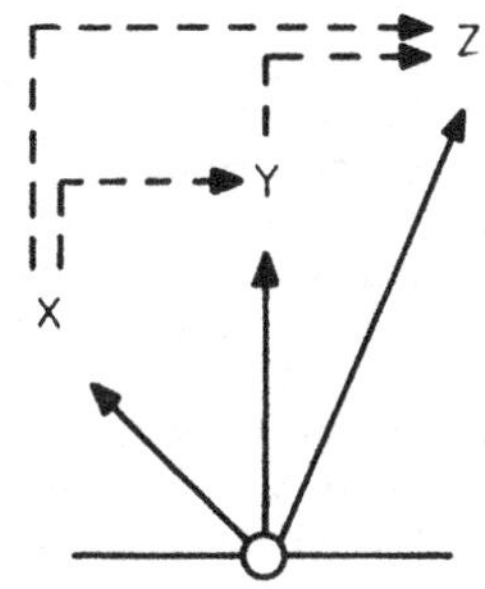

Figure 1. The Foulds concept of multiple pathways in cancer progression. The open circle illustrated incipient neoplasia which is transformed into tumors of graded malignancy x, y, and z. In addition, there can be progression in tumors from lower to higher degrees of malignancy.

tion among a large fraction of the irradiated cells; this generalized alteration is translated into malignant transformation as a rare event by cell proliferation. They postulate that cells respond to DNA damage by the induction of a commonly occurring process that enhances the spontaneous transformation frequency during cell replication. A similar model was proposed by Fernandez and Heidelberger on the basis of the response of cells in culture to chemical carcinogens[12]. Little points out that there are a number of studies in animals that support such a conclusion in rat tracheal epithelial cells[13], mouse mammary cells[14] and thyroid cells[15]. Each of these studies indicate that a large proportion of the carcinogen treated population (5-10%) possess malignant potential in terms of their ability to give rise to either transformed progeny <u>in vitro</u> or malignant tumors <u>in vivo</u> even though only an occasional tumor may arise in situ[16].

As an alternative to the single pathway multistage model, one might consider the combination of the Foulds concept of incipient neoplasia and concurrent with consecutive progression together with the Kennedy and Little evidence highlighting the essential role of cell proliferation in converting incipient neoplasia to neoplastic transformation. This model is described in Figure 2. Here, the normal cell is converted by the carcinogen to a preneoplastic stage characterized by genetic instability which makes it susceptible to neoplastic transformation by proliferation. The degree of malignancy of the induced tumors is indicated as variable with provision for sequential progression to higher degrees of malignancy. The frequency distribution of tumors, according to the degree of malignancy, may vary with target tissue, carcinogen, and dose. For example, at moder-

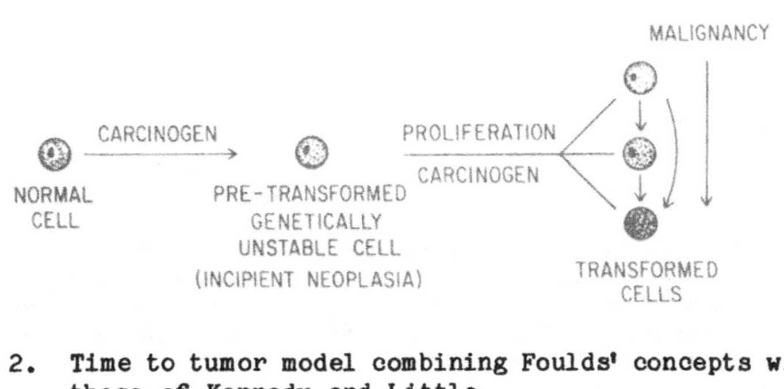

Figure 2. Time to tumor model combining Foulds' concepts with those of Kennedy and Little.

ate doses of ionizing radiation in the rat skin, tumors with a wide variety of growth rates are induced; these growth rates are somewhat higher at high doses but still variable[17]. In the mouse skin with B(a)P, as indicated above, most of the tumors are quite malignant[2]. A possible role of additional carcinogen exposure for converting preneoplastic cells to neoplasia is included in Figure 2 on the grounds that some accelerated progression of papillomas to carcinomas has been observed in the mouse skin with exposure of the papillomas to direct-acting carcinogens[18], and by inference carcinogens may enhance the conversion of preneoplastic to neoplastic cells. However, the main emphasis is on proliferation.

The concept of cell proliferation as the dominant factor that converts incipient neoplasia into neoplasia suggests that the essential feature of carcinogenic response is the time required for tumor development and that the incidence is a secondary consequence of intercurrent mortality which determines the extent to which the carcinogenic process is allowed to be expressed. This biological time-to-tumor model, shown in Figure 3, in terms of the dose-response relationships for chemical DNA adducts, cell proliferation, and time to tumor, can be divided into 3 zones: Zone A is the dose region where adducts increase, shown as linear, and there is little effect on histology and proliferation rate; here, the time to tumor decreases slowly with increasing carcinogen dose. Zone B is the region of cell killing and increasing proliferation which together with increasing levels of DNA adducts rapidly shortens the time-to-tumor with increasing carcinogen dose. Zone C is the region of very high dose where cell proliferation is maximal and cell killing is so high as to kill transformed cells in a therapeutic fashion; in this zone there is no further shortening of time-to-tumor. Thus, Zone A is defined by a normal proliferation rate, Zone B by an increasing proliferation rate. The shape of the time to tumor curve in Zone A is defined by the shape of the dose-adduct or DNA damage curve and the normal rate of proliferation; in Zone B it is defined by the dose-response relationships for both DNA damage and cell proliferation. Swenberg's data on cell proliferation in the nasal mucosa fits this model since in the dose region of rapidly increasing nasal cancer response, there is a linear increase in DNA-protein cross links and a rapid increase in the rate of cell proliferation[19].

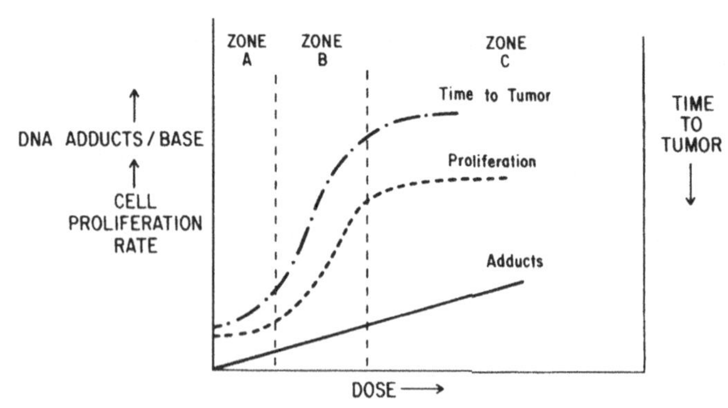

Figure 3. The time to tumor model is shown in relation to the dose response for DNA adducts, cell proliferation and time to tumor.

The above biological time-to-tumor model supports a modified version of the Blum-Druckrey time-to-tumor model for carcinogenesis which was evaluated by Albert and Altshuler in the context of risk assessment[20]. It is one of the cardinal rules of carcinogenic response that, except for special circumstances, the higher the dose, the earlier the occurrence of tumors. With the Blum-Druckrey model, the temporal pattern of tumor occurrence at any chronically administered dose rate can be described by a log-normal distribution. Examples in humans and experimental animals have been shown[20].

The relationship between median time of tumor induction, t, and dose rate, d, is dt^n = constant. The $dt^n = c$ relationship is for the single risk of cancer. Tumor incidence and life shortening are obtained by interacting these log-normal curves with survival. The shape of the dose-incidence curve depends on the value of n with low values of n resembling higher powers of dose and high values of n resembling low powers of dose[20]. It has been pointed out that n is the equivalent of the ratio of the power of time to the power of dose[21].

Biological considerations and experimental observations indicate that the $dt^n = c$ formulation does not hold over the entire range of carcinogen response, namely, that n is not constant at the extremes of response. It is well established that there is a minimum cancer time that cannot be shortened with higher carcinogen exposures and, in fact, increasingly high doses will actually reverse the shortening of the time of appearance. This corresponds to Zone C in the biological time-to-tumor model in Figure 3. Hence, at very high carcinogen dose rates, the value of n rises to infinity. At low dose rates corresponding to Zone A, in Figure 3, the value of n should also be high since the time-to-tumor decreases slowly and the dose-response tends toward linearity. There is experimental evidence suggesting this response pattern. Druckrey's study on liver tumor induction in the rat by diethylnitrosamine (DENA) showed a value for n of 2-3[22]. Peto extended Druckrey's study to lower dose levels where the dose-response was linear[23]. This implies a higher value of n, perhaps, 5-6. Hence, at high and low carcinogen dose rates the value of n tends to be high.

In summary, it might be said that the linearized multistage model is a useful analytic tool for quantitative carcinogen assessment but there is concern about the soundness of its biological basis. Recent developments showing evidence for a preneoplastic state which is converted to neoplasia by cell proliferation raises the possibility that time-to-tumor is the central measure of carcinogenic response and that tumor incidence is a secondary consequence which is dependent on competing causes of death. Further consideration of a modified form of the $dt^n = c$ model, where n is variable, as a basis for quantitative risk assessment may be a step toward finding a more biologically sound basis for the estimation of cancer risks than now exists.

REFERENCES

1. R. E. Albert, R. E. Train, and E. Anderson, Rationale developed by the Environmental Protection Agency for the Assessment of Carcinogenic Risks, J. Natl. Cancer Inst. 58(5): 1537-1541 (May 1977).
2. F. J. Burns and R. E. Albert, Mouse skin papillomas as early stages of carcinogenesis, J. Amer. Coll. Toxicol. 1(1):29-45 (1982).
3. Office of Science and Technology Policy, Chemical carcinogens; A review of the science and its associated principles, February 1985, Part II, Federal Register 50(50):10372-1-442 (March 1985).
4. Environmental Protection Agency, Proposed Guidelines for Carcinogen Risk Assessment; Request for Comments, Part VII, Federal Register 49:46294-46301 (November 1984).
5. C. O. Nordling, A new theory on the cancer inducing mechanism, Br. J. Cancer 7:68-72 (1953).
6. P. Armitage and R. Doll, The age distribution of cancer and a multi-stage theory of the cancer producing mechanism, Br. J. Cancer 7:407-417 (1953).

7. J. C. Fisher, Multiple-mutation theory of carcinogenesis, Nature 181:651-652 (1957).
8. T. Thorslund, Personal communication (June 1985).
9. L. Foulds, "Neoplastic Development, Vol. 1," Academic Press, London and New York (1969).
10. A. R. Kennedy and J. B. Little, Investigation of the mechanism for enhancement of radiation transformation in vitro by 12-0-tetradecanoylphorbol-13-acetate, Carcinogenesis 1:1039-1047 (1980).
11. A. R. Kennedy and J. B. Little, Evidence indicating that the second step in x-ray induced transformation in vitro occurs during cellular proliferation, Radiat. Res. 99:228-248 (1984).
12. A. Fernandez, S. Mondal, and C. Heidelberger, Probabilistic view of the transformation of cultured C3H/10T1/2 mouse embryo fibroblasts by 3-methyl-cholanthrene, Proc. Natl. Acad. Sci. USA 77:7272-7276 (1980).
13. M. Terzaghi and P. Nettesheim, Dynamics of neoplastic development in carcinogen-exposed tracheal mucosa, Cancer Res. 39:4003-4010 (1979).
14. K. H. Clifton, K. Kamiya, R. T. Mulcahy, and M. N. Gould, Radiogenic neoplasia in the thyroid and mammary clonogens: progress, problems and possibilities, in: "Symposium Proceedings, Estimation of Risk from Low Doses of Radiation and Chemicals: A Critical Overview," Brookhaven National Laboratory, New York (1984).
15. R. T. Mulcahy, M. N. Gould, and K. H. Clifton, Radiogenic initiation of thyroid cancer: a common cellular event, Int. J. Radiat. Biol. 45:419-426 (1984).
16. J. B. Little, The radiobiology of in vitro neoplastic transformation, "Radiation Carcinogenesis and DNA Alterations," F. J. Burns, A. C. Upton, and G. Silini, eds., Plenum Press, New York, submitted.
17. R. E. Albert, M. E. Phillips, P. Bennett, F. Burns, and R. Heimbach, The morphology and growth characteristics of radiation-induced epithelial skin tumors in the rat, Cancer Res. 39:658-688 (1969).
18. H. Hennings, R. Shores, M. L. Wenk, E. F. Spangler, R. Tarone, and S. H. Yuspa, Malignant conversion of mouse skin tumours is increased by tumour initiators and unaffected by tumour promoters, Nature 304:67-67 (1983).
19. J. A. Swenberg, E. A. Gross, and H. W. Randall, Localization and quantitation of cell proliferation following exposure to nasal irritants, in: "Toxicity of the Nasal Passages," C. S. Barrow, ed., Hemisphere Publishing Corporation, New York, in press.
20. R. E. Albert and B. Altshuler, Considerations relating to the formulation of limits for unavoidable population exposures to environmental carcinogens, in: "Radionuclide Carcinogenesis," C. L. Sanders, R. H. Busch, J. E. Ballou, and D. D. Mahlum, eds., AEC Symposium Series, CONF-720505, NTIS, Virginia (1973), pp. 233-253.
21. P. Emmelot and E. Scherer, Multi-hit kinetics of tumor formation, with special reference to experimental liver and human lung carcinogenesis and some general conclusions, Cancer Res. 37:1702-1708 (1977).
22. H. Druckrey, Quantitative aspects in chemical carcinogenesis, in: "Potential Carcinogenic Hazards from Drugs. Evaluation of Risks," R. Truhaut, ed., UICC Monograph Series, Vol. 7, Springer-Verlag, Berlin (1967), pp. 60-77.
23. R. Peto, R. Gray, P. Brantom and P. Grasso, Nitrosamine carcinogenesis in 5120 rodents: chronic administration of sixteen different concentrations of NDEA, NDMA, NPYR and NPIP in the water of 4440 inbred rats, with parallel studies on NDEA alone of the effects of age of starting (3,6 or 20 weeks) and of species (rats, mice or hamsters), in: "N-Nitroso Compounds: Occurrence Biological Effects and Relevance to Human Cancer," IARC Publication No. 57 pp. 627-665 (1984).

THE ROLE OF NUTRITIONAL FACTORS IN CELLULAR PROTECTION AGAINST DNA DAMAGE, ALTERED GENE EXPRESSION AND MALIGNANT TRANSFORMATION

Carmia Borek

Departments of Radiology and Pathology
Columbia University
College of Physicians & Surgeons
New York, New York 10032

INTRODUCTION

In recent years data from epidemiological studies and laboratory experiments have revealed numerous links between diet and cancer[1].

The complex role of nutritional factors in modifying cancer incidence may be attributed in part to several modes of action. Dietary agents can act a) as potentiators or promoters of cancer, serving as auxilliary agents to other environmental factors, b) as causes of cancer, or c) as cancer preventive agents[2,3].

While we aim to identify (a) and (b), we try to focus on (c) and determine the roles and mechanisms of dietary factors in cancer prevention.

Such studies can be carried on in vitro, in cell culture systems, where malignant transformation serves as an end point[4-6]. These systems afford us the opportunity to study the direct effect of nutrition in oncogenesis and the role of dietary factors in modulating the frequency and course of neoplastic development in its various stages.

CELLULAR TRANSFORMATION BY NUTRITIONAL DEPRIVATION

Hepatocytes in vitro have served as a model in our studies on the direct role of nutrition and cancer. When rat hepatocytes are subjected to conditions of nutritional stress by growth in exhausted medium they transform into hepatoma-like cells. The transformed cells lose their intercellular communication via permeable junctions, and display aneuploidy and a loss in their ability to synthesize high-molecular weight gangliosides. The cells grow in agar and form tumors in the animal[7,8].

Parallel cultures maintained under optimal nutritional conditions with a bi-weekly change of medium do not transform (Fig. 1) and remain diploid.

Recent studies (Borek unpublished) indicated that the progression of transformation can be suppressed in part by nutritional supplementation within 2 weeks of culture and that there exists a clonal variation in the cellular response to malignant progression.

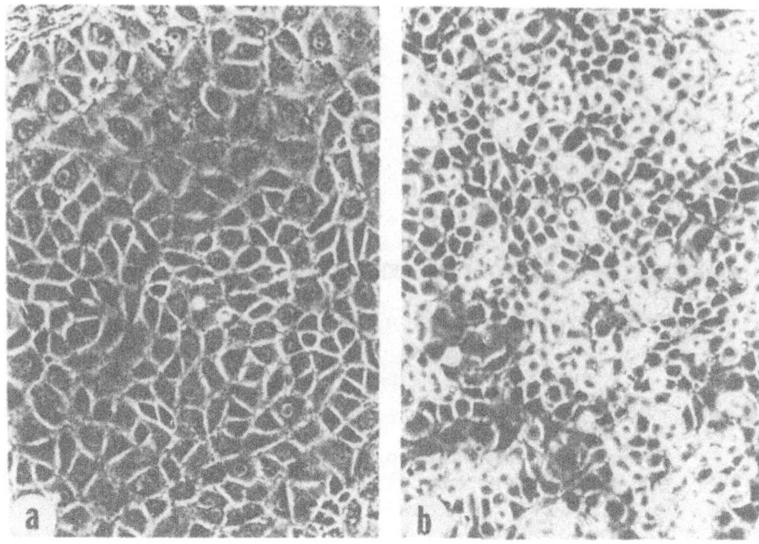

Figure 1. Normal rat liver cells (a) transformed in vitro under nutritional stress (b).

FREE RADICALS, DNA DAMAGE AND MALIGNANT TRANSFORMATION

Cellular malignant transformation which can be studied in epithelial cells or fibroblasts is initiated by hereditary damage to DNA, resulting in abnormal expression of cellular genes. The types of DNA damage may include mutations, gene amplifications, altered methylation or oncogene activation[6].

The sources of damage include radiation, chemical carcinogens and tumor promoters, capable of producing free radicals by interaction with organic matter. The free radicals interact with cellular lipids, protein and nucleic acids, causing lipid peroxidation, protein oxidation and cross linking of DNA. A cascade of free radical-mediated events ensues, resulting in toxic products, oxidant stress, loss in cellular integrity, and altered gene expression, which may culminate in a neoplastic phenotype[5].

Free radicals are inevitable by-products of aerobic metabolism. Thus, reactive oxygen derivatives including the superoxide anion radical or hydrogen peroxide are generated in mitochondria by single electron transfer to molecular oxygen during the process of electron transport in respiration.

Under optimal cellular conditions a balance prevails between the oxidative challenge, a source of DNA damage, and the inherent mechanisms which include enzymes and nutrients that suppress the damaging action of free radicals[9,10]. Some nutritional protective agents can act directly by scavenging the free radicals and preventing peroxidation and autoxidation of macromolecules. Vitamin A, β carotene, non protein thiols (glutathione) and vitamin E come into this category[11]. These nutritional defense sources are complemented by enzymatic factors, such as superoxide dismutase which dismutates oxygen free radicals to hydrogen peroxide, catalase which detoxifies peroxide to water, as well as peroxidases such as the glutathione dependent peroxidase and glutathione transferase which break down peroxidases formed in the course of free radical-mediated lipid peroxidation. These prevent a continued cascade of a variety of free radicals (including lipid radicals) which may damage DNA and protein and alter the genetic apparatus.

We have shown that a variety of retinoids inhibit transformation by radiation (x rays) and chemicals (Benzo(a)Pyrene and the tryptophan pyrolysate carcinogens derived from broiled protein foods). They suppress the synergism between radiation and the pyrolysate in transforming cells and inhibit the promoting action of TPA[2]. Vitamin C, another powerful scavenger of free radicals, is also an inhibitor of transformation _in vitro_[11,12].

Another nutrient important in controlling free radical damage is selenium[2,13,14]. Epidemiological evidence has shown that areas deficient in selenium have an increased cancer incidence. Our work _in vitro_ has enabled us to show that selenium is a radioprotector and a chemoprotector at the cellular level[2,13,14]. We found that its mechanism of action in inhibiting transformation is mediated via induction of cellular thiols, catalase, and glutathione peroxidase. Selenium thus acts by elevating cellular scavenging systems enabling a doubling of peroxide and preventing peroxide-mediated damage to cellular macromolecules[2,13,14]. Our studies also show that vitamin E inhibits radiogenic and chemically induced transformation [14]. While the inhibitory action is additive with that of selenium, vitamin E, a known antioxidant, does not modify the enzymes studied above[14].

BISULFITES AS INHIBITORS OF TRANSFORMATION

Since transformation involves damage to DNA and its metabolism[6,14,15], agents which modify transformation must act either directly or indirectly to prevent DNA damage. Bisulfites have long been used as food additives and as enhancers of flavor in beer. We find that they may actually be beneficial.

In recent studies[16] we have shown that bisulfites inhibit transformation in hamster and mouse cells (Table 2). While we do not know the molecular mechanism underlying their action, they could be acting as scavengers, a possibility which may underlie their action in inhibiting mutagenesis.

PROTEASE INHIBITORS

Protease inhibitors of high and low molecular weight prevail in many protein foods. Earlier studies have shown that a variety of protease inhibitors, including antipain, inhibits radiogenic transformation _in vitro_[6,15].

Our current experiments indicate that low-molecular weight protease inhibitors, but not the high-molecular weight ones, suppress poly(ADP)ribose polymerase, thus having a direct effect on the consequences of DNA damage[17] (Table 3).

Poly(ADP)ribose polymerase action is triggered by DNA damage induced by radiation and a variety of alkylating chemicals[18]. The events that follow result in the transfer of the (ADP)ribose moieties of NAD and their polymerization and covalent binding to a variety of nuclear and non nuclear proteins. The resulting effect is altered gene expression[18-20].

The role of poly(ADP)ribose in transformation has been shown in earlier studies by the fact that benzamides at low concentrations, which specifically suppress poly (ADP)polymerase, also inhibit transformation induced by x ray, UV light[19] and methylating chemical carcinogens[19,20]

Table 1. Transformation, Glutathione Peroxidase (GSH), Catalase and Non-Protein Thiols (NPSH) in Selenium Pre-treated and Untreated C3H 10 T1/2 Cells.

	Untreated	Selenium Treated*
Transformation by 400 rad X-ray	9.8×10^{-4}	1.8×10^{-4}
Transformation by B(a)P 1.2μg/ml	11.6×10^{-4}	1.6×10^{-4}
GSH peroxidase	5.6	17.7
Catalase**	4.2	9.0
NPSH+	10	21

* 2.5μM $NaSeO_3$
**n moles H_2O_2 reduced/min/mg protein
+n moles/mg protein.

From reference 14.

Thus our recent findings that some food products such as low-molecular weight protease inhibitors can directly modify the consequences of DNA damage have important implications in cancer prevention.

FINAL WORD

There is no complete escape from ubiquitous physical and chemical agents which prevail in our environment. These can induce DNA damage directly or indirectly. Some of these damaging processes are mediated via free radicals and oxidant stress which result in cascades of free radical mediated events and progressive deterioration of cellular structure and function. Such damage can trigger oncogenic events. Carcinogens as well as tumor promoters which produce free radicals in the cell can therefore challenge cellular integrity by DNA or protein modification, resulting in abnormal expression of cellular genes.

We see, however, that cells have powerful antioxidant defense mechanisms which under optimal conditions monitor oxidant stress and prevent its consequences. Under excessive oxidant stress, such as exposure to a variety of physical or chemical environmental pollutants

Table 2. Effect of Bisulfite on C_3H 10T1/2 Cell Transformation

Treatment	Transformation/ surviving cell	Transformation/surviving cell ($\pm$ S.E.x10^{-4})
Control	0/37500	0
Bisulfite (5.0 μg/ml)	0/47064	0
400 rad	28/41028	6.8 $\pm$ 0.51
400 rad + bisulfites (5.0 μg/ml)	12/45600	2.6 $\pm$ 0.23
Benzo(a)pyrene (B(a)P)	23/36300	6.3 $\pm$ 0.71
B(a)P + bisulfite (5.0 μg/ml)	10/43624	2.3 $\pm$ 0.21

From[16]

Table 3. Protease Inhibitors which Suppress Poly (ADP)* Ribosylation and Transformation.

Protease Inhibitor	Effect on Transformation	Effect on Poly (ADP) Ribosylation
Antipain	Inhibition	Suppression
Leupeptin	Inhibition	Suppression
Soybean	Inhibition	No Effect
Bowman	Inhibition	No Effect

*From reference 19.

and some dietary carcinogens[3], one can enhance cellular defenses. By intake of appropriate and adequate vitamins and micronutrients -- and, as we find now, low molecular weight protease inhibitors -- one can either inhibit free radical-mediated damage to DNA and other cellular macromolecules or, as in the case of the low-molecular weight protease inhibitors, prevent the consequences of DNA damage if it has already taken place.

This article was supported by a contract from the National Foundation for Cancer Research and Grant no. CA 12536 from the National Cancer Institute.

REFERENCES

1. R. Doll, and R. Petro, The causes of cancer: Quantitative estimates of avoidable risks of cancer in the United Stated today, J. Nat. Cancer Inst. 66:1191-1308 (1981).
2. C. Borek, Vitamins and micronutrients modify carcinogenesis and tumor promotion in vitro, in: Molecular Interrelations of Nutrition and Cancer, M. S. Arnott, J. Van Eys, and Y. M. Wang, eds., pp. 337-350, Raven Press, New York, (1982).
3. B. N. Ames, Dietary carcinogens and anticarcinogens, Oxygen radicals and degenerative diseases, Science 211:1256-1264 (1983).
4. C. Borek, Radiation oncogenesis in cell culture, Adv. Cancer Res. 37:159-232 (1982).
5. C. Borek, In vitro cell cultures as tools in the study of free radicals and free radical modifiers in carcinogenesis, in: Methods in Enzymology, Volume on Oxygen Radicals in Biological Systems, pp. 465-479, C. P. Colowick, N. O. Kaplan, and L. Packer, eds., Academic Press, New York (1984).
6. C. Borek, The induction and control of radiogenic transformation in vitro: Cellular and molecular mechanisms, J. Pharmac. Therap 27:99-142 (1985).
7. C. Borek, Neoplastic transformation in vitro of a clone of adult liver epithelial cells into differentiated hepatoma-like cells under conditions of nutritional stress, Proc. Nat. Acad. Sci. (USA) 69:956-959 (1972).
8. C. Borek, Differentiation, metabolic activation and malignant transformation in cultured liver cells exposed to chemical carcinogens, in: Advances in Modern Environmental Toxicology, Vol. 1, pp. 297-318, N. Mishra, V. Dunkel, and M. A. Mehlman, eds., Senate Press, Princeton, New Jersey (1981).
9. Radioprotectors and Anticarcinogens, O. F. Myaard, and M. G. Simic, eds., Academic Press, New York (1983).

10. W. A. Pryor, The role of free radical reactions in biological systems, in: Free Radicals in Biology, Vol. 1, pp. 1-49, W. A. Pryor, ed., Academic Press, New York (1976).
11. C. Borek, Free radicals dietary antioxidants and mechanisms in cancer prevention; in vitro studies, in: Nutrition and Cancer Proceedings of the Second International Symposium on Modulation and Mediation of Cancer by Vitamins and Micronutrients, Tuscan AZ, F. Meyskens, and K. Prasad, eds., Humana Press, New Jersey (in press) (1985).
12. W. F. Benedict, W. L. Wheatley, and P. A. Jones, Inhibition of chemically induced morphological transformation and reversion of transformed phenotype by ascorbic acid in C3H 10T1/2 cells, Cancer Res. 402796-2801 (1980).
13. C. Borek, and J. E. Biaglow, Factors controlling cellular peroxide breakdown: Relevance to selenium protection against radiation and chemically induced carcinogenesis, Proc. Am. Ass. Cancer Res. 25:125 (abstract), (1984).
14. C. Borek, A. Ong, L. Donahue, and J. E. Biaglow, Selenium and vitamin E inhibit radiogenic and chemically induced transformation in vitro via different mechanisms of action, Pro. Nat. Acad. Sci. (1985).
15. C. Borek, R. Miller, C. Pain, and W. Troll, Conditions for inhibiting and enhancing effects of the protease inhibitor antipain on x-ray-induced neoplastic transformation in hamster and mouse cells. Proc. Nat. Acad. Sci. (USA) 76:1800-1803 (1979).
16. C. Borek, A. Ong, and H. Mason, Sodium bisulfite protects against radiogenic and chemically induced transformation in hamster embryo and mouse 10T 1/2 cells, Toxicol. Indust. Health 1:69-74 (1985).
17. J. E. Cleaver, M. Banda, W. Troll, and C. Borek, Some protease inhibitors are also inhibitors of poly(ADP)ribose, Carcinogenesis (in press) (1985).
18. T. Sugimura, and M. Miwa, Poly (ADP ribose) and cancer research, (in press) (1985).
19. C. Borek, W. F. Morgan, A. Ong, and J. Cleaver, Inhibition of malignant transformation in vitro by inhibitors of poly)ADP-ribose) synthesis, in: Synthesis, Vol. 81, pp. 243-247 (1984).
20. E. Kun, E. Kirsten, G. E. Milo, P. Kurian, and H. L. Kumari, Cell cycle-dependent intervention by benzamide of carcinogen-induced neoplastic transformation and in vitro poly(ADP-ribosylation) of nuclear proteins in human fibroblasts, Proc. Nat. Acad. Sci. (USA) 80:7219-7223 (1983).

THE ROLE OF RISK ASSESSMENT IN FOOD SAFETY POLICY*

Robert J. Scheuplein

Food and Drug Administration
Center for Food Safety and Applied Nutrition
200 C. Street, S. W., Washington, D.C. 20204

INTRODUCTION

Virtually every category of food substance contains, either naturally or through contamination, trace levels of carcinogens. Whether they are traditional foods like grains, cheese, or mushrooms, spices like pepper or nutmeg, migrants from packaging materials, residues from pesticides or animal drugs, or unavoidable impurities from additives in processed foods; it is impossible to purge them completely. As our awareness of the ubiquity of carcinogens has grown, there has been a growing need to distinguish between tolerable and intolerable risk from carcinogens in food. This distinction -- which relies inter alia on the toxicological principle of dose-response, the essence of quantitative risk assessment -- is now regarded as crucial in the regulation of carcinogens, as it has been in the regulation of non-carcinogens. As suggested above this was not always so. Early FDA policy statements and the food safety laws themselves contained the implied promise of absolute safety from carcinogens in food. To understand the current role of carcinogen risk assessment in food safety policy, its relatively slow development, its conservative application, and the legal difficulties that still bar its wider use, it will be helpful to review some of the history of the food laws in the U.S.

HISTORICAL OVERVIEW OF FOOD SAFETY LAW AND POLICY

The first attempt at federal control of food was a bill introduced January 20, 1879, "for preventing the adulteration of articles of food and drink." During this period, Americans were becoming concerned over the safety of their food supply for several reasons, all relating to the fact that much of the food was no longer produced in the home or in the neighborhood, and incidents of adulteration were increasing. Unwholesome imports were being "dumped" in the U.S., and domestic food was being debased by the intentional substitution of inferior substances. Damage was concealed, and some food was simply filthy. These concerns culminated in the 1906 Act which aimed at the elimination of adulteration of otherwise natural wholesome food. Proponents of the 1906 Act were also alarmed by the accumulating number of unfamiliar chemicals that were "added" in order to preserve food and improve its appearance and texture, a problem created

---

*This article contains material that has been published in "Handbook of Carcinogen Testing," H. A. Milman and E. K. Weisburger editors, Noyes Publications, Park Ridge, New Jersey, 1985.

by the necessity to store food longer and transport it farther. The notion of adulteration, originally intended to apply more to international debasement of food for profit, came to embrace the presence of "unsafe chemical additives" which seemed to be added more for the producer's convenience than the consumer's. There was no expressed intent to improve the safety of natural food, as selected and prepared by individuals. If greater health could be had through better nutrition or a better selection of foodstuffs, this was not perceived in 1906 as a matter for federal intervention.

The food safety laws were fundamentally changed in 1938. The 1938 Act and its interpretation by FDA substantially broadened FDA's mission in the direction of protecting the consumer from unsafe food per se, rather than acting solely in response to an act of food adulteration by a food manufacturer or distributor. For the first time, the law contained a provision that prohibited the marketing of even a traditional food that was "ordinarily injurious." No intervening human act of adulteration was required. A distinction was made, however, in the safety standard that was to apply to naturally occurring "poisonous or deleterious substances", as compared to intentionally added substances. Natural constituents in food, such as oxalic acid in rhubarb or solaine in potatoes, were considered adulterants only if the food contained them in sufficient quantity to be considered "ordinarily injurious." FDA has used this provision to regulate very few substances, most notably poisonous mushrooms and Burma beans. FDA has never attempted to restrict the marketing of a traditional food because it contained a "non-added" carcinogen. If FDA wished, for example, to bring the nitrosamines in some natural substances, the hydrazines in mushrooms, or the safroles in spices under the ambit of this provision, the agency would have to demonstrate a real probability (not a mere possibility) of harm to a significant percentage of ordinary consumers.

"Added" constitutents on the other hand were made subject to the far stiffer "may render injurious" standard under the same Section of the Act. Under this provision, FDA need show only the existence of some significant possibility that the food could be injurious to the health of a consumer. The courts have stated, "if ..., it cannot by any possibility, when the facts are reasonably considered, injure the health of any consumer, such food, though having a small addition of poisonous or deleterious ingredients, may not be condemned under the act"[1]. Until relatively recently FDA took the position that the presence of an added carcinogen at any level inherently violated this provision. In the 1950's, two artificial sweeteners, dulcin (4-ethoxyphenylurea) and P-4000 (5-nitro-2 propoxyaniline), were prohibited under Section 402 (a) (1) after studies indicated that they produced tumors in animals.

The 1938 Act authorized FDA, for the first time, to permit the presence of "safe" levels of poisonous and deleterious added substances in certain justifiable instances. If such substances were necessary in the production of food or if they were unavoidable even under the best food manufacturing practices, the Act authorized the FDA to set tolerances limiting the quantity of such substances to the extent found necessary to protect the public health. FDA has taken the position that it may, under this section of the Act, establish tolerances for environmental contaminants that find their way into food. The tolerance is established by considering the contaminant's toxicity, the extent to which its occurrence can be controlled, and the capability of analytical methods to measure the contaminant and enable enforcement of the tolerance. Under both the 1906 and 1938 Acts, the burden of providing needed information remained with FDA. It was up to FDA to become aware of possible toxicological problems, to conduct the appropriate safety studies, to establish the appropriate tolerance, and to provide assurance that this amount would not be exceeded, taking into account the diverse dietary habits of the American public. Limitations in FDA personnel and resources mitigated against the frequent use of this provision. Another major limitation was the lack of sophistication in the art of toxicological testing with regard to cancer and other chronic effects.

The statutory provisions described, above, extensively supplemented by others, still exist and embody the agency's basic food safety requirements. They permit greater risk or less assurance of safety for substances of greater perceived value (traditional food); they permit less risk for inherently unnecessary substances (additives), while tolerating intermediate risk for necessary or unavoidable substances (contaminants).

BASIC (PRE-1958) STATUTORY FOOD SAFETY STANDARDS

FOOD CATEGORIES	STATUTORY SAFETY STANDARDS*
Traditional Foods possibly containing naturally occurring poisonous or deleterious substances (could include carcinogens)	Sec 402 (a) (1) the food is adulterated if the substance renders the food ordinarily injurious to health
Added Substances possibly poisonous and deleterious substances (could include carcinogens)	Sec 402 (a) (1) the food is adulterated if the substance may render the food injurious to health
Unavoidable Added Substances Substances either required in the production of food or unavoidable by good manufacturing practice (could include carcinogens)	Sec 406 The food is adulterated if the amount of substance exceeds the tolerance which attempts to balance the need to protect the public health with the unavoidability or essentially of the substance

*The reader is referred to the quoted sections of the FD&C Act for the precise phraseology of the various provisions.

The original triad of provisions discussed above has been amended several times since 1938, usually to separate out a particular class of added substances for special regulatory attention. For example, Congress required the registration and licensing of pesticides under The Pesticide and Chemical Act of 1954. In 1958, the Food Additives Amendment established a similar preclearance scheme for substances directly added to formulated foods or for substances used in processing or in packaging that could migrate into food or otherwise affect its characteristics. However, the Food Additives Amendment did not apply to all intentionally added ingredients or to all potential migrants. An exception was made for GRAS substances and for substances that were approved either by FDA or the U.S. Department of Agriculture (U.S.D.A.) prior to 1958. GRAS---"generally recognized as safe"---substances included a large class of traditional additives; e.g., sugar, salt...etc. in common use, and others later to be established as GRAS by panels of scientific experts. "Prior sanctioned" substances were in effect "grandfathered" on the basis of previous approvals or advisory statements by U.S.D.A. and/or F.D.A. Despite such exceptions, the Food Additives Amendment reversed the legal burden of proof demanded under the law and fundamentally changed the regulation of food additives in the U.S. FDA was no longer required to show that a marketed food additive might be injurious to consumers; instead the sponsor had to provide experimental evidence to FDA demonstrating the additive to be safe for its intended use prior to marketing approval.

The safety standard for food additives (Sec 402 (a) (1) and Sec 409), was a demonstration of a "reasonable certainty of no harm" under anticipated and reasonably foreseeable levels of exposure. Carcinogenic food additives were in addition banned under the Delaney Clause. The Color Additive Amendments of 1960 established very similar rules for color additives, including a Delaney Clause.

From this brief discussion it is evident that current U.S. food safety requirements are an amalgam of several provisions of the FD&C Act that have been enacted at different times since 1938. The amendments in 1958 and subsequently have tended to produce specialized regulatory requirements for different functional categories of additives at the cost of some internal consistency in the treatment of comparable risks. Each set of category-specific provisions operates essentially without regard to the treatment accorded other classes of food constituents.

THE REGULATION OF CARCINOGENS

Before 1958, as stated above, carcinogens were not legally distinguished from other toxic or poisonous and deleterious substances. When the Delaney Clause was enacted in 1958, career scientists at FDA regarded it as neither particularly helpful nor harmful, but merely redundant. No one then believed that a food additive would be declared safe under the general safety provisions if it was an animal carcinogen. Furthermore, the Delaney Clause was a weaker provision in one important respect. While a suspicion of carcinogenicity could be sufficient to undermine a showing of safety, the Delaney Clause required a frank finding of carcinogenicity. As a result, from 1958 to 1975 many more substances were prohibited as food or color additives under the authorization of the general safety provision than under the Delaney Clause.

But Perceptions of the Carcinogenic Hazard were changing. During the 70's the number of regulatory actions on carcinogens in food and cosmetics increased. Perhaps two scientific trends were primarily responsible: first, the sensitivity of analytical methods improved, and the instrumentation to carry out these analyses became widely available; and second, many more long-term animal bioassays were conducted--notably, by the National Cancer Institute's (NCI) Cancer Bioassay Program. To the average person, the results of this "progress" were constant reminders about the carcinogenic hazards from corn, peanuts, milk, grain, canned food, hamburger, poultry, beef, fish, hot dogs, and even scotch and beer. People eventually began to question the soundness of a regulatory policy that threatened to remove useful and popular articles from commerce on the basis of what in many instances appeared to them to be very low and even trivial levels of risk.

The initial concern over carcinogens in 1958 developed against the background of scientific knowledge of cancer, or more accurately the uncertainty in that knowledge that existed at the time.

During the Color Additive Hearing of 1960, the Secretary's testimony in support of the Delaney Clause stated in part:

"We have no basis for asking Congress to give us discretion to establish a safe tolerance for a substance which definitely has been shown to produce cancer when added to the diet of test animals. We simply have no basis in which such discretion could be exercised because no one can tell us with any assurance at all how to establish a safe dose of any cancer-producing substances"[2].

The House Report accompanying the reported bill summarized the Committee's position:

"In view of the uncertainty surrounding the determination of safe tolerances for carcinogens, the committee decided that the Delaney anti-cancer provision in the reported bill should be retained without change"[3].

The awareness of growing regulatory problems gradually increased, but as late as May 1974, FDA continued officially to support the Delaney Clause, but with some reservations:

"...we are not prepared to state that the Delaney Clause has had a deleterious effect, to date, upon the food supply, nor could we suggest any particular change in the anti-cancer clauses. We believe that the growth of knowledge in carcinogenesis may eventually permit safe levels of carcinogenic additives to be determined, but that day is not yet here. We also believe that the elected representatives of the people may some day be called up to permit exceptions to the clause for additives which may be important to the food supply and yet for which there is evidence of carcinogenicity." Dr. Alexander Schmidt, Commissioner of Food and Drugs[4].

The statement proved prophetic. In November 1977, Congress enacted the Saccharin Study and Labeling Act (PL-95-203) which in effect created an indefinite exception for saccharin. Other problems occurred with lead (in tin cans), lead acetate (in hair dyes), carcinogenic impurities in color additives and packaging materials, and DES and nitrites in meat. Recognizing that such problems were consuming a disproportionate and increasing share of the agency's resources, that the public health benefits resulting from banning very low levels of carcinogens were doubtful at best, that the agency's credibility was being harmed, and that it appeared to be possible to establish crude but nonetheless "safe" tolerances for carcinogens, the FDA, at the 1983 Food Safety Hearings, asked Congress for authorization to use a risk assessment approach for trace amounts of added carcinogens in food[5].

The motivation for this change in regulatory approach was not primarily scientific. No scientific breakthrough had suddenly provided the knowledge that now permitted "safe" tolerances for carcinogens to be determined with certainty. Much had indeed been learned in the 20-odd intervening years about the mechanism of cancer. The existence of at least a two-stage carcinogenic process was substantially verified; the importance of modifications in DNA in the initiation step was discovered, as was the need for metabolic activation to electrophiles. The existence of DNA repair processes was established, new methods of "screening" for carcinogens such as the "Ames test" were being developed and tested, and much more But this knowledge, however valuable and crucial to an eventual understanding and possible control of cancer, was not of compelling regulatory significance. Much more information was needed to be able to establish safe tolerances for human exposure to carcinogens from first principles. The crucial event that occurred during the intervening two decades was not a scientific breakthrough but rather a continuous advance in technology. The improved "sensitivity" of analytical instrumentation and techniques eventually allowed the detection of carcinogens at the ppb and even the ppt level, and virtually assured finding carcinogens both in new and in familiar food and color additives. However, the ability to <u>detect</u> a substance does not always mean that it can be reliably identified and confirmed. The reliability of identifying and quantifying substances decreases exponentially as concentration decreases[6]. Moreover, the vastly improved ability to detect substances, often in artifically ideal conditions, does not automatically mean that comparably precise or reliable chemical measurements are available in practice. Electronic noise in measuring instruments, the leaching of impurities from containers, background contamination, and the difficulties in separating and isolating substances in complex food matrices generally mean that the improvements in analytical science are more effective in finding problems than in enabling regulators to deal with them. In 1958, Congress did not foresee the widespread presence of trace amounts of carcinogens in food, the possibility that common food substances might themselves be shown to contain carcinogens, or the occurrence in humans of carcinogens as essential nutrients or as an outcome of normal metabolism. Congress did not anticipate the extent to which substances then regarded either as absent from foods, or as non-carcinogenic on the basis of weaker technology, would later prove to be present and to be animal carcinogens -- saccharin, nitrites, lead, etc. In short, the acknowledged scientific uncertainty surrounding carcinogenesis was larger and less amenable to the apparently prudent solution they proposed, than they could have foreseen.

RISK ASSESSMENT OF CARCINOGENS

In the early forties, some investigators had proposed quantitative analysis of carcinogen dose-response data[7] and even systems of grading carcinogenic potency[8,9]. By 1970 there were hundreds of animal studies that supported quantitative dose-response behavior at least to the extent that lower doses in the same strain of rodent consistently produced lower tumor incidence. Methods of using the dose-response behavior for assessing human risk from carcinogens at very low exposures were proposed in 1961 by Mantel and Bryan[10] and were soon extended by others[11-15]. However, the two crucial quantitative extrapolation steps: (a) high dose to low dose and (b) animals to humans have not been equally well developed. The use of conservative assumptions was and is generally regarded as a necessary prudence to bridge over gaps in our understanding of the effects of differing species sensitivities, metabolic responses, diets, patterns of consumption, etc., between man and animals, as well as the occurrence of additive or synergistic effects. These all provide opportunities for error aside from the question of the accuracy of the dose-response extrapolation model. Despite these difficulties, support gradually developed for the notion that while scientific advances still could not provide a basis for establishing an absolutely safe threshold, they might provide a basis on which discretion could be exercised in establishing tolerances for carcinogens at the cost of acceptably low and, within bounds, ascertainable risks. Another factor supporting the use of risk assessment was the much larger risk from unregulated exposures to added carcinogens. The estimated (U.S.) dietary intake of nitrosamines is from 1 to 10 μg/day, and it is approximately the same in Western Europe and at least as great in China and Japan[16]. Our exposure to polynuclear aromatic hydrocarbons (PNA's) from a variety of sources is at least this great[17]. We have learned that traditional methods of preparing and preserving food contaminate it with both classes of carcinogens; e.g., char-broiling and smoking with PNA's, and pickling with nitrosamines. When the upper-bound human risks from these exposures are estimated by the same techniques proposed for food additives (linear extrapolation) on the basis of animal data the risk exceeds 10^{-4} (i.e., one in ten thousand in a lifetime). This is just the tip of the iceberg; the spectrum of natural carcinogenic contaminants at low levels in food is far larger than these two examples can suggest. These "added" carcinogens are officially ignored because the exposures are ubiquitous and would be extraordinarily difficult if not impossible to control and regulate. Finally, there is the risk from the diet itself; a risk not necessarily quantitatively attributable to the level of carcinogenic contamination, but possibly more to the promotional character of some major nutrients. A recent report of the National Academy of Sciences on Diet, Nutrition, and Cancer states:[18]

"...Judging from the observed differences in cancer rates among populations with different diets, it is highly likely that the United States will eventually have the option of adopting a diet that reduces its incidence of cancer by approximately one-third,..."

The impact of all this information gradually created a distinction between the regulatory consequences forced by the Delaney prohibition against carcinogenic additives at any level and "safe" tolerances which now seemed feasible under the "reasonable certainty of no harm" standard of the general safety provisions. The focus of the distinction was primarily the low-level cancer risk from small amounts of carcinogenic animal drug residues, migrants from food packaging materials, and contaminants in direct food and color additives. The wisdom of banning these useful additives, which could be expected to produce only miniscule levels of carcinogens in food seemed open to question. Control of these substances appeared manageable through risk-assessment procedures. Under current law, FDA has very limited administrative discretion to deal with low-level cancer risks from additives. Quantitative risk assessment of carcinogens has played a significant role in the development and use of FDA's administrative authority. However, because of the uncertainty in the accuracy of risk-assessment procedures and the legal constraint still imposed by the Delaney Clause, FDA's approach to risk-assessment has been cautious and limited.

FDA first suggested the use of risk-assessment in a 1973 proposed rule dealing with animal drug residues[19]. This was later issued as a regulation in February 1977[20]. Because of a special exemption proviso added to the law as part of the Drug Amendments of 1962, the use of risk-assessment in this instance was not considered to be in violation of the Delaney Clause. That provision, the DES proviso, permitted the use of carcinogenic drugs in animals raised for food production if "no residue" was found in the edible tissue (Sec. 512(d) (1) (H)). This proviso makes the detection of residues of the drug in edible animal tissues rather than its <u>administration</u> to animals or <u>addition</u> to animal feed the critical inquiry[21]. It also means permitting some carcinogenic residue, namely the amount present below the limit of the official method of detection. For some years FDA insisted that the protection of human health from carcinogenic animal drugs required that the most sensitive methods be used as soon as they were available--thus continually reducing the apparent "zero residue" level. This was called the "sensitivity of method" regulation or SOM. Finally in 1979, concluding that the continually moving regulatory target made little sense when cancer risks of that order were coming to be regarded as insignificant, FDA adopted a fixed "insignificant risk" level of 10^{-6}/lifetime (by linear extrapolation) regardless of the capability of analytical techniques to detect residues at or below this level[22].

With the increasing ability of analytical methods to operate at lower levels and the Delaney prohibition against additives at any level, the anticancer provisions were operating like the early models of the SOM regulation. Legally permissible carcinogen levels were being driven down to current analytical limits. This was a special problem for color additives and food packaging materials, many of which can contain potentially carcinogenic contaminants. Time seemed to be running out rapidly for many of these products. If section 409 or 706 of the Food, Drug, and Cosmetic Act were to be literally and fully enforced, it was difficult to imagine how many of them would continue to remain on the market.

In response to the growing problem with low-level carcinogenic impurities, FDA proposed its "Constituents Policy" under which these impurities would be regulated under the general safety provisions of the Act by risk-assessment procedures[23]. The language of the Delaney provision, being silent on the impurity question, lent itself to interpretation on this issue. FDA proposed that the word "additive" in the Delaney phrase ("...no such additive shall be deemed to be safe if it is found to induce cancer...") should be interpreted to mean the "additive as a whole" including the trace constituents that may accompany it. Therefore, if the additive as a whole, when tested in animal feeding studies, is not found to be carcinogenic, it does not come under the ambit of the Delaney Clause, despite the fact that a trace constituent(s) when tested by itself at higher doses is known to be carcinogenic. The constituent is not ignored however; it is regulated by "risk-assessment" under the general safety provision of the Act. FDA has, for the present, used 10^{-6}/lifetime (by linear extrapolation) as the cutoff for a significant risk level. The "Constituents Policy" is currently under review, and no one knows its future. In the absence of a change in the statute, however, some administrative flexibility along these lines appears essential. FDA believes that some use of "dose-response" in the regulation of carcinogens comes closer to preserving the original intent of the Delaney Clause -- that is, of distinguishing between substances that present carcinogenic risks from those that don't -- than an interpretation that would, given current analytical capabilities, result in the indiscriminate banning of virtually all additives regardless of the amount of carcinogen present.

RISK ASSESSMENT - ITS VALUE

Many foods contain carcinogens. Our regulatory task is hopeless without some method of distinguishing between large and small risks. Risk assessment, at least as currently practiced, embodies two implicit axioms. The first is the certainty of some risk from chemicals (everyone now seems to recognize that absolute safety is unattainable), and the second is the uncertainty in its measurement. If carcinogenic risk were avoidable, risk assessment would be unnecessary; if carcinogenic risk could be measured easily, risk assessment would be trivial. The scientific basis of quantitative risk assessment at present is weak, particularly the extrapolation from animals to humans. However, when combined with other evidence and used in a conservative manner chosen in a way that is not likely to understate the risk, it can be a valuable regulatory tool. It provides a measure of orderliness, openess and equity in the regulation of carcinogens. FDA has accepted it for a part but not for the whole of the regulatory process and applies it cautiously and on a limited basis because of its scientific disabilities as well as current legal proscriptions. The law and the need to act promptly require FDA to make decisions usually in a climate of scientific uncertainty[24]. Quantitative risk assessment is not at present a fully adequate procedure, but this cannot be expected when scientific information is lacking. It is, however, the only alternative to inaction presently available.

REFERENCES

1. U.S. v Lexington Mill & Elevator Co. 232 U.S. 399 (1914).
2. Color Additives: Hearings on H R 7624 and S2197. Before the House Commitee on Interstate and Foreign Commerce, 86th Congress, 2nd Session 62 (1960).
3. H. R. Rep. No. 1761 86th 2nd Session (1960).

4. Food and Drug Administration "Study of the Delaney Clause and other Anti-Cancer Clauses:" Hearings on Environmental and Consumer Protection Appropriations for 1975, Before the House Subcommittee, of the Committee on Appropriations, 93rd Congress 2nd Session, May (1974).
5. Oversite of Food Safety, 1983. Hearings before the Senate Committee on Labor and Human Resources, 98th Congress, first session, (June 8-10, 1983).
6. W. Horwitz, Today's Chemical Realities, J. Assoc. Off. Anal. Chem. 66:1295-1301 No. 5 (1983).
7. W. R. Bryan, and M. B. Shimkin, Quantitative analysis of dose-response data obtained with three carcinogenic hydrocarbons in strain C3H male mice, J. Nat. Cancer Inst. 3:503-531 (1943).
8. I. Berenblum, Systems of grading carcinogenic potency, Cancer Res. 5:561-564 (1945).
9. H. Druckrey, Quantitative Grundlagen der Krebsergeugun Klin. Wschr. 22:532-540 (1943).
10. N. Mantel, and W. R. Bryan, "Safety" testing of carcinogenic agents J. Nat. Cancer Inst. 27:455-470 (1961).
11. R. E. Albert and B. Altshuler, Considerations relating to the formulation of limits for unavoidable population exposures to environmental carcinogens, in: "Radionuclide Carcinogenesis", J. E. Ballou, R. H. Busch, D. D. Mahlum and C. L. Sanders, eds., AEC Symposium Series, CONF-72050, NTIS, Springfield, VA, pp. 233-253 (1973).
12. L. Friedman, A proposed procedure for the assessment of health hazards of carcinogens at very low levels of exposures, in: "Assessment of the Carcinogenicity and Mutagenicity of Chemicals", World Health Organization Technical Report #546, 14-19 (1974).
13. P. G. Hoel, D. W. Gaylor, R. L. Kirschstein, U. Saffiotti, and M. A. Schneiderman, Estimation of risks of irreversible delayed toxicity, J. Toxicol. Environ. Health, I, 133-151 (1975).
14. H. Guess, K. Crump, and R. Peto, Uncertainty estimates for low-dose-rate extrapolation of animal carcinogenicity data, Cancer Res. 37:3475-3483 (1977).
15. H. O. Hartley and R. L. Sielken, Estimation of "safe doses" in carcinogenic experiments, Biometrics 33:1-30 (1977).
16. "IARC Monographs on the Evaluation of the Carcinogenic Risk of Chemicals to Humans: Some N-Nitroso Compounds", Vol. 17, World Health Organization (1978).
17. J. W. Howard and T. Fazio, Review of polycyclic aromatic hydrocarbons in foods, J. Assoc. Off. Anal. Chem. 63:1077-1104 (1980).
18. National Research Council, Report of the Committee on Diet, Nutrition, and Cancer, Assembly of Life Sciences, Diet, Nutrition and Cancer, Chapter 2, p. 1-13 (1982).
19. Chemical Compounds in Food-Producing Animals, Federal Register 38:19226-19227 (July 19, 1973).
20. Chemical Compounds in Food-Producing Animals, Federal Register 42:10412-10437 (Feb. 1977).
21. R. A. Merrill, Regulation of Carcinogens in Food: A legislator's guide to the food safety provisions of the Federal Food, Drug, and Cosmetic Act, Mich. Law Rev. 77:171-250 No. 2 (1978).
22. "Chemical Compounds in Food-Producing Amimals", Federal Register 44:17070-17114, (March 1979).
23. D & C Green, No. 6 Federal Register 47:14138-14142 (April 2, 1982).
24. P. B. Hutt, Food regulation, Food Drug Cosmet. Law J. 33:501-558 (1978).

CONTRIBUTORS

INDEX

Made in the USA
Monee, IL
18 October 2026

66036440R10328